W9-BJG-741

SYSTEM OF EXPERIMENTAL DESIGN

SYSTEM OF EXPERIMENTAL DESIGN

Engineering Methods to Optimize Quality and Minimize Costs

GENICHI TAGUCHI

Don Clausing
Technical Editor for the English Edition

English translation by LOUISE WATANABE TUNG

VOLUME TWO

UNIPUB
KRAUS INTERNATIONAL PUBLICATIONS
White Plains, New York

AMERICAN SUPPLIER INSTITUTE, INC.
Dearborn, Michigan

Copublished by:

UNIPUB/Kraus International Publications
(Division of Kraus-Thomson Organization Limited)
One Water Street
White Plains, NY 10601
(914) 761-9600

and

American Supplier Institute, Inc.
Six Parklane Boulevard
Suite 411
Dearborn, MI 48126
(313) 336-8877

First printing 1987
Second printing 1988

Printed in the United States of America

The paper used in this publication meets the minimum requirements of American National Standard for Information Sciences—Permanence of Papers for Printed Library Materials, ANSI Z39.48–1984.
The book is Smyth sewn and casebound in F-grade Library Buckram, which contains no synthetic fibers.

Library of Congress Cataloging-in-Publication Data

Taguchi, Gen'ichi, 1924–
System of experimental design.

Translation of: Jikken keikakuhō.
1. Experimental design. I. Title.
QA279.T3313 1987 001.44'34 87-5005
ISBN 0-527-91621-8 (UNIPUB/Kraus; set)
ISBN 0-941243-00-1 (ASI; set)

Table of Contents

VOLUME ONE

VOLUME TWO

Preface to Volume Two

In the preface to the second edition, I wrote as follows:

> A few persons made the criticism regarding Volume Two of the old edition that, "It's not organized, differing from Volume One. It's confused." It was exactly so. But then, when I finished writing the second volume of the new edition (the second edition), it seemed that the confusion had doubled. Since it was the purpose of the second volume to give as much foundation as possible to the formulas and methods discussed in the first volumes, perhaps it would have been impossible to avoid encyclopedia-like disorder, but could it not be that the primary reason for the confusion was because the foundation of the theory of statistics regarding data analysis has not yet been established. . . .

For the author, who regards the third edition as the final edition, it has been a painful challenge through nearly thirty years to *organize the analysis methods for widely varying experimental data under a single guiding principle.* As I mentioned in the preface to Volume One, I travelled to the United States in January 1982, during the proofreading of the second edition. Coincidentally, the research theme that I tackled at Bell Telephone Laboratories while in America was "Design of Experiments and the Communication Theory" (see Section 40.4). Upon reflection after finishing Volume 2 of the second edition, it seemed that the research that was performed at that time had become the guiding principle of Volume 2 of the present edition.

Just as the total output is resolved into the signal voltage and noise voltage and is resolved into the sum of the powers of the frequency components in communication engineering, in data analysis I decided to use as the unifying principle the so-called resolution of variation, by which the total sum of squares of the data which are the total output of the factorial effects is expressed by the sum of the magnitudes of the individual sources that have influenced these data. The ideas of SN ratio and beta coefficient were born entirely naturally from this method of thinking. I believe I have explained this new method of thinking in some detail in Chapter 16 (in Volume One) and in Chapters 18 to 24 (in Volume Two). I feel that under the new method of thinking, it is now possible to *evaluate the magnitude of the influence of each source and the magnitude of the error toward the target characteristic, without being bound by the probability theory.* As a result, such logical treatment methods as the following have become possible:

1. The mathematical foundation is placed upon quadratic-form calculations or orthogonal resolution which is orthogonal expansion (the Parseval equality) and linear algebra.
2. Testing is no longer performed, but point estimation by the beta coefficient is carried out.
3. It is possible to rationalize expressions of dynamic characteristic values and stability of functions.
4. It is also possible to explain the economic basis of why resolution of the sum of the squares in performed.

Especially, the definition of SN ratio as an expression of dynamic properties, explained, in Chapter 24, will probably result in a fundamental change in the design of experiments on dynamic characteristics, such as the maneuverability of a vehicle, the performance of a machine tool, or the performance of articles used in sports.

Ever since the second edition, it has come about that the design of experiments method is applied in unexpectedly wide-ranging areas, such as calibration, medical and pharmaceutical products, steps of assembly, the marketplace, and in business. Since applications in the respective fields have unique properties, it is to be hoped that complete volumes will appear for each area. Until then, however, this volume provides information on several of these fields: calibration relations are explained in Chapter 22; medical and pharmaceutical product relations are explained in Chapter 26, in Sections 27.1 and 27.2, and in Section 30.4.3; assembled product relations are discussed in Chapter 25; and marketplace relations are explained in Chapter 26. Since a complete volume has already been completed for business relations (See Reference 34), a separate chapter on this topic has not been incorporated in Volume Two. It is regrettable that, although I had wished to write a chapter on design research, it was not possible to provide a chapter especially on this subject, due to insufficiency of examples. I hope that those individuals who are concerned with design will themselves attempt rationalization of design research by tying together Chapter 16 (in Volume One) and Chapter 24 (in Volume Two).

Besides the chapters mentioned above, minute accumulating analysis, discussed in Chapter 32, and the compounding technique, covered in Chapter 35, are important subjects. The contents of these chapters should have been published in Volume One, but for various reasons, they are given in this volume instead. Therefore, I feel that in Volume Two, Chapters 18 to 24, Chapter 32, and Chapter 35 are important for all researchers. Concerning the other chapters, it should suffice if the reader examines only those that are related to his or her problems, or reads them as necessary.

I await criticisms and comments from readers in both the areas of theory and actual practice concerning the new method of thinking present in Volume 2.

GENICHI TAGUCHI

June 24, 1977

18 Resolution of Variation

In this chapter, *linear mathematics* (orthogonal expansion and the quadratic form), which constitutes the foundation of *resolution of the sum of squares* of the experimental values, and the economic basis of why we resolve the sum of squares will be explained. Since the discourse is rather mathematical, beginners in linear mathematics may begin with Section 18.3. However, even if the finer points are unclear, it is to be hoped that the main thread of Sections 18.1 and 18.2 will be read. Please refer to Section 40.4 for somewhat greater detail.

18.1 Resolution of Variation and the Quadratic Form

We denote n observed values by $y_1, y_2, \ldots, y_n$ and we consider the sum of their squares, S_T.

$$S_T = y_1^2 + y_2^2 + \cdots + y_n^2 \tag{18.1}$$

It will be explained in Section 18.2 why we use the sum of squares. Equation (18.1) is a *unit quadratic form,* and the *matrix* for this is the $n \times n$ *unit matrix* $\boldsymbol{E}$. To resolve a unit matrix and to resolve a unit quadratic form are equivalent; we will mainly consider matters in terms of the quadratic form.

As a simple example, when the heights of six Japanese and four American male youths were measured, the following results were obtained (unit: cm).

A_1: Japanese	$y_1=158,\ y_2=162,\ y_3=155,\ y_4=172,\ y_5=160,\ y_6=168$
A_2: American	$y_7=186,\ y_8=172,\ y_9=176,\ y_{10}=180$

The sum of squares of the heights of the ten persons, S_T, is

$$S_T=158^2+162^2+\cdots+180^2=286\,197 \quad (f=10) \tag{18.2}$$

With respect to a quadratic form Q in $y_1, y_2, \ldots, y_n$,

$$Q=\sum_{i,j=1}^{n} a_{ij}y_iy_j \tag{18.3}$$

the *rank* of the matrix $\boldsymbol{M}$ whose elements are the coefficients a_{ij},

$$\boldsymbol{M}=(a_{ij}) \tag{18.4}$$

is termed the *number of degrees of freedom of the quadratic form*. The quadratic forms which appear in analyses of variance are all positive-definite, and their elements are always real numbers. The number of degrees of freedom f of the unit quadratic form of Equation (18.1) is 10.

Let us now consider the problem of resolving the total sum of squares S_T of Equation (18.2) into the following three components.

m = the effect of the cause which is common to the heights of the ten persons; this is a source which is also termed the general mean
A = source termed the difference between Japanese and Americans
e = source termed individual difference, expressed by the initial letter of "error"

In other words, this is the problem of resolving S_T of Equation (18.2) into three components of the quadratic form: S_m, S_A, and S_e. To do this, we consider the following linear equations for source m and source A, and we find the answer by squaring them.

Mean Value

$$L_m=\frac{1}{10}(y_1+y_2+\cdots+y_{10}) \tag{18.5}$$

Difference Between Nationalities

$$L_A=\frac{1}{6}(y_1+y_2+\cdots+y_6)-\frac{1}{4}(y_7+y_8+\cdots+y_{10}) \tag{18.6}$$

L_m of Equation (18.5) expresses the mean height of the ten male youths, and L_A of Equation (18.6) indicates the difference between the mean height of the Japanese and that of the Americans. Inasmuch as these ten data of height are all heights of present-day male youths, they possess a condition in common. If we consider not youths but ten-year-old boys, the mean value will be different. *The effect of a source that is common to n data appears in the mean value.* Anyone can understand that the difference between the mean heights of Japanese and American male youths appears in L_A of Equation (18.6).

In general, the linear equation with constant coefficients of n observed values $y_1, y_2, \ldots, y_n$ is expressed as

$$L=c_1y_1+c_2y_2+\cdots+c_ny_n \tag{18.7}$$

It is to be understood that at least one among the coefficients $c_1, c_2, \ldots, c_n$ is not zero. Then, the quotient obtained by squaring the value of L of Equation (18.7) and dividing the result by the sum of squares of the coefficients, termed the *unit number D,*

$$D=c_1^2+c_2^2+\cdots+c_n^2 \tag{18.8}$$

is a quadratic form of rank 1 or, in other words, one degree of freedom, and this becomes one component of the total sum of squares S_T.

For example, the quadratic forms S_m and S_A for the linear equations L_m and L_A of Equations (18.5) and (18.6) are

$$S_m=\frac{\left[\frac{1}{10}(y_1+y_2+\cdots+y_{10})\right]^2}{\left(\frac{1}{10}\right)^2+\cdots+\left(\frac{1}{10}\right)^2}=\frac{(y_1+y_2+\cdots+y_{10})^2}{10} \tag{18.9}$$

$$=\frac{(158+162+\cdots+180)^2}{10}=\frac{1\,689^2}{10}=285\,272.1 \tag{18.10}$$

$$S_A=\frac{\left[\frac{1}{6}(y_1+y_2+\cdots+y_6)-\frac{1}{4}(y_7+y_8+\cdots+y_{10})\right]^2}{\left(\frac{1}{6}\right)^2\times 6+\left(\frac{-1}{4}\right)^2\times 4}$$

$$=\frac{[2(y_1+y_2+\cdots+y_6)-3(y_7+y_8+\cdots+y_{10})]^2}{2^2\times 6+(-3)^2\times 4} \tag{18.11}$$

$$=\frac{(2\times 975-3\times 714)^2}{60}=614.4 \tag{18.12}$$

The difference in heights among the individuals within the same nationality is termed individual difference, and the quadratic form for this, S_e, is obtained by subtracting S_m and S_A from S_T. S_e is also termed the *residual sum of squares.*

$$S_e=S_T-S_m-S_A$$

$$=286\,197.0-285\,272.1-614.4=310.5 \tag{18.13}$$

The condition under which the quadratic form corresponding to the individual difference S_e may be found by subtracting S_m+S_A from S_T is limited to the case where the matrices corresponding to the quadratic forms S_m and S_A, $\boldsymbol{M}_m$ and $\boldsymbol{M}_A$, are orthogonal in the sense that their product is zero. This is true in the present case.

$$\boldsymbol{M}_m=\begin{bmatrix} \frac{1}{10} & \frac{1}{10} & \cdots & \frac{1}{10} \\ \frac{1}{10} & \frac{1}{10} & \cdots & \frac{1}{10} \\ \vdots & \vdots & & \vdots \\ \frac{1}{10} & \frac{1}{10} & \cdots & \frac{1}{10} \end{bmatrix} \tag{18.14}$$

$$\boldsymbol{M}_A=\begin{bmatrix} \frac{4}{60} & \frac{4}{60} & \cdots & \frac{-6}{60} \\ \frac{4}{60} & \frac{4}{60} & \cdots & \frac{-6}{60} \\ \vdots & \vdots & & \vdots \\ \frac{-6}{60} & \frac{-6}{60} & \cdots & \frac{9}{60} \end{bmatrix} \tag{18.15}$$

$$\boldsymbol{M}_m \cdot \boldsymbol{M}_A=(0) \tag{18.16}$$

The foregoing calculation results can be organized into an analysis of variance table (also written ANOVA), as shown in Table 18.1.

In analysis of variance, the value of the quadratic form for the respective sources if termed their *variation*. Also, the quotient obtained by dividing the variation by its number of degrees of freedom is termed the *variance* (or the mean square), and it is denoted by the symbol V. In the total sum of squares S_T, there are included ten persons' worth of the magnitude of the cause in common, m^2, ten persons' worth of the magnitude of individual difference, σ^2, and 4.8 times the magnitude of difference between nationalities, $\sigma_A{}^2$. However, we wish to define the magnitude of the cause in common per person as m^2 and the magnitude of individual difference per person as σ^2. As to the effect of A, the magnitude per replication of A_1 and A_2 is defined as $\sigma_A{}^2$. We then have

$$S_T=10m^2+4.8\sigma_A{}^2+10\sigma^2 \tag{18.17}$$

As can be seen in the section on $E(V)$, S_m, S_A, and S_e are estimated values of σ^2+10m^2, $\sigma^2+4.8\sigma_A{}^2$, and $8\sigma^2$, respectively. Since the individuals differ among the six Japanese of A_1 and among the four Americans of A_2, individual difference enters the nationality difference L_A. Therefore, one person's worth of individual difference enters the quantity that is obtained by squaring this and by dividing by the number of units. The

Table 18.1 Analysis of Variance Table

Source	f	S	V	$E(V)$	S'	ρ [%]
m	1	285 272.1	285 272.1	σ^2+10m^2	285 233.3	99.66
A	1	614.4	614.4	$\sigma^2+4.8\sigma_A{}^2$	575.6	0.20
e	8	310.5	38.8	σ^2	388.1	0.14
T	10	286 197.0			286 197.0	100.00

variations that are obtained by eliminating the magnitude of individual difference from S_m and S_A are termed net (or pure) variations, and they are written S_m' and S_A'.

$$S_m' = S_m - V_e = 285272.1 - 38.8 = 285233.3 \tag{18.18}$$

$$S_A' = S_A - V_e = 614.4 - 38.8 = 575.6 \tag{18.19}$$

$$S_e' = S_e + 2V_e = 310.5 + 2 \times 38.8 = 388.1 \tag{18.20}$$

The results obtained when these are divided by the total sum of squares and multiplied by 100 are the contribution ratios (or degrees of contribution).

We will explain why the magnitude of the difference between nationalities, σ_A^2, is multiplied by 4.8. The coefficient of σ_A^2 would be 5 in the case of data for a total of ten persons, five Japanese and five American. That five times' worth of difference between nationalities enters in this case is easy to understand. However, if the number of persons differs, it means that the difference between nationalities enters multiplied by the *harmonic mean.* If the harmonic mean (reciprocal of mean of reciprocals) in this case is expressed as $\bar{r}$, we have

$$\frac{1}{\bar{r}} = \frac{1}{2}\left(\frac{1}{6} + \frac{1}{4}\right) = \frac{5}{24} \tag{18.21}$$

Thus, $\bar{r}$ is $\frac{24}{5} = 4.8$.

The reason why the square of the linear equation is divided by the unit number is that the linear equation is not one which has been normalized. In applied fields, orthogonal expansion, which considers approximations of functions in the whole of a certain range, is used more widely than the Taylor expansion (the method whereby the properties of functions in the neighborhood of a certain point are given in terms of the derivatives at that point), which is effective in studies of local properties of functions. Fourier expansion and wave-function expansion are examples of orthogonal expansion.

We will assume that a set of functions which are mutually orthogonal in the interval (a, b), $\phi_0(x)$, $\phi_1(x)$, . . . , form a complete system. When a certain function $f(x)$ has been expanded as

$$f(x) = c_0\varphi_0(x) + c_1\varphi_1(x) + \cdots \tag{18.22}$$

the coefficients c_i can be obtained from the following equation.

$$c_i = \frac{\int_a^b \varphi_i(x) f(x)\,dx}{\int_a^b \varphi_i^2(x)\,dx} \qquad (i = 0, 1, \cdots) \tag{18.23}$$

The magnitude S_i occupied by the ith term in the integral of the square of $f(x)$ (square of norm)

$$S_T = \int_a^b f^2(x)\,dx \tag{18.24}$$

is given by

$$S_i = \frac{\left[\int_a^b \varphi_i(x) f(x) dx\right]^2}{\int_a^b \varphi_i^2(x) dx} \qquad (i=0,\ 1,\ \cdots) \tag{18.25}$$

by Parseval's equality. This is the same as what is obtained by dividing the square of the right side of Equation (18.23) by the unit number D,

$$D = \frac{\int_a^b \varphi_i^2(x) dx}{\left[\int_a^b \varphi_i^2(x) dx\right]^2} = \frac{1}{\int_a^b \varphi_i^2(x) dx} \tag{18.26}$$

Since the integrated value of the denominator of the right side of Equation (18.23) becomes 1 when the functions have been normalized, if one normalizes the linear estimated values L_m and L_A (*i.e.*, if one divides by the square root of the unit number) and obtains

$$L_m' = L_m / \sqrt{D} \qquad \sqrt{D} = \sqrt{1/10} \tag{18.27}$$

$$L_A' = L_A / \sqrt{D} \qquad \sqrt{D} = \sqrt{5/12} \tag{18.28}$$

the unit number becomes 1, and these expressions will become the first term and second term of the normalized complete orthogonal expansion. Since estimated values in general have not been normalized, we divide by the number of units D after squaring. Since analysis of variance involves not a general quadratic form but a positive-definite quadratic form of a finite number of data, the basis of calculation is orthogonal expansion as indicated by the Parseval form. It is orthogonal expansion not in an infinite-dimensional function space but in a finite-dimensional vector space. The most important aspect lies in the fact that the individual researcher must create the quadratic form that corresponds to the information he desires, and this is where the problem is not mathematical.

In ordinary orthogonal expansion, there are differential equations and integral equations that fix naturally the eigenfunctions (or characteristic functions) $\phi_0,\ \phi_1,\ \ldots$ for the eigenvalue series $\lambda_0,\ \lambda_1,\ \ldots$. ϕ_0, $\phi_1,\ \ldots$ must form a complete orthogonal-function system, and what type of orthogonal-function system to create is determined from the solutions of the equations. It is therefore important to find the correct eigenvalues and eigenfunctions. But in data analysis, what sources to consider is a problem for the individual researcher, and it is not a problem of mathematics. Mathematics is necessary only in orthogonalizing and in normalizing. For example, we consider a problem in which the heights of n male youths are expressed as $y_1,\ y_2,\ \ldots,\ y_n$, and these values are influenced by the heights of the fathers, $x_1, x_2,\ \ldots, x_n$ and the heights of the mothers, $z_1,\ z_2,\ \ldots,\ z_n$. The resolution

$$S_T = S_m + S_x + S_z + S_e \tag{18.29}$$

representing the fathers' influences as S_x and the mothers' influences as S_z, does not hold. The researcher would like to have it so, but this is *mathematically impossible.*

To explain the reason for this, since a tall man tends to marry a tall woman and a short man tends to marry a short woman, cause x and cause z are not orthogonal. Mathematically, therefore, a method such as the following becomes necessary.

As a method of orthogonalizing variables that are not orthogonal, we perform so-called *Schmidt's orthogonalization.* In this case, the expansion is as follows. The expression within [ ] is the linear regression between x and z.

$$y=m+b_1(x-\bar{x})+b_2[z-\bar{z}-b_{21}(x-\bar{x})]+e \tag{18.30}$$

The linear expression for the constant term will be called L_m, the linear expression for the terms in x will be L_x, and the linear expression for z will be L_z. e indicates individual difference which cannot be expressed by m, x, and z, and mathematically it is called the residual term.

$$L_m=\sum y_i/n \tag{18.31}$$

$$L_x=\sum(x_i-\bar{x})y_i/\sum(x_i-\bar{x})^2 \tag{18.32}$$

$$L_z=\sum[z_i-\bar{z}-b_{21}(x_i-\bar{x})]y_i/\sum[z_i-\bar{z}-b_{21}(x_i-\bar{x})]^2 \tag{18.33}$$

The three components L_m, L_x, and L_z are orthogonal, but normalization has not been performed. L_m, L_x, and L_z constitute estimated values of m, b_1, and b_2 in Equation (18.30), and these are important numerical values for the researcher. Normalization is no more than a problem of calculation. b_1 indicates by how much the height of the son becomes greater, on the average, when the height of the father becomes taller by 1 cm. That this value is about 0.4 cm at one time evoked the interest of geneticists. Even if the father is 10 cm taller than the mean height of his generation, his son is only taller by a mean of 4 cm relative to the mean height of the son's generation. As is clear from Schmidt's orthogonalization, the effect of the spouse, too, being tall is mixed in this to some degree, but the conclusion holds.

This is actually a proof of the so-called law of general regression in genetics, that "a marked characteristic regresses when it is transmitted to the children and grandchildren." It is the law that proclaims that a reader who is good in mathematics had better give up hope if his son is not as good as he in mathematics. Now, it can be expected that such orthogonalization requires considerable calculating work, but in design of experiments the experiment is designed from the beginning so that the sources are mutually orthogonal. Therefore, Schmidt's orthogonalization is necessary only in limited cases such as regression analysis. Let us assume that for n observed values $y_1, y_2, \ldots, y_n$ there are $k\ (\leq n)$ linear equations,

$$L_l = c_1^{(l)}y_1 + c_2^{(l)}y_2 + \cdots + c_n^{(l)}y_n \qquad (l=1, 2, \cdots, k) \tag{18.34}$$

and that they are orthogonal. If the quadratic forms are

$$S_l = L_l^2/[c_1^{(l)2} + c_2^{(l)2} + \cdots + c_n^{(l)2}] \qquad (l=1, 2, \cdots, k) \tag{18.35}$$

the resolution

$$S_T = S_1 + S_2 + \cdots + S_k + S_e \tag{18.36}$$

holds. In this instance, the number of degrees of freedom of S_e is $(n-k)$, and it can be obtained from

$$S_e = S_T - (S_1 + S_2 + \cdots + S_k) \tag{18.37}$$

If the matrices of $S_1, S_2, \ldots, S_k, S_e$ are expressed as $\boldsymbol{M}_1, \boldsymbol{M}_2, \ldots, \boldsymbol{M}_k$, $\boldsymbol{M}_e$, Equation (18.36) is exactly the same as the following resolution, with the unit matrix as $\boldsymbol{E}$.

$$\boldsymbol{E} = \boldsymbol{M}_1 + \boldsymbol{M}_2 + \cdots + \boldsymbol{M}_k + \boldsymbol{M}_e \tag{18.38}$$

If both sides of Equation (18.38) are multiplied by $\boldsymbol{M}_i$, the relationship

$$\boldsymbol{M}_i = \boldsymbol{M}_i^2 \qquad (i=1, 2, \cdots, k, e) \tag{18.39}$$

holds for the matrices. Multiplication is the same whether on the right or the left. Thus, there is the property that all matrices of quadratic forms which appear in analysis of variance are unchanged if squared.

It is not necessarily the case that the number of degrees of freedom is 1 for quadratic forms in analysis of variance, and this is not only true of the quadratic form for error, S_e. That the number of degrees of freedom is not 1 means the same as that one evaluates by combining several terms together in the orthogonal expansion. For example, if in the Parseval form of Fourier expansion we arrange to use the constant term, the term combining sin x and cos x, the term combining sin $2x$ and cos $2x$, . . . , the numbers of degrees of freedom become 1, 2, 2, ... And in fact, the orthogonal array L_{27} of a three-level system is a method of expression by components of two degrees of freedom which correspond to 13 columns where the number of degrees of freedom is two except for the constant term. For harmonic analysis when a continuum of components exists, Wiener's theory for finding the cumulative spectrum is necessary; frequency components of a certain range are combined and the cumulative power thereof, or in other words the variation, is found. The method of thus combining several components is an everyday affair in analysis of variance; it is generally necessary, no matter what, to express orthogonal expansions by combining groups of components in the Parseval form, rather than expanding the individual terms. This will be explained by a specific example.

Table 18.2 Sales Classed by Day of Week and Branch Store (Unit: 10,000 Yen)

	B_1 (Mon)	B_2 (Tues)	B_3 (Wed)	B_4 (Fri)	B_5 (Sat)	B_6 (Sun)	Total
A_1	92	83	95	74	103	100	547
A_2	174	150	149	131	198	176	978
A_3	123	115	132	96	152	141	759
Total	389	348	376	301	453	417	2 284

Table 18.2 gives the sales classed by day of the week of a certain item at three branch stores, A_1, A_2, and A_3. Thursday is not a day of business. Variables such as A and B in Table 18.2 which do not take continuous values are termed integer-type variables or integer variables. In actual problems, often integer variables and continuous variables (real-number variables) coexist. This is also the reason why expansion of the quadratic form (Parseval's orthogonal expansion) is more important than function expansion itself.

In the case of Table 18.2, the total sum of squares S_T is resolved into the variation of the general mean, S_m, variation for differences of sales depending on branch store, S_A, variation for differences of sales by the day of the week B, S_B, and variation S_e due to other causes which cannot be explained by m, A, and B. Now, if the value 92 for A_1B_1 is denoted by y_1, the value 83 for A_1B_2 by y_2, . . . , the value 141 for A_3B_6 by y_{18}, the respective variations can be found as follows. f is the number of degrees of freedom.

$$S_T=y_1^2+y_2^2+\cdots+y_{18}^2=92^2+83^2+\cdots+141^2=310\,760 \qquad (f=18) \tag{18.40}$$

$$S_m=\frac{(y_1+y_2+\cdots+y_{18})^2}{18}=\frac{2\,284^2}{18}=289\,814.2 \qquad (f=1) \tag{18.41}$$

$$S_A=\frac{(y_1+y_2+\cdots+y_6)^2+\cdots+(y_{13}+y_{14}+\cdots+y_{18})^2}{6}-S_m$$

$$=\frac{547^2+978^2+759^2}{6}-289\,814.2=15\,481.4 \qquad (f=2) \tag{18.42}$$

$$S_B=\frac{(y_1+y_7+y_{13})^2+\cdots+(y_6+y_{12}+y_{18})^2}{3}-S_m$$

$$=\frac{389^2+348^2+\cdots+417^2}{3}-289\,814.2=4685.8 \qquad (f=5) \tag{18.43}$$

$$S_e=S_T-S_m-S_A-S_B=310\,760-289\,814.2-4\,685.8-15\,481.4=778.6 \qquad (f=10) \tag{18.44}$$

Therefore, the analysis of variance table becomes as shown in Table 18.3.

Equations (18.40)-(18.44) are all quadratic forms in $y_1, y_2, \ldots, y_{18}$. However, rank calculations for them and the orthogonality condition calculation are not simple. *In actual practice, it is impossible to find the number of degrees of freedom by a mathematical method or to verify*

Table 18.3 Analysis of Variance Table of Sales

Source	f	S	V	$E(V)$	S'	ρ [%]
m	1	289 814.2	289 814.2	σ^2+18m^2	289 736.34	93.23
A	2	15 481.4	3 096.3	$\sigma^2+6\sigma_A{}^2$	15 325.68	4.94
B	5	4 685.8	2 342.9	$\sigma^2+3\sigma_B{}^2$	4 296.5	1.38
e	10	778.6	77.86	σ^2	1 401.48	0.45
T	18	310 760.0			310 760.00	100.00

the orthogonality conditions. It can easily be proved mathematically that if they are quadratic forms which have been obtained by squaring the linear forms and by dividing by the unit number the rank is 1, and they are mutually orthogonal quadratic forms. But when the number of degrees of freedom is 2 or more, to construct quadratic form matrices is itself considerable work, and rank calculations and verification of orthogonality conditions are even more troublesome. For example, in the case of an electric power company, if one is to analyze data classed by branch, by year, by month, by day of the week, and by hour, assuming intervals of three branches, five years, 12 months, seven days, and 30 minutes, the number of data becomes $3 \times 5 \times 12 \times 7 \times 48 = 60{,}480$, and it is necessary to construct $60{,}480 \times 60{,}480$ matrices and calculate their ranks. By general methods, the calculation cannot be carried out no matter what kind of high-grade computer is used.

When every combination of the respective levels of A and B has been taken with equal frequency, we then judge the orthogonality by assuming that S_A and S_B are orthogonal, or by having the researcher learn such rules. It is therefore necessary to train oneself to be able to judge intuitively the rank of the quadratic form and the orthogonal property through practice. Although this is completely unmathematical, it cannot be helped, and it must be learned through training. Even in mathematics, for example how to draw an adjoint line or perform factorization requires intuition attained through training, so perhaps it is the same here. Now, in many books

$$S_T' = \sum_1^n (y_i - \bar{y})^2 \qquad (f = n-1) \tag{18.45}$$

is used as the total variation; this point will be explained. Depending on the problem, it may be correct to use Equation (18.45).

In the case of the sales figures of Table 18.2, the effect of the general mean, S_m, is not important. Thus, if one is considering what caused the variation in the sales, one need only eliminate S_m from Table 18.3 and revise the total variation S_T as follows:

$$S_T = 310\,760 - S_m \qquad (f = 17) \tag{18.46}$$

Then the total number of degrees of freedom is 17. However, since this becomes messy mathematically, the general mean, too, was incorporated as a source in Table 18.3.

18.2 The Reason Why the Sum of Squares Is Used

Analysis of variance was introduced mainly by R. A. Fisher, but it is generally treated as a *normal regression theory,* as it is termed in mathematical statistics, differing from the approach in the previous section. In spite of this, the normal regression theory is not used in this book; rather, analysis of variance is used in the form of resolution of the quantity of work which various types of sources have performed on the target characteristic. It is believed that this is more closely related to engineering and applied science. In applied science, consideration of the essential nature of the characteristic value itself is important.

A characteristic value has two problem points:

(1) Whether or not it correctly expresses the objective.
(2) Whether or not the effects of the causal systems are additive.

For example, in many cases the economic value is lower when the yield improves from 90% to 99% than when it improves from 80% to 95%. The former has a value which is only 60% of that of the latter. But compared with improving a yield of 80% to 95%, improving a yield of 90% to 99% is more difficult by a factor of about 2.3, based on the omega method. A source or a set of sources whose effect is 2.3 times greater is necessary.

As to how much a certain cause influences the target characteristics, or in other words the result, we use the method of Chapter 16 to estimate the magnitude of the effect by *a deductive theory which ties together the cause and the result.* But to perform such a feat successfully, there must exist a relationship equation connecting the cause and the result. When such a relationship equation does not exist or when only qualitative theories exist, one cannot help resorting to experimentation to investigate quantitatively how much the cause influences the result. Since a theory which ties together the cause and the result is less likely to exist the newer the subject of study, experimental methods and observational methods become important. It is the objective of assignment of the experiment to investigate logically and quantitatively how the cause influences the result which is the target characteristic. However, assignment of the experiment is not the sole content of design of experiments. The method of calculating to what degree the cause influences the result from the data of the result, or in other words the calculation method of analysis of variance, is also an important part of the content. In analysis of variance, we express the magnitude of the influence imparted to the target characteristic by the measure of the square, or in other words the *square of the norm.*

In physics, the *quantity of work* is proportional to the *product of force and time* (for constant velocity),

$$F \times t \propto \text{quantity of work} \tag{18.47}$$

and perhaps this is acceptable in the world of natural phenomena where efficiency is not a great problem. But if, after someone had worked long hours with all his energy, *the result were to be zero,* we would regard the *quantity of work* of that person as being *zero.* In the world of applications, it is clear that *the quantity of work should be measured by the output and must not be measured by the input as in Equation (18.47).*

If the yield has been improved to a considerable degree by the addition of just a bit of an additive, the quantity of work of the additive is great. Physics treats mainly matters centered about quantitative characteristics such as force, energy, mass, momentum, and strength of field, but in engineering it is also necessary to consider the effects of qualitative characteristics. Thus, the quantity of work should be calculated not from force × time but from the standpoint of result, or in other words, output.

The method of calculating the magnitude of the effect of a causal system, or in other words the quantity of work, from the output side was first seen in calculations of the power spectrum in communications engineering.

A set of functions such that an integral of their squares exists in the interval (a, b) is usually written L_2, and it constitutes a *Hilbert space.* A function that belongs to Hilbert space L_2 is expressed as $f(x)$, and we write

$$P=\int_a^b f(x)^2 dx \tag{18.48}$$

P is the *square of the norm* of the element $f(x)$ of the function space L_2. P is decomposed into the square of the norm of the signal, P_S, and the square of the norm of the noise, P_N, and is resolved as follows.

$$P=P_S+P_N \tag{18.49}$$

As is well known, Hilbert space is a direct extension of Euclidean space; or, rather, the n-dimensional *Euclidean space* $R^{(n)}$ can be regarded as a special case of Hilbert space. In the case of Euclidean space $R^{(n)}$, the following equations correspond to Equations (18.48) and (18.49).

$$S_T=y_1^2+y_2^2+\cdots+y_n^2 \tag{18.50}$$

$$S_T=S_m+S_A+\cdots+S_e \tag{18.51}$$

What is the reason why we measure the magnitude of effects by the integral of the squares or the sum of squares? Distance in Euclidean space and the magnitude of a vector in vector space, for example, are defined by the square root of the sum of squares of the components, or in other words the norm (it may also be termed the standard deviation). The result obtained by taking the time-average square root of Equation (18.48) is termed voltage in the field of electrical communications.

It will be understood by anyone acquainted with electrical engineering that the square of the voltage is the energy per unit time, and is a quantity of work. When direct current and alternating current are mixed, or when a number of components of alternating current are mixed, the resulting voltage is given not by the sum of the voltages of the respective components but by the square root of the sum of squares of the voltages of each component. This means that addition does not hold in the case of voltage. However, voltage has the advantage that the unit becomes the same as the unit of the original function. Additivity does not hold, but it is desirable that addition be possible for the magnitudes of the influences of various causes, or in other words the quantity of work, on the target characteristic.

One would wish to say that when cause A performs work S_A toward the output and cause B performs work S_B, the sum of the quantities of work by the two causes is $S_A + S_B$. Although it might be said that something like this is a matter of form, it plays a rather essential role in functional analysis, and it is termed a *Parseval equality*. The general expansion theorem of Weyl-Stone-Titchmarsh-Kodaira for the boundary value problem of a second-order linear differential equation is a wonderful theorem which includes all of Fourier series, Fourier integral, Hermite polynomial, Laguerre polynomial, and Bessel function expansions as special cases, and its expansion format is a Parseval equality.*

In mathematics, when approximating a function $f(x)$ with other functions $f_0(x)$, $f_1(x)$, . . . , one is usually concerned with the maximum value of the absolute value of the difference in interval $[a, b]$.

$$\max|f(x)-[a_0f_0(x)+a_1f_1(x)+\cdots+a_nf_n(x)]| \qquad x\epsilon[a, b] \tag{18.52}$$

Since in the case of mathematics the difference converges to 0 in the limit where n is infinitely large, there is no problem. In the real world, however, it is always necessary to expand in finite terms.

Now, if there exist two sequences of functions, $f_0(x)$, $f_1(x)$, . . . and $g_0(x)$, $g_1(x)$, . . . , then with respect to two approximate expressions for $f(x)$,

$$f(x)\fallingdotseq F_n(x)\equiv a_0f_0(x)+a_1f_1(x)+\cdots+a_nf_n(x) \tag{18.53}$$

$$f(x)\fallingdotseq G_n(x)\equiv b_0g_0(x)+b_1g_1(x)+\cdots+b_ng_n(x) \tag{18.54}$$

which should we say is the better approximate expression? If, for example, we consider the maximum values of error,

$$\varepsilon(f)=\max|f(x)-a_0f_0(x)-a_1f_1(x)-\cdots-a_nf_n(x)| \qquad x\epsilon[a, b]$$

$$\varepsilon(g)=\max|f(x)-b_0g_0(x)-b_1g_1(x)-\cdots-b_ng_n(x)| \qquad x\epsilon[a, b]$$

and

$$\varepsilon(f)<\varepsilon(g) \tag{18.55}$$

*Kōsaku Yoshida: Sekibun Hōteishiki-ron (Theory of Integral Equations), Chapter 5, Iwanami Zensho, 1950.

then is it appropriate to say that $F_n(x)$ of Equation (18.53) is a better approximate expression than $G_n(x)$ of Equation (18.54)? It will be assumed that $G_n(x)$ of Equation (18.54) takes the same value as $f(x)$ at most values of $x \in [a, b]$, and that its error is large at only a single point or for only a part of the x's. If $F_n(x)$ of Equation (18.53) possesses error close to $\epsilon(f)$ even at all of these points x, Equation (18.53) probably could not be regarded as better.

There was the problem at the Japanese National Railways as to where to build a repair plant for engines. This could also be a problem of where to build a cargo collection center or a warehouse for stock.

To use a problem that is nearer at hand, let us assume that heaps of dust as depicted in Figure 18.1 exist at three locations in a narrow and long hallway. The problem is where to collect the dust.

In this case, dust in the quantity 8 exists at point A, dust in the quantity 6 exists at point B, and dust in the quantity 15 exists at point C. The cleaning lady knows well the correct answer: to collect at point C. Persons who halfway understand physics often answer, the position of the center of gravity. As to where in the town one should locate a telephone office, too, it is not at the position of the center of gravity of those subscribing to the telephone service. In the case of dust, incidentally, it is to be understood that the labor of collecting the dust is expressed as Σ(quantity of dust) $\times$ (distance). In other words, it is assumed that loss is proportional to distance.

The center of gravity is the answer when loss is proportional to (quantity of dust) $\times$ (distance)^2. The mean value $\bar{y}$ of n is measured values, y_1, y_2, . . . , y_n, is given by the value of m that minimizes the sum of squares of the differences

$$S_m=(y_1-m)^2+\cdots+(y_n-m)^2 \tag{18.56}$$

The m that minimizes the sum of the absolute values of the differences,

$$S_m'=|y_1-m|+\cdots+|y_n-m| \tag{18.57}$$

gives the median, as in the case of dust discussed before. When y_1, y_2, . . . , y_n are arranged in sequence from the smallest to the greatest value, the median is the $\frac{1}{2}(n + 1)$-th value at the center if n is odd, and it is a value between the $\frac{1}{2}n$-th value and the $(\frac{1}{2}n + 1)$-th value if n is even.

When given n measured values, y_1, y_2, . . . , y_n, there is almost no one who uses the median as the estimated value of their representative

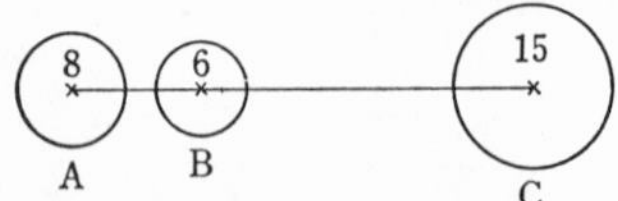

FIGURE 18.1 Quantities of Dust at the Three Points A, B, and C in the Hallway

value or true value. Nearly everyone uses the arithmetic mean. Why is this so? Is it no more than custom? Or is it that the arithmetic mean is used as the best estimated value because the distribution of $y_1, y_2, \ldots, y_n$ is a normal distribution? The arithmetic mean has been used since before various statistical estimation theories became known. Moreover, even persons who are not versed in statistical estimation theories use the arithmetic mean.

Where lies the true reason why the arithmetic mean is used? Is estimation by the method of least squares, which is a generalization of the arithmetic mean, correct estimation? Why isn't the least absolute deviation method used?

The author interprets the reason for this as follows. It is because we wish to find the true value to the extent possible that we estimate or form approximate expressions. *If our estimated value is exactly the same as the true value, we can minimize loss by taking action on it.* Prediction and estimation are necessary in order to take some action. If the proceeds from sales for the coming period are predicted, the most appropriate production plan for this predicted value is then drawn up. If the predicted value differs from actual demand, we will suffer extra loss by this production plan. Put differently, loss becomes minimum when the predicted value matches the true value. Let us assume that when the estimated value is expressed as y and the true value is expressed as m, we have taken the most advantageous action based on y. The loss function (or minus the gain function) is then written as $L(y)$. We expand $L(y)$ about the true value m.

$$L(y) = L(m + y - m)$$

$$= L(m) + \frac{L'(m)}{1!}(y-m) + \frac{L''(m)}{2!}(y-m)^2 + \cdots \qquad (18.58)$$

When predicted value y matches true value m in Equation (18.58), loss becomes minimum. We then have

$$L'(m) = 0$$

Then, by Equation (18.58), the loss that arises from the difference between the estimated value y and the true value m is such that the third term is the initial term and the leading term. If the coefficient of $(y - m)^2$ is written as

$$k = \frac{L''(m)}{2!} \qquad (18.59)$$

then for the loss function L,

$$L \doteqdot k(y-m)^2 \qquad (18.60)$$

holds for the leading term of the loss caused by divergence from the true value.

This shows that *the term that is proportional to the square of the difference between the estimated value y and the realized value m is the leading term of loss as a function of the error of the estimated value y*. In other words, *loss that arises from the error of the estimated value y becomes minimum when the square of the difference between the estimated value y and the realized value is minimized.*

This indicates that the estimation method that minimizes the residual sum of squares is important for practical purposes. There is almost no relationship with what kind of distribution is assumed for error. That there is almost no relationship means that since terms of at least cubic degree exist in the loss function, in certain cases they cannot be disregarded.

This means that when approximating a function $f(x)$ by a linear equation in n functions $f_0(x), f_1(x), \ldots, f_n(x)$, it is a good method to estimate the coefficients, $a_0, a_1, \ldots, a_n$, so as to minimize the sum of squares of the differences,

$$\overline{\varepsilon^2}=\int_a^b |f(x)-a_0f_0(x)-a_1f_1(x)-\cdots-a_nf_n(x)|^2dx \tag{18.61}$$

This is because in this case, too, if the approximate function is expressed as

$$F_n(x)=a_0f_0(x)+a_1f_1(x)+\cdots+a_nf_n(x) \tag{18.62}$$

it can be assumed that the loss becomes minimum when $F_n(x)$ is equal to the true function $f(x)$.

The discussion thus far shows that the method of least squares is not a formal choice of about the same degree of validity as the method of least fourth powers or the least absolute deviation method, but that it is deeply rooted in our daily actions and embodies a substantial, important property. *In terms of calculation, resolution of squares is the simplest method, and yet it also has practical value. That complete harmony of linear mathematics (the quadratic form theory) and analysis (Parseval resolution, orthogonal expansion) is realized in such daily problems can be said to be a marvelous attribute of mathematics.*

Therefore, *if resolution of the sum of squares does not work well, it becomes important to redefine the characteristic value itself.* This is the greatest problem of rationalization of characteristic values, and assignment by orthogonal arrays plays no more than the role of a litmus test for distinguishing the rationality of the characteristic value. Herein lies the reason why the author is stressing that an orthogonal array is *not for solution of the problem but merely plays the role of judging as to whether or not solution is possible.*

The Parseval equality is, in the last analysis, the Pythagorean theorem when a specific orthogonal coordinate system has been taken in a high-dimensional space, and explanation of analysis of variance is possible without passing by way of the normal regression theory. The normal regression theory becomes relevant not to analysis of variance but to the F-test, which questions the significance of the variance ratio. The F table, which has been derived from the normal regression theory, is used when one forms the so-called variance ratio by dividing the variances of other sources by the error variance V_e and judges whether or not the factorial effect is significant. The reader is referred to Chapter 19 and Chapter 26 concerning this problem.

However, the variance ratio is a numerical value possessing meaning even when one is not doing mathematical statistics, which is based on probability theory. For example, let us assume that we have taken the first three terms of the Taylor expansion as an approximation for sin x. The residual term is R_x.

$$\sin x = x - \frac{x^3}{3!} + \frac{x^5}{5!} + R_x \tag{18.63}$$

In mathematics, the maximum error for $0 \leq x \leq 2\pi$ is taken as the error.

$$\max_{0\leq x\leq 2\pi} \left| \sin x - x + \frac{x^3}{3!} - \frac{x^5}{5!} \right| = \max_{0\leq x\leq 2\pi} |R_x| \tag{18.64}$$

In analysis of variance, one divides $0 \leq x \leq 2\pi$ into n equal parts and discusses the sum of squares of the errors between the approximate values and the true values at these division points:

$$S_e = \sum_{l=0}^{n} R_{2\pi l/n}{}^2 = \sum_{l=0}^{n} \left| \sin \frac{2\pi l}{n} - \frac{2\pi l}{n} + \frac{1}{3!}\left(\frac{2\pi l}{n}\right)^3 + \frac{1}{5!}\left(\frac{2\pi l}{n}\right)^5 \right|^2 \tag{18.65}$$

One then seeks the value of the error variance, which is obtained by dividing the sum of squares of error of the equation above, S_e, by the number of degrees of freedom, $n - 2$.

$$V_e = \frac{S_e}{n-2} \tag{18.66}$$

Next, the third term is omitted and one finds the sum of squares of residuals, S_e', at the same $n + 1$ points.

$$S_e' = \sum_{l=0}^{n} \left| \sin \frac{2\pi l}{n} - \frac{2\pi l}{n} + \frac{1}{3!}\left(\frac{2\pi l}{n}\right)^3 \right|^2 \qquad (f = n-1) \tag{18.67}$$

The extent to which S_e' is greater than S_e is a measure of the usefulness of the approximate value of the third term on the right side of Equation (18.63).

$$S_3 = S_e' - S_e \tag{18.68}$$

If S_3 is at most comparable to V_e, it is useless to consider such a term. This is because it does not increase the accuracy of approximation more than the mean of the omitted higher-order terms. Thus, if S_3/V_e is no more than a certain value (usually 2 or less), it is better to omit such a term.

This example suggests that analysis of variance is a method that can be used in all numerical value calculations, not only in mathematical statistics. It is useful especially in the numerical analysis of several variables. The reader is referred to Chapter 16. There, a complicated rational function with 13 variables was approximated by a polynomial with nine variables, and it was proved that there was energy of expression of 99.1%. If, in the function

$$y=f(A, B, \cdots) \tag{18.69}$$

$A, B, \ldots$ are continuous variables and they are changed by an infinitesimal amount, then for the increment of y, dy, it is possible to omit higher-order terms such as $dAdB$ in the equation of expansion,

$$dy=\frac{\partial f}{\partial A}dA+\frac{\partial f}{\partial B}dB+\cdots+\frac{\partial^2 f}{\partial A\partial B}dAdB+\cdots \tag{18.70}$$

However, if there are integer variables among $A, B, \ldots$ or if the change is not small, one needs an evaluation of the magnitude of error when the total difference of y, Δy,

$$\Delta y=\frac{\Delta f}{\Delta A}\Delta' A+\frac{\Delta f}{\Delta B}\Delta' B+\cdots+\frac{\Delta^2 f}{\Delta A\Delta B}\Delta' A\Delta' B+\cdots \tag{18.71}$$

is expressed only by the first-order difference terms $\Delta' A, \Delta' B, \ldots$ at the beginning. It becomes important to evaluate the magnitude of error when the higher-order terms are omitted. It can therefore be claimed that *the method of finding the power of expression of the equation of expansion by the contribution ratio, by finding the magnitude of the error of the approximate expression by the mean of the square of the differences between the accurate value and the value of the approximation equation, is analysis of variance.*

18.3 Rules of Calculation for Resolution of Variation

It will be assumed with regard to factor A at a levels that each of A_1, $A_2, \ldots, A_a$ is a sum of r data. Variation S_A is given by

$$S_A=\frac{A_1^2+A_2^2+\cdots+A_a^2}{r}-\frac{(A_1+A_2+\cdots+A_a)^2}{ar} \qquad (f=a-1) \tag{18.72}$$

If often happens that one wishes to resolve the main effect of A into more finely differentiated components. For example, if we wish to investigate a certain characteristic for a total of five types of electric wire, two types of rubber-coated wire A_1 and A_2, and three types of vinyl-coated wire A_3, A_4, and A_5, we wish in effect to resolve the differences among the A_i as follows.

$$S_A = S(\text{difference between rubber and vinyl}) + S(\text{difference between rubber } A_1 \text{ and } A_2) + S(\text{difference among vinyl } A_3, A_4, \text{ and } A_5) \tag{18.73}$$

To state it differently, we separate unknowns of four degrees of freedom for the main effect of A into one unknown for the difference between rubber and vinyl, one unknown for the difference between the two types of rubber-coated wires, and unknowns of two degrees of freedom for the differences among vinyl-coated wires. As long as the form of the observation equation is not known, analysis of variance must be a resolution with respect to those unknowns which are believed to be the most logical.

If the foregoing three variations are expressed as S_1, S_2, and S_3, we have

$$\left.\begin{aligned} S_1 &= \frac{1}{30r}[3(A_1+A_2)-2(A_3+A_4+A_5)]^2 \\ S_2 &= \frac{1}{2r}(A_1-A_2)^2 \\ S_3 &= \frac{1}{r}(A_3^2+A_4^2+A_5^2)-\frac{(A_3+A_4+A_5)^2}{3r} \end{aligned}\right\} \tag{18.74}$$

and how to calculate these will be explained below.

[RULE 1]

It will be assumed that the total for each of the a levels of factor A is the sum of r data. If, in the *linear equation (first-order equation with constant coefficients)* of $A_1, A_2, \ldots, A_a$,

$$L = c_1A_1 + c_2A_2 + \cdots + c_aA_a$$

the sum of the coefficients is 0,

$$c_1 + c_2 + \cdots + c_a = 0$$

L is termed a *contrast*. If L is a contrast, the quotient obtained by dividing its square by the unit number $D = (c_1^2 + c_2^2 + \cdots + c_a^2)$,

$$S_L = \frac{L^2}{(c_1^2+c_2^2+\cdots+c_a^2)r} = \frac{(c_1A_1+c_2A_2+\cdots+c_aA_a)^2}{(c_1^2+c_2^2+\cdots+c_a^2)r} \tag{18.75}$$

constitutes *a variation of one degree of freedom* and it becomes *one component* of S_A. In this case, $S_{\text{res}} = S_A - S_L$ is a variation of number of degrees of freedom $a - 2$, and it becomes *independent* of S_L. The confidence limits of L and the expected value of variation are as follows.

$$L \pm \sqrt{F \times V_e (c_1^2+c_2^2+\cdots+c_a^2) \times r} \tag{18.76}$$

$$E(S_L) = \sigma^2 + n_e\theta^2$$

θ here is the true value which linear equation L is estimating; n_e is the effective number of replications, and it is given by the following equation.

$$n_e = \frac{1}{(c_1^2 + c_2^2 + \cdots + c_a^2)r} \tag{18.77}$$

Anyone would compare the difference between A_1 and A_2, on the one hand, and A_3, A_4, and A_5, on the other, which is the difference between rubber and vinyl, by

$$L = \frac{A_1 + A_2}{2r} - \frac{A_3 + A_4 + A_5}{3r}$$

Clearly, the equation above is a contrast of

$$c_1 = c_2 = \frac{1}{2r}$$

$$c_3 = c_4 = c_5 = -\frac{1}{3r}$$

Therefore,

$$S_L = \frac{\left(\frac{A_1 + A_2}{2r} - \frac{A_3 + A_4 + A_5}{3r}\right)^2}{\left(\frac{1}{2r}\right)^2 \times 2r + \left(\frac{-1}{3r}\right)^2 \times 3r} = \frac{[3(A_1 + A_2) - 2(A_3 + A_4 + A_5)]^2}{30r} \tag{18.78}$$

is a variation with one degree of freedom which expresses the difference between rubber and vinyl.

And if the true value of the difference between rubber and vinyl is expressed by θ,

$$E(L) = \theta$$

holds; the expected value of variation becomes, using the effective number of replications $n_e = 6r/5$,

$$E(S_L) = \sigma^2 + \frac{6r}{5}\theta^2 \tag{18.79}$$

The confidence limits of L are

$$L \pm \sqrt{F \times V_e \times \frac{5}{6r}} \tag{18.80}$$

since the effective number of replications n_e is

$$\frac{1}{n_e} = \left(\frac{1}{2r}\right)^2 \times 2r + \left(\frac{-1}{3r}\right)^2 \times 3r$$

$$= \frac{1}{2r} + \frac{1}{3r} = \frac{5}{6r} \tag{18.81}$$

The value of variation S_L does not change even if, instead of using L, we multiply its value by constants. In the case of rubber and vinyl, instead of L, one may use

$$L' = 6rL = 3(A_1 + A_2) - 2(A_3 + A_4 + A_5) \tag{18.82}$$

obtained by multiplying it by $6r$, and calculate the variation by

$$S_{L'} = \frac{(L')^2}{3^2 \times 2r + (-2)^2 \times 3r} = \frac{[3(A_1 + A_2) - 2(A_3 + A_4 + A_5)]^2}{30r} \tag{18.83}$$

Calculation is easier by this method. However, for the confidence limits, it is necessary to obtain n_e from the original L.

[RULE 2]

When there are two contrasts,

$$L_1 = c_1 A_1 + c_2 A_2 + \cdots + c_a A_a$$

$$L_2 = c_1' A_1 + c_2' A_2 + \cdots + c_a' A_a$$

and when the *inner product of the coefficients is zero,* or in other words

$$c_1 c_1' + c_2 c_2' + \cdots + c_a c_a' = 0 \tag{18.84}$$

then

$$S_{L_1} = \frac{L_1^2}{r(c_1^2 + c_2^2 + \cdots + c_a^2)} \tag{18.85}$$

$$S_{L_2} = \frac{L_2^2}{r[(c_1')^2 + (c_2')^2 + \cdots + (c_a')^2]} \tag{18.86}$$

are mutually *independent variations of one degree of freedom.* For example, if S_1 and S_2 in Equation (18.74) are multiplied by $6r$ and r, then since the coefficients for A_1, A_2, A_3, A_4, and A_5 are

	A_1	A_2	A_3	A_4	A_5
S_1	3	3	−2	−2	−2
S_2	1	−1	0	0	0

the sum of products of the coefficients is zero and the contrasts are orthogonal. The same can be said concerning this relationship no matter how many contrasts there are. For example, if $L_1, L_2, \ldots, L_k$ are mutually independent contrasts, no matter which two are taken, S_A can be resolved as follows and the number of degrees of freedom of S_{res} becomes $a - 1 - k$.

$$S_A = S_{L_1} + S_{L_2} + \cdots + S_{L_k} + S_{\text{res}}$$

Therefore, if there are $a - 1$ mutually independent contrasts, S_A can be expressed completely by them.

[RULE 3]

When the number of replications r is different for each of A_1, $A_2, \ldots, A_a$, and when, for

$$L=c_1A_1+c_2A_2+\cdots+c_aA_a$$

we have

$$r_1c_1+r_2c_2+\cdots+r_ac_a=0$$

where the numbers of replications of $A_1, A_2, \ldots, A_a$ are $r_1, r_2, \ldots, r_a$, L is a contrast and the variation for L, S_L, is given by

$$S_L=\frac{(c_1A_1+c_2A_2+\cdots+c_aA_a)^2}{r_1c_1^2+r_2c_2^2+\cdots+r_ac_a^2} \tag{18.87}$$

For example, replications differ between A_1 and A_2 when, in the conducting of an experiment of a three-level system, A_3 is the dummy level of A_1 since A is at two levels. If the replication of A_2 is r, that of A_1 becomes $2r$. Therefore, in regard to S_A with one degree of freedom, if

$$L=A_1-2A_2 \quad (c_1=1, c_2=-2)$$

then, since

$$r_1c_1+r_2c_2=2r\times1+r\times(-2)=0$$

we have

$$S_A=S_L=\frac{(A_1-2A_2)^2}{2r\times1^2+r\times(-2)^2}=\frac{(A_1-2A_2)^2}{6r} \tag{18.88}$$

[RULE 4]

Let us assume that factor A at a levels and factor B at b levels have been assigned so as to enable one to find the *interaction* $A\times B$. If the number of data for A_iB_j is r in each case, then

$$S_{A\times B}=\frac{A_1B_1^2+A_1B_2^2+\cdots+A_aB_b^2}{r}-CF-S_A-S_B \tag{18.89}$$

with A_iB_j as the total of the r data for A_iB_j, is the *variation of the interaction* of A and B, and its number of degrees of freedom is $(a-1)(b-1)$. Arbitrary contrasts of the main effects of A and B are expressed as L_A' and L_B'.

$$L_{A'}=c_1A_1+c_2A_2+\cdots+c_aA_a \tag{18.90}$$

$$L_{B'}=c_1'B_1+c_2'B_2+\cdots+c_b'B_b \tag{18.91}$$

Then $L_{A'\times B'}$, which is one component of interaction $A\times B$, is

$$L_{A'\times B'}==[c_1'L_{A'}(B_1)+c_2'L_{A'}(B_2)+\cdots+c_b'L_{A'}(B_b)]ab$$

Here, from Equation (18.90),

$$L_{A'}(B_i)=c_1A_1B_i+c_2A_2B_i+\cdots+c_aA_aB_i \qquad (i=1, 2, \cdots, b)$$

The variation

$$S_{A'\times B'}=\frac{(L_{A'\times B'}/ab)^2}{r(c_1^2+c_2^2+\cdots+c_a^2)[(c_1')^2+(c_2')^2+\cdots+(c_b')^2]} \tag{18.92}$$

is a variation of one degree of freedom and it constitutes one component of interaction $A \times B$. Moreover, if A' and A'' are two contrasts of A that are orthogonal and B' and B'' are two contrasts of B that are orthogonal, $A' \times B'$, $A'' \times B'$, $A' \times B''$, and $A'' \times B''$ are all orthogonal contrasts of interaction.

If B_1, B_2, B_3, and B_4 are four equispaced levels of temperature and if A_1, A_2, A_3, and A_4 are four types of products,

$$L(A_i) = -3A_iB_1 - A_iB_2 + A_iB_3 + 3A_iB_4 \qquad (i=1,\ 2,\ 3,\ 4)$$

is the contrast of the temperature coefficients for each product, so that if the values for A_iB_j have been measured for r test pieces, we use the following variation to test whether or not there are differences of temperature coefficient among these four types of products. Since

$$c_1^2 + c_2^2 + c_3^2 + c_4^2 = (-3)^2 + (-1)^2 + 1^2 + 3^2 = 20$$

we have

$$S_{A\times L} = \frac{L^2(A_1) + L^2(A_2) + \cdots + L^2(A_4)}{20r} - \frac{[L(A_1) + \cdots + L(A_4)]^2}{80r} \tag{18.93}$$

Incidentally, for example when A_1 is the product of one's own company and A_2, A_3, and A_4 are those of other companies, comparisons of the means of A_1 and of A_2, A_3, and A_4 might pose a problem. Then, with A' as a source of comparison between A_1 and $\dfrac{A_2 + A_3 + A_4}{3}$, we have

$$S_{A'\times L} = \frac{[3L(A_1) - L(A_2) - L(A_3) - L(A_4)]^2}{[3^2 + (-1)^2 + (-1)^2 + (-1)^2][(-3)^2 + (-1)^2 + 1^2 + 3^2)]r}$$

$$= \frac{[3L(A_1) - L(A_2) - L(A_3) - L(A_4)]^2}{240r} \tag{18.94}$$

[RULE 5]

If the numbers of replications of $A_1, A_2, \ldots, A_a$ are $r_1, r_2, \ldots, r_a$,

$$S_A = \frac{A_1^2}{r_1} + \frac{A_2^2}{r_2} + \cdots + \frac{A_a^2}{r_a} - \frac{(A_1 + A_2 + \cdots + A_a)^2}{r_1 + r_2 + \cdots + r_a} \qquad (f = a-1) \tag{18.95}$$

For example, when the number of replications of A_1 and A_2 is r and that of A_3 is $2r$,

$$S_A = \frac{A_1^2 + A_2^2}{r} + \frac{A_3^2}{2r} - \frac{(A_1 + A_2 + A_3)^2}{4r} \qquad (f=2) \tag{18.96}$$

Of course, cases where $r_1, r_2, \ldots, r_a$ differ for the various levels of A, as above, are rare, and usually it is because dummies have been introduced for several levels. Actually, it might happen, for example, that for a two-way array of A and B, it was not possible to measure the same number of test pieces for the different A_iB_j. There could be a case such as the following. The numerals are the number of measured values.

	B_1	B_2	B_3	B_4
A_1	3	4	2	4
A_2	5	4	5	4
A_3	5	4	4	5

Please refer to the discussion of missing values in Section 30.2 regarding cases such as this. It is useless to apply the rules above since the levels of B are not combined balanced for the levels of A. Let us try to find the main effects and interactions when, for an experiment with a four-level system, A_1 and B_1 are dummy levels. The number of replications of a combination in which there are no dummies is represented by r.

	B_1	B_2	B_3	Total
A_1	$4r$	$2r$	$2r$	$8r$
A_2	$2r$	r	r	$4r$
A_3	$2r$	r	r	$4r$
Total	$8r$	$4r$	$4r$	$16r$

$$S_A=\frac{A_1^2}{8r}+\frac{A_2^2+A_3^2}{4r}-\frac{T^2}{16r} \qquad (f=2) \tag{18.97}$$

$$S_B=\frac{B_1^2}{8r}+\frac{B_2^2+B_3^2}{4r}-\frac{T^2}{16r} \qquad (f=2) \tag{18.98}$$

$$S_{A\times B}=\frac{A_1B_1^2}{4r}+\frac{A_1B_2^2+A_1B_3^2+A_2B_1^2+A_3B_1^2}{2r}$$
$$+\frac{A_2B_2^2+A_2B_3^2+A_3B_2^2+A_3B_3^2}{r}-\frac{T^2}{16r}-S_A-S_B$$
$$(f=4) \tag{18.99}$$

For judgment of the orthogonality of variations where the number of degrees of freedom is 2 or more, it is necessary to see whether the product of the matrices of the quadratic forms is zero, as was indicated at the beginning of this chapter, but this judgment method is useless for practical purposes. Rather than judging when it is orthogonal, it is more important to assign so that the sources are mutually orthogonal; and thus, it becomes unnecessary to judge the orthogonality each time.

Exercises (18)

(1) Prove that the estimated values of m and β,

$$L_m=\frac{1}{n}(y_1+\cdots+y_n)$$

$$L_\beta=\frac{\sum(x_i-\bar{x})y_i}{\sum(x_i-\bar{x})^2}$$

which have been obtained by the method of least squares from n pairs of data (x_1, y_1), (x_2, y_2), . . . , (x_n, y_n) for the regression equation $y = m + \beta(x - \bar{x})$, are mutually orthogonal linear expressions. Also find their variations and expected values.

(2) With respect to four types of products, A_1 = foreign product, A_2 = product of one's own company, A_3 = product of company α within the country, and A_4 = product of company β within the country, we have taken 2, 5, 4, and 4 products, respectively, and have performed a continuous degradation test lasting 300 hours. The data represent the following percentage of change:

$$y = \frac{\text{value after testing} - \text{initial value}}{\text{initial value}} \times 100\%$$

The following data were obtained by subtracting the working mean 20 from the foregoing y.

A_1	−8	−6			
A_2	0	−2	−8	−3	−5
A_3	6	−1	0	8	
A_4	4	5	−2	2	

(a) Perform an analysis of variance by resolving A into three components: A' = foreign product and Japanese product, A'' = product of one's own company and products of other companies, and A''' = product of company α and product of company β.
(b) Perform an estimation concerning significant sources and show the conclusions.

19 Testing and Estimation

In this chapter, we treat the β coefficient method, and the F test which is an approximation thereto, as measures of the error possessed by estimated values. The reader is also referred to Chapter 33 and Chapter 40 concerning this problem.

19.1 Introduction

As shown by several examples in Chapter 26, it is not unusual for problems of design of experiments and data analysis to be presented that concern the effects of medicines and treatments. When this happens, it is astonishing that there are so many researchers who are overly concerned with testing in evaluating the effects of medicines and of treatment. The author believes that *testing is a preparatory stage for the performing of better estimation*. He believes that testing itself has no value, and that the value of testing is in doing better estimation; and he has studied what is the optimum level of significance for this, as shown in Section 19.8. However, that study is based upon having estimated the form of distribution. Since he has separately devised a method which can be used in general, with no relationship to testing, the fundamental principles of this will be explained.

Not only can it be expected that this method will replace all testing methods in future, but the author also predicts that it will be applied widely in such fields as the design of control systems, as *the only logical method of preventing the hunting phenomenon which has its origin in error.*

19.2 Beta Coefficient*

We will assume that with respect to A_1 and A_2 the means of the experimental values for r replications, $\bar{A}_1$ and $\bar{A}_2$, have been determined. For example, it may be that factor A at two levels has been assigned to orthogonal array L_{16} and the means of A_1 and A_2 have been obtained. The true values of A_1 and A_2 will be called m_1 and m_2. $\bar{A}_1$, the mean value of A_1, is a statistic that distributes about m_1, and its mean value will be m_1; and $\bar{A}_2$ is a statistic that distributes about m_2 and its mean value will be m_2. Thus, $\bar{A}_1$ and $\bar{A}_2$ are unbiased estimates of m_1 and m_2, as they are called in mathematical statistics.

However, $\bar{A}_1 - \bar{A}_2$ is too large an estimate for $m_1 - m_2$. This fact has already been pointed out in reference 35), and at that time there were queries from W. Edwards Deming and many other specialists. The author replied as follows. As a simple example, suppose that two teams, A_1 and A_2, have played n games, and that A_1 has won r times. If the probability that team A_1 wins is expressed as p_1, then for the expected value of r/n,

$$E\left(\frac{r}{n}\right)=p_1 \tag{19.1}$$

holds. But it is before the games are played that Equation (19.1) holds, and we are not speaking of the situation after the games have been played. For example, if one game has been played and team A_1 has won,

$$\frac{r}{n}=\frac{1}{1}=1=100\,\% \tag{19.2}$$

This value is clearly too large an estimate of the probability p_1 that A_1 will win. Before the game is played, since the probability that team A_1 will win and will acquire 1 point is p_1 and the probability that it will lose and acquire 0 points is $(1-p_1)$,

$$E\left(\frac{r}{1}\right)=1\times p_1+0\times(1-p_1)=p_1 \tag{19.3}$$

holds. It is therefore better not to use the term *unbiased estimate* in estimation which is performed after data have been obtained.

* The Japanese term, "waribiki keisū," means discount coefficient.

There is much doubt regarding Bayes' theorem in statistics, but when that method is used the estimate of p_1 when team A_1 has won one game becomes, as is well known,

$$\hat{p}_1=\frac{r+1}{n+2}=\frac{2}{3} \tag{19.4}$$

This is an estimate that fits common sense better. However, Bayes' theorem has not been used very often for the alleged reason that it requires a knowledge of the *a priori* distribution and so its practical utility is small.

In Chapter 11 of reference 36), the author considered an optimum correction method for continuous value control, and he suggested a method such as the following. Suppose that in the control of a certain characteristic value, say the sheet thickness of a certain metal, it was found to be 10 microns thicker than the target value y_0 when actually measured at a certain time t. A person with no experience in control will probably correct the draft of the rolling roller so that the sheet will become exactly 10 microns thinner. But a person with long experience in rolling operations will usually correct by a lesser value than 10 microns, such as about 7 or 8 microns. We will clarify the reason why the latter method is more correct.

The following hypothesis will be used. In general, the actual measured value y of the sheet thickness includes an error, and the magnitude of the error variance will be called σ_e^2. It will be assumed that the actual measured value y of the sheet thickness is distributed with variance σ_e^2 about the true value at that time, m. It will also be assumed that in general the true value m of the sheet thickness varies with time in the same manner on the plus side and the minus side, and that the magnitude of this change is σ_l^2 during l minutes. (It is assumed that time of l minutes is taken in the measurement of the sheet thickness and correction of the draft.) It is assumed that if the true value of the sheet thickness at time t is m_t and the true value of the sheet thickness after l minutes is m_{t+l}, we have

$$\lim_{T\to\infty}\frac{1}{2T}\int_{-T}^{T}(m_{t+l}-m_t)^2dt=\sigma_l^2 \tag{19.5}$$

In other words, σ_l^2 is the value obtained by averaging over time the square of the difference between the true value of the sheet thickness l minutes later and the true value of the sheet thickness at present. When the actual measured value of the sheet thickness is y, then for measurement error we have, statistically,

$$E(y)=m_t \tag{19.6}$$

$$\mathrm{Var}(y)=\sigma_e^2 \tag{19.7}$$

from the earlier assumptions regarding measurement error. Therefore, the actual measured value y of sheet thickness is an unbiased estimate of the true value of the sheet thickness l minutes later, m_{t+l}; and the mean value of the squares of the error becomes as follows.

$$E(y-m_{t+l})^2=E(y-m_t+m_t-m_{t+l})^2$$
$$=E[(y-m_t)^2+2(y-m_t)(m_t-m_{t+l})+(m_t-m_{t+l})^2]$$
$$=\sigma_e^2+0+\sigma_l^2 \quad (19.8)$$
$$=\sigma^2 \qquad (\text{putting } \sigma^2=\sigma_e^2+\sigma_l^2) \quad (19.9)$$

To explain why the middle term is 0 in Equation (19.8), since the measurement error of sheet thickness $(y - m_t)$ and the change in the true value of the sheet thickness in l minutes $(m_t - m_{t+l})$ are clearly independent, the expected value of their product is 0.

If, when the target value of sheet thickness is expressed as y_0, one corrects just by the difference between the actual measured value y and the target value y_0 (it will be assumed that there is no error in the correction operation; refer to Chapter 11 of reference 36) for cases where there is error in the correction operation), the mean of the squares of the difference between the true value after correction and target value y_0, σ_{out}^2, is as follows.

$$\sigma_{out}^2=E[m_{t+l}-(y-y_0)-y_0]^2=\sigma^2 \quad (19.10)$$

As is well known, according to control theory, as long as the difference between the measured value and the target value is correct as is, it never happens that the mean of the square of the difference between the true value and target value of sheet thickness which has been produced, σ_{out}^2, becomes smaller than the expected error variance σ^2, when the expected error variance of m_{t+l} is σ^2 and is not zero, even if the measured value y is an unbiased estimate of the true value after time-lag l minutes, m_{t+l}.

In fact, a better control method exists. Instead of simply correcting the difference between the measured value and the target value, one corrects by discounting this difference. The author has given this method the name *beta coefficient method*. Let us obtain the *optimum beta coefficient, β*. The mean of the squares of the differences between the true value and the target value of the sheet thickness after correction when the beta coefficient β has been used can be obtained as follows.

$$E[m_{t+l}-\beta(y-y_0)-y_0]^2=[m_{t+l}-\beta(m_{t+l}-y_0)-y_0]^2+(-\beta)^2\sigma^2$$
$$=(1-\beta)^2(m_{t+l}-y_0)^2+\beta^2\sigma^2 \quad (19.11)$$

We find the β which minimizes Equation (19.11). If we differentiate with respect to β and set it equal to 0, we get

$$\beta=\frac{(m_{t+l}-y_0)^2}{\sigma^2+(m_{t+l}-y_0)^2}=1-\frac{1}{[\sigma^2+(m_{t+l}-y_0)^2]/\sigma^2} \quad (19.12)$$

If we put

$$F = \frac{\sigma^2 + (m_{t+l} - y_0)^2}{\sigma^2} \tag{19.13}$$

we have

$$\beta = 1 - \frac{1}{F} \tag{19.14}$$

Now, in estimating $\sigma^2 + (m_{t+l} - y_0)^2$, we use

$$E(y - y_0)^2 = \sigma^2 + (m_{t+l} - y_0)^2 \tag{19.15}$$

If $\sigma^2 + (m_{t+l} - y_0)^2$ is replaced by $(y - y_0)^2$, then by putting

$$\hat{F} = F_0 = \frac{(y - y_0)^2}{\sigma^2} \tag{19.16}$$

the beta coefficient β is given by the following equation.

$$\beta = 1 - \frac{1}{F_0} \tag{19.17}$$

By substituting Equation (19.12) into Equation (19.11), we obtain

$$\begin{aligned}\sigma_{\text{out}}^2 &= E[m_{t+l} - \beta(y - y_0) - y_0]^2 \\ &= \left[\frac{\sigma^2}{\sigma^2 + (m_{t+l} - y_0)^2}\right]^2 (m_{t+l} - y_0)^2 + \left[1 - \frac{\sigma^2}{\sigma^2 + (m_{t+l} - y_0)^2}\right]^2 \sigma^2 \\ &= \frac{\sigma^2 (m_{t+l} - y_0)^2}{\sigma^2 + (m_{t+l} - y_0)^2} = \left[1 - \frac{\sigma^2}{\sigma^2 + (m_{t+l} - y_0)^2}\right] \sigma^2 \\ &\fallingdotseq \beta\sigma^2\end{aligned} \tag{19.18}$$

Therefore, when the actual measured value of the sheet thickness y has been found to be different from the target value y_0, the optimum correction quantity is given by

$$-\beta(y - y_0) \tag{19.19}$$

Here,

$$\beta = \begin{cases} 0 & \text{when } F_0 = \dfrac{(y - y_0)^2}{\sigma^2} \leq 1 \\ 1 - \dfrac{1}{F_0} & \text{when } F_0 = \dfrac{(y - y_0)^2}{\sigma^2} > 1 \end{cases} \tag{19.20}$$

Then the mean value of the squares of the differences between the true value and target value y_0 of the thickness after correction, σ_{out}^2, is given as follows, with $\bar{\beta}$ being the mean of β.

$$\sigma_{\text{out}}^2 = \bar{\beta}\sigma^2 \tag{19.21}$$

Since β is always at most 1, its mean value $\bar{\beta}$ is less than 1, which means that the efficiency is better with the use of coefficient β than by correcting with the actual difference.

The author believes that how to render prediction error small is a problem of specialized technique, but that statistics is truly useful in that it teaches us what to do when the error does not become smaller than a certain value σ^2 no matter how one uses that technique. Although there are methods, such as the control chart method, of correcting only by the genuine difference when there is displacement by $3\sigma/\sqrt{n}$, it is a superior method to use the beta coefficient of Equation (19.20). This is because it is clearly strange from a common-sense standpoint that *in statistical testing the difference is trusted 100% when it is greater than a certain critical value but it is totally disregarded when it is less than the critical value.*

19.3 Beta Coefficient for Experimental Data and Its Applications

What has been shown in the previous section holds true in exactly the same way in the case of analysis of experimental data. With the true values for A_1 and A_2 as m_1 and m_2, let us assume that we are to estimate $m_1 - m_2$ by the difference of the means of the experimental data for A_1 and A_2, $\bar{A}_1 - \bar{A}_2$. This time, we consider

$$L=\beta(\bar{A}_1-\bar{A}_2)-(m_1-m_2) \tag{19.22}$$

as the error of the estimated values. We find the mean square of the error of the equation above.

$$E(L^2)=E[\beta(\bar{A}_1-\bar{A}_2)-(m_1-m_2)]^2=(1-\beta)^2(m_1-m_2)^2+\beta^2\frac{2}{n}\sigma^2 \tag{19.23}$$

Here, n is the effective number of replications of $\bar{A}_1$ and $\bar{A}_2$ and σ^2 is the error variance of the experiment. If we differentiate Equation (19.23) with respect to β and put it equal to 0, we find,

$$\beta=\frac{\frac{n}{2}(m_1-m_2)^2}{\sigma^2+\frac{n}{2}(m_1-m_2)^2} \tag{19.24}$$

Since both σ^2 and $(m_1 - m_2)^2$ are unknown, we estimate their values from the data. σ^2 is replaced by V_e, the error variance in the analysis of variance table. If the total values for A_1 and A_2 are denoted by A_1 and A_2, we have

$$S_A=\frac{(A_1-A_2)^2}{2n} \tag{19.25}$$

and the expected value of V_A (S_A divided by 1 degree of freedom) is

$$E(V_A)=\sigma^2+\frac{n}{2}(m_1-m_2)^2 \tag{19.26}$$

Therefore, instead of σ^2 and $\sigma^2 + \frac{n}{2}(m_1 - m_2)^2$, we use V_e and V_A as their estimates.

$$\widehat{\sigma^2} = V_e \tag{19.27}$$

$$\sigma^2 + \frac{n}{2}(m_1 - m_2)^2 \doteqdot V_A \tag{19.28}$$

Equations (19.27) and (19.28) are substituted into Equation (19.24).

$$\beta = 1 - \frac{\sigma^2}{\sigma^2 + \frac{n}{2}(m_1 - m_2)^2}$$

$$\doteqdot 1 - \frac{1}{F_0} \tag{19.29}$$

Here, F_0 is the variance ratio, given by the following equation.

$$F_0 = \frac{V_A}{V_e} \tag{19.30}$$

The value of β is 1 when the variance ratio is ∞ and it approaches 0 as the variance ratio approaches 1. When the variance ratio is less than 1, we regard β as 0 since, of course, the difference is smaller than when the variance ratio is 1. It has thus been demonstrated that the value which minimizes the mean square error in estimating the difference between the true values of A_1 and A_2, m_1 and m_2, is given by the following equation.

Estimated Value of

$$\boldsymbol{m_1 - m_2 = \beta(\bar{A}_1 - \bar{A}_2)} \tag{19.31}$$

where

$$\boldsymbol{\beta} = \begin{cases} \boldsymbol{0} & \text{when } \boldsymbol{F_0 = \frac{V_A}{V_e} \leq 1} \\ \boldsymbol{1 - \frac{1}{F_0}} & \text{when } \boldsymbol{F_0 = \frac{V_A}{V_e} > 1} \end{cases} \tag{19.32}$$

The value of the mean square error when $m_1 - m_2$ has been estimated by using the beta coefficient β will be expressed as σ_{est}^2. By substituting Equation (19.24) into Equation (19.23), σ_{est}^2 is obtained as

$$\sigma_{est}^2 = \left[1 - \frac{\sigma^2}{\sigma^2 + \frac{n}{2}(m_1 - m_2)^2}\right] \times \frac{2}{n}\sigma^2 \tag{19.33}$$

This becomes simple, as follows, by substituting from Equation (19.29).

$$\sigma_{est}^2 = \beta \times \frac{2}{n}\sigma^2 \tag{19.34}$$

Actually, β takes various values, but if its mean is represented as $\bar{\beta}$, Equation (19.34) becomes

$$\sigma_{\mathrm{est}}{}^2 = \bar{\beta} \times \frac{2}{n}\sigma^2 \tag{19.35}$$

Figure 19.1 shows the beta coefficient β for various F_0's. What is clear from Figure 19.1 is that it is advisable to consider that there is no effect of factor A when the variance ratio is 1 or less and to regard one-half the difference which has been obtained by experimentation as the estimated value of the difference when the variance ratio is 2. One might say that this is taking the story at half its face value. Even if there is a high degree of significance, with the variance ratio F_0 as much as 10, it is advisable to take 0.9 times the observed difference as the estimate of the difference. Persons with a practical background should have had the experience that when results obtained by experiment are actually applied, it is rare that an effect greater than the experimental results is otained, and that in most cases less than the expected effect is obtained. This indicates that even with experiments that have been conscientiously carried out, often the effect comes out overly great.

The method in which the variance ratio F_0 is found and is tested for significance by comparing it with the value in the F table can be described as the method in which it is judged that there is an effect when the variance ratio F_0 is greater than the F value of significance level 5% in the F table, and the difference is taken at its full value; while the difference is disregarded if it is smaller. For example, when the 5% level in the F table is 4.75, β is then as indicated by the broken line in Figure 19.1.

To accept the difference totally when the F_0 value is slightly greater than 4.75 and not to accept the difference at all when it is slightly less than 4.75 constitutes a total departure from common sense. It is only natural that many knowledgeable specialists do not use statistical testing in their fields. The degree of difference is approximately the same whether the variance ratio F_0 is slightly greater or slightly smaller than 4.75. Thus, it is more reasonable to round to the nearest whole number when using the F test. However, one should actually use a con-

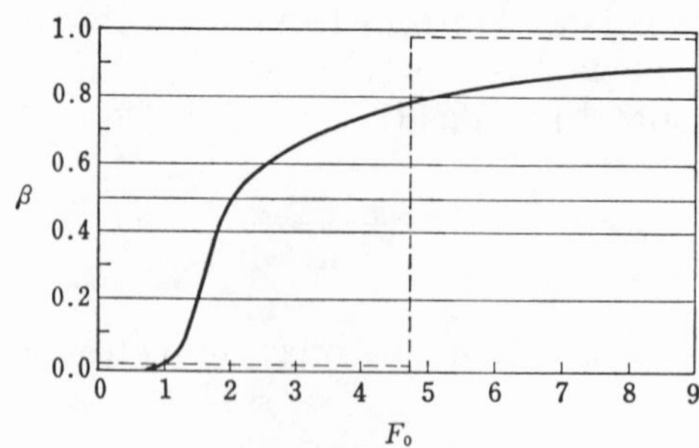

FIGURE 19.1 Beta Coefficient β

tinuous evaluation method for the difference, such as the beta coefficient method.

Incidentally, there are people who assert that it is peculiar to consider that there is no difference in cases where there is a difference, but not a significant one. They claim that whether or not there is a significant difference, one should use the estimate of difference which has been obtained by experiment, as is. But in that case, why did we test at all? Does this not mean that testing itself is a totally useless calculation procedure? The author believes that the purpose of testing is estimation, and that testing should be performed in order to estimate better. But all present-day testing methods judge whether or not the difference is significant. Such a method of arriving at a conclusion might be good for simplification, but it is too superficial and narrow as a mode of thinking. The author believes that to estimate and to evaluate the magnitude of the difference is, indeed, the important task. And if the existence of a statistical method is to be justified, it must be one that provides us with a method of solving, even if only to some degree, the various problems that arise because of error. Thus, *he believes that testing is no more than a rough means of approximating the beta coefficient method as a tool for estimation.* With respect to the correlation coefficient, too, whether or not there is correlation is not the problem, but the degree of magnitude of correlation is the problem. Also, when a certain scientific theory has been advanced as a hypothesis, the degree of accuracy of that theory (the magnitude of error possessed by the theory) is the problem. In the example of side-effects of medicines, too, whether or not there are side-effects is not the problem; the difference between the social loss from the side-effects and the social gain from the effect of the medicine constitutes the problem. Such examples indicate that estimation is, indeed, important. Experiments and tests to prove to the extent possible that there are no side-effects are themselves meaningless, and only contribute needless confusion and loss to society.

19.4 Reason Why Mean Square Error Is Used

Often in mathematical statistics, bias and dispersion are treated separately. The reason for this is probably that the moduli of the normal distribution are m and σ, and parameter estimation is regarded as the problem because of a preoccupation with the form of the distribution. However, this is very questionable for practical purposes. Whether in an experiment or in surveying or measuring, we wish to estimate some true value. The estimated value will be represented by y and the true value by m.

Then, when the estimated value y is equal to the true value m, loss becomes minimum, as was also explained in the preceding chapter. The loss function $L(y)$ becomes minimum when y is equal to m. If $L(y)$ can be differentiated in the neighborhood of m, then in the equation expanding $L(y)$ about m,

$$L(y)=L(m)+\frac{L'(m)}{1!}(y-m)+\frac{L''(m)}{2!}(y-m)^2+\cdots \tag{19.36}$$

$L'(m)$ is zero. Also, since loss increases as y becomes smaller than m and also increases as it becomes greater, in many cases $L''(m)$ is convex in the neighborhood of m. This means that $L''(m) = 0$ is difficult to conceive of.

Then, the loss function $L(y)$ is given approximately by

$$L(y)=L(m)+\frac{L''(m)}{2!}(y-m)^2 \tag{19.37}$$

in the neighborhood of m. The term that is proportional to the square of the difference between the estimated value y and the true value m has become the leading term of the loss due to error. This is *the reason why mean square error is used irrespective of the distribution,* and it is probably the reason why the arithmetic mean and orthogonal expansion have long been used more commonly than the median and Taylor expansion.

We will not discuss here the case where it is not possible to differentiate the loss function in the neighborhood of m, but even in that case, if one evaluates the error by a method other than the mean square, probably a reason to justify doing so is necessary. It is advisable to measure the magnitude of error by the mean of the squares of the difference. This again means that the form of distribution is not important but that it is best to examine the goodness or badness of the estimated value by the mean square error.

Already, at a number of companies, testing has been stopped and only beta coefficients are being used. The author predicts that in the future, testing will hardly ever be used. The F test of the present is not such a bad method as long as it is regarded as a rough approximation of the beta coefficient method. It is therefore felt that *it does not matter if the F test is used as a simplification of the beta coefficient method.* This stand has been taken in Volume 1 of this book. But it is not particularly difficult to rewrite the calculations of Volume 1 completely by the beta coefficient method.

19.5 Example of Application of Beta Coefficient (1): The Case of a Two-Level System

Let us try to apply the method of the preceding section to the experiment example of Section 17.3. The data comprised the results of experimentation by the use of orthogonal array L_{16} to find ten two-level factors, A, B, C, D, E, F, G, H, I, and J, and four two-factor interactions, $A \times B$, $B \times D$,

$B \times E$, and $D \times E$, with the purpose of improving the yield of a lubricant and the coagulation temperature. The analysis of variance table for the yield data is given again in Table 19.1.

The optimum conditions, for heavy crude oil A_2 (factor A is an indicative factor and it is not possible to select levels) are $B_1C_1D_2E_2F_1G_2H_1I_1J_1$. Let us estimate the process average under these conditions by three methods: (1) the method using the difference as is, (2) the method using the beta coefficient, and (3) the method using the F test.

(1) *Method Using Difference As Is*

$\bar{A}_1$, $\bar{T}$, and $\overline{(A \times B)}_2$ represent the mean of A_1, the total mean, and the mean of the second level of interaction $A \times B$. The estimated value of the process average $\hat{\mu}$ is given by the following equation.

$$\begin{aligned}\hat{\mu} &= \bar{T}+(\bar{A}_2-\bar{T})+(\bar{B}_1-\bar{T})+[\overline{(A\times B)}_2-\bar{T}]+(\bar{C}_1-\bar{T})+(\bar{D}_2-\bar{T})+[\overline{(B\times D)}_2-\bar{T}]\\ &\quad+(\bar{E}_2-\bar{T})+[\overline{(B\times E)}_2-\bar{T}]+[\overline{(D\times E)}_1-\bar{T}]+(\bar{F}_1-\bar{T})+(\bar{G}_2-\bar{T})\\ &\quad+(\bar{H}_1-\bar{T})+(\bar{I}_1-\bar{T})+(\bar{J}_1-\bar{T})\\ &=\bar{A}_2+\bar{B}_1+\overline{(A\times B)}_2+\bar{C}_1+\bar{D}_2+\overline{(B\times D)}_2+\bar{E}_2+\overline{(B\times E)}_2+\overline{(D\times E)}_1\\ &\quad+\bar{F}_1+\bar{G}_2+\bar{H}_1+\bar{I}_1+\bar{J}_1-13\bar{T}\\ &=\frac{7.6}{8}+\frac{25.1}{8}+\frac{14.9}{8}+\frac{27.7}{8}+\frac{37.9}{8}+\frac{30.8}{8}+\frac{20.1}{8}+\frac{26.0}{8}+\frac{23.2}{8}+\frac{26.0}{8}+\frac{30.2}{8}\\ &\quad+\frac{34.9}{8}+\frac{26.2}{8}+\frac{18.1}{8}-13\times\frac{47.4}{16}=5.08\doteqdot 5.1\end{aligned} \tag{19.38}$$

Table 19.1 Analysis of Variance Table for Yield (Difference from Initial Yield)

Source	f	S	V	F_0	β
A	1	64.80	64.80	37.8**	0.974
B	1	0.49	0.49	—	0
$A\times B$	1	19.36	19.36	11.3*	0.912
C	1	4.00	4.00	2.3	0.565
D	1	51.12	51.12	30.0**	0.967
$B\times D$	1	12.60	12.60	7.4*	0.865
E	1	3.24	3.24	1.9	0.474
$B\times E$	1	1.32	1.32	—	0
$D\times E$	1	0.06	0.06	—	0
F	1	1.32	1.32	—	0
G	1	10.56	10.56	6.2*	0.839
H	1	31.36	31.36	18.3**	0.945
I	1	1.56	1.56	—	0
J	1	7.84	7.84	4.6	0.783
(e)	(7)	(11.99)	(1.71)		
T	14†	209.63			

† The number of degrees of freedom has become 14 instead of 15 because of missing values.

Table 19.2 Various Prediction Methods and Their Error

Method of estimating process average	Predicted value	Actually realized value	Difference
(1) Method using difference	5.1	4.1	+1.0
(2) Method using beta coefficient	4.0	4.1	−0.1
(3) Method using F test	4.7	4.1	+0.6

(2) *Method Using Beta Coefficient*

$\beta(A_2)$, etc., are the beta coefficients for the main effects of A, etc.

$$\hat{\mu}=\overline{T}+\beta(A_2)\times(\overline{A}_2-\overline{T})+\beta(B_1)\times(\overline{B}_1-\overline{T})+\beta(A\times B)[\overline{(A\times B)}_2-\overline{T}]$$
$$+\cdots+\beta(J_1)(\overline{J}_1-\overline{T})$$
$$=\frac{47.4}{16}+0.974\left(\frac{7.6}{8}-\frac{47.4}{16}\right)+0\left(\frac{25.1}{8}-\frac{47.4}{16}\right)+0.912\left(\frac{14.9}{8}-\frac{47.4}{16}\right)$$
$$+0.565\left(\frac{27.7}{8}-\frac{47.4}{16}\right)+0.967\left(\frac{37.9}{8}-\frac{47.4}{16}\right)+0.865\left(\frac{30.8}{8}-\frac{47.4}{16}\right)$$
$$+0.474\left(\frac{20\cdot1}{8}-\frac{47.4}{16}\right)+0\left(\frac{26.0}{8}-\frac{47.4}{16}\right)+0\left(\frac{23.2}{8}-\frac{47.4}{16}\right)+0\left(\frac{26.0}{8}-\frac{47.4}{16}\right)$$
$$+0.839\left(\frac{30.2}{8}-\frac{47.4}{16}\right)+0.945\left(\frac{34.9}{8}-\frac{47.4}{16}\right)+0\left(\frac{26.2}{8}-\frac{47.4}{16}\right)$$
$$+0.783\left(\frac{18.1}{8}-\frac{47.4}{16}\right)=2.962-1.960+0-1.003+0.282+1.716+0.768-0.213$$
$$+0+0+0+0.681+1.323+0-0.548=4.008\doteqdot 4.0 \qquad (19.39)$$

(3) *Method Using F Test*

In the analysis of variance table of Table 19.1, the significant sources are A, $A\times B$, D, $B\times D$, G, and H. We therefore have

$$\hat{\mu}=\overline{A_2B_1}+\overline{B_1D_2}-2\overline{B}_1+\overline{G}_2+\overline{H}_1-\overline{T}$$
$$=\frac{0.1}{4}+\frac{23.2}{4}-2\times\frac{25.1}{8}+\frac{30.2}{8}+\frac{34.9}{8}-\frac{47.4}{16}$$
$$=4.72\doteqdot 4.7 \qquad (19.40)$$

Now, the mean value of the increase in yield when a confirmatory trial was carried out under the optimum conditions was 4.1%. The differences between the three types of prediction methods and the confirmatory trial values were as given in Table 19.2.

The results of Table 19.2 indicate that the error is least by the beta coefficient method. Of course, it is necessary to study the reproducibility of the process averages and the factorial effects by a larger number of actual examples. The author would welcome reports from many experimenters regarding the beta coefficient method.

19.6 The Case of Three or More Levels

We have explained the beta coefficient for the case where all of the factors are at two levels; we will now explain the case of three or more levels. Let us assume that A is at three levels. They will be represented as A_1, A_2, and A_3. The totals for A_1, A_2, and A_3 will be expressed by the same symbols, A_1, A_2, and A_3, and the means will be expressed by $\bar{A}_1, \bar{A}_2$, and $\bar{A}_3$. The estimate considering the effect of A when an arbitrary level of A_1, A_2, and A_3, say A_3, has been selected is given by

$$\hat{\mu}=\bar{T}+(\bar{A}_3-\bar{T}) \tag{19.41}$$

with the number of replications of $\bar{A}_1, \bar{A}_2$, and $\bar{A}_3$ as n and the mean for the whole experiment as $\bar{T}$. By substituting

$$\bar{T}=\frac{1}{3}(\bar{A}_1+\bar{A}_2+\bar{A}_3) \tag{19.42}$$

into the second term on the right side of Equation (19.41), we get

$$\hat{\mu}=\bar{T}+\left(\bar{A}_3-\frac{\bar{A}_1+\bar{A}_2+\bar{A}_3}{3}\right)=\bar{T}+\frac{2\bar{A}_3-\bar{A}_1-\bar{A}_2}{3} \tag{19.43}$$

as the usual estimation formula for A_3.

However, as explained before, generally speaking the second term in Equation (19.41) is an excessive estimate. We multiply it by a beta coefficient and choose it so that the mean square error becomes minimum.

$$\hat{\mu}=\bar{T}+\beta(\bar{A}_3-\bar{T}) \tag{19.44}$$

Since T and $(\bar{A}_3-\bar{T})$ are orthogonal, the mean square error of the second term in Equation (19.44), with $E(\bar{T})=m$ and $E(\bar{A}_3)=m+a_3$, is given by the following equation.

$$\begin{aligned}E[\beta(\bar{A}_3-\bar{T})-a_3]^2&=\{E[\beta(\bar{A}_3-\bar{T})-a_3]\}^2+\mathrm{Var}[\beta(\bar{A}_3-\bar{T})]\\&=[\beta(m+a_3-m)-a_3]^2+\mathrm{Var}\left(\beta\frac{2\bar{A}_3-\bar{A}_1-\bar{A}_2}{3}\right)\\&=(\beta-1)^2a_3{}^2+\frac{\beta^2}{9}\left(\frac{4}{n}+\frac{1}{n}+\frac{1}{n}\right)\sigma^2\\&=(\beta-1)^2a_3{}^2+\frac{6\beta^2}{9n}\sigma^2\end{aligned} \tag{19.45}$$

If we differentiate Equation (19.45) with respect to β and put it equal to 0 and then solve with respect to β, we have

$$\beta=\frac{a_3{}^2}{a_3{}^2+\dfrac{2}{3n}\sigma^2} \tag{19.46}$$

Now if, in comparing A_1, A_2, and A_3, the contrast of $3(A_3 - \overline{T})$, or in other words A_3 and the other two levels, is expressed as L, we have

$$L = 2A_3 - A_1 - A_2$$

If the variation of this is expressed as S_L,

$$S_L = \frac{(2A_3 - A_1 - A_2)^2}{6n} \tag{19.47}$$

$$E(L) = [2(m + a_3) - (m + a_1) - (m + a_2)]n = (2a_3 - a_1 - a_2)n \tag{19.48}$$

Since $a_1 + a_2 + a_3 = 0$, we substitute $a_1 + a_2 = -a_3$ into Equation (19.48) and obtain

$$E(L) = (2a_3 + a_3)n = 3na_3 \tag{19.49}$$

Also, since the variance of L is

$$\mathrm{Var}(L) = \{2^2 \times n + [(-1)^2 + (-1)^2]n\}\sigma^2 = 6n\sigma^2 \tag{19.50}$$

we have

$$E(S_L) = \sigma^2 + \frac{3}{2}na_3^2 \tag{19.51}$$

Therefore, if V_e is the error variance, we have

$$\hat{a}_3^2 = (S_L - V_e)\frac{2}{3n} \tag{19.52}$$

If this is substituted into Equation (19.46), we have

$$\beta \doteqdot \frac{(S_L - V_e)\dfrac{2}{3n}}{(S_L - V_e)\dfrac{2}{3n} + \dfrac{2}{3n}V_e} \doteqdot \frac{S_L - V_e}{S_L} = 1 - \frac{1}{F_L} \tag{19.53}$$

F_L is the variance ratio of contrast L,

$$F_L = \frac{S_L}{V_e} \tag{19.54}$$

In general, in the case of a levels, we assume that level A_i has been selected, with the sums of n replications written as $A_1, A_2, \ldots, A_a$.

$$\hat{\mu} = \overline{T} + \beta(\overline{A}_i - \overline{T}) \tag{19.55}$$

The mean square error is

$$E(\hat{\mu} - \mu)^2 = E[\overline{T} + \beta(\overline{A}_i - \overline{T}) - (m + a_i)]^2 = (\beta - 1)^2 a_i^2 + \left(\frac{1}{an} + \frac{a-1}{an}\beta^2\right)\sigma^2 \tag{19.56}$$

If we differentiate with respect to β so as to minimize Equation (19.56) and put it equal to 0, we obtain

$$(\beta - 1)a_i^2 + \frac{a-1}{an}\beta\sigma^2 = 0 \tag{19.57}$$

When this is solved with respect to β, we get

$$\beta=\frac{a_i^2}{a_i^2+\frac{a-1}{an}\sigma^2} \tag{19.58}$$

If the contrast of A_i and the means of the other levels is expressed as L, we have

$$L=(a-1)A_i-(A_1+\cdots+A_{i-1}+A_{i+1}+\cdots+A_a)$$

$$E(L)=(a-1)n(m+a_i)-\{(m+a_1)+\cdots+(m+a_{i-1})+(m+a_{i+1})+\cdots+(m+a_a)\}n \tag{19.59}$$

Now, if

$$a_1+\cdots+a_{i-1}+a_{i+1}+\cdots+a_a=-a_i \tag{19.60}$$

is substituted in, we have

$$E(L)=ana_i \tag{19.61}$$

Also,

$$\mathrm{Var}(L)=a(a-1)n\sigma^2 \tag{19.62}$$

From this we obtain

$$E(L^2)=(ana_i)^2+a(a-1)n\sigma^2 \tag{19.63}$$

Therefore, the variation of L, S_L, becomes as follows.

$$S_L=\frac{L^2}{(a-1)^2n+(-1)^2n(a-1)}=\frac{L^2}{a(a-1)n} \tag{19.64}$$

For the expected value of S_L, we substitute the result of Equation (19.63) into Equation (19.64) and obtain

$$E(S_L)=\sigma^2+\frac{an}{(a-1)}a_i^2 \tag{19.65}$$

Thus, since the estimated value of the error variance σ^2 is V_e, the estimated value of a_i^2 becomes as follows.

$$\hat{a}_i^2=\frac{(a-1)}{an}(S_L-V_e) \tag{19.66}$$

$$\hat{\sigma}^2=V_e \tag{19.67}$$

By substituting these into Equation (19.58), we obtain

$$\beta=\frac{a_i^2}{a_i^2+\frac{a-1}{an}\sigma^2}\doteqdot\frac{\frac{a-1}{an}(S_L-V_e)}{\frac{a-1}{an}(S_L-V_e)+\frac{a-1}{an}V_e}$$

From this we have

$$\beta=1-\frac{1}{F_L} \tag{19.68}$$

F_L is the variance ratio of S_L and V_e. When the variance ratio is 1 or less, β is to be 0.

$$F_L=\frac{S_L}{V_e} \tag{19.69}$$

19.7 Example of Application of Beta Coefficient (2): The Case of a Three-Level System

The example of Table 19.3 is that of carbon powder, given in Section 8.3. For three factors at two levels, A, B, and E, and two factors at three levels, C and D,

A:	material	A_1 = current, A_2 = new
B:	washing	B_1 = done, B_1 = not done
C:	temperature	C_1 = 1000°C, C_2 = 1100°C, C_3 = 1200°C
D:	treatment time	D_1 = 1h, D_2 = 2h, D_3 = 3h
E:	additive	E_1 = none, E_2 = used

were assigned to orthogonal array L_9. A and B were combined, and dummies were inserted into E. Regarding the electric resistance values of carbon powders, the initial-period value was expressed as y_0 and the value after testing for 1000 hours was expressed as y_t, and the data give the percentage change. Since $y_t - y_0$ was small compared with y_0, $(y_t - y_0)/y_0$ was used instead of the logarithmic value, $\log(y_t/y_0)$. For,

$$y=\log\frac{y_t}{y_0} \tag{19.70}$$

$$=\log\frac{y_0+\varepsilon_t}{y_0} \qquad (\varepsilon_t=y_t-y_0)$$

$$\doteqdot\frac{\varepsilon_t}{y_0} \tag{19.71}$$

Table 19.3 Assignment and Data

No.	(AB) 1	C 2	D 3	E 4	Change (%)		Total
1	11	1	1	1	+11	+ 9	20
2	11	2	2	2	+ 5	+ 7	12
3	11	3	3	3	+ 5	+ 3	8
4	21	1	2	3	+22	+25	47
5	21	2	3	1	+17	+14	31
6	21	3	1	2	+10	+10	20
7	12	1	3	2	+ 8	+11	19
8	12	2	1	3	+ 9	+ 8	17
9	12	3	2	1	+ 5	+ 5	10
					Total		184

In cases such as this, of percentage change or quantity abraded, the general mean too becomes a target of testing. For the general mean, of course, one must not calculate by subtracting the working mean. In the present case,

$$S_m = \frac{184^2}{18} = 1\,880.9 \tag{19.72}$$

The analysis of variance table is given by Table 19.4, and estimates of factorial effects are shown in Table 19.5. A single error, e, was created by pooling e_1, D, and B with e_2. The effect of F_0 is considered even if it is not significant, but it is disregarded when F_0 is 1 or less. This means that we disregard only the effect of B which has a value for V less than $V_e = 3.7$. Then only A enters column 1 of Table 19.3.

Therefore, the combination for which the change becomes minimum is A_1, C_3, D_1, and E_2. In

$$\hat{\mu} = \bar{T} + \beta_A(\bar{A}_1 - \bar{T}) + \beta_C(\bar{C}_3 - \bar{T}) + \beta_D(\bar{D}_1 - \bar{T}) + \beta_E(\bar{E}_2 - \bar{T}) \tag{19.73}$$

the beta coefficients β_A, β_C, β_D, and β_E are as follows.

$$\beta_A = \left(1 - \frac{1}{F_{A_1}}\right) \tag{19.74}$$

$$\beta_C = \left(1 - \frac{1}{F_{C_3}}\right) \tag{19.75}$$

$$\beta_D = \left(1 - \frac{1}{F_{D_1}}\right) \tag{19.76}$$

$$\beta_E = \left(1 - \frac{1}{F_{E_2}}\right) \tag{19.77}$$

Now, F_{A_1}, etc., are the variance ratios obtained by dividing the variation of the contrast for the difference between A_1 and $\bar{T}$ by V_e.

$$L(A_1) = \bar{A}_1 - \bar{T} = \frac{A_1}{12} - \frac{A_1 + A_2}{18} = \frac{A_1 - 2A_2}{36} \tag{19.78}$$

Table 19.4 Analysis of Variance Table

Source	f	S	V
m	1	1 880.9	1 880.9
(AB)	2	339.1	169.6
[A	1	280.3	280.3
[B	1	3.0	3.0
C	2	192.4	96.2
D	2	14.8	7.4
E	1	26.7	26.7
e_1	1	10.1	10.1
e_2	9	20.0	20.0
T	18	2 484.0	
(e)	(13)	(47.9)	(3.7)

Table 19.5 Estimation of Factorial Effects

T	$\bar{T}=\frac{184}{18}=10.2$	D	$\bar{D}_1=\frac{57}{6}=9.5$
A	$\bar{A}_1=\frac{86}{12}=7.2$		$\bar{D}_2=\frac{69}{6}=11.5$
	$\bar{A}_2=\frac{98}{6}=16.3$		$\bar{D}_3=\frac{58}{6}=9.7$
C	$\bar{C}_1=\frac{86}{6}=14.3$	E	$\bar{E}_1=\frac{133}{12}=11.1$
	$\bar{C}_2=\frac{60}{6}=10.0$		$\bar{E}_2=\frac{51}{6}=8.5$
	$\bar{C}_3=\frac{38}{6}=6.3$		

From this we have

$$S_L=\frac{(A_1-2A_2)^2}{12+(-2)^2\times 6}=\frac{(A_1-2A_2)^2}{36}=\frac{(86-2\times 98)^2}{36}=336.1 \tag{19.79}$$

Therefore,

$$F_{A_1}=\frac{S_{L(A_1)}}{V_e}=\frac{336.1}{3.7}=90.8 \tag{19.80}$$

From this we find

$$\beta_A=\left(1-\frac{1}{F_{A_1}}\right)=\left(1-\frac{1}{90.8}\right)=0.989 \tag{19.81}$$

In a similar manner, we get

$$L(C_3)=3C_3-(C_1+C_2+C_3)=2C_3-C_1-C_2$$

$$S_L=\frac{(2C_3-C_1-C_2)^2}{4\times 6+(-1)^2\times 6+(-1)^2\times 6}=\frac{(2\times 38-86-60)^2}{36}=136.1$$

$$F_{C_3}=\frac{136.1}{3.7}=36.8$$

$$\beta_C=1-\frac{1}{36.8}=0.973 \tag{19.82}$$

$$L(D_1)=\bar{D}_1-\bar{T}=\frac{D_1}{6}-\frac{D_1+D_2+D_3}{18}$$

$$S_L=\frac{(2D_1-D_2-D_3)^2}{36}=\frac{(2\times 57-69-58)^2}{36}=4.7$$

$$F_{D_1}=\frac{4.7}{3.7}=1.3$$

$$\beta_D=1-\frac{1}{F_{D_1}}=0.231 \tag{19.83}$$

$$L(E_2)=\frac{E_2}{6}-\frac{E_1+E_2}{18}$$

$$S_L=\frac{(2E_2-E_1)^2}{24+(-1)^2\times 12}=\frac{(2\times 51-133)^2}{36}=26.7$$

$$F_{E_2}=\frac{26.7}{3.7}=7.2$$

$$\beta_E=1-\frac{1}{F_0}=1-\frac{1}{7.2}=0.861 \qquad (19.84)$$

Therefore, we substitute in Equation (19.73)

$$\beta_A=0.989,\ \beta_C=0.973,\ \beta_D=0.231,\ \beta_E=0.861$$

and obtain

$$\begin{aligned}\hat{\mu}&=\bar{T}+\beta_A(\bar{A}_1-\bar{T})+\beta_C(\bar{C}_3-\bar{T})+\beta_D(\bar{D}_1-\bar{T})+\beta_E(\bar{E}_2-\bar{T})\\&=10.2+0.989(7.2-10.2)+0.973(6.3-10.2)+0.231(9.5-10.2)\\&\quad+0.861(8.5-10.2)\\&=10.2-2.97-3.79-0.16-1.46\\&=1.8\end{aligned} \qquad (19.85)$$

For the confidence limits, usually one makes a prediction for the range of one run of supplementary experimental values. Speaking approximately, it is probably all right to regard this as being the range of the individual values. Since in the formula,

$$\pm\sqrt{F\times V_e\left(\frac{1}{n_e}+\frac{1}{1}\right)} \qquad (19.86)$$

the number of degrees of freedom of the error variance is 13, we have

$$F_{13}^{1}=4.67$$

$$V_e=3.7$$

$$\begin{aligned}n_e&=\frac{18}{\left[1+\beta_A{}^2\left(\frac{3}{2}-1\right)+\beta_C{}^2(3-1)+\beta_D{}^2(3-1)+\beta_E{}^2(3-1)\right]}\\&=\frac{18}{(1+0.989^2\times 0.5+0.973^2\times 2+0.231^2\times 2+0.861^2\times 2)}=3.6\end{aligned} \qquad (19.87)$$

Where dummies have been inserted, as in the cases of A and E, the effective number of degrees of freedom was regarded as

$$\left(\frac{3}{2}-1\right)=0.5$$

when the dummy level was selected; and when a level that was not a dummy was selected, its effective number of degrees of freedom was taken as 2. Therefore, by substituting into Equation (19.86), we get

$$\pm\sqrt{4.67\times3.7\times\left(\frac{1}{3.6}+\frac{1}{1}\right)}=\pm4.7 \tag{19.88}$$

Thus, the range of the individual values x is

$$x=1.8\pm4.7 \tag{19.89}$$

Since calculation by Equation (19.87) is tedious in many instances, one may use the approximate expression

$$F\left(\frac{1}{n_e}+\frac{1}{1}\right)\doteqdot8 \tag{19.90}$$

and find the confidence limits from

$$x=\hat{\mu}\pm\sqrt{8V_e} \tag{19.91}$$

If this is done, we have

$$x=1.8\pm\sqrt{8\times3.7}=1.8\pm5.4 \tag{19.92}$$

For the general mean, too, with its variance ratio F_0 as

$$F_0=\frac{S_m}{V_e}=\frac{1\,880.9}{3.7}=508.4 \tag{19.93}$$

by using

$$\beta_m=1-\frac{1}{F_0}=1-\frac{1}{508.4}=0.998 \tag{19.94}$$

we may calculate alternatively that

$$\begin{aligned}x&=0.998\times\bar{T}+\beta_A(\bar{A}_1-\bar{T})+\beta_C(\bar{C}_3-\bar{T})+\beta_D(\bar{D}_1-\bar{T})+\beta_E(\bar{E}_2-\bar{T})\\&=1.7\pm4.7\end{aligned} \tag{19.95}$$

19.8 How to Determine the Level of Significance α of Testing Theory

Although the method using the beta coefficient is the most logical, as an approximate method it is also possible to use the method of estimating only those differences which are significant by performing the F test to determine whether to recognize the difference or not. From this standpoint, let us study what is the optimum *level of significance*. It is a shortcoming of the testing theory that accurate calculation is impossible unless the form of distribution is determined. Let us calculate here assuming a normal distribution. Also, although we will treat the problem of testing of interactions here, it is exactly the same for main effects.

Let us investigate *how to decide on the level of significance* α in the F test. To simplify the explanation, we will assume that we have experimented by means of a two-way array for the two factors A at a levels and

B at b levels with r replications $V_1, V_2, \ldots, V_r$. We will study the problem of performing the F test of level of significance α with regard to interaction $A \times B$ and of estimating the population mean, μ_{ij}, for A_iB_j. It is assumed that

(1) If $A \times B$ is not significant, we estimate μ_{ij} by

$$\hat{\mu}_{ij} = \bar{A}_i + \bar{B}_j - \bar{T} \tag{19.96}$$

(2) If $A \times B$ is significant, we estimate by

$$\hat{\mu}_{ij} = \overline{A_iB_j} \tag{19.97}$$

If the probability that $A \times B$ *does not emerge as significant* at level of significance α *is expressed as* β, then W, which is the mean square of the error of the estimated value when estimation is done as explained above, is, from references 7) and 19),

$$\begin{aligned} W &= (1-\beta)\mathrm{Var}(\overline{A_iB_j}) + \beta[I_{ij}^2 + \mathrm{Var}(\bar{A}_i + \bar{B}_j - \bar{T})] \\ &= (1-\beta)\frac{\sigma^2}{r} + \beta\left(I_{ij}^2 + \frac{a+b-1}{abr}\sigma^2\right) \\ &= \frac{\sigma^2}{r} + \beta\left[I_{ij}^2 - \frac{(a-1)(b-1)}{abr}\sigma^2\right] \end{aligned} \tag{19.98}$$

I_{ij} is the value of the component of the interaction in A_iB_j, and β is an *error of the second kind* and is given by

$$\begin{aligned} \beta &= e^{-\lambda} \sum_{j=0}^{\infty} \frac{\lambda^j}{j!} I_{x_\alpha}\left[\frac{(a-1)(b-1)}{2} + j, \frac{(r-1)(ab-1)}{2}\right] \\ \lambda &= \frac{r\sigma_{A\times B}^2}{2\sigma^2} = \frac{\psi^2}{2} \\ x_\alpha &= \frac{F_\alpha}{1+F_\alpha} \\ F_\alpha &= \alpha\% \text{ point of } F \text{ test} \end{aligned} \tag{19.99}$$

ψ^2 in the equation for β is termed the *mean eccentricity,* and it is obtained by subtracting 1 from the result of dividing the expected value of the mean square by the population variance σ^2. Since, in this case,

$$E[S_{A\times B}/(a-1)(b-1)] = \sigma^2 + r\sigma^2{}_{A\times B}$$

it becomes

$$\psi^2 = \frac{\sigma^2 + r\sigma^2{}_{A\times B}}{\sigma^2} - 1 = r\frac{\sigma^2{}_{A\times B}}{\sigma^2} \tag{19.100}$$

I_{x_α} is the *incomplete beta function,* and $I_x(n, m)$ is defined by

$$I_x(n, m) = \frac{1}{I(n, m)} \int_0^x t^{n-1}(1-t)^{m-1} dt$$

We are to treat the problem of minimizing W in Equation (19.98) as much as possible, but since the unknown I_{ij} is involved, this problem is similar to that in Chapter 33. The most desirable level of significance is $\alpha = 0\%$ *if* I_{ij} *is 0* and $\alpha = 100\%$ *if* I_{ij} *is large*. Assuming that I_{ij} takes various values, approximately what α is the most reasonable? In order to solve this problem by the method of Chapter 33, it is necessary further to decide on the relationship between I_{ij} and $\sigma^2{}_{A\times B}$. The following *inequality** holds between $I_{ij}{}^2$ and $\sigma^2_{A\times B}$.

$$\max_{ij} I_{ij}{}^2 \leq \frac{(a-1)^2(b-1)^2}{ab}\sigma^2{}_{A\times B}$$

However, since it is unclear which A_iB_j will be chosen, this does not necessarily mean that the maximum $|I_{ij}|$ will be chosen. It would seem better to consider the mean value for I_{ij}. When this is done, we take $\sigma^2_{A\times B}/ab$ as $I_{ij}{}^2$. From the discussions above we have

$$W = \frac{\sigma^2}{r} + \beta\left[\frac{\sigma^2{}_{A\times B}}{ab} - \frac{(a-1)(b-1)}{abr}\sigma^2\right]$$
$$= \frac{\sigma^2}{r}\left\{1 + \beta\left[\frac{\psi^2}{ab} - \frac{(a-1)(b-1)}{ab}\right]\right\}$$

Since this means that it is unnecessary to consider σ^2/r for the maximum and minimum of W, in the last analysis we need only examine the maximum or minimum of

$$W = 1 + \beta\left[\frac{\psi^2}{ab} - \frac{(a-1)(b-1)}{ab}\right] \tag{19.101}$$

β is given by Equation (19.99).

Let us draw a graph of Equation (19.101) assuming that

$$a=3, \qquad b=3, \qquad r=3$$

in order to obtain a graphical relationship with L_{27}. In this case, we have

$$W = 1 + \beta\left(\frac{\psi^2}{9} - \frac{4}{9}\right) = 1 + \frac{\beta}{9}(\psi^2 - 4) \tag{19.102}$$

The values of β for various values of α and ψ, incidentally, become as given by Table 19.6.

If we substitute the values above into W of Equation (19.100) and if we then calculate what happens to the mean square error for various α and ψ values and show this graphically, we get something like Figure 19.2.

What is evident from Figure 19.2 is that the estimation error for $\alpha = 0$, or in other words the test leading to the conclusion that $A \times B$ does not exist whatever the case, becomes greater as $A \times B$ becomes greater. However, if ψ is smaller, in effect the precision of estimation is higher when

* See reference 10, pp. 90–94 for a general discussion on these inequalities.

Table 19.6 Value of Error of the Second Kind, β

α \ ψ	0.0	0.5	1.0	1.5	2.0	2.5	3.0	3.5	4.0	5.0	6.0
0	1.00	1.00	1.00	1.00	1.00	1.00	1.00	1.00	1.00	1.00	1.00
0.01	0.99	0.98	0.98	0.96	0.91	0.85	0.75	0.61	0.47	0.18	0.07
0.05	0.95	0.94	0.91	0.85	0.75	0.63	0.48	0.33	0.20	0.06	0.02
0.10	0.90	0.88	0.84	0.75	0.63	0.49	0.33	0.20	0.11	0.04	0.01
0.25	0.75	0.72	0.65	0.53	0.39	0.25	0.14	0.07	0.03	0.00	0.00
0.50	0.50	0.47	0.39	0.29	0.18	0.10	0.05	0.02	0.01	0.00	0.00
1.00	0.00	0.00	0.00	0.00	0.00	0.00	0.00	0.00	0.00	0.00	0.00

$A \times B$ is disregarded even if it exists; therefore, if it is certain that $A \times B$ is rather small, it is the most advantageous to perform a test with $\alpha = 0$. The level of significance that minimizes error in the worst case is $\alpha = 1.00$, in other words the case where one arrives at the decision that $A \times B$ always exists, but this is probably by no means a good method if it happens frequently that the interaction is small.

If we find the point giving $\psi = 2$ from

$$\psi^2 = \frac{r\sigma^2_{A\times B}}{\sigma^2}$$

it is

$$\sigma_{A\times B} = \frac{2}{\sqrt{r}} \times \sigma = \frac{2}{\sqrt{3}}\sigma \doteqdot 1.15\sigma$$

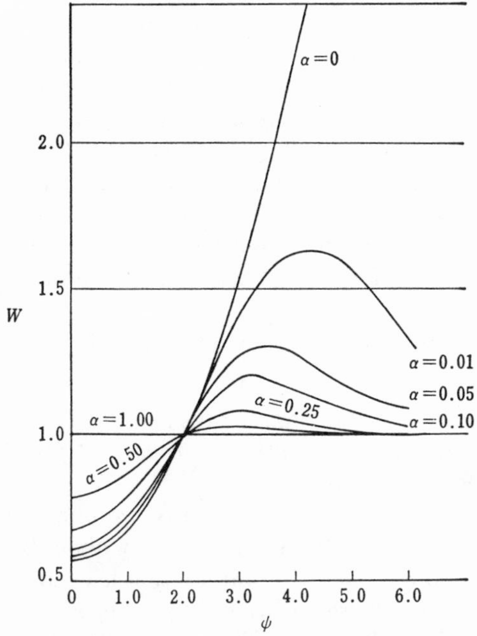

FIGURE 19.2 Mean Square Error for Various α Values

Thus, since it is common that $\sigma_{A\times B}$ is less than 1.15 times the standard deviation of error, $\alpha = 1.00$ cannot be regarded as a particularly good level of significance. It seems to the author that if it happens only about once in 10 times that $\sigma^2_{A\times B}$ is greater than σ^2, the best level of significance is in the range 0.01 – 0.10, from the graph. Probably *about 5% is the best in this case.*

The graph differs in the case of main effects since the number of degrees of freedom differs, but the difference is not very great. If it is assumed that two factors C and D at three levels have been inserted in place of $A \times B$, we obtain

$$\sigma^2{}_{A\times B} = \frac{3}{2}(\sigma_C{}^2 + \sigma_D{}^2)$$

Therefore, we have

$$\psi^2 = \frac{9}{2\sigma^2}(\sigma_C{}^2 + \sigma_D{}^2) \qquad (19.103)$$

But the likelihood is far greater that main effects exist than that interactions exist. Moreover, since the coefficient in ψ^2 becomes $\frac{9}{2}$ and the ratio of σ^2 is 1.5, it is not possible to achieve a direct analogy with interactions. Since the main effect becomes $\sqrt{\frac{3}{2}}$ times the interaction in its influence on ψ, if the level of significance is the same, detection precision of the main effect becomes higher. This fact means that the greater the likelihood of existence and the greater the number of replications, the more one should consider centering about the point where ψ is great.

Therefore, if the matter is *considered impartially for the whole range of* ψ, it would seem that *the best level of significance is in the range from* $\alpha = 0.05$ *to* $\alpha = 0.25$. This means that aside from a value depending on psychological effect as to the reliability of the conclusions, the most logical level of significance for a main effect in an experiment of about 10 replications (therefore cases such as L_{27} and L_{16}) is about 10%. Even if psychological effect is stressed, it will probably not become smaller than 5%.

The most important conclusions will be given, based on the foregoing discussions.

(1) For the cases of orthogonal arrays such as L_{27} and L_{16}, *the best level of significance is believed to be 10% for the main effect and 5% for an interaction.*

(2) In the case of an experiment with few replications, the level of significance should be relaxed. Probably about 25% for the main effect and about 10% for the interaction is good.

(3) For cases such as those of L_{81} and L_{64}, it is best when the level of significance is taken as 5% for the main effect and 1% for interaction.

(4) When one wishes to affix a * sign to emphasize psychological effect, it is believed better to test with a level of significance that is one step smaller than the levels of significance of (1), (2), and (3).

In this book, as a rule, we have been using 5% and 1%, which are the values commonly used, but in actual cases it is probably better to use the foregoing results. To be even more accurate, it is best to use beta coefficients.

Exercises (19)

(1) In regard to the tile experiment of Chapter 6, predict $\hat{\mu}$ for the optimum conditions when β coefficients are used and the realized value of fraction defective x when a confirmatory trial is run on 10,000 tiles under these conditions.

(2) Estimate the coefficient b of linear regression by the beta coefficient method.

20 Applications of Analysis of Variance

The range of application of resolution of variation, or in other words the method of analysis of variance, discussed in Chapter 18, is very wide. Here, its applications to *periodic analysis* and *analysis of cause,* which are important in quality control, for example, will be shown. Please refer to Chapters 22–24 for applications to measurement methods and dynamic characteristics.

20.1 Application to Quality Control

20.1.1 Control, Prediction, and Correction of Continuous Values

We consider the problem that a target value has been given for a continuous quantity, such as the thickness of a wire, the thickness of a metal sheet, the dimensions of a part, the viscosity of a liquid, the amount packed into a bag by an automatic bag-packing machine, temperature, flow, or generated power, and *the deviation from the target value is corrected by a specific signal factor*. This constitutes process correction rather than process adjustment. In process adjustment, we search for the cause and take measures to change it, but in process correction we do not search for the cause but merely bring about mutual cancellation by the signal factor, or in other words carry out correction treatment (reference 36). For example, if the thickness of a metal sheet is 10 microns greater than the target value, regardless of whether the cause lies in the ingot components or temperature change or wear of the rolling mill, we take

Table 20.1 Measured Values of Sheet Thickness (μm) (Difference from Target Value Is Given)

Consecutive No. A	Coil R	Position V	K_1 Left	K_2 Center	K_3 Right	5th-order unit total	4th-order unit total	3rd-order unit total	2nd-order unit total	1st-order unit total
1	1	1	−3	−3	−5	−11				
2	1	2	2	0	−3	−1	−12			
3	1	3	3	4	4	11				
4	1	4	8	7	4	19	30	18		
5	2	1	−8	−3	−5	−16				
6	2	2	−4	−4	−5	−13	−29			
7	2	3	−2	1	1	0				
8	2	4	3	7	8	18	18	−11	7	
9	3	1	2	5	4	11				
10	3	2	3	6	7	16	27			
11	3	3	8	10	7	25				
12	3	4	11	19	18	48	73	100		
13	4	1	−11	−10	−14	−35				
14	4	2	−10	−11	−12	−33	−68			
15	4	3	−7	−4	−5	−16				
16	4	4	−2	−2	−6	−10	−26	−94	6	13
17	5	1	−14	−16	−22	−52				
18	5	2	−9	−8	−15	−32	−84			
19	5	3	−7	−5	−7	−19				
20	5	4	−4	−2	−6	−12	−31	−115		
21	6	1	−14	−17	−16	−47				
22	6	2	−14	−12	−16	−42	−89			
23	6	3	−12	−8	−13	−33				
24	6	4	−8	−7	−8	−23	−56	−145	−260	
25	7	1	−4	−1	0	−5				
26	7	2	3	6	4	13	8			
27	7	3	3	7	5	15				
28	7	4	11	14	12	37	52	60		
29	8	1	−10	−7	−8	−25				
30	8	2	−10	−8	−12	−30	−55			
31	8	3	−5	−6	−7	−18				
32	8	4	1	4	0	5	−13	−68	−8	−268
Total			−100	−44	−111					−255

measures involving the variable that directly affects the result, in this case the draft of the rolling mill. This is a method of bringing a quantity close to its target value by changing the values of a specific factor whose level can be changed (this is termed a *signal factor*), rather than by finding the cause and correcting that.

When the sheet thickness is too large by 10 microns, we shift the draft M of the rolling mill, which is a signal factor, by a certain number of graduations so that the thickness will become 10 microns thinner, but since a certain time is necessary for such treatment a time-lag l develops. In practice, feedback cannot take place, and the fact that we correct the difference between the target value and the actually measured value means that we are predicting that if we do nothing the difference between the target value and the actually measured value will remain the same. Since we are predicting that the future value will be equal to the present one, this means that we are using a very simple method of prediction.

We will explain how to determine the optimum measurement interval for the case where a certain characteristic value is measured from time to time, and where we correct the difference from the target value by a signal factor in order to minimize this difference. The example of sheet thickness will be used for calculation.

At present, the rolled sheet thickness of a certain metal is being measured at the beginning of each coil and the deviation from the target value is being corrected. Let us consider as our problem whether it is better also to measure the sheet thickness midway in the coil or whether only one measurement and one correction of the coil is sufficient. In a case such as this, we try measuring the thickness of a number of coils at intervals, for example, of every $\frac{1}{4}$ of the coil. Let us assume that one ingot has been rolled and that it ultimately becomes a thin-sheet coil of length 1200 m.

The target value of the thickness at the final stage is 1.000 mm, and let us assume that we wish to bring the dispersion of the thickness to within 1.000 ± 0.020 mm. A product whose thickness does not fall in this range must be scrapped, and let us assume that there is a loss of 8000 yen per coil.

For eight coils $R_1, R_2, \ldots, R_8$, when the sheet thickness was measured at the three points of the left side, center, and right side of the width of the coil for each of the four positions

V_1 = position at 150 m from the starting point of the coil
V_2 = position at 450 m from the starting point of the coil
V_3 = position at 750 m from the starting point of the coil
V_4 = position at 1050 m from the starting point of the coil

the data obtained were as given in Table 20.1.

Factor K is the difference from the target value of the sheet thickness at the three points in the coil width direction:

$$K_1 = \text{left side},\quad K_2 = \text{center},\quad K_3 = \text{right side}$$

The 5th-order unit total in Table 20.1 is the total of the three data for left, center, and right of the sheet thickness and the 4th-order unit total is the sum of the totals of two consecutive 5th-order unit totals in the longitudinal direction. The sum of consecutive 4th-order unit totals is the 3rd-order unit total. Therefore, the 3rd-order unit total is the total of 12 data per coil. The 2nd-order unit total is the total of the data for two consecutive coils. The 1st-order unit total is the total of the data for the first four coils or the second four coils, and the 0th-order unit total is the total of all data.

20.1.2 Resolution of Variation

By using these data for the totals, we find the following sums of squares (variations).

0th-Order Unit Variation S_0

$$S_0=\frac{(-255)^2}{96}=677.34 \qquad (f=1) \tag{20.1}$$

1st-Order Unit Variation S_1

$$S_1=\frac{13^2+(-268)^2}{48}=1\,499.85 \qquad (f=2) \tag{20.2}$$

2nd-Order Unit Variation S_2

$$S_2=\frac{7^2+6^2+(-260)^2+(-8)^2}{24}=2\,822.88 \qquad (f=4) \tag{20.3}$$

3rd-Order Unit Variation S_3

$$S_3=\frac{18^2+(-11)^2+\cdots+(-68)^2}{12}=5\,146.25 \qquad (f=8) \tag{20.4}$$

4th-Order Unit Variation S_4

$$S_4=\frac{(-12)^2+30^2+\cdots+(-13)^2}{6}=6\,433.83 \qquad (f=16) \tag{20.5}$$

5th-Order Unit Variation S_5

$$S_5=\frac{(-11)^2+(-1)^2+\cdots+5^2}{3}=6\,938.33 \qquad (f=32) \tag{20.6}$$

Variation Among Left, Center, and Right S_K

$$S_K=\frac{(-100)^2+(-44)^2+(-111)^2}{32}-S_0=80.69 \qquad (f=2) \tag{20.7}$$

Total Variation S_6

$$S_6=(-3)^2+(-3)^2+\cdots+0^2=7229 \qquad (f=96) \tag{20.8}$$

From these variations, we find the error variations as follows.

0th-Order Unit Error Variation S_{e_0}

$$S_{e_0}=S_0=677.34 \qquad (f=1) \tag{20.9}$$

1st-Order Unit Error Variation S_{e_1}

$$S_{e_1}=S_1-S_0=1499.85-677.34=822.51 \qquad (f=1) \tag{20.10}$$

2nd-Order Unit Error Variation S_{e_2}

$$S_{e_2}=S_2-S_1=2822.88-1499.85=1323.03 \qquad (f=2) \tag{20.11}$$

3rd-Order Unit Error Variation S_{e_3}

$$S_{e_3}=S_3-S_2=5146.25-2822.88=2323.37 \qquad (f=4) \tag{20.12}$$

4th-Order Unit Error Variation S_{e_4}

$$S_{e_4}=S_4-S_3=6433.83-5146.25=1287.58 \qquad (f=8) \tag{20.13}$$

5th-Order Unit Error Variation S_{e_5}

$$S_{e_5}=S_5-S_4=6938.33-6433.83=504.50 \qquad (f=16) \tag{20.14}$$

6th-Order Unit Error Variation S_{e_6}

$$S_{e_6}=S_6-S_5-S_K=7229-6938.33-80.69=209.98 \qquad (f=62) \tag{20.15}$$

The 3rd-order unit error variation expresses the dispersion between successive coils, and the 4th-order unit error variation expresses the dispersion of the sheet thickness between the first half of a coil and the second half. The 5th-order unit error variation expresses the dispersion of the sheet thickness corresponding to changes by $\frac{1}{4}$ unit. The variation S_K expresses the magnitude of the systematic difference among the

Table 20.2 Analysis of Variance Table for Dispersions

Source	f	S	V	$E(V)$
e_0	1	677.34	677.34	$\sigma_6^2+3\sigma_5^2+6\sigma_4^2+12\sigma_3^2+24\sigma_2^2+48\sigma_1^2+96\sigma_0^2$
e_1	1	822.51	822.51	$\sigma_6^2+3\sigma_5^2+6\sigma_4^2+12\sigma_3^2+24\sigma_2^2+48\sigma_1^2$
e_2	2	1 323.03	661.52	$\sigma_6^2+3\sigma_5^2+6\sigma_4^2+12\sigma_3^2+24\sigma_2^2$
e_3	4	2 323.37	580.84	$\sigma_6^2+3\sigma_5^2+6\sigma_4^2+12\sigma_3^2$
e_4	8	1 287.58	160.95**	$\sigma_6^2+3\sigma_5^2+6\sigma_4^2$
e_5	16	504.50	31.53**	$\sigma_6^2+3\sigma_5^2$
K	2	80.69	40.34**	$\sigma_6^2+32\sigma_K^2$
e_6	62	209.98	3.39	σ_6^2
T	96	7 229.00		

thicknesses at the left, center, and right in the coil width direction. Therefore, if the sheet thickness is measured at the beginning of each coil and if correction of the draft of the rolling mill and number of passes is performed once per coil, it can be expected that the dispersions among coils, the 3rd-order unit error variation S_{e_3}, 2nd-order unit error variation S_{e_2}, 1st-order unit error variation S_{e_1}, and 0th-order unit error variation S_{e_0}, will become zero (to be accurate, when these error variations are tested against the 4th-order unit error variation, they are no longer significant).

Such a method of performing resolution of error variation is absolutely necessary in order to decide whether correction of sheet thickness should be performed once for each coil or once per $\frac{1}{2}$ coil, or whether it suffices to perform it once very two coils. Actually, it is easy to understand if one constructs an analysis of variance table such as Table 20.2.

When testing, we test e_5 and K by e_6, e_4 by e_5, e_3 by e_4, e_2 by e_3, e_1 by e_2, and e_0 by e_1. However, as is evident at a glance, there is hardly any difference among e_0, e_1, e_2, and e_3. This indicates that with about eight coils, change of thickness per coil is the main thing, and there is no long-wave change in units such as two coils or four coils. The effect of K is great, but if, for example, it is desirable to cause the center part to be about 3 microns thicker than the left and right, we recalculate S_K and the total variation S_T as follows.

$$S_K=\frac{(-100)^2+(-44-3\times 32)^2+(-111)^2}{32}-\frac{(-255-3\times 32)^2}{96}$$

$$=26.69 \tag{20.16}$$

$$S_T=S_T \text{ of Table 20.2} - (S_K \text{ of Table 20.2} - S_K \text{ of equation above})$$

$$=7229.00-(80.69-26.69)=7175.00 \tag{20.17}$$

The S_K and S_T in this case are the same as though, when the sheet thickness was measured at the beginning, we had separately subtracted different target thicknesses from the left, center, and right and then per-

Table 20.3 Readjusted Analysis of Variance Table

Source	f	S	V	$E(V)$	F_0
e_3(inter-coil)	8	5 146.25	643.28	$\sigma_6^2+3\sigma_5^2+6\sigma_4^2+12\sigma_3^2$	4.00*
e_4(inter-$\frac{1}{2}$ coil)	8	1 287.58	160.95	$\sigma_6^2+3\sigma_5^2+6\sigma_4^2$	5.10**
e_5(inter-$\frac{1}{4}$ coil)	16	504.50	31.53	$\sigma_6^2+3\sigma_5^2$	9.30**
K(left, center, right)	2	26.69	13.34	$\sigma_6^2+32\sigma_K^2$	3.94*
e_6	62	209.98	3.39	σ_6^2	
T	96	7 175.00			

formed the calculations of Table 20.2. If we pool e_0, e_1, e_2, and e_3 and substitute in Equations (20.16) and (20.17), we obtain an analysis of variance table such as Table 20.3.

Since, according to the results of Table 20.3, K is slightly significant, it is indicated that in the coil width direction the center has become a little under 3 microns thicker than the two sides. It can be understood that the dispersion of thickness among coils is large and the dispersion of thickness within a coil is also large. When a systematic thickness change is conceivable in the rolling direction of the coil, it is best to consider such a systematic difference as a source before e_4 and e_5. Such tendencies exist in this example but analysis will be omitted because of space limitations.

20.1.3 Estimation of Variance and Optimum Adjustment Interval

We estimate the components of the respective variances.

$$\hat{\sigma}_6^2 = V_6 = \frac{209.98}{62} = 3.39, \qquad \sigma_K^2 = \frac{13.34-3.39}{32} = 0.31 \tag{20.18}$$

$$\hat{\sigma}_5^2 = \frac{1}{3}(V_5 - V_6) = \frac{1}{3}(31.53-3.39) = 9.38 \tag{20.19}$$

$$\hat{\sigma}_4^2 = \frac{1}{6}(V_4 - V_5) = \frac{1}{6}(160.95-31.53) = 21.57 \tag{20.20}$$

$$\hat{\sigma}_3^2 = \frac{1}{12}(V_3 - V_4) = \frac{1}{12}(643.28-160.95) = 40.19 \tag{20.21}$$

These are the estimated values of the dispersions of the independent thicknesses which develop within $\frac{1}{4}$ coil, between $\frac{1}{4}$ coils, between $\frac{1}{2}$ coils, and between coils, respectively. Therefore, if measurement and proper correction have been performed for each coil, the variance of thickness is

$$\hat{\sigma}_6^2 + \hat{\sigma}_5^2 + \hat{\sigma}_4^2 + \hat{\sigma}_K^2 \tag{20.22}$$

The variance of thickness when the thickness has been checked midway in the coil, as well, and proper correction has been performed, is

$$\hat{\sigma}_6^2 + \hat{\sigma}_5^2 + \hat{\sigma}_K^2 \tag{20.23}$$

However, when the thickness is measured at the beginning of each coil and the draft of the rolling mill is adjusted, prediction error enters the picture since we use the initial data of the coil to estimate the mean value of the thickness of this coil. Let us assume that we find the thickness at the beginning of the coil at the three points in the width direction (left, center, right), that we estimate the thickness of the whole coil by its mean value, and that we then adjust the coil thickness. The predicted error in this case is given approximately by

$$\frac{1}{3}\sigma_6^2+\sigma_4^2 \tag{20.24}$$

Therefore, in this instance, if we measure the three points at the very beginning of each coil, then average these values and adjust the deviation from the target value, the dispersion of the thickness of the product becomes

$$\left(\frac{1}{3}\sigma_6^2+\sigma_4^2\right)+(\sigma_6^2+\sigma_5^2+\sigma_4^2+\sigma_K^2) \tag{20.25}$$

The part within the first parentheses constitutes the error variance and the part within the second parentheses constitutes the error variance within the lot. No matter what prediction is being made, one need only find the error variance of the prediction by a regression equation and substitute it in.

Similarly, the variance of the dispersion of the thickness when measurement and correction are performed once per ½ coil and once per eight coils are given by the following equations.

Once Per 8 Coils

$$\left(\frac{1}{3}\sigma_6^2+\sigma_3^2\right)+(\sigma_6^2+\sigma_5^2+\sigma_4^2+\sigma_3^2+\sigma_K^2) \tag{20.26}$$

Once Per Coil

$$\left(\frac{1}{3}\sigma_6^2+\sigma_4^2\right)+(\sigma_6^2+\sigma_5^2+\sigma_4^2+\sigma_K^2) \tag{20.27}$$

Once Per ½ Coil

$$\left(\frac{1}{3}\sigma_6^2+\sigma_5^2\right)+(\sigma_6^2+\sigma_5^2+\sigma_K^2) \tag{20.28}$$

By substituting the estimated values into these, we obtain

$$\left(\frac{3.39}{3}+40.19\right)+(3.39+9.38+21.57+40.18+0.31)=116.15 \tag{20.29}$$

$$\left(\frac{3.39}{3}+21.57\right)+(3.39+9.38+21.57+0.31)=57.35 \tag{20.30}$$

$$\left(\frac{3.39}{3}+9.38\right)+(3.39+9.38+0.31)=23.59 \tag{20.31}$$

If the cost of a single correction is B, the correction interval is u, the variance at that time is σ^2, and the loss proportional to the variance (Chapter 18) is $k\sigma^2$, the *loss function* L is given by

$$L=\frac{B}{u}+k\sigma^2 \tag{20.32}$$

Since the loss when coils are scrapped is 8000 yen per coil,

$$k=\frac{\text{loss from having become a defective}}{(\text{tolerance})^2}=\frac{8000}{20^2}$$

$$=20 \text{ yen} \tag{20.33}$$

If $B = 800$ yen, L becomes as follows.

Once Per 8 Coils

$$L=\frac{800}{8}+20\times 116.16=2\,423.0 \quad \text{yen} \tag{20.34}$$

Once Per Coil

$$L=\frac{800}{1}+20\times 57.35=1\,947.0 \quad \text{yen} \tag{20.35}$$

Once Per $\frac{1}{2}$ Coil

$$L=\frac{800}{0.5}+20\times 23.59=2\,071.8 \quad \text{yen} \tag{20.36}$$

Therefore, measurement and correction once per coil, as at present, is the optimum.

20.2 Application to Search for Cause

20.2.1 Survey Data

In the case of trouble which it is not possible to correct, such as blow-hole change, strength change, and appearance change, a method of searching for the cause and removing it becomes important. The data of Table 20.4 give the number of rolls examined per processed lot in the production of a certain film, and the number of second-grade products among them. These are data obtained by finding the percentage of second-grade products and taking the subtotal per raw material batch unit and the sub-sub-total per product type. The total of the whole was 16.138.

Table 20.4 Data

Lot (C)	Product type (A)	Raw material batch (B)	Number of samples (n)	Number of 2nd-grade samples (r)	Percent	Sub-subtotal	Subtotal
1	1	1	15	0	0.000		
2	1	1	13	2	0.154	0.194	
3	1	1	25	1	0.040		
4	1	2	15	3	0.200	0.200	
5	1	3	23	4	0.174	0.174	
6	1	4	7	1	0.143	0.143	1.355
7	1	5	6	2	0.333	0.333	
8	1	5	12	0	0.000		
9	1	6	25	2	0.080	0.311	
10	1	6	13	3	0.231		
11	2	7	8	6	0.750	1.750	
12	2	7	7	7	1.000		
13	2	8	12	7	0.583		4.325
14	2	8	8	5	0.625		
15	2	8	6	4	0.667	2.575	
16	2	8	10	7	0.700		
17	3	9	18	16	0.889	0.889	0.889
18	4	10	15	5	0.333		
19	4	10	12	3	0.250		
20	4	10	10	4	0.400	0.983	
21	4	10	5	0	0.000		
22	4	11	8	5	0.625		
23	4	11	10	6	0.600	2.003	4.580
24	4	11	9	7	0.778		
25	4	12	8	2	0.250		
26	4	12	6	3	0.500	1.594	
27	4	12	10	4	0.400		
28	4	12	9	4	0.444		
29	5	13	25	12	0.480		
30	5	13	18	13	0.722	2.084	2.084
31	5	13	17	15	0.882		
32	6	14	10	6	0.600		
33	6	14	39	28	0.718	1.703	
34	6	14	13	5	0.385		
35	6	15	5	0	0.000		
36	6	15	10	0	0.000	0.133	
37	6	15	15	2	0.133		
38	6	15	2	0	0.000		2.905
39	6	16	18	4	0.222	0.222	
40	6	17	3	0	0.000	0.000	
41	6	18	15	0	0.000		
42	6	18	8	1	0.125		
43	6	18	18	4	0.222	0.847	
44	6	18	10	3	0.300		
45	6	18	5	1	0.200		
Total			556	207			16.138

20.2.2 Analysis of Variance

Calculation of variation is performed as follows.

$$S_m = \frac{16.138^2}{45} = 5.787445 \qquad (f=1) \tag{20.37}$$

$$S_{T_1}=\frac{1.355^2}{10}+\frac{4.325^2}{6}+\frac{0.889^2}{1}+\cdots+\frac{2.905^2}{14}$$
$$=8.048946 \qquad (f=6) \tag{20.38}$$

$$S_{T_2}=\frac{0.194^2}{3}+\frac{0.200^2}{1}+\frac{0.174^2}{1}+\cdots+\frac{0.847^2}{5}$$
$$=9.012030 \qquad (f=18) \tag{20.39}$$

$$S_{T_3}=0.000^2+0.154^2+0.040^2+\cdots+0.200^2$$
$$=9.479088 \qquad (f=45) \tag{20.40}$$

From this, we obtain

Variation Between Product Types

$$S_A=S_{T_1}-S_m=8.048946-5.787445=2.261501 \qquad (f=5) \tag{20.41}$$

Variation Between Batches

$$S_B=S_{T_2}-S_{T_1}=0.963084 \qquad (f=12) \tag{20.42}$$

Variation Between Lots

$$S_C=S_{T_3}-S_{T_2}=0.467058 \qquad (f=27) \tag{20.43}$$

The Variation Between Samples, S_e, Is Found as Follows

$$S_e=\frac{1}{\bar{r}}S_e' \qquad (f=556-45=511) \tag{20.44}$$

Here, $\bar{r}$ is the harmonic mean, and we have

$$\frac{1}{\bar{r}}=\frac{1}{45}\left(\frac{1}{15}+\frac{1}{13}+\frac{1}{25}+\cdots+\frac{1}{5}\right)=0.1122342 \tag{20.45}$$

$$S_e'=\Sigma\left(r_i-\frac{r_i^2}{n_i}\right)=\Sigma r_i-\Sigma\frac{r_i^2}{n_i}$$
$$=207-\left(\frac{0^2}{15}+\frac{2^2}{13}+\frac{1^2}{25}+\cdots+\frac{1^2}{5}\right)=84.554444 \tag{20.46}$$

Therefore, we have

$$S_e=0.1122342\times 84.554444=9.489900 \qquad (f=511) \tag{20.47}$$

The analysis of variance table, Table 20.5, was obtained from this. The net variation S' was found as follows.

$$S_m'=S_m-0.452300$$
$$S_A'=S_A+0.452300-6\times 0.080257$$

Table 20.5 Analysis of Variance Table

Source		f	S	V	S'	ρ[%]
General mean m		1	5.787445	5.787445*	5.335145	28.12
Product type A		5	2.261501	0.452300**	2.232259	11.77
Batch	B	12	0.963084	0.080257**	1.111500	5.86
Lot	C	27	0.467058	0.017298		
Sample	e	511	9.489900	0.018571		
Total	T	556	18.968988		18.968988	100.00
$(C+e)$	(e)	538	9.956958	0.018507	10.290084	54.25

$$S_B'=S_B+6\times 0.080257-18\times 0.018507$$

$$S_e'=S_e+S_C+18\times 0.018507$$

The influence of the cause of dispersion between processed lots is not seen; the effect of the cause of dispersion between raw material batches is 5.86%, that of the cause between product types is 11.77%, and that of the cause in common, m, is 28.12%. Furthermore, individual differences among samples account for 54.25% and are the greatest source. To examine the influence of the cause for the individuals, one by one, in effect we search for the cause which changes per individual.

When cause A and B are confounded, we find the combined effect of (AB). In other words, the ordered-pair combinations of (AB) are used as levels of the factor. If one wishes to examine whether or not defectives are especially numerous in the first single lot at the time of switching of product type, it is best to perform an analysis of covariance by using the additional factor x. If it is the first lot, x is to be 1; and if not, it is to be 0.

The reader is also referred to Chapter 25 and Chapter 31.

Exercise (20)

(1) A certain product is bag-packed with an automatic weighing machine, worked eight hours a day. The target value of content quantity is 100 g, but since the dispersion is large, the product packed every day at the beginning of work is weighed by a separate scale, and correction (calibration) of the automatic weighing machine is performed. In order to investigate to what degree the error becomes smaller if different procedures are followed in making the correction every day at the beginning of work, bags were weighed every hour for four days and the error in weight was measured, whereupon the following data were obtained.

0	5	8	2	4	2	2	4	−3	−2	−1	0	−1	0	−3	−2
4	6	7	9	−3	−1	−2	−5	−3	−6	−2	−5	0	2	4	2

Resolve the sum of squares of the difference from the target value, S_T, into periods, and estimate to what degree the error variance (mean of squares of difference between target value and actual content) becomes small when the correction once a day is done by various methods.

21 Fitting of Functions

The method of proving formulas which appear in the physical sciences, engineering, etc., by observation and experimentation will be discussed in this chapter. Please also refer to Chapters 15 and 16 of Volume 1 for more complicated cases.

21.1 A Simple Example

Because of the development of coaxial cables, random current (noise current) coming from leakage current and induced current in communication cables has become nearly zero, but thermal noise arising because the cable itself is not at zero absolute temperature still exists, and this is the principal noise in coaxial cables at present. It is not possible to design multiplex communication systems unless the magnitude of the noise is understood. We therefore consider the following problem.

[Problem] We wish to investigate whether or not the following law regarding the magnitude of random current which derives from thermal motion of electrons within the circuit,

$$\text{Power of random current } y = k\frac{T}{C} \tag{21.1}$$

holds. How should one experiment, and how should one perform data analysis? One expects that T is the absolute temperature, C is the capacity of the circuit, and k is Boltzmann's constant.

If one completely trusts that Formula (21.1) is correct, one should not conduct follow-up testing or the like; if one does not, one should believe his own experimental results. If one discusses how to perform the experiment for the problem above, *the design of the experiment must be debated only on the premise that the formula above is probably correct but cannot be trusted 100%.*

But when the problem above was presented, only a few persons gave the answers desired by the author regarding the following two points.

(1) Method of assignment (nearly all persons gave wrong answers)
(2) Analysis method (wrong answers were especially numerous among persons acquainted with statistics)

(1) *Problem Point Concerning Assignment*

It is inconceivable that the formula given above holds at all absolute temperatures. For example, since generally speaking the superconduction phenomenon occurs in the neighborhood of absolute zero, it can be assumed that the formula does not hold there; and since at temperatures of several thousand degrees or more the existence of the circuit itself becomes doubtful, the validity of the formula at high temperatures must also be questioned. If one wishes to determine the limits within which such a formula holds, an experiment with the levels focused in the neighborhood of absolute zero or several hundreds or several thousands of degrees of absolute temperature T becomes necessary.

When there is progress in supermultiplexing by coaxial cables, as recently, thermal noise as the principal component of noise comes to be a problem. Thus, if one wishes to obtain several thousands or several tens of thousands of channels with a coaxial cable, it is not possible to design an adequate carrier system unless one investigates the magnitude of thermal noise. So when one wishes to examine the validity of the formula above for practical design, as here, one takes as the range of the absolute temperatures in which to experiment such levels as

$$T_1 = -30°\text{C} = 243°\text{K}$$

$$T_2 = 20°\text{C} = 293°\text{K}$$

$$T_3 = 70°\text{C} = 343°\text{K}$$

representing the values at both ends and at the center of the temperature range in which the coaxial cable is actually used.

Table 21.1 Two-Way Array of *T* and *C*

	C_1	C_2	C_3
T_1	y_{11}	y_{12}	y_{13}
T_2	y_{21}	y_{22}	y_{23}
T_3	y_{31}	y_{32}	y_{33}

Although the experiment has been designated as one for coaxial cables, there is no difference in the discussion below even if the temperature level has been focused at absolute zero. It will be assumed that, similarly, we have taken for example the three levels C_1, C_2, and C_3 for the circuit capacity. We then might perform experiments of all of the nine combinations of T and C by randomizing the sequence of the experiments. In other words, we would carry out the experiments of the two-way array of T and C as given in Table 21.1.

An answer such as the foregoing was given by most of the people. But in this problem, we are debating an experiment by which to investigate whether or not the formula above holds. *To assume that it is adequate to experiment by a two-way layout, taking suitable levels of T and C within the range in which one wishes to experiment, is to accept without question what has been given and merely to imitate others.*

The formula above means that the random thermal current is proportional to the absolute temperature and is inversely proportional to the circuit capacity, and that moreover the proportionality constant is Boltzmann's constant. The assumption that the proportionality constant is an absolute constant means that the random current is determined only by the absolute temperature and the capacity, and that *it is unrelated to other factors, such as the types of circuit elements, resistance values, inductances, etc.* Besides T and C, which are in the formula, one should also take, for example,

A: structure of the circuit
B: resistance values of the circuit
D: inductances

at several levels, being careful not to change the capacity and temperature information, and should consider them in the assignment of the experiment along with T and C. In other words, *the number of factors that should be considered in the experiment is not two, but far more.* Of course, factors A, B, and D have been selected in order to show that their effects do not exist, but whatever the case, the number of factors is much more than two; and it is hoped that the reader will understand that it is by no means a two-way layout problem and that even for problems such as this, an experiment by orthogonal array is often necessary.

Table 21.2 Two-Way Array and Data

	C_1	C_2	C_3
T_1	y_{11}	y_{12}	y_{13}
T_2	y_{21}	y_{22}	y_{23}
T_3	y_{31}	y_{32}	y_{33}

In a situation such as this, one should consider matters from the standpoint of the question, what experiment should one have performed if he had originally thought of the formula above, or in other words *from the standpoint of the originator*. And it is important to nurture a spirit that enables one to grapple with original problems.

(2) *Problem in Terms of Data Analysis*

Let us assume that the experiment has been designed without taking up other factors $A, B, \ldots$. The discussion below applies in exactly the same way even if there are other factors. Data such as those given in Table 21.2 are obtained in the case of the two factors T and C. Data analysis such as the following is performed on these data. The levels of T and of C are chosen beforehand so as to be equispaced with respect to T and with respect to $1/C$ since the table of orthogonal polynomials will be used. When this is done, an analysis of variance such as is indicated in Table 21.3 becomes possible by the use of the orthogonal polynomials of the next section.

Data analysis such as this *is performed especially often by persons who have a superficial knowledge of statistical methods.* Calculations such as the following should naturally be performed in a case such as the above. With the observed value at T_iC_j as y_{ij}, one divides y_{ij} by T_i and multiplies by C_j so that the expected effects of T and C (it does not matter if erroneous) become cancelled from the data.

$$y_{ij}' = y_{ij} \times \frac{C_j}{T_i} \tag{21.2}$$

Table 21.3 Analysis of Variance Table

Source	f
T linear	1
T quadratic	1
C linear	1
C quadratic	1
T (linear) $\times$ C (linear)	1
e	3
Total	8

Table 21.4 Data After Performing Variable Conversion

$$y'' = y \times \frac{C}{T} - k$$

	C_1	C_2	C_3	Total
T_1	y_{11}''	y_{12}''	y_{13}''	T_1
T_2	y_{21}''	y_{22}''	y_{23}''	T_2
T_3	y_{31}''	y_{32}''	y_{33}''	T_3
Total	C_1	C_2	C_3	Total

These data will be expressed as y_{ij}'. If y is proportional to T and is inversely proportional to C, y_{ij}' no longer contains the effects of T and C. And if the proportionality constant is Boltzmann's constant k, the value above is equal to k within the range of error. Therefore, if we subtract k and put

$$y_{ij}'' = y_{ij} \times \frac{C_j}{T_i} - k \tag{21.3}$$

then y_{ij}'' should become zero within the range of experimental error if Formula (21.1) holds. Thus, analysis of variance should be performed on Table 21.4, where all of the data have been converted into y_{ij}''. Regarding C, one may take its levels at arithmetically equal intervals, or one may also take them so that their reciprocals are equispaced. Analysis of variance should be performed here for the first time (Table 21.5).

If T is not significant for the data y'', which have been obtained by dividing y by T, it means that *y is proportional to T within the range of error.* And *if C is not significant for these data, which have been obtained by multiplying y by C,* it means that *y is inversely proportional to capacity C.* And *if the general mean m is not significant,* it means that *the proportionality constant agrees with Boltzmann's constant within the range of error.*

Also, if T is significant, the effect of T remains even after dividing by T, which means that the random current is not proportional to T. In this case

Table 21.5 Analysis of Variance Table for y''

Source	f	S
m	1	S_m
T	2	S_T
C	2	S_C
e	4	S_e
Total	9	S_{Total}

the degree to which T deviates from the linear is, for example, expanded into the orthogonal polynomial of the next section.

If none of the main effects of m, T, and C are significant, y'' falls within the following range.

$$y''=0\pm\varepsilon, \qquad \varepsilon=\sqrt{F\times V_e}, \qquad V_e=\frac{S_{\text{Total}}}{9} \tag{21.4}$$

From this, we obtain the following formula.

$$y=(k\pm\varepsilon)\frac{T}{C} \tag{21.5}$$

It is necessary to consider first, when performing data analysis, that if experimental evidence for the formula one wishes to prove already exists beforehand, one should trust this formula and perform variable conversion so that the factorial effects will be cancelled. Such conversion is carried out because the formula is trusted, but if the formula is not trusted 100%, one tests whether or not there are factorial effects after conversion.

This is exactly the same even when one has an expectation or hunch regarding several sources, even if it is not a case of proving a formula. *To place 100% reliance on expectation or hunch is a basis for failure, but it is not a proper method to treat it as totally unreliable, either.* One should perform analysis thoroughly using technical expectations derived from the specific technology, and should not be entirely at the mercy of formulas.

The reader is referred to Chapter 15 for the analysis method when, in the problem given above, the expected functional form is

$$y=\alpha\times\frac{\log(1+\delta T^{\varepsilon})}{1+\gamma C^{\beta}} \tag{21.6}$$

and when α, β, γ, δ, and ϵ are unknown.

21.2 Orthogonal Polynomials

First, let us consider a case where we have experimented at *a equispaced points* x: $x_0, x_0+h, x_0+2h, \ldots, x_0+(a-1)h$. This is, for example, a case where, for factor A in the following table, we have experimented at the four equispaced levels 0, 10, 20, and 30°C. The number of replications for each temperature is denoted by r. The number of replications is a number that indicates the total number of times experiments were conducted in the form such that the main effect of A can be analyzed, no matter what common-factor effects are also included. To word it differently, the number of replications can be found if one knows, when one has found the sums for the respective levels of temperature A, how many measured values are included in each sum. The number of replications in

		B_1	B_2	B_3	B_4	B_5	Total
Working mean 80 has been subtracted	A_1 0	−5	−2	0	1	−3	$A_1=-9$
	A_2 10	−3	1	3	3	−1	$A_2=3$
	A_3 20	−1	2	4	3	1	$A_3=9$
	A_4 30	−6	−4	0	−1	−6	$A_4=-17$

the table above is 5. Also, for simplicity's sake, it will be understood that A_1, A_2, A_3, A_4 also mean the *sums* for the four levels of temperature A_1, A_2, A_3, and A_4, respectively.

Please refer to Section 21.5 regarding the analysis method *when the experiment has not been performed with equispaced levels.* Also, please refer to Section 21.6 in the case of *the general functional form.*

For simplicity's sake, let us also express this variable by A instead of x. It will be assumed that total values $A_1, A_2, \ldots, A_a$ of number of replications r have been obtained by a equispaced points $A, A_1 = A_0, A_2 = A_0 + h, \ldots, A_a = A_0 + (a-1)h$. Then, usually we write the following expression in representing the population mean y of the experimental values for variable A by a polynomial in A.

$$y = b_0' + b_1'A + b_2'A^2 + \cdots + b_{a-1}'A^{a-1} \tag{21.7}$$

But it becomes very troublesome to estimate or test b_i' by this equation as it stands. The reason for this is that *the estimated values* of b_1', b_2', . . . *are not mutually independent.* To avoid this difficulty, one need only revise Equation (21.7) into *an expansion in an orthogonal polynomial,* as follows.

$$\begin{aligned} y = {} & b_0 + b_1(A-\bar{A}) + b_2\left[(A-\bar{A})^2 - \frac{a^2-1}{12}h^2\right] \\ & + b_3\left[(A-\bar{A})^3 - \frac{3a^2-7}{20}(A-\bar{A})h^2\right] \\ & + b_4\left[(A-\bar{A})^4 - \frac{3a^2-13}{14}(A-\bar{A})^2h^2 + \frac{3(a^2-1)(a^2-9)}{560}h^4\right] \\ & + b_5\left[(A-\bar{A})^5 - \frac{5(a^2-7)}{18}(A-\bar{A})^3h^2 + \frac{15a^4-230a^2+407}{1008}(A-\bar{A})h^4\right] \\ & + \cdots \end{aligned} \tag{21.8}$$

$\bar{A}$ is the mean of the level values of a levels. It is a characteristic possessed by such an expansion that if the functions multiplying the coefficients $b_0, b_1, b_2, \ldots$ are expressed as $\phi_0(A), \phi_1(A), \phi_2(A), \ldots$, the orthogonality relationship

$$\sum_{\nu=1}^{a} \varphi_i(A_\nu)\varphi_j(A_\nu) = 0 \qquad (i \neq j;\ i, j = 0, 1, 2, \cdots) \tag{21.9}$$

is satisfied. To put it differently, the expansion is such that when the observation equation

$$b_0 + b_1\varphi_1(A_\nu) + b_2\varphi_2(A_\nu) + b_3\varphi_3(A_\nu) + \cdots = A_\nu/r \qquad (\nu = 1, 2, \cdots, a) \tag{21.10}$$

is solved by the method of least squares, *all terms other than those on the diagonal of the normal equation become zero.* The A_i in $\phi(A_i)$ is the level number of A, and A_i alone means the total of the data for A_i. Thus, if we multiply both sides of Equation (21.10) by $\phi_0(A) = 1$ and then add, only the term in b_0 remains and we have

$$ab_0 = \frac{1}{r}\sum_{\nu=1}^{a} A_\nu \tag{21.11}$$

Therefore, the estimated value of b_0, $\hat{b}_0$, is

$$\hat{b}_0 = \frac{1}{ar}\sum_{\nu=1}^{a} A_\nu = \text{mean of all measured values} \tag{21.12}$$

If we similarly add after multiplying by $\phi_1(A_\nu)$, only the term in b_1 remains and we have

$$\left[\sum_{\nu=1}^{a} \varphi_1{}^2(A_\nu)\right] b_1 = \frac{1}{r}\sum_{\nu=1}^{a} \varphi_1(A_\nu)A_\nu$$

Therefore, the estimated value of b_1, $\hat{b}_1$, becomes

$$\hat{b}_1 = \frac{\frac{1}{r}\sum \varphi_1(A_\nu)A_\nu}{\sum_{\nu=1}^{a} \varphi_1{}^2(A_\nu)} \tag{21.13}$$

By treating $b_2, b_3, \ldots$ similarly, one obtains their estimated values. To find $\phi(A)$ and $\Sigma\phi^2(A)$, as well, usually requires fairly considerable time and work, but actually for such values, a table of orthogonal polynomials, Table 21.6 (up to $a = 7$; see Appendix Table 3 of Volume 1 for up to 20 levels), is available, and we need but use it. Table 21.6 is to be used as follows.

W and λS of Table 21.6 are

$$\left.\begin{aligned} \lambda\varphi(A_\nu) &= W_\nu \\ \frac{1}{\lambda}\Sigma\varphi^2(A_\nu) &= \lambda S \end{aligned}\right\} \tag{21.14}$$

$$\left.\begin{aligned} \hat{b}_0 &= \frac{1}{ar}(A_1 + A_2 + \cdots + A_a) \\ \hat{b}_1 &= \frac{1}{r(\lambda S)h}(W_1A_1 + W_2A_2 + \cdots + W_aA_a) \\ \hat{b}_2 &= \frac{1}{r(\lambda S)h^2}(W_1A_1 + W_2A_2 + \cdots + W_aA_a) \\ \hat{b}_3 &= \frac{1}{r(\lambda S)h^3}(W_1A_1 + W_2A_2 + \cdots + W_aA_a) \\ &\vdots \qquad\qquad \vdots \end{aligned}\right\} \tag{21.15}$$

Table 21.6 Orthogonal Polynomials When Equispaced

b_1, b_2, . . . are coefficients of linear, quadratic, . . . terms. Estimation of the level a i-th order coefficient b_i is performed by using the coefficients W_1, W_2, . . . in the b_i column.

Estimation $\hat{b}_i = \dfrac{W_1A_1 + W_2A_2 + \cdots + W_aA_a}{r(\lambda S)h^i}$

Variation $S_{\hat{b}_i} = \dfrac{(W_1A_1 + W_2A_2 + \cdots + W_aA_a)^2}{r(\lambda^2 S)}$

Unit number unit number of $\hat{b}_i = \dfrac{1}{rSh^{2i}}$

Level	a=2	a=3		a=4			a=5			
	b_1	b_1	b_2	b_1	b_2	b_3	b_1	b_2	b_3	b_4
W_1	−1	−1	1	−3	1	−1	−2	2	−1	1
W_2	1	0	−2	−1	−1	3	−1	−1	2	−4
W_3		1	1	1	−1	−3	0	−2	0	6
W_4				3	1	1	1	−1	−2	−4
W_5							2	2	1	1
$\lambda^2 S$	2	2	6	20	4	20	10	14	10	70
λS	1	2	2	10	4	6	10	14	12	24
S	$\frac{1}{2}$	2	$\frac{2}{3}$	5	4	$\frac{9}{5}$	10	14	$\frac{72}{5}$	$\frac{288}{35}$
λ	2	1	3	2	1	$\frac{10}{3}$	1	1	$\frac{5}{6}$	$\frac{35}{12}$

Level	a=6					a=7				
	b_1	b_2	b_3	b_4	b_5	b_1	b_2	b_3	b_4	b_5
W_1	−5	5	−5	1	−1	−3	5	−1	3	−1
W_2	−3	−1	7	−3	5	−2	0	1	−7	4
W_3	−1	−4	4	2	−10	−1	−3	1	1	−5
W_4	1	−4	−4	2	10	0	−4	0	6	0
W_5	3	−1	−7	−3	−5	1	−3	−1	1	5
W_6	5	5	5	1	1	2	0	−1	−7	−4
W_7						3	5	1	3	1
$\lambda^2 S$	70	84	180	28	252	28	84	6	154	84
λS	35	56	108	48	120	28	84	36	264	240
S	$\frac{35}{2}$	$\frac{112}{3}$	$\frac{324}{5}$	$\frac{576}{7}$	$\frac{400}{7}$	28	84	216	$\frac{3\,168}{7}$	$\frac{4\,800}{7}$
λ	2	$\frac{3}{2}$	$\frac{5}{3}$	$\frac{7}{12}$	$\frac{21}{10}$	1	1	$\frac{1}{6}$	$\frac{7}{12}$	$\frac{7}{20}$

Here, r is the number of replications, λS is, for the respective number of levels a, and b_i, the value of λS in the section at the bottom of the table (it is multiplied by λ so that W_1, . . . , W_a will become integers), h is the interval of A, and $W_1, W_2, W_3, \ldots, W_a$ are the coefficients given in the table for the respective a and b_i.

For example, when r data have been obtained for each of four equispaced levels A_1, A_2, A_3, and A_4, the table of orthogonal polynomials for b_1, b_2, and b_3 constitutes the section for $a = 4$ in Table 21.6.

Therefore, b_0 is obtained from the first equation of Equations (21.15) and b_1, b_2, and b_3 are obtained from Table 21.6:

$$\hat{b}_0 = \frac{1}{r \times 4}(A_1 + A_2 + A_3 + A_4) \tag{21.16}$$

$$\left.\begin{aligned} \hat{b}_1 &= \frac{1}{r \times 10 \times h}(-3A_1 - A_2 + A_3 + 3A_4) \\ \hat{b}_2 &= \frac{1}{r \times 4 \times h^2}(A_1 - A_2 - A_3 + A_4) \\ \hat{b}_3 &= \frac{1}{r \times 6 \times h^3}(-A_1 + 3A_2 - 3A_3 + A_4) \end{aligned}\right\} \tag{21.17}$$

These equations can be summarized as

$$\hat{b}_i = \frac{1}{r(\lambda S)h^i}(W_1A_1 + W_2A_2 + \cdots + W_aA_a) \tag{21.18}$$

For $i = 0$, λS is to be a and $W_1, W_2, \ldots, W_a$ are all 1. The confidence limits of b_i at 95% reliability are given by the following equation.

$$b_i = \hat{b}_i \pm \sqrt{\frac{F \times V_e}{n_e}} \tag{21.19}$$

Here, V_e is the error variance for number of degrees of freedom f_2; F is the 5% point of the F distribution where the number of degrees of freedom of the numerator is 1 and the number of degrees of freedom of the denominator is that of the error variance f_2; and n_e is the effective number of replications of $\hat{b}$, given by the following equation.

$$n_e = rSh^{2i} \tag{21.20}$$

For example, when A_1, A_2, A_3, and A_4 are four levels of temperature at 10°C intervals, and when the tensile strength of a certain product has been measured at A_1, A_2, A_3, and A_4 with four replications, *the confidence limits at 95% reliability* of the estimated value b_1 of the coefficient of linear tendency b_1 of the tensile strength as it depends on temperature become, from Table 21.6,

$$\hat{b}_1 = \frac{1}{r(\lambda S)h}(W_1A_1 + W_2A_2 + W_3A_3 + W_4A_4)$$

$$= \frac{1}{4 \times 10 \times 10}(-3A_1 - A_2 + A_3 + 3A_4)$$

$$b_1 = \hat{b}_1 \pm \sqrt{\frac{F^1_{f_2} \times V_e}{n_e}}; \quad n_e = rSh^{2i} = 4 \times 5 \times 10^2 = 2000$$

$$=\hat{b}_1 \pm \sqrt{\frac{F^1_{f_2} \times V_e}{2\,000}}$$

And *the sum of squares* S_A *for the main effect of* A can be resolved as follows.

$$S_A = S_{\hat{b}_1} + S_{\hat{b}_2} + \cdots + S_{\hat{b}_{a-1}} \tag{21.21}$$

Where

$$S_{\hat{b}_i} = \frac{1}{r(\lambda^2 S)}(W_1A_1 + W_2A_2 + \cdots + W_aA_a)^2 \qquad (i=1,\,2,\,\cdots,\,a-1) \tag{21.22}$$

The number of degrees of freedom of these components is in every case 1. Also, $\lambda^2 S = W_1{}^2 + W_2{}^2 + \cdots + W_a{}^2$. For example, as to whether or not linear tendency is significant, $\hat{b}_1$ above, one need but test the following.

$$S_{\hat{b}_1} = \frac{1}{r(\lambda^2 S)}(W_1A_1 + W_2A_2 + W_3A_3 + W_4A_4)^2$$

$$= \frac{1}{r \times 20}(-3A_1 - A_2 + A_3 + 3A_4)^2$$

Since high-order terms are not particularly important, not infrequently the $\hat{b}_p$ from a certain $p < a$ onward are combined in a residual term.

$$S_A = S_{\hat{b}_1} + \cdots + S_{\hat{b}_{p-1}} + S_{\text{res}} \tag{21.23}$$

The number of degrees of freedom up to $S_{\hat{b}_{p-1}}$ is 1 in each case and the number of degrees of freedom of S_{res} is $(a-1)-(p-1)=a-p$.

Source		f	S
A	Linear	1	$S_{\hat{b}_1}$
	Quadratic	1	$S_{\hat{b}_2}$
	⋮	⋮	⋮
	$(p-1)$th order	1	$S_{\hat{b}_{p-1}}$
	The rest(res)	$a-p$	S_{res}
Total		$a-1$	S_A

Of course,

$$S_{\text{res}} = S_A - (S_{\hat{b}_1} + \cdots + S_{\hat{b}_{p-1}}) \tag{21.24}$$

When only the linear and quadratic terms are significant and the others can be omitted as a result of analysis of variance, the quadratic equation is fitted, and the confidence limits of this fitted quadratic equation are as follows. Since $\hat{b}_0$, $\hat{b}_1$, and $\hat{b}_2$ are mutually independent, we have

$$\hat{f}(x) = \hat{b}_0 + \hat{b}_1(A-\bar{A}) + \hat{b}_2\left[(A-\bar{A})^2 - \frac{a^2-1}{12}h^2\right] \pm \sqrt{\frac{F \times V_e}{n_e}} \tag{21.25}$$

where

$$\frac{1}{n_e} = \frac{1}{ar} + \frac{(A-\bar{A})^2}{rSh^2} + \frac{\left[(A-\bar{A})^2 - \frac{a^2-1}{12}h^2\right]^2}{rSh^4} \tag{21.26}$$

When it is expected that the data can be expanded into a polynomial in $1/T$, it is possible to use the formula above, as is, if the experiment is conducted so the levels are *equispaced with respect to 1/T. In certain cases, the number of replications r is not necessarily a positive integer*. Nor is it rare that *r is 1*. When *r* is 1, often *one combines the effects of higher-order terms* which one wishes to omit in view of common sense, *and with this as the error variation, one tests the lower-order terms such as the linear, quadratic, and cubic.*

21.3 When There Are Two or More Variables

The orthogonal expansion when there are two or more variables is the product of two orthogonal expansions of one variable each, provided the experiments are equispaced in each case. Thus, when one has experimented so as to be able to find the interaction of A and B with A at a levels and B at b levels, the expansion is as follows.

$$\begin{aligned} y=&b_{00}+b_{10}(A-\bar{A})+b_{01}(B-\bar{B}) \\ &+b_{20}\left[(A-\bar{A})^2-\frac{a^2-1}{12}h_A{}^2\right]+b_{11}(A-\bar{A})(B-\bar{B})+b_{02}\left[(B-\bar{B})^2-\frac{b^2-1}{12}h_B{}^2\right] \\ &+b_{30}\left[(A-\bar{A})^3-\frac{3a^2-7}{20}(A-\bar{A})h_A{}^2\right]+b_{21}\left[(A-\bar{A})^2-\frac{a^2-1}{12}h_A{}^2\right](B-\bar{B}) \\ &+b_{12}(A-\bar{A})\left[(B-\bar{B})^2-\frac{b^2-1}{12}h_B{}^2\right]+b_{03}\left[(B-\bar{B})^3-\frac{3b^2-7}{20}(B-\bar{B})h_B{}^2\right]+\cdots \end{aligned} \tag{21.27}$$

Here, h_A and h_B are the intervals between the levels of factors A and B.

The estimated values of b_{11}, etc., and their sums of squares and variances are as follows.

$$\begin{aligned} \hat{b}_{11}=&\frac{W_1[W_1'(A_1B_1)+W_2'(A_1B_2)+\cdots+W_b'(A_1B_b)]}{r(\lambda S)_A h_A\times(\lambda S)_B h_B} \\ &\frac{+W_2[W_1'(A_2B_1)+W_2'(A_2B_2)+\cdots+W_b'(A_2B_b)]+\cdots}{} \\ &\frac{+W_a[(W_1'(A_aB_1)+W_2'(W_aB_2)+\cdots+W_b'(A_aB_b)]}{} \end{aligned} \tag{21.28}$$

$$S_{\hat{b}_{11}}=\frac{\left\{\sum_{\nu=1}^{a}W_\nu\left[\sum_{n=1}^{b}W_n'(A_\nu B_n)\right]\right\}^2}{r(\lambda^2S)_A(\lambda^2S)_B} \tag{21.29}$$

$$\sigma^2(\hat{b}_{11})=\frac{\sigma^2}{rS_Ah_A{}^2S_Bh_B{}^2} \tag{21.30}$$

W_1', W_2', . . . , W_b' are coefficients by which the total values of B_1, B_2, . . . , B_b for each level of A are multiplied, and can be found in Table 21.6 as the coefficient of b_1 of level number b. W_1, W_2, . . . , W_a are *the coefficients of the linear effect of A by which the values of the linear effect of B obtained for each level* of A are multiplied, and they can similarly be found in the table as the coefficients of b_1 of level number a.

$(\lambda S)_A$ and $(\lambda S)_B$ and so forth are the λS of A and B, respectively. $A_\nu B_n$ is the sum of r experimental values for the combination $A_\nu B_n$. In general, for coefficient b_{ij} of the i-th-order term of A and the j-th-order term of B, we have

$$\hat{b}_{ij}=\frac{\sum_{\nu=1}^{a} W_\nu\left[\sum_{n=1}^{b} W_n'(A_\nu B_n)\right]}{r(\lambda S)_A h_A{}^i(\lambda S)_B h_B{}^j} \tag{21.31}$$

$$S_{\hat{b}_{ij}}=\frac{\left[\sum_{\nu=1}^{a} W_\nu \sum_{n=1}^{b} W_n'(A_\nu B_n)\right]^2}{r(\lambda^2 S)_A(\lambda^2 S)_B} \tag{21.32}$$

$$\sigma^2(\hat{b}_{ij})=\frac{\sigma^2}{rS_A h_A{}^{2i} S_B h_B{}^{2j}} \tag{21.33}$$

Also, since the estimated values of these coefficients are independent, the confidence limits of the function when, for example, we have taken A up to the quadratic term, B up to the linear term, and the product of A and B up to the term in $(A-\bar{A})(B-\bar{B})$, are given by the following equation.

$$\pm\sqrt{\frac{F_{f_2}^1 \times V_e}{n_e}} \tag{21.34}$$

$$\frac{1}{n_e}=\frac{1}{abr}+\frac{(A-\bar{A})^2}{rbS_A h_A{}^2}+\frac{\left[(A-\bar{A})^2-\frac{a^2-1}{12}h_A{}^2\right]^2}{rbS_A h_A{}^4}$$
$$+\frac{(B-\bar{B})^2}{raS_B h_B{}^2}+\frac{(A-\bar{A})^2(B-\bar{B})^2}{rS_A h_A{}^2 S_B h_B{}^2} \tag{21.35}$$

21.4 An Example

To investigate how the tensile strength of a certain synthetic resin changes depending on temperature A, the value of this quantity was measured for five test pieces at the four points $A_1=5$, $A_2=20$, $A_3=35$, and $A_4=50$°C, whereupon the results of Table 21.7 were obtained.

Table 21.7 Data

A_1	43	47	45	43	45	223
A_2	43	41	45	41	39	209
A_3	37	36	39	40	38	190
A_4	34	32	36	35	35	172
Total						794

Table 21.8 Analysis of Variance Table

Source		f	S	V
A	l	1	296**	
	q	1	1	
	c	1	0	
	e	16	51	3.2

In order to see what order of polynomial should be fitted for the main effect A, let us try testing by resolving A into its linear, quadratic, and cubic components.

$$S_{A_l}=\frac{(W_1A_1+W_2A_2+W_3A_3+W_4A_4)^2}{r(\lambda^2S)}=\frac{[-3(223)-209+190+3(172)]^2}{5\times 20}$$

$$=\frac{(-172)^2}{100}=296 \tag{21.36}$$

$$S_{A_q}=\frac{(223-209-190+172)^2}{5\times 4}=\frac{(-4)^2}{20}=1 \tag{21.37}$$

$$S_{A_c}=\frac{[-223+3(209)-3(190)+172]^2}{5\times 20}=\frac{6^2}{100}=0 \tag{21.38}$$

$$S_e=43^2+47^2+\cdots+35^2-CF-S_{A_l}-S_{A_q}-S_{A_c}=51 \tag{21.39}$$

Since only the linear term is significant for A, it may be considered that within the temperature range of the experiment, the relationship between temperature A and tensile strength y *is linear*. The estimated value of its temperature coefficient b_1 is

$$\hat{b}_1=\frac{-3(223)-209+190+3(172)}{r(\lambda S)h}=\frac{-172}{5\times 10\times 15}=-0.23 \tag{21.40}$$

For the confidence limits, by using the pooled error variance 2.9 ($n_e = rSh^2$), we have

$$b=\hat{b}\pm\sqrt{\frac{F\times V_e}{n_e}}=-0.23\pm\sqrt{\frac{4.41\times 2.9}{5\times 5\times 15^2}}$$

$$=-0.23\pm 0.05 \tag{21.41}$$

The tensile strength y of this synthetic resin for an arbitrary temperature A is

$$y=\frac{\text{total}}{20}+\hat{b}_1(A-\bar{A})=\frac{794}{20}-0.23(A-27.5)$$

$$=39.70-0.23(A-27.5) \tag{21.42}$$

And since its confidence limits are

$$\frac{1}{n_e}=\frac{1}{20}+\frac{(A-27.5)^2}{rSh^2}=\frac{1}{20}+\frac{(A-27.5)^2}{5\times 5\times 15^2}$$

$$=0.05+0.00018(A-27.5)^2$$

we have

$$\pm\sqrt{\frac{4.41\times 2.9}{n_e}} = \pm 3.6\times\sqrt{0.05+0.00018(A-27.5)^2} \tag{21.43}$$

As to the manner in which the error varies according to the value of A, the reader is requested to calculate the confidence limits by assuming, for example, that $A = 0$, 20, and 50.

21.5 Orthogonal Polynomial for the Case of Unequal Spacing

Even when the spacing in an experiment is unequal with respect to a variable—when, for example, the spacing is equal if the logarithms are taken or if the reciprocals are taken—it is possible to use the formulas for the equispaced case directly. In general, if $f(A)$ is a known function of A and, moreover, varies monotonically, and if $A_1, A_2, \ldots, A_a$ are so selected that $f(A_i)-f(A_{i-1})$ equals h in the case of a polynomial-type regression, it is possible to apply the formulas for the equispaced case to $f(A)$ as they stand. The case of $f(A) = \log A$ is an example. Since $\log A$ becomes equispaced if A is taken as 1, 2, 4, 8, . . . , it is possible to use the formulas for the equispaced case by putting $x = \log A$. Also, when the levels of A are approximately equispaced, as in the case of $A_1 = 50$, $A_2 = 61$, $A_3 = 70$, and $A_4 = 79$, there will probably be hardly any error even if one uses the formulas for the equispaced case. In most cases, a factor such as temperature cannot be neatly controlled to the scheduled points; and if the variation is within about 20 percent compared with the spacing, one should use the formulas for the equispaced case. When the levels are equispaced but *the number of replications r differs* for each level, although as a basic rule it is necessary to use the formulas for unequal spacing explained in this section, when each number of replications *comes within roughly a factor of 2 of the mean number of replications, it is more realistic to use the formulas for equal spacing by treating the case much as though there were missing data equivalent to the insufficient part of the number of replications.* For example, if there are five replications for A_1, ten replications for A_2, and five replications for A_3, we take the mean value for A_1, A_2, and A_3 and calculate as though there were replications by the harmonic mean $\bar{r}$ at each level; and we use the formulas for the equispaced case of $a = 3$. In the case of *unequal spacing in general,* with the experimental values of A as x_1, $x_2, \ldots, x_n$, we find the mean, $\bar{x}$. It does not matter if, in this case, there are several having equal values among $x_1, x_2, \ldots, x_n$. This is because, for example when $r_1, r_2, \ldots, r_k$ measurements have been made for x_1,

$x_2, \ldots, x_k$, the result is the same as if one applies consecutive numbering to

$$\overbrace{x_1, x_1, \cdots, x_1}^{r_1} \overbrace{x_2, x_2, \cdots, x_2}^{r_2} \cdots \overbrace{x_k, x_k, \cdots, x_k}^{r_k}$$

and uses $x_1, x_2, \ldots, x_n$ $(n = r_1 + r_2 + \cdots + r_k)$. The value obtained by subtracting the mean $\bar{x}$ from x_i is expressed as z_i, and $z = x - \bar{x}$.

Then, expansion by orthogonal polynomial in a case of unequal spacing becomes

$$\begin{aligned} y=&b_0+b_1z+b_2(-M_2^2-M_3z+M_2z^2) \\ &+b_3[(M_3^3-2M_2M_3M_4+M_2^2M_5)+(-M_4^2+M_2^2M_4-M_2M_3^2+M_3M_5)z \\ &+(M_2^2M_3+M_3M_4-M_2M_5)z^2+(M_2M_4-M_2^3-M_3^2)z^3]+\cdots \end{aligned} \tag{21.44}$$

$$M_i=\frac{1}{n}(z_1{}^i+z_2{}^i+\cdots+z_n{}^i) \qquad (i=1,\ 2,\ \cdots)$$

It is apparent from the equations above that although calculation to the quadratic term does not seem to be very troublesome, the cubic term is rather troublesome. As long as $x_1, x_2, \ldots, x_n$ becomes determined, inasmuch as M_i is a constant, the coefficient of b_2 is the quadratic equation of z and the coefficient of b_3 is the cubic equation of z, but it cannot be said that calculation is very easy; thus, when A has four or more levels, it is suggested that the total of the effects of the terms from the cubic on, $S_{A_{res}}$, be found as

$$S_{A\ \mathrm{res}}=S_A-S_{A_l}-S_{A_q}$$

and that one calculate S_{A_c}, etc., only when $S_{A_{res}}$ is large. When A is at three levels, even S_{A_q} is more simply found by subtracting S_{A_l} from S_A. If Equation (21.44) is abbreviated as

$$y=b_0+b_1f_1(z)+b_2f_2(z)+b_3f_3(z)+\cdots \tag{21.45}$$

$f_1(z) = z$ and $f_2(z)$ are linear and quadratic equations in z and $f_3(z)$ is a cubic equation in z. Then, the estimated values $\hat{b}_i$ of the coefficients b_i, the variances $\sigma^2(\hat{b}_i)$, $S_{\hat{b}_i}$, etc., are given by the following equations.

$$\hat{b}_0=\frac{1}{rn}(A_1+A_2+\cdots+A_n) \tag{21.46}$$

$$\hat{b}_i=\frac{1}{rS}(W_1A_1+W_2A_2+\cdots+W_nA_n) \qquad (i=1,\ 2,\ \cdots) \tag{21.47}$$

$$\sigma^2(\hat{b}_0)=\frac{1}{rn}\sigma^2 \tag{21.48}$$

$$\sigma^2(\hat{b}_i)=\frac{1}{rS}\sigma^2 \qquad (i=1,\ 2,\ \cdots) \tag{21.49}$$

$$S_{\hat{b}_i}=\frac{1}{rS}(W_1A_1+W_2A_2+\cdots+W_nA_n)^2 \qquad (i=1,\ 2,\ \cdots) \tag{21.50}$$

Here,

$$S = W_1^2 + W_2^2 + \cdots + W_n^2 \tag{21.51}$$

and W_ν is obtained by substituting $z = z_\nu$ in the coefficient $f_i(z)$ of the respective b_i in Equation (21.44). Thus,

$$W_\nu = f_i(z_\nu) \text{ when estimating } b_i \ (\nu = 1, 2, \ldots) \tag{21.52}$$

For example, since in the case of $r = 5$ and $A_1 = 0, A_2 = 10, A_3 = 30$, and $A_4 = 50$, we have

$$\bar{x} = \frac{1}{4}(0+10+30+50) = 22.5$$

the z_ν become

$$z_1 = 0 - 22.5 = -22.5$$

$$z_2 = 10 - 22.5 = -12.5$$

$$z_3 = 30 - 22.5 = 7.5$$

$$z_4 = 50 - 22.5 = 27.5$$

so that

$$M_2 = \frac{1}{4}[(-22.5)^2 + (-12.5)^2 + 7.5^2 + 27.5^2] = 368.75$$

$$M_3 = \frac{1}{4}[(-22.5)^3 + (-12.5)^3 + 7.5^3 + 27.5^3) = 1968.75$$

and W_1, W_2, W_3, and W_4 for b_1 and b_2 become as follows.

$$W_1 = -22.5, \quad W_2 = -12.5, \quad W_3 = 7.5, \quad W_4 = 27.5 \quad (\text{for } b_1)$$

$$\left.\begin{aligned} W_1 &= -M_2^2 - M_3(-22.5) + M_2(-22.5)^2 = 95\,000 \\ W_2 &= -M_2^2 - M_3(-12.5) + M_2(-12.5)^2 = -53\,750 \\ W_3 &= -M_2^2 - M_3(7.5) + M_2(7.5)^2 = -130\,000 \\ W_4 &= -M_2^2 - M_3(27.5) + M_2(27.5)^2 = 88\,750 \end{aligned}\right\} (\text{for } b_2)$$

It is simple to find $\hat{b}_i$ ($i = 1, 2$), etc., from this.

21.6 The Case of General Linear Functions

A Linear Function in General:

$$y = a_0 g_0(x) + a_1 g_1(x) + \cdots \tag{21.53}$$

We will explain orthogonal expansion when there is no unknown parameter in $g_\nu(x)$ and when we wish to impart a priority sequence to its order. An example would be a case where it can be expected that the relationship between reaction constant y and absolute temperature T will be expanded in the form

$$y=a_0+a_1\frac{1}{T}+a_2 \log T+\cdots \tag{21.54}$$

When it can be assumed that the power series expansion begins with the linear term (*i.e.*, it is a polynomial that passes through the origin), if the quotient obtained by dividing y by x is redefined as y, it becomes possible to use the expansion formula which begins with the constant term, as is. When an expansion such as Equation (21.53) can be used, one might also divide both sides by $g_0(x)$ and, with $y/g_0(x)$ as a new function of x, estimate the constant term and terms from the second term onward.

For example, when given the reaction system

$$2\,NO+O_2=2\,NO_2$$

if we wish to find out whether the differential equation

$$y=\frac{dx}{dt}=k(a-x)^2\left(b-\frac{x}{2}\right)$$

holds or not, we can just regard the quotient obtained by dividing y by $g_0(x)=(a-x)^2\left(b-\dfrac{x}{2}\right)$, as a new function of x and test whether or not terms other than the correction term are significant.

If one wishes to expand Equation (21.53) into an orthogonal function, then to obtain the *new expansion*,

$$y=b_0f_0(x)+b_1f_1(x)+b_2f_2(x)+\cdots \tag{21.55}$$

one need only put $f_i(x)$ in the *linear form* of $g_0(x)$, $g_1(x)$, . . . , with k levels of x as $x_1, x_2, \ldots, x_k$, so that

$$\sum_{\nu=1}^{k} f_i(x_\nu)\times f_j(x_\nu)=0 \qquad (i\neq j;\ i,\ j=0,\ 1,\ 2,\ \cdots)$$

$$\sum_{\nu=1}^{k} f_i^2(x_\nu)=S_i \qquad (i=0,\ 1,\ 2,\ \cdots) \tag{21.56}$$

holds. For this, one chooses $f_0(x)$, $f_1(x)$, . . . as follows.

$$f_0(x)=g_0(x) \tag{21.57}$$

$$f_1(x)=(0,\ 1)g_0(x)-(0,\ 0)g_1(x) \tag{21.58}$$

$$\begin{aligned} f_2(x)=&[(1,\ 1)(0,\ 2)-(0,\ 1)(1,\ 2)]g_0(x)+[(0,\ 0)(1,\ 2)\\ &-(0,\ 1)(0,\ 2)]g_1(x)+[(0,\ 1)^2-(0,\ 0)(1,\ 1)]g_2(x) \end{aligned} \tag{21.59}$$

$$\vdots \qquad\qquad \vdots$$

where

$$(i,j)=\sum_{\nu=1}^{k} g_i(x_\nu)g_j(x_\nu) \qquad (i,j=0,1,2,\cdots) \tag{21.60}$$

If the orthogonal expansion of Equation (21.55) is carried out, we obtain the following as the estimated value $\hat{b}_i$ for the respective b_i, with the sum of r experimental values for x_ν expressed as A:

$$\hat{b}_i=\frac{f_i(x_1)A_1+f_i(x_2)A_2+\cdots+f_i(x_k)A_k}{[f_i^2(x_1)+f_i^2(x_2)+\cdots+f_i^2(x_k)]r} \tag{21.61}$$

Please refer to Chapter 15 for the most general case where the function expansion is not the linear function of unknowns $b_0, b_1, b_2, \ldots$ shown in Equation (21.55).

21.7 Orthogonal Expansion of Regression Function with Two or More Variables

To simplify the explanation, we will assume that there are two descriptive variables, x and y, and that g is a characteristic value. We write the regression equation as follows.

$$g=m+ax+by \tag{21.62}$$

The method of least squares is used to estimate the values of m, a, and b from n classes of data for (g, x, y). But an estimated value by the method of least squares harbors the following problem point. Even if m, a, and b are found by solving Equation (21.62), this does not by any means constitute having found proportionality coefficients for x and y. According to C. R. Rao (Indian Statistical Institute), when the quantity of rice consumed per household, g, was expressed by a linear equation of the number of adult males in the household, x, the number of adult females, y, and the number of children, z,

$$g=ax+by+cz \tag{21.63}$$

and *when a, b, and c were found by the method of least squares, coefficient b turned out negative*. This is because there are correlations among the number of adult men x, the number of children z, and the number of adult women y, and therefore the estimation precision of a, b, and c worsens. To solve a problem such as this, it becomes necessary either to use the method of Chapter 15 or to perform a Schmidt-type expansion as in this section.

Let us consider the problem of expressing the target characteristic g by a linear equation in the descriptive variables x and y. First, we consider the linear regression equation,

$$g=m+a(x-\bar{x}) \tag{21.64}$$

between x and g, and what we wish to do is to use the same method as in the preceding section of fitting the linear term in y to the difference between the right side and g. The interpretation is that there is a relationship between x and y, and that all of this part of y that influences g is regarded as the effect of x. A relationship such as this where the effects of all other variables that have been found to be related with oneself have been incorporated into one's own effect is termed *a single regression relationship.* Coefficient a of the single regression is given, from Equation (21.64), as

$$a=\frac{(xg)}{(xx)} \tag{21.65}$$

Here, (xx) and (xg) mean the following.

$$(xx)=\sum(x-\bar{x})^2$$

$$(xg)=\sum(x-\bar{x})(g-\bar{g})=\sum(x-\bar{x})g=\sum xg-\frac{(\sum x)(\sum g)}{n} \tag{21.66}$$

Such inner-product symbols will be used from here on.

But, now, regarding the residue of Equation (21.64), how would we best find the relationship with y? For this, we find a coefficient termed the *partial regression coefficient;* but without using knowledge of matrices, here we will find this quantity by orthogonal expansion.

This method is the same as though, in an expansion of the following form,

$$g=m+a(x-\bar{x})+b[(y-\bar{y})-b'(x-\bar{x})] \tag{21.67}$$

b' is chosen so that all of the sums of products of the coefficients of unknowns m, a, and b will become zero. Since the sums of the products of m and of a and b become zero at

$$\sum(x-\bar{x})=0$$

$$\sum(y-\bar{y})-b'\sum(x-\bar{x})=0$$

we have, putting the sum of products of coefficients a and b as zero,

$$\sum(x-\bar{x})(y-\bar{y})-b'\sum(x-\bar{x})^2=0$$

$$b'=\frac{(xy)}{(xx)} \tag{21.68}$$

By substituting into Equation (21.64), we obtain the orthogonal expansion

$$g=m+a(x-\bar{x})+b\left[y-\bar{y}-\frac{(xy)}{(xx)}(x-\bar{x})\right] \tag{21.69}$$

Estimation of coefficient b in the equation above indicates the relationship between y and g after having completely eliminated the effect of x related with g either directly or through y. m, a, and b are all orthogonal, and

$$\hat{m}=\frac{1}{n}(g_1+\cdots+g_n) \tag{21.70}$$

$$\hat{a}=\frac{(xg)}{(xx)} \tag{21.71}$$

$$\hat{b}=\frac{(gy)-\frac{(xy)}{(xx)}(gx)}{\sum\left[(y-\bar{y})-\frac{(xy)}{(xx)}(x-\bar{x})\right]^2}=\frac{(xx)(gy)-(xy)(gx)}{(xx)(yy)-(xy)^2} \tag{21.72}$$

The error variation S_e when g is expressed by Equation (21.69) is as follows, since $m=\bar{g}$:

$$\begin{aligned}S_e&=\sum\{g-m-a(x-\bar{x})-b[(y-\bar{y})-b'(x-\bar{x})]\}^2\\&=S_g-S_x-S_{y/x}\end{aligned} \tag{21.73}$$

where

$$S_g=(gg) \tag{21.74}$$

$$S_x=\frac{(xg)^2}{(xx)} \tag{21.75}$$

$$S_{y/x}=S_{\hat{b}}=\frac{[(xx)(gy)-(xy)(gx)]^2}{(xx)^2(yy)-(xy)^2(xx)} \tag{21.76}$$

In general, when expressing the target characteristic value g in terms of $x, y, z, \ldots$, frequently the following *successive orthogonal expansion* is used.

$$\begin{aligned}g=&m+a(x-\bar{x})+b[y-\bar{y}-b'(x-\bar{x})]\\&+c\{z-\bar{z}-c'(x-\bar{x})-c''[y-\bar{y}-b'(x-\bar{x})]\}+\cdots\end{aligned} \tag{21.77}$$

$$\begin{aligned}V_e=&V_g-a^2V_x-b^2[V_y-(b')^2V_x]\\&-c^2\{V_z-(c')^2V_x-(c'')^2[V_y-(b')^2V_x]\}-\cdots\end{aligned} \tag{21.78}$$

The second, third, fourth, . . . term of the right side is obtained by dividing the square of the numerator by the denominator with regard to the estimated values of a, b, and c, and they are $S_a, S_b, S_c, \ldots$. As to the contribution ratios, we obtain approximately the following.

$$\rho_x=\frac{S_a}{(gg)} \tag{21.79}$$

$$\rho_{y/x}=\frac{S_b}{(gg)} \tag{21.80}$$

$$\rho_{z/x,y}=\frac{S_c}{(gg)} \tag{21.81}$$

$$\vdots$$

$$\rho_e=\frac{S_e}{(gg)} \tag{21.82}$$

21.8 A Worked Example

According to reports by the technical committee on survey of the build of Japanese persons (Japanese Standards Association, 1970, 1973), the mean, standard deviation, and correlation coefficient of the slacks length x, waist y, hip z, and crotch-to-waist measurement g of women about 20 years old are as given in Tables 21.9.

First, we investigate whether it is possible to adequately express hip z in terms of x and y; and if it is possible, we investigate whether it is possible to express g in terms of x and y. If it is not possible to adequately express z in terms of x and y, we consider the problem of expressing g in terms of x, y, and z. That it is possible to adequately express, here, means that the proportion of women falling within the tolerance width for the hip of 3 cm is 95%. [Tolerance width is a measurement of a pair of slacks: when it is allowed to be ± 1.5 cm (or $+2$ cm or -1 cm) from the exactly right measurement for the hip, the difference, 3 cm, is termed the tolerance width. If $\pm 2\times$ (standard deviation of error of the representation equation) is within one-half the tolerance width, it can be expected that at least 95% of women will not complain about the hip measurement.] First, we construct the table of variance and covariance, Table 21.10. Variance V is the square of the standard deviation, and the covariance is found, for example in the case of x and y, by $V_{xy} = r_{xy} s_x s_y$.

Then, since we are considering z instead of g in Equation (21.69), we have

$$m = \text{mean of } z = 88.36 \tag{21.83}$$

$$a = \frac{(xz)}{(xx)} = \frac{V_{xz}}{V_x} = \frac{5.96}{13.40} = 0.445 \tag{21.84}$$

Tables 21.9 (a) Fundamental statistics

	x	y	z	g
Mean (m)	88.66	60.54	88.36	25.94
Standard deviation (s)	3.66	3.63	4.02	2.09

(b) Correlation coefficient (r)

	x	y	z	g
x	1.000	0.248	0.405	0.467
y		1.000	0.694	0.266
z			1.000	0.355
g				1.000

Table 21.10 Table of Variance and Covariance

	x	y	z	g
x	13.40	3.29	5.96	3.57
y		13.18	10.13	2.02
z			16.16	2.98
g				4.37

13.40÷3.66²
3.29÷3.66×3.63×0.248

$$b' = \frac{(xy)}{(xx)} = \frac{V_{xy}}{V_x} = \frac{3.29}{13.40} = 0.246 \qquad (21.85)$$

$$b = \frac{(xx)(zy) - (xy)(zx)}{(xx)(yy) - (xy)^2} = \frac{V_x V_{zy} - V_{xy} V_{zx}}{V_x V_y - {V_{xy}}^2}$$
$$= \frac{13.40 \times 10.13 - 3.29 \times 5.96}{13.40 \times 13.18 - 3.29^2} = \frac{116.13}{165.79} = 0.700 \qquad (21.86)$$

Therefore, we have

$$z = 88.36 + 0.445(x - 88.66) + 0.700[y - 60.54 - 0.246(x - 88.66)] \qquad (21.87)$$

$$V_e = V_z - a^2 V_x - b^2[V_y - (b')^2 V_x]$$
$$= 16.16 - 0.445^2 \times 13.40 - 0.700^2(13.18 - 0.246^2 \times 13.40)$$
$$= 16.16 - 2.65 - 6.06 = 7.45 \qquad (21.88)$$

$$s_e = \sqrt{V_e} = 2.73 \qquad (21.89)$$

Therefore, for one-half the tolerance width, we have

$$k = \frac{1.50}{2.73} = 0.55 \qquad (21.90)$$

From the simplified normal distribution table, Table 21.11, we subtract the percentage for $k = 0.55$, 0.2912, from 0.5 and double the answer. This covers no more than about 42% of the people. Therefore, for the slacks measurements, it is necessary to use the hip as an independent variable besides the slacks length and waist. Table 21.12 is obtained when we show ready-to-wear measurements with at least 1% covering rate, with the tolerance width of the slacks length as 3 cm, the tolerance width of the waist as 3 cm, and the tolerance width of the hip as 3 cm.

Although not shown here, the standard deviations of the crotch-to-waist and the maximum thigh girth are 1.81 cm and 2.21 cm, when expressed by x, y, and z as the target variables.

Table 21.11 Simplified Normal Distribution Table (Proportion from k to ∞)

k	.00	.01	.02	.03	.04	.05	.06	.07	.08	.09
0.0	.5000	.4960	.4920	.4880	.4840	.4801	.4761	.4721	.4681	.4641
0.1	.4602	.4562	.4522	.4483	.4443	.4404	.4364	.4325	.4286	.4247
0.2	.4207	.4168	.4129	.4090	.4052	.4013	.3974	.3936	.3897	.3859
0.3	.3821	.3783	.3745	.3707	.3669	.3632	.3594	.3557	.3520	.3483
0.4	.3446	.3409	.3372	.3336	.3300	.3264	.3228	.3192	.3156	.3121
0.5	.3085	.3050	.3015	.2981	.2946	.2912	.2877	.2843	.2810	.2776
0.6	.2743	.2709	.2676	.2643	.2611	.2578	.2546	.2514	.2483	.2451
0.7	.2420	.2389	.2358	.2327	.2296	.2266	.2236	.2206	.2177	.2148
0.8	.2119	.2090	.2061	.2033	.2005	.1977	.1949	.1922	.1894	.1867
0.9	.1841	.1814	.1788	.1762	.1736	.1711	.1685	.1660	.1635	.1611
1.0	.1587	.1562	.1539	.1515	.1492	.1469	.1446	.1423	.1401	.1379
1.1	.1357	.1335	.1314	.1292	.1271	.1251	.1230	.1210	.1190	.1170
1.2	.1151	.1131	.1112	.1093	.1075	.1056	.1038	.1020	.1003	.0985
1.3	.0968	.0951	.0934	.0918	.0901	.0885	.0869	.0853	.0838	.0823
1.4	.0808	.0793	.0778	.0764	.0749	.0735	.0721	.0708	.0694	.0681
1.5	.0668	.0655	.0643	.0630	.0618	.0606	.0594	.0582	.0571	.0559
1.6	.0548	.0537	.0526	.0516	.0505	.0495	.0485	.0475	.0465	.0455
1.7	.0446	.0436	.0427	.0418	.0409	.0401	.0392	.0384	.0375	.0367
1.8	.0359	.0351	.0344	.0336	.0329	.0322	.0314	.0307	.0301	.0294
1.9	.0287	.0281	.0274	.0268	.0262	.0256	.0250	.0244	.0239	.0233
2.0	.0228	.0222	.0217	.0212	.0207	.0202	.0197	.0192	.0188	.0183
2.1	.0179	.0174	.0170	.0166	.0162	.0158	.0154	.0150	.0146	.0143
2.2	.0139	.0136	.0132	.0129	.0125	.0122	.0119	.0116	.0113	.0110
2.3	.0107	.0104	.0102	.0099	.0096	.0094	.0091	.0089	.0087	.0084
2.4	.0082	.0080	.0078	.0075	.0073	.0071	.0069	.0068	.0066	.0064
2.5	.0062	.0060	.0059	.0057	.0055	.0054	.0052	.0051	.0049	.0048
2.6	.0047	.0045	.0044	.0043	.0041	.0040	.0039	.0038	.0037	.0036
2.7	.0035	.0034	.0033	.0032	.0031	.0030	.0029	.0028	.0027	.0026
2.8	.0026	.0025	.0024	.0023	.0023	.0022	.0021	.0021	.0020	.0019
2.9	.0019	.0018	.0018	.0017	.0016	.0016	.0015	.0015	.0014	.0014
3.0	.0013	.0013	.0013	.0012	.0012	.0011	.0011	.0011	.0010	.0010

Exercises (21)

(1) In order to improve the rate of bleaching of sugar syrup, two factors, A and B, were each taken at the four levels

A = quantity of active carbon used
0.3, 0.5, 0.7, 0.9%

B = bleaching treatment temperature
60, 70, 80, 90°C

Table 21.12 Body Measurements for Ready-to-Wear Slacks (For Products of Comprehensive Covering Rate at Least 1%)

Region	Slacks length x	Waist y	Hip z	Crotch to waist g_1	Maximum thigh girth g_2	Comprehensive covering rate (%)
Tolerance widths	±1.5	±1.5	±1.5	±1.5	±3.0	
Standard deviation of error				1.81	2.21	
Covering rate by item				59	83	
1	87.0	57.0	87.0	25.3	49.8	1.1
2	87.0	60.0	87.0	25.4	50.2	1.1
3	90.0	57.0	87.0	26.0	49.5	2.4
4	90.0	57.0	90.0	26.2	51.3	2.0
5	90.0	60.0	87.0	26.1	49.8	2.1
6	90.0	60.0	90.0	26.4	51.7	4.0
7	90.0	60.0	93.0	26.6	53.5	1.8
8	90.0	63.0	90.0	26.5	51.7	2.4
9	90.0	63.0	93.0	26.7	53.9	2.4
10	90.0	66.0	96.0	27.7	55.8	1.3
11	93.0	60.0	90.0	27.0	51.4	1.4
12	93.0	60.0	93.0	27.3	53.2	2.0
13	93.0	63.0	93.0	27.4	53.6	2.3
14	93.0	63.0	96.0	27.6	55.4	1.7
				Total		28.0

and the following data were obtained when the color intensity was read off in terms of the Lumetron photoelectric colorimetric index.

	B_1	B_2	B_3	B_4
A_1	57	50	57	48
A_2	56	36	25	21
A_3	43	34	21	13
A_4	15	30	13	18

(a) Resolve A and B into the components of an orthogonal polynomial, and perform an analysis of variance considering $A_l \times B_l$ as well.

(b) We wish to bring the color index to 18 or less. Find the ranges of A and B that make this possible.

(2) The following data were obtained (unit, m) when three factors, A, B, and C, were taken as

A: speed $A_1 = 30, A_2 = 55, A_3 = 80$ km/h
B: vehicle type $B_1 =$ Morris Minicoupe
$B_2 =$ Corona 1600 GT
C: road surface $C_1 =$ dry, $C_2 =$ wet

and the braking distance was measured twice each, replicated, having attached new radial tires by the three-way layout method.

A	B	C	Data		A	B	C	Data	
1	1	1	3.25	3.35	2	2	1	10.80	10.75
1	1	2	5.15	4.95	2	2	2	19.25	20.70
1	2	1	3.72	3.82	3	1	1	26.55	27.50
1	2	2	6.35	6.08	3	1	2	49.85	51.25
2	1	1	9.45	9.40	3	2	1	30.08	28.05
2	1	2	16.30	16.20	3	2	2	60.52	60.25

(a) Resolve A into the components of an orthogonal polynomial and perform an analysis of variance.
(b) State the conclusions.

22 Design of Experiments for Instrumentation Method and SN Ratio

In this chapter, the problem of measurement error and the use of the SN ratio for comparison of measurement errors will be explained. This method is also a preparation for the expression of dynamic characteristics by the SN ratio, in Chapter 24.

22.1 Calibration Method and SN Ratio

22.1.1 Calibration of Hardness Tester and SN Ratio

We will show a method of preparing the calibration table for a certain hardness tester.

The readings when three persons, R_1, R_2, and R_3, measured the hardness of five levels of hardness reference pieces, $M_1 = 20$, $M_2 = 30$, $M_3 = 40$, $M_4 = 50$, and $M_5 = 60$, twice each with this hardness tester were as given in Table 22.1.

It will be assumed that there was no error in the indicated values for the reference pieces. It will be assumed that even if there was error, it was only from the second decimal place on, and that for practical purposes the value could be regarded as being the true value.

Table 22.1 Readings by Hardness Tester

	M_1	M_2	M_3	M_4	M_5	Total
R_1	21.0	30.6	40.5	50.25	59.8	
	21.0	30.5	40.4	50.3	59.8	
R_2	21.2	30.5	40.4	50.2	59.9	
	21.1	30.5	40.4	50.1	59.9	
R_3	21.1	30.6	40.4	50.4	59.9	
	21.0	30.4	40.4	50.2	59.9	
Total	126.4	183.1	242.5	301.45	359.2	1 212.65

We apply the following linear equation to the relationship between the reading y and the values of the hardness reference pieces M.

$$y=m+\beta(M-\overline{M})+e \tag{22.1}$$

Here $\overline{M}$ is the mean hardness of the five reference pieces.

$$\overline{M}=\frac{1}{5}(M_1+M_2+M_3+M_4+M_5)$$

$$=\frac{1}{5}(20+30+40+50+60)=40 \tag{22.2}$$

Also, e is error. The estimated values of m and β by the method of least squares, $\hat{m}$ and $\hat{\beta}$, are given by the following equations.

$$\hat{m} = \text{mean of all of the readings}$$

$$= \tfrac{1}{30}(126.40 + 183.10 + 242.50 + 301.45 + 359.20)$$

$$= \frac{1212.65}{30} = 40.42 \tag{22.3}$$

$$\hat{\beta}=\frac{\text{covariation of true values } M \text{ and readings } y}{\text{total variation of true values } M} \tag{22.4}$$

Here, total variation of true values M means 6 times

$$(M_1-\overline{M})^2+(M_2-\overline{M})^2+\cdots+(M_5-\overline{M})^2 \tag{22.5}$$

When 30 readings all have different M values, the total variation of the true values M is given by

$$\sum_{i=1}^{30}(M_i-\overline{M})^2 \tag{22.6}$$

and in this case, in effect, $M_1, M_2, \ldots, M_5$, have each been measured six times. Therefore, essentially the value which corresponds to Equation (22.6) is 6 times the value of Equation (22.5).

$$\text{total variation of true values } M=6[(M_1-\overline{M})^2+(M_2-\overline{M})^2+\cdots+(M_5-\overline{M})^2]$$

$$=6[(20-40)^2+(30-40)^2+\cdots+(60-40)^2]=6000 \tag{22.7}$$

Next, the sum of the products of the deviations of the true values M and readings y (termed covariation) is given by the following equation if all 30 of the values of the true values M are different and if the value of y has been obtained for each of them.

$$(M_1-\bar{M})y_1+(M_2-\bar{M})y_2+\cdots+(M_{30}-\bar{M})y_{30} \tag{22.8}$$

Since only the five values $M_1, M_2, \ldots, M_5$ are taken for the M in this case, however, Equation (22.8) can be written as follows.

$$\begin{aligned}&(M_1-\bar{M})(\text{total of values of readings } y \text{ from measurements on reference } M_1\\&\quad+(M_2-\bar{M})(\text{total of values of readings } y \text{ from measurements on reference } M_2+\cdots\\&\quad+(M_5-\bar{M})(\text{total of values of readings } y \text{ from measurements on reference } M_5\\&=(20-40)\times 126.40+(30-40)\times 183.10+\cdots+(60-40)\times 359.20\\&=5\,839.5\end{aligned} \tag{22.9}$$

Therefore, $\hat{\beta}$ in Equation (22.4) is

$$\hat{\beta}=\frac{5\,839.5}{6\,000}=0.9732 \tag{22.10}$$

and

$$y=40.42+0.9732(M-40) \tag{22.11}$$

In general, when the totals of the r_0 readings y for each of $M_1, M_2, \ldots, M_k$, $Y_1, Y_2, \ldots, Y_k$, have already been obtained, $\hat{m}$ and $\hat{\beta}$ can be obtained by

$$\hat{m}=\frac{Y_1+Y_2+\cdots+Y_k}{kr_0} \tag{22.12}$$

$$\hat{\beta}=\frac{(M_1-\bar{M})Y_1+(M_2-\bar{M})Y_2+\cdots+(M_k-\bar{M})Y_k}{r_0[(M_1-\bar{M})^2+(M_2-\bar{M})^2+\cdots+(M_k-\bar{M})^2]} \tag{22.13}$$

If Equation (22.11) is solved for M, one can obtain the equation for calculating the estimated value $\hat{M}$ of the true hardness M from the readings y.

$$\hat{M}=\frac{y-40.42}{0.9732}+40=-1.53+1.0275y \tag{22.14}$$

Thus, if the reading of the hardness of a certain object with this hardness tester is y, the estimated values of the true hardness can be obtained from Equation (22.14). This is termed *linear equation calibration.* In certain cases, one need only prepare beforehand a table that shows the estimated values of M by giving readings y at 1.0 intervals, or in other words a calibration table. In this case, such a table would be Table 22.2 provided the range of the readings y was from 20.0 to 62.0. For example, if the reading y of the hardness tester in question is 23.0, the estimated

Table 22.2 Calibration Table

reading	value of M
20.0	19.02
21.0	20.04
22.0	21.08
23.0	22.10
⋮	⋮
62.0	62.18

proportional part table	
0.1	0.10
0.2	0.21
0.3	0.31
0.4	0.41
0.5	0.51
0.6	0.62
0.7	0.72
0.8	0.82
0.9	0.92
1.0	1.03

value of the true hardness is 22.10 according to Table 22.2. If the reading of the hardness tester 23.3, one should add the proportional part for 0.3, 0.31, to the value in the calibration table for 23.0, 22.10, and 22.41 would be the resulting estimated value of the hardness after correction.

If the study described above took place in the developing stages of the hardness tester, one would insert a temporary scale into the tester and would switch later to the scale of M. If Equation (22.11) is the relationship equation for such a temporary scale y and the true values M, in effect one would graduate the two points

$$\begin{aligned} M_0 &= \text{value of } y \text{ corresponding to } 20 \\ &= 40.42 + 0.9732(20 - 40) = 20.956 \end{aligned} \tag{22.15}$$

$$\begin{aligned} M_1 &= \text{value of } y \text{ corresponding to } 60 \\ &= 40.42 + 0.9732(60 - 40) = 59.884 \end{aligned} \tag{22.16}$$

and would calibrate by dividing the difference between them into 40 equal parts.

22.1.2 Error Variance When Linear Equation Calibration Has Been Performed, and the SN Ratio as Its Reciprocal

Unless we state the contrary, we will seek the error variance when linear equation calibration has been performed. This is because in the stages of research and development of instrumentation methods, an arbitrary temporary scale is usually inserted at the beginning and only after obtaining the relationship Equation (22.14) and calibrating does one insert the new scale. Therefore, one need only concern oneself with the problem of the magnitude of the error variance after calibration by the linear equation.

It will be assumed that r_0 readings y_i have been obtained for each of the k levels of true values, $M_1, M_2, \ldots, M_k$. It does not matter if r_0 is 1. The totals of the r_0 for each of the readings for $M_1, M_2, \ldots, M_k$ will be expressed as $y_1, y_2, \ldots, y_k$. Each of $y_1, y_2, \ldots, y_k$ includes a total of r_0 readings. Therefore, in the linear equation

$$y = m + \beta(M - \overline{M}) + e \tag{22.17}$$

the estimated values of m and β, $\hat{m}$ and $\hat{\beta}$, are given by

$$\hat{m} = \frac{y_1 + y_2 + \cdots + y_k}{r_0 k} \tag{22.18}$$

$$\hat{\beta} = \frac{(M_1 - \overline{M})y_1 + (M_2 - \overline{M})y_2 + \cdots + (M_k - \overline{M})y_k}{r_0[(M_1 - \overline{M})^2 + (M_2 - \overline{M})^2 + \cdots + (M_k - \overline{M})^2]} \tag{22.19}$$

The term e is error.

If Equation (22.17) is solved for M,

$$M = \frac{1}{\beta}(y - m + \beta\overline{M}) - \frac{e}{\beta} \tag{22.20}$$

The left side of the equation above is the true values M, and the first term in the right side is the estimated values $\hat{M}$ of the true values M which have been obtained by readings y. Therefore, the value of the difference when the true values M are subtracted from the estimated values of the true value $\hat{M}$ is given by

$$\frac{y - m + \beta\overline{M}}{\beta} - M = \frac{e}{\beta} \tag{22.21}$$

The mean of the square of the difference on the left side of Equation (22.21) is the error variance, and this is equal to the mean of the square of the right side. The mean of the square of the right side, or in other words the error variance of the estimated value, is defined by the following equation.

$$\text{error variance of estimated value} = \text{mean of}\left(\frac{e}{\beta}\right)^2 = \frac{\text{error variance of regression equation}}{\beta^2} \tag{22.22}$$

It is simple to estimate the error variance of the regression equation, and it can be obtained from the following equation which consists of the error variation of the regression equation divided by the number of degrees of freedom.

$$V_e = \frac{\text{error variation of regression equation } S_e}{\text{degrees of freedom of error variation}}$$

$$= \frac{1}{kr_0 - 2}\left(\text{total variation of readings} - \text{variation of 1st-order term}\right) \tag{22.23}$$

Here,

$$\text{total variation of readings } S_T = \text{sum of squares of } kr_0 \text{ readings } y - \frac{\left(\begin{array}{c}\text{total of } kr_0 \\ \text{readings } y\end{array}\right)^2}{kr_0} \tag{22.24}$$

$$\text{variation of 1st-order term } S_\beta = \frac{[(M_1-\overline{M})y_1+(M_2-\overline{M})y_2+\cdots+(M_k-\overline{M})y_k]^2}{r_0[(M_1-\overline{M})^2+(M_2-\overline{M})^2+\cdots+(M_k-\overline{M})^2]} \tag{22.25}$$

It would seem advisable to estimate the β^2 in the denominator of Equation (22.22) by squaring the value of $\hat{\beta}$ in Equation (22.19). However, the squared $\hat{\beta}$ of Equation (22.19) is a somewhat excessive estimate of β^2. This is because $\hat{\beta}$ is an estimated value of β and there is error.

We will therefore use S_β in Equation (22.25) and will estimate $\hat{\beta}$ from the following equation.

$$\text{estimated value of } \beta^2 = \frac{S_\beta - V_e}{r_0[(M_1-\overline{M})^2+(M_2-\overline{M})^2+\cdots+(M_k-\overline{M})^2]} \tag{22.26}$$

The denominator of Equation (22.26) is termed the *effective number of replications* r. It is also expressed as n_e.

$$r = r_0[(M_1-\overline{M})^2+(M_2-\overline{M})^2+\cdots+(M_k-\overline{M})^2] \tag{22.27}$$

Actually, S_β is the estimate of the following value.

$$E(S_\beta) = \sigma^2 + r_0[(M_1-\overline{M})^2+(M_2-\overline{M})^2+\cdots+(M_k-\overline{M})^2]\beta^2 = \sigma^2 + r\beta^2 \tag{22.28}$$

Therefore, if the error variance V_e, which is the estimated value of σ^2, is subtracted from S_β, the resulting quantity is the estimated value of $r\beta^2$; and if this is divided by the number of replications r, the result is the estimated value of β^2.

Therefore, the estimated value of the error variance after performing linear equation calibration, V_e', can be expressed by the following equation.

$$V_e' = \frac{V_e}{\frac{1}{r}(S_\beta - V_e)} \tag{22.29}$$

Instead of using the error variance V_e', obtained by the equation above, which indicates the degree of badness, it is often more convenient to express the degree of goodness by using its reciprocal. We will term this the *SN ratio,* imitating the practice in communications. SN ratio is a term derived from the initials of the English term *signal-to-noise ratio*. The SN ratio will be *expressed symbolically by using the Greek letter* η. Thus, the SN ratio η is defined by the following equation.

$$\eta = \frac{\frac{1}{r}(S_\beta - V_e)}{V_e} \tag{22.30}$$

The reciprocal of the SN ratio η is none other than the value of the error variance, whose units are the square of the units of the signal factor M, and it is the error variance which indicates the mean of the square of the difference between the true value and the estimated value after performing linear equation calibration.

In the case of the hardness tester of the previous section, the total variation S_T, the variation of the linear term S_β, and the error variation S_e can be obtained as follows.

$$S_T = 21.0^2 + 21.0^2 + 21.2^2 + \cdots + 59.9^2 - \frac{1212.65^2}{30} = 5683.728 \quad (f=29) \tag{22.31}$$

$$S_\beta = \frac{[-20\times 126.40 + (-10)\times 183.10 + 10\times 301.45 + 20\times 359.20]^2}{6[(-20)^2 + (-10)^2 + 10^2 + 20^2]}$$
$$= 5683.293 \quad (f=1) \tag{22.32}$$

$$S_e = S_T - S_\beta = 0.435 \quad (f = 29 - 1 = 28) \tag{22.33}$$

From these, the error variance V_e and the number of replications r are

$$V_e = \frac{0.435}{28} = 0.01554 \tag{22.34}$$

$$r = 6[(20-40)^2 + (30-40)^2 + \cdots + (60-40)^2] = 6000 \tag{22.35}$$

The SN ratio η is then

$$\eta = \frac{\frac{1}{r}(S_\beta - V_e)}{V_e} = \frac{\frac{1}{6000}(5683.293 - 0.01554)}{0.01554} = 60.95 \tag{22.36}$$

The reciprocal of this, V_e', is

$$V_e' = \frac{1}{\eta} = \frac{1}{60.95} = 0.01640 \tag{22.37}$$

Thus, if the true hardness M is estimated from the readings y of this hardness tester by using the calibration table of Table 22.2, in effect its *error variance* is 0.01640. From this, the standard deviation of error, s', becomes as follows.

$$s' = \sqrt{0.01640} = 0.128 \tag{22.38}$$

If it is permissible to assume that the measurement error comes more or less within $\pm 3s'$, the limits of such error become as follows.

$$\pm 3 \times 0.128 = \pm 0.384 \tag{22.39}$$

Calculation of the SN ratio is easy to understand in practice if one approaches it by preparing an analysis of variance table such as Table 22.3.

Table 22.3 Analysis of Variance Table

Source	f	S	V	$E(V)$
β (linear term)	1	5 683.293	5 683.293	$\sigma^2+r\beta^2$
e (error term)	28	0.435	0.01554	σ^2
T	29	5 683.728		

σ^2 = true error variance of regression equation

$V_e = 0.01554$ is its estimated value (22.40)

$$r=r_0[(M_1-\overline{M})^2+(M_2-\overline{M})^2+\cdots+(M_k-\overline{M})^2]$$
$$=6[(-20)^2+(-10)^2+\cdots+20^2]=6000 \quad (22.41)$$

22.1.3 How to Obtain the SN Ratio

In order to obtain the SN ratio, it is necessary to calculate, for example, the total variation S_T, the variation of the linear term (also called the regression variation) S_β, the error variation S_e, and the number of replications r. The readings at 240-minute intervals of a certain alarm clock were as given in Table 22.4. We will find the SN ratio of this clock.

We calculate as follows. In this case we have $r_0 = 1$.

$$S_T=2.0^2+243.0^2+\cdots+1449.5^2-\frac{5078.5^2}{7}=1629644.00 \quad (f=6) \quad (22.42)$$

$$\overline{M}=\frac{1}{7}(0+240+\cdots+1440)=720 \quad (22.43)$$

$$M_1-\overline{M}=0-720=-720,\ M_2-\overline{M}=-480,\ \cdots,\ M_7-\overline{M}=720$$

From this,

$$r=r_0[(M_1-\overline{M})^2+(M_2-\overline{M})^2+\cdots+(M_7-\overline{M})^2]$$
$$=1[(-720)^2+(-480)^2+\cdots+720^2]=1612800 \quad (22.44)$$

Table 22.4 True Time and Read Time (Elapsed Time in Minutes)

	M_1	M_2	M_3	M_4	M_5	M_6	M_7	Total
Value of M	0	240	480	720	960	1 200	1 440	
y	2.0	243.0	484.0	725.5	966.5	1 208.0	1 449.5	5 078.5

Therefore,

$$S_\beta=\frac{[(M_1-\overline{M})y_1+(M_2-\overline{M})y_2+\cdots+(M_7-\overline{M})y_7]^2}{(M_1-\overline{M})^2+(M_2-\overline{M})^2+\cdots+(M_7-\overline{M})^2}$$

$$=\frac{(-720\times 2.0-480\times 243.0+\cdots+720\times 1449.5)^2}{1\,612\,800}=\frac{1\,621\,200^2}{1\,612\,800}$$

$$=1\,629\,643.75 \tag{22.45}$$

$$S_e=S_T-S_\beta=1\,629\,644.00-1\,629\,643.75=0.25 \tag{22.46}$$

Thus, the analysis of variance table is as shown in Table 22.5.

$$\eta=\frac{\frac{1}{1\,612\,800}(1\,629\,643.75-0.050)}{0.050}=20.2 \tag{22.47}$$

The reciprocal of this is

$$V_e'=\frac{1}{20.2}=0.0495 \tag{22.48}$$

One can also obtain the SN ratio η of Equation (22.47) by using the $k=7$ of Appendix Table 7, (1) with $r_0=1$, $h=240$.

Essentially, calculation of the SN ratio is the method of obtaining the error variance after calibration without allowing the incorporation of the error for the calibration, doing away with the need to obtain the mean of the sum of squares of the error after obtaining the value after calibration for each reading.

The value of the SN ratio η itself, given in Equation (22.47), indicates the inverse of the error variance of this measurement method after performing accurate calibration. Therefore, if one merely seeks the measurement error, the "raw" value of the SN ratio suffices, but if one wishes to compare the SN ratios of various instrumentation methods, rather than using the raw value, it is more convenient to use the *decibel value,* which is 10 times its common logarithm.

In order to obtain the decibel value of the SN ratio η of Equation (22.47), we obtain the decibel value of 2.02 as 3.05 db from Appendix Table 6; since η is 10 times 2.02, we add 10 db and obtain 13.05 db.

$$\text{decibel value of } 20.2 = 10 \text{ db} + \text{decibel value of } 2.02$$

$$= 10.00 + 3.05 = 13.05 \text{ db} \tag{22.49}$$

Table 22.5 Analysis of Variance Table for Readings of Alarm Clock

Source	f	S	V	$E(V)$
β	1	1 629 643.75	1 629 643.75	$\sigma^2+1\,612\,800\,\beta^2$
e	5	0.25	0.050	σ^2
T	6	1 629 644.00		

When actually calculating the SN ratio, the values of S_β, V_e, and r are needed. The calculation methods for these differ according to how the data are acquired in experimentation or surveys. It also differs according to whether the values of the levels of signal factors are known. This matter will be discussed in the next section.

Determination of the *confidence limit of the SN ratio* η is carried out by using Appendix Table 8. To use Appendix Table 8, the *variance ratio* F_0 and the *degrees of freedom of the error variance*, f_2, are necessary. The number of degrees of freedom of the error variance f_2 is 5, and

$$F_0=\frac{V_\beta}{V_e}=\frac{1\,629\,643.75}{0.05}\doteqdot\infty$$

Therefore, from Appendix Table 8, with $f_2 = 5$ and $F_0 = \infty$,

$$\eta=13.05^{-7.8}_{+4.1}\ \text{[db]} \qquad (22.50)$$

The method of finding the SN ratio differs according to the experimental conditions and method, and is divided between the cases where the linear effect of the signal, S_β, can be obtained and those where this is not possible. Cases where the linear effect of the signal can be obtained are:

(1) The case where the values of the levels of the signal factor are known with sufficient accuracy, as in the examples of the previous section and the section before that. The case where the values for k objects which have been obtained by precision measurement with the error being smaller than at least one decimal place is also such a case. The case where there are several standard samples of small error is such a case, as well.

(2) The case where the values of the levels of the signal factor are unknown but the intervals are known, such as the case where the quantity x of a certain component in the sample is unknown but four levels of substances of known concentration have been added so that the additional concentration of that component has become $+h$, $+2h$, and $+3h$, *viz.*, the samples of x, $x+h$, $x+2h$, and $x+3h$ are used as signal factors. Unequal intervals are acceptable.

(3) The case where both the values of the levels of the signal factors and the interval are unknown but the interval ratios can be determined. For example, when testing levels which measure flatness, if we assume that one takes a thin metal shim and, M_1 = one inserts it from the right, M_2 = one does not insert it, and M_3 = one inserts it from the left, onto the surface plate, we have three levels of equal intervals; although the true values of the angle and the intervals are unknown, the intervals are equal. If a certain analysis sample is diluted to two times, four times, and eight times, samples whose concentration ratios are 1, 0.5, 0.25, and 0.125

are obtained. Again, when one wishes to study the measurement method for lengths, if one brings one block gauge and leaves it standing at k levels of room temperature, $M_1, M_2, \ldots, M_k$, it can be assumed that the gauge intervals change linearly with temperature. Room temperatures M_1, $M_2, \ldots, M_k$ are the values of the levels of the signal factor, and the ratios of the differences of the gauge intervals become equal to the ratios of the temperature differences. *If there is a variable M which linearly influences the characteristic value that one wishes to measure, it is possible to vary the levels of the characteristic value one wishes to measure by taking M as the signal factor.* Although the true value of the characteristic value one wishes to measure is unknown, the ratios of the differences between levels become equal to the ratios of the differences between the levels of M. This method is a very important one for creating useful signal factors.

As to case (2), the formula for case (1) can be used as it is if the first level x is taken as 0 or as its expected value x_0, and if we let $M_1 = 0$ (or x_0), $M_2 - h_1$ (or $x_0 + h_1$), $\ldots$, $M_k = h_k$ (or $x_0 + h_k$).

As to case (3), one can use the same formula as in case (1) if the interval ratios or their linear expression are expressed as $M_1, M_2, \ldots, M_k$. For example, since it is equal intervals in the case of the level, one can represent them as $M_1 = -1, M_2 = 0, M_3 = 1$ or as $M_1 = 0, M_2 = 1, M_3 = 2$. In the case of the block gauge, the room temperature itself is to be M_1, $M_2, \ldots, M_k$.

For the reasons given above, we use only case (1) for the formula for the SN ratio. This is Appendix Table 8. It is separated into two parts: the case of linear equation calibration and the case of proportional equation calibration. The proportional equation calibration differs only in the respect that the proportional equation $y = \beta M + e$ is considered as the relationship between the reading y and the signal factor M. General equations will be given for case (1).

(1) *Linear Equation Calibration*

For the signal factor level values $M_1, M_2, \ldots, M_k$, when the sums of r_1, $r_2, \ldots, r_k$ readings $y_1, y_2, \ldots, y_k$ have been given,

$$S_\beta = \frac{\left[M_1 y_1 + \cdots + M_k y_k - \dfrac{(r_1 M_1 + \cdots + r_k M_k)(y_1 + \cdots + y_k)}{r_1 + \cdots + r_k}\right]^2}{r} \tag{22.51}$$

$$r = r_1 M_1^2 + \cdots + r_k M_k^2 - \frac{(r_1 M_1 + \cdots + r_k M_k)^2}{r_1 + \cdots + r_k}$$

$$\eta = \frac{\dfrac{1}{r}(S_\beta - V_e)}{V_e}$$

The error variance is obtained by finding the error variation and dividing this by the number of degrees of freedom. The total variation is the difference obtained by subtracting the correction term from the sum of squares of the readings.

(2) *Proportional Equation Calibration*

$$S_\beta = \frac{(M_1y_1 + \cdots + M_ky_k)^2}{r} \tag{22.52}$$

$$r = r_1M_1^2 + \cdots + r_kM_k^2$$

$$\eta = \frac{\frac{1}{r}(S_\beta - V_e)}{V_e}$$

where the total variation used is the sum of squares of the readings without subtracting the correction term. Whichever calibration equation is used, one should not only subtract S_β from the total variation S_T but in certain cases also subtract effects that are counted as neither error nor signal. Since, however, this problem differs according to the individual experimental design, only the methods of obtaining S_β and r have been shown in Appendix Table 7.

22.2 Instrumentation Method and Its Error

22.2.1 Measurement of Moment of Inertia

The value of the moment of inertia of a rotating body can be obtained with sufficient accuracy by a formula in which the diameter, mass, etc., are used when the shape of the rotating body is simple. For example, the moment of inertia I_g of a disc (to be accurate, a cylinder) can be obtained, if

D = diameter of disc (m)
W = mass of disc (kg)

by the formula

$$I_g = \frac{W \times D^2}{8} \tag{22.53}$$

For example, the moment of inertia I_g of a disc whose D is (diameter 100 mm) and whose weight is 5 kg is as follows if the unit is kg · m².

$$I_g = \frac{5 \times 0.1^2}{8} = 0.00625 \text{ [kg·m}^2\text{]} \tag{22.54}$$

However, if the shape is somewhat complicated, such as the blade of a fan, it becomes difficult to obtain the moment of inertia from a formula.

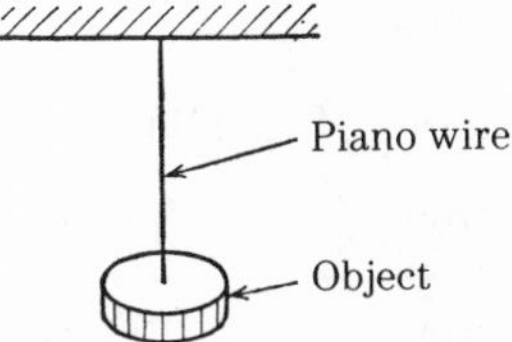

FIGURE 22.1 Moment of Inertia Measurement Apparatus

In a case such as this, therefore, a method is used in which the object whose moment of inertia is being sought is hung on a piano wire, as shown in Figure 22.1, a torque is applied to the object, and the moment of inertia is obtained by measuring the period of the rotational oscillation of the object.

It is known that the period y of the rotational oscillation when torque has been applied to an object as illustrated in Figure 22.1 is given by an equation such as the following.

$$y = 2\pi\sqrt{\frac{I_g}{C}} \tag{22.55}$$

where

I_g = moment of inertia of object (kg · m^2)
C = rigidity of piano wire (kg · m^2/s^2)

Let us consider the method of obtaining the moment of inertia I_g of an object by using Formula (22.55). The relationship between the observed value y of the period of oscillation and the square root M of the moment of inertia I_g of the object becomes as follows.

$$y = \beta M \tag{22.56}$$

where β is a constant which is determined by the piano wire. In this case,

$$\beta = \frac{2\pi}{\sqrt{C}} \tag{22.57}$$

If one hangs an object whose moment of inertia can be calculated, such as a cylindrical object, on a piano wire, and if one applies a torque and measures the period y of the oscillation, one can obtain the constant of proportionality β of Equation (22.56) therefrom. Once the constant β is obtained, if one next hangs an object whose moment of inertia cannot be calculated on the same piano wire, then applies torque and measures the period of the oscillation one can divide the value obtained by the proportionality constant β and obtain the square root of the moment of inertia, M, so that one can obtain the moment of inertia I_g by squaring this value. We will explain the measurement method for the moment of inertia by an actual example.

22.2.2 Calculation of SN Ratio

Four objects whose moment of inertia could be calculated,

M_1 = cylinder made of iron	200.0ϕ	7.250 kg
M_2 = cylinder made of iron	180.0ϕ	2.155 kg
M_3 = cylinder made of iron	100.0ϕ	0.950 kg
M_4 = cylinder made of aluminum	100.0ϕ	0.212 kg

were taken and, using a piano wire of nominal value 0.6ϕ of a certain length, three persons, R_1, R_2, and R_3, read the period of oscillation, and the data obtained were as shown in Table 22.6.

In general, if one wishes to measure a certain quantity, it is necessary to prepare several objects whose true value of this quantity varies. Since the objective is to measure the moment of inertia in this case, we prepare four objects, M_1, M_2, M_3, and M_4, whose true value of the moment of inertia, in other words the adequately accurate value, is known. As the levels of this signal factor M, it is advisable to gather objects whose level has been varied sufficiently broadly as to cover the ranges of the moment of inertia that we wish to measure. It is also important that they be so chosen that it is possible to study the errors when M has been obtained from Equation (22.56), no matter what the material type and no matter what the height of the cylinder, by varying the material type, cylinder height, etc., as much as possible.

The square root values of the moments of inertia of the four objects can be obtained from the following equations.

$$M_1=\sqrt{\frac{W}{8}}\times D=\sqrt{\frac{7.250}{8}}\times 0.20=0.19039 \tag{22.58}$$

$$M_2=\sqrt{\frac{2.155}{8}}\times 0.18=0.09342 \tag{22.59}$$

$$M_3=\sqrt{\frac{0.950}{8}}\times 0.10=0.03446 \tag{22.60}$$

$$M_4=\sqrt{\frac{0.212}{8}}\times 0.10=0.01628 \tag{22.61}$$

Table 22.6 Readings of Period of Rotation Vibration (Unit: s) (Nippondensō, Shōgo Yamamoto)

	R_1	R_2	R_3	Total
M_1	27.3	27.1	26.9	81.3
M_2	13.2	13.2	13.4	39.8
M_3	5.0	5.0	5.0	15.0
M_4	2.4	2.3	2.3	7.0

From these we obtain the variations in the following manner. Since there is no constant term in Equation (22.56) in the present case, the total variation becomes the sum of squares of the data, and one must not subtract the correction term. In this example we are seeking the SN ratio for the case of proportional equation calibration.

$$S_T=27.3^2+27.1^2+\cdots+2.3^2=2822.69 \quad (f=12) \tag{22.62}$$

Also, the variation of the linear term S_β is obtained from the following equation.

$$\begin{aligned} S_\beta &= \frac{(M_1\times y_1+M_2\times y_2+M_3\times y_3+M_4\times y_4)^2}{3(M_1^2+M_2^2+M_3^2+M_4^2)} \\ &= \frac{(0.19039\times 81.3+0.09342\times 39.8+0.03446\times 15.0+0.01628\times 7.0)^2}{3\times(0.19039^2+0.09342^2+0.03446^2+0.01628^2)} \\ &= \frac{19.827683^2}{3\times 0.0464281785}=2822.55 \end{aligned} \tag{22.63}$$

From this, the error variation S_e is

$$S_e=S_T-S_{M_l}=2822.69-2822.55=0.14 \tag{22.64}$$

Therefore, we obtain the table of analysis of variance of Table 22.7. From this, the SN ratio η can be obtained as follows.

$$\eta=\frac{\frac{1}{3\times 0.0464281785}(2822.55-0.013)}{0.013}=1558810=61.93\ [\text{db}] \tag{22.65}$$

The SN ratio η is the reciprocal of the error variance when M, which is the square root of the moment of inertia, was obtained. Thus, the error variance when the square root of the moment of inertia, M, was obtained by dividing the reading y of the oscillation period by the proportionality constant β is

$$\frac{1}{\eta}=\frac{1}{1558810}=0.000000642 \tag{22.66}$$

Therefore, if it is allowable to regard three times the standard deviation of error as the error of the estimated value of the signal factor M, its value is

$$\pm 3\times\sqrt{0.000000642}=\pm 0.0024 \tag{22.67}$$

Table 22.7 Analysis of Variance Table

Source	f	S	V	$E(V)$
β	1	2822.55	2822.55	$\sigma^2+3(M_1^2+M_2^2+M_3^2+M_4^2)\beta^2$
e	11	0.14	0.013	σ^2
T	12	2822.69		

Therefore, the error of the estimated value of the moment of inertia I_g is

$$I_g=(M\pm0.0024)^2\fallingdotseq M^2\pm0.0048\times M \tag{22.68}$$

22.2.3 How to Obtain the Moment of Inertia

The SN ratio is merely the reciprocal of the error variance of the given measurement method. For the person who is in charge of calibration, once a good measurement method has been selected by using the SN ratio, it is important that he first establish a concrete method of measurement on the basis of this measurement method and the error range of the measured values.

In order to obtain the square root of the moment of inertia, M, by using Equation (22.56), one need only obtain the estimated value $\hat{\beta}$ of the first-order coefficient β, then use the following equation.

$$M=\frac{y}{\hat{\beta}}\pm\sqrt{F\times\frac{1}{\eta}\left[1+\frac{1}{3(M_1^2+\cdots+M_4^2)}\left(\frac{y}{\hat{\beta}}\right)^2\right]} \tag{22.69}$$

Here,

$$\hat{\beta}=\frac{M_1\times y_1+M_2\times y_2+M_3\times y_3+M_4\times y_4}{3(M_1^2+M_2^2+M_3^2+M_4^2)}=\frac{19.827683}{3\times0.0464281785}=142.35 \tag{22.70}$$

F = the 5% value of the F table when the number of degrees of freedom of the numerator is 1 and that of the denominator is the same as that of the error variance V_e when the SN ratio was calculated

$$= 4.84 \tag{22.71}$$

Therefore,

$$M=\frac{y}{142.35}\pm\sqrt{4.84\times\frac{1}{1\,559\,000}\left[1+\frac{1}{3\times0.0464281785}\times\left(\frac{y}{142.35}\right)^2\right]}$$

$$=0.007025y\pm\sqrt{0.000003105(1+0.0003543y^2)} \tag{22.72}$$

If we assume, for example, that when the period of oscillation of a certain object was measured with the piano wire in question it was

$$y=8.4\ [\mathrm{s}] \tag{22.73}$$

then the square root M of the moment of inertia of that object is

$$M=\frac{8.4}{142.35}\pm\sqrt{0.000003105(1+0.0003543\times8.4^2)}$$

$$=0.0590\pm\sqrt{0.000003105(1+0.025)}=0.0590\pm0.0018 \tag{22.74}$$

Therefore, the moment of inertia I_g is

$$I_g=(M\pm0.0018)^2=(0.0590\pm0.0018)^2=0.00348\pm0.00021 \tag{22.75}$$

In many cases, one can also use the following *simplified formula* for the confidence limits of the level value of signal factor M:

$$M = M \pm 3 \times \frac{1}{\sqrt{\eta}} \tag{22.76}$$

When the equation above is used,

$$M = 0.0590 \pm 0.0024 \tag{22.77}$$

Therefore, for the moment of inertia I_g, by squaring the equation above we get

$$I_g = 0.00348 \pm 0.00028 \tag{22.78}$$

We recommend the use of the simple formula when reasonable standards have been used.

22.2.4 When Only One Standard Product Is Used

Thus far, when estimating the moment of inertia of a product by using a piano wire of 0.6ϕ, we used four sample items whose moments of inertia could be obtained; what is more, we conducted three measurements in every case and obtained the analysis of variance and the first-order coefficient β. However, during the use of this piano wire, perhaps the value of the initially determined coefficient $\hat{\beta}$ gets thrown off a little at a time. In such a case, it would be far too much trouble to use as many as four objects each time, to find the period of the oscillation three times for each, and to prepare a table of the analysis of variance or recalculate the linear coefficient β. Generally speaking, therefore, one would wish to calibrate by conducting estimation of β by using only one object whose moment of inertia is known.

It will be assumed that for a certain object whose moment of inertia is known, the square root of this quantity is M_1. Then, if the reading of the period of oscillation when torque has been applied to this object is expressed as y_1, we obtain

$$\hat{\beta} = \frac{M_1 \times y_1}{M_1^2} = \frac{y_1}{M_1} \tag{22.79}$$

Therefore, if the reading of the period of oscillation when torque has been applied to an object of unknown moment of inertia, found by the use of $\hat{\beta}$ in Equation (22.79), is y, the value of the square root of the moment of inertia M of this object can be estimated by the following equation.

$$M = M_1 \times \frac{y}{y_1} \pm \sqrt{F \times \frac{1}{\eta} \times \left[1 + \left(\frac{y}{y_1}\right)^2\right]} \tag{22.80}$$

where, for the SN ratio η we use the value which was obtained when the SN ratio η was calculated. In this case, we use

$$F = 4.84, \qquad \eta = 1\,558\,810$$

For example, if the object for calibration is a cylinder made of iron of 180ϕ and 2.155 kg,

$$M_1=\sqrt{\frac{W}{8}}\times D=\sqrt{\frac{2.155}{8}}\times 0.180=0.09342 \tag{22.81}$$

Therefore, as an example it will be assumed that the reading of the period of oscillation y_1 when torque has been applied to this object (termed the standard sample or reference product or the like) was 13.2.

Next, we bring an object for which calculation of the moment of inertia is difficult and hang this onto the same piano wire; it will be assumed that when torque is applied, the reading of the period of oscillation y is 8.4.

From Formula (22.80), therefore, if the moment of inertia of this object is I_g,

$$\begin{aligned}\sqrt{I_g}&=M_1\times\frac{y}{y_1}\pm\sqrt{F\times\frac{1}{\eta}\left[1+\left(\frac{y}{y_1}\right)^2\right]}\\&=0.09342\times\frac{8.4}{13.2}\pm\sqrt{4.84\times\frac{1}{1558810}\left[1+\left(\frac{8.4}{13.2}\right)^2\right]}\\&=0.0594\pm 0.0021\end{aligned} \tag{22.82}$$

By squaring both sides,

$$I_g=0.00353\pm 0.00025 \tag{22.83}$$

In actual practice, one uses Equation (22.76). Thus, in

$$M=M_1\times\frac{y}{y_1}\pm 3\times\frac{1}{\sqrt{\eta}} \tag{22.84}$$

we substitute

$$y=8.4$$

$$\frac{y_1}{M_1}=\frac{13.2}{0.09342}=141.3$$

$$\eta=1558810$$

and we obtain

$$\sqrt{I_g}=\frac{8.4}{141.3}\pm 3\times\frac{1}{\sqrt{1558810}}=0.0594\pm 0.0024 \tag{22.85}$$

We square and get

$$I_g=0.00353\pm 0.00029 \tag{22.86}$$

It is necessary to use a precision formula for the confidence limits of the estimated value of the signal factor level value M only when conducting a very bad calibration. In the present case, an instance where calibration is performed using a very light object corresponds to this. When a disc of aluminum is used as M_1, we obtain

$$M_1 = 0.01628 \tag{22.87}$$

so that, when

$$y_1 = 2.4 \qquad y = 8.4$$

the precision formula of Equation (22.82) becomes as follows.

$$\sqrt{I_g} = \frac{0.01628}{2.4} \times 8.4 \pm \sqrt{4.84 \times \frac{1}{1558810}\left[1 + \left(\frac{8.4}{2.4}\right)^2\right]} = 0.0570 \pm 0.0064$$

$$I_g \doteqdot 0.00325 \pm 0.00073 \tag{22.88}$$

On the other hand, when the simple formula of Equation (22.84) is used, we have

$$\hat{\beta} = \frac{2.4}{0.01628} = 147.4$$

and therefore

$$\sqrt{I_g} = \frac{8.4}{147.4} \pm 3\sqrt{\frac{1}{\eta}} = 0.0570 \pm 0.0024$$

$$I_g \doteqdot 0.00325 \pm 0.00028 \tag{22.89}$$

Equations (22.88) and (22.89) give confidence limits that differ by a wide margin. Since the correct equation is the former, this means that it is not possible to use Formula (22.84), which is the approximation method for the confidence limit. However, Equation (22.82) is a very cumbersome formula to use. To be able to use Formula (22.84), which is desirable for practical purposes, it is advisable to conduct calibration logically. We mentioned that there are linear equation calibration and proportional equation calibration, but in any of the following cases:

(1) One calibrates at the zero point and at a reference point which is close to the maximum.
(2) One uses a reference point which is near the mean and a reference point which is near either the minimum or the maximum.
(3) In the case of proportional equation calibration, one calibrates with a standard which is near the maximum (at least greater than a value close to the mean).
(4) One calibrates at two or more reference points which are suitably far apart.

it is permissible to use

$$M = \hat{M} \pm 3 \times \frac{1}{\sqrt{\eta}} \tag{22.90}$$

for the confidence limits.

One uses the precision formula in the case of estimation which is regarded as being especially important and in a case where the conditions mentioned above are not satisfied.

22.2.5 Formula for Measurement in a General Case of Linear Equation Calibration

Before continuing our discussions, we will show the formula to use when measuring by using k objects whose true values $M_1, M_2, \ldots, M_k$, are known. This is the formula to be used when estimating M from the true value M and reading y in the following linear relationship equation:

$$y = m + \beta(M - \overline{M})$$

The SN ratio of this measurement method will be represented by η.

It will be assumed that m and β have been estimated by measuring M_1, $M_2, \ldots, M_k$ r_0 times each, and that calibration has been performed thereby. It will be assumed that by using this calibration equation, the value of y has been read for an object of unknown true value M.

We then estimate M from the following equation.

$$M = \hat{M} \pm \sqrt{F \times \frac{1}{\eta}\left[1 + \frac{1}{kr_0} + \frac{1}{r_0\sum(M_i - \overline{M})^2}\left(\frac{y - \hat{m}}{\hat{\beta}}\right)^2\right]} \quad (22.91)$$

for practical purposes, we use

$$M = \hat{M} \pm 3 \times \frac{1}{\sqrt{\eta}} \quad (22.92)$$

where

$$\hat{M} = \frac{y - \hat{m}}{\hat{\beta}} + \overline{M}$$

$$\overline{M} = \frac{1}{k}(M_1 + M_2 + \cdots + M_k)$$

$\hat{m}$ = mean y of kr_0 y values

$$\hat{\beta} = \frac{\Sigma(M_i - \overline{M})y_i}{r_0\Sigma(M_i - \overline{M})^2}$$

y_i = total of r_0 readings for M_i objects

F = the 5% value of the F table when the number of degrees of freedom of the numerator is 1 and that of the denominator is the same as that of the error variance V_e when the SN ratio was calculated.

$\eta = SN$

Especially when conducting calibration by using values y_1 and y_2 which are readings taken once each on two standard samples, we have

$$M = \hat{M} \pm \sqrt{F \times \frac{1}{\eta}\left[1 + \frac{1}{2} + \frac{2}{(M_1 - M_2)^2} \times \left(\frac{y - \hat{m}}{\hat{\beta}}\right)^2\right]} \quad (22.93)$$

$$\hat{M} = \overline{M} + \frac{y - \hat{m}}{\hat{\beta}}$$

$$\hat{m} = \frac{y_1 + y_2}{2}$$

$$\hat{\beta}=\frac{M_1y_1+M_2y_2-(M_1+M_2)(y_1+y_2)/2}{(M_1-M_2)^2/2}$$

The approximate expression for practical purposes is given by the following equation.

$$M=\hat{M}\pm\frac{3}{\sqrt{\eta}} \tag{22.94}$$

Here, $\hat{M}$ is the same as in Equation (22.93).

22.2.6 When Calibrating with the Zero Point and One Standard

The formula for carrying out calibration by using the reading at the zero point y_1 and the reading at a certain standard (its true value is to be M_1) y_2, and for estimating the true value M from the reading y of a certain object (its true value is to be M) is as follows.

Precision Formula

$$M=\hat{M}\pm\sqrt{F\times\frac{1}{\eta}\left[\frac{3}{2}+2\left(\frac{y-y_1}{y_2-y_1}-\frac{1}{2}\right)^2\right]} \tag{22.95}$$

where

$$\hat{M}=M_1\times\frac{y-y_1}{y_2-y_1}$$

y_2-y_1 = reading when standard is measured − reading at zero point

$y-y_1$ = reading when object is measured − reading at zero point

Simplified Formula

$$M=M_1\times\frac{y-y_1}{y_2-y_1}\pm\frac{3}{\sqrt{\eta}} \tag{22.96}$$

22.2.7 Calculation Example

The SN ratio η in the quantitative analysis of alcohol in soy sauce was

$$\eta=22.4\ \text{[db]}$$

with alcohol content 1% as the unit increment and with the number of degrees of freedom of error variance being 6. We wish to perform calibration by using standard samples whose alcohol content is 0.0% and 1.0%, and to determine the alcohol in a certain soy sauce product. The readings for the alcohol components of the standard products and the soy sauce when read by gas chromatography were as follows.

Standard sample of alcohol content 0.0% (0) = 0.08 ($=y_1$)
Standard sample of alcohol content 1.0% (M_1) = 2.12 ($=y_2$)
Soy sauce sample of unknown content (M) = 1.05 ($=y$)

From formula (22.95),

$$y_2 - y_1 = 2.12 - 0.08 = 2.04$$

$$y - y_1 = 1.05 - 0.08 = 0.97$$

$$\hat{M} = M_1 \times \frac{y - y_1}{y_2 - y_1} = 1.0 \times \frac{0.97}{2.04} = 0.48 \tag{22.97}$$

Since the SN ratio η of 22.4 db is 174.7, and since F for number of degrees of freedom of error variance 6 is 5.99, we have

$$M = \hat{M} \pm \sqrt{F \times \frac{1}{\eta}\left[\frac{3}{2} + 2\left(\frac{y - y_1}{y_2 - y_1} - \frac{1}{2}\right)^2\right]}$$

$$= 0.48 \pm \sqrt{5.99 \times \frac{1}{174.7}\left[\frac{3}{2} + 2\left(\frac{0.97}{2.04} - \frac{1}{2}\right)^2\right]} = 0.48 \pm 0.23 \tag{22.98}$$

By the simplified formula,

$$M = M_1 \times \frac{y - y_1}{y_2 - y_1} \pm \frac{3}{\sqrt{\eta}} = 1.0 \times \frac{0.97}{2.04} \pm \frac{3}{\sqrt{174.7}} = 0.48 \pm 0.23 \tag{22.99}$$

22.3 Comparison of Measurement Methods (1): Example of Chemical Analysis

22.3.1 Introduction

In the preceding section we showed a measurement method for finding the moment of inertia by using a piano wire, but actually prior thereto, studies of the optimum measurement parameters such as the thickness of the piano wire, the length of the piano wire, and the number of oscillations measured had been carried out. Generally speaking, when studying improvements of a measurement method it is necessary to distinguish the following three types of factors.

(1) *Control Factors.* These are factors for the improvement of the SN ratio; for example, the type of measurement instrument, the manner of acquisition of the samples, the steps of the procedure of measurement, and the number of replications are all control factors. When several persons are selected to do the measuring, except in the case when people who are adept at measuring are selected, the persons are an error factor. When discussing whether or not to measure with the room temperature constant, it is possible to treat the room temperature as an error factor. In general, environmental conditions during measurement are error factors. Of course, it is possible to express the level when measurement is conducted in a constant temperature, constant humidity room as A_1 and the level when it is conducted in a given ordinary room as A_2, and to treat A as a control factor.

(2) *Signal Factors.* Signal factors are for ascertaining the magnitude of the signal, S, of the SN ratio. As to how to choose the signal factor level, it is advisable to classify the possible cases as follows.

a. *When the Linear Effect of the Signal Factor Can Be Obtained*

When the true value of each of the levels of $M_1, M_2, \ldots, M_k$ is known, when the true value of the intervals can be determined, or when the true value of the interval ratio can be determined, the linear effect can be obtained. From Appendix Table 8 we have

$$\eta=\frac{\frac{1}{r}(V_\beta-V_e)}{V_e} \quad (r=r_0Sh^2) \tag{22.100}$$

It does not matter in this case if error of up to several percent exists in the estimated value of interval h. This is because fundamentally there is that much error in the calculated SN ratio.

b. *When the True Value of the Interval Ratio Between the Signal Factor Levels Is Not Known with Sufficient Precision*

With the total values of r data for each of the signal factor levels M_1, $M_2, \ldots, M_k$ as $y_1, y_2, \ldots, y_k$, and

$$V_M=\frac{1}{k-1}\left[\frac{y_1^2+y_2^2+\cdots+y_k^2}{r}-\frac{(y_1+y_2+\cdots+y_k)^2}{kr}\right] \tag{22.101}$$

we find the SN ratio η'.

$$\eta'=\frac{\frac{1}{r}(V_M-V_e)}{V_e} \tag{22.102}$$

We assume that the approximate values of the true values of M_1, $M_2, \ldots, M_k$ are known, and these values will be expressed as M_1, $M_2, \ldots, M_k$. We calculate the following σ_M^2.

$$\sigma_M^2=\frac{1}{k-1}\left[M_1^2+M_2^2+\cdots+M_k^2-\frac{(M_1+M_2+\cdots+M_k)^2}{k}\right] \tag{22.103}$$

The ultimate SN ratio η is given by the following equation.

$$\eta=\eta'\times\frac{1}{\sigma_M^2} \tag{22.104}$$

For the true values of $M_1, M_2, \ldots, M_k$, we assume that only the linear term is significant. Since such an assumption is sometimes not valid when differences of $M_1, M_2, \ldots, M_k$ are large, in such a case one divides the levels of M into several classes and finds the SN ratio of Equation (22.104) for each class.

It does not matter if there is error of several percent in σ_M^2. This is because fundamentally there is error of about 20% in the SN ratio.

(3) *Error Factors.* This refers to all variables of the environmental conditions that affect the reading of the characteristic value y. Degradation of the measuring instrument does not necessarily constitute an error. The reason for this is that even if the sensitivity of a certain instrument drops and the zero point or slope β changes, as long as there is a standard it is possible to calibrate with it. The amount of change of the error factor that one should incorporate in an experiment is a difficult problem. This is because that part of the measured values coming from the error factors which can be corrected by means of the standard does not constitute error.

Often, the error factors are taken at three levels, taking into consideration, for example, changes in levels within the calibration period. With the error factor as K and the mean of its distribution as m, and the standard deviation as σ, it is best to take the three levels as

$$\left.\begin{aligned} K_1 &= m - \sqrt{\frac{3}{2}}\sigma \\ K_2 &= m \\ K_3 &= m + \sqrt{\frac{3}{2}}\sigma \end{aligned}\right\} \qquad (22.105)$$

This is because, in this case, the effect of these error factors enters S_e.

22.3.2 Measurement of Gel Fraction of Crosslinked Polyethylene (Tatsuta Electric Wire and Cable Co., Ltd., Shirō Katayama)

A method that is being used widely at present for measuring the degree of crosslinking of crosslinked polyethylene is gel fraction measurement, where the sample is extracted with a solvent and its gel content is measured. The following experiment was conducted in order to find the optimum conditions for gel fraction measurement.

(1) *Control Factors*

A: solvent type (influences of dissolving power of solvent and differences of boiling point)

A_1 = xylene (boiling point 141°C)

A_2 = toluene (boiling point 110.6°C)

B: extraction temperature

	When A_1 (xylene) was used	When A_2 (toluene) was used
B_1	120°C	110.6°C
B_2	80°C	80°C

C: extraction time
$C_1 = 4$ hours, $C_2 = 8$ hours, $C_3 = 24$ hours

(2) *Signal Factors* In order to prepare samples with different degrees of crosslinking, the quantity M of DCP, which influences the degree of crosslinking, was varied and we prepared three types of samples with different degrees of crosslinking.

M_1 = DCP optimum quantity
M_2 = DCP medium quantity
M_3 = DCP small quantity

The steps of the procedure of sample preparation and gel fraction measurement were as follows.

(i) The base polyethylene and the DCP are kneaded with a test roller at $120 \pm 2°C$, and the product is formed into a sheet of thickness about 3 mm.
(ii) The sheet is crosslinked under specified conditions and is finished to a sheet of 1 mm thickness.
(iii) A sample in the amount of 0.1 g is weighed to 0.1 mg and is placed in a test tube, then 20 cc of solvent is added and the test tube is sealed.
(iv) This test tube is heated for a prescribed period of time in a vessel held at a prescribed temperature.
(v) When the solvent is toluene and the extraction temperature is the boiling point, the sample and solvent are placed in a flask, then a cooling tube is attached. The procedure thereafter follows steps (vi) to (viii).
(vi) After heating, the test tube is removed; a suitable filter is placed upon a beaker and the liquid in the test tube is poured into this.
(vii) Since the extracted sample remains on the filter, this is placed in a weighing bottle and it is dried for 15 hours in a constant-temperature vessel at 120°C.
(viii) After drying, it is left to cool in a desiccator, then the sample is removed and is precisely weighed.

$$\text{Gel fraction} = \frac{\text{weight after extraction}}{\text{initial weight of sample}} \times 100\%$$

(3) *Error Factor*
R = number of replications

22.3.3 Assignment and Data

It is necessary to assign the control factors, to assign the signal factors and error factor together, and to obtain the data of all combinations of both assignments. In this case, the control factors were assigned as a three-

Table 22.8 Data of Gel Fractions and Omega-Transformed Values

No.	A	B	C	y' = gel fraction						$y=-10\log\left(\frac{1}{y'}-1\right)$					
				M_1		M_2		M_3		M_1		M_2		M_3	
1	1	1	1	94.0	94.0	89.5	88.2	81.1	79.7	11.95	11.95	9.31	8.73	6.32	5.94
2	1	1	2	93.8	94.2	87.6	85.8	69.6	72.7	11.80	12.11	8.49	7.81	3.60	4.25
3	1	1	3	91.1	90.5	77.1	77.2	58.9	63.5	10.10	9.79	5.28	5.30	1.56	2.40
4	1	2	1	95.3	95.1	91.5	91.3	84.0	84.4	13.07	12.88	10.32	10.21	7.20	7.33
5	1	2	2	94.9	95.3	89.7	88.5	80.4	78.9	12.70	13.07	9.40	8.86	6.13	5.73
6	1	2	3	92.5	92.4	82.1	84.4	67.9	67.2	10.91	10.85	6.62	7.33	3.25	3.12
7	2	1	1	94.6	94.5	87.2	86.6	74.2	78.3	12.43	12.35	8.33	8.10	4.59	5.57
8	2	1	2	93.6	94.1	83.8	84.0	78.3	85.0	11.65	12.03	7.13	7.20	5.57	7.53
9	2	1	3	91.1	91.0	77.1	77.2	65.7	65.7	10.10	10.05	5.28	5.30	2.81	2.81
10	2	2	1	95.4	95.4	90.1	90.9	84.2	83.3	13.10	13.17	9.89	10.00	7.27	6.98
11	2	2	2	95.6	96.0	89.1	88.0	79.7	77.7	13.37	13.80	9.12	8.65	6.00	5.42
12	2	2	3	92.1	92.1	81.2	83.1	66.0	66.3	10.67	10.67	6.35	6.92	2.88	2.94

way layout and the signal factors were assigned as a one-way layout, and we performed experiments of combinations of the two.

The data were as given in Table 22.8. Only values between 0 and 100% were used for the fractions. In a case such as this, it is better to calculate by using omega-transformed data. The results of omega transformation according to Appendix Table 5 are given in the right columns of the table.

22.3.4 Calculation of SN Ratios

(1) *Comparison of A_1 and A_2.* Analysis of variance tables are prepared for A_1 and A_2, and the SN ratio is calculated.

a. The Case of A_1.

$$CF=\frac{767^2}{36}=16\,341 \qquad (f=1)$$

$$S_T=395^2+395^2+\cdots+(-488)^2-CF=3\,860\,267-16\,341=3\,843\,926 \qquad (f=35)$$

Table 22.9 Data of A_1: $(y-8)\times 100$

	M_1		M_2		M_3		M_1	M_2	M_3	Total
B_1C_1	395	395	131	73	−168	−206	790	204	−374	620
B_1C_2	380	411	49	−19	−440	−375	791	30	−815	6
B_1C_3	210	179	−272	−270	−644	−560	389	−542	−1 204	−1 357
B_2C_1	507	488	232	221	−80	−67	995	453	−147	1 301
B_2C_2	470	507	140	86	−187	−227	977	226	−414	789
B_2C_3	291	285	−138	−67	−475	−488	576	−205	−963	−592
					Total		4518	166	−3 917	767

Table 22.10 Analysis of Variance Table for A_1

Source		f	S	V	$E(V)$
Signal source	M	2	2 965 556	1 482 778	$\sigma^2+12\sigma_M{}^2$
Junction source	row	5	798 904		
	$M\times$ row	10	62 256		
Error	e	18	17 210	956	σ^2
Total	T	35	3 843 926		

$$S_{\text{row}}=\frac{620^2+\cdots+(-592)^2}{6}-CF=798\,904 \qquad (f=5)$$

$$S_M=\frac{4\,518^2+166^2+(-3\,917)^2}{12}-CF=2\,981\,897-16\,341=2\,965\,556$$
$$(f=2)$$

$$S_{M\times\text{row}}=\frac{790^2+204^2+\cdots+(-963)^2}{2}-CF-S_{\text{row}}-S_M=62\,256 \qquad (f=10)$$

$$S_e=S_T-S_{\text{row}}-S_M-S_{M\times\text{row}}=17210 \qquad (f=18)$$

Therefore, the analysis of variance table for A_1 is as given in Table 22.10. Therefore, the SN ratio of A_1 is

$$\eta(A_1)=\frac{\frac{1}{12}(1\,482\,778-956)}{956}=129.2\rightarrow 21.11\ [\text{db}] \qquad (22.106)$$

b. The Case of A_2.

If calculated by the same method as for A_1, the analysis of variance table is as shown in Table 22.11.

$$\eta(A_2)=\frac{\frac{1}{12}(1\,461\,554-1\,718)}{1\,718}=70.81=18.50\ [\text{db}] \qquad (22.107)$$

(2) *SN Ratios of B_1, B_2, C_1, C_2, C_3.* If analysis of variance is conducted as with A for factors B and C, we obtain Tables 22.12–Table 22.16.

Table 22.11 Analysis of Variance Table for A_2

Source	f	S	V	$E(V)$
M	2	2 923 108	1 461 554	$\sigma^2+12\sigma_M{}^2$
Row	5	712 014		
$M\times$ row	10	87 219		
e	18	30 927	1 718	σ^2
T	35	3 753 268		

Table 22.12 Analysis of Variance Table for B_1

Source	f	S	V	$E(V)$
M	2	2 934 293	1 467 146	$\sigma^2+12\sigma_M{}^2$
Row	5	588 971		
$M\times$ row	10	108 540		
e	18	36 387	2 022	σ^2
T	35	3 668 191		

$$\eta(B_1)=\frac{\frac{1}{12}(1\,467\,146-2\,022)}{2022}=60.38=17.81\text{ [db]} \tag{22.108}$$

Table 22.13 Analysis of Variance Table for B_2

Source	f	S	V	$E(V)$
M	2	2 944 414	1 472 207	$\sigma^2+12\sigma_M{}^2$
Row	5	692 351		
$M\times$ row	10	50 890		
e	18	11 750	653	σ^2
T	35	3 699 405		

$$\eta(B_2)=\frac{\frac{1}{12}(1\,472\,207-653)}{653}=187.79=22.73\text{ [db]} \tag{22.109}$$

Table 22.14 Analysis of Variance Table for C_1

Source	f	S	V	$E(V)$
M	2	1 544 928	772 464	$\sigma^2+8\sigma_M{}^2$
Row	3	111 651		
$M\times$ row	6	19 746		
e	12	8 334	694	σ^2
T	23	1 684 659		

$$\eta(C_1)=\frac{\frac{1}{8}(772\,464-694)}{694}=139.01=21.43\text{ [db]} \tag{22.110}$$

Table 22.15 Analysis of Variance Table for C_2

Source	f	S	V	$E(V)$
M	2	2 008 323	1 004 161	$\sigma^2+8\sigma_M{}^2$
Row	3	79 223		
$M\times$ row	6	84 516		
e	12	31 513	2 626	σ^2
T	23	2 203 575		

$$\eta(C_2)=\frac{\frac{1}{8}(1\,004\,161-2\,626)}{2\,626}=47.67=16.78\ [\text{db}] \tag{22.111}$$

Table 22.16 Analysis of Variance Table for C_3

Source	f	S	V	$E(V)$
M	2	2 367 761	1 183 881	$\sigma^2+8\sigma_M{}^2$
Row	3	62 671		
$M\times$ row	6	12 864		
e	12	8 291	691	σ^2
T	23	2 451 587		

$$\eta(C_3)=\frac{\frac{1}{8}(1\,183\,881-691)}{691}=214.03=23.30\ [\text{db}] \tag{22.112}$$

22.3.5 Conclusions

When the foregoing results are organized, we obtain Table 22.17.

Therefore, compared with the present analysis methods A_1, B_1, and C_1, A_2 shows a loss of 2.6 db compared to A_1, B_2 shows a gain of 4.9 db, and C_3 shows a gain of 1.9 db. The optimum conditions $A_1B_2C_3$ yield a gain of $4.9 + 1.9 = 6.8$ db compared with the analysis method $A_1B_1C_1$, and in terms of the true values this means a gain of 4.8 times.

Table 22.17 At-a-Glance Table of SN Ratio (Unit: db)

A_1	21.1	B_1	17.8	C_1	21.4
A_2	18.5	B_2	22.7	C_2	16.8
				C_3	23.3

22.3.6 SN Ratio of Omega Transformation

Mr. Shirō Katayama of Tatsuta Electric Wire and Cable Company used the SN ratio for the two types of data of Table 22.8, the gel fraction data as they stand and the omega-transformed data, and indicated that the omega-transformed data were better. Here we give only the results. The SN ratios η when significant factors are all regarded as signals, from Table 22.18 and from Table 22.19, are:

Gel Fraction, as Is

$$\eta=\frac{94.6}{5.4}=17.5=12.4\ [\text{db}] \tag{22.113}$$

Omega-Transformed

$$\eta=\frac{96.6}{3.4}=28.4=14.5\ [\text{db}] \tag{22.114}$$

Table 22.18 Analysis of Variance Table for Gel Fraction, as Is

Source	f	S	V	ρ[%]
M	2	4 356.20	2 178.10**	67.9
A	1	0.20	0.20	
B	1	172.98	172.98**	2.6
C	2	1 226.31	613.16**	19.0
$M\times A$	2	19.84	9.92	
$M\times B$	2	29.20	14.60	
$M\times C$	4	356.32	89.08**	5.1
$A\times B$	1	7.48	7.48	
$A\times C$	2	15.59	7.80	
$B\times C$	2	2.27	1.14	
e_1	16	136.77	8.55**	5.4 (e_1, e_2 combined)
e_2	36	60.51	1.68	
T	71	6 383.67		100.0

This means that precision has been improved by 2.1 db, or 1.6 times, by carrying out omega transformation.

However, the author believes that the fact that there was interaction means that the expression of the data was clumsy, and that actually it is better to include the interaction in error. In this case, the SN ratio of the data as gel fraction, untransformed, becomes as follows.

$$\eta=\frac{89.5}{10.5}=8.52=9.3\,[\text{db}] \tag{25.115}$$

Table 22.19 Analysis of Variance Table After Omega Transformation

Source	f	S	V	F_0	ρ[%]
M	2	586.9320	293.7208	479.28**	77.1
A	1	0.0018	0.0018	0.00	
B	1	23.0407	23.0407	37.63**	3.0
C	2	125.9128	62.9565	102.81**	16.5
$M\times A$	2	1.9344	0.9672	1.58	
$M\times B$	2	0.9294	0.4646	0.75	
$M\times C$	4	5.1015	1.2753	2.08	
$A\times B$	1	0.2058	0.2058	0.33	
$A\times C$	2	0.6701	0.3350	0.54	
$B\times C$	2	0.3812	0.1906	0.31	
e_1	16	9.7974	0.6123	4.58**	
e_2	36	4.8138	0.1337		
Remainder					3.4
T	71	759.7209			100.0

Note: e_1 was tested with e_2, and since it was 1% significant, the factors were tested with e_1.

Therefore, in this case we have a difference of 5.2 db and 3.3 times, and it is more advantageous to omega-transform.

For the reasons given above, we have used data after omega transformation for comparisons of SN ratios such as the percentage.

In general, variable conversion and goodness or badness of quantification are expressed by the SN ratio. The reader is referred to Chapter 39, as well, regarding this problem.

22.4 Comparison of Instrumentation Methods (2)

22.4.1 Experiment on Dynamic Balance of Drive Shaft*

If there is great *residual unbalance* of the drive shaft of an automobile, it constitutes a cause of noise and vibration. Therefore, to minimize noise and vibration which derive from the drive shaft, the practice at present is to measure the magnitude of the residual unbalance of the drive shaft, to attach balance weights of a mass which compensates this and so to effect dynamic balancing. However, when there is error in the measurement of the unbalance, erroneous corrections are made; the residual unbalance does not become sufficiently small and the noise and vibration are not reduced to a desirable level. Therefore, decreasing the error in measurement of the unbalance can be said to be one of the important quality control problems for the maker of drive shafts.

22.4.2 Factors and Levels

Control factors

A:	testing machine	$A_1 = \text{new}$, $A_2 = \text{old}$
B:	master rotor	$B_1 = \#1$, $B_2 = \#2$, $B_3 = \#3$, $B_4 = \#4$
C:	number of rotations at handling time	$C_1 = \text{current}$, $C_2 = \text{new}$
D:	number of rotations at measurement	$D_1 = \text{current}$, $D_2 = \text{new}$
E:	signal sensitivity at handling time	$E_1 = 10$, $E_2 = 20$ $E_3 = 30$, $E_4 = 40$
F:	sequence of correction of unbalance	$F_1 = \text{current}$, $F_2 = \text{reverse}$, $F_3 = \text{new} - 1$, $F_4 = \text{new} - 2$
G:	unbalance correction location	$G_1 = \text{current}$, $G_2 = \text{new}$

* Kenichi Ueno: QCRG Data, 1969.

Signal factors

M':	product	$M_1' =$ good product
		$M_2' =$ good product
		$M_3' =$ bad product
M:	amount of unbalance	$M_1 = 0, M_2 = +10,$
		$M_3 = +20, M_4 = +30$ g

Indicative factor

K	$K_1 =$ flange side
	$K_2 =$ sleeve side

Indicative factors are factors that are neither signal factors nor error factors. These can be assigned together with the control factors if one wishes.

22.4.3 Assignment and Data

The control factors were assigned to orthogonal array L_{16} as shown in Table 22.20. For each of three drive shafts of different unbalance quantities, M_1', M_2', and M_3', the signal factor M was taken at the four levels:

$M_1 =$ measured as is
$M_2 =$ a 10 g balance weight is attached to the deficient-mass side
$M_3 =$ a 20 g balance weight is attached to the deficient-mass side
$M_4 =$ a 30 g balance weight is attached to the deficient-mass side,

and the readings of the unbalance quantity by the balancing machine were as in the data columns of Table 22.20. Since there were readings for the six combinations of M' and K, the total number of data was

$$16 \times 3 \times 2 \times 4 = 384$$

22.4.4 Calculation of SN Ratio

The SN ratio is found for each combination of the experiment numbers No. 1–No. 16 in the orthogonal array and the levels of the indicative factor K. The three drive shafts, M_1', M_2', and M_3', are the three levels of the signal factor, but their accurate unbalance quantities are unknown. Therefore, in effect, the signal effect is obtained from the main effects of four levels of equal-interval factors, M_1, M_2, M_3, and M_4, as variations of the linear term. Since the unbalance quantity differs at each level of M', differences in readings within M' are not error. That is, the main effect of M' should not be included in the signal or the error.

Table 22.20 Assignment and Data

No.	A	G	e	C	B	F	E	D	(E)	(B)	(F)	$C \times D$	(F)	(E)	(B)
	1	2	3	4	5	6	7	8	9	10	11	12	13	14	15
1	1	1	1	1	1	1	1	1				1			
2	1	1	1	1	2	2	2	2				2			
3	1	1	1	2	3	3	3	1				2			
4	1	1	1	2	4	4	4	2				1			
5	1	2	2	1	2	4	3	1				1			
6	1	2	2	1	1	3	4	2				2			
7	1	2	2	2	4	2	1	1				2			
8	1	2	2	2	3	1	2	2				1			
9	2	1	2	1	3	2	4	1				1			
10	2	1	2	1	4	1	3	2				2			
11	2	1	2	2	1	4	2	1				2			
12	2	1	2	2	2	3	1	2				1			
13	2	2	1	1	4	3	2	1				1			
14	2	2	1	1	3	4	1	2				2			
15	2	2	1	2	2	1	4	1				2			
16	2	2	1	2	1	2	3	2				1			

No.	K_1 (flange side)											
	M_1'				M_2'				M_3'			
	M_1	M_2	M_3	M_4	M_1	M_2	M_3	M_4	M_1	M_2	M_3	M_4
1	−4	6	18	27	−4	6	15	25	−20	−10	2	14
2	−7	10	23	42	−3	15	32	46	−34	−20	4	16
3	−4	9	22	34	−7	6	18	30	−26	−15	0	12
4	−2	10	22	36	−4	8	22	34	−30	−18	−7	10
5	−6	6	16	28	−5	6	16	27	−21	−12	3	13
6	−7	13	32	50	−7	14	31	48	−45	−27	−8	13
7	−13	10	30	52	−8	10	30	50	−37	−18	7	26
8	−19	8	27	48	−14	7	29	52	−42	−25	−2	22
9	−10	4	16	29	−8	6	16	26	−29	−20	−14	−14
10	−14	11	32	51	−18	4	25	46	−44	−26	−11	16
11	−3	2	10	16	−4	2	17	13	−13	−8	−5	7
12	−5	5	16	25	−7	3	12	22	−22	−14	−8	12
13	−4	6	18	30	−8	2	16	28	−23	−15	−10	18
14	−6	16	38	62	−16	6	32	55	−44	−25	−6	21
15	−4	2	7	14	−4	2	6	12	−10	−6	−3	7
16	−5	6	16	30	−7	4	16	27	−25	−14	−6	14

No.	K_2 (sleeve side)											
	M_1'				M_2'				M_3'			
	M_1	M_2	M_3	M_4	M_1	M_2	M_3	M_4	M_1	M_2	M_3	M_4
1	−13	0	12	28	−12	2	9	20	−22	−5	9	21
2	−10	14	24	41	−7	14	30	46	−6	13	36	55
3	−14	3	18	32	−17	16	20	34	−30	−5	16	36
4	−10	−4	6	15	−8	−1	8	18	−10	8	16	26
5	−14	2	10	23	−9	2	10	18	−16	−5	5	16
6	−13	8	12	22	−6	4	11	18	−14	10	14	24
7	−10	2	12	23	−8	2	10	19	−15	−5	7	17
8	−10	10	24	42	−9	7	21	40	−19	−8	14	28
9	−10	−4	5	11	−12	−7	0	8	−6	4	13	21
10	−34	−14	16	32	−17	−5	20	35	−27	−17	−9	16
11	−5	2	7	12	−4	3	17	12	−7	−2	4	10
12	−9	4	12	20	−8	3	12	20	−11	−4	7	14
13	−8	3	12	22	−8	3	12	22	−17	−8	4	14
14	−14	7	24	41	−19	4	22	41	−32	−15	6	24
15	−4	4	8	15	−5	2	7	12	−12	−6	2	7
16	−13	2	10	20	−10	2	13	21	−13	−10	3	20

Calculation of SN Ratio of K_1 in No. 1

In No. 1, the data for K_1 number 12 in all. Therefore,

$$S_T=(-4)^2+6^2+18^2+\cdots+14^2-\frac{75^2}{12}=2707-468.75$$

$$=2238.25 \qquad (f=11) \tag{22.116}$$

The totals for M_1, M_2, M_3, and M_4 are

$$M_1=(-4)+(-4)+(-20)=-28$$

$$M_2=6+6+(-10)=2$$

$$M_3=18+15+2=35$$

$$M_4=27+25+14=66$$

and therefore their linear effect, S_{M_l} is

$$S_{M_l}=\frac{(-3M_1-M_2+M_3+3M_4)^2}{3\times 20}=\frac{[-3\times(-28)-2+35+3\times 66]^2}{60}$$

$$=1653.75 \qquad (f=1) \tag{22.117}$$

Also, the variation of M' which is not included in signal or error, $S_{M'}$, is

$$S_{M'}=\frac{47^2+42^2+(-14)^2}{4}-\frac{75^2}{12}=573.50 \qquad (f=2) \tag{22.118}$$

$$S_e=S_T-S_{M_l}-S_{M'}=2238.25-1653.75-573.50=11.00 \qquad (f=8) \tag{22.119}$$

Table 22.21 Values of SN Ratio of Control Factor

No.	1 Raw value K_1	K_2	2 Decibel value K_1	K_2	3 Total
1	0.802	0.204	−0.96	−6.90	−7.86
2	0.542	0.278	−2.66	−5.56	−8.22
3	3.018	0.080	4.80	−10.97	−6.17
4	1.162	0.124	0.64	−9.07	−8.43
5	0.966	0.242	−0.15	−6.16	−6.31
6	2.504	0.044	3.98	−13.57	−9.59
7	1.442	0.558	1.58	−2.53	−0.95
8	1.300	0.652	1.14	−1.86	−0.72
9	0.038	0.236	−14.15	−6.72	−20.87
10	0.642	0.076	−1.92	−11.19	−13.11
11	0.040	0.036	−13.98	−14.44	−28.42
12	0.154	0.438	−8.12	−3.59	−11.71
13	0.082	1.990	−10.86	2.99	−7.87
14	1.058	1.458	0.25	1.64	1.89
15	0.160	0.396	−7.96	−4.02	−11.98
16	0.286	0.138	−5.44	−8.60	−14.04

$$V_e = \frac{S_e}{8} = \frac{11.00}{8} = 1.375 \tag{22.120}$$

$$\eta = \frac{\frac{1}{rSh^2}(S_{M_l} - V_e)}{V_e} = \frac{\frac{1}{3\times5\times10^2}(1653.75 - 1.375)}{1.375} = 0.801 \tag{22.121}$$

$$= -0.96 \text{ [db]} \tag{22.122}$$

Similarly, we calculate the SN ratios for K_1 for experiments No. 2 to No. 16. We calculate in the same way for the sleeve side K_2, as well. The results are as given in Table 22.21.

22.4.5 Analysis of Variance

We carry out ordinary analysis of variance with the decibel values of the SN ratio values in Table 22.21 as the target characteristic. We prepare the auxiliary table, Table 22.22.

$$CF = \frac{(-154.36)^2}{32} = 744.5941 \qquad (f=1)$$

$$S_A = \frac{(-48.25 + 106.11)^2}{32} = 104.6181 \qquad (f=1)$$

$$S_B = \frac{(-59.91)^2 + (-38.22)^2 + (-25.87)^2 + (-30.36)^2}{8} - CF$$

$$= 85.5263 \qquad (f=3)$$

$$S_C = \frac{(-71.94 + 82.42)^2}{32} = 3.4322 \qquad (f=1)$$

$$S_D = 21.9453 \qquad (f=1)$$

$$S_{C\times D} = 0.0496 \qquad (f=1)$$

Table 22.22 Auxiliary Table

	K_1	K_2	Total		K_1	K_2	Total
A_1	8.37	−56.62	−48.25	E_1	−7.25	−11.38	−18.63
A_2	−62.18	−43.93	−106.11	E_2	−26.36	−18.87	−45.23
B_1	−16.40	−43.51	−59.91	E_3	−2.71	−36.92	−39.63
B_2	−18.89	−19.33	−38.22	E_4	−17.49	−33.38	−50.87
B_3	−7.96	−17.91	−25.87	F_1	−9.70	−23.97	−33.67
B_4	−10.56	−19.80	−30.36	F_2	−20.67	−23.41	−44.08
C_1	−26.47	−45.47	−71.94	F_3	−10.20	−25.14	−35.34
C_2	−27.34	−55.08	−82.42	F_4	−13.24	−28.03	−41.27
D_1	−41.68	−48.75	−90.43	G_1	−36.35	−68.44	−104.79
D_2	−12.13	−51.80	−63.93	G_2	−17.46	−32.11	−49.57
$(C\times D)_1$	−37.90	−39.91	−77.81	計	−53.81	−100.55	−154.36
$(C\times D)_2$	−15.91	−60.64	−76.55				

$$S_E=74.2964 \qquad (f=3)$$

$$S_F=9.0114 \qquad (f=3)$$

$$S_G=95.2890 \qquad (f=1)$$

$$S_K=\frac{(-53.81+100.55)^2}{32}=68.2696$$

$$S_{A\times K}=\frac{8.37^2+(-56.62)^2+(-62.18)^2+(-43.93)^2}{8}-CF-S_A-S_K$$

$$=216.5280 \qquad (f=1)$$

$$S_{B\times K}=\frac{(-16.40)^2+(-43.51)^2+\cdots+(-19.80)^2}{4}-CF-S_B-S_K$$

$$=46.6711 \qquad (f=3)$$

$$S_{C\times K}=2.3871 \qquad (f=1)$$

$$S_{D\times K}=33.2112 \qquad (f=1)$$

$$S_{C\times D\times K}=57.0312 \qquad (f=1)$$

$$S_{E\times K}=118.7270 \qquad (f=3)$$

$$S_{F\times K}=13.3664 \qquad (f=3)$$

$$S_{G\times K}=9.5048 \qquad (f=1)$$

$$S_{T_1}=\frac{1}{2}[(-7.86)^2+(-8.22)^2+\cdots+(-14.04)^2]-CF$$

$$=420.4496 \qquad (f=15)$$

$$S_{e_1}\doteq S_{T_1}-(S_A+S_B+\cdots+S_G)=26.2812 \qquad (f=1)$$

$$S_T=(-0.96)^2+(-6.90)^2+\cdots+(-8.60)^2-CF$$

$$=989.3594 \qquad (f=31)$$

$$S_{e_2}=S_T-(S_{T_1}+S_K+S_{A\times K}+\cdots+S_{G\times K})=3.2131 \qquad (f=1)$$

Therefore, the analysis of variance table is as given in Table 22.23. There is no need to distinguish between primary error and secondary error. This is because even among the primary factors, there are several with about the same degree of error variance as secondary factors. Error variance was drawn up by pooling the factors marked O as being in the neighborhood of 15.5 degrees of freedom, one-half the total number of degrees of freedom, 31.

E and $D\times K$ are not significant, but their effects will not be disregarded since they are close to 5% significant.

22.4.6 Estimation of Significant Sources

Since the interactions $A\times K$, $C\times D\times K$, and $E\times K$ are significant, we prepare the two-way array of AK, the three-way array of CDK, and the two-way array of EK, and we estimate the main effects of B and G.

Table 22.23 Analysis of Variance Table

Source	f	S	V
A	1	104.6181	104.6181**
B	3	85.5263	28.5088*
C	1	3.4322	3.4322 o
D	1	21.9453	21.9453
$C\times D$	1	0.0496	0.0496 o
E	3	74.2964	24.7655(*)
F	3	9.0114	3.0038 o
G	1	95.2890	95.2890**
e_1	1	26.2812	26.2812 o
K	1	68.2696	68.2696**
$A\times K$	1	216.5280	216.5280**
$B\times K$	3	46.6711	15.5570 o
$C\times K$	1	2.3871	2.3871 o
$D\times K$	1	33.2112	33.2112(*)
$C\times D\times K$	1	57.0312	57.0312*
$E\times K$	3	118.7270	39.5757*
$F\times K$	3	13.3664	4.4555 o
$G\times K$	1	9.5048	9.5048 o
e_2	1	3.2131	3.2131 o
O symbols pooled (e)	(15)	(113.9169)	7.5945
T	31	989.3594	

As to the confidence limits, those of the two-way array of AK and the main effect of B are

$$\pm\sqrt{\frac{4.54\times7.5945}{8}}=\pm2.08$$

Those of the three-way array of CDK and the two-way array of EK are

$$\pm\sqrt{\frac{4.54\times7.5945}{4}}=\pm2.94$$

The confidence limits of the main effect of G are one-half these, ±1.47. From Table 22.24, one possible set of optimum conditions is

Table 22.24 Estimation of Source Effect

	K_1	K_2
A_1	1.05	−7.08
A_2	−7.77	−5.49

	K_1	K_2
C_1D_1	−6.53	−4.20
C_1D_2	−0.09	−7.17
C_2D_1	−3.89	−7.99
C_2D_2	−2.94	−5.78

	K_1	K_2
E_1	−1.81	−2.84
E_2	−6.59	−4.72
E_3	−0.68	−9.23
E_4	−4.37	−8.34

B_1	−7.49
B_2	−4.78
B_3	−3.23
B_4	−3.80

G_1	−6.55
G_2	−3.10

$A_1B_3C_1D_2E_1G_2$. We estimate the process average for K_1 and K_2. The factors which are regarded as being significant are the main effects of A, B, E, G, and K and the interactions $A \times K$, $D \times K$, $C \times D \times K$, and $E \times K$.

For K_1,

$$\begin{aligned}\hat{\mu} &= \overline{A_1K_1} + \overline{C_1D_2K_1} - \overline{C_1K_1} - \overline{C_1D_2} + \overline{E_1K_1} + \bar{B}_3 + \bar{C}_1 + \bar{G}_2 - \bar{K}_1 - 2\bar{T} \\ &= 1.05 + (-0.09) - (-3.31) - (-3.63) + (-1.81) + (-3.23) + (-4.50) \\ &\quad + (-3.10) - (-3.36) - 2 \times (-4.8) \\ &= 8.26 \end{aligned} \tag{22.123}$$

For K_2,

$$\begin{aligned}\hat{\mu} &= \overline{A_1K_2} + \overline{C_1D_2K_2} - \overline{C_1K_2} - \overline{C_1D_2} + \overline{E_1K_2} + \bar{B}_3 + \bar{C}_1 + \bar{G}_2 - \bar{K}_2 - 2\bar{T} \\ &= (-7.08) + (-7.17) - (-5.70) - (-3.63) + (-2.84) + (-3.23) \\ &\quad + (-4.50) + (-3.10) - (-6.28) - 2 \times (-4.82) \\ &= -2.67 \end{aligned} \tag{22.124}$$

Accordingly, the results

Flange Side

$$\eta = 6.70 \tag{22.125}$$

Sleeve Side

$$\eta = 0.541 \tag{22.126}$$

were obtained. This means that the error variance is

Flange Side

$$\sigma^2 = \frac{1}{6.70} = 0.1493$$

Sleeve Side

$$\sigma^2 = \frac{1}{0.541} = 1.848$$

Therefore, if it is allowed that three times the standard deviation can be regarded as measurement error, the error becomes as follows.

Flange Side

$$\pm 3 \times \sqrt{0.1493} = \pm 1.2 \text{ [g]} \tag{22.127}$$

Sleeve Side

$$\pm 3 \times \sqrt{1.848} = \pm 4.1 \text{ [g]} \tag{22.128}$$

These results indicate that the unbalance quantity can be obtained accurately only for the flange side. Therefore, if one wishes to reduce the measurement error of the sleeve side further even though this means somewhat increasing the measurement error of the flange side, one should stress the K_2 side of Table 22.24 and use as the optimum conditions $A_1B_3C_1D_1E_1G_2$. In this case, the process averages are

Flange Side

$$\begin{aligned}\hat{\mu} &= \overline{A_1K_1} + \overline{C_1D_1K_1} - \overline{C_1K_1} - \overline{C_1D_1} + \overline{E_1K_1} + \bar{B}_3 + \bar{C}_1 + \bar{G}_2 - \bar{K}_1 - 2\bar{T} \\ &= 1.05 + (-6.53) - (-3.31) - (-5.36) + (-3.23) + (-1.81) \\ &\quad + (-4.50) + (-3.10) - (-3.36) - 2(-4.82) \\ &= 3.55 \text{ [db]}\end{aligned} \tag{22.129}$$

Sleeve Side

$$\begin{aligned}\hat{\mu} &= (-7.08) + (-4.20) - (-5.68) - (-5.36) + (-2.84) \\ &\quad + (-3.23) + (-4.50) + (-3.10) - (-6.28) - 2(-4.82) \\ &= 2.01 \text{ [db]}\end{aligned} \tag{22.130}$$

Therefore, we obtain the predicted results given in Table 22.25.

Therefore, $A_1B_3C_1D_1E_1G_2$, which is better on balance, can be recommended as the optimum conditions.

22.4.7 Interpretation

In order to achieve dynamic balance of a rotating device, it is necessary to estimate the unbalance quantity. The magnitude of the unbalance quantity and the precision of estimation of the angular position of the mass which is to correct the unbalance become the problems. As in this example, it is best to collect about three objects:

M_1' drive shaft whose unbalance quantity is small
M_2' drive shaft whose unbalance quantity is of medium degree
M_3' drive shaft whose unbalance quantity is large

Table 22.25 Two Choices for Optimum Conditions

		Flange side		Sleeve side	
	Optimum conditions	η	Error	η	Error
First set	$A_1B_3C_1D_2E_1G_2$	6.70	±1.2 g	0.541	±4.1 g
Second set	$A_1B_3C_1D_1E_1G_2$	2.26	±2.0 g	1.59	±2.4 g

(It does not matter if the true values of the unbalance quantities are unknown.)

and to create levels of the signal factor M in the direction where the unbalance is greatest. For example,

$$M_1=0$$
$$M_2=+h$$
$$M_3=+2h$$
$$M_4=+3h$$

If the unbalance direction is not clear, the direction which is believed to be the direction of unbalance (*i.e.*, which has been measured by the testing machine in question; it does not matter if it is wrong, but it will be assumed that the error is within about $\pm 30°$) is expressed as W_2, and one creates levels such as

$$W_1=W_2-30°$$
$$W_2=W_2$$
$$W_3=W_2+30°$$

One obtains data such as are shown in Table 22.26 for each of the three objects M_1', M_2', and M_3'.

The unbalance quantity should become maximum when the direction of unbalance is correct. We use the data of Table 22.26 and perform orthogonal polynomial expansion.

$$y=m+a_1(W-\overline{W})+a_2\left[(W-\overline{W})^2-\frac{3^2-1}{12}h_W{}^2\right]+b_1(M-\overline{M})$$
$$+b_2\left[(M-\overline{M})^2-\frac{4^2-1}{12}h_M{}^2\right]+\cdots+C_{11}(W-\overline{W})(M-\overline{M})+\cdots \qquad (22.131)$$

Here, h_W and h_M are the intervals of W and M, and in this case they are 30 and 10.

$$m=\frac{T}{12} \qquad (22.132)$$

$$a_1=\frac{-W_1+W_3}{4\times 2\times 30} \qquad (22.133)$$

Table 22.26 Readings for M_i'

M_i'	M_1	M_2	M_3	M_4	Total	
W_1	y_1	y_2	y_3	y_4	W_1	$L_1=-3y_1-y_2+y_3+3y_4$
W_2	y_5	y_6	y_7	y_8	W_2	$L_2=-3y_5-y_6+y_7+3y_8$
W_3	y_9	y_{10}	y_{11}	y_{12}	W_3	$L_3=-3y_9-y_{10}+y_{11}+3y_{12}$
Total	M_1	M_2	M_3	M_4	T	

$$a_2 = \frac{W_1 - 2W_2 + W_3}{4 \times 2 \times 30^2} \tag{22.134}$$

$$b_1 = \frac{-3M_1 - M_2 + M_3 + 3M_4}{3 \times 10 \times 10} \tag{22.135}$$

$$\vdots$$

$$C_{11} = \frac{-L_1 + L_3}{1 \times 10 \times 10 \times 2 \times 30} \tag{22.136}$$

The main effect term for W in Equation (22.131) is differentiated with respect to W and we set the derivative equal to 0.

$$\frac{\partial y}{\partial W} = a_1 + 2a_2(W - \overline{W}) = 0$$

Therefore,

$$W = \overline{W} - \frac{a_1}{2a_2} \tag{22.137}$$

In essence, the signal factors M_1, M_2, M_3, and M_4 are newly attached to the direction of the angle obtained by Equation (22.137). Even if there is some mistake in the direction of unbalance, it is not too important for comparisons of the SN ratio since it affects every number of the experiment equally, but there is a problem in accurately obtaining the measurement error from the SN ratio of the process average of the optimum conditions. In certain cases, therefore, it is possible to first obtain the optimum conditions of the SN ratio and then to recalculate the SN ratio for several product articles by taking data classed by the signal factor levels M_1, M_2, M_3, and M_4, and to obtain the measurement error by

$$\pm \frac{3}{\sqrt{\eta}}$$

Next, let us consider control factors. As experiments for the user of the balancing machine, it might be considered that experiments of the above degree suffice, but in the case of a maker of balancing machines, he should search for a design method in which the SN ratio is large by taking a greater number of factors in designing the machine. He should include dispersions and degradation of the parts which are used in the machine among the design conditions. To begin with, it should be the social role of the testing machine maker to design a balancing machine whose SN ratio is large. As studies on the user side, there are only selection of a testing machine whose SN ratio is large, and studies toward its use. As factors of the method of use, B, C, D, E, F, and G were selected in this case, but it is the level of the master rotor B, the signal sensitivity E, and the unbalance correction position G that have a great effect, and what is overwhelmingly great is the term related with the choice of testing machine, A.

Error factors are extremely important when the user is conducting experiments on testing methods. As error factors, although this has not been especially treated in any way here, it would seem best to select, for example,

L = person testing 4 persons
Q = electric source voltage 3 levels
R = days 4 days

and the like. The object M' should also be taken at four levels, and one assigns L, Q, R, M', and M to orthogonal array L_{16}. For example, if one takes Q_1, Q_2, Q_3, in effect one performs the analysis of variance of Table 22.27 with each of the orthogonal-array L_{16} control factors assigned.

The SN ratio η is obtained from this by

$$\eta = \frac{\frac{1}{4 \times 5 \times 10^2}(S_{Ml} - V_e)}{V_e} \tag{22.138}$$

and when it has been converted to decibel value it is applied to analysis of variance as shown previously. Factors such as

K_1 = flange side K_2 = sleeve side

are such that it is necessary to obtain the SN ratios for K_1 and K_2, and these are termed indicative factors. Indicative factors are sometimes assigned to the same orthogonal arrays to which the control factors have been assigned, but if possible it is advisable to take every combination with the control-factor arrays, in other words to perform direct-sum experiments. Therefore, if the error-factor part is eliminated, the experimental design that has been introduced here can be recommended as an example of a logical design of experiment.

22.5 When There Are Two or More Different Types of Signal Factors

22.5.1 When There Are Two or More Signal Factors

In the case of a quantitative analysis where two or more chemical components influence a given reading, at least two different types of readings

Table 22.27 Analysis of Variance Table Per Experiment Number of Control Factor

Source	f	S
M'	3	$S_{M'}$
M_l	1	S_{Ml}
e	11	S_e
T	15	

are necessary. In general, if one wishes to estimate at most k types of signal factors from k types of readings, it is necessary to solve simultaneous equations. Here, we will simply explain an example where analysis of two signal factors is conducted from two types of readings, but it is simple to expand this to k variables. If it becomes too laborious for practical purposes to solve the simultaneous equations, one can treat the effect of the signal factor whose influence is less as an error. Since in this case there is one less signal factor, in effect one reduces the signal factors to fewer than k.

22.5.2 The Problem of Simultaneous Determination of Total Quantities of Nitrogen and Salt

In determining the total nitrogen quantity M and salt (sodium chloride) quantity N in a certain food product, one wishes to perform this determination by using the data of viscosity y in Bé (Baumé) and sugar content z, in the step prior to bottling.

The total nitrogen M and salt quantity N were taken at equal intervals and an experiment by two-way layout was carried out, and the data obtained when viscosity y and sugar content z were measured were as given in Table 22.28.

From the data of viscosity y (upper row) and sugar content z (lower row) in Table 12.25, we seek the variation of y, the variation of z, and the covariation of y and z.

Table 22.28 Data of Two-Way Layout (Upper Row Is y Values, Lower Row Is z Values)

M \ N	N_1 17.3	N_2 17.5	N_3 17.7	N_4 17.9	N_5 18.1	N_6 18.3	N_7 18.5	Total
M_1 1.36	21.00	21.30	21.40	21.50	21.80	21.90	22.05	150.95
	31.6	31.7	32.0	32.0	32.5	32.6	32.7	225.1
M_2 1.38	21.10	21.40	21.50	21.60	21.90	21.90	22.15	151.55
	31.5	31.9	32.1	32.3	32.6	32.8	32.9	226.1
M_3 1.40	21.20	21.50	21.60	21.70	21.90	22.00	22.25	152.15
	31.6	32.1	32.2	32.4	32.6	32.8	33.0	226.7
M_4 1.42	21.25	21.60	21.70	21.80	22.00	22.10	22.35	152.80
	32.0	32.2	32.5	32.5	32.8	33.0	33.2	228.2
M_5 1.44	21.35	21.65	21.80	21.90	22.10	22.25	22.45	153.50
	32.0	32.4	32.6	32.6	32.9	33.0	33.4	228.9
M_6 1.46	21.50	21.65	21.90	22.00	22.15	22.35	22.55	154.10
	32.0	32.6	32.7	32.8	33.0	33.1	33.5	229.7
Total	127.40	129.10	129.90	130.50	131.85	132.50	133.80	915.05
	190.7	192.9	194.1	194.6	196.4	197.3	198.7	1364.7

$$S_m(yy)=\frac{915.05^2}{42}=19936.1072 \qquad (f=1)$$

$$S_m(yz)=\frac{915.05\times 1364.7}{42}=29732.589 \qquad (f=1)$$

$$S_m(zz)=\frac{1364.7^2}{42}=44343.002 \qquad (f=1)$$

$$S_T(yy)=21.00^2+21.30^2+\cdots+22.55^2-S_m(yy)=5.7653 \qquad (f=41)$$

$$S_T(yz)=21.00\times 31.6+21.30\times 31.7+\cdots+22.55\times 33.5-S_m(zy)$$
$$=7.411 \qquad (f=41)$$

$$S_T(zz)=31.6^2+31.7^2+\cdots+33.5^2-S_m(zz)=9.908 \qquad (f=41)$$

$$S_{M_l}(yy)=\frac{(-5\times 150.95-3\times 151.55-152.15+152.80+3\times 153.50+5\times 154.10)^2}{7\times 70}$$
$$=\frac{22.25^2}{490}=1.0103 \qquad (f=1) \tag{22.139}$$

$$S_{M_l}(yz)=\frac{22.25\times 32.9}{7\times 70}=1.494 \qquad (f=1) \tag{22.140}$$

$$S_{M_l}(zz)=\frac{(-5\times 225.1-3\times 226.1-226.7+228.2+3\times 228.9+5\times 229.7)^2}{7\times 70}$$
$$=\frac{32.9^2}{490}=2.209 \qquad (f=1) \tag{22.141}$$

$$S_{N_l}(yy)=\frac{(-3\times 127.40-2\times 129.10-129.90+131.85+2\times 132.50+3\times 133.80)^2}{6\times 28}$$
$$=\frac{27.95^2}{168}=4.6500 \qquad (f=1) \tag{22.142}$$

$$S_{N_l}(yz)=\frac{27.95\times 35.1}{168}=5.840 \qquad (f=1) \tag{22.143}$$

$$S_{N_l}(zz)=\frac{(-3\times 190.7-2\times 192.9-194.1+196.4+2\times 197.3+3\times 198.7)^2}{6\times 28}$$
$$=\frac{35.1^2}{168}=7.333 \qquad (f=1) \tag{22.144}$$

$$S_e(yy)=S_T(yy)-S_{M_l}(yy)-S_{N_l}(yy)=5.7653-1.0103-4.6500$$
$$=0.1050 \qquad (f=39) \tag{22.145}$$

$$S_e(yz)=S_T(yz)-S_{M_l}(yz)-S_{N_l}(yz)=7.411-1.494-5.840$$
$$=0.077 \qquad (f=39) \tag{22.146}$$

$$S_e(zz)=S_T(zz)-S_{M_l}(zz)-S_{N_l}(zz)=9.908-2.209-7.333$$
$$=0.366 \qquad (f=39) \tag{22.147}$$

Thus, we obtain the table of variations, covariations, variances, and covariances of Table 22.29. We now estimate M_l and N_l.

For y, we use the values of the numerators of Equation (22.139) and Equation (22.142), and

Table 22.29 Variation, Covariation, Variance, and Covariance

Source	f	Variation (yy)	Variation (yz)	Variation (zz)	Variance V_y	Variance V_{yz}	Variance V_z
M_l	1	1.0103	1.494	2.209	1.0103	1.494	2.209
N_l	1	4.6500	5.840	7.333	4.6500	5.840	7.333
e	39	0.1050	0.077	0.366	0.00269	0.00197	0.00938
T	41	5.7653	7.411	9.908			

$$\hat{\beta}(M_l)=\frac{22.25}{7\times 35\times 0.02}\pm\sqrt{\frac{4.09\times 0.00269}{7\times\frac{35}{2}\times 0.02^2}}=4.54\pm 0.47 \tag{22.148}$$

$$\hat{\beta}(N_l)=\frac{27.95}{6\times 28\times 0.2}\pm\sqrt{\frac{4.09\times 0.00269}{6\times 28\times 0.2^2}}=0.832\pm 0.040 \tag{22.149}$$

For z, we use the values of the numerators of Equation (22.141) and Equation (22.143), and

$$\hat{\beta}(M_l)=\frac{32.9}{7\times 35\times 0.02}\pm\sqrt{\frac{4.09\times 0.00938}{7\times\frac{35}{2}\times 0.02^2}}=6.71\pm 0.88 \tag{22.150}$$

$$\hat{\beta}(N_l)=\frac{35.1}{6\times 28\times 0.2}\pm\sqrt{\frac{4.09\times 0.00938}{6\times 28\times 0.2^2}}=1.045\pm 0.076 \tag{22.151}$$

Also, since

$$\bar{y}=\frac{915.05}{42}=21.79$$

$$\bar{z}=\frac{1364.7}{42}=32.49$$

the relationship equations of M, N, and y and of M, N, and z become as follows.

$$y=\bar{y}+\hat{\beta}(M_l)(M-\bar{M})+\hat{\beta}(N_l)(N-\bar{N})$$

$$y=21.79+4.54(M-1.41)+0.832(N-17.9) \tag{22.152}$$

$$z=32.49+6.71(M-1.41)+1.045(N-17.9) \tag{22.153}$$

If we solve the simultaneous Equations (22.152) and (22.153) for M and N, we get

$$M=1.41-1.244(y-21.79)+0.99(z-32.49) \tag{22.154}$$

$$N=17.9+7.989(y-21.79)-5.41(z-32.49) \tag{22.155}$$

Therefore, when the errors of $\bar{y}$, $\bar{z}$, $\hat{\beta}(M_l)$, $\hat{\beta}(N_l)$, etc., are sufficiently small,

$$\begin{aligned}\text{Var}(M) &\fallingdotseq (-1.244)^2\sigma_y^2+2\times(-1.244)\times 0.99\sigma_{xy}+0.99^2\sigma_z^2\\ &\fallingdotseq (-1.244)^2\times 0.00269+2\times(-1.244)\times 0.99\times 0.00197+0.99^2\times 0.00938\\ &=0.00850\end{aligned} \tag{22.156}$$

$$\text{Var}(N)=7.989^2\times 0.00269+2\times 7.989\times(-5.41)\times 0.00197+(-5.41)^2\times 0.00938$$
$$=0.2759 \qquad (22.157)$$

From this, when $\pm 3\sigma$ is regarded as error,

$$M=1.41-1.244(y-21.79)+0.99(z-32.49)\pm 0.28 \qquad (22.158)$$

$$N=17.9+7.989(y-21.79)+0.99(z-32.49)\pm 1.58 \qquad (22.159)$$

These results indicate that the precision is too poor by these methods to allow satisfactory use. Instead of the variance of Equations (22.156) and (22.157), one can also find the SN ratios $\eta(M)$ and $\eta(N)$ by the following equations.

$$\eta(M)=\frac{1}{0.00850}=117.6 \qquad (22.160)$$

$$\eta(N)=\frac{1}{0.2759}=3.62 \qquad (22.161)$$

Exercises (22)

(1) In electrically measuring the water content x% (x is unknown) of a certain product, water was added so that the water content would become $(x+1)$%, $(x+2)$%, and $(x+3)$% and measurement was conducted with the gain of the measuring instrument set at the two levels

$$A_1 = 7 \text{ db}$$
$$A_2 = 5 \text{ db},$$

and the following results were obtained. R_1 and R_2 are two persons, and these are incorporated in the error.

	A_1		A_2	
Water	R_1	R_2	R_1	R_2
$M_1=x$	1.1	1.4	0.8	1.0
$M_2=(x+1)$	2.3	2.4	2.1	2.2
$M_3=(x+2)$	4.9	4.6	3.5	3.5
$M_4=(x+3)$	7.2	6.8	5.0	4.8

(a) Compare the SN ratios of A_1 and A_2, and calculate to find which is better and by how many decibels. Use Appendix Table 9 and also test for significant difference.

(b) Estimate x by the measurement method with the better SN ratio. It should be assumed that there is no zero-point error. One need only solve $\bar{y}+\bar{\beta}(M-x-1.5)=0$.

(2) We wished to measure the volume of a certain powder by using the relationship of atmospheric pressure and volume. Taking care that air would not leak, with the mass of the product in the neighborhood of $M_1' = 100$ g and in the neighborhood of $M_2' = 300$ g, the signal factor M was taken as

$$M_1 = 0, \quad M_2 = 20, \quad M_3 = 40 \text{ g}$$

Readings of atmospheric pressure were taken and the data were as follows. There were two repetitions.

	M_1	M_2	M_3
M_1'	2.65	3.16	3.70
	2.78	3.29	3.84
M_2'	7.75	8.34	8.93
	7.58	8.10	8.81

Obtain the SN ratio of this method. Assume that it has been predicted that the volume of 100 g of this article is about 82 cm^3. Also, obtain the range of error when the volume is obtained by a single run of measurement.

(3) Experiment with Magnetic Powder Deep Flaw Detection Method

We consider a testing method for discovering flaws that are so small as to be impossible to see with the naked eye, for example in a metal rod, by immersion in a magnetic powder. Three factors, A, B, and C, were each taken at two levels, and testing was conducted under the conditions of all eight combinations on three objects whose depth of flaws differed, M_1, M_2, and M_3. Testing was carried out one time each on three days, R_1, R_2, and R_3. The data were obtained by reading the width of the magnetic powder with a microscope in 10-micron units.

			M_1			M_2			M_3		
A	B	C	R_1	R_2	R_3	R_1	R_2	R_3	R_1	R_2	R_3
1	1	1	7	9	10	19	25	21	70	39	60
1	1	2	2	1	2	4	3	2	5	8	10
1	2	1	20	16	10	36	40	30	130	160	140
1	2	2	2	3	5	8	6	4	9	10	12
2	1	1	5	8	7	8	14	13	42	45	36
2	1	2	1	2	2	2	3	5	8	7	7
2	2	1	3	5	8	15	21	23	82	69	93
2	2	2	4	4	3	4	7	10	15	10	13

The value may be regarded as 0 when there is no flaw. (Hint: For example, from the nine data of $A_1B_1C_1$, assume that

$$S_M = \frac{26^2 + 65^2 + 169^2}{3}, \quad S_T = 7^2 + 9^2 + \cdots + 60^2$$

and after finding the SN ratio, carry out analysis of variance.)

(4) For improvement of the Sb microanalysis method in H_3Sb gas generation, each of the factors reduction method D, gas generation time A, acid concentration C, quantity of acid added E, and air quantity B, was taken at two levels (1st level is the current method and 2nd level is the new plan), and this was assigned to orthogonal array L_8 so that all principal effects and the interactions $A \times D$ and $C \times D$ could be obtained. Signal factor M was taken as

$$M_1 = x, \qquad M_2 = x + h, \qquad M_3 = x + 2h$$

with x and h both as unknowns. h was about 0.5 ppm. The data were obtained in two days, R_1 and R_2.

(a) Perform data analysis and find the optimum conditions, and determine how many decibels of gain there will be compared with the present method.

(b) Find the analysis error when a single run of the experiment is carried out by this method.

	D	A	$A \times D$	C	$D \times C$	E	B	R_1			R_2		
No.	1	2	3	4	5	6	7	M_1	M_2	M_3	M_1	M_2	M_3
1	1	1	1	1	1	1	1	0.7	1.4	1.5	0.4	1.3	1.9
2	1	1	1	2	2	2	2	1.5	2.6	3.7	1.4	2.5	3.6
3	1	2	2	1	1	2	2	0.9	1.7	2.8	0.4	1.9	3.0
4	1	2	2	2	2	1	1	0.8	1.8	2.8	0.7	1.8	3.0
5	2	1	2	1	2	1	2	1.1	2.4	3.1	0.6	1.9	2.2
6	2	1	2	2	1	2	1	1.2	2.4	3.6	1.2	2.6	3.7
7	2	2	1	1	2	2	1	0.8	1.5	2.0	0.9	1.6	2.4
8	2	2	1	2	1	1	2	0.9	2.0	3.0	1.2	2.0	3.1

23 Reliability Test, Sensory Test, Qualitative Analysis, and SN Ratio

Here we will discuss the problems of how to conduct *reliability testing* when there is only *one required action,* and of testing of *sensory judgments.* Complicated treatment such as that used for dynamic characteristics in Chapter 24 is not necessary, since in both cases either calibration is useless or subjectivity enters the calibration. The reader is referred to Section 24.1 concerning reliability tests for cases such as digital communications methods. For dynamic characteristics when there is hysteresis, usually it is possible to handle the two curves of increase and decrease separately, and when this is done, the problem becomes one of Section 24.4. Hysteresis is absent in photographic characteristics, copying characteristics, and magnetic tape characteristics, but since these are continuous data they are the problem of Section 24.4 and not the problem of this chapter.

23.1 Design and Analysis of Reliability Test

23.1.1 Continuous Data and Discrete Data

The problem of controllability will be treated further in Chapter 24; here we will discuss a simple case. To select levels of signal factors depending on what value of the target characteristic one desires, is control. If the value of the target characteristic is influenced not only by the signal factors but by many noise factors (error factors), no matter how ingeniously one selects the levels of the signal factors, it is not possible to cause the value of the object characteristic to match perfectly with the target value.

Thus, *there is a limit to control if, to begin with, the dynamic characteristics and the hardware of the control system are not good.* The dynamic characteristics of hardware and the degree of effectiveness of the control system can be compared by the ratio of the magnitude of the effect of the signal factor affecting the output to the magnitude of the effect of the error factor affecting the output. This is the SN ratio. The *SN ratio is a measure which is effective for comparison of the dynamic characteristics of the hardware of control systems.* Unless a machine tool has a good SN ratio, in other words unless it has process capability, there is a limit to one's control of the quality of the product no matter how ingeniously one controls.

Now, in processes where the objective is attained by signals the signals do not necessarily have continuous values. For example, an automatic vending machine is a device that causes objects such as tickets to emerge in response to a signal that money has been inserted. Again, a relay is a device that closes or opens a circuit depending on the availability or unavailability of electric current. Machines and control systems where the value assumed by the desired characteristic is one of a finite number of possibilities, such as one or two, as in these examples, exist in large numbers. Actually, computers and control systems are changing from the analog type to such a digital type. In the case of a digital system, the values or states which are required of the desired characteristics are discrete values, in other words they are of the so-called *integer type.*

In this chapter, we will proceed to discuss the SN ratio for the integer type, and its applications. We will explain *the case where the values which the desired characteristic takes are only 1 or 0,* which is the simplest among integer-type SN ratios. However, in the case of digital systems where hysteresis is used, such as bimetals and parametrons, it becomes the problem of this chapter if the two types of input are analyzed separately.

The problem of distinguishing between 0 and 1 depending on whether there are radio waves or not, as in PCM communications, is dealt with according to Section 24.2. The method of this chapter cannot be used when distinguishing between 0 and 1 by a continuous function.

23.1.2 Simple Reliability and SN Ratio

About the time when automatic vending machines were being developed, when one went to the laboratory of a manufacturer, the technicians would be inserting coins as many times as 3000 or 10,000 times, and they were taking the fraction of times the article emerged properly as the measure of reliability. Here, a variable defined when the signal of feeding a coin is given n times and the variable takes the value 1 if the product emerges and 0 if it does not emerge will be expressed as y. Such a series of length n can be expressed as follows:

$$y_1, y_2, y_3, \cdots, y_n \tag{23.1}$$

The reliability p from the observed data of length n of Equation (23.1) can be calculated by the following equation.

$$p=\frac{y_1+y_2+y_3+\cdots+y_n}{n} \tag{23.2}$$

For simplicity, the total in the numerator of Equation (23.2) will be called r.

$$r=y_1+y_2+y_3+\cdots+y_n \tag{23.3}$$

Since there is only the one type of work involved in bringing forth the product each time, if this amount of work is taken as the *unit of quantity of work,* in effect r in Equation (23.3) becomes the total quantity of work which that system has performed for n signals. It will be assumed here that when there is no signal, in other words when no money is inserted, a ticket (or other product) does not emerge. In Section 24.1 we will discuss cases where this assumption is not possible.

If the total quantity of work is expressed as S_T, S_T is given by the following equation.

$$S_T=y_1+y_2+\cdots+y_n \tag{23.4}$$

This is also equal to the sum of squares. That is because $y_1, y_2, \ldots, y_n$ take only the values 0 or 1; and the first power and second power of the values 0 and 1 are equal.

$$S_T=y_1^2+y_2^2+\cdots+y_n^2 \tag{23.5}$$

The proportion, p, of 1's in $y_1, y_2, \ldots, y_n$ is, as indicated by Equation (23.2), a linear expression with constant coefficients in $y_1, y_2, \ldots, y_n$. (In this case, the coefficient is always $1/n$). As to what magnitude an arbitrary linear expression in the observed values $y_1, y_2, \ldots, y_n$,

$$L=c_1y_1+c_2y_2+\cdots+c_ny_n \tag{23.6}$$

occupies in the total sum of squares (this is the total quantity of work and the total output), one need only divide the square of the linear equation by the sum of squares of the coefficients, which is said to be the unit number.

$$S_L=\frac{L^2}{c_1^2+c_2^2+\cdots+c_n^2} \tag{23.7}$$

If this formula is used, the variation S of p (p is called the reliability) is

$$S=\frac{p^2}{\left(\frac{1}{n}\right)^2+\left(\frac{1}{n}\right)^2+\cdots+\left(\frac{1}{n}\right)^2}=\frac{\left(\frac{y_1+y_2+\cdots+y_n}{n}\right)^2}{\left(\frac{1}{n}\right)^2\times n}$$

$$=\frac{(y_1+y_2+\cdots+y_n)^2}{n} \tag{23.8}$$

This is also n times p^2, which indicates the magnitude of the signal per unit.

$$S=p^2\times n=\frac{(y_1+y_2+\cdots+y_n)^2}{n^2}\times n=\frac{(y_1+y_2+\cdots+y_n)^2}{n} \tag{23.9}$$

The value obtained when the output of signal is subtracted from the total output S_T is the output of noise. If this is expressed as S_e,

$$S_e=S_T-S=y_1+y_2+\cdots+y_n-\frac{(y_1+y_2+\cdots+y_n)^2}{n}=r-\frac{r^2}{n}=np-\frac{(np)^2}{n}$$

$$=np(1-p) \tag{23.10}$$

This is, again, equal to n times the binomial error variance $p(1-p)$. S_e also agrees with the value which is obtained by subtracting one-nth the square of $y_1+y_2+\cdots+y_n$, which is the correction term, from the sum of squares of $y_1, y_2, \ldots, y_n$.

$$S_e=y_1^2+y_2^2+\cdots+y_n^2-\frac{(y_1+y_2+\cdots+y_n)^2}{n} \tag{23.11}$$

When these are organized into an analysis of variance table, we obtain Table 23.1.

The expected value of variance here was obtained as follows. With the mean of y_i as p',

$$E(y_1+y_2+\cdots+y_n)=np'$$

$$E(y_1^2+y_2^2+\cdots+y_n^2)=np'$$

$$E(y_1+y_2+\cdots+y_n)^2=(np')^2+np'(1-p')$$

$$E\left[\frac{(y_1+y_2+\cdots+y_n)^2}{n}\right]=p'(1-p')+np'^2 \tag{23.12}$$

Table 23.1 Analysis of Variance Table

Source	f	S	V	$E(V)$
Signal	1	S	V	$p'(1-p')+np'^2$
Error	$n-1$	S_e	V_e	$p'(1-p')$
T	n	S_T		

$$E\left[y_1+y_2+\cdots+y_n-\frac{(y_1+y_2+\cdots+y_n)^2}{n}\right]=np'-p'(1-p')-np'^2$$
$$=(n-1)p'(1-p') \qquad (23.13)$$

Therefore, if we use straightforwardly the formula for the SN ratio η in the case of discrete data,

$$\eta=\frac{p'^2}{p'(1-p')}\doteqdot\frac{\frac{1}{n}(V-V_e)}{V_e}=\frac{\frac{1}{n}\left[\frac{(y_1+y_2+\cdots+y_n)^2}{n}-\frac{1}{n-1}S_e\right]}{\frac{1}{n-1}S_e}$$
$$=\frac{n-1}{n}\times\frac{\left[\frac{(np)^2}{n}-\frac{n}{n-1}p(1-p)\right]}{np(1-p)}=\frac{n-1}{n}\times\frac{p}{1-p}\times\left(1-\frac{1}{n-1}\right)$$
$$=\frac{n-2}{n}\times\frac{p}{1-p} \qquad (23.14)$$

Since usually the frequency of testing n in the case of a digital system is far greater than 1, Equation (23.14) can be approximated by the following equation, with $n-2 \doteqdot n$.

$$\eta=\frac{p}{1-p} \qquad (23.15)$$

Ten times the common logarithm of this is the value in decibel units, and

$$\eta=10\log\frac{p}{1-p}=-\mathbf{10}\ \mathbf{log}\left(\frac{\mathbf{1}}{\boldsymbol{p}}-\mathbf{1}\right) \qquad (23.16)$$

Therefore, the SN ratio η when there is only one type of signal and the desired action is performed r times with n signals is given approximately by Equation (23.16), where

$$p=\frac{r}{n} \qquad (23.17)$$

The reader is referred to Appendix Table 5 for the decibel values of Equation (23.16). Also, when $p=0$ and $p=1$, substitute

$$p=\frac{1}{2n} \text{ and } p=1-\frac{1}{2n}$$

respectively, as approximations.

23.2 Application to Simulated Test

Why did we use the SN ratio as in Equation (23.16) in place of p, which already gives a value for the reliability? It is because *one can predict arithmetic additivity* for the *SN ratio (unit db)* of Equation (23.16).

Table 23.2 Simulated Testing Data

	When not used	When used
Countermeasure *A*	90%	99%
Countermeasure *B*	95	98
Countermeasure *C*	80	95

We will explain here an efficient method of reliability testing for machinery and systems where there is only one required action by using the SN ratio which is given by Equation (23.16).

The reliability of automatic vending machines at present (the proportion of correct operation when money is inserted) is *0.994 in the marketplace.* Three proposals, *A*, *B*, and *C*, were advanced for the improvement of this reliability.

A: change part *A* to a new one
B: use a new circuit
C: attach a dust shield

In order to examine the effects of these three proposals efficiently, the electric power used was reduced to $\frac{1}{3}$ and testing was carried out while applying vibrations and dust to the machine. The data in Table 23.2 were all obtained by testing the countermeasures 100 times each separately under these *simulated conditons.*

The effects of countermeasure *A*, countermeasure *B*, and countermeasure *C* are expressed in the decibel units of Equation (23.16). From Appendix Table 5, the decibel value of 90%, for example, is 9.54 db. Table 23.3 is thus obtained.

Therefore, the gain when all three countermeasures are used is 21.30 db. Since the reliability in the marketplace, 0.994, is 22.19 db in decibel units, the decibel value when all three countermeasures are used is

$$21.30+22.19=43.49 \text{ [db]} \tag{23.18}$$

Therefore, if

$$-10\log\left(\frac{1}{p}-1\right)=43.49 \tag{23.19}$$

Table 23.3 Data Expressed in Decibel Units

	When not used	When used	Gain
Countermeasure *A*	9.54 db	19.96 db	+10.42 db
Countermeasure *B*	12.79	16.90	+ 4.11
Countermeasure *C*	6.02	12.79	+ 6.77
Total			+21.30

is solved by using the table of logarithms, we obtain finally,

$$p=0.999955 \tag{23.20}$$

The reliability improves from 99.4% to 99.9955%.

To have an inspector insert coins 3000 or 1000 times for testing is inefficient. To raise the efficiency, one should reduce the signal (in this case the electric power was reduced to $\frac{1}{3}$) or increase the noise (error factor). This is because the reason why the object does not emerge reliably when money is inserted is that there is noise or the signal becomes distorted. In such a case, vibrations, dust, rust, and moisture, for example, are noise sources and distort the signal. What noise to choose is the most important consideration for rationalization of testing.

In 1952, CCIF (international communications consultative committee) decided on the method which uses *articulation* for measuring the *quality of telephone systems*. In this method, first two standard telephones (such as telephones which are at present being used in the United States), X_0 and X_0', are brought and testing is performed as shown in Figure 23.1.

When a girl speaks 100 randomly-listed words of Esperanto (at the Electrical Communication Laboratory in Japan, random listings of the Japanese syllabary, sonants, p-sounds, and contracted sounds are being used) using telephone X_0, another girl, listening with telephone X_0', counts the number of errors. In this case, if an ordinary line is used, testing must be performed several tens of thousands or hundreds of thousands of times to be certain that there are nearly no errors, and it becomes a great deal of trouble.

Therefore, testing is conducted by impressing white noise onto the line that connects the two telephones, and by increasing the error rate by reducing the signal electric power as well. *CCIF recommends that the signal be reduced and white noise be applied so that the error rate becomes about 20% and articulation becomes about 80%.* We will assume here that the articulation is 80%.

Then, under this simulated test, the SN ratio is

$$-10\log\left(\frac{1}{0.8}-1\right)=6.02\ [\text{db}] \tag{23.21}$$

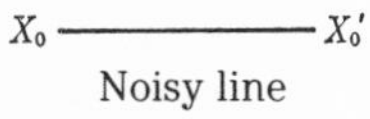

X_0 ———————— X_0'

Noisy line

FIGURE 23.1
Test of Telephone System

Next, with this noisy line left as it is, similar testing is conducted by switching the telephones at both ends with the telephones to be tested (new products or telephones being developed). So that the learning effect will not affect the data, testing is performed, for example, with a different list of words or the two sequences are assigned as levels to an $L_4(2^3)$, orthogonal array. Whatever the case, we will assume that articulation has become 95% with the telephones being tested. The SN ratio of this is, from Appendix Table 5,

$$-10 \log\left(\frac{1}{0.95}-1\right)=12.79 \text{ [db]} \tag{23.22}$$

Therefore, the telephones for testing have the gain

$$12.79-6.02=6.77 \text{ [db]} \tag{23.23}$$

CCIF uses

$$\eta=500+50 \log\left(\frac{1}{p}-1\right) \tag{23.24}$$

instead of Equation (23.16), but there is no fundamental difference. We believed that Equation (23.24) was used because the relationship between analysis of variance and the SN ratio was not understood clearly at the time.

What is very important is that Equation (23.16) *cannot be used for electrical transmission of information between terminals and a mainframe computer*. This is because the data consist of two signals, 0 and 1, as with relays and switches. Therefore, the SN ratio is defined by a completely different method. We will explain this in Section 24.2.

23.3 Applications in the Chemical Field

The method which has been shown here can also be applied to experimental data for the yield when reverse reaction and reaction do not occur too much, or to experiments for improvement of purity or insecticidal rate or the like. It cannot be used, for example, for the removal rate when removing harmful components from industrial waste water or raw materials, or for the extraction rate when extracting useful components from ores. The reason for this is that substances other than the desired component are mixed with the product of removal or extraction. For example, in the process of copper smelting, when copper is extracted from copper ore the purity of the copper in the ingot is not 100% but only about 98%–99%.

As an example of the method, the number of flaws in carbon fiber crystals can be calculated by the decibel value which is shown here (n is the total number of crystals or the total number of molecules of carbon). However, when n is very large, we recommend the use of

$$-10 \log \frac{p}{1-p} \doteqdot -10 \log p \tag{23.25}$$

since

$$\frac{p}{1-p} \doteqdot p$$

(Please refer to the data of gel fraction in Section 22.3.2 for the analysis method.)

23.4 Problem Points of Sensory Test

Many tests that rely on the senses, such as those for goodness or badness of taste, are difficult to do by means of instruments. The author considers the reason for this to be as follows. Even concerning the simple problem of gloss, for example, he regards this as being the *dispersion of the intensity of light* to *the light-sensitive cell units* of the retina. Therefore, there is a problem as to the definition of brilliance according to the Munsell system. It is the author's prediction that the magnitude of gloss is proportional to (the square of) the *spatial light amplitude* and the magnitude of differences of stimuli between cell units such that certain light-sensitive cells sense strong light and adjoining cells sense weak light. It should suffice if one creates a light distribution meter, similar to a surface roughness meter.

Nor is it correct that flavors are determined by the proportions of several components such as sweetness, sourness, saltiness, and bitterness. At Company N in Shinjuku (in Tokyo), the components of rice curry were ground very finely and were eaten, whereupon it was discovered that the mixture was no longer tasty at all. Although the components were exactly the same, the flavor became bad. The author believes that unbalance between cell units on the tongue or the surface of the mouth cavity plays an important role in flavor; for example, certain taste bud cells taste the flavor of curry powder, while adjoining taste bud cells taste the flavor of meat. This is the reason why the flavor of beer is improved by the addition of tasteless and odorless carbon dioxide gas.

No matter how much one measures the stiffness of fibers, the feeling does not change. This is said to be a strange characteristic. Each single cell at the surface of the skin of a human being is a sensory organ, and it is important to measure the dispersion of the sensations among cell units. For timewise changes, instruments have been developed for measurement in terms of sound pressure or voltage, but the development of measurement methods for *spatial differences (by cell unit)* is a major problem which remains for measurement in scientific areas.

Therefore, until new measuring instruments are developed, direct measurement by human beings, or so-called sensory testing, is necessary. Calculations and analyses of experiments when conducting taste tests and comparing gloss by using sensory test data are as indicated in Chapter 2 and Chapter 3, and these are no different from other experiments. Since human beings perform sensory tests, we will explain in this chapter how to investigate what persons are suitable for such tests. The method indicated here is an experimental planning and data analysis method for the finding of optimum conditions for sensory tests and qualitative analyses.

23.5 Chord Discriminating Capability

The following experiment was performed by Masaki Ogawa (1975); and the ability to distinguish three types of chords was tested with four persons, A_1, A_2, A_3, and A_4. If one is concerned with discrimination among the sounds of do, re, mi rather than chords, one analyzes by the accumulation method since between the case where do is mistaken for re and the case where do is mistaken for mi, the mistake is greater in the latter case. In the present case, we analyze by *frequency analysis.*

A_1 = person who began learning the piano before entering elementary school
A_2 = person who began learning the piano after entering elementary school
A_3 = person who learned the piano as a hobby in junior college
A_4 = person who never especially learned a musical instrument

(A_1–A_3 are not learning piano at present. A_2 played the guitar as a hobby but is not playing much at present. The other persons never played the guitar.)

M_1 = chord of do mi so
M_2 = chord of do fa la
M_3 = chord of ti re so

The test was repeated ten times each for M_1, M_2, and M_3 randomly.

Table 23.4 gives the results of organization of the answers when the sounds M_1, M_2, and M_3 were played randomly.

SN Ratio for the Case of A_1

$$W_1=\frac{30^2}{8(30-8)}=5.11 \qquad W_2=\frac{30^2}{10(30-10)}=4.50 \qquad W_3=\frac{30^2}{12(30-12)}=4.17 \tag{23.26}$$

$$CF=\frac{8^2\times 5.11+10^2\times 4.50+12^2\times 4.17}{30}=45.91 \qquad (f=2) \tag{23.27}$$

Table 23.4 Organized Data

Data for A_1

Actual \ Answer	M_1	M_2	M_3	Total
M_1	8	0	2	10
M_2	0	10	0	10
M_3	0	0	10	10
Total	8	10	12	30

Data for A_2

Actual \ Answer	M_1	M_2	M_3	Total
M_1	8	1	1	10
M_2	0	9	1	10
M_3	4	1	5	10
Total	12	11	7	30

Data for A_3

Actual \ Answer	M_1	M_2	M_3	Total
M_1	6	0	4	10
M_2	3	7	0	10
M_3	2	3	5	10
Total	11	10	9	30

Data for A_4

Actual \ Answer	M_1	M_2	M_3	Total
M_1	6	2	2	10
M_2	0	8	2	10
M_3	4	3	3	10
Total	10	13	7	30

$$S_M=\frac{8^2+0^2+0^2}{10}\times 5.11+\frac{0^2+10^2+0^2}{10}\times 4.50+\frac{2^2+0^2+10^2}{10}\times 4.17-CF$$

$$=75.16 \qquad (f=4) \tag{23.28}$$

$$S_T=90 \qquad (f=58) \tag{23.29}$$

$$S_e=S_T-S_M=90-75.16=14.84 \qquad (f=54) \tag{23.30}$$

Therefore, we obtain the analysis of variance table of Table 23.5.

$$\eta=\frac{\frac{1}{10}(18.79-0.27)}{0.27}=6.85=8.36\ [\text{db}] \tag{23.31}$$

Similarly,

$$\eta(A_2)=\frac{\frac{1}{10}(9.43-0.97)}{0.97}=0.872=-0.60\ [\text{db}] \tag{23.32}$$

$$\eta(A_3)=\frac{\frac{1}{10}(5.38-1.27)}{1.27}=0.324=-4.89\ [\text{db}] \tag{23.33}$$

Table 23.5 Table of Analysis of Variance of A_1

Source	f	S	V	$E(V)$
M	4	75.16	18.79	$\sigma^2+10\sigma_M{}^2$
e	54	14.84	0.27	σ^2
T	58	90.00		

Table 23.6 Comparison of Discriminating Capability

	η	Decibel value	Gain	Multiplier
A_1	6.85	8.36 db	+14.98 db	31.5 Times
A_2	0.872	−0.60	+6.02	4.0
A_3	0.324	−4.89	+1.73	1.5
A_4	0.218	−6.62	Standard	

$$\eta(A_4)=\frac{\frac{1}{10}(4.30-1.35)}{1.35}=0.218=-6.62\text{ [db]} \tag{23.34}$$

Therefore, the conclusions of Table 23.6 are obtained.

There is a considerable difference in the SN ratio between A_1 and A_2, and A_1 possesses a discriminating capability which is 8.96 db greater than or 7.9 times that of A_2.

23.6 Comparison of Discriminating Capability for Beer Brand by Smoker and Nonsmoker

23.6.1 Problem and Data

It is said that the discriminating capability of persons who smoke is lower in taste tests. For research regarding this, eight smokers and five nonsmokers were selected and the experiment of having them judge the brands of three types of beer in two trials was conducted.

(1) Control factors A_1 = smoker
A_2 = nonsmoker

Table 23.7 Data

Data for A_1

R_1	M_1	M_2	M_3	Total
M_1	4	4	1	9
M_2	4	3	2	9
M_2	2	0	4	6
Total	10	7	7	24

R_2	M_1	M_2	M_3	Total
M_1	5	4	1	10
M_2	2	4	1	7
M_3	1	0	6	7
Total	8	8	8	24

R_1+R_2	M_1	M_2	M_3	Total
M_1	9	8	2	19
M_2	6	7	3	16
M_3	3	0	10	13
Total	18	15	15	48

Data for A_2

R_1	M_1	M_2	M_3	Total
M_1	4	0	0	4
M_2	1	3	0	4
M_3	0	0	7	7
Total	5	3	7	15

R_2	M_1	M_2	M_3	Total
M_1	2	1	0	3
M_2	1	5	0	6
M_3	2	1	3	6
Total	5	7	3	15

R_1+R_2	M_1	M_2	M_3	Total
M_1	6	1	0	7
M_2	2	8	0	10
M_3	2	1	10	13
Total	10	10	10	30

(2) Signal factors M_1 = Asahi
M_2 = Kirin
M_3 = Suntory
(3) Error factors R_1 = first time
R_2 = second time

After each person was first informed of the brands, the persons were asked to drink M_1, M_2, and M_3. After having rinsed their mouths, they were requested to drink three types of beer (actually, maybe all of the three types were of the same brand and maybe all were different), and data were obtained by having them guess the brands (Table 23.7).

23.6.2 Calculation of SN Ratio

We calculate by frequency analysis.

(1) SN Ratio of A_1

We find weights W_1, W_2, and W_3.

$$W_1=\frac{48^2}{18\times(48-18)}=4.267 \quad W_2=\frac{48^2}{15\times 33}=4.655 \quad W_3=\frac{48^2}{15\times 33}=4.655$$

The total variation S_T is

$$S_T=48\times 3=144 \qquad (f=94) \tag{23.35}$$

The variation among products S_M is

$$S_M=\left(\frac{9^2}{19}+\frac{6^2}{16}+\frac{3^2}{13}-\frac{18^2}{48}\right)\times W_1+\left(\frac{8^2}{19}+\frac{7^2}{16}+\frac{0^2}{13}-\frac{15^2}{48}\right)\times W_2$$

$$+\left(\frac{2^2}{19}+\frac{3^2}{16}+\frac{10^2}{13}-\frac{15^2}{48}\right)\times W_3$$

$$=0.453\times 4.267+1.743\times 4.655+3.779\times 4.655$$

$$=27.64 \qquad (f=4) \tag{23.36}$$

$$S_e=S_T-S_M=144-27.64=116.36 \qquad (f=90) \tag{23.37}$$

From this, the analysis of variance table of A_1 becomes as given in Table 23.8. The number $\bar{r}$ below is the harmonic mean of the different row totals, 19, 16, and 13, and this is obtained by the following equation.

Table 23.8 Analysis of Variance Table of A_1

Source	S	f	V	$E(V)$
M	27.64	4	6.91	$\sigma^2+\bar{r}\sigma_M{}^2$
e	116.36	90	1.29	σ^2
T	144.00	94		

$$\frac{1}{\bar{r}}=\frac{1}{3}\left(\frac{1}{19}+\frac{1}{16}+\frac{1}{13}\right)=0.0640 \tag{23.38}$$

$$\eta=\frac{\frac{1}{\bar{r}}(V_M-V_e)}{V_e}$$

$$=\frac{0.0640(6.91-1.29)}{1.29}$$

$$=0.278=-5.5\,[\text{db}] \tag{23.39}$$

(2) SN Ratio of A_2

$$W_1=\frac{30^2}{10\times 20}=4.5 \qquad W_2=\frac{30^2}{10\times 20}=4.5 \qquad W_3=\frac{30^2}{10\times 20}=4.5$$

$$S_T=90 \qquad (f=94) \tag{23.40}$$

$$S_M=\left(\frac{6^2}{7}+\frac{2^2}{10}+\frac{2^2}{13}-\frac{10^2}{30}\right)\times W_1+\left(\frac{1^2}{7}+\frac{8^2}{10}+\frac{1^2}{13}-\frac{10^2}{30}\right)\times W_2$$

$$+\left(\frac{0^2}{7}+\frac{0^2}{10}+\frac{10^2}{13}-\frac{10^2}{30}\right)\times W_3$$

$$=2.518\times 4.5+3.287\times 4.5+4.359\times 4.5$$

$$=45.74 \qquad (f=4) \tag{23.41}$$

$$S_e=90.00-45.74=44.26 \qquad (f=54) \tag{23.42}$$

$$\frac{1}{\bar{r}}=\frac{1}{3}\left(\frac{1}{7}+\frac{1}{10}+\frac{1}{13}\right)=0.1066 \tag{23.43}$$

$$\eta=\frac{\frac{1}{\bar{r}}(V_M-V_e)}{V_e}=\frac{0.1066(11.44-0.820)}{0.820}$$

$$=1.38=1.4\,[\text{db}] \tag{23.44}$$

23.6.3 Conclusions

From the previous section, the SN ratios of smokers and nonsmokers were obtained as follows.

	η	Decibel value	Gain
A_1	0.278	−5.5 db	−6.9 db
A_2	1.38	1.4	Standard

Table 23.9 Analysis of Variance Table of A_2

Source	S	f	V	$E(V)$
M	45.74	4	11.44	$\sigma^2+\bar{r}\sigma_M{}^2$
e	44.26	54	0.820	σ^2
T	90.00	58		

This means that smokers have a discriminating capability of −6.9 db, or in other words about one-fifth that of nonsmokers.

Exercises (23)

(1) In actual use, the reliability of an automatic vending machine (the proportion of times that the product will emerge when a coin is inserted) is estimated to be 99.2%. In order to improve this reliability, five countermeasures, A, B, C, D, and E, were experimented with by orthogonal array L_9.

A:	part A	A_1 = current A_2 = new
B:	part B	B_1 = current B_2 = new
C:	circuit	C_1 = current C_2 = new 1 C_3 = new 2
D:	power used	$D_1 = D_2 \times 0.8$ $D_2 = \text{current} \times \frac{1}{3}$ $D_3 = D_2 \times 1.2$
E:	dustproofing method	E_1 = current E_2 = new dustproofing device used

The experimental data were obtained by lowering the power used to $\frac{1}{3}$ the normal power and imparting vibrations to the dust in order to raise testing efficiency.

	(AB) 1	C 2	D 3	E 4	Data (number of correct actions in 100 trials)
1	11	1	1	1	37
2	11	2	2	2	94
3	11	3	3	1′	60
4	12	1	2	1′	58
5	12	2	3	1	98
6	12	3	1	2	60
7	21	1	3	2	83
8	21	2	1	1′	88
9	21	3	2	1	62
Total					640

A, B is assigned by the combination method, and a dummy was introduced into E_1.

(a) Convert each data item to the decibel value of the SN ratio and perform an analysis of variance.
(b) Compare with the current conditions and estimate the gain by the optimum conditions as well as the reliability under the actual conditions of use, approximately.

(2) In predicting tomorrow's weather, we wish to compare the prediction capabilities of

A_1 = the Meterological Agency
A_2 = a patient with neuritis

by classifying the weather as

M_1 = clear
M_2 = sometimes clear and sometimes cloudy
M_3 = cloudy
M_4 = clouding and occasional rain
M_5 = rain

The data for two weeks between November 10 and November 23, 1975, were as follows.

Data for A_1

Day	1	2	3	4	5	6	7	8	9	10	11	12	13	14
Actual weather	3	3	4	1	1	3	2	4	3	1	1	2	1	2
Predicted weather	3	4	4	2	1	1	2	4	4	1	2	2	1	2

Data for A_2

Day	1	2	3	4	5	6	7	8	9	10	11	12	13	14
Actual weather	3	3	4	1	1	3	2	4	3	1	1	2	1	2
Predicted weather	4	3	3	3	1	3	1	5	4	3	1	2	1	1

Find the SN ratio and calculate by how many decibels the prediction capability of the Meteorological Agency is better than that of the patient with neuritis. Use the accumulating method for analysis.

24 Dynamic Characteristics and SN Ratio

In this chapter we will discuss the methods of assignment and performance evaluation for experiments on dynamic functions, such as the maneuverability of automobiles and aircraft, and the performance of machine tools, sporting equipment, and control systems. Digital communication systems, selection of particles, and diagnosis also involve dynamic characteristics and can be described by the SN ratio. These applications, which are not testing, are the most important fields of application of *analysis of variance* in the experimental design method.

24.1 What Are Dynamic Characteristics?

24.1.1 Definition

Many characteristics such as the maneuvering capability of an automobile, the performance of skis and golf balls, the rolling properties of raw materials and rolling mills, the effectiveness of various types of control systems, and the sports ability of human beings can be said to be dynamic characteristics.

Nearly all of these have the graphic formula:

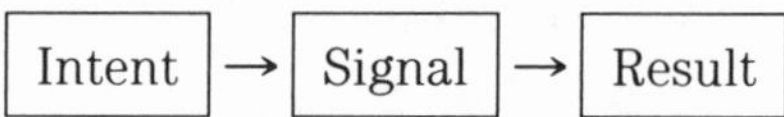

When a human being does not mediate, intent is, in effect, replaced by the target value.

In the case of skiing, depending on whether one wishes to advance straight ahead or to turn right, one places the center of gravity of the body at the center or deflects it to the right. *If there is intent, or a target that one wishes to achieve in a certain way, it is necessary to put forth changeable variables, or in other words a signal, in order to accomplish this.* In this case, since there are targets that one wishes to assume various states, *e.g.*, one wishes to move straight ahead or to turn right by a certain amount, it is also necessary to vary the amount of the signal, such as changing the center of gravity, depending on the target. This is why we use the term dynamic characteristic.

Even in the case of skiing, if only the capability for advancing straight forward, as in jumping, is the problem, it is enough to improve only this property by putting several channels in the skis. Essentially, in this case, dynamic performance is unnecessary.

However, a *material whose rollability is good must be one that can be rolled to various thicknesses.* It is required of it that it become 1 mm when one wishes to roll it to 1 mm thickness, and that it become 0.1 mm when one wishes to roll it to 0.1 mm thickness. The requirement differs according to the case. If the material always fits these differing requirements well, one claims that the dynamic performance of that material is good.

Differences of requirements and objectives, such as that one time one wishes to fabricate the material to 1 mm thickness and the next time one wishes to fabricate it to 0.1 mm thickness, are realized by changing the signal, as by changing the rolling pressure and the number of passes. *Our intent to produce a certain thickness is initially expressed as a specific signal, such as the reduction ratio or the number of passes.*

In the case of the turning capability of skis, signals would be the application of the body weight and the use of the edge of the skis. In the case of the maneuvering performance of an automobile, the steering wheel is the medium for the expression of the intent, and the angle of its turning is the signal. In the case of the intensity of color of a given dye, capability means that it is possible to dye with this dye to the desired intensity. When one varies the quantity of dye to change the color intensity of the finish-dyed product, the quantity of dye used is the signal factor. When one uses the same quantity, if the dyed intensity of the finished product varies with the yarn count number and thickness of the material being dyed, or with differences in the temperature when dyeing or differences in the stirring conditions, this means that the dyed intensity varies with factors other than the signal factor, and it means that the dye affinity is not good. That dye affinity is good means that the dyed intensity changes very accurately if the level of the signal factor is changed but is influenced as little as possible by factors other than the signal factor (these are termed error factors or noise factors). The former is the *signal factor effect* and the latter, the *noise factor effect.* The value that is obtained when the magnitude of the former factor is divided by the magnitude of the latter factor is the SN ratio.

One might raise the question: If the effect of the signal were too great, and the intensity after dyeing were changed greatly by very slight differences in the quantity used, would this not be undesirable in a dye? In other words, since it is undesirable if the effect of the signal (the sensitivity) is too great, it is true that dye affinity, which also includes ease of use, cannot be represented only by the SN ratio.

If the intensity after dyeing changes too much when the quantity used changes very little, it would be troublesome if one mistook the quantity only a little, and minute care became necessary in measuring it. In other words, this would be a dye which would be very difficult to use. When the quantity used, M, is the signal factor, the fact that the intensity of color after dyeing, y, differs greatly with small changes in the quantity of use means that the function

$$y = f(M) \tag{24.1}$$

has a derivative,

$$y' = f'(M) \tag{24.2}$$

which is large. $[f'(M)]^2$, which is the magnitude of $f'(M)$, is termed the magnitude of sensitivity, but no matter what value this may have, it can be changed simply. For example, it is possible to reduce the magnitude of this sensitivity as much as one wishes by diluting the dye or by introducing a suitable filler.

Although *it is undesirable that the sensitivity be too great,* the sensitivity that is necessary to obtain the optimum in terms of human operation is determined merely by observation, and it is not essentially a problem of improvement of technique. Therefore, it is a good sequence of procedure to determine what to do with the sensitivity after finding a case where the SN ratio is maximum.

In the case of maneuvering performance of an automobile, too, the steering system is bad if, by a slight turning of the steering wheel, the path of the vehicle is turned by a large amount. However, since it is possible to change the sensitivity as much as one wishes simply by changing the ratio of the gears, large sensitivity does not cause any difficulty. In many cases, improvement of the SN ratio is the only substantial technical improvement needed, and is essential. In the present case, the signal is the turning angle of the steering wheel, and noise factors are, for example, the quality and the hills and valleys of the road or the kind of wind that is blowing. Please refer to the following sections for specific methods of experimentation and calculation.

24.1.2 How to Obtain Dynamic Characteristics

To study improvement of dynamic properties, it is important to consider the following four types of factors, exactly as in studies of improvement of testing and measurement methods.

(1) *Control Factors.* These are factors that have been chosen in order to improve dynamic performance; the superiority or inferiority of their levels is compared by finding the SN ratio of each level. For example, type of dye, production conditions, steering mechanism of an automobile, type of material to be rolled, and type of rolling mill are all control factors. All control factors possess levels and they have been chosen in order to select a good level.

(2) *Indicative Factors.* There are factors such as the conditions of use where selection of level is not possible but one wishes to find the SN ratios of the levels. These are termed indicative factors. The interaction of an indicative factor and a control factor is a control factor.

(3) *Signal Factors.* These are factors that are chosen in order to express intent, or because one wishes to obtain results in accordance with the target, and these too must necessarily possess levels. For example, quantity of dye, turning angle of a steering wheel, how body weight is applied, and reduction ratio and number of passes of a rolling mill are all signal factors. Let us assume that when dyeing with a given dye, the finished concentration of this dye varies greatly depending on the temperature while dyeing. In a case where one adjusts the temperature while dyeing in order to control the finished dyed intensity, the dyeing temperature too is a signal factor. Therefore, *a signal factor is a factor by which intent or the desire to keep values near to the target is expressed, and the researcher himself decides what to use as the signal factor.* What to use as a signal factor and what to use as a noise factor are freely decided on by the researcher himself, and it is not automatically determined. Also, one often combines the use of two types of signal factors, *signal factors for substantial adjustment* and *signal factors for fine adjustment.*

(4) *Noise Factors.* This term refers to all factors other than the control factors and signal factors which affect the result. In the case of skiing, it refers for example to the hills and valleys on the ski slope and the quality of the snow. If levels have not been chosen for these factors,

one finds the magnitude of the effect of the noise factors from the sum of squares of the residuals.

We will assume that we have selected two types of materials A_1 and A_2 as control factors and that we wish to study which has the better rolling property. For the two types of materials in this case, the reduction ratio M of a given rolling mill is used as the signal factor and one rolls by varying the reduction ratio at equal intervals according to its scale.

As the rolling property, up to now, it has been the practice to determine the ultimate thinness value, that is, how thin the material can be rolled, but there are many problems in so doing. Once, to compare the image-resolving power of a Japanese film and a Kodak film, a Japanese film company conducted tests to determine how many stripes of a black-and-white zebra pattern could be distinguished in 1 inch. Test charts similar to a television test pattern had been prepared and superiority or inferiority was compared by the value of the greatest number of stripes one could distinguish in 1 inch. It is said that when this was done, a greater number of stripes could be distinguished with the Japanese film than with the Kodak film. In spite of this, when viewed as a whole photograph, the Kodak film appeared clearer. There was consternation as to where the essential nature of the problem lay.

This is the same problem as the Hertz range over which it is possible to reproduce vocal sounds in a sound apparatus such as a stereo set. We will assume that as limiting values, a stereo set of one's own company could reproduce vocal sounds up to 20,000 Hertz and one of another company was able to reproduce only up to 18,000 Hertz. It will be assumed that, as shown in Figure 24.1, the frequency characteristics of the product of the other company are flat and it reproduces from 30 Hertz to 18,000 Hertz with about the same sensitivity, and that the product of one's own company possesses a sensitivity curve that traces a gentle arc from 20 Hertz to 20,000 Hertz. In this case, the probability that the sound reproduction characteristics of the product of one's own company are bad is high.

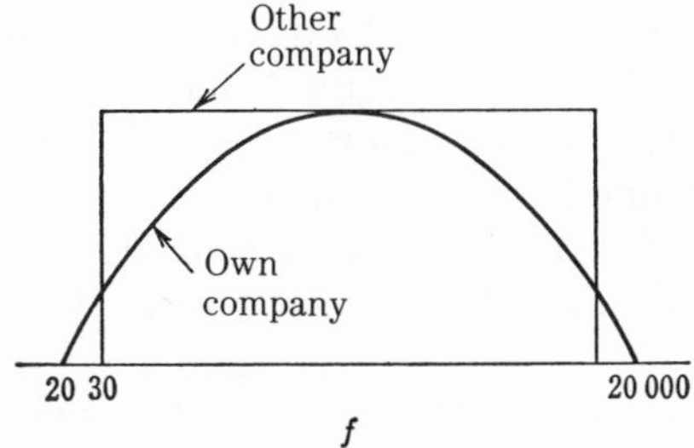

FIGURE 24.1 Two Types of Frequency Characteristics

Even if the Japanese film is better in terms of limiting value, if the frequency characteristics are not flat the image-resolving power will be reduced. Therefore, if there is a black-and-white zebra pattern with a variety of numbers of stripes per inch, *one will not have a true comparison unless one gets quantitative information with the degree of image-resolving power versus the number of stripes as the frequency characteristic.*

24.1.3 Dynamic Characteristics (The Case of Rollability)

When comparing materials, *finding the limiting values of thinness that it is ultimately possible to roll down to does not essentially indicate the true rollability.* Rather, it is necessary to compare the rollability when it is given for various thicknesses.

We will assume that for two types of materials A_1 and A_2, the rolled thickness was measured while varying the reduction ratio of the rolling mill over the four levels M_1, M_2, M_3, and M_4. It will be assumed that M_1, M_2, M_3, and M_4 are four levels of an equal-interval signal factor when the scale of the reduction ratio has been changed by one graduation at a time. Thus, in effect M is a signal factor which has been taken at equal intervals and this interval is 1.

A good rolling material must be one such that even if the temperature of the material were to be displaced slightly or if the components of the material differed slightly, the dispersion of the thickness after rolling would be small. For it is troublesome to change the reduction ratio every time simply because the temperature of the material differs somewhat. If the material components and temperature differ slightly at different locations in the material, it is not possible to vary the rolling pressure for each location. It might be said that the reason the roll of the rolling mill has not been designed as a perfect cylinder is that averaged temperature diferences, etc., depending on location have been considered and it has been corrected accordingly, but it is not really possible to correct localized differences.

Consider a product which has been made by rolling under the conditions of two levels: R_1 as the coil immediately after removal from the furnace and R_2 as the coil after being left standing for a certain period of time. It will be assumed that the data obtained when the thickness was measured at two locations, center and edge, are given in the upper and lower rows of Table 24.1. The working mean, 1000 μm has been subtracted.

In this case, R is in effect the noise factor, with two levels.

What one does is to conduct an analysis of variance for A_1 and for A_2, to find the SN ratios, and to perform a comparison of A_1 and A_2 quantita-

Table 24.1 Experimental Data for Comparison of Rollability (Unit: μm)

		M_1	M_2	M_3	M_4
A_1	R_1	102	86	58	34
		103	86	58	37
	R_2	110	88	64	40
		98	73	56	40
A_2	R_1	85	77	69	65
		85	77	72	64
	R_2	86	78	70	63
		86	79	70	62

tively. First, we prepare auxiliary Table 24.2 classed by A_1 and A_2 in which, for both A_1 and A_2, the working mean 70 has further been subtracted.

(1) *SN Ratio of* A_1. We calculate the total variation, signal variation, and error variation for the case of A_1. First, the total variation S_T is given by the following equation from the theory of analysis of variance.

$$S_T = 32^2 + 16^2 + \cdots + (-30)^2 - \frac{13^2}{16} = 10036 \qquad (f = 15) \qquad (24.3)$$

As the effect of the signal, one calculates only the linear effect of M. The reason for this is that unless the change of thickness with respect to the change of reduction ratio is linear, it is extremely difficult to achieve the target value for thickness. It does not matter if linearity does not hold in

Table 24.2 Supplementary Table

Data of A_1

	M_1	M_2	M_3	M_4	Total
R_1	32	16	−12	−36	0
	33	16	−12	−33	4
R_2	40	18	−6	−30	12
	28	3	−14	−30	−13
Total	133	53	−44	−129	13

Data of A_2

	M_1	M_2	M_3	M_4	Total
R_1	15	7	−1	−5	16
	15	7	2	−6	18
R_2	16	8	0	−7	17
	16	9	0	−8	17
Total	62	31	1	−26	68

the entire range where the reduction ratio differs greatly, but in such a case one would seek the influence on the thickness per unit scale reading separately for each small range and would seek the SN characteristics as in the case of the frequency characteristics, as with a weighing scale. Here, this means that changes in this range are approximated by a linear relationship, and that deviations from this will be included in error. The linear effect of M will be expressed as M_l. We seek its variation S_{M_l} by the following equation, using the coefficients of the orthogonal polynomial.

$$S_{M_l}=\frac{[-3(133)-53+(-44)+3(-129)]^2}{80}=9746 \qquad (f=1) \tag{24.4}$$

Therefore, the error variation S_e is

$$S_e=S_T-S_{M_l}=10036-9746=290 \qquad (f=14) \tag{24.5}$$

From this, the analysis of variance table becomes as given in Table 24.3.

The SN ratio is the value obtained by dividing the magnitude of change per unit scale reading β^2 by the error variance per unit, σ^2. The estimated value of σ^2 is the error variance, $V_e = 20.7$, and the variance of M_l, 9746, is the estimated value of $\sigma^2 + 20\beta^2$. Therefore, the SN ratio η can be estimated by the following equation.

$$\eta=\frac{\beta^2}{\sigma^2}\doteqdot\frac{\frac{1}{20}(V_{M_l}-V_e)}{V_e}$$

$$=\frac{\frac{1}{20}(9746-20.7)}{20.7}=23.5 \longrightarrow 13.7\text{ [db]} \tag{24.6}$$

Appendix Table 8 is used to obtain the confidence limit of this.

We use the variance ratio,

$$F_0=\frac{9746}{20.7}=470.8 \tag{24.7}$$

$$f_2=14 \tag{24.8}$$

and from Appendix Table 8,

$$\eta=13.7^{-4.1}_{+2.7}\text{ [db]} \tag{24.9}$$

Table 24.3 Analysis of Variance Table for A_1

Source	f	S	V	$E(V)$
M_l	1	9 746	9 746	$\sigma^2+20\beta^2$
e	14	290	20.7	σ^2
T	15	10 036		

Table 24.4 Analysis of Variance Table for A_2

Source	f	S	V	$E(V)$
M_l	1	1 080.4	1 080.4	$\sigma^2+20\beta^2$
e	14	14.6	1.04	σ^2
T	15	1 095.0		

(2) *SN Ratio of A_2.*

$$S_T=15^2+7^2+\cdots+(-8)^2-\frac{68^2}{16}=1\,095 \qquad (f=15)$$

$$S_{M_l}=\frac{[-3\times 62-31+1+3(-26)]^2}{80}=1080.4 \qquad (f=1)$$

$$S_e=1095-1080.4=14.6 \qquad (f=14)$$

$$\eta=\frac{\frac{1}{20}(1\,080.4-1.04)}{1.04}=51.9=17.2 \text{ [db]} \tag{24.10}$$

We find the confidence limits as in the case of A_1. By using

$$F_0=1039$$

	η	Gain
A_1	13.7	Standard
A_2	17.2	+3.5

$$\eta=17.2^{-4.0}_{+2.7} \text{ [db]} \tag{24.11}$$

A comparison of A_1 and A_2 is given in the table at the left. It shows that compared with A_1, A_2 gives +3.5 db and is a material whose rollability is better by 2.2 times. This means that regardless of thickness, the dispersion of thickness of A_2 becomes $\frac{1}{2.2}$ that of A_1 in terms of variance and about $\frac{1}{1.5}$ in terms of standard deviation. To examine whether the difference between A_1 and A_2, of +3.5 db is significant, one need only compare the square roots of the sum of squares of the width of the plus-side confidence limit in the equation for confidence limits, Equation (24.9), +2.7, and of the width of the minus-side confidence limit in Equation (24.11), −4.0,

$$\sqrt{2.7^2+(-4.0)^2}=4.8 \tag{24.12}$$

with the +3.5 given earlier. Since the gain of +3.5 in this case is smaller than 4.8, essentially one sees no significant difference.

It is necessary to use the significant difference as a reference, but the scale of this experiment was somewhat small; and if experimentation had been conducted with the number of degrees of freedom of error at least 20 or so, it is believed that a more clear-cut conclusion would have been arrived at.

24.2 When There Are Two Types of Error

24.2.1 Error in Digital Systems

A relay, an ON-OFF switch, a digital computer and the like are all machines that function by two states, 0 and 1, or by combinations of these. For example, the alphabet, numerals, *kana* (Japanese letters), and *kanji* (Chinese characters) are expressed (in information transmission and information processing) by a certain number of digits in the binary system of 0 and 1, and the information proceeds from the computer terminal to the mainframe and, after having undergone information processing at the mainframe, is transmitted to the memory and terminal.

The author has been consulted several times by researchers and designers working for makers of terminals as to what to do since they were not succeeding in studies for improvement of the reliability of the terminals. At these companies, reliability was calculated by subtracting the erroneous transmission rate per 10,000 characters from 1. However, this will not do.

This measure of reliability should be discontinued when studying digital transmissions, for the sake of efficiency in research. First, we will explain the reason for this.

24.2.2 Output When There Are Two Types of Error

If two states are assumed by a system and a 0 state is changed to the state 1 by one signal, while the next arrival of exactly the same signal causes the state 1 to revert to the state 0, if a mistake develops once in the sequence the mistake will continue in effect unless another mistake occurs or one becomes aware of it. Such an information system is an extremely ineffectual digital system. However, since in this case there is only one type of signal, one tests by the method of Section 23.1. We will consider here a case where there are two states, and one of them is assigned the value 0 and the other the value 1. These two states may be such that the value is 1 when a given characteristic value is greater than a certain value and 0 when it is less. For example, in the temperature control of an electric foot warmer, once the temperature becomes higher than a certain value, the bimetal reverses and the switch turns off. Conversely, when the temperature becomes lower than a certain value, the bimetal reverses and the switch turns on.

The signal factor in this case is the temperature. The switch turns on or off depending on the value of the signal factor. Thus, the output assumes two states. Generally, the temperature when a switch that is on is turned off and the temperature when a switch that is off is turned on are not the same. There is so-called hysteresis.

Table 24.5 Input-Output Data When There Are Two Types of Error

Input \ Output	0	1	Total
0	n_{00}	n_{01}	n_0
1	n_{10}	n_{11}	n_1
Total	r_0	r_1	n

When there is hysteresis, as in this case, one performs analysis of variance by varying the levels of signal as, for example, M_1, M_2, and M_3, separately to obtain a curve for each action. One performs analysis of variance of the data of 0, 1 with the linear effects of M_1, M_2, and M_3 as the signals. Therefore, in terms of calculation one experiments or analyzes separately by the same method as in Section 24.2 or Section 22.1. In xerography there is a situation which is similar to this. A xerographic copier is designed to output as 1 when the intensity is greater than a certain value and as 0 (white) below a certain intensity. Thus, when a slightly soiled original is copied by xerography, the background blemishes disappear and often a copy which is cleaner than the original is obtained because the output system is digital.

We will discuss the SN ratio for the case where the signals are of two types and there are two types of output depending on their magnitude, as in digital communications. We will assume that the output was as given in Table 24.5 for the two input signals 0 and 1.

Table 24.5 means that when the two signal states, 0 and 1, were each independently tested n_0 times and n_1 times, for the signal 0, the number of correct responses was n_{00} and the number of erroneous responses was n_{01}. Also, for the signal 1, the number of correct responses was n_{11} and the number of erroneous responses was n_{10}.

In his famous thesis [reference 40], C. Shannon of the Bell Laboratories clarified that an output which is effective for information is judged by the degree to which 0 and 1 are mixed, and he defined the so-called information quantity by *entropy*. If the total entropy in Table 24.5 is expressed as E,

$$E = -n[p \log p + (1-p)\log(1-p)] \tag{24.13}$$

where

$$p = \frac{r_1}{n} \tag{24.14}$$

It does not matter whether the logarithm is the natural logarithm or the common logarithm or logarithm with base 2. As long as it is decided which, it does not matter. Entropy, incidentally, is also used in physics and chemistry and it is an enticing definition, but in actual calculations it has two defects, as follows.

(1) *Entropy cannot be calculated unless the shape of the distribution is known. It is impossible to know the distribution shape from a few observed values, and it is not practical to proceed with calculations based on a hypothesis for the distribution shape the validity of which is unclear.*
(2) *Even if the total entropy of the output has been obtained, when resolving this into components of various causal systems* (although there are only two, the information source and noise, in the case of communication theory, there are many factors in the case of experimental data), *calculation is difficult and it is not practical if the causal systems are numerous.*

As a measure with properties similar to entropy, then, the author has been using *variance* for a long time. The total variance S_T of the output of Table 24.5 is given by

$$S_T = np(1-p) \tag{24.15}$$

Both the total variance S_T and the total entropy E become 0 at $p = 0, p = 1$ and become maximum at $p = \frac{1}{2}$. Both express the degree of mixing of 0 and 1, the degree of dispersion, the magnitude of deviation from orderliness, and the magnitude of change undergone.

Since, however, the variance of Equation (24.15) is a measure of the square (so-called square of norm), Parseval's orthogonal decomposition, as it is termed in mathematics, is valid. Therefore, as long as the causal systems are orthogonal it is simple to resolve the total variance into the sum of the effects of the individual causes.

24.2.3 Resolution of Total Output (Total Variation)

If total variation is defined by Equation (24.15), the value of the total output of Table 24.5 can be calculated in the following manner.

$$S_T = np(1-p) = n\frac{r_1}{n}\left(1-\frac{r_1}{n}\right) = r_1 - \frac{r_1^2}{n} \tag{24.16}$$

$$= \frac{r_1(n-r_1)}{n} = \frac{r_0 r_1}{n} \tag{24.17}$$

On the other hand, as to how accurately the input-side signals 0 and 1 have appeared on the output side in Table 24.5, one can determine this by contrasting 0 and 1. If this contrast is expressed as L, L can be obtained by the following equation.

$$L = \frac{n_{11}}{n_1} - \frac{n_{01}}{n_0} \tag{24.18}$$

The variation of Equation (24.18), or in other words the magnitude of the signal on the output side, S, can be obtained by the following equation from the formula for variation of contrasts.

$$S = \frac{L^2}{\text{number of units}} = \frac{\left(\dfrac{n_{11}}{n_1} - \dfrac{n_{01}}{n_0}\right)^2}{\left(\dfrac{1}{n_1}\right)^2 \times n_1 + \left(\dfrac{-1}{n_0}\right)^2 \times n_0}$$

$$= \frac{(n_0 n_{11} - n_1 n_{01})^2}{n_0^2 \times n_1 + n_1^2 \times n_0} = \frac{(n_{00} n_{11} - n_{10} n_{01})^2}{n_0 n_1 n} \tag{24.19}$$

Equation (24.19) can also be derived from the calculation equation which determines the variation S between the two levels 0 and 1 for numbers of repetitions n_0 and n_1.

Specifically,

$$S = \frac{n_{01}^2}{n_0} + \frac{n_{11}^2}{n_1} - \frac{(n_{01} + n_{11})^2}{n} \tag{24.20}$$

$$= \frac{n n_1 n_{01}^2 + n n_0 n_{11}^2 - n_0 n_1 (n_{01} + n_{11})^2}{n_0 n_1 n}$$

$$= \frac{(n_0 + n_1) n_1 n_{01}^2 + (n_0 + n_1) n_0 n_{11}^2 - n_0 n_1 (n_{01}^2 + 2 n_{01} n_{11} + n_{11}^2)}{n_0 n_1 n}$$

$$= \frac{(n_1 n_{01} - n_0 n_{11})^2}{n_0 n_1 n} = \frac{[(n_{10} + n_{11}) n_{01} - (n_{00} + n_{01}) n_{11}]^2}{n_0 n_1 n}$$

$$= \frac{(n_{10} n_{01} - n_{00} n_{11})^2}{n_0 n_1 n} \tag{24.21}$$

The magnitude of the error on the output side (noise), S_e, becomes as follows from the sum of the error for 0 and the error for 1.

$$S_e = n_0 \left(\frac{n_{01}}{n_0}\right)\left(1 - \frac{n_{01}}{n_0}\right) + n_1 \left(\frac{n_{10}}{n_1}\right)\left(1 - \frac{n_{10}}{n_1}\right)$$

$$= \left(n_{01} - \frac{n_{01}^2}{n_0}\right) + \left(n_{11} - \frac{n_{11}^2}{n_1}\right) \tag{24.22}$$

By adding the signal output S and the error output S_e, we obtain the total output S_T.

$$S + S_e = \left[\frac{n_{01}^2}{n_0} + \frac{n_{11}^2}{n_1} - \frac{(n_{01} + n_{11})^2}{n}\right] + \left(n_{01} - \frac{n_{01}^2}{n_0}\right) + \left(n_{11} - \frac{n_{11}^2}{n_1}\right)$$

$$= n_{01} + n_{11} - \frac{(n_{01} + n_{11})^2}{n} = r_1 - \frac{r_1^2}{n} = S_T \tag{24.23}$$

Equation (24.23) is *Parseval's equality* (extension of the Pythagorean theorem).

The proportion of the signals in the total output is termed the *contribution ratio ρ*.

$$\rho = \frac{S}{S_T} = \frac{(n_{00} n_{11} - n_{10} n_{01})^2 / n_0 n_1 n}{r_0 r_1 / n} = \frac{(n_{00} n_{11} - n_{01} n_{10})^2}{r_0 r_1 n_0 n_1} \tag{24.24}$$

Although fundamentally one should use the value that is obtained by subtracting the error variance

$$V_e = \frac{1}{n-2} S_e \tag{24.25}$$

from S, we will omit this since the expression in Equation (24.25) is small in comparison with S. Henceforth we will omit sample errors since they are not important. This is where our method differs from mathematical statistics. *In actuality, there cannot be cases where it is uncertain whether there is a signal or not. In the actual world, the problem is what is the magnitude of the signal.*

From this, the SN ratio η is

$$\eta = \frac{\rho}{1-\rho} \tag{24.26}$$

In decibel units, this is given by the following equation.

$$10 \log \eta = -10 \log\left(\frac{1}{\rho} - 1\right) \quad \text{[db]} \tag{24.27}$$

It can be seen that the contribution ratio ρ corresponds to the simple reliability p of Section 23.1.

24.2.4 Application to ○× Test

There is such a thing as a ○× test. It will be assumed that, given 20 problems that are correctly solved if answered ○ and 10 problems that are correctly solved if answered ×, the answers of a certain person were as given in Table 24.6. Here we do not question the value of the ○× test itself, but we discuss the rationalizing of the marking of the ○× test.

Since this person has correctly solved 16 of 20 problems where ○ is the correct answer and 8 of 10 problems where × is the correct answer, or a total of 24 problems correctly out of a total of 30 problems, an ordinary instructor would give a mark of 80. The author, however, *would mark this 33*.

The reason is that even if ○ or × is decided by a roll of dice, about one-half the questions would be correctly answered and the mark would be 50. Also, a person who wrote ○ for all, feeling that there seem to be more of ○, would obtain a mark of 67. A person who has decided with dice, a person who has written ○ for everything, and a person who has

Table 24.6 ○× Test Data

Input \ Output	Answer ○	Answer ×	Total
○	16	4	20
×	2	8	10
Total	18	12	30

written × for everything should all be given a zero mark. A person who cannot distinguish between O and × deserves a zero mark.

Even for the problem of finding in which year of the Western calendar a certain incident occurred, a random answer will be correct at a probability of about $\frac{1}{2000}$. When a right answer has been correctly chosen from among an infinite set of possible answers, it is proper to mark by multiplying the ratio of correct answers to total answers by 100.

The ordinary marking method is not good when selecting from a set of two (in the case O and ×) or a finite number. Of course, in the case, for example, of which year in the Western calendar did the incident in question happen, the ordinary marking is probably acceptable approximately since even though the set is finite it is fairly large, but strictly speaking this is mistaken. The proof of a mathematical equation, for example, might be said to be the selection of words and equations from sets that are nearly infinite. Probably many persons will recall the famous story of the person who calculated the probability that when a monkey uses a typewriter, a passage in Shakespeare will be obtained accidentally.

Even assuming that the mark will be about 50 if a O× test is answered randomly, if 50 is subtracted from an ordinary mark, even a person making 100 will get 50. The mark should be arrived at by multiplying the ρ of Equation (24.24) by 100.

In that case, for the mark for the answerer of Table 24.6,

$$\rho=\frac{(n_{00}n_{11}-n_{01}n_{10})^2}{r_0r_1n_0n_1}=\frac{(16\times 8-4\times 2)^2}{18\times 12\times 20\times 10}=0.33 \qquad (24.28)$$

Multiplying this by 100, it becomes a mark of 33.

There is one problem point regarding ρ in Equation (24.28). This is that it becomes 100 if all O's are answered as ×'s and all ×'s are answered as O's. This is the same as in the case of a negative and positive of a photograph. Whether the correlation coefficient is 1 or -1, both are complete information. Still, if one wishes to distinguish by affixing a sign one should use the following value.

$$\begin{aligned}\text{marking} &= 100\ \text{sign}(n_{00}n_{11}-n_{01}n_{10})\times\rho \\ &= 100\times\frac{|n_{00}n_{11}-n_{01}n_{10}|\times(n_{00}n_{11}-n_{01}n_{10})}{r_0r_1n_0n_1}\end{aligned} \qquad (24.29)$$

Here, sign(x) is a function which indicates the sign of x.

A weather forecast that never comes true has value. One just doesn't bring an umbrella if that forecaster says it will rain. But weather forecasts that sometimes predict accurately and sometimes do not predict accurately are a problem. Thus, in the world of information, whether the sign is positive or negative is not important. Therefore, the author wishes to stress that one should use Equation (24.24) for the contribution ratio without considering the sign.

(In telling fortunes, there is correct reading and misreading, but the purpose of fortune telling is not necessarily correct reading. If a fortune teller were to say, You will die on March 31 of next year, and if this were reliable, it would be troubling. It is the actual objective of fortune telling to cheer a person up or to cause him to change his ways by saying, for example, You will not die if you change your name or change your ways. However, weather forecasting that does not often come true is useless.) The above contentions suggest that it is best to calculate the SN ratio when there are two types of error by Equation (24.27).

It was *the author's great error that he thought the calculation problems of the SN ratio were solved by Equation (24.27). Consideration of the calibration problem, or in other words a biased* threshold, was lacking in Equation (24.27). Research on the calibration problem for a digital system and the establishment of a better calculation method for the SN ratio when there are two types of error was conducted for the first time at the Bell Telephone Laboratories in 1962.

24.2.5 SN Ratio of Hardware and SN Ratio of Operating Conditions

In a case where there are two types of error, when there are the two errors of mistaking 1 for 0 and mistaking 0 for 1, it is not possible to define the SN ratio by the method of Equation (24.27). For the input-output table of Table 24.5, we use the error rate for the signal 0,

$$p=\frac{n_{01}}{n_0} \tag{24.30}$$

and the error rate for the signal 1,

$$q=\frac{n_{10}}{n_1} \tag{24.31}$$

and we express the input-output table as in Table 24.7.

When $p = 0$, $q = 0$, we substitute

$$p=\frac{1}{2n_0}, \qquad q=\frac{1}{2n_1}$$

in order to prevent the SN ratio from becoming ∞.

Table 24.7 Input-Output Table Indicated by Two Types of Error Rate

Input \ Output	0	1	Total
0	$1-p$	p	1
1	q	$1-q$	1
Total	$1-p+q$	$1+p-q$	2

If, in Table 24.7, $p = q = 0$, this is a perfect digital communication system or a perfect relay. This is not so when p and q are not 0. The *contribution ratio* ρ of Table 24.7 is

$$\rho = \frac{(1-p-q)^2}{(1-p+q)(1+p-q)} \tag{24.32}$$

Therefore, the SN ratio η is

$$\eta = -10 \log\left(\frac{1}{\rho} - 1\right) \tag{24.33}$$

In the SN ratio given above, however, the loss for the reason that the two types of error of erroneously regarding 0 as 1 and erroneously regarding 1 as 0 are not equal is also included. The loss that is caused by the difference between the two types of error depends on the device.

In cases such as the digital communication system, it is most desirable to *adjust the communication system* so as to obtain $p = q$. Although it differs from this problem since there is hysteresis, in a case such as the ON-OFF switch of an electric foot warmer, if it does not turn ON when it should become ON, one is merely cold, but if it does not turn OFF when it should be OFF, there is the possibility of a fire hazard.

As to *how to handle the ratio of* p *and* q, in general, *it is best to perform leveling (adjustment) so that* p *and* q *become inversely proportional to the losses of the two errors in question.* However, in the research of test-construction stage it is difficult to design a test so as to obtain a constant $p:q$ for all products. When the ratio of p and q differs, it is inadvisable to compare by the SN ratio values as they are. *It is necessary to resolve the SN ratio into the SN ratio of hardware and the SN ratio of operating conditions (lack of leveling).*

$$-10 \log\left(\frac{1}{\rho} - 1\right) = \text{(SN ratio of hardware)} + \text{(SN radio of operating conditions)} \tag{24.34}$$

Here, only the SN ratio of hardware indicates the SN ratio of the instrument itself, and the SN ratio of operating conditions indicates the ingeniousness of use of the machine (in this case, it indicates the goodness or badness of leveling or adjustment; leveling and adjustment are calibration operations). *The SN ratio of operating conditions becomes 0 db when that machine has been calibrated or adjusted perfectly. Any other case gives negative decibels.*

To find the SN ratio assuming $p = q$, or in other words the standardized SN ratio, from data where p and q are different, one uses the omega method. If the error rate of regarding 0 as 1, p, is improved by just K db, the error rate of regarding 1 as 0, q, necessarily becomes worse by just K db when the states are of two types and there is no hysteresis or the like. Thus, one need only solve the following equation to find K and cause the error rate of interpreting 0 as 1 and the error rate of interpreting 1 as 0 to have the equal value p'.

$$-10\log\left(\frac{1}{p}-1\right)+K=-10\log\left(\frac{1}{q}-1\right)-K \tag{24.35}$$

If this is solved for K,

$$K=-10\log\sqrt{\left(\frac{1}{q}-1\right)\Big/\left(\frac{1}{p}-1\right)}$$

Substituting into Equation (24.35), with the error rate in common as p', we obtain

$$\begin{aligned}-10\log\left(\frac{1}{p'}-1\right)&=-10\log\left(\frac{1}{p}-1\right)-10\log\sqrt{\left(\frac{1}{q}-1\right)}+10\log\sqrt{\left(\frac{1}{p}-1\right)}\\&=-10\log\sqrt{\left(\frac{1}{p}-1\right)\left(\frac{1}{q}-1\right)}\\&=\frac{1}{2}\left[-10\log\left(\frac{1}{p}-1\right)-10\log\left(\frac{1}{q}-1\right)\right]\\&=\frac{1}{2}(\text{decibel value of } p+\text{decibel value of } q)\end{aligned} \tag{24.36}$$

Thus, the decibel value of p' is the mean of the decibel value of p and the decibel value of q.

24.3 Application to Simulated Testing

24.3.1 Simulated Test

In a given digital communication system, the error rate in the present state of affairs is $p \doteqdot q \doteqdot 0.000\,01$. In order to decrease this error rate,

Countermeasure (1) Change part A to a new part
Countermeasure (2) Use AGC (automatic gain control)

were tried and the test of the digital system was carried out separately with the signal power at $\frac{1}{5}$ and the amplification on the receiving side increased to 5 times; and the resulting data were as given in Table 24.8.

24.3.2 Standardized SN Ratio

Clearly, for countermeasure (1) and countermeasure (2), the error rates for the total output data are

$$\frac{205}{1\,000}=0.205 \tag{24.37}$$

$$\frac{222}{1\,000}=0.222 \tag{24.38}$$

Table 24.8 Data Under Simulated Conditions

Current conditions

I \ O	0	1	Total
0	170	30	200
1	30	170	200
Total	200	200	400

Data as percent

I \ O	0	1	Total
0	0.850	0.150	1.000
1	0.150	0.850	1.000
Total	1.00	1.00	2.000

Change part *A* to a new one

I \ O	0	1	Total
0	495	5	500
1	200	300	500
Total	695	305	1 000

Data as percent

I \ O	0	1	Total
0	0.990	0.010	1.000
1	0.400	0.600	1.000
Total	1.390	0.610	2.000

Use AGC

I \ O	0	1	Total
0	498	2	500
1	220	280	500
Total	718	282	1 000

Data as percent

I \ O	0	1	Total
0	0.996	0.004	1.000
1	0.440	0.560	1.000
Total	1.436	0.564	2.000

and the error rate is greater than the 0.150 of the current conditions. The SN ratios, if the former is expressed as η_1 and the latter as η_2, are

$$\rho_1 = \frac{(0.990 \times 0.600 - 0.010 \times 0.400)^2}{1.390 \times 0.610} = 0.411$$

$$\eta_1 = -10 \log\left(\frac{1}{\rho_1} - 1\right) = -1.56 \quad \text{[db]} \tag{24.39}$$

$$\rho_2 = \frac{(0.996 \times 0.560 - 0.004 \times 0.440)^2}{1.436 \times 0.564} = 0.382$$

$$\eta_2 = -\log\left(\frac{1}{\rho_2} - 1\right) = -2.09 \quad \text{[db]} \tag{24.40}$$

On the other hand, if the contribution ratio in the case of the current conditions is expressed as ρ_0 and the SN ratio as η_0,

$$\rho_0 = \frac{(0.850^2 - 0.150^2)^2}{1.000 \times 1.000} = 0.49$$

$$\eta_0 = -0.17 \quad \text{[db]} \tag{24.41}$$

Thus, by η_1 there is loss of -1.39 db relative to η_0 and by η_2 there is loss of -1.92 db relative to η_0. Therefore, it appears as though both the new part and AGC cause the error to become large, but actually this is not so. This is because for both cases, the levelling is unbalanced.

We find the standardized SN ratio of η_1.

$$\text{decibel value of } p' = \tfrac{1}{2}(\text{decibel value of p} + \text{decibel value of q})$$
$$= \tfrac{1}{2}(\text{decibel value of } 0.010 + \text{decibel value of } 0.400)$$
$$= \tfrac{1}{2}[(-19.96) + (-1.76)] = -10.86 \text{ db} \tag{24.42}$$

Therefore, from Appendix Table 5,

$$p' = 0.076$$

This indicates that if experimentation is done after levelling so that p and q become equal, the test data of (2) in Table 24.8 will become as given in Table 24.9.

This indicates that when part A is replaced the error rate becomes about one-half that of the current state. Let us find the SN ratios of Table 24.9.

$$\rho = \frac{(0.924^2 - 0.076^2)^2}{1.000 \times 1.000} = 0.719 \tag{24.43}$$

$$\eta = 4.08 \quad [\text{db}] \tag{24.44}$$

From this,

$$\text{decibel value of } \eta_1 = (\text{decibel value of hardware}) + (\text{decibel value of operating conditions})$$
$$-1.56 = 4.08 + (-5.65) \text{ db} \tag{24.45}$$

Thus, using a new part A means a gain of

$$4.08 - (-0.17) = 4.25 \quad [\text{db}] \tag{24.46}$$

Similarly, we find the standardized SN ratio when AGC is used. The error rate p' when the error of interpreting 0 as 1 and the error of interpreting 1 as 0 are taken as equal is

$$\text{decibel value of } p' = \tfrac{1}{2}(\text{decibel value of } 0.004 + \text{decibel value of } 0.440)$$
$$= -12.50 \text{ db} \tag{24.47}$$
$$p' = 0.053 \tag{24.48}$$
$$\rho = (1 - 2p')^2 = 0.799$$
$$\eta = 5.99 \text{ db} \tag{24.49}$$

Table 24.9 Error Rate in Standardized State (After Calibration) (When Part *A* Has Been Replaced)

I \ O	0	1	Total
0	0.924	0.076	1.000
1	0.076	0.924	1.000
Total	1.000	1.000	2.000

decibel value of η_2 = (decibel value of hardware − decibel value of operating conditions)

$$-2.09 = 5.99 + (-8.08) \text{ db} \tag{24.50}$$

From this, the gain when AGC is used is

$$5.99-(-0.17)=6.16 \quad [\text{db}] \tag{24.51}$$

Therefore, the gain if a new part A is used and AGC is also used is $4.25 + 6.16 = 10.41$ db. The SN ratio under the current conditions is, by using the current error rate 0.000 01,

$$\eta_0=\frac{(1-2p_0)^2}{1-(1-2p_0)^2}=\frac{(1-0.00002)^2}{1-(1-0.00002)^2}=24999$$

$$=43.98 \quad [\text{db}] \tag{24.52}$$

If the two countermeasures, new part A and AGC, are used, gain 10.41 db adds on and we have

$$43.98+10.41=54.39 \quad [\text{db}] \tag{24.53}$$

Therefore, the error rate p that will be achieved is obtained by solving

$$10 \log\frac{(1-2p)^2}{1-(1-2p)^2}=54.39$$

$$p=9.1\times 10^{-7} \tag{24.54}$$

In other words, if both of the two countermeasures are used, it can be expected that the error rate will decrease from once in 100,000 times to 0.91 times in 1 million times.

24.4 Application to the Chemical Field

24.4.1 Application to Chemical Disciplines

The problem of two signals (or components) 0 and 1 that are distinguished at the output side appears also in component extraction and component separation in chemistry. Although this has decreased recently, when one went to a metal smelting plant in the old days, the following data, termed the yield, were used in laboratories and in process control in plants.

$$y=\frac{(\text{\% of metal in product})\times(\text{mass})}{(\text{\% of metal in raw material})\times(\text{mass of raw material})}\times 100\% \tag{24.55}$$

The denominator of the equation above is the total quantity of the metal in the raw material, and the numerator is the total quantity of that metal in the product, and y appears to be a logical characteristic value.

However, the value of y becomes 100% not only in the case where the smelting efficiency actually is 100%, but even when the raw material emerges unchanged without having been processed.

When the yield y has been compared in the laboratory under varying conditions, even if y appears large under certain conditions, they might be at a level which is close to a method that produces the product as the raw material, unchanged. Thus, the only cases where it is allowable to use the yield of Equation (24.55) as the data are those in which the purity of the product is of a certain specific value.

If the smelting researchers had conducted experiments by orthogonal arrays, they would undoubtedly have understood the awkwardness of the characteristic value of the equation above since the experiment of the orthogonal arrays would not have worked well. If this had happened, to revert to studies of the individual factors as in the past would not have constituted a solution of the problem, but *improvement of the characteristic values so that the experiment by the orthogonal array would work well would have been necessary.* To remove a certain component, distinguishing it from other substances, is the problem of component extraction in chemistry. The simplest such problem is that of removing only one component from a mixture wherein two components are mixed. The most typical among such problems is that of concentration of uranium.

24.4.2 Uranium Concentration Problem

There are two types of natural uranium, U_{235} and U_{238}. How to concentrate and remove U_{235}, in which nuclear fission occurs, from a mixture of the two is the uranium concentration problem. It is said that more than 30 methods have been suggested for the concentration of uranium. In order to indicate clearly the superiority or inferiority of these methods, a characteristic value which clarifies their superiority or inferiority in terms of numerical values is necessary.

If U_{235} is expressed as 0 and U_{238} as 1, how to distinguish 0 and 1 from a mixture of 0 and 1 and how to concentrate the former in the product and the latter in the slag is the problem. This is exactly the same problem as the transmission of the information of a 0-and-1 series over time, and the separation of this on the receiving side was the problem in the case of digital communication. In the case of communication the information is sent moment by moment, but in the case of uranium concentration the mixture of U_{235} and U_{238} is treated all at once.

The proportion of U_{235} in the input will be expressed as p_0 and the quantity of the whole will be expressed as m. It will be assumed that the quantity of product which was obtained by one run of a certain concentrating method was mq_0 and that the slag quantity was $m(1 - q_0)$. Also,

Table 24.10 Input-Ouput Table When Concentrating Uranium

Input \ Output	Product	Slag	Total
U_{235}	$m[p_0-(1-q_0)p']$	$m(1-q_0)p'$	mp_0
U_{238}	$m[q_0-p_0+(1-q_0)p']$	$m(1-q_0)(1-p')$	$m(1-p_0)$
	mq_0	$m(1-q_0)$	m

the concentration of U_{235} in the slag will be expressed as p'. Then we obtain an input-output table such as shown in Table 24.10. Table 24.10 has the same form as the input-output table of Table 24.5. The values in each column of the table will be rewritten using new variables, p and q, as follows. We write

$$p=\frac{m(1-q_0)p'}{mp_0} \tag{24.56}$$

$$q=\frac{m[q_0-p_0-(1-q_0)p']}{m(1-p_0)} \tag{24.57}$$

Then Table 24.10 can be represented as Table 24.11.

Then the SN ratio η of Table 24.11 is given by the following equation according to the theory of the previous section.

$$\eta=\frac{\rho}{1-\rho},\ \rho=\frac{(1-p-q)^2}{(1-p+q)(1+p-q)} \tag{24.58}$$

Also, the *standardized SN ratio* and the levelling loss are obtained by resolving the SN ratio (db units) as follows.

$$10\log\eta=(\text{decibel value of standardized SN ratio } \eta_0)+(\text{levelling loss}) \tag{24.59}$$

where

$$\eta_0=-10\log\left(\frac{1}{\rho_0}-1\right) \tag{24.60}$$

$$\rho_0=(1-2p_0')^2 \tag{24.61}$$

$$p_0'=\frac{1}{1+\sqrt{\left(\frac{1}{p}-1\right)\left(\frac{1}{q}-1\right)}} \tag{24.62}$$

These are exactly the same as in the previous section.

Table 24.11 Input-Ouput Table Expressed by Two Types of Error Rate, *p* and *q*

Input \ Output	Product	Slag	Total
U_{235}	$1-p$	p	1
U_{238}	q	$1-q$	1
Total	$1-p+q$	$1+p-q$	2

However, the optimum levelling is obtained with the following equation, from the loss per unit quantity that occurs when U_{235} enters the slag, C_p, and the loss per unit quantity that occurs when U_{238} enters the product, C_q.

$$p \times C_p = q \times C_q \tag{24.63}$$

However, whether one is deciding the ratio of p and q in such a manner that Equation (24.63) hold true or whether one is deciding so as to obtain $p = q$, comparison of the various concentrating methods is exactly the same numerically, so that even in a case where one wishes to decide $p:q$ in such a manner that Equation (24.63) holds true, it is possible to compare by using the standardized SN ratio calculated for $p = q$.

24.4.3 How to Find the Standardized SN Ratio

In the concentrating of uranium, the data for two methods, A_1 and A_2, were as given in the following table.
Let us try solving problems such as the following with this.

Problem (1)

Which of A_1 and A_2 is the better concentrating method, and by how many times (how many decibels).

Problem (2)

Assuming that one run of concentration time of both A_1 and A_2 is 24 hours and that the cost of this is 40,000 yen, what is the total number of days required by the better concentration method and how many yen will it

Table 24.12 Hypothetical Data of Uranium Concentration, The Case for A_1 (Unit: g)

Input \ Output	Product	Slag	Total
U_{235}	1 025	3 975	5 000
U_{238}	38 975	156 025	195 000
Total	40 000	160 000	200 000

The Case for A_2 (Unit: g)

Input \ Output	Product	Slag	Total
U_{235}	1 018	3 782	4 800
U_{238}	38 982	156 218	195 200
Total	40 000	160 000	200 000

cost to produce 1 kg of concentrated uranium of concentration 20% from a uranium raw material whose U_{235} concentration is 1.0%. It will be assumed that only 1 kg per run can be handled by the concentration process.

We will begin with Problem (1). First, beginning with the data of A_1, we correct the data given in Table 24.13. From this, the two types of error rate, p and q, become as follows.

$$p=0.79500, \quad q=0.19987$$

The error rate p_0 when levelling has been performed so that p and q become equal can be obtained from the following equation.

$$\frac{1}{p_0}-1=\sqrt{\left(\frac{1}{p}-1\right)\left(\frac{1}{q}-1\right)}=\sqrt{\left(\frac{1}{0.79500}-1\right)\left(\frac{1}{0.19987}-1\right)}$$

$$=1.01604 \tag{24.64}$$

$$p_0=\frac{1}{2.01604}=0.49602 \tag{24.65}$$

Therefore, the contribution ratio ρ is

$$\rho=(1-2p_0)^2=0.00006336 \tag{24.66}$$

The standardized SN ratio η_s is

$$\eta_s=-10\log\left(\frac{1}{\rho}-1\right)=-10\log\left(\frac{1}{0.00006336}-1\right)=-41.981 \text{ [db]} \tag{24.67}$$

On the other hand, the SN ratio η and contribution ratio ρ_1 of Table 24.13, without levelling, are

$$\rho_1=\frac{(0.20500\times 0.80013-0.79500\times 0.19987)^2}{1\times 1\times 0.40487\times 1.59513}=0.00004075 \tag{24.68}$$

$$\eta_1=-10\log\left(\frac{1}{0.00004075}-1\right)=-43.899 \text{ [db]} \tag{24.69}$$

Therefore, the SN ratio of the concentration method A_1 can be resolved as follows.

total SN ratio = (SN ratio of hardware) + (SN ratio of operating conditions)

$$-43.899=(-41.981)+(-1.918) \text{ db} \tag{24.70}$$

Table 24.13 Data as Percentages of Total for A_1

Input \ Output	Product	Slag	Total
U_{235}	0.20500	0.79500	1.00000
U_{238}	0.19987	0.80013	1.00000
Total	0.40487	1.59513	2.00000

The SN ratio of operating conditions involves levelling such as the following. For example, let us consider the furnace temperature when smelting copper. If the temperature is raised, the copper component in the furnace dissolves well and little copper enters the slag. In other words, if the temperature is raised, we have $p \to 0$. On the other hand, components other than copper also dissolve well and mix into the melt; in other words, the proportion of impurity in the product, q, becomes large. Thus, it can be considered that the temperature in copper smelting is not adjusted to raise the smelting efficiency but is a factor that changes the proportion of p and q, or in other words is a factor for levelling. Factors of this sort might be the residence time of the raw material in the separating apparatus or the set position of the filter in the case of uranium concentration. Thus, the levelling can be changed simply. Therefore, for example, in the smelting of metal, uranium concentration, and component extraction, only the standardized SN ratio is a measure which is related to fundamental improvement.

We calculate similarly for the case of A_2.

The SN ratio for Table 24.14, as is, can be obtained from the contribution ratio ρ_2.

$$\rho_2 = \frac{(0.21208 \times 0.80030 - 0.78792 \times 0.19970)^2}{0.41178 \times 1.58822 \times 1 \times 1} = 0.00023435 \tag{24.71}$$

$$\eta_2 = -10 \log\left(\frac{1}{0.00023435} - 1\right) = -36.299 \quad \text{[db]} \tag{24.72}$$

For the case of A_2, too, we find the standardized SN ratio.

$$p_0' = 0.49053 \tag{24.73}$$

$$\rho = 0.00035872 \tag{24.74}$$

$$\eta_s = -10 \log\left(\frac{1}{0.00035872} - 1\right) = -34.449 \tag{24.75}$$

Therefore,

Total SN ratio = (standardized SN ratio) + (SN ratio of operating conditions)

$$(-36.299) = (-34.449) + (-1.850) \text{ db} \tag{24.76}$$

From Equations (24.70) and (24.76), A_2 is a concentration method which is better than A_1 by the SN ratio

Table 24.14 Data as Percentages of Total for A_2

Input \ Output	Product	Slag	Total
U_{235}	0.21208	0.78792	1.00000
U_{238}	0.19970	0.80030	1.00000
Total	0.41178	1.58822	2.00000

$$(-34.449) - (-41.981) = 7.532 \text{ db} \quad (24.77)$$

Since the true value corresponding to 7.532 db is 5.66, this means that the concentration effect per run by A_2 is 5.66 times better than by A_1. In other words, it is possible to lower the concentration processing cost per run to $\frac{1}{5.66}$.

It is said that during the Second World War, the United States *created about 3000 plants for uranium concentration.* If there had been a method whose concentrating efficiency was 5.66 times as great, it should have been possible to decrease the number of such plants by the square of this, or to $\frac{1}{32}$.

At present, more than 30 methods have been suggested for uranium concentration. Could it not be that the reason for this is that the efficiency of these various methods is not clearly known as indicated above, and because numerical comparisons of the various methods are being treated rather vaguely?

The problem of accurately calculating the cost of uranium concentration is rather difficult. Regarding the case of Problem (2), we will merely show the solution for the following simple case.

We will assume that by a certain concentrating method, in one run of the concentrating process 1 kg of raw material is converted into 500 g of product and 500 g of slag, and that the concentration gain of this, 10 log $(\frac{1}{p_0} - 1)$, is 0.0154 db. Let us calculate the number of runs of concentration needed in order to convert a uranium raw material of 1.0% U_{235} into concentrated uranium of 20% U_{235}.

$$\text{Decibel value of } 1.0\% = -19.955 \text{ db} \quad (24.78)$$

$$\text{Decibel value of } 20.0\% = -6.020 \text{ db} \quad (24.79)$$

Therefore, the gain when concentrating a product of 1.0% to one of 20.0% is

$$-6.020-(-19.955)=13.935 \text{ [db]} \quad (24.80)$$

The quotient n_0 which is obtained by dividing this by the standardized SN ratio in this case, 0.0154 db, is

$$n_0 = \frac{13.935}{0.0154} = 905 \text{ times}$$

Admittedly, if 1 kg of raw material uranium of 1% U_{235} is concentrated 905 times, one will obtain concentrated uranium of 20%, but the quantity which is obtained in this case becomes

$$\left(\frac{1}{2}\right)^{905} \text{ [kg]}$$

To obtain 1 kg of concentrated uranium, even if the U_{235} concentration in the slag is 0.5%, a raw material of 1% U_{235} is required in the amount of

39.0 kg, as shown next. For, assuming that the U_{235} concentration in the slag is 0.5%,

$$1\,[\text{kg}]\times 20.0\,[\%]+(x-1)\,[\text{kg}]\times 0.5\,[\%]=x\,[\text{kg}]\times 1\,[\%]$$

Solving this,

$$x=\frac{20.0-0.5}{1.0-0.5}=39.0\,[\text{kg}] \tag{24.81}$$

Now, if we concentrate 1 kg of raw material uranium once, by hypothesis we obtain 500 g of a product of gain 0.0154 db and 500 g of slag of loss 0.0154 db. If these are concentrated separately, we obtain

Product of gain 0.0308 db	250 g
Raw material the same as the original	500 g
Product of loss 0.0308 db	250 g

Therefore, to concentrate again the 500 g of original raw material and change all of it into a product of gain 0.0308 db and slag of loss 0.0308 db, that much further work will be necessary. If, now, the quantity of work which is necessary to concentrate 500 g of a product of gain $i\times 0.0154$ db and slag of loss $i\times 0.0154$ db and obtain 500 g of a product of gain $(i+1)\times 0.0154$ slag of loss $(i+1)\times 0.0154$ db is expressed as n_i, with n_1 as the unit of work quantity,

$$n_1=1$$

$$n_2=1+\frac{1}{2}(n_1+n_2)$$

$$n_3=1+\frac{1}{2}(n_2+n_3)$$

$$\vdots$$

$$n_{i+1}=1+\frac{1}{2}(n_i+n_{i+1})$$

By solving this,

$$n_i=2i-1 \tag{24.82}$$

The loss of slag of 0.5% is 197×0.0154 db. The quantity of work which is necessary when 39.0 kg of uranium raw material is concentrated and 19.0 kg of product of gain 197×0.0154 db and slag of loss 197×0.0154 db are obtained is, with the quantity of work which is required to concentrate 1 kg of uranium once as the unit, 39.0 times

$$n_1+n_2+\cdots+n_{197}=\sum_{i=1}^{197}(2i-1)=197^2 \tag{24.83}$$

In general, for n steps of concentration, n^2 times the quantity of work is necessary. Therefore, the work quantity which is necessary in order to obtain 19.5 kg of an intermediate product of 197×0.0154 db is

$$39.0 \times 197^2 = 1\,513\,551 \tag{24.84}$$

Next, the quantity of work to obtain 9.75 kg of concentrated uranium by 394 steps by this method from 19.5 kg of concentrated uranium of gain 197 × 0.0154 db is

$$19.5 \times 394^2 = 3\,027\,102 \tag{24.85}$$

Next, we add all of the quantities of work of obtaining 4.875 kg of concentrated uranium by concentrating 9.75 kg of intermediate product, and of obtaining 2.4375 kg of concentrated uranium by concentrating this, and we obtain

$$39.0(1+2+9+16+25) \times 197^2$$

$$\fallingdotseq 83\,245\,305 \tag{24.86}$$

assuming, in this case, that in order to produce 1 kg of concentrated uranium of just 20%, 38 kg of slag is formed. Thus, it is necessary to perform the operation of concentrating 1 kg of uranium raw material or intermediate-concentration raw material a total of 83 million times. Assuming that one run of concentration can be performed 1 kg at a time in one day, this will take 227,000 years. In other words, at least 220,000 concentrating apparatuses will be necessary and assembly-line operation will be necessary.

Problem (2) is more difficult than this, but it is easy to obtain an approximate solution by proceeding by the method explained above. Therefore, when one has obtained a concentrating method whose standardized SN ratio is good, in effect the efficiency increases as the square of the SN ratio.

24.5 Experiment on Maneuverability

24.5.1 Experiment on Maneuverability of Automobiles

In an automobile maneuverability experiment, it seems that in many cases the steering wheel is turned to a specific angle and fixed, then the speed is gradually increased and the path of the vehicle is determined. In such a case, one obtains a path such as is shown in Figure 24.2.

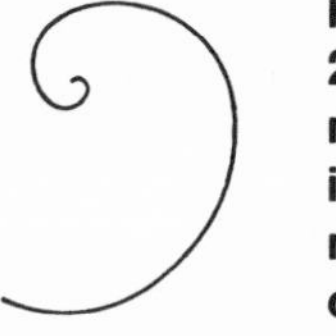

FIGURE 24.2 Maneuverability Experiment (Path of Vehicle)

However, it is known that the path of the vehicle differs according to whether the road is of asphalt or concrete or gravel. Fundamentally, not only is it not clear what shape of the curve of Figure 24.2 is optimum, but the problem is how to make a judgment taking into consideration the fact that such curves vary under varying road conditions.

In other words, unless it is possible to represent maneuverability by a certain scalar quantity, the superiority or inferiority of various designs of an automobile does not become clear. The maneuverability of an automobile means the ability to freely realize two attributes, the speed of the vehicle and the direction of its motion, appropriately to the targets. The maneuverability of the automobile is high when it is possible always to match both of these attributes to the target values as desired.

Factors that might be useful in improving the maneuverability of an automobile, in other words control factors, will be expressed as A, B, ... For example, factor A comprises four levels of rigidity of the steering system and factor B consists of the four types of tires, such as bias tire and radial tire.

These are factors which have been chosen in order to select optimum combinations of levels related to maneuverability later.

It will be assumed that we have assigned the factors to orthogonal array L_{16} (Table 24.15) so as to be able to obtain all the main effects and the interaction $D \times E$ for two factors of four levels, A and B; one of three levels, C; and four of two levels, D, E, F, and G.

Table 24.15 Assignment of Control Factors

	A	B	C	D	F	(B)	G	(C)	E	(B)	$D \times E$	e	(C)
	1 2 3	4	5	6	7	8	9	10	11	12	13	14	15
1	1	1	1	1	1	1	1	1	1	1	1	1	1
2	1	2	2	1	1	2	2	2	2	2	2	2	2
3	1	3	3	2	2	1	1	1	1	2	2	2	2
4	1	4	1′	2	2	2	2	2	2	1	1	1	1
5	2	1	2	2	2	1	1	2	2	1	1	2	2
6	2	2	1	2	2	2	2	1	1	2	2	1	1
7	2	3	1′	1	1	1	1	2	2	2	2	1	1
8	2	4	3	1	1	2	2	1	1	1	1	2	2
9	3	1	3	1	2	1	2	1	2	1	2	1	2
10	3	2	1′	1	2	2	1	2	1	2	1	2	1
11	3	3	1	2	1	1	2	1	2	2	1	2	1
12	3	4	2	2	1	2	1	2	1	1	2	1	2
13	4	1	1′	2	1	1	2	2	1	1	2	2	1
14	4	2	3	2	1	2	1	1	2	2	1	1	2
15	4	3	2	1	2	1	2	2	1	2	1	1	2
16	4	4	1	1	2	2	1	1	2	1	2	2	1

Next, the *signal factor* is the factor for concrete realization of the wish to advance the vehicle in a certain direction, and in this case it is the turning angle of the steering wheel, the so-called steering angle. For the levels of the steering angle, we use the three levels

$M_1' = 100°$ (speed 90 km)
$M_2' = 300°$ (speed 30 km)
$M_3' = 500°$ (speed 10 km)

and it is best to find the SN ratios separately in the neighborhoods of M_1', M_2', and M_3'. Similar to the way in which small weights are used when seeking the SN ratio of a chemical balance, *one changes the steering angle minutely* in the respective neighborhoods of M_1', M_2', and M_3'. We will assume that as the minute changes, three levels, $\frac{1}{1.1}$ of M', M' itself, and 1.1 times M' have been taken as in the following table.

	M_1'	M_2'	M_3'
M_1	91	273	455
M_2	100	300	500
M_3	110	330	550

In this case, since it is necessary to find the SN ratio classed by the level of M' (this is confounded with the speed of the vehicle), it must be assigned as an outer factor of the orthogonal array where the control factor M' has been assigned. If M' is taken as an outer factor of orthogonal array L_{16}, experiments of an orthogonal array where error factors and signal factors have been assigned under a total of 48 sets of conditions become necessary.

Error factors are all factors (variables) other than signal factors and control factors that influence the direction of travel of the vehicle. For example, they are

K	road surface:	concrete, asphalt, gravel
L	road surface:	dry, wet
N	position of center of gravity:	front, center, rear
O	tire air pressure:	left and right both 6, left 3 right 6, left and right both 3 kg/cm^2
Q	3 levels:	left–right difference of load
R	3 levels:	3 drivers

Of course, as the road conditions, there are many more cases, such as snow or ice. Since snowy and icy roads are special conditions, it is probably better to experiment separately from the above.

Error factors and the signal factor M (M' is separate) can be assigned to the same orthogonal array. Now, regarding the error factors, it will be assumed that K, N, O, Q, and R are at three levels and L is at two levels, as explained above. It will be assumed that these have been assigned to orthogonal array L_{18} together with signal factor M, as shown in Table 24.16.

In effect, data of the turning radius are obtained for every combination of L_{16} and M' and L_{18}, or under 864 sets of conditions.

In actuality, it is a problem to determine whether it is easier to obtain successively $3 \times 18 = 54$ sets of data under the respective conditions of L_{16}, that is, 16 times or whether it is better to experiment by reclassifying the factors whose level-change it is difficult to perform among the 864 sets and to experiment by split-unit design, so that one can proceed the most efficiently. If, among the error factors, there are two factors whose level change is especially difficult, and if these are K and L, the best procedure is to make K correspond to the second column of L_{18}, L correspond to the first column of L_{18}, and factors from M on to correspond to the third column and on. And in effect, for the combinations which are created by the six sets of the first column and second column of orthogonal array L_{16} and orthogonal array L_{18}, or the respective conditions of $16 \times 6 = 96$ sets, the nine conditions of the signal factors and error factors are experimented on successively.

As the data, one measures the constant turning radius of the circle obtained when the vehicle is run under the above-mentioned $96 \times 9 = 864$ sets of conditions. The vehicle is designed with reference to the 16

Table 24.16 Assignment of Signal Factors and Error Factors

	L 1	K 2	M 3	N 4	O 5	Q 6	R 7	8
1	1	1	1	1	1	1	1	1
2	1	1	2	2	2	2	2	2
3	1	1	3	3	3	3	3	3
4	1	2	1	1	2	2	3	3
5	1	2	2	2	3	3	1	1
6	1	2	3	3	1	1	2	2
7	1	3	1	2	1	3	2	3
8	1	3	2	3	2	1	3	1
9	1	3	3	1	3	2	1	2
10	2	1	1	3	3	2	2	1
11	2	1	2	1	1	3	3	2
12	2	1	3	2	2	1	1	3
13	2	2	1	2	3	1	3	2
14	2	2	2	3	1	2	2	3
15	2	2	3	1	2	3	1	1
16	2	3	1	3	2	3	1	2
17	2	3	2	1	3	1	2	3
18	2	3	3	2	1	2	3	1

sets of conditions of L_{16}; a part of each road surface B_1, B_2, and B_3 is caused to be in a dry condition and a part is caused to be in a wet condition, and the automobile is run. When, in experimenting under the conditions of M_1' and M_2', the experiment of completely turning the vehicle on the road surface is not possible because of space limitations, it should suffice to estimate the turning radius after having turned, for example, $\frac{1}{4}$ circle. In the case of such a large experiment, it is rarely necessary to measure the value of the turning radius very precisely. This is because even if there is observation error, the precision can be regarded as sufficiently high when the data of all of the experiments are analyzed.

Thus, although it is necessary to strive to match the predetermined conditions, it probably does not matter if, for example, the speed at the time of experimentation changes within about 10% of the indicated value. This is because such an error will become averaged since it is, after all, a vast experiment to measure as many as 864 turning radius values.

If, when performing the experiments on maneuverability, the predetermined six types of road conditions mentioned above cannot be selected satisfactorily, it is possible to create six new levels of road conditions, as for example,

K_1: asphalt in dry condition
K_2: asphalt in wet condition
K_3: concrete in dry condition
K_4: school yard
K_5: gravel path
$K_6 = K_3$

where a dummy level is used as K_6 and these levels are caused to correspond with the six sets of combinations of the first column and second column of L_{18}. If the road conditions are used for the combinations of these columns, it suffices to select the experimental roads one after another in six locations where it is easy to experiment.

Also, it is correct to confound M_1', M_2', and M_3' with the speed of the automobile. One studies about how many km per hour of speed is frequent with M_1, or in other words with a steering angle of about one hundred degrees and how many km per hour is frequent with a steering angle of about M_2', and one may suppose that these are 90 km/hr, 30 km/hr, and 10 km/hr. Essentially, the three levels of signal factor M in the neighborhood of M_1', 91°, 100°, and 110°, are experimented on only with the vehicle speed 90 km/hr; the three levels of signal factor M in the neighborhood of M_2', 273°, 300°, and 330°, are experimented on only with the vehicle speed 30 km/hr; and the three levels of signal factor M in the neighborhood of M_3', 455°, 500°, and 550°, are experimented on only with the vehicle speed 10 km/hr. In other words, M' and the vehicle velocity are confounded, and we have the three levels

Table 24.17 Table of Analysis of Variance for SN Ratio

Source	f	$E(V)$
M_l	1	$\sigma^2+12\beta^2$
e	16	σ^2
T	17	

$$M_1' \ (90),\ M_2' \ (30),\ M_3' \ (10)$$

It will be assumed that a vehicle has been designed according to the conditions of No. 1 among the 16 design conditions indicated by the L_{16} orthogonal array, that data have been acquired on the road surface of K_1 in Table 24.16 by varying the conditions of M', M, N, O, Q, and R nine ways, and that the data of L_1 and L_2 have been obtained. That is, it will be assumed that the two locations L_1 and L_2 have been created on road surface K_1, and L_1 and L_2 have been experimented on consecutively.

And for the combinations of orthogonal array L_{16} and the three levels of M', from the data of each of the 18 rows of L_{18}, the analysis of variance table of Table 24.17 is drawn up for each of the 48 combinations. We analyze by obtaining the logarithmic values of the turning radius. From

Table 24.18 Table of Analysis of Variance for Decibel Value of SN Ratio

Source	f
A	3
B	3
C	2
D	1
E	1
$D\times E$	1
F	1
G	1
e_1	2
M'	2
$A\times M'$	6
$B\times M'$	6
$C\times M'$	4
$D\times M'$	2
$E\times M'$	2
$D\times E\times M'$	2
$F\times M'$	2
$G\times M'$	2
e_2	4
T	47

this we find the SN ratio η from the following equation, with log 1.1 as the unit of difference and $r = 6 \times 2 = 12$.

$$\eta = \frac{(V_\beta - V_e)/12}{V_e} \tag{24.87}$$

Therefore, since we obtain 48 decibel values of the SN ratio, we perform an analysis of variance as in Table 24.18 from them. In Table 24.18, the value of the SN ratio itself is the characteristic value and ordinary analysis of variance is performed.

If one can find a case where the primary factor is significant and there is no interaction with M' from Table 24.18, this indicates that there is a design level for the vehicle by which maneuverability becomes good under any of the conditions M_1', M_2', and M_3'.

If there is a control factor whose interaction with M' is also great, this means that there is a level of the factor such that maneuverability is good if the steering angle is large but is bad if it is small, and that there is a level such that maneuverability is bad if the steering angle is large but is good if the steering angle is small, or that there is nearly no difference in maneuverability no matter what the steering angle. When interaction with M' is great, it is necessary to select the better level by studying which of the two cases, the case where the steering angle is large and the case where the steering angle is small, is more important for the maneuverability of the vehicle. However, only a source whose interaction with M' is small but whose main effect is large is very useful for improvement of the steering property at any speed.

24.5.2 Assignment and Data*

For two control factors,

A:	stiffness of front spring	3 levels
B:	steering geometry	2 levels

we wished to investigate the influence on the maneuverability of a truck in the left and right directions. With the levels of an indicative factor as M_1' and M_2', we selected three levels of speed for each. An at-a-glance table of all of the factors is given in Table 24.19. Since it was easy to change the levels of speed, M' and C, we decided to calculate the SN ratio for all of the combinations of these with the two-way layout of A and B.

Factor C and also factor M take different levels for M_1' and M_2'. M' and C are also indicative factors. Therefore, by combination with the 2×3 two-way layout of indicative factors, in effect the SN ratio was found under 36 conditions. It would seem that it would have sufficed to take M' at three levels and fix the level of C. Signal factor M and error factors G, H, D, E, and F were assigned to orthogonal array L_{18} as shown in Table 24.20.

* Takeshi Ineoi, Isuzu Jidōsha Kōgyō (Isuzu Motors Ltd.), 1974.

Table 24.19 Factors and Levels

Classification of factor	Factor	Symbol	No. of levels	Content of level
Control factor	Front spring	A	3	A_1 = soft, A_2 = standard, A_3 = hard
	Steering geometry	B	2	B_1 = Ackermann, B_2 = parallel
Indicative factor	Speed of vehicle	C	3	$C_1 = 15$, $C_2 = 20$, $C_3 = 25$ km/h (M_1')
		C	3	$C_1 = 5$, $C_2 = 10$, $C_3 = 15$ km/h (M_2')
Signal factor	Steering angle	M	3	$M_1 = 200°$, $M_2 = 250°$, $M_3 = 300°$ (M_1')
		M	3	$M_1 = 560°$, $M_2 = 700°$, $M_3 = 840°$ (M_2')
Error factor	Turning direction	G	2	G_1 = right turn, G_2 = left turn
	Condition of road surface	H	3	H_1 = wet asphalt, H_2 = dry asphalt, H_3 = dry concrete
	Position of load	D	3	D_1 = front load, D_2 = standard, D_3 = rear load
	Tire type	E	3	E_1 = rig tire, E_2 = lug tire, E_3 = radial tire
	Air pressure of front-wheel tire	F	3	F_1 = left, right both 6, F_2 = right 3, left 6, F_3 = left, right both 3 kg/cm^2

Table 24.20 Assignment (Unit: m) and Data (A_1B_1) of Signal Factors and Error Factors

No.	G 1	H 2	M 3	e 4	D 5	E 6	F 7	e 8	Turning radius M_1' C_1	C_2	C_3	Turning radius M_2' C_1	C_2	C_3
1	1	1	1	1	1	1	1	1	41.9	44.0	47.6	14.4	14.8	15.1
2	1	1	2	2	2	2	2	2	33.7	34.2	34.9	11.3	11.5	11.4
3	1	1	3	3	3	3	3	3	27.2	27.4	26.7	9.2	9.2	9.0
4	1	2	1	1	2	2	3	3	42.2	43.7	45.2	13.9	14.5	14.6
5	1	2	2	2	3	3	1	1	31.4	31.9	31.8	10.9	10.9	10.7
6	1	2	3	3	1	1	2	2	30.1	31.8	34.5	9.9	10.2	10.9
7	1	3	1	2	1	3	2	3	44.4	46.6	49.4	15.1	15.6	16.5
8	1	3	2	3	2	1	3	1	33.6	35.5	37.2	11.8	11.7	12.4
9	1	3	3	1	3	2	1	2	27.0	27.3	26.9	8.4	9.0	9.0
10	2	1	1	3	3	2	2	1	37.4	37.2	37.1	12.8	13.2	13.0
11	2	1	2	1	1	3	3	2	36.2	42.8	45.0	11.3	12.7	12.7
12	2	1	3	2	2	1	1	3	25.5	25.5	26.7	8.5	8.8	9.0
13	2	2	1	2	3	1	3	2	38.1	38.2	37.0	12.9	13.3	13.2
14	2	2	2	3	1	2	1	3	31.8	33.7	36.2	10.5	11.4	11.8
15	2	2	3	1	2	3	2	1	26.7	27.6	29.2	9.0	9.1	9.6
16	2	3	1	3	2	3	1	2	38.1	39.5	40.6	13.3	13.3	13.6
17	2	3	2	1	3	1	2	3	27.6	28.1	27.5	9.9	10.0	10.2
18	2	3	3	2	1	2	3	1	30.2	34.6	39.0	9.5	9.9	10.5

Vehicles were designed by the six combinations of control factors A and B, and with each of them we experimented under the 18 experimental conditions No. 1 to No. 18 in Table 24.20. Essentially, the turning radius was measured by the three levels of M_1' of C_1, C_2, and C_3, or in other words 15, 20, and 25 km/h. Subsequently, the turning radius was measured by C_1, C_2, and C_3 of M_2'. The data of Table 24.20 are the data under the conditions A_1B_1. Data were taken in exactly the same way for the other five combinations.

Since the steering angle and the turning radius y have an approximately linear relationship if one uses the logarithmic values, common logarithm table values were used for both the turning radius and the steering angle. Since in this case the logarithmic values of the steering angle are

$$\log 200 = 2.3010, \quad \log 250 = 2.3979, \quad \log 300 = 2.4771$$

$$\log 560 = 2.7482, \quad \log 700 = 2.8451, \quad \log 840 = 2.9243$$

and the intervals are no longer equal, it is necessary to use the formula for unequal intervals when calculating the primary effect of the signal, S_β.

In calculating the analysis of variance after logarithmic conversion, the working mean 1 was subtracted. Here we will show the calculation method for the SN ratio from the 18 data of $A_1B_1M_1'C_1$. Since the values that are obtained by subtracting 1 from the common logarithm table values of 41.9, 33.7, . . . , 30.2 are 0.622, 0.528, . . . , 0.480,

$$S_T = 0.622^2 + 0.528^2 + \cdots + 0.480^2 - \frac{(0.622 + 0.528 + \cdots + 0.480)^2}{18}$$

$$= 0.096349 \tag{24.88}$$

For the linear effect of the signal factor, the orthogonal polynomial table cannot be used since the logarithmic values of the steering angle are not equispaced. We find the covariation $S(My)$ between the levels of the signal factor $S(MM)$ and the total of the signal factor levels and M_1, M_2, and M_3.

$$S(MM) = 6\left[2.3010^2 + 2.3979^2 + 2.4771^2 - \frac{(2.3010 + 2.3979 + 2.4771)^2}{3}\right]$$

$$= 0.093347 \tag{24.89}$$

The sum of M_1, M_2, and M_3 in orthogonal array L_{18} is

$$M_1 = 0.622 + 0.625 + \cdots + 0.581 = 3.629$$

$$M_2 = 0.528 + 0.497 + \cdots + 0.441 = 3.053$$

$$M_3 = 0.435 + 0.479 + \cdots + 0.480 = 2.659$$

Therefore, the covariation is

$$S(My)=2.3010\times 3.629+\cdots+2.4771\times 2.659$$
$$-\frac{(2.3010+\cdots+2.4771)(3.629+\cdots+2.849)}{3}$$
$$=-0.0859454 \tag{24.90}$$

From this, S_β is

$$S_\beta=\frac{(-0.0859454)^2}{0.093347}=0.079131 \qquad (f=1) \tag{24.91}$$

$$S_e=0.096349-0.079131=0.017218 \qquad (f=16) \tag{24.92}$$

From this, the error variance V_e is

$$V_e=\frac{0.017218}{16}=0.001076 \tag{24.93}$$

Also, since

$$E(S_\beta)=\sigma^2+S(MM)\beta^2 \tag{24.94}$$

we have

$$\eta=\frac{\frac{1}{0.093347}(0.079131-0.001076)}{0.001076}=777$$
$$=28.9 \quad [\text{db}] \tag{24.95}$$

The SN ratios were then found for all 36 combinations of A, B, M', and C, and when these were converted to decibel value we obtained Table 24.21.

Table 24.21 Data of SN Ratio (Four-Way Layout) (Unit: db)

steering angle	front spring	steering geometry	vehicle speed		
			C_1	C_2	C_3
M_1'	A_1	B_1	28.9	25.1	22.2
		B_2	25.8	24.8	22.0
	A_2	B_1	27.4	24.0	22.0
		B_2	28.3	25.4	22.2
	A_3	B_1	27.8	25.2	22.2
		B_2	29.4	24.6	21.1
M_2'	A_1	B_1	31.5	30.6	28.4
		B_2	32.4	30.7	28.3
	A_2	B_1	33.4	30.5	27.9
		B_2	31.8	28.6	26.6
	A_3	B_1	35.0	31.6	28.7
		B_2	29.6	28.2	26.1

Table 24.22 Analysis of Variance Table of SN Ratio

Source	f	S	V	ρ[%]
A	2	0.30	0.15 ○	
B	1	7.45	7.45**	1.6
$A\times B$	2	4.44	2.22 ○	
M'	1	232.77	232.77**	52.1
$A\times M'$	2	0.77	0.38 ○	
$B\times M'$	1	5.73	5.73*	1.1
$A\times B\times M'$	2	11.28	5.64**	3.3
C	2	168.39	84.20**	37.5
$A\times C$	4	3.05	0.76 ○	
$B\times C$	2	0.16	0.08 ○	
$A\times B\times C$	4	0.50	0.12 ○	
$M'\times C$	2	3.04	1.52 ○	
$A\times M'\times C$	4	0.73	0.18 ○	
$B\times M'\times C$	2	0.45	0.22 ○	
e	4	6.33	1.58 ○	
○ symbols pooled (e)	(28)	(19.77)	(0.706)	4.4
T	35	445.40		100.0

24.5.3 Analysis of SN Ratio

With the SN ratio (decibel value) of Table 24.21 as the simple continuous value, we perform the analysis of variance of an ordinary four-way layout.

$$S_T = 28.9^2 + 25.1^2 + \cdots + 26.1^2 - CF = 445.40 \qquad (f = 35) \qquad (24.96)$$

It is the same for the other sources. The analysis of variance table is as given in Table 24.22.

Table 24.23 was obtained when the process average was estimated by using the significant factors, $m, B, M', B\times M', A\times B\times M'$, and C, and the results obtained, drawn as graphs, are given in Figure 24.3. Estimation of the process average for the combination $A_iB_jM_k'C_l$ was done according to the following formula.

$$\hat{\mu} = \overline{A_iB_jM_k'} - \overline{A_iB_j} - \overline{A_iM_k'} + \bar{A}_i + \bar{B}_j + \bar{M}_k' + \bar{C}_l - 2\bar{T} \qquad (24.97)$$

Table 24.23 Estimated Values of Process Average (Unit: db)

A	B	M_1'			M_2'		
		C_1	C_2	C_3	C_1	C_2	C_3
1	1	28.4	25.7	23.1	32.8	30.1	27.4
1	2	26.7	24.1	21.4	32.6	29.9	27.3
2	1	27.4	24.7	22.1	33.7	31.0	28.4
2	2	27.7	25.0	22.4	31.6	28.9	26.3
3	1	27.1	24.4	21.8	34.1	31.4	28.8
3	2	28.1	25.4	22.8	31.3	28.6	26.0

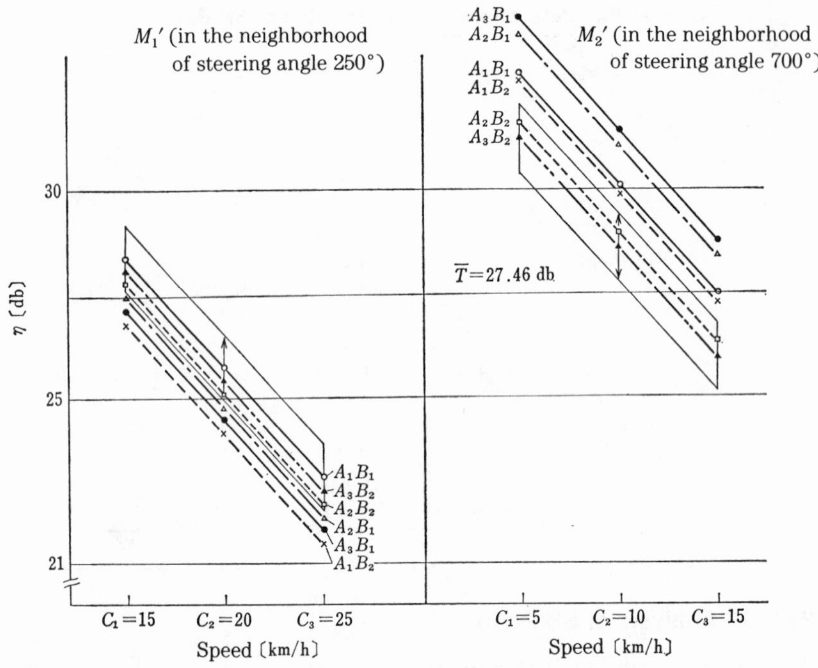

FIGURE 24.3 Relationship Between Vehicle Speed and SN Ratio

This means that, compared with M_1', the SN ratio of M_2' is such that the variation of the level of the signal factor comes out higher by about

$$10\log\left(\frac{700}{250}\right)^2 = 8.9 \quad [\text{db}] \tag{24.98}$$

This means that we consider matters by adding 8.9 db to the data of M_1' in Figure 24.2.

From the table of analysis of variance and Figure 24.3, the design group at Isuzu Motors Ltd. have derived the following conclusions [reference 38].

(1) Influence of Range of Steering Angle. For M_1' and M_2', a comparatively gentle curve and a sharp curve have been hypothesized, and according to the analysis results, the SN ratio is higher in the neighborhood of steering angle 250° than in the neighborhood of 700° (from the results of Figure 24.3). That the SN ratio is high here means that the turning radius is not easily influenced by the five elements, direction of turning, road surface, position of load, tire type, and air pressure of the front-wheel tires, that were selected as error factors, and by sources that were not specifically selected as error but that could become error. When the numerator and denominator of the SN ratio are examined in detail,

we see that the sensitivity of M_1' is better by about 6.3 times than that of M_2' but that the error is only 1.5 times as large.

(2) Influence of Vehicle Speed. The result obtained was that regarding the vehicle speed, whether in the case of M_1' or in the case of M_2', the SN ratio became better the lower the speed. If the denominator and numerator of the SN ratio are again compared separately, we see that the magnitude of the signal does not change much with vehicle speed but the influence of error is large.

(3) Steering Geometry. In the range where the steering angle is small (neighborhood of 250°), the effect of the control factor is only 1.0 db to less than 2.0 db. It is believed that with a turning radius and vehicle speed of this amount, the automobile is in a sufficiently stable state and differences in specification do not appear. On the other hand, in the range where the steering angle is large (neighborhood of 700°), an effect of 3.5 to 4.0 db can be seen. The highest combination is A_3B_1 and the lowest is A_3B_2. This difference is believed to be mainly due to steering geometry.

B_1 is the Ackermann mechanism and B_2 is the parallel mechanism, and although details will be omitted, with the Ackermann mechanism there is less sideslipping of the tire and it is possible to turn a circle more smoothly. The parallel mechanism is a link system where the angle of attack of the inner wheel and outer wheel become the same, and the tire becomes gouged during turning and there is much sideslipping.

Actually, there are other merits and one cannot necessarily always assert that the parallel mechanism is unconditionally bad. It can be supposed that since the interaction between the tires and the road surface at low speed was more unstable than with the Ackermann mechanism in this test, it was affected more readily by the influences of the road surface, load, and air pressure. These are results which are quite understandable in view of past experience and theories.

(4) Spring Constant of Front Spring. The influence of the front spring (A) appeared as an interaction with the steering geometry (B). In the relationship between the SN ratio and the vehicle speed in Figure 24.3, when we direct our attention to the neighborhood of steering angle 700°, the best SN ratio is obtained with A_3B_1 and the worst with A_3B_2. A_1B_1 and A_1B_2 rank intermediately, and from this it can be seen that there is no difference between geometries when the front spring is soft but Ackermann is better when the spring is hard.

Although the tire was used as an error factor in this experiment, when a tire maker is doing the research the tire is a control factor and the vehicle type becomes an error factor. Also, since the turning radius was logarithmically converted, the steering angle should have been taken as a geometric series. If this were done, it would be possible to use an orthogonal polynomial.

Exercises (24)

(1) These are data of comparison of the goodness of two methods, A_1 and A_2, for refining a certain metal X. In each case, the proportion of production of product and slag for a given ore and the quality of each are indicated. Compare the standardized SN ratios of A_1 and A_2 and draw conclusions.

A_1	Product	Slag	Total
Weight (%)	3.2	96.8	100.0
Quality of X (%)	98.0	0.28	

A_2	Product	Slag
Weight (%)	3.0	97.0
Quality of X (%)	99.3	0.91

(2) It is required that dynamite for blasting be delay-blasted in accordance with a specified time lag which depends on the charge quantity. The following data (unit: ms) were obtained when, as the signal factor, the charge quantity M was taken at the two levels, $M_1 = 75$ mg and $M_2 = 150$ mg, then four factors of two levels, $A, B, C,$ and D (in all cases, the first level is a new method and the second level is the current method), were assigned to orthogonal array L_8, and the time to blasting was investigated over five days.

(a) Perform appropriate data analysis. Assume there is no time lag in the case of charge quantity 0.
(b) Find the charge quantity for which the time lag becomes 0 ms, 40 ms, or 60 ms under optimum conditions.

No.	A	B	$A \times B$	C	$A \times C$	$B \times C$	D	M_1					M_2				
	1	2	3	4	5	6	7	R_1	R_2	R_3	R_4	R_5	R_1	R_2	R_3	R_4	R_5
1	1	1	1	1	1	1	1	18	19	20	20	20	40	40	41	39	40
2	1	1	1	2	2	2	2	13	18	18	19	18	34	36	36	38	42
3	1	2	2	1	1	2	2	20	21	20	18	23	40	43	44	36	40
4	1	2	2	2	2	1	1	18	19	19	21	19	35	40	33	36	44
5	2	1	2	1	2	1	2	20	20	19	19	20	41	40	42	40	40
6	2	1	2	2	1	2	1	21	19	20	16	22	39	38	40	37	36
7	2	2	1	1	2	2	1	19	18	19	18	18	38	42	38	42	44
8	2	2	1	2	1	1	2	19	17	19	17	20	37	42	38	37	30

(3) In a certain transmission system, the received levels of pulses in the current state are as follows. Bad space and bad mark mean that there is no mistake but the shape of the space and mark is degraded.

Input \ Reception	Good space	Bad space	Bad mark	Good mark	Total (%)
Space	99.8735	0.1221	0.0044	0.0000	100.0000
Mark	0.0000	0.0030	0.0845	99.9125	100.0000
Total [%]	99.8735	0.1251	0.0889	99.9125	200.0000

In order to improve this, seven factors at two levels were taken and were assigned to orthogonal array L_8, and the state of reception when 10,000 each of spaces and marks were transmitted in each experiment was as follows.

(a) For Group I, Group II, and Group III of the cumulative frequency, find the standardized SN ratios and perform an analysis of variance with these as the data of outer factors K_1, K_2, and K_3 at the respective levels.

(b) Find the significant factors and obtain the SN ratios under optimum conditions classed by levels of K_1, K_2, and K_3.

(c) Perform an analysis by the accumulation method with H regarded as the outer factors, with $H_1 = S$ and $H_2 = M$, and compare with the results of (a) and (b).

(4) We wished to experiment on an apparatus that converts the number of revolutions of an engine to voltage, and that emits the signal ON if this value becomes lower than the reference level and emits the signal OFF if it becomes higher. For 12 parts, we selected the levels of above and below the following central values in each case.

Exp. No.	Assignment							Input pulse	Received pulse type				Total
	A 1	B 2	C 3	D 4	E 5	F 6	G 7		Good space	Bad space	Bad mark	Good mark	
1	1	1	1	1	1	1	1	S	9 678	310	12	0	10 000
								M	0	6	238	9 756	10 000
2	1	1	1	2	2	2	2	S	9 892	104	4	0	10 000
								M	0	3	72	9 925	10 000
3	1	2	2	1	1	2	2	S	9 971	28	1	0	10 000
								M	0	1	42	9 957	10 000
4	1	2	2	2	2	1	1	S	9 009	956	35	0	10 000
								M	0	52	1 426	8 522	10 000
5	2	1	2	1	2	1	2	S	9 874	119	7	0	10 000
								M	0	5	85	9 910	10 000
6	2	1	2	2	1	2	1	S	9 957	42	1	0	10 000
								M	0	0	32	9 968	10 000
7	2	2	1	1	2	2	1	S	9 963	35	2	0	10 000
								M	0	3	49	9 948	10 000
8	2	2	1	2	1	1	2	S	7 942	1 980	78	0	10 000
								M	0	110	2 820	7 070	10 000

$A=2.2\,\text{k}\Omega,\ B=470\,\Omega,\ C=0.68\,\mu\text{F},\ D=100\,\text{k}\Omega,\ E=10\,\mu\text{F},\ F=180\,h_{fe},$

$G=10\,\text{k}\Omega,\ H=1.5\,\text{k}\Omega,\ I=10\,\text{k}\Omega,\ J=10\,\text{k}\Omega,\ K=180h_{fe},\ L=6.6\,\text{V}$

where A, B, C, D, G, H, I, and J are $\pm 5\%$, E is $\pm 20\%$, F and K are $\pm 50\%$, and L is ± 2 V each thus being at two levels, and these were assigned to orthogonal array L_{16} (Nippondensō, Shōgo Yamamoto, 1976). The standards of the ON value and OFF value are

ON value 600 ± 100 r.p.m.
OFF value Less than 125 r.p.m.

What analysis should be performed in order to cause N rather than SN ratio to be small? Decide on a method of decreasing dispersion by minimizing the tolerance of the error factors, regarding $A, B, \ldots, L$ as being the error factors. Please also refer to Chapter 16.

No.	A 1	B 2	C 3	D 4	E 5	F 6	G 7	H 8	I 9	J 10	K 11	L 12	e 13	e 14	e 15	Data ON value	Data OFF value
1																523	74
2																430	66
3						Levels omitted										674	101
4																572	90
5																609	90
6																534	80
7																578	95
8																527	88
9																605	97
10																707	92
11																541	87
12																669	85
13																430	78
14																480	72
15																578	89
16																668	85

25 Design of Experiments for Assembled Products

The design for determining which part is responsible for defectiveness of an assembled product, and which part causes degradation and variance of the product, will be discussed in this chapter. In the case of a chemical product, this is the problem of determining which batch of raw material constitutes the cause of variance. The question, whether dispersion of color prints originates in the film batch or in differences of the cameras or of the developing and printing concerns, is also a problem of the same type. The methods of Section 25.1 and Section 25.2 can be used to substitute for length-of-life tests when defectives and discrepancies are abundant; Section 25.3 can be used for the same purpose in the case of rare defectives, and it is the most important part of the chapter. It is therefore recommended that the reader begin with Section 25.3.

25.1 When the Parts Are of Two Types or Two Sets

When we wish to investigate whether the body or the needle has the greater influence on a certain characteristic of the nozzle of a diesel engine, we bring over from the work site four bodies and four needles chosen randomly. These will be denoted by A_1, A_2, A_3, and A_4; and B_1, B_2,

B_3, and B_4. A nozzle is constructed by combining body A_1 and needle B_1 and its characteristic y_{11} is studied. Afterward, it is possible to disassemble it, and then body A_1 and needle B_2 can be assembled and the nozzle characteristic can again be measured.

It is exactly the same when, in studying the variance of a characteristic of an electric fan (such as the air pressure, noise, or vibrations), we perform an assembling experiment to determine whether the greater influence is that of the motor or that of the blade, and disassembling and reassembling are possible as many times as one wishes. In the case of an assembled product where caulking or welding or gluing is used, it is not possible to disassemble or reassemble freely; please refer to Section 25.6 regarding the design of experiments in such a case.

Let us assume that we have brought over four bodies and four needles from the work site and have assigned them by a *two-way layout* such as is depicted by Table 25.1.

When using an assignment such as this, it is best also to use the sequence of assembling, V, as a factor, as shown by Table 25.2. The levels of V are ordered by an assignment known as the Latin square, but in practice this is an assignment where, in orthogonal array $L_{16}(4^5)$ of Appendix Table 12, A has been placed in the first column, B in the second column, and V in the third column. The combinations for V_1, *viz.*, A_1B_1, A_2B_2, A_3B_3, and A_4B_4, are assembled the first time, and the characteristic is measured. Next, those of V_2, *viz.*, A_1B_2, A_2B_1, A_3B_4, and A_4B_3, are assembled and the characteristic is again measured. When one proceeds in this manner, there is no pair of parts which is assembled twice, and by virtue of the effect of V it becomes clear how the characteristic changes if one disassembles and assembles several times.

Of course, *variances of the system cannot be thoroughly understood from only four units; as the number of units, it is better to use at least about 20*. Moreover, if possible, it is probably best to bring over several from those produced each day, for a total of 20.

Let us assume that we have performed an experiment where we have brought over 20 each of bodies and needles, then we have randomly divided them into five sets of four each, then we have assigned each set to a two-way layout of A and B and have inserted the sequence factor V as

Table 25.1 Assignment of Two Factors

	B_1	B_2	B_3	B_4	Total
A_1	y_{11}	y_{12}	y_{13}	y_{14}	A_1
A_2	y_{21}	y_{22}	y_{23}	y_{24}	A_2
A_3	y_{31}	y_{32}	y_{33}	y_{34}	A_3
A_4	y_{41}	y_{42}	y_{43}	y_{44}	A_4
Total	B_1	B_2	B_3	B_4	T

Table 25.2 Sequence of Experiment

	B_1	B_2	B_3	B_4
A_1	V_1	V_2	V_3	V_4
A_2	V_2	V_1	V_4	V_3
A_3	V_3	V_4	V_1	V_2
A_4	V_4	V_3	V_2	V_1

indicated in Table 25.2. Let us assume that the results of Table 25.3 were obtained when, *in conducting a jetting experiment with the nozzles, the jet shapes were distinguished as good, middling, and bad.* If the data are nearly the same when replicated, it is meaningless to replicate. Even in such a case, however, it is desirable to vary the conditions of jetting two ways and to categorize the jet shape for each as good, middling, or bad.

We analyze the data of Table 25.3 by *accumulating analysis.* First, we render the data into cumulative frequency.

CLASS I = bad
CLASS II = bad + middling
CLASS III = bad + middling + good

The data, rendered into cumulative frequency, become as given in Table 25.4. Supplementary tables such as Tables 25.5 are constructed for calculation of the factorial variations.

Table 25.3 Data of Nozzle Assembling Experiment

R_1	B_1	B_2	B_3	B_4
A_1	Gd	Md	Bad	Gd
A_2	Md	Md	Bad	Bad
A_3	Gd	Gd	Md	Gd
A_4	Bad	Bad	Md	Bad

R_2	B_5	B_6	B_7	B_8
A_5	Bad	Md	Bad	Bad
A_6	Gd	Md	Gd	Gd
A_7	Md	Gd	Gd	Md
A_8	Md	Gd	Gd	Gd

R_3	B_9	B_{10}	B_{11}	B_{12}
A_9	Bad	Md	Bad	Bad
A_{10}	Md	Md	Bad	Md
A_{11}	Gd	Gd	Md	Gd
A_{12}	Gd	Gd	Gd	Gd

R_4	B_{13}	B_{14}	B_{15}	B_{16}
A_{13}	Md	Bad	Md	Md
A_{14}	Gd	Gd	Gd	Gd
A_{15}	Gd	Gd	Md	Md
A_{16}	Md	Bad	Bad	Bad

R_5	B_{17}	B_{18}	B_{19}	B_{20}
A_{17}	Md	Md	Gd	Md
A_{18}	Gd	Md	Md	Md
A_{19}	Md	Bad	Md	Bad
A_{20}	Bad	Bad	Bad	Bad

Gd: Good
Md: Middling
Bad: Bad

Table 25.4 Data Rendered Cumulative

(R_1)

A	B	V	Bad	Md	Gd	I	II	III
1	1	1	0	0	1	0	0	1
1	2	2	0	1	0	0	1	1
1	3	3	1	0	0	1	1	1
1	4	4	0	0	1	0	0	1
2	1	2	0	1	0	0	1	1
2	2	1	0	1	0	0	1	1
2	3	4	1	0	0	1	1	1
2	4	3	1	0	0	1	1	1
3	1	3	0	0	1	0	0	1
3	2	4	0	0	1	0	0	1
3	3	1	0	1	0	0	1	1
3	4	2	0	0	1	0	0	1
4	1	4	1	0	0	1	1	1
4	2	3	1	0	0	1	1	1
4	3	2	0	1	0	0	1	1
4	4	1	1	0	0	1	1	1

...

(R_5)

A	B	V	Bad	Md	Gd	I	II	III
17	17	1	0	1	0	0	1	1
17	18	2	0	1	0	0	1	1
17	19	3	0	0	1	0	0	1
17	20	4	0	1	0	0	1	1
18	17	2	0	0	1	0	0	1
18	18	1	0	1	0	0	1	1
18	19	4	0	1	0	0	1	1
18	20	3	0	1	0	0	1	1
19	17	3	0	1	0	0	1	1
19	18	4	1	0	0	1	1	1
19	19	1	0	1	0	0	1	1
19	20	2	1	0	0	1	1	1
20	17	4	1	0	0	1	1	1
20	18	3	1	0	0	1	1	1
20	19	2	1	0	0	1	1	1
20	20	1	1	0	0	1	1	1

Tables 25.5 Supplementary Tables

R_1	I	II	III	R_2	I	II	III	R_3	I	II	III	R_4	I	II	III	R_5	I	II	III
A_1	1	2	4	A_5	3	4	4	A_9	3	4	4	A_{13}	1	4	4	A_{17}	0	3	4
A_2	2	4	4	A_6	0	1	4	A_{10}	1	4	4	A_{14}	0	0	4	A_{18}	0	3	4
A_3	0	1	4	A_7	0	2	4	A_{11}	0	1	4	A_{15}	0	2	4	A_{19}	2	4	4
A_4	3	4	4	A_8	0	1	4	A_{12}	0	0	4	A_{16}	3	4	4	A_{20}	4	4	4
Total	6	11	16	Total	3	8	16	Total	4	9	16	Total	4	10	16	Total	6	14	16

R_1	I	II	III	R_2	I	II	III	R_3	I	II	III	R_4	I	II	III	R_5	I	II	III
B_1	1	2	4	B_5	1	3	4	B_9	1	2	4	B_{13}	0	2	4	B_{17}	1	3	4
B_2	1	3	4	B_6	0	2	4	B_{10}	0	2	4	B_{14}	2	2	4	B_{18}	2	4	4
B_3	2	4	4	B_7	1	1	4	B_{11}	2	3	4	B_{15}	1	3	4	B_{19}	1	3	4
B_4	2	2	4	B_8	1	2	4	B_{12}	1	2	4	B_{16}	1	3	4	B_{20}	2	4	4
Total	6	11	16	Total	3	8	16	Total	4	9	16	Total	4	10	16	Total	6	14	16

	I	II	III
V_1	5	15	20
V_2	4	13	20
V_3	7	12	20
V_4	7	12	20
Total	23	52	80

The calculations of analysis of variance are as follows. Since these are fixed marginal enumerative values, we first calculate the *weights* W of class I and class II.

$$W_1=\frac{80^2}{23\times 57}=4.88 \qquad W_2=\frac{80^2}{52\times 28}=4.40 \tag{25.1}$$

$$CF=\frac{529\times 4.88+2704\times 4.40}{80}=180.99 \qquad (f=2) \tag{25.2}$$

$$S_R=\frac{(6^2+3^2+4^2+4^2+6^2)\times 4.88+(11^2+8^2+9^2+10^2+14^2)\times 4.40}{16}-CF$$

$$=8.03 \qquad (f=8) \tag{25.3}$$

$$S_V=\frac{(5^2+4^2+7^2+7^2)\times 4.88+(15^2+13^2+12^2+12^2)\times 4.40}{20}-CF$$

$$=2.97 \qquad (f=6) \tag{25.4}$$

$$S_A=S_{A_1\sim A_4}+S_{A_5\sim A_8}+\cdots+S_{A_{17}\sim A_{20}}$$

$$=\left[\frac{(1^2+2^2+0^2+3^2)\times W_1+(2^2+4^2+1^2+4^2)\times W_2}{4}-\frac{6^2\times W_1+11^2\times W_2}{16}\right]+\cdots$$

$$+\left[\frac{(0^2+0^2+2^2+4^2)\times W_1+(3^2+3^2+4^2+4^2)\times W_2}{4}-\frac{6^2\times W_1+14^2\times W_2}{16}\right]$$

$$=83.64 \qquad (f=30) \tag{25.5}$$

$$S_B=16.48 \qquad (f=30) \tag{25.6}$$

$$S_T=80+80=160 \qquad (f=158) \tag{25.7}$$

Therefore, the analysis of variance table becomes as given by Table 25.6.

Only A is significant in this case. Although fundamentally the effects of both A and B have entered the effect of R, it may be considered that only the effect of A has entered, not only because the effect of B is not significant but also because the variance ratio F_0 is at most 1 and it can therefore be disregarded. Although usually the effect of R, too, would be significant if A is significant, it is not so here since the number of degrees of freedom is small. However, if the variance ratio F_0 is as much as 1.73, it

Table 25.6 Analysis of Variance Table

Source	f	S	V	F_0
R	8	8.03	1.004	1.73(△)
A	30	83.64	2.788	4.79**
B	30	16.48	0.549	—
V	6	2.97	0.495	—
e	84	48.88	0.582	
T	158	160.00		

Table 25.7 Adjusted Analysis of Variance Table

Source	f	S	V	F_0	S'	ρ[%]
A	38	91.67	2.412	4.24**	70.05	43.7
e	120	68.33	0.569		89.95	56.3
T	158	160.00			160.00	100.0

should be judged that the effect of A has entered. In the last analysis, it is reasonable to regard the effects of sources other than A as error in the total variation S_T. Therefore, the *adjusted analysis of variance table* becomes as shown in Table 25.7.

These results mean that *the contribution ratio of part A is 43.7% and that the contribution ratio of part B may be disregarded. Although it appears as though the contribution ratio of error,* ρ_e, *is large, in effect variances of measurement have also entered the error. If one wishes to see the influence of part A in the variances of the products themselves, having eliminated the variances of measurement, it is necessary to replicate measurements in the experiment.*

25.2 When the Factors Are Numerous

25.2.1 Assignment the First Time

If there are three or more types of parts and, moreover, assembling and disassembling are possible as many times as one wishes with any of the types, we assign as follows by using an orthogonal array. It is best to vary the orthogonal array according to the number of types of parts. For example, if there are ten types of parts, probably L_{27} can be used. But since it is better to disassemble and assemble many parts a few times rather than to assemble and disassemble the same parts many times, replicating to the extent that is possible, probably a four-level system is the easiest to use except when this is impossible. If there are four levels, however, it is possible to experiment only on up to five types of parts even by $L_{16}(4^5)$, and four levels would be a problem when there are ten types of parts. Even in such a case, we can go about it as follows.

We designate these ten types of parts $A, B, C, D, E, F, G, H, I$, and J; and we take the most important part as A, the parts which are next in importance as B and C, and regard the remainder as being not particularly important. Assuming that we will be bringing over $4 \times 5 = 20$ of each of the parts, we create four compounding factors as follows from the above factors.

$A' = A$
$B' =$ 20 sets of B and C
$C' =$ 20 sets of D, E, and F
$D' =$ 20 sets of G, H, I, and J

Here, 20 levels of factor A mean 20 levels of part A. *The 20 levels of compounding factor B' are 20 sets of part B and part C, obtained by randomly combining 20 of part B and 20 of part C.* Since these 20 sets can be used for an assembling experiment anytime as pairs, we are in effect regarding the pair of part B and part C as a single part B'.

Similarly, we are treating the 20 sets obtained by randomly combining 20 each of D, E, and F as though they were the levels of a single part C'. And the 20 levels of D' are 20 sets in which G, H, I, and J have been randomly combined.

If we proceed thus, in effect ten factors at 20 levels may be regarded as four factors at 20 levels, A', B', C', and D'. There is no difference in this even if the sequence of assembly is such that D is attached to A and subsequently B, C, E, F, G, H, I, and J are attached in sequence. However, if assembly-disassembly of A and D is especially difficult compared with assembly-disassembly of the other parts, one might create a factor A' compounded of A and D. In this case, if we have assembled 20 types of subassembled products of A and D beforehand, it will be unnecessary to disassemble them. When creating levels by compounding factors, it is best to take into consideration not only the importance of the factors but also the difficulty of disassembly-assembly.

The four factors at 20 levels, A', B', C', and D' and the assembling sequence factor V are then assigned five sets of $L_{16}(4^5)$, as in Table 25.8.

Table 25.8 Assignment by Five Classes of L_{16}

R_1	A_1' ⋮ A_4' 1	B_1' ⋮ B_4' 2	C_1' ⋮ C_4' 3	D_1' ⋮ D_4' 4	V 5	Data		R_5	A_{17}' ⋮ A_{20}' 1	B_{17}' ⋮ B_{20}' 2	C_{17}' ⋮ C_{20}' 3	D_{17}' ⋮ D_{20}' 4	V 5	Data
1	1	1	1	1	1	y_{11}		1	1	1	1	1	1	y_{51}
2	1	2	2	2	2	y_{12}		2	1	2	2	2	2	y_{52}
3	1	3	3	3	3	y_{13}		3	1	3	3	3	3	y_{53}
4	1	4	4	4	4	y_{14}		4	1	4	4	4	4	y_{54}
5	2	1	2	3	4	y_{15}		5	2	1	2	3	4	y_{55}
6	2	2	1	4	3	y_{16}	···	6	2	2	1	4	3	y_{56}
7	2	3	4	1	2	y_{17}		7	2	3	4	1	2	y_{57}
8	2	4	3	2	1	y_{18}		8	2	4	3	2	1	y_{58}
9	3	1	3	4	2	y_{19}		9	3	1	3	4	2	y_{59}
10	3	2	4	3	1	y_{110}		10	3	2	4	3	1	y_{510}
11	3	3	1	2	4	y_{111}		11	3	3	1	2	4	y_{511}
12	3	4	2	1	3	y_{112}		12	3	4	2	1	3	y_{512}
13	4	1	4	2	3	y_{113}		13	4	1	4	2	3	y_{513}
14	4	2	3	1	4	y_{114}		14	4	2	3	1	4	y_{514}
15	4	3	2	4	1	y_{115}		15	4	3	2	4	1	y_{515}
16	4	4	1	3	2	y_{116}		16	4	4	1	3	2	y_{516}

Calculations of the variations are the same as before. For example, we have

$$S_R=\frac{R_1^2+R_2^3+\cdots+R_5^2}{16}-CF \tag{25.8}$$

$$S_{A'}=\frac{(A_1')^2+(A_2')^2+\cdots+(A_{20}')^2}{4}-CF-S_R \tag{25.9}$$

$$S_{B'}=\frac{(B_1')^2+(B_2')^2+\cdots+(B_{20}')^2}{4}-CF-S_R \tag{25.10}$$

...

$$S_V=\frac{V_1^2+V_2^2+\cdots+V_4^2}{20}-CF \tag{25.11}$$

$$S_e=S_T-(S_R+S_{A'}+S_{B'}+S_{C'}+S_{D'}+S_V) \tag{25.12}$$

Therefore, the analysis of variance table becomes as shown in Table 25.9.

In this case, *even if it is clear that source B′ has effect, it is unclear whether the effect is due to B or to C or both.* In such a case, it is necessary to separate the effects of B and C by the next experiment. However, if it emerges that B' is not significant, it is probably safe to assume that there is no effect of B or C. This is because it is nearly inconceivable when there are as many as 20 levels, as here, that the effects of B and C would cancel so that there is no resultant effect of B'.

This is true because, in fact, the following relationship holds when the levels of B and of C are randomly combined:

$$\sigma_{B'}^2=\sigma_{BC}^2=\sigma_B^2+\sigma_C^2 \tag{25.13}$$

Similarly, if a compounded factor has a larger number of levels, we can say the following:

$$\sigma_{C'}^2=\sigma_{DEF}^2=\sigma_D^2+\sigma_E^2+\sigma_F^2 \tag{25.14}$$

$$\sigma_{D'}^2=\sigma_{GHIJ}^2=\sigma_G^2+\sigma_H^2+\sigma_I^2+\sigma_J^2 \tag{25.15}$$

Table 25.9 Analysis of Variance Table

Source	f	$E(V)$
R	4	$\sigma^2+4(\sigma_{A'}^2+\sigma_{B'}^2+\sigma_{C'}^2+\sigma_{D'}^2)$
$A'(A)$	15	$\sigma^2+4\sigma_A^2$
$B'(BC)$	15	$\sigma^2+4\sigma_{BC}^2$
$C'(DEF)$	15	$\sigma^2+4\sigma_{DEF}^2$
$D'(GHIJ)$	15	$\sigma^2+4\sigma_{GHIJ}^2$
V	3	$\sigma^2+20\sigma_V^2$
e	12	σ^2
T	79	

This means that if it turns out that D' is not significant, we may conclude that G, H, I, and J, all have no effect. Therefore, if, in the analysis of variance table, Table 25.9, it emerges that R', A', B', C', and V are significant and D' is not significant, the second experiment becomes as described in 25.2.2. If the contribution ratio $\rho_{D'}$ is small even if D' is significant, we may probably consider the effects of G, H, I, and J as small.

The contribution ratios can be calculated in the following manner. If it is found that

$$\rho_A = \frac{S_A - 15 \times V_e}{S_T} \times 100 \tag{25.16}$$

$$\rho_{B'} = \frac{S_{B'} - 15 \times V_e}{S_T} \times 100 \tag{25.17}$$

$$\rho_{C'} = \frac{S_{C'} - 15 \times V_e}{S_T} \times 100 \tag{25.18}$$

we find the contribution ratios after correction, ρ'_A, ρ'_{BC}, and ρ'_{DEF}, as follows. By distributing the effect of R, we get

$$\rho'_A = \rho_A \times \frac{100}{100 - \rho_R} \tag{25.19}$$

$$\rho'_{BC} = \rho_{B'} \times \frac{100}{100 - \rho_R} \tag{25.20}$$

$$\rho'_{DEF} = \rho_{C'} \times \frac{100}{100 - \rho_R} \tag{25.21}$$

Table 25.10 Assignment the Second Time

R_1	$(BC)_1 \ldots (BC)_4$ 1	$D_1 \ldots D_4$ 2	$E_1 \ldots E_4$ 3	$F_1 \ldots F_4$ 4	$(A\cdots)_1 \ldots (A\cdots)_4$ 5
1	1	1	1	1	1
2	1	2	2	2	2
3	1	3	3	3	3
4	1	4	4	4	4
5	2	1	2	3	4
6	2	2	1	4	3
7	2	3	4	1	2
8	2	4	3	2	1
9	3	1	3	4	2
10	3	2	4	3	1
11	3	3	1	2	4
12	3	4	2	1	3
13	4	1	4	2	3
14	4	2	3	1	4
15	4	3	2	4	1
16	4	4	1	3	2

R_5	$(BC)_{17} \ldots (BC)_{20}$ 1	$D_{17} \ldots D_{20}$ 2	$E_{17} \ldots E_{20}$ 3	$F_{17} \ldots F_{20}$ 4	$(A\cdots)_{17} \ldots (A\cdots)_{20}$ 5
1	Omitted				
2					
3					
4					
5					
6					
7					
8					
9					
10					
11					
12					
13					
14					
15					
16					

25.2.2 Assignment the Second Time

In order to separate the compounded sources B and C from D, E, and F, although in general we resort to an experiment by the compounding method, in this case it is not possible to solve the problem neatly merely by performing the assembling-disassembling experiment of five L_{16}'s once more. This is because of the fact that there are four levels in L_{16} and the fact that three factors are compounded. We therefore re-choose the levels as follows for the experiment.

B 20 levels
C 20 levels
D 20 levels
E 20 levels
F 20 levels
A, G, H, I, J 20 levels
V 4 levels

A – J are to be at 20 levels; then, further, it is possible to compound V with them. If (BC) is not significant, one need only assign $D, E, F, (ABCGHIJ)$, and V the five columns of L_{16}. This means that we assign D to the first column of $L_{16}(4^5)$, E to the second column, F to the third column, the compounded factor for A, B, C, G, H, I, and J, $(ABCGHIJ)$ to the fourth column, and V to the fifth column.

The analysis of variance table is as shown in Table 25.11. If D is significant, we assume that,

$$\rho_D = \frac{S_D - 15V_e}{S_T} \times 100 \tag{25.22}$$

and we need only take $\rho_{D'}$ after correction as

$$\rho_{D'} = \rho_D \times \frac{100}{100 - \rho_R} \tag{25.23}$$

It is exactly the same for the other sources also.

Table 25.11 Analysis of Variance Table for the Second Experiment

Source	f	$E(V)$
R	4	$\sigma^2+4(\sigma_D^2+\sigma_E^2+\sigma_F^2+\sigma_A^2)$
D	15	$\sigma^2+4\sigma_D^2$
E	15	$\sigma^2+4\sigma_E^2$
F	15	$\sigma^2+4\sigma_F^2$
$(ABCGHIJ)$	15	$\sigma^2+4\sigma^2_{(ABCGHIJ)}$
V	3	$\sigma^2+4\sigma^2_{(ABCGHIJ)}+20\sigma_V^2$
e	12	σ^2
T	79	

When it is necessary to separate the effects of B and C, as here, this may be possible by the use of orthogonal array $L_{32}(4^9 \times 2^1)$ but it is difficult with L_{16}. If we wish, no matter what, to experiment by using $L_{16}(4^5)$, there are several methods, including the following.

(1) Estimations for B_1', B_2', B_3', and B_4' are performed from the experiment R_1.

$$\bar{B}_1' = \frac{B_1'}{4} \quad \bar{B}_2' = \frac{B_2'}{4} \quad \bar{B}_3' = \frac{B_3'}{4} \quad \bar{B}_4' = \frac{B_4'}{4}$$

We then select from among $\bar{B}_1'$, $\bar{B}_2'$, $\bar{B}_3'$, and $\bar{B}_4'$ the one whose value is the greatest and the one whose value is the smallest. We will assume that these are $\bar{B}_1'$ and $\bar{B}_3'$. We will suppose that, in this case, B_1' is the combination B_1C_1 and that B_3' is B_3C_3. Similarly, for the data of R_2 we perform estimations for B_5', B_6', B_7', and B_8' and we select the classes of BC whose values are the greatest and the smallest. We do the same for R_3, R_4, and R_5. This means that 10 classes have been selected for part B and part C. We renumber the levels of B and C for these 10 classes of products consecutively as $B_1, B_2, \ldots, B_{10}$ and $C_1, C_2, \ldots, C_{10}$.

When this is done, the factors ultimately become as follows.

B 10 levels
C 10 levels
D 20 levels
E 20 levels
F 20 levels
A, G, H, I, J 20 classes (20 levels)
V 4 levels

This means that, with regard to B and C, we have halved the numbers of levels without losing too much information. We combine B and C as follows.

$$(BC)_1 = B_1C_1 \quad (BC)_5 = B_3C_3 \quad \cdots \quad (BC)_{17} = B_9C_9$$

$$(BC)_2 = B_1C_2 \quad (BC)_6 = B_3C_4 \quad \cdots \quad (BC)_{18} = B_9C_{10}$$

$$(BC)_3 = B_2C_1 \quad (BC)_7 = B_4C_3 \quad \cdots \quad (BC)_{19} = B_{10}C_9$$

$$(BC)_4 = B_2C_2 \quad (BC)_8 = B_4C_4 \quad \cdots \quad (BC)_{20} = B_{10}C_{10}$$

When this is done, B and C can be regarded as one factor at 20 levels. A method such as this can also be used when the numbers of parts B and C cannot be taken very large.

The combination factor (BC) is now assigned to the first column; then, the fifth column, the 20 levels of the combination factor $(AGHIJ)$ and assembly sequence V are assigned confounded.

$$S_{(BC)} = \frac{(BC)_1^2 + \cdots + (BC)_{20}^2}{4} - S_R \qquad (f = 15) \tag{25.24}$$

can be resolved into S_B, S_C, and $S_{B \times C}$.

$$S_B = \frac{(B_1 - B_2)^2}{16} + \frac{(B_3 - B_4)^2}{16} + \cdots + \frac{(B_9 - B_{10})^2}{16} \qquad (f=5) \qquad (25.25)$$

$$S_C = \frac{(C_1 - C_2)^2}{16} + \frac{(C_3 - C_4)^2}{16} + \cdots + \frac{(C_9 - C_{10})^2}{16} \qquad (f=5) \qquad (25.26)$$

$$S_{B\times C} = \frac{[(BC)_1 + (BC)_4 - (BC)_2 - (BC)_3]^2}{16} + \cdots$$

$$+ \frac{[(BC)_{17} + (BC)_{20} - (BC)_{18} - (BC)_{19}]^2}{16} \qquad (f=5) \qquad (25.27)$$

$$S_{(A\cdots)V} = \frac{(A\cdots)_1^2 + \cdots + (A\cdots)_{20}^2}{4} - S_R \qquad (f=15) \qquad (25.28)$$

It is exactly the same for the other sources. We find the contribution ratios by assuming that

$$\rho_B = \frac{S_B - 5V_e}{S_T} \qquad (25.29)$$

In a case where E, F, G, and H have also been found significant, it is insufficient to follow up with five sets of experiments by L_{16}, but for the three factors (DEF), sometimes the method of rendering them into three factors at 10 levels is useful.

For the experiment R_1, we perform estimation for each level $(DEF)_1, \ldots, (DEF)_4$. Those parts D, E, and F for which these values turn out to be the greatest are redesignated D_1, E_1, and F_1. The levels of the parts D, E, and F in the class whose estimated values were the smallest are redesignated D_2, E_2, and F_2. Then the first four levels of a new combination factor C' are defined by

$$C_1' = D_1\ E_1\ F_1$$

$$C_2' = D_1\ E_2\ F_2$$

$$C_3' = D_2\ E_1\ F_2$$

$$C_4' = D_2\ E_2\ F_1$$

From $R_2, \ldots, C_5', C_6', C_7', C_8', \ldots$ are similarly determined, and all these are regarded as one factor C' at 20 levels.

(2) The confounding method is used. In this method, B, C, D, E, and F are assigned to the 5th column of L_{16}; A, G, H, I, J and V are caused to correspond randomly, and with respect to $R_1, R_2, \ldots, R_5$, they are confounded with B, C, D, E, and F. Details of calculation in such a case should be clear even without explanation.

(3) The levels of B and C are shifted by one place relative to each other and factors with new levels are thus created. For example:

$$B_1' = B_1C_2 \qquad B_5' = B_5C_6 \ \cdots \ B_{17}' = B_{17}C_{18}$$

$$B_2' = B_2C_1 \qquad B_6' = B_6C_5 \ \cdots \ B_{18}' = B_{18}C_{17}$$

$$B_3' = B_3C_4 \qquad B_7' = B_7C_8 \cdots B_{19}' = B_{19}C_{20}$$

$$B_4' = B_4C_3 \qquad B_8' = B_8C_7 \cdots B_{20}' = B_{20}C_{19}$$

Thus, by compounding B and C at 20 levels, the new compounding factor B' is created; then this is assigned together with C, D, and E to the 1st, 2nd, 3rd, and 4th columns of L_{16}. Since the calculation in this case is of a cyclic nature and becomes rather laborious, it will be omitted because of space limitations.

25.3 The Case of Rare Defectives

25.3.1 When Disassembling and Reassembling Are Possible

One of the most important and also interesting problems in searching for causes is searching for the causes of rare troubles. For example, let us suppose that deafening noise in a passenger vehicle occurs in one car among several hundred. Even if only one car in 500 cars is defective in this way, if as many as 50,000 are produced in one month, as has been the case recently, it means that there are as many as 100 cars in a month that are defective. In a case such as this, even if we perform the assembling experiment explained in Sections 25.1 and 25.2 by arbitrarily bringing over several tens of parts from the work site, most likely it will turn out that all of the assembled products are nondefectives, and one will obtain no useful information at all. In such a situation, it is important to create factor levels in the following manner.

We collect, for example, 16 vehicles that are defective assemblies (or products that have been used for a long period of time) and we disassemble them. We consider parts (or subassembled products consisting of several assembled parts) that seem to influence the target characteristic. Let us assume that these parts can be disassembled and reassembled. We will also assume that the number of parts and units that can be disassembled and reassembled, which constitute these assembled products, amounts to 10 types. These are designated A, B, C, D, E, F, G, H, I, and J. $A, B, C, \ldots, J$ are still in production, and we bring over 16 of each of these. *Since there are few defectives among the assembled products, even if we randomly bring over currently produced parts, we may assume that the products of assembling parts belong among the nondefectives.* Of course, in certain cases it is probably all right to disassemble assembled products that were nondefective and to bring over 16 each of $A, B, \ldots, J$ from those.

For the ten factors A–J, this means that levels have been created as follows.

A_1 = ordinary part in production A_2 = part taken from defective assembled product
B_1 = ordinary part in production B_2 = part taken from defective assembled product
C_1 = ordinary part in production C_2 = part taken from defective assembled product
D_1 = ordinary part in production D_2 = part taken from defective assembled product
E_1 = ordinary part in production E_2 = part taken from defective assembled product
F_1 = ordinary part in production F_2 = part taken from defective assembled product
G_1 = ordinary part in production G_2 = part taken from defective assembled product
H_1 = ordinary part in production H_2 = part taken from defective assembled product
I_1 = ordinary part in production I_2 = part taken from defective assembled product
J_1 = ordinary part in production J_2 = part taken from defective assembled product

Thus, ten factors at two levels have been born. Of course, if part A does not influence the target characteristic, there should be no difference between A_1 and A_2, and this means there is no difference between an assembled product constructed with A_1 and one constructed with A_2. We can regard the ten factors above as a 2^{10}-type experiment and assign it to an orthogonal array.

Even when disassembling-reassembling is possible, when the product becomes large in size it may well be too much trouble to disassemble and reassemble it many times. In such a case, once the factors have been created as above, one sometimes wishes to only reassemble it once. This means that parts $A-J$ are sent around for reassembling just one additional time and the target characteristic is measured. In this case, regarding the important factors of $A-J$ as being A, B, and C, assignment to orthogonal array L_{16} (whereby it is also possible to find $A \times B, A \times C$, and $B \times C$) is as shown in Table 25.12. The reader is referred to Exercise Problem (3) of Chapter 7 for actual examples.

In the experiments of L_{16}, there are two replications. The analysis of variance table is as given by Table 25.13.

$$S_T = \text{sum of squares of the individual values of the 32 items} - CF \tag{25.30}$$

$$S_A = \frac{(A_1 - A_2)^2}{32} \tag{25.31}$$

$$\vdots$$

$$S_{e_1} = S_{\text{column 14}} + S_{\text{column 15}} \tag{25.32}$$

$$S_{e_2} = S_T - S_{T_1} \tag{25.33}$$

Table 25.12 Assignment for First Reassembling

	A	B	$A \times B$	C	$A \times C$	$B \times C$	D	E	F	G	H	I	J	e	e	Data (replicated twice)
	1	2	3	4	5	6	7	8	9	10	11	12	13	14	15	
1	1	1	1	1	1	1	1	1	1	1	1	1	1	1	1	–
2	1	1	1	1	1	1	1	2	2	2	2	2	2	2	2	–
3	1	1	1	2	2	2	2	1	1	1	1	2	2	2	2	–
4	1	1	1	2	2	2	2	2	2	2	2	1	1	1	1	–
5	1	2	2	1	1	2	2	1	1	2	2	1	1	2	2	–
6	1	2	2	1	1	2	2	2	2	1	1	2	2	1	1	–
7	1	2	2	2	2	1	1	1	1	2	2	2	2	1	1	–
8	1	2	2	2	2	1	1	2	2	1	1	1	1	2	2	–
9	2	1	2	1	2	1	2	1	2	1	2	1	2	1	2	–
10	2	1	2	1	2	1	2	2	1	2	1	2	1	2	1	–
11	2	1	2	2	1	2	1	1	2	1	2	2	1	2	1	–
12	2	1	2	2	1	2	1	2	1	2	1	1	2	1	2	–
13	2	2	1	1	2	2	1	1	2	2	1	1	2	2	1	–
14	2	2	1	1	2	2	1	2	1	1	2	2	1	1	2	–
15	2	2	1	2	1	1	2	1	2	2	1	2	1	1	2	–
16	2	2	1	2	1	1	2	2	1	1	2	1	2	2	1	–

In this case, with regard to part A, individual differences in A_1 and individual differences in A_2 are included in σ_2^2. For $B, C, \ldots$, too, the effect of variances within the respective levels is included in σ_2^2. In this sense, it is not possible to find completely the influence of individual differences of part A by an experiment such as this. However, it is clear

Table 25.13 Analysis of Variance Table

Source	f	$E(V)$
A	1	$\sigma_2^2+2\sigma_1^2+16\sigma_A^2$
B	1	〃 $+16\sigma_B^2$
$A\times B$	1	〃 $+8\sigma_{A\times B}^2$
C	1	〃 $+16\sigma_C^2$
$A\times C$	1	〃 $+8\sigma_{A\times C}^2$
$B\times C$	1	〃 $+8\sigma_{B\times C}^2$
D	1	〃 $+16\sigma_D^2$
E	1	〃 $+16\sigma_E^2$
F	1	〃 $+16\sigma_F^2$
G	1	〃 $+16\sigma_G^2$
H	1	〃 $+16\sigma_H^2$
I	1	〃 $+16\sigma_I^2$
J	1	〃 $+16\sigma_J^2$
e_1	2	$\sigma_2^2+2\sigma_1^2$
e_2	16	σ_2^2
T	31	

that for testing of the difference between A_1 and A_2, it is necessary to use the error which includes the variances of A within A_1 and within A_2. Since interactions between A_1 and A_2 and between D_1 and D_2 are also included in the first-order error variance σ_1^2, this must be distinguished from the second-order error variance σ_2^2.

Therefore, when e_1 and many of the first-order sources (about one-half in terms of the total number of degrees of freedom) are not significant when tested by e_2, we test by pooling them with e_2, but if this is not the case, we test by pooling the smaller of the first-order sources (the total of the numbers of degrees of freedom is about 6–9) with e_1.

25.4 When There Are Categorization Factors

Let us consider a case where we bring over 20 of part A and 20 of part B and conduct an experiment by two-way layout where disassembly-reassembly is possible, taking A at 20 levels $A_1, A_2, \ldots, A_{20}$ and B at 20 levels $B_1, B_2, \ldots, B_{20}$.

Although the contribution ratios of part A and part B, ρ_A and ρ_B, were found in Section 25.1, if it emerges that ρ_A is large this does not constitute a solution as to which of the characteristic values of part A is having effect. Even if it turns out that A is great in Section 25.1, it is unclear whether this is attributable to the roundness of body A or to the difference in tip shape.

Let us assume that for part A the most important characteristic, A', is measurable before the experiment. We measure the value of A' for part A, and depending on the value of A' we categorize part A into four classes:

A_1' = five whose value of A' is the smallest
A_2' = five whose value of A' is the next smaller
A_3' = five whose value of A' is the next greater
A_4' = five whose value of A' is the greatest

Similarly, for part B, if there is a characteristic value which we are concerned with that is measurable, we measure this value and divide B into four classes.

Sometimes such measurement is impossible but it is nevertheless possible to arrange the parts in sequence; or by matching them with another part or with a gauge one may categorize them, for example, as

B_1' = five whose fit is loose
B_2' = five whose fit is somewhat loose
B_3' = five whose fit is just right
B_4' = five whose fit is tight

Table 25.14 Two-Way Layout Method When Categorized (Numbers Are the Levels of the Factor of Sequence of Experiment, *V*)

R_1	B_1'	B_2'	B_3'	B_4'
A_1'	1	2	3	4
A_2'	2	1	4	3
A_3'	3	4	1	2
A_4'	4	3	2	1

We then bring over one part from each of A_1', A_2', A_3', and A_4' and one part from each of B_1', B_2', B_3', and B_4' and, by a two-way layout of the first class, we perform an experiment with an assignment such as that of Table 25.14. Similarly, experiments R_2, R_3, R_4, and R_5 are carried out.

Analysis of variance is performed as follows.

$$CF=\frac{T^2}{80} \qquad (f=1) \tag{25.34}$$

$$S_R=\frac{R_1^2+\cdots+R_5^2}{16}-CF \qquad (f=1) \tag{25.35}$$

$$S_{A'}=\frac{(A_1')^2+\cdots+(A_4')^2}{20}-CF \qquad (f=3) \tag{25.36}$$

$$S_{\text{within }A}=\frac{A_1^2+A_2^2+\cdots+A_{20}^2}{4}-CF-S_R-S_{A'} \qquad (f=12) \tag{25.37}$$

$S_{\text{within }A}$ is also the interaction of A' and R by the two-way array

$r=4$	R_1	R_2	R_3	R_4	R_5	Total
A_1'	—	—	—	—	—	A_1'
A_2'	—	—	—	—	—	A_2'
A_3'	—	—	—	—	—	A_3'
A_4'	—	—	—	—	—	A_4'
Total	R_1	R_2	R_3	R_4	R_5	T

Similarly, we have

$$S_{B'}=\frac{(B_1')^2+\cdots+(B_4')^2}{20}-CF \qquad (f=3) \tag{25.38}$$

$$S_{\text{within }B}=\frac{B_1^2+B_2^2+\cdots+B_{20}^2}{4}-CF-S_{B'}-S_R \qquad (f=12) \tag{25.39}$$

$$S_V=\frac{V_1^2+V_2^2+V_3^2+V_4^2}{20}-CF \qquad (f=3) \tag{25.40}$$

$$S_T=\text{sum of squares of the individual values of the 80 items}-CF \quad (f=79) \tag{25.41}$$

$$S_e=S_T-(S_R+S_{A'}+S_{\text{within }A}+S_{B'}+S_{\text{within }B}+S_V) \quad (f=42) \tag{25.42}$$

Table 25.15 Analysis of Variance Table

Source	f	$E(V)$
R	4	$\sigma^2+4(\sigma_{A内}{}^2+\sigma_{B内}{}^2)$
A'	3	$\sigma^2+\sigma_{A内}{}^2+20\sigma_{A}{}'^2$
Within A	12	$\sigma^2+4\sigma_{A内}{}^2$
B'	3	$\sigma^2+4\sigma_{B内}{}^2+20\sigma_{B}{}'^2$
Within B	12	$\sigma^2+4\sigma_{B内}{}^2$
V	3	$\sigma^2+20\sigma_V{}^2$
e	42	σ^2
T	79	

Therefore the analysis of variance table becomes as given in Table 25.15.

If there are two categorization items each for part A and part B—A' and A'', and B' and B''—we take 16 each of part A and part B. Then we successively categorize $A_1, A_2, \ldots, A_{16}$ by A' and A''. Once we have measured the values of A' and A'' for $A_1, A_2, \ldots, A_{16}$, we first categorize into four classes of four each, A_1', A_2', A_3', and A_4', according to the sequence of the values of A'. With respect to the four parts within A_1', we categorize into A_1'', A_2'', A_3'', and A_4'' by the magnitude of the values of A''. For A_2', too, we categorize into the four classes A_1'', A_2'', A_3'', and A_4'' in the sequence of the values of A'' for the four parts in A_2'. It does not matter if there is a difference between the values of A_1'' of A_1' and A_1'' of A_2'. The values of A'' within A_1' and within A_2' are categorized by the relative sequence. Assignment is possible by orthogonal array L_{16} or L_{64}, and the analysis of variance table is given by Table 25.16.

25.5 The Case of Split Form

With seven types of parts A, B, C, D, E, F, and G, let us consider a case where not very many of parts A, B, and C can be used but it does not matter if we bring over many pieces of D, E, F, and G. In this case, we

Table 25.16 Analysis of Variance Table

Source	f	$E(V)$
R	4	$\sigma^2 + 4(\sigma^2_{\text{within } A} + \sigma^2_{\text{within } B})$
A'	3	$\sigma^2 + 4\sigma^2_{\text{within } A} + 16\sigma_{A'^2}$
A''	3	$\sigma^2 + 4\sigma^2_{\text{within } A} + 16\sigma_{A''^2}$
Within A	6	$\sigma^2 + 4\sigma^2_{\text{within } A}$
B'	3	$\sigma^2 + 4\sigma^2_{\text{within } B} + 16\sigma_{B'^2}$
B''	3	$\sigma^2 + 4\sigma^2_{\text{within } B} + 16\sigma_{B''^2}$
Within B	6	$\sigma^2 + 4\sigma^2_{\text{within } B}$
V	3	$\sigma^2 + 16\sigma_V{}^2$
e	36	σ^2
T	63	

decide that *A, B, and C are first-order parts and D, E, F, and G are second-order parts.*

We take

First-order parts	A, B, C	8 apiece
Second-order parts	D, E, F, G	16 apiece

Thus, we have an assignment with

A	8 levels
B	8 levels
C	8 levels
D	16 levels
E	16 levels
F	16 levels
G	16 levels

Assignment in this case is to be as follows. A_1, A_2; B_1, B_2; and C_1, C_2 are assigned to L_4. A_3, A_4; B_3, B_4; and C_3, C_4 are assigned to a different L_4. If we continue in this way, the assignment becomes as given by Table 25.17.

Next, we assign $D_1 - D_4$, $E_1 - E_4$, $F_1 - F_4$, and $G_1 - G_4$ to the second column, third column, fourth column, and fifth column of $L_{16}(4^5)$. The four levels

$$(ABC)_1 = A_1B_1C_1$$

$$(ABC)_2 = A_1B_2C_2$$

$$(ABC)_3 = A_2B_1C_2$$

$$(ABC)_4 = A_2B_2C_1$$

are assigned to the four levels of the first column of the $L_{16}(4^5)$. Therefore, if the assignment of this class is redesignated R_1, it becomes the class R_1 of Table 25.18.

For R_2, we assign the four levels

$$(ABC)_5 = A_3B_3C_3$$

$$(ABC)_6 = A_3B_4C_4$$

$$(ABC)_7 = A_4B_3C_4$$

$$(ABC)_8 = A_4B_4C_3$$

Table 25.17 Assignment of First-Order Parts

R_1	A_1 / A_2 / 1	B_1 / B_2 / 2	C_1 / C_2 / 3		R_4	A_7 / A_8 / 1	B_7 / B_8 / 2	C_7 / C_8 / 3
1	1	1	1	···	1	1	1	1
2	1	2	2		2	1	2	2
3	2	1	2		3	2	1	2
4	2	2	1		4	2	2	1

Table 25.18 Assignment with Split

R_1	(ABC) 1	D 2	E 3	F 4	G 5	R_2	(ABC) 1	D 2	E 3	F 4	G 5		R_4	(ABC) 1	D 2	E 3	F 4	G 5
1	(111)	1	1	1	1		(333)	5	5	5	5			(777)	13	13	13	13
2	(111)	2	2	2	2		(333)	6	6	6	6			(777)	14	14	14	14
3	(111)	3	3	3	3		(333)	7	7	7	7			(777)	15	15	15	15
4	(111)	4	4	4	4		(333)	8	8	8	8			(777)	16	16	16	16
5	(122)	1	2	3	4		(344)	5	6	7	8			(788)	13	14	15	16
6	(122)	2	1	4	3		(344)	6	5	8	7			(788)	14	13	16	15
7	(122)	3	4	1	2		(344)	7	8	5	6			(788)	15	16	13	14
8	(122)	4	3	2	1		(344)	8	7	6	5	...		(788)	16	15	14	13
9	(212)	1	3	4	2		(434)	5	7	8	6			(878)	13	15	16	14
10	(212)	2	4	3	1		(434)	6	8	7	5			(878)	14	16	15	13
11	(212)	3	1	2	4		(434)	7	5	6	8			(878)	15	13	14	16
12	(212)	4	2	1	3		(434)	8	6	5	7			(878)	16	14	13	15
13	(221)	1	4	2	3		(443)	5	8	6	7			(887)	13	16	14	15
14	(221)	2	3	1	4		(443)	6	7	5	8			(887)	14	15	13	16
15	(221)	3	1	4	1		(443)	7	6	8	5			(887)	15	14	16	13
16	(221)	4	2	3	2		(443)	8	5	7	6			(887)	16	13	15	14

and D_5–D_8, E_5–E_8, F_5–F_8, and G_5–G_8 in the same way. Therefore, the assigned array becomes the class R_2 of Table 25.18. By proceeding similarly, we obtain the rest of the assignment of Table 25.18.

Calculations of analysis of variance are as follows.

$$S_R=\frac{R_1^2+R_2^2+R_3^2+R_4^2}{16}-CF \qquad (f=3) \tag{25.43}$$

$$S_A=\frac{1}{16}[(A_1-A_2)^2+(A_3-A_4)^2+(A_5-A_6)^2+(A_7-A_8)^2] \qquad (f=4) \tag{25.44}$$

A_1-A_2 is the difference of the totals of A_1 and A_2 of the experimental data of R_1.

$$S_B=\frac{1}{16}[(B_1-B_2)^2+(B_3-B_4)^2+(B_5-B_6)^2+(B_7-B_8)^2] \qquad (f=4) \tag{25.45}$$

$$S_C=\frac{1}{16}[(C_1-C_2)^2+(C_3-C_4)^2+(C_5-C_6)^2+(C_7-C_8)^2] \qquad (f=4) \tag{25.46}$$

$$S_D=\frac{D_1^2+D_2^2+\cdots+D_{16}^2}{4}-\frac{R_1^2+\cdots+R_4^2}{16} \qquad (f=12) \tag{25.47}$$

$$S_E=\frac{E_1^2+E_2^2+\cdots+E_{16}^2}{4}-\frac{R_1^2+\cdots+R_4^2}{16} \qquad (f=12) \tag{25.48}$$

$$S_F=\frac{F_1^2+F_2^2+\cdots+F_{16}^2}{4}-\frac{R_1^2+\cdots+R_4^2}{16} \qquad (f=12) \tag{25.49}$$

$$S_G=\frac{G_1^2+G_2^2+\cdots+G_{16}^2}{4}-\frac{R_1^2+\cdots+R_4^2}{16} \qquad (f=12) \tag{25.50}$$

$$S_{e_1}=S_{T_1}-(S_R+S_A+S_B+S_C) \qquad (f=0) \tag{25.51}$$

Table 25.19 Distribution of Number of Degrees of Freedom

Source	f	$E(V)$
R	3	$\sigma_2^2+4\sigma_1^2+$ confounded sources
A	4	$\sigma_2^2+4\sigma_1^2+8\sigma_A^2$
B	4	$\sigma_2^2+4\sigma_1^2+8\sigma_B^2$
C	4	$\sigma_2^2+4\sigma_1^2+8\sigma_C^2$
e_1	0	$(\sigma_2^2+4\sigma_1^2)$
D	12	$\sigma_2^2+4\sigma_D^2$
E	12	$\sigma_2^2+4\sigma_E^2$
F	12	$\sigma_2^2+4\sigma_F^2$
G	12	$\sigma_2^2+4\sigma_G^2$
e_2	0	(σ_2^2)
T	63	

where

$$S_{T_1}=\frac{(A_1B_1C_1)^2+\cdots+(A_8B_8C_8)^2}{4}-CF \qquad (f=12) \tag{25.52}$$

Since interactions between A, B, and C are included in the first-order error variance e_1, the variance e_1 was calculated separately, but in certain cases it is probably not very important to distinguish between first-order and second-order error variance.

Total variation S_T is found by subtracting the correction factor CF from the sum of squares of the 64 individual values. Therefore, second-order error variance S_{e_2} is

$$S_{e_2}=S_T-(S_R+S_A+\cdots+S_G+S_{e_1}) \qquad (f=0) \tag{25.53}$$

Therefore the analysis of variance table is as given by Table 25.19.

25.6 When There Are Parts That Cannot Be Disassembled and Reassembled

Sometimes, in an experiment using assembled products, *parts that cannot be disassembled and reassembled are included.* This is because, in the cases of assembled parts where caulking or welding or the like is necessary, or of packing materials, etc., the part breaks if disassembled and cannot be used again for reassembly, or even if it can be used, its performance changes altogether.

Let us assume that A and B are parts that can be disassembled and reassembled, and that C is a part that cannot be disassembled and reassembled. We bring over twelve A's and twelve B's and designate them A_1, A_2, . . . , A_{12} and B_1, B_2, . . . B_{12}. Assuming that we are to use three classes of two-way layouts with A and B, the assignment array is as shown in Table 25.20.

Table 25.20 Assignment of First-Order Parts *A* and *B*

R_1	B_1 B_2 B_3 B_4	R_2	B_5 B_6 B_7 B_8	R_3	B_9 B_{10} B_{11} B_{12}
A_1		A_5		A_9	
A_2		A_6		A_{10}	
A_3		A_7		A_{11}	
A_4		A_8		A_{12}	

Since one part C is necessary for each assembly, in the case of Table 25.20, this means that $16 \times 3 = 48$ are necessary. We bring over 48 part C's and designate them $C_1, C_2, \ldots, C_{48}$. However, if a certain characteristic of C can be measured, it is desirable to measure it. For example, if C is a packing material and it is possible to measure the thickness C' and weight C'' of this material, we measure these characteristics for the products from C_1 through C_{48}.

Once the values of C' and C'' have been measured for the 48 part C's, C is categorized in the order of the magnitude of C'. Three to four levels is the most favorable number of levels for categorization. Since we have a four-level orthogonal array in the present case, we categorize into four levels. Thus, we categorize into the four sets of 12 whose value of C' is the smallest, the next 12, the 12 which are the next greater, and the 12 which are the greatest. These are designated C_1', C_2', C_3', and C_4'. Next, within C_1' we categorize into four levels in the sequence of the magnitude of C''. The three whose C'' value is the smallest are expressed as C_1'', the next three as C_2'', the three which are next greater as C_3'', and the three which are the greatest as C_4''. If the third, fourth, and fifth of those whose value of C'' is small have equal values, one among them is randomly incorporated into the class C_1''. There must be exactly three parts in each of C_1'', C_2'', C_3'', and C_4''.

Next, regarding the 12 of C_2', too, we categorize them into the four classes

C_1'' = three whose C'' is the smallest
C_2'' = three whose C'' value is next smallest
C_3'' = three whose C'' value is next greater
C_4'' = three whose C'' value is the greatest

in the sequence of the magnitude of the value of C''. It does not matter if the ranges of the values of C_1'' of C_1' and C_1'' of C_2' differ. We categorize by the magnitude of the relative sequence within C_2'.

This means that part C has been *successively categorized* by four levels each of its two characteristics C' and C''.

It now becomes possible to assign A, B, C', C'', and V (sequence of assembly) to orthogonal array $L_{16}(4^5)$. For example, assignment of R_1 becomes as shown by Table 25.21. Parts A_1–A_4 and B_1–B_4 are used in

Table 25.21 Assignment of *A*, *B*, *C′*, *C″*, and *V*

R_1	A 1	B 2	C' 3	C'' 4	V 5	Data
1	1	1	1	1	1	y_1
2	1	2	2	2	2	y_2
3	1	3	3	3	3	y_3
4	1	4	4	4	4	y_4
5	2	1	2	3	4	y_5
6	2	2	1	4	3	y_6
7	2	3	4	1	2	y_7
8	2	4	3	2	1	y_8
9	3	1	3	4	2	y_9
10	3	2	4	3	1	y_{10}
11	3	3	1	2	4	y_{11}
12	3	4	2	1	3	y_{12}
13	4	1	4	2	3	y_{13}
14	4	2	3	1	4	y_{14}
15	4	3	2	4	1	y_{15}
16	4	4	1	3	2	y_{16}

each of V_1, V_2, V_3, and V_4, but in the case of part C, one from each of the 16 classes of categorization is used for assembly each time.

In this case, it is not possible to calculate the total effect of part C, and S_C is resolved as

$$S_C = S_{C'} + S_{C''} + S_{e(C)} \tag{25.54}$$

$S_{C'}$ expresses the effect of characteristic C' in the influence of part C, and $S_{C''}$ expresses the magnitude of the effect of characteristic C''. To be accurate, $S_{C'}$ also includes the effects of other characteristics of part C that are related with C'. $S_{C'}$ corresponds to the so-called *single regression variation* of C', and $S_{C''}$ to the *partial regression variation* of C''.

$$S_{C'} = \frac{C_1'+C_2'+C_3'+C_4'}{12} - CF \qquad (f=3) \tag{25.55}$$

$$S_{C''} = \frac{C_1''+C_2''+C_3''+C_4''}{12} - CF \qquad (f=3) \tag{25.56}$$

The effect of those characteristics in the influence of part C which cannot be expressed by characteristics C' and C'' comes into $S_{e(C)}$. In analysis of variance, in effect this gets mixed into error variance S_e.

Therefore, the analysis of variance table is as given by Table 25.22.

The fundamental error,

$$\sigma_{e(C)}^2 = \sigma_C^2 - (\sigma_{C'}^2 + \sigma_{C''}^2) \tag{25.57}$$

is included in error variance σ^2.

The number of parts was small in the example above; what should we do if the number of parts becomes large? Let us consider the following

Table 25.22 Analysis of Variance Table

Source	f	$E(V)$
R	2	$\sigma^2+4\sigma_A{}^2+4\sigma_B{}^2$
A	9	$\sigma^2+4\sigma_A{}^2$
B	9	$\sigma^2+4\sigma_B{}^2$
C'	3	$\sigma^2+12\sigma_{C'}{}^2$
C''	3	$\sigma^2+12\sigma_{C''}{}^2$
V	3	$\sigma^2+12\sigma_V{}^2$
e	18	σ^2
T	47	

case, for example. Let us assume an assembling experiment for:

Parts that can be disassembled and reassembled A, B, C
Parts that cannot be disassembled and reassembled d, e, f, g

First, for A, B, and C, as before, we bring over 16 each.

A 16 levels
B 16 levels
C 16 levels

Next, for parts $d, e, f,$ and g, it is not possible to find the total contribution ratios of the individual differences of these parts. For these parts we find measurable items of categorization and obtain the contribution ratios of these items. We will assume

Part d successively categorized by the two characteristics D and E
Part e successively categorized by the three characteristics F, G, and H
Part f categorized by the characteristic I
Part g there is no characteristic for categorization

Then, for parts $d, e, f,$ and g, we bring over 64 apiece and we have

Part d	D_1E_1	4
	D_1E_2	4
	·	·
	·	·
	·	·
	D_4E_4	4
Part e	$F_1G_1H_1$	1
	$F_1G_1H_2$	1
	$F_1G_1H_3$	1
	$F_1G_1H_4$	1
	$F_1G_2H_1$	1
	·	·
	·	·
	·	·
	$F_4G_4H_4$	1

Part f	I_1	16
	I_2	16
	I_3	16
	I_4	16

Part g has no categorization, and numbering is at random

Next, let us express the sequence of assembling by V.

$$V = 4 \text{ levels}$$

The method of assignment by orthogonal array L_{64} is left to the reader.

26 Design and Analysis for Clinical Experiments (Especially, Effectiveness Tests for Medicines)

In this chapter, criticism of the χ^2 test, which is often used for medicinal effect tests, will be aired, and accumulating analysis will be advocated to replace it. Also, the method explained in this chapter can often be applied as well to survey data, as at plants and in the marketplace. Please also refer to Chapter 27 concerning analysis of more general time series data.

26.1 Problem Points of χ^2 Test

The χ^2 test is often used in analyzing clinical data other than continuous values. The χ^2 method is a statistical testing method which was developed by the English statistician K. Pearson at the beginning of this century (Phil. Mag. *50,* 1900, p. 157). However, there was no research on statistical testing theories in those times, and since this is a method which was developed without consideration of *power of test* (research related to errors of the second kind: of inflicting great damage on society by judging medical drugs by their beneficial effects and regarding side-effects as being of no consequence), there is much question regarding the practicality of this method because of the following three points. The u test, too, has the defects of points (2) and (3).

(1) When data have been categorized into three or more classes, such as markedly effective, effective, and without effect, the power of test by the χ^2 method drops considerably (Section 26.4).

(2) When cross-over design has been used for data of curative treatment effect or they have been categorized by the attributes of the patients, the power of test by the χ^2 method becomes poor (Section 26.5, Section 26.6).

(3) If the χ^2 test is used in a case where there are observed time series data for a given patient, error of the first kind becomes large and the result appears too significant. (Section 26.7, Section 26.8).

26.2 The Case of 2 × 2 Contingency Table

With A_1 representing the control and A_2 representing the medicine, let us assume that the data of curative treatment effect for n_1 patients and n_2 patients, respectively, after a specific period of time were as given by Table 26.1. The + sign indicates cured and the − sign not cured.

We assume that the data of Table 26.1 were obtained by randomly selecting n_1 persons for the control, A_1, and n_2 persons for the medicine, A_2, from among an infinite population of patients, and that x_1 persons were cured in the former case and x_2 in the latter case. In actual practice, n_1 persons are randomly selected from among n persons who have come to the hospital and A_1 is given to these patients, while the remaining patients are treated with A_2. In this case, there develop the problems of dropout patients and correlation between x_1 and x_2, and it becomes more troublesome. Please refer to the Note at the end of the chapter.

If the cure rate of A_1 in an infinite population is expressed as p_1 and that of A_2 is expressed as p_2, the probability $L(x_1, x_2)$ that we will obtain the data in Table 26.1 is given by the following equation as the product of two binomial distributions.

$$L(x_1, x_2) = \binom{n_1}{x_1} p_1^{x_1}(1-p_1)^{n_1-x_1} \binom{n_2}{x_2} p_2^{x_2}(1-p_2)^{n_2-x_2} \tag{26.1}$$

Although p_1 and p_2 are unknown, the total number of combinations in which x_1, x_2 can occur for an arbitrary pair of values of p_1, p_2 is $(n_1+1)(n_2+1)$.

Table 26.1 2 × 2 Contingency Table

	−	+	Total
A_1	n_1-x_1	x_1	n_1
A_2	n_2-x_2	x_2	n_2
Total	$n-r_1$	r_1	n

The $(n_1 + 1)(n_2 + 1)$ combinations of x_1 and x_2 are split into two sets, *viz.*, the rejection (or critical) region R and the acceptance region Q, and if the value of (x_1, x_2) falls into the rejection region R, the null hypothesis

$H_0: p_1 = p_2$, or in other words there is no difference in the cure rate for A_1 and A_2

is denied, while if the value of (x_1, x_2) falls into the acceptance region Q, the null hypothesis is not denied, or in other words we question the effect of the treatment. Since to deny the hypothesis H_0 means to judge that there is a significant difference of cure rate between A_1 and A_2, naturally values of the class (x_1, x_2) such that the observed difference

$$S = \frac{x_1}{n_1} - \frac{x_2}{n_2} \tag{26.2}$$

is large constitute the rejection region.

The purpose of the statistical hypothesis testing theory is none other than to find a logical statistic that will separate the class of $(n_1 + 1)(n_2 + 1)$ values of (x_1, x_2), which is the sample space, into two parts, the rejection region R and the acceptance region Q. In the case of the 2 × 2 contingency table, the statistics which have been studied up to now include, for example,

(1) χ^2 statistic
(2) The direct probability calculation method
(3) Variance ratio
(4) Likelihood ratio
(5) Entropy

We will discuss only (1), (2), and (3) because of space limitations.

We will discuss the matter by the simple example of Table 26.1 in order to carry out actual numerical calculations. The rejection regions by the respective methods of (1), (2), and (3) in this case are determined as follows. It will be assumed that the level of significance is 5%.

(1) Rejection Region by the χ^2 Method

$$\chi^2 = \frac{n[(n_1 - x_1) \times x_2 - (n_2 - x_2) \times x_1]^2}{n_1 \times n_2 \times (n - x_1 - x_2) \times (x_1 + x_2)} \geq \chi^2(0.05) \tag{26.3}$$

(2) The Direct Probability Calculation Method

$$P = \sum \frac{n_1!n_2!(n-r_1)!r_1!}{n!} \times \frac{1}{(n_1-x_1)!x_1!(r_1-x_1)!(n-r_1-x_1)!} \tag{26.4}$$

The sum Σ in the equation above indicates that one draws up the sum of probabilities of all values which are more biased than the value x_2 which has been obtained for those who have been cured among the patients given A_2. If this sum is less than 0.025, we judge it to be significant.

(3) Rejection Region by Variance Ratio

According to this method, we consider binary representation, *viz.*, 1 when the patient has been cured, and 0 if not; and the same analysis of variance as in the case of continuous values is performed according to Note 3.1 of Chapter 3. First, the variation of effect due to difference of treatment method, S_A, is

$$S_A = \frac{x_1^2}{n_1} + \frac{x_2^2}{n_2} - \frac{(x_1+x_2)^2}{n} \qquad (f=1) \tag{26.5}$$

The sum of variations due to individual differences between the n_1 patients and n_2 patients for the respective treatment methods A_1 and A_2, S_e, is given by

$$S_e = n_1 \times \frac{x_1}{n_1}\left(1-\frac{x_1}{n_1}\right) + n_2 \times \frac{x_2}{n_2}\left(1-\frac{x_2}{n_2}\right) = (x_1+x_2) - \left(\frac{x_1^2}{n_1} + \frac{x_2^2}{n_2}\right) \qquad (f=n-2) \tag{26.6}$$

The analysis of variance table, Table 26.2, is constructed from this.

Although in testing for significance by variance ratio F_0, *to be accurate one should calculate the correct level of significance, in practice we use the F value of level of significance 5% for 1 degree of freedom of the numerator and n − 2 degrees of freedom of the denominator in the F table, so named in commemoration of Fisher,* $[F^1_{n-2}(0.05)]$. Thus, by using the value in the F table, the rejection region R is given by the set of x_1, x_2 which satisfies

$$F_0 = \frac{V_A}{V_e} \geq F^1_{n-2}(0.05) \tag{26.7}$$

If one uses the value in the F table, which has been derived for continuous values and, moreover, for the case of a normal distribution, in judging whether or not the variance ratio F_0 is significant, of course there arises the question of whether or not the level of significance of *error of the first kind* becomes false. But the level of significance of an error of the first kind is questionable not only in the χ^2 method but also in the direct probability calculation method, which is regarded as correct. Before going into *error of the second kind,* we will study what happens to an error of the first kind in the cases of the three statistics mentioned above.

Table 26.2 Analysis of Variance Table

Source	f	S	V	F_0
A (between treatment methods)	1	S_A	V_A	V_A/V_e
e (indiv. difference)	$n-2$	S_e	V_e	
T	$n-1$	S_T		

26.3 Error of the First Kind, Comparisons Regarding Significance Level α

Once the data of Table 26.1 have been obtained, we find the correct significance level for the three testing methods by using Equation (26.1).

We will show the results of numerical value calculations for the following two cases, without going into general considerations.

Case (a) $n_1 = 10, \quad n_2 = 10$
Case (b) $n_1 = 5, \quad n_2 = 15$

The rejection region, R, by the χ^2 method, the direct probability calculation method, and the variance ratio method is found in each case as the values of (x_1, x_2) which satisfies the following equation. Only the equation for case (a) will be given.

(1) *χ^2 Method.* Since the value of the 5% significance level of the χ^2 distribution of 1 degree of freedom is 3.841, we have

$$\chi_A{}^2 = \frac{20[(10-x_1)x_2-(10-x_2)x_1 \pm 10]^2}{10 \times 10 \times (20-x_1-x_2)(x_1+x_2)} \geq 3.841 \tag{26.8}$$

± 10 is a term that is necessary only when performing Yates' correction, and the sign is decided so that the square of the numerator becomes small.

(2) *Direct probability calculation method.* Here we have taken only the range of (x_1, x_2) such that probability P is at most 2.5%. This is because it is necessary to halve the level of significance for the direct probability calculation method in order to compare it with other methods since it is one-sided testing, as mentioned in reference 37.

$$P = \frac{10!10!(20 - x_1 - x_2)!(x_1 + x_2)!}{20!}\left[\frac{1}{x_1!x_2!(10 - x_1)!(10 - x_2)!}\right]$$

$$+ \text{[sum of similar values when more biased than this]}$$

$$\geq 0.025 \tag{26.9}$$

(3) *Variance ratio method.* Since the 5% value in the F table for 1 degree of freedom for the numerator and 18 degrees of freedom for the denominator is 4.41, we have

$$V_A = \frac{x_1{}^2}{10} + \frac{x_2{}^2}{10} - \frac{(x_1+x_2)^2}{20} = \frac{(x_1-x_2)^2}{20}$$

$$V_e = \frac{1}{18}\left[\left(x_1 - \frac{x_1{}^2}{10}\right) + \left(x_2 - \frac{x_2{}^2}{10}\right)\right]$$

$$F_0 = \frac{V_A}{V_e} \geq 4.41 \tag{26.10}$$

When the ranges of (x_1, x_2) that satisfy Equations (26.8), (26.9), and (26.10) are found by calculating, we obtain Tables 26.3.

Table 26.3 Classes of Values of x_1 and x_2 Which Become Significant at 5% Significance Level

Case (a) (The base of $n_1 = n_2 = 10$)

x_1 \ x_2	0	1	2	3	4	5	6	7	8	9	10
0					○ *	◎ * +	◎ * +	◎ * +	◎ * +	◎ * +	◎ * +
1							○ *	◎ * +	◎ * +	◎ * +	◎ * +
2								○ *	◎ * +	◎ * +	◎ * +
3									○ *	◎ * +	◎ * +
4	○ *									○ *	◎ * +
5	◎ *										◎ * +
6	◎ *	○ *									○ *
7	◎ *	◎ *	○ *								
8	◎ *	◎ *	◎ *	○ *							
9	◎ *	◎ *	◎ *	◎ *	○ *						
10	◎ *	◎ *	◎ *	◎ *	◎ *	◎ *	○ *				

Case (b) (The case of $n_1 = 5$, $n_2 = 15$)

x_2 \ x_1	0	1	2	3	4	5
0			○ *	◎ * +	◎ * +	◎ * +
1				○ *	◎ * +	◎ * +
2				○ *	◎ * +	◎ * +
3					○ *	◎ * +
4					○ *	◎ * +
5						◎ * +
6						○ *
7						○ *
8	○ *					
9	○ *					
10	◎ * +					
11	◎ * +	○ *				
12	◎ * +	○ *				
13	◎ * +	◎ * +	○ *			
14	◎ * +	◎ * +	○ *			
15	◎ * +	◎ * +	◎ * +	○ *		

◎ χ^2 method with Yates' correction
○ χ^2 method without Yates' correction
* Variance ratio method (when the F table is used)
\+ Direct probability calculation method

In this example, the rejection region coincides for the χ^2 method and the variance ratio method, and the rejection region also coincides for the χ^2 method with Yates' correction and the direct probability calculation method. Generally, the rejection region is wider by the χ^2 method than by the variance ratio method.

The sum of probabilities of the ranges where the statistics become significant as indicated by Tables 26.3 is calculated by using Equation (26.1). For case (a), we have

Table 26.4 Values of Probability for Various Combinations of (x_1, x_2) ($\times 10^{-6}$)
Case (a), $p_1 = p_2 = 0.5$

x_1 \ x_2	0	1	2	3	4	5	6	7	8	9	10
0	1	10	43	114	200	240	200	114	43	10	1
1	10	95	429	1 144	2 003	2 403	2 003	1 144	429	95	10
2	43	429	1 931	5 150	9 012	10 815	9 012	5 150	1 931	429	43
3	114	1 144	5 150	13 733	24 033	28 839	24 033	13 733	5 150	1 144	114
4	200	2 003	9 012	24 033	42 057	50 468	42 057	24 033	9 012	2 003	200
5	240	2 403	10 815	28 839	50 468	60 562	50 468	28 839	10 815	2 403	240
6	200	2 003	9 012	24 033	42 057	50 468	42 057	24 033	9 012	2 003	200
7	114	1 144	5 150	13 733	24 033	28 839	24 033	13 733	5 150	1 144	114
8	43	429	1 931	5 150	9 012	10 815	9 012	5 150	1 931	429	43
9	10	95	429	1 144	2 003	2 403	2 003	1 144	429	95	10
10	1	10	43	114	200	240	200	114	43	10	1

$$L(x_1, x_2) = \binom{10}{x_1} p_1^{x_1}(1-p_1)^{10-x_1} \binom{10}{x_2} p_2^{x_2}(1-p_2)^{10-x_2} \tag{26.11}$$

The hypothesis under consideration, H_0, is

$$H_0 : p_1 = p_2 = p \tag{26.12}$$

Table 26.4 is obtained when, with $p_1 = p_2 = 0.5$, the probabilities in Equation (26.11) are found for all combinations of x_1 and x_2. The sums of the probabilities for the respective combinations of the ◎ symbol, (○ + ◎) symbols, * symbol, and + symbol in Table 26.3, are the values for the largest value of $p = 0.5$ in Table 26.5.

The other values of Table 26.5 were obtained similarly when p was taken as 0.4, 0.3, 0.2, and 0.1, and the correct values of the error of the first kind were found for both case (a) and case (b).

According to Table 26.5, the results obtained by the direct probability calculation method are exactly the same levels of significance as the results of the χ^2 method with Yates' correction. That the results of the

Table 26.5 Values of Correct Level of Significance for 5% Rejection Region (%)
[Upper level is for case (a), lower level is for case (b)]

Type of statistic	Value of $p_1 = p_2 = p$				
	0.1 (0.9)	0.2 (0.8)	0.3 (0.7)	0.4 (0.6)	0.5
χ^2 (without Yates' correction)	0.90 2.21	3.04 2.29	3.71 4.38	3.72 5.02	4.22 5.21
χ^2 (with Yates' correction)	0.11 0.20	0.76 0.46	1.19 0.60	1.26 0.63	1.28 0.60
Direct probability calculation method	0.11 0.20	0.76 0.46	1.19 0.60	1.26 0.63	1.28 0.60
Variance ratio F_0	0.90 2.21	3.04 2.29	3.71 4.38	3.72 5.02	4.22 5.21

direct probability calculation method, which should be absolutely accurate values, differ greatly from the 2.5% value indicates that the direct probability calculation method cannot be used for comparisons of treatment methods for an infinite population. The accurate level of significance obtains when 20 patients are randomly divided into 10 persons each and, moreover, when it is taken into account that some patients who are cured by A_1 are also cured by A_2. But when the sets of patients who are cured by A_1 and by A_2 are not considered to overlap, a finite population correction is necessary and the structure of the probability changes. Please refer to the Note at the end of the chapter.

In this example, the χ^2 method without Yates' correction and the cumulative method (or accumulation method) coincided perfectly, but the level of significance obtained by the χ^2 method has a value which is too large in the general case. *Since the χ^2 method is the same as the F test with the number of degrees of freedom of the numerator as 1 and with the number of degrees of freedom of the error variance as* ∞, if the number of degrees of freedom of the error is small, the level of significance becomes too high compared with that in the variance ratio method. The variance ratio method cannot be used when $n_1 + n_2 < 6$, but otherwise the level of significance becomes somewhat large in only a few cases ($n_1 = 5$, $n_2 = 15$; $p = 0.5$, $p = 0.4$). But this effect is only of the order of the level of significance 5% becoming 5.21% or 5.02% or so; it does not cause substantial harm, and thus, among the four statistics mentioned above, the variance ratio can probably be regarded as being the best, together with χ^2.

The variance ratio method cannot be used when $n_1 + n_2$ is 5 or less. But at $n_1 + n_2 \leq 5$, no matter what the data, it will not become statistically significant. Henceforth, therefore, it will be a premise that when categorizing into the two classes 0 and 1, the total number of patients is at least six. Even fewer persons will suffice in the case of continuous values or categorization into more than two classes.

Table 26.6 shows the levels of significance of an error of the first kind for the case of $p_1 = p_2 = 0.5$ with a greater number of values of n_1 and n_2. When $p_1 = p_2 \neq 0.5$, it is even more on the safe side. It is surprising that even by the use of the F table, which has been derived from a normal distribution, the level of significance by the variance ratio test for data that differ completely from a normal distribution is close to 5%. As mentioned in Chapter 19 and Chapter 33, the author does not recognize the importance of testing itself. This is because *even if a medicine with no effect is judged as having effect, there is no social loss other than that of cost*. Contrasted with this, *an error of the second kind, such as that of overlooking a medicine with therapeutic effect as having no significant effect, or of using a medicine with side-effects as though that made no significant difference, is more important for human life*. If it is one's intent that errors of the first kind should be regarded as not

Table 26.6 Values of Correct Level of Significance α (When $p_1 = p_2 = 0.5$)

n_1	n_2	5% Test			1% Test		
		Variance ratio	χ^2	χ^2 Cor-rec'n	Variance ratio	χ^2	χ^2 Cor-rec'n
5	5	6.05	6.05	0.19	2.15	2.15	0.00
5	10	4.80	4.80	1.47	1.47	0.74	0.10
5	15	5.21	5.21	1.06	1.09	1.09	0.12
5	20	5.56	5.56	1.01	1.01	0.86	0.04
5	25	5.81	5.81	1.39	0.78	0.78	0.05
5	30	4.25	5.98	1.21	0.64	0.65	0.05
10	10	4.22	4.22	1.28	1.28	1.28	0.26
10	15	4.80	6.14	1.84	1.06	1.02	0.22
10	20	5.20	5.20	2.01	1.34	1.26	0.24
10	25	4.92	4.92	2.03	1.21	1.08	0.25
10	30	5.42	5.42	2.11	1.21	0.75	0.26
15	15	4.56	4.56	1.65	1.61	1.27	0.29
15	20	4.79	4.79	2.14	0.91	0.91	0.37
15	25	5.50	5.50	2.42	0.99	0.94	0.36
15	30	4.58	4.58	2.58	0.97	0.97	0.38
20	20	4.25	4.25	2.04	0.95	0.95	0.29
20	25	4.90	4.90	2.55	1.01	1.01	0.34
20	30	4.88	5.63	2.46	1.05	0.93	0.44
25	25	6.49	6.49	3.28	1.32	0.98	0.35
25	30	5.41	5.41	2.84	1.11	0.94	0.45
30	30	5.19	5.19	2.74	1.35	1.35	0.62

mattering and one wishes to minimize errors of the second kind, instead of the F table currently in use it is necessary to construct a table of F values that are smaller the less the number of degrees of freedom. This means that we must regard testing as only a reference point, and *a reference point which is not important,* at that. Rather, what one should regard as important is evaluation of the magnitude of the factorial effects by analysis of variance.

26.4 When There Are Three or More Classes of Categorization

The situation is as presented in the preceding section regarding an error of the first kind (the mistake of erroneously discarding the null hypothesis when it is correct), and the variance ratio method is by no means especially superior in that case. It is where error of the second kind is small that we may regard the variance method as outstanding. When one considers the loss incurred by erroneously deciding that it cannot be said that there is curative effect when the null hypothesis H_0 is not correct, that an error of the second kind is large means that *by erroneously discarding a treatment method or a medicine which is more useful than the control as being without effect, or by delaying the time of sending it forth into society, one shortens the lives of patients who might have been saved by that method, or one increases the social loss of prolonging pain and agony.* Or, it means that *one commits the error of sending a harmful medicine into society believing that there are no significant consequences when in fact there are side-effects.*

As to which is more important between the social loss generated by performing an ineffectual treatment method or using an ineffective medicine by an error of the first kind, on the one hand, or the social loss which derives from an error of the second kind, on the other hand, the relative importance differs in individual cases, but the present state of statistical testing methods, where only errors of the first kind are regarded as important, is a matter of concern.

However, the variance ratio method is by no means a method for minimizing an error of the second kind by ignoring an error of the first kind. It *offers a method by which an error of the second kind is far smaller than by the χ^2 method, etc., after one has fixed the error of the first kind (level of significance α) at approximately the correct value.* Specifically, the loss from overlooking treatment methods and medicines that are actually effective, or harmful treatment method and medicines, becomes far smaller than with the χ^2 method, etc.

That the power of test by the χ^2 method is poor can be readily understood from the mistaken conclusion drawn by the χ^2 test regarding the data of Tables 26.7, even without resorting to mathematical calculations.

Since the number of patients is equal for the control and for the medicine, χ^2 can be found by the following equations.

$$\chi_A{}^2=\frac{(40-24)^2}{64}+\frac{(24-40)^2}{64}+\frac{(10-10)^2}{20}+\frac{(6-6)^2}{12}=8.00^* \quad (f=3) \qquad (26.13)$$

$$\chi_B{}^2=\frac{(40-24)^2}{64}+\frac{(24-29)^2}{53}+\frac{(10-16)^2}{26}+\frac{(6-11)^2}{17}=7.33 \quad (f=3) \qquad (26.14)$$

The 5% point in the χ^2 table where the number of degrees of freedom f is 3 is 7.815. As is apparent from Tables 26.7, in spite of the fact that the effect of B is clearly greater than that of A, the effect of medicine A is

Table 26.7 Data of Curative Effects of Two Types of Medicines, *A* and *B*

The case of medicine A

	−	+	++	+++	Total
A_1 (Control)	40	24	10	6	80
A_2 (A used)	24	40	10	6	80
Total	64	64	20	12	160

The case of medicine B

	−	+	++	+++	Total
B_1 (Control)	40	24	10	6	80
B_2 (B used)	24	29	16	11	80
Total	64	53	26	17	160

− (no effect), + (somewhat effective), ++ (effective), +++ (markedly effective)

Table 26.8 Data Converted Into Cumulative Frequency

The case of A

	I	II	III	IV
A_1	40	64	74	80
A_2	24	64	74	80
Total	64	128	148	160

The case of B

	I	II	III	IV
B_1	40	64	74	80
B_2	24	53	69	80
Total	64	117	143	160

significant and the effect of medicine B is not significant. This indicates that it is inappropriate to use the χ^2 method when there is a sequence to the treatment data, such as ⧻, #, +, and −.

We analyze the data of Tables 26.7 as follows by the accumulating analysis. First, we construct the cumulative frequencies of Tables 26.8. Here,

CLASS I = (number of −)
CLASS II = (number of −) + (number of +)
CLASS III = (number of −) + (number of +) + (number of #)
CLASS IV = (number of −) + (number of +) + (number of #) + (number of ⧻)

(1) *Test of effect of medicine A*. In the accumulating analysis, after having constructed the cumulative frequencies, we eliminate the last class and perform a comprehensive analysis of variance on the remaining classes (in this case, on classes I, II, and III). For this, separately for class I, class II, and class III, we form the variation of the variable with value 1 if it is in the class and 0 if it is not, as explained in Chapter 3, then we summarize by multiplying these by weights W_1, W_2, and W_3. First, when we direct our attention to class I, those among the 80 persons of A_1 and the 80 persons of A_2 who appear in class I (those whose symptoms have not changed or have worsened) number 40 and 24, respectively. When we consider, for the individual patients, the data given the value 1 if the patient is in class I and 0 if he is not, in the case of A_1, 1 is the value for 40 persons and 0 the value for 40 persons, and in the case of A_2, 1 is the value for 24 persons and 0 is the value for 56 persons. Therefore S_A, the variation of the difference between A_1 and A_2, which constitutes the effect of medicine A, is given by

$$S_A(\mathrm{I}) = \frac{(\text{total of } A_1)^2}{\text{number of patients given } A_1} + \frac{(\text{total of } A_2)^2}{\text{number of patients given } A_2} - \frac{(\text{total of } A_1 + \text{total of } A_2)^2}{\text{total number of patients}}$$

$$= \frac{40^2}{80} + \frac{24^2}{80} - \frac{64^2}{160} = 1.600 \qquad (f = 1) \qquad (26.15)$$

The variation of individual difference as to whether a patient belongs in class I or does not, S_e, is given by

$$S_e(\mathrm{I}) = \begin{pmatrix}\text{variation of individual}\\ \text{differences of patients}\\ \text{given } A_1\end{pmatrix} + \begin{pmatrix}\text{variation of individual}\\ \text{differences of patients}\\ \text{given } A_2\end{pmatrix}$$

$$= \left(1^2 \times 40 + 0^2 \times 40 - \frac{40^2}{80}\right) + \left(1^2 \times 24 + 0^2 \times 56 - \frac{24^2}{80}\right)$$

$$= \left(40 - \frac{40^2}{80}\right) + \left(24 - \frac{24^2}{80}\right)$$

$$= 36.800 \qquad (f = 79 + 79 = 158) \tag{26.16}$$

This value can also be found by subtracting the variation between medicines, $S_A(\mathrm{I})$, from the total variation among all of the 160 patients, $S_T(\mathrm{I})$.

$$S_T(\mathrm{I}) = (\text{sum of squares of 0's and 1's which are the individual data of the 160 patients}) - (\text{correction factor})$$

$$= 0^2 \times 96 + 1^2 \times 64 - \frac{64^2}{160} = 38.400 \qquad (f = 157) \tag{26.17}$$

One may also find $S_e(\mathrm{I})$ by subtracting $S_A(\mathrm{I})$ from this.

$$S_e(\mathrm{I}) = S_T(\mathrm{I}) - S_A(\mathrm{I}) = 38.400 - 1.600 = 36.800 \qquad (f = 156) \tag{26.18}$$

The variations of class II and class III are found similarly.

$$S_A(\mathrm{II}) = \frac{(64-64)^2}{160} = 0 \qquad (f = 1) \tag{26.19}$$

$$S_T(\mathrm{II}) = 128 - \frac{128^2}{160} = 25.600 \qquad (f = 159) \tag{26.20}$$

$$S_e(\mathrm{II}) = S_T(\mathrm{II}) - S_A(\mathrm{II}) = 25.600 \qquad (f = 158) \tag{26.21}$$

$$S_A(\mathrm{III}) = \frac{(74-74)^2}{160} = 0 \qquad (f = 1) \tag{26.22}$$

$$S_T(\mathrm{III}) = 148 - \frac{148^2}{160} = 11.100 \qquad (f = 159) \tag{26.23}$$

$$S_e(\mathrm{III}) = S_T(\mathrm{III}) - S_A(\mathrm{III}) = 11.100 \qquad (f = 158) \tag{26.24}$$

In accumulating analysis, we draw up the summarized variation by multiplying the variations of class I, class II, and class III by the following weights W_1, W_2, and W_3. These weights relate to the effects of the respective classes I, II, and III, and take into consideration that the scales measuring their effects differ. Although it may not be particularly difficult to improve a cure rate of 50% to 60%, it is difficult to improve a cure rate of 90% to 99% or 100%. As was indicated in Note 3.2 and Note 3.5, this means that if the cure rate is expressed as p, we are assuming that the difficulty of changing p by just dp is proportional to $p(1-p)$. We assume that the factorial effects of class I, class II, and class III are proportional to the quantities $p_1(1-p_1)$, $p_2(1-p_2)$, and $p_3(1-p_3)$, and we define the weights W_1, W_2, and W_3 by their reciprocals. Please refer to Note 3.2 of Chapter 3.

$$W_1=\frac{160^2}{64\times(160-64)}=4.17 \tag{26.25}$$

$$W_2=\frac{160^2}{128\times(160-128)}=6.25 \tag{26.26}$$

$$W_3=\frac{160^2}{148\times(160-148)}=14.41 \tag{26.27}$$

We find the effect of the medicine using the summarized classes I, II, and III, S_A, and the individual difference variation, S_e. The variation is the total after having multiplied by the weights, and the number of degrees of freedom is the simple sum.

$$\begin{aligned} S_A &= S_A(\text{I})\times W_1+S_A(\text{II})\times W_2+S_A(\text{III})\times W_3 \\ &=1.600\times 4.17+0.000\times 6.25+0.000\times 14.41 \\ &=6.67 \qquad (f=1\times 3=3) \end{aligned} \tag{26.28}$$

$$\begin{aligned} S_e &= S_e(\text{I})\times W_1+S_e(\text{II})\times W_2+S_e(\text{III})\times W_3 \\ &=36.800\times 4.17+25.600\times 6.25+11.100\times 14.41 \\ &=473.33 \qquad (f=158\times 3=474) \end{aligned} \tag{26.29}$$

Similarly, for the effect of medicine B we find S_B, S_T, and S_e by using the following three weights.

$$W_1=\frac{160^2}{64\times(160-64)}=4.17 \tag{26.30}$$

$$W_2=\frac{160^2}{117\times(160-117)}=5.09 \tag{26.31}$$

$$W_3=\frac{160^2}{143\times(160-143)}=10.53 \tag{26.32}$$

$$\begin{aligned} S_B &= \frac{(40-24)^2\times 4.17+(64-53)^2\times 5.09+(74-69)^2\times 10.53}{160} \\ &=12.17 \qquad (f=1\times 3=3) \end{aligned} \tag{26.33}$$

$$\begin{aligned} S_T &= \left(64-\frac{64^2}{160}\right)\times\frac{160^2}{64(160-64)}+\left(117-\frac{117^2}{160}\right) \\ &\quad\times\frac{160^2}{117(160-117)}+\left(143-\frac{143^2}{160}\right)\times\frac{160^2}{143(160-143)} \\ &=160+160+160=480 \qquad (f=159\times 3=477) \end{aligned} \tag{26.34}$$

As can be seen from Equation (26.34), in the case of accumulating analysis the total variation S_T can be found simply as (total number of data) ×

Table 26.9 Analysis of Variance Tables

The case of medicine A

Source	f	S	V	F
A	3	6.67	2.22	2.22*
e	474	473.33	0.999	
T	477	480.00		

The case of medicine B

Source	f	S	V	F
B	3	12.17	4.06	4.11**
e	474	467.83	0.987	
T	477	480.00		

(number of classes being analyzed). From this, the individual difference variation, S_e, is

$$S_e = S_T - S_B = 467.83 \tag{26.35}$$

Therefore, the analysis of variance tables shown in Tables 26.9 are obtained.

When tested by variance ratio F, the effect of medicine B has a ratio which is about twice as large as that of medicine A.

As explained above, if the χ^2 test is performed when there are data on the effects of a treatment method, one gets a large error in the evaluation of the magnitude of its effect. When there is no effect, the χ^2 test gives the correct level of significance about as well as the variance ratio method, but it yields a totally erroneous result regarding the degree of significance when there is an effect. This is one of the reasons why the χ^2 test cannot be used.

26.5 When There Are Other Sources

When there are other sources besides treatment, if one uses the variance ratio method one finds the error variation having eliminated these other factorial effects and tests by drawing up the variance ratio. Since there is no method of eliminating other sources, if they exist, by the χ^2 method, other factorial effects intrude into the error and the result is that the power of test of the treatment effect worsens.

For example, let us assume that in the case of a certain disease, there is 90% cure if the patient is young but an elderly person is hardly ever cured.

Table 26.10 Data of Clinical Experiment Categorized by Age Groups B_1 and B_2

Conditions	−	+	Total
A_1B_1	1	9	10
A_1B_2	5	0	5
A_2B_1	0	10	10
A_2B_2	2	3	5
Total	8	22	30

Table 26.11 Data with Age Groups Pooled

Condition	−	+	Total
A_1	6	9	15
A_2	2	13	15
Total	8	22	30

Let us assume that the data of Table 26.10 were obtained when a clinical experiment was performed with placebo A_1 and medicine A_2 on 15 patients each, differentiating the age groups as the young age group B_1 and the old age group B_2.

If the data are drawn up without distinction of age but taking all patients together, they are as presented by Table 26.11.

When the χ^2 test is performed on the basis of Table 26.11, it is not significant.

$$\chi_A{}^2=\frac{30(6\times 13-2\times 9)^2}{15\times 15\times 8\times 22}=2.73 \tag{26.36}$$

However, if the data are only for old age, it becomes significant.

$$\chi_A{}^2=\frac{10(5\times 3-2\times 0)^2}{5\times 5\times 7\times 3}=4.29^{*} \tag{26.37}$$

It is a problem, from the standpoint of common sense, if it is significant when only a part of the data are used but not significant when all are used. Actually, this has happened by the χ^2 test because in regard to the cure rate, the influence of age is greater than that of treatment. The data categorized by age are as given by Table 26.12.

If the effect of B is tested using Table 26.12, we have

$$\chi_B{}^2=\frac{30(3-7\times 19)^2}{20\times 10\times 8\times 22}=14.40^{**} \tag{26.38}$$

Generally, when there are large sources other than treatment, the precision of the χ^2 test becomes completely bad regarding the treatment effect.

By the variance ratio method, we find the total variation S_T of the data of all of the 30 patients counting those who have been cured as 1 and those who have not been cured as 0.

$$S_T=0^2\times 8+1^2\times 22-\frac{(0\times 8+1\times 22)^2}{30}=22-\frac{22^2}{30}=5.87 \qquad (f=29) \tag{26.39}$$

We resolve into the variation for the treatment effect, S_A, the variation for the difference of cure rate by age group, S_B, the variation for the interaction indicating whether or not the effects by treatments A_1 and A_2 differ depending on the age group, $S_{A\times B}$, and the sum of individual difference variations within the four types of combinations of the same age group and same treatment, S_e.

Table 26.12 Data Categorized by Age

Condition	−	+	Total
B_1	1	19	20
B_2	7	3	10
Total	8	22	30

$$S_A = \frac{(\text{total of } A_1)^2 + (\text{total of } A_2)^2}{15} - CF = \frac{9^2+13^2}{15} - \frac{22^2}{30}$$

$$= 0.53 \qquad (f=1) \tag{26.40}$$

$$S_B = \frac{19^2}{20} + \frac{3^2}{10} - \frac{22^2}{30} = 2.82 \qquad (f=1) \tag{26.41}$$

$$S_{A\times B} = \text{variation between combinations of } AB - S_A - S_B$$

$$= \frac{9^2+10^2}{10} + \frac{0^2+3^2}{5} - \frac{22^2}{30} - 0.53 - 2.82 = 0.42 \qquad (f=1) \tag{26.42}$$

$$S_e = S_T - (S_A + S_B + S_{A\times B}) = 2.10 \qquad (f = 29-1-1-1=26) \tag{26.43}$$

The analysis of variance table, Table 26.13, is obtained from these. S' is the net variation and ρ is the contribution ratio of each source.

When categorizing by age or by sex or by symptom, often the categorization is carried out after obtaining the data. In such a case of post-sorting, there are differences in the number of patients depending on the level combinations of the factors, such as A_1B_1, A_1B_2, A_2B_1, and A_2B_2. When using analysis of variance, as long as there is at least one patient, whichever the combination, it is possible to resolve the total variation; and by eliminating factorial effects other than treatment, analysis of variance of high precision regarding treatment effect becomes possible.

26.6 Clinical Data Where the Number of Replications Is Nonuniform

With A_1 as the control and A_2 as the medicine being tested by the double blind method on 105 patients and 90 patients, respectively, with a stomach ailment (some of the patients dropped out), post-sorting was conducted by age B and symptom C as factors which were recorded in their charts.

B: age $\qquad B_1$ = at most 39 years old

$\qquad\qquad B_2$ = at least 40 years old

Table 26.13 Analysis of Variance Table

Source	f	S	V	F_0	S'	ρ[%]
A (Treatment)	1	0.53	0.53	6.54*	0.449	7.6
B (Age group)	1	2.82	2.82	34.81**	2.739	46.7
$A \times B$ (Interaction)	1	0.42	0.42	5.06*	0.339	5.8
e (Individual difference)	26	2.10	0.081		2.343	39.9
T	29	5.87			5.870	100.0

C: degree of symptom C_1 = weak

C_2 = medium

C_3 = strong

The data are: − (ineffective), + (somewhat effective), ⧺ (effective), and ⧻ (markedly effective).

Various methods have been considered for treating dropout patients, but this topic will be omitted here.

In accumulating analysis, after having constructed tables of cumulative frequencies I, II, III, and IV, we analyze as follows after revising the data into percent falling into the respective classes I, II, III, and IV. In a case such as this one, where there are at least two factors (variables) and the number of persons tested differs for the various combinations, we find, after having drawn up the cumulative frequencies, the proportions of persons who are in the respective classes I, II, III, and IV. The results are as given by Table 26.15.

For the 12 combinations of A, B, and C, the mean proportion in class I is

$$p_1 = \frac{2.526}{12.000} \tag{26.44}$$

The weight of class I, W_1, is

$$W_1 = \frac{1}{p_1(1-p_1)} = \frac{12.000^2}{2.526(12.000-2.526)} = 6.02 \tag{26.45}$$

Similarly, we have

$$W_2 = \frac{12^2}{5.307(12-5.307)} = 4.05 \tag{26.46}$$

Table 26.14 Curative Effect for Stomach Disease

Conditions	−	+	⧺	⧻	Total	Cumulative frequency I	II	III	IV
A_1 B_1 C_1	4	2	9	3	18	4	6	15	18
A_1 B_1 C_2	5	4	4	2	15	5	9	13	15
A_1 B_1 C_3	4	9	3	1	17	4	13	16	17
A_1 B_2 C_1	3	6	6	6	21	3	9	15	21
A_1 B_2 C_2	2	6	5	7	20	2	8	13	20
A_1 B_2 C_3	5	3	3	3	14	5	8	11	14
A_2 B_1 C_1	1	6	4	5	16	1	7	11	16
A_2 B_1 C_2	3	0	3	4	10	3	3	6	10
A_2 B_1 C_3	7	6	5	2	20	7	13	18	20
A_2 B_2 C_1	4	3	6	4	17	4	7	13	17
A_2 B_2 C_2	0	1	2	8	11	0	1	3	11
A_2 B_2 C_3	3	2	9	2	16	3	5	14	16
Total	41	48	59	47	195	41	89	148	195

Table 26.15 Data Rendered Into Percent

A	B	C	I	II	III	IV
1	1	1	0.222	0.333	0.833	1.000
1	1	2	0.333	0.600	0.867	1.000
1	1	3	0.235	0.765	0.941	1.000
1	2	1	0.143	0.429	0.714	1.000
1	2	2	0.100	0.400	0.650	1.000
1	2	3	0.357	0.571	0.786	1.000
2	1	1	0.063	0.438	0.688	1.000
2	1	2	0.300	0.300	0.600	1.000
2	1	3	0.350	0.650	0.900	1.000
2	2	1	0.235	0.418	0.765	1.000
2	2	2	0.000	0.091	0.273	1.000
2	2	3	0.188	0.312	0.875	1.000
Total			2.526	5.307	8.892	12.000

$$W_3=\frac{12^2}{8.892(12-8.892)}=5.21 \tag{26.47}$$

For the correction factor S_m' we get, after summarizing the correction factors of classes I, II, and III by W_1, W_2, and W_3,

$$S_m'=\frac{2.526^2}{12}\times W_1+\frac{5.307^2}{12}\times W_2+\frac{8.892^2}{12}\times W_3$$

$$=47.03495 \qquad (f=3) \tag{26.48}$$

Since S_m' in the equation above constitutes the variation of the value obtained by dividing by the number of patients for each set of conditions (this is termed the number of replications), it is too small by this number of replications, and therefore it is necessary to multiply it by the number of replications. When the number of replications is different for the 12 combinations of conditions, we designate their harmonic mean (reciprocal of mean of reciprocals) as $\bar{r}$, and we treat the data as though the number of replications were $\bar{r}$ everywhere. We solve

$$\frac{1}{\bar{r}}=\frac{1}{12}\left(\frac{1}{18}+\frac{1}{15}+\cdots+\frac{1}{16}\right) \tag{26.49}$$

and obtain

$$\bar{r}=15.4873 \tag{26.50}$$

From this, the correction term which has been multiplied by the harmonic mean, S_m', is

$$S_m=\bar{r}\times S_m'=15.4873\times 47.03495=728.4444 \qquad (f=3) \tag{26.51}$$

Similarly, we have

$$S_A=\bar{r}\left[\frac{(1.390^2+1.136^2)\times W_1+(3.098^2+2.209^2)\times W_2+(4.791^2+4.101^2)\times W_3}{6}\right]$$
$$-S_m=7.8335 \qquad (f=3) \tag{26.52}$$

$$S_B=15.4873\left[\frac{(1.503^2+1.023^2)\times 6.02+(3.086^2+2.221^2)\times 4.05+(4.829^2+4.063^2)\times 5.21}{6}\right]$$
$$-S_m=9.6464 \qquad (f=3) \tag{26.53}$$

$$S_{A\times B}=15.4873\left[\frac{(0.790^2+\cdots+0.423^2)\times 6.02+\cdots+(2.641^2+\cdots+1.913^2)\times 5.21}{3}\right]$$
$$-S_m-S_A-S_B=0.7696 \qquad (f=3) \tag{26.54}$$

$$S_C=15.4873\left[\frac{(0.633^2+\cdots+1.130^2)\times 6.02+\cdots+(3.000^2+\cdots+3.502^2)\times 5.21}{4}\right]$$
$$-S_m=22.4545 \qquad (f=6) \tag{26.55}$$

$$S_{A\times C}$$
$$=15.4873\left[\frac{(0.365^2++\cdots+0.538^2)\times 6.02+\cdots+(1.547^2+\cdots+17.75^2)\times 5.21}{2}\right]$$
$$-S_m-S_A-S_C=9.4900 \qquad (f=6) \tag{26.56}$$

$$S_{B\times C}=15.4873\left[\frac{(0.285^2+\cdots+0.545^2)\times 6.02+\cdots+(1.521^2+\cdots+1.661^2)\times 5.21}{2}\right]$$
$$-S_m-S_B-S_C=11.0247 \qquad (f=6) \tag{26.57}$$

The three-factor interaction, $A\times B\times C$, is found by subtracting the main-effect two-factor interaction effects given above from the variation among the 12 combinations, S_{T_1}.

$$S_{A\times B\times C}=15.4873[(0.222^2+\cdots+0.188^2)\times 6.02+\cdots+(0.833^2+\cdots+0.875^2)$$
$$\times 5.21]-S_m-S_A-S_B-S_C-S_{A\times B}-S_{A\times C}-S_{B\times C}$$
$$=4.5777 \qquad (f=6) \tag{26.58}$$

Next, we find the individual difference variation S_e. The individual difference variation of class I is

$$S_e(\mathrm{I})=\left(4-\frac{4^2}{18}\right)+\cdots+\left(3-\frac{3^2}{16}\right) \tag{26.59}$$

Similarly, we find S_e(II) and S_e(III), and by summarizing with the weights, we find S_e.

$$S_e=\left[\left(4-\frac{4^2}{18}\right)+\cdots+\left(3-\frac{3^2}{16}\right)\right]\times W_1+\cdots+\left[\left(15-\frac{15^2}{18}\right)+\cdots+\left(14-\frac{14^2}{16}\right)\right]\times W_3$$
$$=610.2090 \qquad (f=183\times 3=549) \tag{26.60}$$

The analysis of variance table, Table 26.16, is thereby obtained.

According to Table 26.16, inter-medicine, inter-age-group, and degree of subjective symptoms are all significant for the cure rate. But since the

Table 26.16 Analysis of Variance Table

Source	f	S	V	F_0
A (Control and medicine)	3	7.8335	2.6112.	2.35*
B (Age group)	3	9.6464	3.2155	2.89*
C (Subjective symptom difference)	6	22.4545	3.7424	3.37**
$A\times B$	3	0.7696	0.2565	0.23
$A\times C$	6	9.4900	1.5817	1.42
$B\times C$	6	11.0247	1.8374	1.65
$A\times B\times C$	6	4.5777	0.7629	0.69
e (Individual difference)	549	610.2090	1.1115	
T	582	676.0054		

interactions with the medicine are not significant in all cases, we understand that the medicine has similar effects both for young persons and for middle-aged and elderly persons; and that it has similar effects for the group whose subjective symptoms are strong and the group whose subjective symptoms are weak. If we thus categorize by other factors and determine, for example, that there is no interaction with those factors, or that it is small, when investigating the effects and side-effects of a medicine, the reliability of the investigation is improved.

Based on the results of Table 26.16, sources which are not significant are pooled into the individual difference, and we construct an adjusted analysis of variance table, Table 26.17.

Here, S' is the net variation; it is found by eliminating from the variations of the sources which have become significant the individual difference variations included therein. The net variations are found as follows.

$$S_A'=S_A-3\times V_e=7.8335-3\times 1.1159=4.4858 \tag{26.61}$$

$$S_B'=9.6464-3\times 1.1159=6.2987 \tag{26.62}$$

$$S_C'=22.4545-6\times 1.1159=15.7591 \tag{26.63}$$

$$S_e'=S_e+12\times V_e=649.4618 \tag{26.64}$$

The contribution ratio is the value obtained by dividing the net variation by the total variation S_T and multiplying by 100. If it is 100% − for the control and 100% + for the medicine, ρ_A becomes 100%. That it is − in

Table 26.17 Adjusted Analysis of Variance Table

Source	f	S	V	F_0	S'	ρ[%]
A (Intermedicine)	3	7.8335	2.6112	2.34*	4.4858	0.66
B (Interage group)	3	9.6464	3.2155	2.88*	6.2987	0.93
C (Inter-subject. symptom)	6	22.4545	3.7424	3.35**	15.7591	2.33
e (Individual difference)	570	636.0710	1.1159		649.4618	96.08
T	582	676.0054			676.0054	100.00

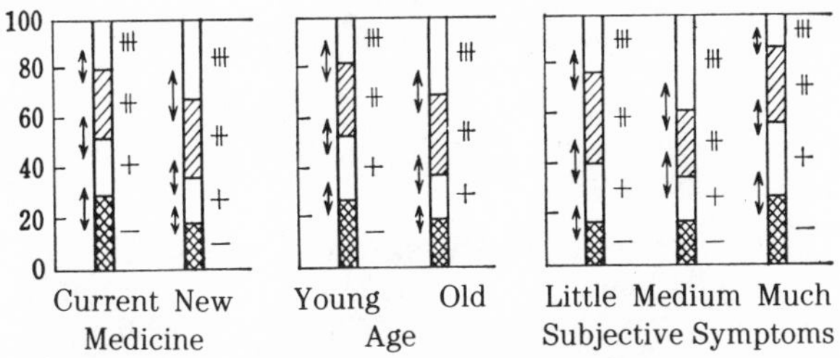

FIGURE 26.1 Graphs of Factorial Effects

certain cases and +, ++, and +++ in certain other cases even though the same medicine was given to the same age groups and the same subjective symptom groups means that even if it is the same group with respect to age and the subjective symptoms, other individual differences still have a contribution ratio of 96.08%. Besides individual difference, differences in the degree of subjective symptoms have the greatest influence on the cure rate; and it can be seen from the contribution ratios that this is followed by the age group, then by difference between medicines.

By χ^2 test,

$$\chi_A^2=\frac{\left(25-47\times\frac{90}{195}\right)^2}{47\times\frac{90}{195}}+\cdots+\frac{\left(23-41\times\frac{105}{195}\right)^2}{41\times\frac{105}{195}}=2.70 \qquad (f=3) \qquad (26.65)$$

and the effect of medicine is not significant.

We estimate as shown by Table 26.18 for the sources which are significant in Table 26.17. If necessary, graphs such as shown by Figures 26.1 are also drawn.

The confidence limits of the estimated values are given by the following formula. Confidence limits can be obtained only for data of cumulative frequency.

Table 26.18 Estimations of Significant Sources (%)

	Cumulative frequency				Density frequency				
	I	II	III	IV	−	+	++	+++	Total
A_1	23.2	51.6	80.0	100.0	23.2	28.4	28.4	20.0	100.0
A_2	18.9	36.8	68.4	100.0	18.9	17.9	31.6	31.6	100.0
B_1	25.0	51.4	80.5	100.0	25.0	26.4	29.1	19.5	100.0
B_2	17.0	36.8	67.7	100.0	17.0	19.8	30.9	32.3	100.0
C_1	16.6	40.4	75.0	100.0	16.6	23.8	34.6	25.0	100.0
C_2	18.3	34.8	59.8	100.0	18.3	16.5	25.0	40.2	100.0
C_3	28.2	57.4	87.8	100.0	28.2	29.2	30.4	12.2	100.0
Mean	21.0	44.2	74.1	100.0	21.0	23.2	29.9	25.9	100.0

$$\hat{\mu} \pm \sqrt{F \times V_e \times \hat{\mu}(1-\hat{\mu}) \times \frac{1}{n_e}} \qquad (26.66)$$

where

$\hat{\mu}$ = estimated value

F = 5% value in the F table where the number of degrees of freedom of the numerator is 1 (this is always so) and the number of degrees of freedom of the denominator is the number of degrees of freedom of the error variance in the analysis of variance table, V_e

V_e = value of error variance in analysis of variance table

$\hat{\mu}(1-\hat{\mu})$ = In the case of fixed marginal enumerative values, the error variance is found by multiplying by the weight W. Therefore, V_e in the analysis of variance table is the mean of the constants which are proportional to the binomial error variance $p(1-p)$. Therefore, to obtain the error variance in the neighborhood of $\hat{\mu}$, it is necessary to multiply V_e by binomial error variance $\hat{\mu}(1-\hat{\mu})$.

n_e = effective number of replications; this refers to the number of individual items as to the mean of how many the estimated value is. The mean of A_i is regarded as $6 \times \bar{r}$ in this case.

For example, let us try finding the confidence limits of the estimated value of 23.2% of class I of A_1 in Table 26.18.

$\hat{\mu} = 23.2\% = 0.232$

F = 5% value in F table for 1 degree of freedom of numerator and 570 degrees of freedom of denominator = 3.86

$$V_e = 1.1159$$

$$n_e = 6 \times \bar{r} = 6 \times 15.873 \fallingdotseq 92.9$$

$$\mu = 0.232 \pm \sqrt{3.86 \times 1.1159 \times 0.232 \times 0.768 \times \frac{1}{92.9}}$$

$$= 0.232 \pm 0.091 = 23.2 \pm 9.1 \,[\%] \qquad (26.67)$$

Furthermore, in practice, we would wish to predict the state of cure after the treatment for arbitrary combinations of B and C. For example, in the case of a person who is B_2, thus being at least 40 years old, who has many subjective symptoms, the cure rate differs between the control, A_1, and the tested medicine, A_2, as follows. Detailed explanation regarding the prediction equation given here will be omitted, but it has been obtained by assuming that the effect is proportional to the binomial error variance $p(1-p)$. In foreign countries, too, this is being used for estimation of the cure rate, as logit transformation. From Appendix Table 5, we have, in decibel values,

$$\hat{\mu}_{\text{I}} = \bar{A}_1 + \bar{B}_2 + \bar{C}_3 - 2\bar{T} = (-5.20) + (-6.88) + (-4.06) - 2(-5.75)$$

$$= -4.64 \quad [\text{db}] \qquad (26.68)$$

Therefore, $\hat{\mu}_I$ for -4.64 db is

$$\hat{\mu}_I = 25.6 \,[\%] \qquad (26.69)$$

Similarly, we have

$$\hat{\mu}_{II} = 57.1 \,[\%] \qquad (26.70)$$

$$\hat{\mu}_{III} = 88.1 \,[\%] \qquad (26.71)$$

And when A_2 is used, we have

$$\hat{\mu}_I = 21.0 \,[\%] \qquad (26.72)$$

$$\hat{\mu}_{II} = 42.1 \,[\%] \qquad (26.73)$$

$$\hat{\mu}_{III} = 80.0 \,[\%] \qquad (26.74)$$

The predicted values in Table 26.19 are obtained from these.

Actually, it is best to estimate for every combination of B and C, not merely B_2C_3. The confidence limits in this case are obtained as follows, for example for class I of A_2. The numerator 5 of the last term is the sum of the number of degrees of freedom of m, A, B, and C.

$$\hat{\mu}_I = 0.210 \pm \sqrt{3.86 \times 1.1159 \times 0.210 \times 0.790 \times \frac{5}{12 \times 15.4873}}$$

$$= 0.210 \pm 0.139 = 21.0 \pm 13.9 [\%] \qquad (26.75)$$

26.7 The Case of Time Series Data (1)

We wish to compare two types of medicine for allergy, A_1 = control (product currently in use) and A_2 = new medicine. The data of Table 26.20 were obtained when 12 allergy patients were randomly divided into two classes, and A_1 was given to one-half, $R_1, R_2, \ldots, R_6$ and A_2 was given to the remaining six persons, and when the allergy symptoms were observed for ten days, $K_1, K_2, \ldots, K_{10}$.

With class I as the class of $-$, class II as the sum of the class of $-$ and the class of $+$, and class III as the sum of the whole of the class of $-$, the class of $+$, and the class of ++, we show time data as in Table 26.21. Since class III was 1 in every case, this has been omitted.

$$W_1 = \frac{120^2}{33 \times 87} = 5.02 \qquad W_2 = \frac{120^2}{81 \times 39} = 4.56 \qquad (26.76)$$

$$S_m = \frac{33^2 \times 5.02 + 81^2 \times 4.56}{120} = 294.88 \qquad (f = 2) \qquad (26.77)$$

Table 26.19 Predicted Values of Cure Rate

	I	II	III	IV	−	+	++	+++	Total
A_1	25.6	57.1	88.1	100.0	25.6	31.5	31.0	11.9	100.0
A_2	21.0	42.1	80.0	100.0	21.0	21.1	37.9	20.0	100.0

Table 26.20 Data of Allergy Symptoms (Hypothetical Example)
(− severe, + mild, ╫ barely any)

		K_1	K_2	K_3	K_4	K_5	K_6	K_7	K_8	K_9	K_{10}
A_1	R_1	−	+	−	−	+	−	−	−	−	−
	R_2	+	╫	╫	╫	+	+	+	╫	╫	+
	R_3	+	+	+	−	+	+	+	+	+	−
	R_4	−	+	−	−	╫	−	−	+	−	−
	R_5	−	╫	╫	−	−	+	+	+	−	−
	R_6	−	+	╫	+	+	+	╫	╫	╫	╫
A_2	R_7	−	+	−	−	+	+	−	+	+	+
	R_8	╫	+	╫	╫	╫	╫	╫	+	╫	+
	R_9	╫	╫	╫	╫	╫	+	+	╫	╫	╫
	R_{10}	−	╫	+	+	+	+	+	╫	╫	+
	R_{11}	−	−	+	+	+	−	+	+	−	−
	R_{12}	+	╫	╫	╫	╫	╫	╫	+	+	╫

$$S_A=\frac{(23^2+10^2)\times 5.02+(47^2+34^2)\times 4.56}{60}-294.88=13.49 \quad (f=2) \tag{26.78}$$

The variation among the 12 patients, $S_{R(A)}$, is

$$S_{R(A)}=\frac{(8^2+0^2+\cdots+0^2)\times 5.02+(10^2+5^2+\cdots+3^2)\times 4.56}{10}-294.88$$
$$=100.32 \qquad (f=22) \tag{26.79}$$

The individual difference variation after having eliminated treatment effect from $S_{R(A)}$, S_{e_1}, is

$$S_{e_1}=S_{R(A)}-S_A=86.83 \qquad (f=20) \tag{26.80}$$

$$S_K=\frac{1}{12}[(7^2+1^2+\cdots+5^2)\times 5.02+(10^2+7^2+\cdots+9^2)\times 4.56]-294.88$$
$$=20.00 \qquad (f=18) \tag{26.81}$$

$$S_{A\times K}=\frac{1}{6}[(4^2+0^2+\cdots+1^2)\times 5.02+(6^2+4^2+\cdots+4^3)\times 4.56]-294.88-S_A-S_K$$
$$=8.14 \qquad (f=18) \tag{26.82}$$

$$S_T=(1^2+0^2+0^2+\cdots+1^2+0^2)\times W_1+(1^2+1^2+1^2+\cdots+1^2+0^2)\times W_2-294.88$$
$$=240.00 \qquad (f=238) \tag{26.83}$$

$$S_{e_2}=S_T-(S_A+S_{e_1}+S_K+S_{A\times K})=111.99 \qquad (f=18) \tag{26.84}$$

The analysis of variance table, Table 26.22, is obtained from these.

K and $A\times K$ are tested by e_2. Variation between days, K, is significant. This indicates that there are days on which allergy symptoms tend to manifest themselves and days on which they tend not to occur. Although it is unclear whether this is a result of weather conditions or, for example,

Table 26.21 Data Converted Into Cumulative Frequency

	K_1		K_2		K_3		K_4		K_5		K_6		K_7		K_8		K_9		K_{10}		Total	
	I	II	I	II	I	II	I	II	I	II	I	II	I	II	I	II	I	II	I	II	I	II
A_1 R_1	1	1	0	1	1	1	1	1	0	1	1	1	1	1	1	1	1	1	1	1	8	10
R_2	0	1	0	0	0	0	0	0	0	1	0	1	0	1	0	0	0	0	0	1	0	5
R_3	0	1	0	1	0	1	1	1	0	1	0	1	0	1	0	1	0	1	1	1	2	10
R_4	1	1	0	1	1	1	1	1	0	0	1	1	1	1	0	1	1	1	1	1	7	9
R_5	1	1	0	0	0	0	1	1	1	1	0	1	0	1	0	1	1	1	1	1	5	8
R_6	1	1	0	1	0	0	0	1	0	1	0	1	0	0	0	0	0	0	0	0	1	5
Sub-total	4	6	0	4	2	3	4	5	1	5	2	6	2	5	1	4	3	4	4	5	23	47
A_2 R_7	1	1	0	1	1	1	1	1	0	1	0	1	1	1	0	1	0	1	0	1	4	10
R_8	0	0	0	1	0	0	0	0	0	0	0	0	0	0	0	1	0	0	0	1	0	3
R_9	0	0	0	0	0	0	0	0	0	0	0	0	0	1	0	0	0	0	0	0	0	1
R_{10}	1	1	0	0	0	1	0	1	0	1	0	1	0	1	0	0	0	0	0	1	1	7
R_{11}	1	1	1	1	0	1	0	1	0	1	1	1	0	1	0	1	1	1	1	1	5	10
R_{12}	0	1	0	0	0	0	0	0	0	0	0	0	0	0	0	1	0	1	0	0	0	3
Sub-total	3	4	1	3	1	3	1	3	0	3	1	3	1	4	0	4	1	3	1	4	10	34
Comb. total	7	10	1	7	3	6	5	8	1	8	3	9	3	9	1	8	4	7	5	9	33	81

of the food which is served to everyone at the hospital, the interday variation K indicates that there are differences in the way the symptoms appear depending on the day.

The effect of the medicines must be tested by e_1, which constitutes the difference between patients, since the patients differ between A_1 and A_2. Of course, if e_1, which is the difference among patients, is not significant when tested by e_2, either there is no individual difference among the patients or it is small, and therefore it is permissible to test the difference between the medicines with the error obtained by pooling e_1 with e_2. The e_1 in this case is significant when tested by e_2.

$$F_0 = \frac{V_{e_1}}{V_{e_2}} = \frac{4.34}{0.622} = 6.98^{**} \tag{26.85}$$

It is therefore not possible to pool e_1 and e_2 together. We test the effect of A by the difference among patients, e_1. The variance ratio is

Table 26.22 Analysis of Variance Table

Source	f	S	V	F_0
A (Intermedicine)	2	13.49	6.74	1.55
e_1 (Interpatient)	20	86.83	4.34	6.98**
K (Interday)	18	20.00	1.11	1.78*
$A \times K$	18	8.14	0.45	—
e_2 (Error)	180	111.99	0.622	
T	238	240.00		

Table 26.23 Density Frequency for χ^2 Test

	−	+	++	Total
A_1	23	24	13	60
A_2	10	24	26	60
Total	33	48	39	120

$$F_0=\frac{6.74}{4.34}=1.55 \tag{26.86}$$

and is not significant. A significant difference between the two types of medicine, A_1 and A_2, cannot be seen for improvement of the allergy symptoms.

Let us try the χ^2 test. There is no problem if we are testing only on a specific day, but the precision drops markedly. If tested by all of the data, the level of significance becomes 1%, as shown by Table 26.23, and it becomes significant.

$$\chi_A{}^2=\frac{(23-10)^2}{33}+\frac{(24-24)^2}{48}+\frac{(13-26)^2}{39}=9.45^{**} \qquad (f=2) \tag{26.87}$$

This is because individual differences among patients have been disregarded. Since the χ^2 test is unable to distinguish between first-order error variance e_1 and second-order error variance e_2, it cannot be used for comprehensive analysis when there is correlation, as in the case of time series data.

26.8 The Case of Time Series Data (2)

Patients with a liver disease were randomly assigned to two types of treatment methods (A_1, A_2); their serum bilirubin quantity was measured after the 1st–4th weeks (K_1–K_4), and the amounts of decrease in this quantity were compared. However, since the value before treatment differed rather considerably from patient to patient, the percentage of drop was found for each patient by comparing the measured value each week with the value before treatment, and the amount of drop was categorized into four classes (Table 26.24).

When cumulative frequencies I (sum of 1), II (sum of 1 and 2), III (sum of 1 and 2 and 3), and IV (sum of the whole) were obtained, the supplementary tables, Tables 26.25, were obtained.

$$W_1=\frac{176^2}{58\times(176-58)}=4.526 \qquad W_2=\frac{176^2}{122\times(176-122)}=4.702$$

Table 26.24 Data of Total Serum Bilirubin
(1: marked drop; 2: drop; 3: some drop; 4: unchanged)

	K_1	K_2	K_3	K_4		K_1	K_2	K_3	K_4
$A1$ $R1$	3	1	1	1	$A2$ $R26$	4	2	1	1
$R2$	4	2	1	1	$R27$	3	2	1	1
$R3$	4	4	4	4	$R28$	1	1	1	1
$R4$	4	3	3	2	$R29$	4	2	2	1
$R5$	3	2	1	1	$R30$	4	1	1	2
$R6$	3	2	1	1	$R31$	4	2	2	1
$R7$	2	1	1	1	$R32$	2	1	1	1
$R8$	4	2	1	2	$R33$	3	2	2	1
$R9$	4	2	2	2	$R34$	3	2	2	1
$R10$	2	1	1	1	$R35$	3	2	1	1
$R11$	2	2	1	1	$R36$	2	1	1	1
$R12$	4	2	1	1	$R37$	2	1	1	1
$R13$	2	1	1	1	$R38$	3	2	2	1
$R14$	3	2	2	2	$R39$	3	2	2	2
$R15$	4	3	3	2	$R40$	2	2	2	2
$R16$	4	3	2	3	$R41$	4	2	3	2
$R17$	4	4	4	2	$R42$	3	1	1	1
$R18$	3	1	1	2	$R43$	2	2	1	1
$R19$	4	3	2	2	$R44$	4	3	2	1
$R20$	3	2	2	1					
$R21$	2	2	2	2					
$R22$	4	3	3	3					
$R23$	4	3	3	3					
$R24$	3	2	2	2					
$R25$	3	3	2	2					

$$W_3=\frac{176^2}{153\times(176-153)}=8.802$$

$$S_m=\frac{58^2\times W_1+122^2\times W_2+153^2\times W_3}{176}=1654.9$$

$$S_A=\left(\frac{26^2}{100}+\frac{32^2}{76}\right)\times 4.526+\left(\frac{61^2}{100}+\frac{61^2}{76}\right)\times 4.702+\left(\frac{83^2}{100}+\frac{70^2}{76}\right)\times 8.802-S_m$$

$$=15.8 \qquad (f=3) \tag{26.88}$$

$$S_{R(A)}=\frac{1}{4}[(3^2+2^2+\cdots+1^2)\times 4.526+(3^2+3^2+\cdots+2^2)\times 4.702$$

$$+(4^2+3^2+\cdots+3^2)\times 8.802]-S_m=201.2 \qquad (f=129) \tag{26.89}$$

$$S_{e_1}=S_{R(A)}-S_A=185.4 \qquad (f=126) \tag{26.90}$$

$$S_K=\frac{1}{44}[(1^2+11^2+\cdots+25^2)\times 4.526+(11^2+34^2+\cdots+40^2)\times 4.702$$

$$+(26^2+42^2+\cdots+43^2)\times 8.802]-S_m=131.9 \qquad (f=9) \tag{26.91}$$

Tables 26.25 Supplementary Tables

	I	II	III	IV
A_1 K_1	0	5	13	25
A_1 K_2	5	16	23	25
A_1 K_3	11	19	23	25
A_1 K_4	10	21	24	25
Sub-total	26	61	83	100
A_2 K_1	1	6	13	19
A_2 K_2	6	18	19	19
A_2 K_3	10	18	19	19
A_2 K_4	15	19	19	19
Sub-total	32	61	70	76
K_1	1	11	26	44
K_2	11	34	42	44
K_3	21	37	42	44
K_4	25	40	43	44
Total	58.	122	153	176

	I	II	III	IV
R_1	3	3	4	4
R_2	2	3	3	4
R_3	0	0	0	4
⋮		⋮		
R_{25}	0	2	4	4
R_{26}	2	3	3	4
⋮		⋮		
R_{44}	1	2	3	4
Total	58	122	153	176

$$S_{A\times K}=\left(\frac{0^2+5^2+11^2+10^2}{25}+\frac{1^2+6^2+10^2+15^2}{19}\right)\times 4.526$$

$$+\left(\frac{5^2+16^2+19^2+21^2}{25}+\frac{6^2+18^2+18^2+19^2}{19}\right)\times 4.702$$

$$+\left(\frac{13^2+23^2+23^2+24^2}{25}+\frac{13^2+19^2+19^2+19^2}{19}\right)\times 8.802-S_m-S_A-S_K$$

$$=5.3 \qquad (f=9) \tag{26.92}$$

$$S_{e_2}=S_T-(S_A+S_{e_1}+S_K+S_{A\times K})=189.6 \qquad (f=378) \tag{26.93}$$

Therefore, the analysis of variance table becomes as given by Table 26.26.

When the effects of the significant factors A and K are estimated, Table 26.27 is obtained.

Estimations for combinations of A_1 and A_2 versus K and how to find their confidence limits, etc., will be omitted.

Table 26.26 Analysis of Variance Table

Source	f	S	V	F_0
A	3	15.8	5.27	3.59*
e_1	126	185.4	1.47	2.93**
K	9	131.9	14.66	29.19**
$A\times K$	9	5.3	0.59	1.18
e_2	378	189.6	0.502	
T	525	528.0		

Table 26.27 Estimation of Significant Sources

	1. Dropped markedly	2. Dropped	3. Dropped somewhat	4. Unchanged
A_1 Control	26.00	35.00	22.00	17.00
A_2 Tested drug	42.11	38.15	11.85	7.89
K_1 Week 1	2.27	22.73	34.09	40.91
K_2 Week 2	25.00	52.27	18.19	4.54
K_3 Week 3	47.73	36.36	11.37	4.54
K_4 Week 4	56.82	34.09	6.82	2.27

Exercise (26)

(1) When the therapeutic performance for 16 patients ($R_1, R_2, \ldots,$ R_{16}) and 12 patients ($R_{17}, R_{18}, \ldots, R_{28}$) with oral cavity cancer when treated by the two respective levels $A_1 =$ only X-rays and $A_2 =$ Bleomycin in combination with X-rays was studied, the results were as follows. Here, $B_1 =$ one week later, $B_2 =$ two weeks later, and $B_3 =$ three weeks later, and the data were obtained by judging with respect to five classes: ineffective (0), somewhat effective (1), effective (2), markedly effective (3), and disappeared (4). Perform the appropriate analysis of variance and draw conclusions.

A_1

R	B_1	B_2	B_3
1	0	1	2
2	1	1	2
3	2	3	4
4	1	3	3
5	0	2	2
6	1	1	2
7	1	2	4
8	0	2	3
9	0	1	3
10	0	2	3
11	1	3	3
12	0	1	2
13	2	3	4
14	1	3	4
15	2	2	4
16	1	3	3

A_2

R	B_1	B_2	B_3
17	0	1	2
18	1	1	2
19	0	1	1
20	0	2	3
21	1	1	2
22	0	1	1
23	1	1	2
24	0	0	1
25	0	1	3
26	0	0	1
27	2	2	3
28	2	2	3

Chapter 26 Note

We consider the test data for two types of medicine categorized into the two classes Does not cure and Cures, A_1 and A_2. We assume that among n

patients the number who are not cured by either of the medicines is r_1, the number who are cured by A_1 but not by A_2 is r_2, the number who are cured by A_2 but not by A_1 is r_3, and the number who are cured by either is r_4. The number of patients cured among n_1 persons randomly selected from among n patients and given medicine A_1 is represented by x_1 and the number of patients who have been cured among the remaining n_2 persons upon having been given medicine A_2 is represented by x_2. Since n persons is the population, the distribution of x_1, x_2 is given by the following equation.

$$P(x_1, x_2)=\sum \frac{n_1!n_2!}{n!} \times \frac{r_1!}{z_1!(r_1-z_1)!} \times \frac{r_2!}{z_2!(r_2-z_2)!} \times \frac{r_3!}{z_3!(r_3-z_3)!} \times \frac{r_4!}{z_4!(r_4-z_4)!} \qquad (*\ 26.1)$$

Σ, here, is the sum over every combination of z_1, z_2, z_3, and z_4 which satisfies the following conditions.

$$\left.\begin{aligned} &z_1+z_2+z_3+z_4=n \\ &z_2+z_4=x_1 \\ &z_3+z_4=x_2 \end{aligned}\right\} \qquad (*\ 26.2)$$

Also, if the rejection region of a certain statistic is R, the value of the true level of significance α for this statistic is given by the following equation under the zero hypothesis $r_2 = r_3$.

$$\alpha=\sum P(x_1, x_2) \qquad (*\ 26.3)$$

Σ here is the sum of $(x_1, x_2)\, \varepsilon R$.

According to the variance ratio method, when $n_1 = n_2$, generally the level of significance of Equation (* 26.3) becomes maximum when $r_2 = r_3 = 0$. If when $n_1 = n_2 = 10$, $r_1 = 10$, $r_2 = 0$, $r_3 = 0$, and $r_4 = 10$, the rejection region R is where the values of (x_1, x_2) are, from Table 26.3, (0, 10), (1, 9), (2, 8), (10, 0), (9, 1), and (8, 2), and the level of significance is 2.3%.

$$\alpha=\frac{10!10!}{20!}\left[\frac{10!}{0!10!}\times1\times1\times\frac{10!}{10!0!}+\frac{10!}{1!9!}\times1\times1\times\frac{10!}{1!9!}+\frac{10!}{2!8!}\times1\times1\times\frac{10!}{8!2!}\right]\times2=0.023 \qquad (*\ 26.4)$$

27 Analysis Method for Time Series Data

Often, survey data are time series data. In this chapter, the design and analysis method for drug effectiveness tests and for surveys done when performing market experiments will be explained. Persons concerned with medical drugs should read Chapter 26, Section 27.1, and Section 27.2, and persons concerned with marketing should read Section 27.3 and Section 27.4. Persons concerned with technical matters are also requested to refer to Chapter 32.

27.1 Data Analysis of Medical Drug Effectiveness Test

27.1.1 Data of Medical Drug Test

The design of experiments for drug tests and related medical matters, and the data analysis method for these, have become subjects of contention. It is a characteristic of such data that, as with market data, continuous values and qualitative data are found for a certain number of patients or animals in a certain time interval. These are a type of time series data.

For example, the information shown in Table 27.1 tells us how a

Table 27.1 Data of Materia Medica Test

	B_1' (Before administration)				B_2' (During administration and after administration)							
	B_1	B_2	B_3	Sub-total	B_4	B_5	B_6	B_7	B_8	B_9	Sub-total	Total
A_1	9	6	7	22	11	12	5	10	11	3	52	74
A_1' A_2	12	12	11	35	9	10	6	6	6	5	42	77
A_3	7	4	4	15	8	10	12	8	9	9	56	71
A_4	14	11	11	36	15	15	16	19	13	13	91	127
Subtotal	42	33	33	108	43	47	39	43	39	30	241	349
A_5	6	9	3	18	2	41	34	29	27	7	140	158
A_2' A_6	19	18	18	55	27	32	43	44	26	18	190	245
A_7	7	6	6	19	21	26	41	41	38	29	196	215
A_8	18	11	13	42	27	41	29	38	17	9	161	203
Subtotal	50	44	40	134	77	140	147	152	108	63	687	821
Total	92	77	73	242	120	187	186	195	147	93	928	1170

certain pharmaceutical product influences a certain component of the blood, in comparison with a placebo.

A_1' placebo (fake medicine, a type of control)
A_2' medicine being tested

For both the placebo A_1' and the tested pharmaceutical product A_2', an experiment was performed on four patients each. Patients tested by placebo A_1' will be denoted by A_1, A_2, A_3, and A_4 and patients tested by the pharmaceutical product A_2' will be denoted by A_5, A_6, A_7, and A_8. With the three days before administering the drug as B_1, B_2, and B_3, and the six days during and after administration as B_4, B_5, B_6, B_7, B_8, and B_9, measurement was performed once a day at a specific time on each patient. The total number of data therefore was $8 \times 9 = 72$.

An important point when analyzing such data by a two-way layout is that there are individual differences in the quantity of the component in the blood from person to person; for example, in the case of two persons among those tested by the placebo, A_2 and A_4, we note the tendency that the quantity of this component was greater than in A_1 and A_3. And with regard to two persons, A_6 and A_8, among the four persons given the test medicine, too, we see the tendency that the content of this component was greater than in A_5 or A_7.

When individual difference is large, there is a method one can use in which one finds the mean value for each of the persons before dispensing the medicine (for both the placebo and the test medicine), and then divides the individual values measured after the drug is administered by this mean value and finds the percentage of change. For example, for patient A_1, the data of days $B_4, B_5, \ldots, B_9$ become as follows.

$$11 \times \frac{3}{22} \times 100 = 150 \ [\%]$$

$$12 \times \frac{3}{22} \times 100 = 164 \ [\%]$$

$$5 \times \frac{3}{22} \times 100 = 68 \ [\%]$$

$$10 \times \frac{3}{22} \times 100 = 136 \ [\%]$$

$$11 \times \frac{3}{22} \times 100 = 150 \ [\%]$$

$$3 \times \frac{3}{22} \times 100 = 41 \ [\%]$$

Sometimes the data on the other patients, too, are first converted into percentages of change relative to the mean value, and one calculates with them in the form of a two-way layout of the eight levels of A and the six levels B_4, B_5, B_6, B_7, B_8, and B_9. This is often a good method in the sense that it simplifies the problem, making it more easily understood. Especially for an after-effect, as of polio, when one wishes to examine the effect of drugs or rehabilitation which might improve it, it is a good method to express the degree of improvement or change by use of such a drug or method compared with the condition before administration or treatment, for each person and moreover for each motor function, for example as

(1) Worsens
(2) No difference
(3) Improves somewhat
(4) Improves fairly considerably
(5) Improves markedly

thereby converting the data into fixed marginal enumerative values. However, for the method of analyzing data of comparison with the situation before treatment, such as these, refer to Section 26.8. Here, we will show a method of analysis using the original data of Table 27.1.

Next, as to factor B (day or time), assuming that blood is tested once a day at a specific time, whether or not the patients show a common effect depending on the day before administration of the medicine becomes something of a problem. If all of the eight patients have been taking similar meals at the same hospital and the content of the meals influences the blood component in question, it means that there are fairly similar tendencies in all of the patients for B_1, B_2, and B_3.

Even when the eight patients are commuting to the hospital and each person is taking meals as he or she likes, if the characteristic values are influenced by meteorological conditions it is not possible to disregard differences among B_1, B_2, and B_3 which are similarly in common. When it is clear that the differences of B_1, B_2, and B_3 do not in any way have a common influence on the eight patients $A_1, A_2, \ldots, A_8$, the differences among B_1, B_2, and B_3 are treated as error from the beginning. But even in a case where there is no difference among B_1, B_2, and B_3, it is actually easier to understand in terms of the calculations if one finds separately the variation of this difference and pools it with the error variation upon confirming that there is no difference; therefore, whether there is no difference among B_1, B_2, and B_3 or it is unclear whether there is or not, it is best to find the difference among B_1, B_2, and B_3 separately and to compare it with the error variance, and then pool it with the error variation when it is not significant. In the calculations to follow, therefore, as a basic rule the differences among B_1, B_2, and B_3 will be calculated.

Next, regarding the data during or after administration, it can be assumed that both the placebo group (persons given the placebo) and the treated group (persons given the test medicine) show an effect common to the data of the six days $B_4, B_5, \ldots, B_9$. This is because if the medicine or placebo has been administered only once, during the morning of B_4, information as to how many days or hours its effect endures will appear in the curve of $B_4, B_5, \ldots, B_9$, while if it is administered continuously every day, the information as to what becomes of the cumulative effect of administration and approximately when it reaches a steady state will appear in the change of the mean value between $B_4, B_5, \ldots, B_9$.

In the case of the example of Table 27.1, continuous administration was performed for the three days B_4, B_5, and B_6, and even after discontinuing administration, measurement was conducted for three days, B_7, B_8, and B_9. Therefore, if the enduring effect of this pharmaceutical product is about 1-2 days, matters will revert to the original state on days B_8 and B_9. For this, we need only examine how the curve of the mean values for $B_4, B_5, \ldots, B_9$ changes.

Therefore, analysis of variance is carried out as in the next section.

27.1.2 Calculation of Variation

In the case of such data in a two-way layout, the variation of patients A_1–A_8 is resolved into three parts: the main effect of factor A' for the difference between the placebo and the medicine being tested, the individual difference among the four persons of A_1', A_1–A_4, and the individual difference among the four persons of A_2', A_5–A_8. This is a resolution of main effects of A_1–A_8, at eight levels.

$$S_A = S_{A'} + S_{(A_1-A_4)} + S_{(A_5-A_8)} \tag{27.1}$$

$$\text{Number of degrees of freedom } 7 = 1 + 3 + 3 \tag{27.2}$$

Next, we resolve the time factor B, also, into the main effect of factor B' at two levels which have been categorized first into before administration of medicine, B_1', and during and after administration, B_2', and then into the effects of B_1–B_3 within B_1' and B_4–B_9 within B_2'.

By such resolution, this two-way layout becomes split into the numbers of degrees of freedom given in Table 27.2.

Since there are data for each combination of patient and day/time, fundamentally such data of a time series are data of a two-way layout. Since various types of treatment are performed on different groups of individuals, in effect difference of treatment (in this case, the difference between the placebo and the tested product) includes individual difference. If one wishes to eliminate such individual difference, one should find the difference between before treatment (B_1') and after treatment (B_2') for each individual and examine whether or not this difference differs for each type of treatment. That is, one finds the difference between B_1' and B_2' for each individual and determines how $B_1' - B_2'$ differs for $A_1, A_2, \ldots, A_8$.

The sources to be considered are

(1) $B' \times (A_1$–$A_4)$: how the difference of B' varies among A_1–A_4
(2) $B' \times A'$: how the difference of B' varies between A_1' and A_2'
(3) $B' \times (A_5$–$A_8)$: how the difference of B' varies among A_5–A_8
(4) $A' \times (B_4$–$B_9)$: how the mean curve of B_4–B_9 varies between A_1' and A_2'

But since $A' \times (B_1$–$B_3)$, $(A_1$–$A_4) \times (B_1$–$B_3)$, and $(A_5$–$A_8) \times (B_1$–$B_3)$ constitute changes with time of the individual difference before administration of the medicine, these are all errors.

Table 27.2 Two-Way Layout Method with Resolution

Source		Number of degrees of freedom
A	A'	1
	A_1–A_4	3
	A_5–A_8	3
B	B'	1
	B_1–B_3	2
	B_4–B_9	5
$A \times B$	$A' \times B'$	1
	$A' \times (B_1$–$B_3)$	$2 \rightarrow e$ (error variation)
	$A' \times (B_4$–$B_9)$	5
	$(A_1$–$A_4) \times B'$	3
	$(A_1$–$A_4) \times (B_1$–$B_3)$	$6 \rightarrow e$ (error variation)
	$(A_1$–$A_4) \times (B_4$–$B_9)$	$15 \rightarrow e'$ (dispersion of placebo group)
	$(A_5$–$A_8) \times B'$	3
	$(A_5$–$A_8) \times (B_1$–$B_3)$	$6 \rightarrow e$ (error variation)
	$(A_5$–$A_8) \times (B_4$–$B_9)$	$15 \rightarrow e''$ (dispersion of treated group)
T		71

Also, the interaction $(A_1-A_4)\times(B_4-B_9)$ is the change in individual difference after the placebo has been given; and whether it should be regarded as an error or as an interaction depends on the problem. Since in the case of the product being tested the interaction $(A_5-A_8)\times(B_4-B_9)$ constitutes the individual difference of the curve of change of the characteristic value during and after administration of the medicine, it is best to examine its magnitude separately from the error, regarding it as the dispersion of the individual difference of reaction after administration.

When A_1' is not a placebo but is, for example, a product currently in use, it is best to find the interaction $(A_1-A_4)\times(B_4-B_9)$, too, as in the case of $(A_5-A_8)\times(B_4-B_9)$, as the dispersion of individual difference after administration of the medicine. This is because the reactions of individuals to the medicine vary widely and it supplies information by which to examine the magnitude of the deviation from the mean effect curve among the whole group of individuals. Therefore, in the present instance, only $A'\times(B_1-B_3)$, $(A_1-A_4)\times(B_1-B_3)$, and $(A_5-A_8)\times(B_1-B_3)$ become pure errors.

Let us carry out the calculations of analysis of variance according to Table 27.2.

$$CF=\frac{1170^2}{72}=19012 \qquad (f=1) \tag{27.3}$$

$$S_T=9^2+6^2+7^2+\cdots+9^2-CF=9551 \qquad (f=71) \tag{27.4}$$

$$S_{A'}=\frac{(349-821)^2}{72}=3094 \qquad (f=1) \tag{27.5}$$

$$S_{(A_1-A_4)}=\frac{74^2+77^2+71^2+127^2}{9}-\frac{349^2}{36}=236 \qquad (f=3) \tag{27.6}$$

$$S_{(A_5-A_8)}=\frac{158^2+245^2+215^2+203^2}{9}-\frac{821^2}{36}=435 \qquad (f=3) \tag{27.7}$$

$$S_{B'}=\frac{(2B_1'-B_2')^2}{2^2\times 24+(-1)^2\times 48}=\frac{(2\times 242-928)^2}{144}=1369 \qquad (f=1) \tag{27.8}$$

$$S_{(B_1-B_3)}=\frac{92^2+77^2+73^2}{8}-\frac{242^2}{24}=25 \qquad (f=2) \tag{27.9}$$

$$S_{(B_4-B_9)}=\frac{120^2+187^2+\cdots+93^2}{8}-\frac{928^2}{48}=1090 \qquad (f=5) \tag{27.10}$$

$$S_{A'\times B'}=\frac{[(687-2\times 134)-(241-2\times 108)]^2}{24+(-2)^2\times 12+24+(-2)^2\times 12}=1078 \qquad (f=1) \tag{27.11}$$

Interaction $A'\times B'$ is one of the most important sources. In explaining this, the difference between the mean after administration for the placebo patients and the mean before administration will be expressed as $L(A_1')$. We have

$$L(A_1')=\frac{241}{24}-\frac{108}{12} \tag{27.12}$$

Similarly, the difference between the mean after administration for the group given the test medicine and the mean before administration is $L(A_2')$.

$$L(A_2')=\frac{687}{24}-\frac{134}{12} \tag{27.13}$$

The difference between $L(A_2')$ and $L(A_1')$ is the *interaction* $A' \times B'$.

$$L(A'\times B')=L(A_2')-L(A_1')=\left(\frac{687}{24}-\frac{134}{12}\right)-\left(\frac{241}{24}-\frac{108}{12}\right) \tag{27.14}$$

Variation $S_{A'\times B'}$ is found by dividing the square of the value of Equation (27.14) by the unit number. Since 687 is the total of 24 observed values, the number $\frac{687}{24}$ means that the coefficient, $\frac{1}{24}$, has been multiplied by the 24 individual observed values. Therefore, the sum of squares of these coefficients is

$$\left(\frac{1}{24}\right)^2\times 24=\frac{1}{24}$$

Similarly, $-\frac{134}{12}$ means that the coefficient $-\frac{1}{12}$ has been multiplied by the sum of the 12 observed values. Since it is the same for the others also, the unit number of Equation (27.14) is

$$\left(\frac{1}{24}\right)^2\times 24+\left(\frac{-1}{12}\right)^2\times 12+\left(\frac{-1}{24}\right)^2\times 24+\left(\frac{1}{12}\right)^2\times 12=\frac{6}{24} \tag{27.15}$$

Therefore, by dividing the square of Equation (27.14) by its unit number, $\frac{6}{24}$, we obtain

$$S_{A'\times B'}=\frac{\left[\left(\frac{687}{24}-\frac{134}{12}\right)-\left(\frac{241}{24}-\frac{108}{12}\right)\right]^2}{\frac{6}{24}}$$

Multiplying the denominator and numerator by 24^2, we have,

$$S_{A'\times B'}=\frac{[(687-2\times 134)-(241-2\times 108)]^2}{24\times 6}$$

This is the same as Equation (27.11). This result can also be found as follows.

$$S_{A'\times B'}=\frac{(A_1'B_1')^2+(A_2'B_1')^2}{12}+\frac{(A_1'B_2')^2+(A_2'B_2')^2}{24}-\frac{T^2}{72}-S_{A'}-S_{B'}$$

$$=\frac{108^2+134^2}{12}+\frac{241^2+687^2}{24}-\frac{1170^2}{72}-S_{A'}-S_{B'} \tag{27.16}$$

Next, let us find the variation of source $A' \times (B_1-B_3)$. This is the interaction in the two-way layout of the two levels A_1' and A_2' with B_1, B_2, and B_3; it indicates how the difference between the class A_1' and the class A_2' before administration of the drug varies among B_1, B_2, and B_3, and together with the variation of the difference within A_1' among B_1, B_2, and B_3 and the variation of the difference within A_2' among B_1, B_2, and B_3, it forms the error variance S_e.

$$S_{A'\times(B_1-B_3)}=\frac{42^2+33^2+33^2+50^2+44^2+40^2}{4}-\frac{242^2}{24}-\left(\frac{108^2+134^2}{12}-\frac{242^2}{24}\right)$$

$$-\left(\frac{92^2+77^2+73^2}{8}-\frac{242^2}{24}\right)$$

$$=2494-2440-28-25=1 \qquad (f=2) \tag{27.17}$$

Next, we find the variation of source $A' \times (B_4-B_9)$. This is the variation that tells whether or not the mean curve at (B_4-B_9) after administration differs between the placebo and the test medicine, and it is the most important source in data of this type.

$$S_{A' \times (B_4-B_9)} = \frac{43^2+47^2+\cdots+63^2}{4} - \frac{928^2}{48} - \left(\frac{241^2+687^2}{24} - \frac{928^2}{48}\right) - S_{(B_4-B_9)}$$

$$= 756 \qquad (f=5) \tag{27.18}$$

Since A' is at two levels, this can also be found as follows. It is done by finding the difference between A_1' and A_2' classed by the levels of B_4, B_5, . . . , B_9, by then dividing its sum of squares by the unit number 8, and by subtracting $S_{A'}$ at B_2'.

$$S_{A' \times (B_4-B_9)} = \frac{1}{8}[(43-77)^2+(47-140)^2+\cdots+(30-63)^2] - \frac{(241-687)^2}{48}$$

$$= 4\,900 - 4\,144 = 756 \tag{27.19}$$

Next, source $(A_1-A_4) \times B'$ indicates whether or not the individual difference among (A_1-A_4) differs from before administration to after administration. Since this is the placebo group, if the placebo has no influence this becomes an error, but usually it does not happen so.

$$S_{(A_1-A_4) \times B'} = \frac{22^2+35^2+15^2+36^2}{3} + \frac{52^2+42^2+56^2+91^2}{6} - \frac{349^2}{36}$$

$$- S_{(A_1 \sim A_4)} - \left(\frac{108^2}{12} + \frac{241^2}{24} - \frac{349^2}{36}\right)$$

$$= \frac{3\,230}{3} + \frac{15\,885}{6} - 3\,383 - 236 - 9 = 96 \qquad (f=3) \tag{27.20}$$

$$S_{(A_1-A_4) \times (B_1-B_3)} = 9^2+6^2+7^2+\cdots+11^2 - \frac{108^2}{12} - \left(\frac{22^2+35^2+15^2+36^2}{3} - \frac{108^2}{12}\right)$$

$$- \left(\frac{42^2+33^2+33^2}{4} - \frac{108^2}{12}\right)$$

$$= 1\,094 - 972 - 105 - 14 = 3 \qquad (f=6) \tag{27.21}$$

$$S_{(A_1-A_4) \times (B_4-B_9)} = 11^2+12^2+\cdots+13^2 - \frac{241^2}{24} - \left(\frac{52^2+42^2+56^2+91^2}{6} - \frac{241^2}{24}\right)$$

$$- \left(\frac{43^2+47^2+\cdots+30^2}{4} - \frac{241^2}{24}\right)$$

$$= 2\,773 - 2\,420 - 228 - 42 = 83 \qquad (f=15) \tag{27.22}$$

$$S_{(A_5-A_8) \times B'} = \frac{18^2+55^2+19^2+42^2}{3} + \frac{140^2+190^2+\cdots+161^2}{6} - \frac{821^2}{36}$$

$$- \left(\frac{134^2}{12} + \frac{687^2}{24} - \frac{821^2}{36}\right) - \left(\frac{158^2+245^2+215^2+203^2}{9} - \frac{821^2}{36}\right)$$

$$= 21\,831 - 18\,723 - 2\,438 - 435 = 235 \qquad (f=3) \tag{27.23}$$

$$S_{(A_5-A_8)\times(B_1-B_3)}=6^2+9^2+\cdots+13^2-\frac{134^2}{12}-\left(\frac{50^2+44^2+40^2}{4}-\frac{134^2}{12}\right)$$

$$-\left(\frac{18^2+55^2+19^2+42^2}{3}-\frac{134^2}{12}\right)$$

$$=1870-1496-13-329=32 \qquad (f=6) \qquad (27.24)$$

$$S_{(A_5-A_8)\times(B_4-B_9)}=2^2+41^2+\cdots+9^2-\frac{687^2}{24}-\left(\frac{140^2+190^2+196^2+161^2}{6}-\frac{687^2}{24}\right)$$

$$-\left(\frac{77^2+140^2+\cdots+63^2}{4}-\frac{687^2}{24}\right)$$

$$=22827-19665-341-1804=1017 \qquad (f=15) \qquad (27.25)$$

By organizing these results we obtain Table 27.3.

27.1.3 Analysis of Variance

Error (e) is created by pooling $(A_1-A_4)\times(B_4-B_9)$, which is not significant, with error e.

In Table 27.3, we have given the contribution ratios of the sources that were found significant when tested by error e. In order to test the main effect of A', which is a comparison of the placebo and the tested drug, inasmuch as the individual differs between A_1' and A_2', we do not test with the error variance $V_e = 2.6$ (or pooled $V_{e'} = 4.1$) but we test by the main effect among A_1-A_4 and A_5-A_8.

$$S_e(A)=236+435=671 \qquad (f=6) \qquad (27.26)$$

Table 27.3 Analysis of Variance Table

Source		f	S	V	S'	ρ[%]
A	A'	1	3 094	3 094	3 090	32.4
	A_1-A_4	3	236	79	224	2.3
	A_5-A_8	3	435	145	423	4.4
B	B'	1	1 369	1 369	1 365	14.3
	B_1-B_3	2	25	12	17	0.2
	B_4-B_9	5	1 090	363	1 069	11.2
$A\times B$	$A'\times B'$	1	1 078	1 078	1 074	11.2
	$A'\times(B_4-B_9)$	5	756	151	735	7.7
	$(A_1-A_4)\times B'$	3	96	32	84	0.9
	$(A_1-A_4)\times(B_4-B_9)$	15	83	6	—	—
	$(A_5-A_8)\times B'$	3	235	78	223	2.3
	$(A_5-A_8)\times(B_4-B_9)$	15	1 017	68	955	10.0
e		14	37	2.6		
(e)		(29)	(120)	(4.1)	292	3.2
T		71	9 551		9 551	100.0

From this, we have

$$V_e(A)=\frac{671}{6}=112 \tag{27.27}$$

$$F=\frac{3\,094}{112}=27.6^{**} \tag{27.28}$$

Although it is significant in this case, often it is not significant if individual difference is large.

As to testing of the difference between B_1' and B_2' for class A_2',

$$L=\frac{687}{24}-\frac{134}{12} \tag{27.29}$$

we test by the error between (B_1-B_3) of A_2' and between (B_4-B_9) of A_2'.

$$S_e'(B)=\frac{13+1\,804}{2+5}=259 \tag{27.30}$$

$$S_L=\frac{(687-2\times 134)^2}{24+4\times 12}=2\,438 \tag{27.31}$$

$$F=\frac{2\,438}{259}=9.4^{*} \tag{27.32}$$

However, the precision of this test is not particularly good and it is at most 5% significant.

Generally, in the case of time series data where there are individual differences, it is advisable to test A' and B' by forming two types of first-order error variance (between persons and between days), *viz.*,

$$S_{e_1}(A)=S_{(A_1-A_4)}+S_{(A_5-A_8)}$$

$$S_{e_1}(B)=S_{(B_1-B_3)}+S_{(B_4-B_9)}$$

The same can be said for $A'\times B'$. That the interaction $(A_5-A_8)\times(B_4-B_9)$ was significant in the analysis of variance table indicates that the change of the characteristic value of the four persons given the test medicine, after administration of this medicine, differed from person to person. Thus, we tested by error variance V_e in order to determine the degree to which the change of characteristic value indicating the effect of a certain medicine differed from person to person.

27.1.4 Estimations

As to the degree to which this tested medicine causes this characteristic value to become plus, information regarding the mean value is given by the estimated value of the interaction of Equation (27.14) and by its variation.

$$L=\left(\frac{687}{24}-\frac{134}{12}\right)-\left(\frac{241}{24}-\frac{108}{12}\right)=16.4 \tag{27.33}$$

For the confidence limits of this, only the term $\frac{687}{24}$ among the four terms of the right side of Equation (27.33) possesses a large error variance.

The expression within the square root sign of the confidence limits consists of two terms.

$$L \pm \sqrt{F^1_{15} \times V_e' \times \frac{1}{24} + F^1_{29} \times V_e \times \frac{5}{24}} \tag{27.34}$$

V_e within the square root sign is the error variance which is found from interaction $(A_5 - A_8) \times (B_4 - B_9)$. We use

$$V_e' = 68$$

$$V_e = 4.1$$

$$F^1_{15} = 4.54$$

$$F^1_{29} = 4.18$$

and obtain

$$\pm \sqrt{4.54 \times \frac{68}{24} + 4.18 \times 4.1 \times \frac{5}{24}} = \pm 4.1 \tag{27.35}$$

Therefore, the estimated value of L and the confidence limits are given by

$$16.4 \pm 4.1 \tag{27.36}$$

Instead of using Equation (27.36) one might also pool all of the interactions of the four quadrants, $A_1'B_1'$, $A_1'B_2'$, $A_2'B_1'$, and $A_2'B_1'$, and by using

$$S_e = S_{(A_1-A_4)\times(B_1-B_3)} + S_{(A_1-A_4)\times(B_4-B_9)} + S_{(A_5-A_8)\times(B_1-B_3)} + S_{(A_5-A_8)\times(B_4-B_9)}$$
$$+ S_{A'\times(B_1-B_3)}$$
$$= 83 + 1017 + 37 = 1137$$

$$V_e = \frac{1137}{44} = 25.8$$

obtain

$$L \pm \sqrt{F^1_{44} \times V_e \times \frac{6}{24}} = 16.4 \pm \sqrt{4.06 \times 25.8 \times \frac{1}{4}} = 16.4 \pm 5.1 \tag{27.37}$$

and use Equation (27.37) instead of Equation (27.36). As it turns out, the interaction $(A_5 - A_8) \times (B_4 - B_9)$ was not needed in order to evaluate Equation (27.34) but only in order to determine whether or not its magnitude is large compared with that for $(A_1 - A_4) \times (B_4 - B_9)$ and the error variance before administration of the medicine.

Next, let us find the confidence limits of the curves of A_1' and A_2'. Although in this case dispersions due to the main effect of individual difference are included in these curves, the main effect of individuals is not included in the curve of $A' \times (B_4 - B_9)$.

Therefore, we estimate as follows.

	B_1'	B_4	B_5	B_6	B_7	B_8	B_9
A_1'	9.0±1.2	10.8	11.8	9.8	10.8	9.8	7.5
A_2'	11.2±1.2	19.2	35.0	36.8	38.0	27.0	15.8

In this case, for the confidence limits of the mean value for $A_1'B_1'$ and $A_2'B_1'$, by using $V_e = 4.5$ of 29 degrees of freedom we get

$$\pm\sqrt{\frac{4.18\times 4.1}{12}}=\pm 1.2 \tag{27.38}$$

And the confidence limits of the mean value of A_1' for B_4–B_9 are

$$\pm\sqrt{\frac{4.18\times 4.1}{4}}=\pm 2.1 \tag{27.39}$$

But for the confidence limits of the mean value of A_2' for B_4–B_9, the variance of the interaction $(A_5-A_8)\times(B_4-B_9)$ is used as the error.

$$V_e'=\frac{S_{(A_5-A_8)\times(B_4-B_9)}}{14}=68 \qquad (f=15) \tag{27.40}$$

$$\pm\sqrt{\frac{4.54\times 68}{4}}=\pm 8.8 \tag{27.41}$$

When these quantities are plotted in a graph, Figure 27.1 is obtained.

To determine from this graph whether or not the characteristic value has returned to the same level as before administration at B_9, which is the third day after interruption of administration to the tested group, we need but examine the significant difference between the mean of $A_2'B_1'$,

$$11.2\pm 1.2 \tag{27.42}$$

and the mean of $A_2'B_9$,

$$15.8\pm 8.8 \tag{27.43}$$

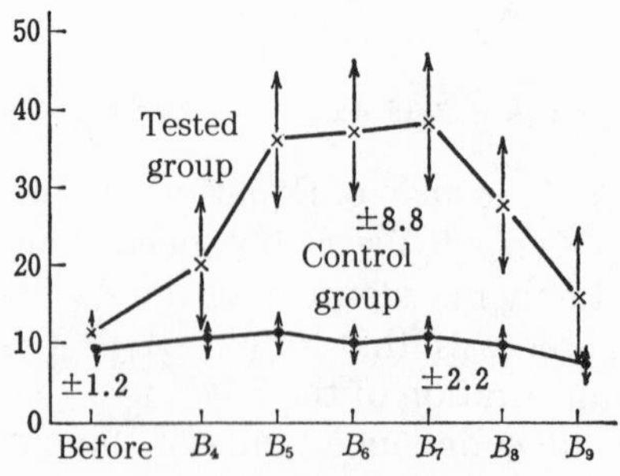

FIGURE 27.1 Effect Curves

In order to test the difference between the two mean values, we merely need to compare the square root of the sum of squares of the two confidence limits,

$$\sqrt{1.2^2+8.8^2}=8.9 \tag{27.44}$$

with the difference between the two mean values,

$$15.8-11.2=4.6 \tag{27.45}$$

Since 4.6 is smaller than 8.9 in this case, we may assume that the effect has reverted to its original at B_9.

27.2 When an Orthogonal Array Is Used

27.2.1 Assignment and Data

The analysis method for a simple example of an experiment of two levels, for comparison between a tested medicine and a placebo, was shown in the previous section; here, we will give a somewhat more complicated worked example of an animal test, using an orthogonal array.

Let us assume that we have taken the levels as follows for four factors *A, B, C,* and *D.*

$$A \begin{cases} A_0 \text{ placebo} \\ A_1 \text{ tested medicine in the amount of 1 mg} \\ A_2 \text{ tested medicine in the amount of 2 mg} \\ A_3 \text{ tested medicine in the amount of 4 mg} \\ A_4 \text{ tested medicine in the amount of 8 mg} \end{cases}$$

B:	method of administering	2 levels
C:	distinction by sex	2 levels
D:	distinction by age class	2 levels

The four levels of A_1, A_2, A_3, and A_4 and factors *B, C,* and *D* were assigned to orthogonal array L_8, and for A_0 an experiment by the partial supplementing design was carried out under the same conditions as for A_4. The experiments on A_4 consisted of the two types No. 7 and No. 8, but in this case experiments No. 7 and No. 8 were also performed with A_0 inserted in place of A_4. Therefore, the total number of animals tested was ten.

For comparison of treatments of the ten types in Table 27.4, to cancel differences of the individual animals the data of the first two days, K_1 and K_2, were measured before the treatment; those of the next two days, K_3 and K_4, were measured during administration; and those for K_5 and K_6 were measured two days after discontinuing administration. The data were as shown in Table 27.5.

Table 27.4 Assignment of Treatments

No.	A (1 2 3)	B (4)	e (5)	C (6)	D (7)	Animal
1	1	1	1	1	1	R_1
2	1	2	2	2	2	R_2
3	2	1	1	2	2	R_3
4	2	2	2	1	1	R_4
5	3	1	2	1	2	R_5
6	3	2	1	2	1	R_6
7	4	1	2	2	1	R_7
8	4	2	1	1	2	R_8
9	0	1	2	2	1	R_9
10	0	2	1	1	2	R_{10}

The eight types of treatment by the tested medicine are called R_1, R_2, . . . , R_8, and are collectively denoted by R_1'. The eight levels R_1, R_2, . . . , R_8 within R_1' are resolved into A at four levels, B at two levels, C at two levels, D at two levels, and the error between individuals, e_1.

Variation	$S_{\text{within } R_1'} = S_A + S_B + S_C + S_D + S_{e_1}$
Number of degrees of freedom	$7 = 3 + 1 + 1 + 1 + 1$

R_9 and R_{10} are the treatments by the placebo, and these are expressed as R_2'. For R_9 and R_{10} of R_2', we find the variation for the difference between R_9 and R_{10}.

Next, the time factor K is resolved into the three large levels $K_1'(K_1, K_2)$ before administration of the medicine, $K_2'(K_3, K_4)$ during administration, and $K_3'(K_5, K_6)$ after discontinuation of administration, in other words three levels of so-called major categorization. The difference between the

Table 27.5 Time Series Data

Treatment \ Day		K_1	K_2	K_3	K_4	K_5	K_6	Total
Tested medicine	1	9	7	29	28	20	9	102
	2	3	5	22	25	13	3	71
	3	8	11	24	32	22	11	108
	4	9	13	35	33	22	11	123
	5	22	19	40	46	19	20	166
	6	11	7	27	27	18	11	101
	7	11	15	34	33	26	12	131
	8	20	17	36	41	29	17	160
Control	9	5	3	10	14	7	6	45
	10	9	5	12	15	12	9	62
Total		107	102	269	294	188	109	1069

two levels K_1 and K_2 within K_1' constitutes the diurnal variation which is common to the ten animals, and when all ten are raised similarly it indicates the common difference of tendency between days, due to such causes as the daily meals, the meteorological conditions, and the life rhythm. Next, the difference within K_2', in other words between K_3 and K_4, is a difference that is common to the ten animals during administration of the drug. The difference between K_5 and K_6 within K_3' indicates the difference between the first day and the second day after discontinuation of the drug. If this drug were to have a lasting effect which continues for about one day longer, it is possible that a great difference will be demonstrated between K_5 and K_6, while if it has continuing effect of at least two days, it is possible that K_3' will demonstrate results that do not differ greatly from K_2'. In view of the foregoing, the main effect of factor K, S_K, can be resolved as follows.

Variation

$$\begin{aligned} S_K = S_{K'} &+ S_{\text{within } K_1'} \\ &+ S_{\text{within } K_2'} \\ &+ S_{\text{within } K_3'} \end{aligned}$$

Number of Degrees of Freedom

$$5 = 2 + 1 + 1 + 1$$

Thus, the main effects of R and K can be resolved as in Table 27.6. Once we know the resolutions of the main effects of R and K, as a matter of form it is possible to calculate all of the interactions between them, but often the number of such combinations becomes huge. Even in a simple case such as this example, the components of these interactions number $7 \times 4 = 28$, including errors.

Recently, there have been developed programs (such as DANEX, Nippon Telephone and Telegraph Public Corporation) by which, as long as one indicates the coefficients of resolution of the main effects, their interactions can be found automatically. However, it is still rather troublesome if one wishes to perform all these calculations with a desktop calculator or the like. If it is felt that the calculations are too much trouble, one can find only those among the 28 interactions which are believed to be important, and pool the remainder into the second-order error variance e_2. When doing so, one should consider the following points.

(1) Interactions with error all become second-order error variances. It is therefore unnecessary to find, for example, $e_1 \times K'$, $e_1 \times$ within K_1', $e_1 \times$ within K_2', and $e_1 \times$ within K_3'.

(2) Where a source can be regarded as a difference between normal

Table 27.6 Distribution of Number of Degrees of Freedom

	Source	Degrees of freedom
R	R'	1
	A	3
	B	1
	C	1
	D	1
	e_1	1
	R_9R_{10}	1
K	K'	2
	within K_1'	1
	within K_2'	1
	within K_3'	1
$R \times K$	$R' \times K'$	2
	$R' \times$ within K_2'	1
	$R' \times$ within K_3'	1
	$A \times K'$	6
	$A \times$ within K_2'	3
	$A \times$ within K_3'	3
	$B \times K'$	2
	$B \times$ within K_2'	1
	$B \times$ within K_3'	1
	$C \times K'$	2
	$C \times$ within K_2'	1
	$C \times$ within K_3'	1
	$D \times K'$	2
	$D \times$ within K_2'	1
	$D \times$ within K_3'	1
	$R_9R_{10} \times K'$	1
	e_2	16
	T	59

states, as within K_1', its interactions with others should be regarded as error and it is unnecessary to find them.

(3) Interaction with individual difference of animals treated by the placebo should be found only with K', and the others are incorporated into error.

Thus, the interactions to be found become narrowed to the following 16:

$R' \times K'$, $R' \times$ within K_2', $R' \times$ within K_3'
$A \times K'$, $A \times$ within K_2', $A \times$ within K_3'
$B \times K'$, $B \times$ within K_2', $B \times$ within K_3'
$C \times K'$, $C \times$ within K_2', $C \times$ within K_3'
$D \times K'$, $D \times$ within K_2', $D \times$ within K_3'
$R_9R_{10} \times K'$

Therefore the sources in the analysis of variance table become as given by Table 27.6.

Table 27.7 Supplementary Table

	K_1'		Subtotal	K_2'		Subtotal	K_3'		Subtotal	Total
	K_1	K_2		K_3	K_4		K_5	K_6		
A_1	12	12	24	51	53	104	33	12	45	173
A_2	17	24	41	59	65	124	44	22	66	231
A_3	33	26	59	67	73	140	37	31	68	267
A_4	31	32	63	70	74	144	55	29	84	291
Subtotal	93	94	187	247	265	512	169	94	263	962
A_0	14	8	22	22	29	51	19	15	34	107
Total	107	102	209	269	294	563	188	109	297	1 069
B_1	50	52	102	127	139	266	87	52	139	507
B_2	43	42	85	120	126	246	82	42	124	455
Total	93	94	187	247	265	512	169	94	263	962
$(e)_1$	48	42	90	116	128	244	89	48	137	471
$(e)_2$	45	52	97	131	137	268	80	46	126	491
Total	93	94	187	247	265	512	169	94	263	962
C_1	60	56	116	140	148	288	90	57	147	551
C_2	33	38	71	107	117	224	79	37	116	411
Total	93	94	187	247	265	512	169	94	263	962
D_1	40	42	82	125	121	246	86	43	129	457
D_2	53	52	105	122	144	266	83	51	134	505
Total	93	94	187	247	265	512	169	94	263	962
R_9	5	3	8	10	14	24	7	6	13	45
R_{10}	9	5	14	12	15	27	12	9	21	62
Total	14	8	22	22	29	51	19	15	34	107

27.2.2 Analysis of Variance

To perform the calculations for Table 27.6, we construct a supplementary table such as Table 27.7.

From the supplementary table, the respective variations can be calculated as follows.

$$S_m = \frac{1\,069^2}{60} = 19\,046 \qquad (f=1) \tag{27.46}$$

$$S_T = 9^2 + 7^2 + \cdots + 12^2 + 9^2 - S_m = 6\,787 \qquad (f=59) \tag{27.47}$$

$$S_{A'} = \frac{(A_1')^2}{48} + \frac{(A_2')^2}{12} - S_m = \frac{962^2}{48} + \frac{107^2}{12} - 19\,046 = 1\,188 \quad (f=1) \tag{27.48}$$

The variation among the eight treatments from No. 1 to No. 8 is the total variation for orthogonal array L_8, S_{T_1}, and it is found as follows.

$$S_{T_1}=\frac{1}{6}(102^2+71^2+\cdots+160^2)-\frac{962^2}{48}=1179 \qquad (f=7) \tag{27.49}$$

S_{T_1} is resolved into S_A, S_B, S_C, S_D, and S_{e_1}. Also, the difference between R_9 and R_{10} by placebo A_0 is found as follows.

$$S_{R_9R_{10}}=\frac{(45-62)^2}{12}=25 \qquad (f=1) \tag{27.50}$$

The variation S_K among the six levels $K_1, K_2, \ldots, K_6$ is found by the following equation.

$$S_K=\frac{1}{10}(177^2+102^2+\cdots+109^2)-S_m=3\,741 \qquad (f=5) \tag{27.51}$$

This S_K is resolved into the sum of the variation among the three levels, before administration of the drug (K_1'), during administration (K_2'), and after administration (K_3), $S_{K'}$, and the three variations of one degree of freedom within K_1', within K_2', and within K_3'. Within K_1' is also written K_1, K_2.

$$S_{K'}=\frac{1}{20}(209^2+563^2+297^2)-S_m=3\,397 \qquad (f=2) \tag{27.52}$$

$$S_{K_1K_2}=S_{\text{within } K_1'}=\frac{(107-102)^2}{20}=1 \qquad (f=1) \tag{27.53}$$

$$S_{\text{within } K_2'}=\frac{(269-294)^2}{20}=31 \qquad (f=1) \tag{27.54}$$

$$S_{\text{within } K_3'}=\frac{(188-109)^2}{20}=312 \qquad (f=1) \tag{27.55}$$

This can be checked by the fact that the sum that is obtained by adding these agrees with S_K: $3397+1+31+312=3741$.

Next, let us resolve S_{T_1}.

$$S_A=\frac{A_1^2+A_2^2+A_3^2+A_4^2}{12}-\frac{(A_1+\cdots+A_4)^2}{48}=\frac{173^2+231^2+267^2+291^2}{12}-\frac{962^2}{48}$$
$$=658 \qquad (f=3) \tag{27.56}$$

$$S_{A\times K'}=\frac{24^2+41^2+\cdots+84^2}{4}-19\,280-S_A-S_{K'}(1-8) \qquad (f=6) \tag{27.57}$$

Here, we separately find $S_{K'}$ (1–8) for No. 1–No. 8, not for all of No. 1–No. 10.

$$S_{K'}(1-8)=\frac{187^2+512^2+263^2}{16}-19\,280=3\,613 \qquad (f=2) \tag{27.58}$$

$$S_{A\times K'}=\frac{94\,296}{4}-19\,280-658-3\,613=23 \qquad (f=6) \tag{27.59}$$

$$S_{A\times K_3K_4}=\frac{1}{4}[(51-53)^2+(59-65)^2+\cdots+(70-74)^2]-\frac{(247-265)^2}{16}$$
$$=3 \qquad (f=3) \tag{27.60}$$

$$S_{A\times K_5K_6}=\frac{1}{4}[(33-12)^2+(44-22)^2+\cdots+(55-29)^2]-\frac{(169-94)^2}{16}$$
$$=58 \qquad (f=3) \tag{27.61}$$

$$S_B = \frac{(507-455)^2}{48} = 56 \qquad (f=1) \tag{27.62}$$

$$S_{B\times K'} = \frac{1}{16}[(102-85)^2+(266-246)^2+(139-124)^2] - S_B = 1 \quad (f=2) \tag{27.63}$$

$$S_{B\times K_3K_4} = \frac{(253-259)^2}{16} = 2 \qquad (f=1) \tag{27.64}$$

$$S_{B\times K_5K_6} = \frac{(129-134)^2}{16} = 2 \qquad (f=1) \tag{27.65}$$

We similarly have

$$S_C = \frac{(551-411)^2}{48} = 408 \qquad (f=1) \tag{27.66}$$

$$S_{C\times K'} = 34 \qquad (f=2) \tag{27.67}$$

$$S_{C\times K_3K_4} = 0 \qquad (f=1) \tag{27.68}$$

$$S_{C\times K_5K_6} = 5 \qquad (f=1) \tag{27.69}$$

$$S_D = \frac{(457-505)^2}{48} = 48 \qquad (f=1) \tag{27.70}$$

$$S_{D\times K'} = 12 \qquad (f=2) \tag{27.71}$$

$$S_{D\times K_3K_4} = 42 \qquad (f=1) \tag{27.72}$$

$$S_{D\times K_5K_6} = 8 \qquad (f=1) \tag{27.73}$$

$$S_{e_1} = \frac{(471-491)^2}{48} = 8 \qquad (f=1) \tag{27.74}$$

With the tested medicine as A_1' and the placebo as $A_2' \equiv A_0$, interaction with A' is important.

$$S_{A'\times K'} = \frac{187^2+512^2+263^2}{16} + \frac{22^2+51^2+34^2}{4} - S_m - S_{A'} - S_{K'}$$
$$= 22893 + 1060 - 19046 - 1188 - 3397 = 322 \qquad (f=2) \tag{27.75}$$

Since there is no difference between the values of A_1' for K_1 and K_2, this becomes an error. For interaction $A' \times K_3K_4$, we find the difference between K_3 and K_4 for A_1' and for A_2':

$$L(A_1') = \frac{1}{8}(247-265)$$

$$L(A_2') = \frac{1}{2}(22-29)$$

and we form the variation by finding the difference of these.

$$S_{A'\times K_3K_4} = \frac{\left[\frac{1}{8}(247-265) - \frac{1}{2}(22-29)\right]^2}{\left(\frac{1}{8}\right)^2 \times 16 + \left(-\frac{1}{2}\right)^2 \times 4} = \frac{[-18-4\times(-7)]^2}{16+(-4)^2\times 4}$$
$$= 1 \qquad (f=1) \tag{27.76}$$

$$S_{A'\times K_5K_6} = \frac{[(169-94)-4(19-15)]^2}{80} = 44 \qquad (f=1) \tag{27.77}$$

Table 27.8 Analysis of Variance Table

	Source	f	S	V	
R	A'	1	1 188	1 188	
	A	3	658	219	
	B	1	56	56	
	C	1	408	408	
	D	1	48	48	
	e_1	1	8	8	○
	R_9R_{10}	1	25	25	○
K	K'	2	3 397	1 698	
	within K_1'	1	1	1	○
	within K_2'	1	31	31	○
	within K_3'	1	312	312	
$A' \times K'$		2	322	161	
$A' \times$ within K_2'		1	1	1	○
$A' \times$ within K_3'		1	44	44	
$A \times K'$		6	23	4	○
$A \times$ within K_2'		3	3	1	○
$A \times$ within K_3'		3	58	19	○
$B \times K'$		2	1	0	○
$B \times$ within K_2'		1	2	2	○
$B \times$ within K_3'		1	2	2	○
$C \times K'$		2	34	17	○
$C \times$ within K_2'		1	0	0	○
$C \times$ within K_3'		1	5	5	○
$D \times K'$		2	12	6	○
$D \times$ within K_2'		1	42	42	
$D \times$ within K_3'		1	8	8	○
e		17	98	5.8	○
T		59	6 787		

Error variation S_e is found by subtracting the total of the above factorial variations from the total variation.

Table 27.9 is obtained when sources that are not significant when tested by error variance V_e, according to Table 27.8, are pooled and a new error variance created, and the result organized. Sources in R that are not significant are given separately as individual differences. For the main effect of K, sources that are not significant are shown for reference.

Differences among the ten animals include individual difference and treatment difference. Although e_1, which does not include treatment difference, or in other words is individual difference, is not significant, inasmuch as the number of degrees of freedom is only 1, this is not proof that individual difference is small. Difference between R_9 and R_{10}, is also small, but since it has only two degrees of freedom even including this, this is probably not sufficient guarantee that the variance of individual difference is small. Thus, in this experiment, the information concerning

Table 27.9 Adjusted Analysis of Variance Table

	Source	f	S	V	S'	ρ [%]
R	A'	1	1 188	1 188	1 181	17.4
	A	3	658	219	637	9.4
	B	1	56	56	49	0.7
	C	1	408	408	401	5.9
	D	1	48	48	41	0.6
	(e)	(2)	(33)	(16)	(19)	(0.2)
K	K'	2	3 397	1 698	3 383	49.8
	within K_3'	1	312	312	305	4.5
	(e)	(2)	(32)	(16)	(17)	(0.2)
$R \times K$	$A' \times K'$	2	322	161	304	4.5
	$A' \times$ within K_3'	1	44	44	34	0.5
	$D \times$ within K_2'	1	42	42	35	0.5
Pooled e		45	312	6.9	414	6.2
T		59	6 787			100.0

individual difference is somewhat insufficient. In order to increase the information, one should either experiment on two animals under each of the ten types of conditions or, instead of having R_9 and R_{10} with the placebo be partially supplementing, one should conduct an experiment by L_8, changing all the levels A_1, A_2, A_3, and A_4 with the placebo. In that case, in the last analysis, the form of the experiment is that of orthogonal array L_{16}, but since most of the sources within the placebo group are small, they are pooled and one can obtain sufficient information concerning individual difference. Since (e) in R is not significant here, there is no definite proof, but we will continue our discourse by tentatively assuming that individual difference is not significant.

Next, as to the effect of factor K, the effect among K_1', K_2', and K_3' is large but the difference between K_1 and K_2 and the difference between K_3 and K_4 is small in each case. The difference between K_5 and K_6 has a contribution ratio of 4.5%, but this is because the residual effect of administration up to that time has remained in K_5, and this is believed to be so because such a residual effect no longer exists by the time of K_6. Therefore, we compare K_6 with the mean of K_1 and K_2.

$$L = \frac{K_6}{10} - \frac{K_1 + K_2}{20}$$

$$S_L = \frac{(2K_6 - K_1 - K_2)^2}{2^2 \times 10 + (-1)^2 \times 20} = \frac{(2 \times 109 - 107 - 102)^2}{60} = 1 \qquad (f=1) \tag{27.78}$$

This is not significant by pooled error variance $V_e = 7.0$. This means it may be assumed that the value at K_6 has reverted to the mean before administration.

Sources which are significant among the interactions number only three. This indicates that A, the quantity administered, and B and C possess similar effects whether during administration or after administration. That $A' \times K'$ is significant means that source A', which indicates the difference between the control and the tested medicine, differs significantly between K_1', K_2', and K_3'. Although there is no difference between A_1' and A_2' at K_1', there is a great difference at K_2' and at K_3'. It is also indicated that there is some slight difference of A' within K_3', or in other words between K_5 and K_6. However, since its contribution ratio is 0.5%, the difference is not a particularly large one.

That the interaction $D \times$ within K_2' is significant indicates that the difference between D_1 and D_2 differs somewhat between K_3 and K_4 within K_2'.

Therefore, the estimations are carried out as follows.

27.2.3 Estimation of Significant Sources

Since the main effects of A' and K' and the interactions $A' \times K'$ and $A' \times$ within K_3' are significant, we form the means of A_1' (tested medicine) and of A_2' (placebo) for K_1', K_2', K_5, and K_6 (Table 27.10). The confidence limits were found as follows. For example, for $A_1'K_3$ and A_1' they are

$$\pm\sqrt{\frac{4.06\times 7.0}{16}}=1.3$$

$$\pm\sqrt{\frac{4.06\times 7.0}{8}}=1.9 \qquad (27.79)$$

The effect of D is found separately for K_1', K_3, K_4, and K_3', as shown by Table 27.11. The remainder are sources only of main effects.

$$\bar{A}_1=14.4\pm1.5$$
$$\bar{A}_2=19.2 \quad ''$$
$$\bar{A}_3=22.2 \quad ''$$
$$\bar{A}_4=24.2 \quad ''$$
$$\bar{B}_1=21.1\pm1.1$$
$$\bar{B}_2=19.0 \quad ''$$
$$\bar{C}_1=23.0\pm1.1$$

It is advisable to draw these on a graph, as was done with A'. For A_1, A_2, A_3, and A_4 we draw graphs of exactly the same shape as for A_1', with only

Table 27.10 Effect of *A*

	K_1 K_2	K_3 K_4	K_5	K_6
A_1'	10.3±1.2	32.0±1.3	21.1±1.9	11.8±1.9
A_2'		12.8±2.6	8.5±2.6	8.5±2.6

Table 27.11 Effect of *D*

	K_1 K_2	K_3	K_4	K_5 K_6
D_1	10.2±1.9	31.7±2.6	30.2±2.6	16.1±1.9
D_2	13.1	30.5	36.0	16.8

the mean values differing. For example, in the case of A_1, with $\bar{A}_1'$ as the mean of $\bar{A}_1$-$\bar{A}_4$, we draw four curves by algebraically adding

$$\bar{A}_1-\bar{A}_1'=14.4-20.0=-5.6$$

$$\bar{A}_2-\bar{A}_1'=19.2-20.0=-0.8$$

$$\bar{A}_3-\bar{A}_1'=22.2-20.0=2.2$$

$$\bar{A}_4-\bar{A}_1'=24.2-20.0=4.2$$

to the values of the six points in the graph of the tested medicine in Figure 27.2. Only the graph for A_4 has been drawn for Figure 27.2. This is where the × marks have been connected with a broken line.

27.3 Experiment on Advertisement Publicity Effect

27.3.1 Introduction

The maintenance and expansion of market share is one of the greatest concerns of all manufacturers, and it is worrisome to watch one's market share decrease gradually. It is the state of affairs now that in order to maintain and expand market share, new products and improved products are placed on the market and products are incessantly advertised and publicized. It is the objective of this section to provide a specific estimation method for the advertising effect in such a case, and the method of finding the confidence limits of the estimated value.

There are a variety of methods of advertising, for example, using mass communication media such as television, newspapers, and magazines; selling with prizes or discount sales, with supermarkets and retail stores

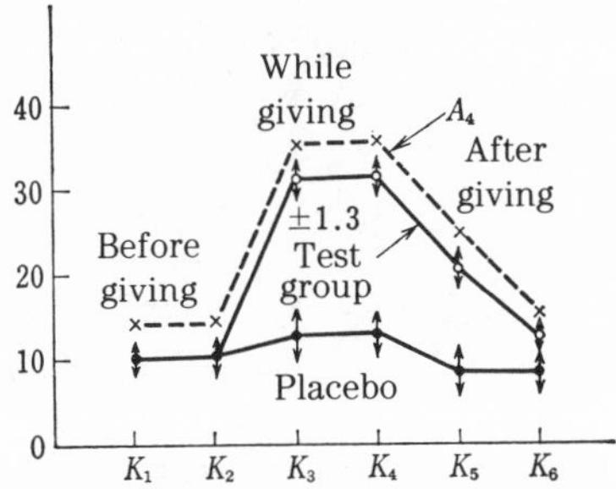

FIGURE 27.2 Adjusted Graph

as the media; advertising directed to individuals by direct mail and the like; and visits to households by salesmen. Estimation of the effect is regarded as difficult in all of these cases. Here we will discuss the estimation method for the effect of mass communication, which is regarded as the most difficult among them.

27.3.2 Estimation Problem for Advertising Effect

The most difficult point when estimating advertising effect is the fact that the economy is a complicated function of time and it is impossible to postulate a steady state. Therefore, theories such as that of the stochastic process assuming steadiness cannot be used straightforwardly in market research. Indeed, changes from month to month, changes depending on the day of the week, changes due to economic policies such as "so-and-so shock," and the emergence of competitive products and their advertising, and so forth, influence the sales of the merchandise in question.

Therefore, when presenting television advertising on a certain network, it is necessary to estimate the correct effect of the advertising having eliminated the influence of such changes. For example, assuming that the sales in June, July, and August were 1.23, 1.45, and 1.38 billion yen, respectively, even if the sales after having broadcast a television advertisement in September became 1.25, 1.1, and 1.21 billion yen in September, October, and November, *this does not mean that the television publicity lowered the sales.* This is because the sales in September, October, and November might have been 1.02, 1.03, and 1.08 billion yen had the television advertisement not been presented. If this were the case, if it is possible to consider only this period of time, the effect of television advertising would have had the effect of increasing sales by

$$(12.5 - 10.2) + (11.0 - 10.3) + (12.1 - 10.3) = 430 \text{ million yen}$$

Thus, in order to grasp correctly the effect of a television advertisement which has begun with a certain month, it is absolutely necessary to know the changes of sales in districts where this television advertisement is not being broadcast. This is because it is necessary to correct for the effects of seasonal changes and shock-type external conditions such as economic prosperity or recession by using the changes in the sales in such geographic districts.

Moreover, when stock is abundant at the wholesalers and retailers, special care is necessary when conducting a survey of the sales in a case such as this. This is because usually, when a certain stratum of the marketing structure judges that recession is nearing, it decreases purchases on the basis of concern about moving its own stock, and as a result, the amount shipped from the manufacturers decreases more than commensurately. In order to avoid influences on the estimated value of the advertisement effect from such increase or decrease of intermediate stock, it is necessary to select a certain number of retail stores from districts where the television commercials were presented and from districts where the television advertisement was not broadcast and to survey the sales in each retail store.

Therefore, in order to examine the effect of the advertisement, having eliminated the influence of such stock correction, when it is run, say, for three months beginning with July, it is necessary to acquire for example data such as given in Table 27.12. Here, A_1 represents the three months before the television advertising, April, May, and June; A_2 represents the three months during the television advertising, July, August, and September; and A_3 represents the three months after the television advertising, October, November, and December. April, May, . . . , December are expressed by symbols as C_1, C_2, . . . , C_9 in the table.

Ten retail stores were selected from geographic region B_1, where this television advertisement was being presented, and five were selected from geographic region B_2, where the advertisement was not presented, and the sales of the merchandise in question every month were studied. In every case, the scale of the retailer was ranked from class 1 to class 5, based on the sales of the product of one's own company, and from each class, two retailers were selected for B_1 and one store was selected for B_2.

In order to estimate the effect of the television advertising from data such as those of Table 27.12, we find for region B_1, where the television advertisement was presented, the difference L_1 between the mean of the sales before the television advertisement and the mean of the sales after the advertisement began.

Table 27.12 Sales Before and After Television Advertisement (y_{ij} Represents Proceeds from Sales)

Store \ Month			A_1 (Before)			A_2 (During)			A_3 (After)		
			C_1	C_2	C_3	C_4	C_5	C_6	C_7	C_8	C_9
B_1	D_1'	D_1	y_{11}	y_{12}	y_{13}	y_{14}	y_{15}	y_{16}	y_{17}	y_{18}	y_{19}
		D_2	y_{21}	y_{22}	y_{23}	y_{24}	y_{25}	y_{26}	y_{27}	y_{28}	y_{29}
	D_2'	D_3	y_{31}	y_{32}	y_{33}	y_{34}	y_{35}	y_{36}	y_{37}	y_{38}	y_{39}
		D_4	y_{41}	y_{42}	y_{43}	y_{44}	y_{45}	y_{46}	y_{47}	y_{48}	y_{49}
	⋮	⋮	⋮	⋮	⋮	⋮	⋮	⋮	⋮	⋮	⋮
	D_5'	D_9	y_{91}	y_{92}	y_{93}	y_{94}	y_{95}	y_{96}	y_{97}	y_{98}	y_{99}
		D_{10}	$y_{10,1}$	$y_{10,2}$	$y_{10,3}$	$y_{10,4}$	$y_{10,5}$	$y_{10,6}$	$y_{10,7}$	$y_{10,8}$	$y_{10,9}$
B_2	D_1'	D_{11}	$y_{11,1}$	$y_{11,2}$	$y_{11,3}$	$y_{11,4}$	$y_{11,5}$	$y_{11,6}$	$y_{11,7}$	$y_{11,8}$	$y_{11,9}$
	D_2'	D_{12}	$y_{12,1}$	$y_{12,2}$	$y_{12,3}$	$y_{12,4}$	$y_{12,5}$	$y_{12,6}$	$y_{12,7}$	$y_{12,8}$	$y_{12,9}$
	⋮	⋮	⋮	⋮	⋮	⋮	⋮	⋮	⋮	⋮	⋮
	D_5'	D_{15}	$y_{15,1}$	$y_{15,2}$	$y_{15,3}$	$y_{15,4}$	$y_{15,5}$	$y_{15,6}$	$y_{15,7}$	$y_{15,8}$	$y_{15,9}$

If we put

$$A_1B_1 = y_{11}+y_{12}+y_{13}+\cdots+y_{10,1}+y_{10,2}+y_{10,3}$$

$$A_2B_1+A_3B_1 = (y_{14}+y_{15}+\cdots+y_{19})+\cdots+(y_{10,4}+\cdots+y_{10,9})$$

L_1 is given by the following equation.

$$L_1 = \frac{A_1B_1}{30} - \frac{A_2B_1+A_3B_1}{60} \tag{27.80}$$

We similarly find the difference L_2 between the mean of the sales in months C_1–C_3 in the region where television advertisement was not presented and the mean of the sales in months C_4–C_9.

$$L_2 = \frac{A_1B_2}{15} - \frac{A_2B_2+A_3B_2}{30} \tag{27.81}$$

The difference between L_1 and L_2 is the difference in the mean sales per month for the case where television advertising was done and the case where it was not done. Therefore, the estimated value is given by the following equation.

$$L = L_1 - L_2 \tag{27.82}$$

The problem lies in the degree to which the value of L, found by Equation (27.82), is reliable. To determine this, it is necessary to test L and to find the confidence limits. These calculations are possible only by an ingenious application of analysis of variance. Since it is much too troublesome to write the general calculation equations for analysis of variance of data of this type, we will show the calculation method by a simpler hypothetical example incorporating numerical values. Please refer to Section 27.4 regarding the analysis method for Table 27.12.

27.3.3 A Worked Example of a Case of Television Advertising

Television advertising was presented from July of this year in the six geographic districts D_1–D_6 of business establishments throughout the country. This television advertising was not aired in the three districts D_7, D_8, and D_9. It is to be understood that C_1 = January, C_2 = February, . . . , C_{12} = December. The data constitute the growth of the sales every month this year relative to the same month last year in each business district. For example, if the sales in January of last year in district D_1 were 12 million yen and the sales in January of this year were 12.61 million yen, the growth of sales y is given by the following equation.

$$y = \frac{1261-1200}{1200} \times 100 \doteqdot 5.1 \ [\%] \tag{27.83}$$

Table 27.13 gives such data of growth.

For region B_1, where television advertising was presented, we find the difference L_1 between the mean value of the growth of sales before the television publicity and its mean value after the television publicity began.

Table 27.13 Data of Growth of Sales

		A_1 (Before television advertising)							A_2 (During and after television advertising)							
		C_1	C_2	C_3	C_4	C_5	C_6	Sub-total	C_7	C_8	C_9	C_{10}	C_{11}	C_{12}	Sub-total	Total
B_1	D_1	5.1	3.2	2.0	6.0	5.1	4.5	25.9	3.4	4.2	2.5	−2.5	0.2	−1.0	6.8	32.7
	D_2	3.8	4.1	2.4	5.6	3.2	4.1	23.2	1.3	2.2	0.1	−4.7	−1.0	−1.8	−3.9	19.3
	D_3	1.0	−0.3	−0.4	−1.2	3.0	4.0	6.1	0.5	3.0	−3.5	−5.8	−6.3	−3.0	−15.1	−9.0
	D_4	8.2	4.0	5.1	3.5	4.8	7.4	33.0	4.3	1.8	3.1	−3.5	−1.5	−0.2	4.0	37.0
	D_5	6.2	2.8	4.1	3.6	5.0	4.0	25.7	3.4	2.2	1.9	−3.7	−5.1	−0.5	−1.8	23.9
	D_6	5.4	6.3	2.5	4.1	4.1	3.0	25.4	1.2	1.4	−1.0	−0.4	0.0	1.2	2.4	27.8
	Sub-total	29.7	20.1	15.7	21.6	25.2	27.0	139.3	14.1	14.8	3.1	−20.6	−13.7	−5.3	−7.6	131.7
B_2	D_7	7.0	4.8	5.1	5.4	5.5	7.1	34.9	−5.2	−4.4	−8.8	−4.9	−4.1	1.2	−26.2	8.7
	D_8	1.2	3.0	−0.8	−0.9	4.3	0.7	7.5	−9.9	−8.2	−12.9	−10.8	−8.9	−2.5	−53.2	−45.7
	D_9	4.6	3.4	4.5	5.1	4.1	4.6	26.3	−6.9	−3.7	−13.8	−4.6	−7.0	−0.9	−36.9	−10.6
	Sub-total	12.8	11.2	8.8	9.6	13.9	12.4	68.7	−22.0	−16.3	−35.5	−20.3	−20.0	−2.2	−116.3	−47.6
Total		42.5	31.3	24.5	31.2	39.1	39.4	208.0	−7.9	−1.5	−32.4	−40.9	−33.7	−7.5	−123.9	84.1

$$L_1 = \frac{-7.6}{36} - \frac{139.3}{36} = -4.08 \tag{27.84}$$

Next, we find the difference L_2 between the mean value of the growth of sales between January and June for the districts where television advertising was not presented and the mean value from July to December.

$$L_2 = \frac{-116.3}{18} - \frac{68.7}{18} = -10.28 \tag{27.85}$$

Although according to Equation (27.84) *the sales decreased 4.08% when the television advertisements were presented, inasmuch as they decreased 10.28% in the districts where the television advertisements were not shown,* this means that the sales were increased 6.2% by the presenting of the television commercials compared with the case where they were not presented.

However, it is not possible by such simple calculations to understand clearly the effect of television advertising. It is important to perform an analysis of variance on the data of Table 27.13 and to conduct research regarding the correctness of the estimated value of the effect of the television advertising, which has prevented a large decrease in sales.

27.3.4 Analysis of Variance

Analysis of variance in this instance involves resolution of the total variation of the 108 items of data, S_T, into the sum of the influences of various causes believed to have influenced it. What can be considered as causes in this case are, for example,

(1) Difference of growth depending on whether there was television advertising or not in geographic region B_1.
(2) Difference of growth depending on differences of districts D_1–D_9.
(3) Difference of growth depending on months, between C_1–C_6 and C_6–C_{12}.
(4) Difference of growth classed by district during the television advertising and after the advertising in geographic region B_2.
(5) Difference of growth classed by month after the television publicity began in region B_1.

Let us find the influences of these sources one by one and proceed to evaluate them correctly. We start by finding the total variation of growth, S_T.

$$S_T = 5.1^2 + 3.2^2 + \cdots + (-0.9)^2 - \frac{84.1^2}{108} = 2156.98 \qquad (f = 107) \tag{27.86}$$

The effect of television advertising is found by comparing the difference between the mean of A_2 and the mean of A_1 in regions B_1 and B_2. For this, we resolve the variations between A_1B_1, A_1B_2, A_2B_1, and A_2B_2 into the main effect of A, the main effect of B, and the interaction $A \times B$.

$$S_A=\frac{1}{108}[208.0-(-123.9)]^2=1019.98 \qquad (f=1) \tag{27.87}$$

$$S_B=\frac{131.7^2}{72}+\frac{(-47.6)^2}{36}-\frac{84.1^2}{108}=238.35 \qquad (f=1) \tag{27.88}$$

$$S_{A\times B}=\left[\frac{139.3^2+(-7.6)^2}{36}+\frac{68.7^2+(-116.3)^2}{18}-\frac{84.1^2}{108}\right]-S_A-S_B$$

$$=1488.76-1019.98-238.35=230.43 \qquad (f=1) \tag{27.89}$$

This variation $S_{A\times B}$ is the same as the variation of the estimated value of television advertising effect L, which is the difference between Equations (27.84) and (27.85). This is because, with A_1B_1 etc. as the total of the data of A_1B_1, we may modify Equation (27.89) and obtain

$$S_{A\times B}=\frac{A_1B_1{}^2+A_2B_1{}^2}{36}+\frac{A_1B_2{}^2+A_2B_2{}^2}{18}-\frac{(A_1B_1+\cdots+A_2B_2)^2}{108}$$

$$-\left[\frac{A_1{}^2+A_2{}^2}{54}-\frac{(A_1+A_2)^2}{108}\right]-\left[\frac{B_1{}^2}{72}+\frac{B_2{}^2}{36}-\frac{(B_1+B_2)^2}{108}\right]$$

$$=\frac{[(A_1B_1-A_2B_1)-2(A_1B_2-A_2B_2)]^2}{216} \qquad (f=1) \tag{27.90}$$

This value is equal to the variation of the difference between L_1 and L_2. L, which is the difference between

$$L_1=\frac{A_2B_1}{36}-\frac{A_1B_1}{36}$$

$$L_2=\frac{A_2B_2}{18}-\frac{A_1B_2}{18}$$

is

$$L=L_1-L_2=\frac{(A_2B_1-A_1B_1)}{36}-\frac{(A_2B_2-A_1B_2)}{18} \tag{27.91}$$

To obtain the variation S_L of this, we need merely divide the square of L by the sum of squares of the coefficients of L. The sum of squares of the coefficients is as follows.

$$\left(\frac{1}{36}\right)^2\times 36+\left(\frac{-1}{36}\right)^2\times 36+\left(\frac{-1}{18}\right)^2\times 18+\left(\frac{1}{18}\right)^2\times 18=\frac{2}{36}+\frac{2}{18}$$

From this, the variation S_L of L becomes as follows.

$$S_L=\frac{\left(\dfrac{A_2B_1-A_1B_1}{36}-\dfrac{A_2B_2-A_1B_2}{18}\right)^2}{\dfrac{2}{36}+\dfrac{2}{18}}$$

$$=\frac{[(A_1B_1-A_2B_1)-2(A_1B_2-A_2B_2)]^2}{216} \qquad (f=1) \tag{27.92}$$

We therefore have

$$S_{A\times B}=S_L \tag{27.93}$$

$S_{e_1}(B)$ is the variation which indicates about how much difference there is in the growth of sales between the nine districts $D_1, D_2, \ldots, D_9$, and we find this by dividing it into two parts, the difference of the growth among the districts D_1–D_6 before the television advertising and the difference of the growth among D_7–D_9 during the entire period, and this quantity expresses the difference in growth among the districts when there was no television advertising. Since difference of district is confounded with the difference between B_1 and B_2, S_B must be tested by this variation. It is because it is the error for testing of the main effect of B that it is abbreviated $S_{e_1}(B)$. $S_{e_1}(B)$ is also termed the first-order error variance of B.

$$S_{e_1}(B) = \left(\frac{25.9^2+23.2^2+\cdots+25.4^2}{6} - \frac{139.3^2}{36}\right) + \left[\frac{8.7^2+(-45.7)^2+(-10.6)^2}{12} - \frac{(-47.6)^2}{36}\right] = 194.58 \qquad (f=7) \tag{27.94}$$

Inter-month difference $S_{e_1}(A)$ is the difference among months when there was no television advertising, and this is given by the sum of the variations among $C_1, C_2, \ldots, C_6$ during A_1 and the variations among $C_7, C_8, \ldots, C_{12}$ in region B_2.

$$S_{e_1}(A) = \left(\frac{42.5^2+31.3^2+\cdots+39.4^2}{9} - \frac{208.0^2}{54}\right) + \left[\frac{(-22.0)^2+(-16.3)^2+\cdots+(-2.2)^2}{3} - \frac{(-116.3)^2}{18}\right] = 216.43 \qquad (f=10) \tag{27.95}$$

It is usual that a large change occurs in the growth of sales at the time of television advertising from July to December and as a result of the subsequent residual effect. The effect in the residual time is especially important. Even if there were large sales during the six-month period thanks to mass communication media or special sales, if it turns out that the total sales proceeds decreased thereafter by reaction and the total sales were nearly the same as without advertising, such an expanded sales technique has served only to hasten sales, and in the last analysis it is a waste of the cost of the advertising. Therefore, when there is reason to believe that the effect of the advertising has great residual effect, it is necessary to survey the sales for a sufficiently long time afterward. But if the number of new consumers increases as a result of such advertising and if part of these consumers feel that the merchandise is better than earlier products and become long-term customers, such a residual effect will last permanently. However, residual effect which comes from market expansion is related to the problem of the market value of the merchandise, and estimation for this must be performed by a different method. It is unnecessary to include this in the residual effect as the term is used here. The residual effect we are discussing here means that sales continue to be abnormal; and the effect after sales reach a steady state must be interpreted as the effect due to market expansion.

The effect of the tendency curve of the growth of sales due to this television advertising, averaged over the districts, can be found by the following equation.

$$S(\text{Difference among } C_7, C_8, \ldots, C_{12} \text{ within } B_1) = \frac{14.1^2+14.8^2+\cdots+(-5.3)^2}{6} - \frac{(-7.6)^2}{36} = 176.33 \quad (f=5) \tag{27.96}$$

Next, the difference in the shapes of the curves of the growth of sales during and after the television advertising in the districts D_1-D_6, where the advertising was presented, is given by the interaction within $B_1 \times$ within A_2.

$$S(\text{within } B_1 \times \text{within } A_2) = 3.4^2+4.2^2+\cdots+1.2^2 - \frac{(-7.6)^2}{36} - \left[\frac{6.8^2+\cdots+2.4^2}{6} - \frac{(-7.6)^2}{36}\right] - \left[\frac{14.1^2+\cdots+(-5.3)^2}{6} - \frac{(-7.6)^2}{36}\right] = 62.54 \quad (f=25) \tag{27.97}$$

Lastly, we must find the variation that compares what has happened to the growth of sales from July to December as a result of the television advertising as it depends on the districts $D_1, D_2, \ldots, D_6$. This is the variation that compares the mean values of the growth of sales from July to December in the respective districts $D_1, D_2, \ldots, D_6$, and this is abbreviated as, for example, the variation of D_1-D_6 within A_2 or $S[B_1$ within $A_2]$.

$$S(B_1 \text{ within } A_2) = \frac{6.8^2+(-3.9)^2+\cdots+2.4^2}{6} - \frac{(-7.6)^2}{36} = 50.81 \quad (f=5) \tag{27.98}$$

With the foregoing, all calculations for the so-called sources have been finished. What remains is the error variation; this can be found by subtracting the total of these source variations from the total variation S_T. However, calculations of the various types of variations as explained above are a lot of trouble, and in order to find errors in the calculating equations and the calculations themselves it is best to find the error variation directly, then to calculate the total with the source variations and to check by comparing with the total variation.

For this example, the error variation is found by adding three components. The first is the error $S(\text{within } B_1 \times \text{within } A_1)$ which the districts D_1-D_6 in which the television advertisement was presented possessed during the six months $C_1, C_2, \ldots, C_6$ before the television advertisement.

$$S(\text{within } B_1 \times \text{within } A_1) = 5.1^2 + 3.2^2 + \cdots + 3.0^2 - \frac{139.3^2}{36}$$
$$- \left(\frac{29.7^2 + 20.1^2 + \cdots + 27.0^2}{6} - \frac{139.3^2}{36} \right) - \left(\frac{25.9^2 + \cdots + 25.4^2}{6} - \frac{139.3^2}{36} \right)$$
$$= 51.35 \qquad (f = 25) \qquad (27.99)$$

The interaction between districts D_7, D_8, and D_9 where the television advertisement was not shown and the months $C_1, C_2, \ldots, C_{12}$ becomes an error. This is abbreviated as: within $B_2 \times C$.

$$S(\text{within } B_2 \times C) = 7.0^2 + 4.8^2 + \cdots + (-0.9)^2 - \frac{(-47.6)^2}{36}$$
$$- \left[\frac{(-26.2)^2 + \cdots + (-36.9)^2}{12} - \frac{(-47.6)^2}{36} \right] - \left[\frac{12.8^2 + \cdots + (-2.2)^2}{3} \right.$$
$$\left. - \frac{(-47.6)^2}{36} \right] = 40.58 \qquad (f = 22) \qquad (27.100)$$

Next, the interaction of $C_1, C_2, \ldots, C_6$ during A_1, and B_1 and B_2 is an error. This is termed: within $A_1 \times B$.

$$S(\text{within } A_1 \times B) = \left(\frac{29.7^2 + \cdots + 27.0^2}{6} + \frac{12.8^2 + \cdots + 12.4^2}{3} - \frac{208.2^2}{54} \right)$$
$$- \left(\frac{42.5^2 + \cdots + 39.4^2}{9} - \frac{208.0^2}{54} \right) - \left(\frac{139.3^2}{36} + \frac{68.7^2}{18} - \frac{208.0^2}{54} \right)$$
$$= 2.36 \qquad (f = 5) \qquad (27.101)$$

Therefore, by adding these error variations, we find error variation S_{e_2} with 52 degrees of freedom.

$$S_{e_2} = S(\text{within } B_1 \times \text{within } A_1) + S(\text{within } B_2 \times C)$$
$$+ S(\text{within } A_1 \times B)$$
$$= 51.35 + 40.58 + 2.36 = 94.29 \qquad (f = 52) \qquad (27.102)$$

The analysis of variance table, Table 27.14, is obtained from the foregoing calculations.

Since $e_1(A)$ and $e_1(B)$ are significant when tested with second-order error variance e_2, A and B were tested with $e_1(A)$ and $e_1(B)$. Source within $A_2 \times$ within B_2 is not significant by e_2. Thus, it may be concluded that the change in growth of sales after the television advertisement was about the same in all of the districts. This means that we may express the curve of change of growth after the advertisement by a single curve. Within $A_2 \times$ within B_1 was pooled with error e_2 and the pooled error (e_2) was created, and error variance $V_2 = 2.04$ was found.

Calculations of the contribution ratios were performed as follows. Clearly, effect classed by month on the growth is mixed in with the effect of A. Therefore, the net variation of A, S_A' is

Table 27.14 Analysis of Variance Table

Source	f	S	V	F_0	S'	ρ[%]
A	1	1 019.98	1 019.98	47.1**	998.34	43.7
$e_1(A)$	10	216.43	21.64	12.0**	215.63	9.4
B	1	238.35	238.35	8.6*	210.55	9.2
$e_1(B)$	7	194.58	27.80	13.6**	206.06	9.0
$A\times B$	1	230.43	230.43	127.3**	228.39	10.0
A_2 within B_1	5	176.33	35.27	19.4**	166.13	7.3
B_1 within A_2	5	50.81	10.16	5.6**	40.61	1.8
within $A_2\times$ within B_1	25	62.54	2.50	1.38		
e_2	52	94.29	1.81			
(e_2)	(77)	(156.83)	(2.04)		218.03	9.6
T	107	2 283.74			2 283.74	100.0

$$S_A' = S_A - (\text{number of degrees of freedom of } A) \times V(e_1(A)) = 1019.98 - 1 \times 21.64 = 998.34 \qquad (27.103)$$

The net variation of $e_1(A)$ is found by adding one error variance $V(e_1(A))$ which has been subtracted from the S_A above and subtracting the part of the second-order error variance which is mixed in with it.

$$S'(e_1(A)) = S(e_1(A)) \times \frac{11}{10} - 11 \times V_2 = 21.643 \times \frac{11}{10} - 11 \times 2.04 = 215.63 \qquad (27.104)$$

The reason why it has been multiplied by $\frac{11}{10}$ in the equation above is that we have added error variance of one degree of freedom $V(e_1(A))$ which has been subtracted from S_A. The other net variations are found similarly.

$$S_B' = S_B - 1 \times 27.80 = 210.55$$

$$S'(e_1(B)) = 194.58 + 27.80 - 8 \times 2.04 = 206.06$$

$$S_{A\times B}' = S_{A\times B} - 1 \times 2.04 = 228.39$$

$$\cdots$$

$$S_{e_2}' = 156.83 + 30 \times 2.04 = 218.03$$

According to the analysis of variance table, Table 27.14, the 108 items of data of the growth of sales generally possess different values one to the next; 43.7% of this variation is caused by the difference of presenting or not presenting television advertising and of economic variation between the earlier half and the later half, and troublesome calculations become necessary in order to separate the part of this that was due to the effect of television advertisement and the part that was due to the economic variation. Also, the total of the variation classed by month of the earlier half and the variation classed by month of the later half shows a contribution ratio of 10.6%. The major objective of such an analysis of variance is the information on $A \times B$, and further, the information on A_2 within B_1 and B_1 within A_2 and within $A_2 \times$ within B_2. The detailed meaning of these will be explained in the next section.

27.3.5 Estimation of Advertising Effect

The interaction of A and B is large and significant, and its contribution ratio also is large, being 10.0%. $A \times B$ indicates how much difference there was in the mean values of the growth of sales in districts where television advertising was used and in districts where it was not used. Its value L is, from Equations (27.84) and (27.85),

$$L = L_1 - L_2 = -4.08 - (-10.28) = 6.20 \pm 1.16 [\%] \tag{27.105}$$

Here the confidence limits were found from the following equation.

$$\sqrt{F \times V_e \times \text{unit number}} = \sqrt{3.97 \times 2.04 \left(\frac{2}{36} + \frac{2}{18}\right)} = \pm 1.16\% \tag{27.106}$$

These numbers indicate that when television advertising is presented, compared with when it is not, the growth of the proceeds from sales becomes greater by at least 5.04% and, on the higher side, by about 7.36%. Just because the growth of the proceeds from sales has increased by only 1.32% in a certain district in spite of the fact that this television advertisement was shown, that does not contradict this conclusion. Had the growth of sales when the television advertisement was not presented been -5.00%, it would have meant that the proceeds from sales had grown 6.32% because of the television advertisement. Thus, *in regard to the effect of television advertising, only comparison with the case when it has not been used is important, and it does not constitute an estimation of the absolute value of the growth of sales.* Worded differently, it means that since there are, for example, abrupt changes of economic conditions, one should not find the effect of television advertising from the difference between the mean before the advertising, A_1, and the mean after the advertising, A_2.

Next, the curves of change of the growth during the television advertising and afterward can be found as follows. We compare the differences of growth for group B_1 and group B_2. We find these for each of the months $C_7, C_8, \ldots, C_{12}$. The results are as given in Table 27.15.

Differences between the districts of group B_1 and group B_2 are clearly confounded in these values of the differences. It is therefore necessary to further correct the geographic region difference B. For this, we estimate the value of L', which is the difference of the mean values before television advertising

Table 27.15 Growth by Month

	C_7	C_8	C_9	C_{10}	C_{11}	C_{12}
B_1	2.35	2.47	0.52	−3.43	−2.28	−0.88
B_2	−7.33	−5.43	−11.83	−6.77	−6.33	−0.73
Difference	9.68	7.90	12.35	3.34	4.05	−0.15

$$L' = \bar{B}_1 - \bar{B}_2 = \frac{139.3}{36} - \frac{68.7}{18} = 0.05 \tag{27.107}$$

Since this difference is very small, correction is actually unnecessary, but let us try correcting here. We need merely further subtract 0.05 from the differences given in Table 27.15. The results become as given by Table 27.16.

The confidence limits were found as follows.

$$\sqrt{3.97 \times 2.04\left(\frac{1}{6} + \frac{1}{3} + \frac{1}{36} + \frac{1}{18}\right)} = 2.17 \tag{27.108}$$

The unit numbers within the parentheses in the equation above are as follows: The first is for the mean of district B_1 per month, the next is for the mean of district B_2 per month, the third is for the mean of district B_1 before the television advertising, and the last is for the mean of district B_2 before the advertising.

We estimate the profit from this television advertising as follows. We will assume that the proceeds of sales in the latter half of last year in the respective districts $D_1, D_2, \ldots, D_6$ were 1500, 860, 620, 180, 340, and 500 million yen, respectively, that the gross profit ratio was 40%, and that the proportional expense of selling constituted about one-half of this. If it is assumed that there is no effect of this television advertising after December, the amount of profit due to increase of sales during this period can be found as follows.

$$(1500 + 860 + 620 + 180 + 340 + 500) \times 0.20 \times 0.0620 = 49.6 \text{ million yen} \tag{27.109}$$

Table 27.16 Effect Classed by Month of Growth of Sales Proceeds by Television Advertisement

Month	Growth difference	
C_7	9.63	±2.17
C_8	7.85	〃
C_9	12.30	〃
C_{10}	3.29	〃
C_{11}	4.00	〃
C_{12}	−0.20	〃

Therefore, if the expense of this television advertisement had been 50 million yen, this would mean a deficit of about 400,000 yen. If a certain fraction of the buying public who increased their purchases here remain as customers for the product in the future, it is necessary to calculate this as well. For example, if 10% of the new consumers were to continue to buy this merchandise, the effect would be, in one-half year, 49.6 million yen $\times$ 0.1 = 4.96 million yen. If the intra-business interest is 15%, this is the same as though one had gained a further income of (4.96 million yen $\div$ 0.15) $\times$ 2 = 66 million yen. Thus, in the last analysis, by an investment of 50 million yen, the effect of 49.6 + 66.2 = 115.8 million yen has been generated, and a profit of 65.9 million yen has been gained.

27.4 When Surveying Small Businesses and Consumer Strata

27.4.1 Objective

In the previous section, we have shown the analysis method when we took as our data the growth of sales in each geographic district. Adequate calculations are possible by the method of the previous section, for example, in the cases of sales at stores, deposits and contract amounts at banks and insurance companies, electric power use, leisuretime travel, color television service contracts, and services such as delivery of perishable foods where intermediate stock is useless. In the case of products that can be stocked, however, sometimes the increase or decrease of sales of company divisions actually means the increase or decrease of stock at wholesalers and retailers. Demand from one's own company sometimes increases or decreases temporarily for reasons such as speculation or high-pressure sales. In such a case, *it is not possible to understand correctly the effects of advertising and of expansion-of-sales counter measures unless one surveys whether or not the ultimate consumer has actually bought.*

Thus, as explained in the previous section, it is necessary to examine the sales by small retail store units. How many retail stores should be investigated by any company unit differs somewhat depending on the type of merchandise, but whatever the case, it is enough to arrange that the number of items bought by the consumer amounts to from several hundred to several thousand for the whole of the survey. Let us assume that when investigating the sales every week by retail store units for five weeks before newspaper advertising and ten weeks after television advertising, it is expected that the mean number of this merchandise sold is about 0.5 in one week. For each retail store, sales of about 0.5 $\times$ 15 = 7.5 can be expected. If we take 60 retail stores, in this case, the total number sold will be 450, and it is a generally good figure.

If the sale per store is much more, as in the case of chewing gum or instant noodles, one should survey so that the total number sold will be several thousand pieces. For the data themselves, one should obtain not only the number sold but the proceeds of the sales, but one should decide on the scale of the survey by the number of items.

27.4.2 Analysis of Variance

We will discuss analysis for a case where a survey such as given by Table 27.12 has been carried out. In a case such as this, one should first distribute the numbers of degrees of freedom in the analysis of variance table. Distribution of the degrees of freedom becomes as given by Table 27.17.

A survey such as Table 27.12 is a *typical market survey*. Instead of the two levels of B, there might be two types of sales-expansion countermeasures, or a situation where two types of direct mail were sent during the months C_4–C_6. In such a case, however, it is better if there is information from somewhat earlier months. The calculation equations for the sources shown in Table 27.17 will be given. T is the total of the whole, A_1 is the total of the data of A_1, and A_1B_1 means A_1 and also the total of the data of B_1, and so forth. The variations are found as follows.

$$CF=\frac{T^2}{135} \qquad (f=1) \tag{27.110}$$

$$S(A_1 \text{ and } A_2, A_3)=\frac{A_1^2}{45}+\frac{(A_2+A_3)^2}{90}-CF \qquad (f=1) \tag{27.111}$$

$$S(A_2, A_3)=\frac{(A_2-A_3)^2}{90} \qquad (f=1) \tag{27.112}$$

$$S[e_1(A)]=\left(\frac{C_1^2+C_2^2+C_3^2}{15}-\frac{A_1^2}{45}\right)+\left(\frac{B_2C_4^2+\cdots+B_2C_6^2}{5}-\frac{B_2A_2^2}{15}\right)$$
$$+\left(\frac{B_2C_7^2+\cdots+B_2C_9^2}{5}-\frac{B_2A_3^2}{15}\right) \qquad (f=6) \tag{27.113}$$

Table 27.17 Distribution of Numbers of Degrees of Freedom When Retail Stores Were Surveyed

Source		f	Source		f
A	A_1 and A_2, A_3	1	$A\times B\times D'$	$(A_1$ and $A_2, A_3)\times B\times D'$	4
	A_2, A_3	1		$(A_2, A_3)\times B\times D'$	4
$e_1(A)$		6	$A\times$ within D'	$(A_1$ and $A_2, A_3)\times$ within D'	5
B		1		$(A_2, A_3)\times$ within D'	5
D'		4			
$B\times D'$		4	B_1 within A_2, A_3		4
$e_1(B)=$ within D'		5	within $B_1\times$ (within A_2, A_3)		36
$A\times B$	$(A_1$ and $A_2, A_3)\times B$	1	e_2		44
	$(A_2, A_3)\times B$	1			
$A\times D'$	$(A_1$ and $A_2, A_3)\times D'$	4	T		134
	$(A_2, A_3)\times D'$	4			

$$S_B=\frac{(B_1-2B_2)^2}{270} \qquad (f=1) \tag{27.114}$$

$$S_{D'}=\frac{(D_1')^2+\cdots+(D_5')^2}{27}-CF \qquad (f=4) \tag{27.115}$$

$$S_{B\times D'}=\frac{(B_1D_1')^2+\cdots+(B_1D_5')^2}{18}+\frac{(B_2D_1')^2+\cdots+(B_2D_5')^2}{9}$$

$$-CF-S_B-S_{D'} \qquad (f=4) \tag{27.116}$$

$$S[e_1(B)]=\frac{1}{18}[(D_1-D_2)^2+(D_3-D_4)^2+\cdots+(D_9-D_{10})^2] \qquad (f=5) \tag{27.117}$$

$$S[(A_1 \text{ and } A_2,\ A_3)\times B]=\frac{(A_1B_1)^2}{30}+\frac{(A_2B_1+A_3B_1)^2}{60}+\frac{(A_1B_2)^2}{15}+\frac{(A_2B_2+A_3B_2)^2}{30}$$

$$-CF-S(A_1 \text{ と } A_2,\ A_3)-S_B \qquad (f=1) \tag{27.118}$$

$$S[(A_2,\ A_3)\times B]=\frac{(A_2B_1)^2+(A_3B_1)^2}{30}+\frac{(A_2B_2)^2+(A_3B_2)^2}{15}-\frac{(A_2B_1+A_3B_1)^2}{60}$$

$$-\frac{(A_2B_2+A_3B_2)^2}{30}-S(A_2,\ A_3) \qquad (f=1) \tag{27.119}$$

$$S[(A_1 \text{ and } A_2,\ A_3)\times D']=\frac{(A_1D_1')^2+\cdots+(A_1D_5')^2}{9}$$

$$+\frac{(A_2D_1'+A_3D_1')^2+\cdots+(A_2D_5'+A_3D_5')^2}{18}$$

$$-CF-S_{D'}-S(A_1 \text{ and } A_2,\ A_3) \qquad (f=4) \tag{27.120}$$

$$S[(A_2,\ A_3)\times D']=\frac{(A_2D_1')^2+\cdots+(A_3D_5')^2}{9}$$

$$-\frac{(A_2D_1'+A_3D_1')^2+\cdots+(A_2D_5'+A_3D_5')^2}{18}-S(A_2,\ A_3) \quad (f=4) \tag{27.121}$$

$$S[(A_1 \text{ and } A_2,\ A_3)\times B\times D']=\frac{(A_1B_1D_1')^2+\cdots+(A_1B_1D_5')^2}{6}$$

$$+\frac{(A_2B_1D_1'+A_3B_1D_1')^2+\cdots+(A_2B_1D_5'+A_3B_1D_5')^2}{12}$$

$$+\frac{(A_1B_2D_1')^2+\cdots+(A_1B_2D_5')^2}{3}$$

$$+\frac{(A_2B_2D_1'+A_3B_2D_1')^2+\cdots+(A_2B_2D_5'+A_3B_2D_5')^2}{6}$$

$$-CF-S(A_1 \text{ and } A_2,\ A_3)-S_B-S_{D'}-S[(A_1 \text{ and } A_2,\ A_3)\times B]$$

$$-S[(A_1 \text{ and } A_2,\ A_3)\times D']-S_{B\times D'} \qquad (f=4) \tag{27.122}$$

$$S[(A_2,\ A_3)\times B\times D']=\frac{(A_2B_1D_1')^2+\cdots+(A_3B_1D_5')^2}{6}$$

$$+\frac{(A_2B_2D_1')^2+\cdots+(A_3B_2D_5')^2}{3}-CF'-S(A_2,\ A_3)-S_B'-S_{D'}'-S_{B\times D'}'$$

$$-S[(A_2,\ A_3)\times B]-S[(A_2,\ A_3)\times D'] \qquad (f=4) \tag{27.123}$$

Here, CF$'$, S_B', $S_{D'}'$, and $S_{B\times D'}'$ are the variations of CF, S_B, S_D, and $S_{B\times D}$, obtained only from the data of A_2 and A_3.

$$S[(A_1 \text{ and } A_2, A_3) \times \text{within } D'] = \frac{1}{36}\{[(2A_1D_1 - A_2D_1 - A_3D_1) - (2A_1D_2 - A_2D_2 - A_3D_2)]^2 + \cdots + [(2A_1D_9 - A_2D_9 - A_3D_9) - (2A_1D_{10} - A_2D_{10} - A_3D_{10})]^2\} \quad (f=5) \tag{27.124}$$

$$S[(A_2, A_3) \times \text{within } D'] = \frac{1}{12}\{[A_2D_1 - A_3D_1) - (A_2D_2 - A_3D_2)]^2 + \cdots + [(A_2D_9 - A_3D_9) - (A_2D_{10} - A_3D_{10})]^2\} \quad (f=5) \tag{27.125}$$

It is necessary to find the main effect of within A_2, A_3 from the period during the television advertising to afterward from the data of the group of retail stores of B_1.

$$S(B_1 \text{ within } A_2, A_3) = \left(\frac{B_1C_4^2 + \cdots + B_1C_6^2}{10} - \frac{A_2B_1^2}{30}\right) + \left(\frac{B_1C_7^2 + \cdots + B_1C_9^2}{10} - \frac{A_3B_1^2}{30}\right) \quad (f=4) \tag{27.126}$$

S[within $B_1 \times$ (within A_2, A_3)] is the difference of the effects of within B_1 within A_2, A_3.

$$S[(\text{within } A_2, A_3) \times \text{within } B_1] = \left[y_{14}^2 + y_{15}^2 + \cdots + y_{10,6}^2 - \frac{A_2B_1^2}{30} - S(\text{within } A_1 \text{ of } B_1)\right] - S(\text{within } B_1 \text{ of } A_2)] + \left[y_{17}^2 + y_{18}^2 + \cdots + y_{10,9}^2 - \frac{A_3B_1^2}{30} - S(\text{within } A_3 \text{ of } B_1)\right] - S(\text{within } B_1 \text{ of } A_3) \; (f = 36) \tag{27.127}$$

In certain cases, this effect is also often resolved as

$$S[(\text{within } A_2, A_3) \times \text{within } B_1] = S[(\text{within } A_2, A_3) \times D' \text{ of within } B_1] + S[(\text{within } A_2, A_3) \times \text{within } D'] \tag{27.128}$$

The error variation can be found by subtracting the sum of the above source variations from the total variation, but orthogonal resolution is difficult and calculation errors are abundant unless one is rather used to calculations of analysis of variance. Therefore, in order to examine whether or not the respective variations have been resolved correctly, it is absolutely necessary to check by finding the error variation directly. In order to examine mathematically the orthogonality of the respective source variations, it is necessary to draw up 135×135 matrices in the quadratic form for 17 variations and to prove that all the product matrices are zero matrices, and such calculations are nearly impossible even with the newest computers. Therefore, *it is best to examine the orthogo-*

nality by checking. In this case, the error variation consists of three components. We need but find them separately, then add them.

(1) within $B_1 \times$ within A_1 is found from within A_1
(2) within $B_2 \times$ (within A_1, A_2, A_3) is found from within B_2
(3) within $A_1 \times B$

It should be clear that these are all components of error.

$$S(\text{within } B_1 \times \text{within } A_1) = y_{11}{}^2 + y_{12}{}^2 + \cdots + y_{10,3}{}^2 - \frac{A_1B_1{}^2}{30}$$
$$- S(\text{main effects of } C_1, C_2, \text{and } C_3 \text{ within } B_1)$$
$$- S(\text{main effects of } D_1\text{–}D_{10} \text{ within } A_1)\ (f = 18) \qquad (27.129)$$

$$S[\text{within } B_2 \times (\text{within } A_1, A_2, A_3)] = [y_{11,1}{}^2 + y_{11,2}{}^2 + \cdots + y_{15,9}{}^2 - \frac{B_2^2}{45} - S(\text{within } B_2)$$
$$- S(\text{main effects of } C_1, C_2, \ldots, C_9 \text{ at } B_2)$$
$$- S(D' \times A \text{ at } B_2)]\ (f = 24) \qquad (27.130)$$

$$S(\text{within } A_1 \times B) = \frac{1}{30}[(B_1C_1 - 2B_2C_1)^2 + \cdots + (B_1C_3 - 2B_2C_3)^2]$$
$$- \frac{1}{90}[(A_1B_1 - 2A_1B_2)^2] \qquad (f = 2) \qquad (27.131)$$

By adding these, we get

$$S_{e_2} = S(\text{within } B_1 \times \text{within } A_1) + S[\text{within } b_2 \times (\text{within}$$
$$A_1, A_2, A_3)] + S(\text{within } A_1 \times B)\ (f = 44) \qquad (27.132)$$

Here ends the explanation of the calculation formulas.

27.4.3 Market Development and Product Quality Problem

Market expansion and market development are major problems for every business. If the quality of the product and service is bad in proportion to the price, no matter how ingeniously one advertises and performs market development, ultimately only a single item is bought and the market share does not increase. However, it is also a fact that if it is not possible to cause the consumer to buy the first one, the market share does not increase no matter how good a product or service is sent to the market in proportion to the price.

When mass communication media are used, it is not possible to vary the conditions easily, as it is when developing a product, in studying the maintainance and expansion of the market share. Often, for data of mass communication media such as are used in advertising, there is no other way but to devote all one's energy to the data analysis method. Sections 27.3 and 27.4 have explained the method of applying the most exhaustive data analysis conceivable at present, through simple examples. It would probably not be difficult to apply these methods by expanding them to more complicated assignments, as by multi-way layouts and orthogonal arrays.

Exercise (27)

(1) For the purpose of investigating the influence of a certain medicine on heartbeat, with A_1 = placebo and A_2 = medicine, six persons, R_1, $R_2, R_3, \ldots, R_6$, were randomly divided into two classes, and the following data were obtained for the number of heartbeats B_1 = before exercise, B_2 = immediately after exercise, B_3 = 10 minutes after exercise, and B_4 = 20 minutes after exercise. Perform appropriate analysis and draw conclusions.

	B_1	B_2	B_3	B_4
A_1R_1	74	119	97	77
R_2	64	114	85	62
R_3	77	127	98	72
A_2R_4	78	106	84	76
R_5	72	102	78	67
R_6	68	96	76	65

28 Analysis of Covariance (Experiment When There Are Supplementary Measured Values)

The method of performing factorial analysis when it is not possible to fix the levels or to categorize data beforehand, but when there are factors whose states and intermediate experimental values can be recorded, will be explained. The method also takes into account the influences of these supplementary variables on the target characteristic. The method of Chapter 15 is useful when there are only supplementary variables. Here we give an analysis method for cases where supplementary variables are few but assigned sources are numerous.

28.1 Experiment on Insecticides

The following data are those of an experiment on insecticidal effect for threadworms, which was performed at the Rothamsted Agricultural Experiment Station in 1935. For four types of insecticide,

$$A_1 = \text{CN}, A_2 = \text{CS}, A_3 = \text{CM}, A_4 = \text{CK}$$

each of the 12 combinations of treatment conditions using these at the three levels

$$B_1 = 0, B_2 = \text{standard quantity}, B_3 = \text{twice the standard quantity}$$

(actually, since there is no difference among A_1, A_2, A_3, and A_4 for B_1, there are nine types) was assigned to four blocks, R_1, R_2, R_3, and R_4.

Table 28.1 Assignment and Data (Middle Level Is x, Bottom Level Is y)

R_1				R_2			
0	2 CK	1 CN	1 CN	2 CM	2 CS	2 CK	0
269	283	252	212	59	127	80	134
466	280	398	386	199	166	142	592
1 CS	0	0	2 CM	1 CK	1 CN	1 CM	0
138	100	197	263	107	89	41	74
194	219	421	379	236	332	176	137
2 CS	1 CK	0	2 CN	0	0	2 CN	1 CS
282	230	216	145	88	25	42	62
372	256	708	304	356	212	308	221

R_3				R_4			
1 CK	0	1 CS	2 CK	2 CK	0	1CK	1 CM
124	211	194	222	193	209	109	153
268	505	433	408	292	352	132	454
0	2 CN	2 CS	1 CN	0	2 CN	2 CS	0
102	193	128	42	29	9	17	19
363	561	311	222	254	92	28	106
2 CM	0	1 CM	0	1 CS	1 CN	0	2 CM
162	191	107	67	23	19	44	48
365	563	415	338	80	114	268	298

Before the treatment, each block was divided into 12 geographic sectors and, after tilling the soil, 400 g of soil was taken from each and the number of threadworms x was determined. The insecticides were ploughed in, then oats were planted; after harvesting, 400 g of soil was again taken from each sector and the number of threadworms y was ascertained. Table 28.1 gives the number of worms x, indicating the conditions before treatment (termed a *supplementary measured value*), and the number of worms y after harvest beside the data of the treatment conditions. Here, 0 CN, 1 CN, and 2 CN represent zero quantity, standard quantity, and double quantity of CN.

In this experiment, the 12 modes of treatment were randomly distributed among the 12 sectors within each block, but from the standpoint of Design of Experiments this is an unskillful method. It would have been better to categorize the sectors into three classes by the number of threadworms x in the 12 sectors of each block (V_1 = four sectors with few worms, V_2 = four middle sectors, V_3 = four sectors with numerous worms), to assign to L_9 by the three levels of B and A_1, A_2, and A_3, and to partially supplement with A_4.

According to reference 6), Table 28.1 was accepted tentatively and assignment was made so that *analysis of covariance* could be performed with x as a supplementary measured value. The results were as follows.

Table 28.2 Analysis of Covariance Table (After Cochran and Cox)

Source	f	S	V	ρ[%]
$B\left\{\begin{matrix} l \\ q \end{matrix}\right.$	1 1	106 090 10 664	106 090 10 664	10.7 0.4
$A\times B_l$	3	120 566	40 189	10.0
$A\times B_q$	3	19 711	6 570	—
Block	3	110 066	36 689	9.0
x	1	374 876	374 876	37.1
e	35	249 592	7 131	32.8
T	47	991 565		100.0

Except for the confounded parts, the contribution ratios of the sources amounted to 21.1%. Therefore, the SN ratio η which indicates the *analysis efficiency* is

$$\eta=\frac{21.1}{32.8}=0.64 \tag{28.1}$$

When studying changes of data which assume arbitrary positive values, often the following method of variable transformation is desirable. The data after transformation are shown in Table 28.3.

$$10 \log \frac{\text{value after treatment}}{\text{value before treatment}} = 10 \log \frac{y}{x} \text{ db} \tag{28.2}$$

The supplementary tables, Tables 28.4, are constructed from this.

$$CF=\frac{213.15^2}{48}=946.5192$$

$$S_T=2.39^2+(-0.05)^2+\cdots+7.93^2-946.5192$$

$$=311.3501 \qquad (f=47) \tag{28.3}$$

$$S_{B_l}=\frac{(-86.73+60.18)^2}{32}=22.0282 \qquad (f=1) \tag{28.4}$$

$$S_{B_q}=\frac{(86.73-2\times 66.24+60.18)^2}{96}=2.1690 \qquad (f=1) \tag{28.5}$$

Table 28.3 Data After Transformation

R_1	2.39 1.48 1.20	−0.05 3.40 0.47	1.98 3.30 5.16	2.60 1.59 3.22	R_3	3.35 5.51 3.53	3.79 4.63 4.69	3.49 3.86 5.89	2.64 7.23 7.03
R_2	5.28 3.44 6.07	1.16 5.72 9.28	2.49 6.33 8.65	6.44 2.67 5.52	R_4	1.80 9.42 5.41	2.26 10.08 7.78	0.83 2.17 7.85	4.72 7.47 7.93

Tables 28.4 Supplementary Tables

	Total
R_1	26.74
R_2	63.05
R_3	55.64
R_4	67.72
Total	213.15

	B_1	B_2	B_3	B_2+2B_3	$2B_2-B_3$
A_1	86.73	22.71	26.58	75.87	18.84
A_2		15.90	8.39	32.68	23.41
A_3		19.54	18.33	56.20	20.75
A_4		8.09	6.88	21.85	9.30
Total	86.73	66.24	60.18	186.60	72.30

The difference of the linear effect relative to the quantities of use of A_1, A_2, A_3, and A_4 is given as the variation between $B_2 + 2B_3$.

$$S_{A\times B_l}=\frac{75.87^2+32.68^2+56.20^2+21.85^2}{20}-\frac{186.60^2}{80}$$

$$=87.7606 \qquad (f=3) \tag{28.6}$$

$$S_{A\times B_q}=\frac{18.84^2+23.41^2+20.75^2+9.30^2}{20}-\frac{72.30^2}{80}$$

$$=5.6602 \qquad (f=3) \tag{28.7}$$

The value obtained by adding Equations (28.4)–(28.7) agrees with the variation among the nine modes of treatment, S_{AB}.

$$S_{(AB)}=\frac{86.73^2}{16}+\frac{22.71^2+26.58^2+\cdots+6.88^2}{4}-CF$$

$$=117.6180 \qquad (f=8) \tag{28.8}$$

$$S_R=\frac{26.74^2+\cdots+67.72^2}{12}-946.5192=84.4923 \tag{28.9}$$

Therefore, the analysis of variance table is as given by Table 28.5.

When this is compared with Table 28.2, we see that the ratio of the total of the contribution ratios of the target sources, B_l and $B_l \times A$, to the error has increased from 64% to 71%. Calculation is simpler by using Table 28.5 than by troublesome analysis of covariance. Thus, in this case, it is *more desirable not to perform analysis of covariance.* Moreover, it is be-

Table 28.5 Analysis of Variance Table

Source	f	S	V	ρ[%]
R	3	84.4923	28.1641**	24.2
B_l	1	22.0282	22.0282**	6.1
B_q	1	2.1690	2.1690	
$A\times B_l$	3	87.7606	29.2535**	25.3
$A\times B_q$	3	5.6602	1.8867	
e	36	109.2398	3.0344	44.4
T	47	311.3501		100.0

lieved that this would have become an experiment of even higher precision had categorizing been performed by the number of threadworms, as mentioned at the beginning.

Thus, the cases where analysis of covariance becomes necessary are limited to those where the following two conditions are satisfied with respect to the supplementary factor.

(1) *Pre-sorting and post-sorting are not possible.*
(2) *It is not possible to correct its effect by ratio or known relationships.*

We analyzed in the example given above by taking the logarithm of the ratio, but it is also possible to perform an analysis of covariance with the logarithmic value of the ratio itself as the target characteristic and by using x as a supplementary variable. This will be left to the reader. Also, when the supplementary variable x is influenced by sources, combined use of analysis of covariance together with experimental regression analysis (Chapter 15) becomes important if nonlinear unknown constants are included between the supplementary variable and y. In this method, temporary numerical values are taken at three levels for the unknown constants, then, for the data corrected for the effect of x, ordinary analysis of variance is performed. This means that sequential approximation is conducted while re-taking the three levels so as to minimize error variation (often interactions too are included). By this method, however, calculations are very difficult unless a computer can be used. Please refer to Chapter 15 and reference 33).

28.2 How to Take Supplementary Measured Values

Sometimes, when designing an experiment, since it is not possible to take levels freely, we take the step of recording the state of a factor that influences the experiment results. This is termed a *supplementary factor*. An indicative factor (sometimes it is a block factor) which has been randomized by necessity is a supplementary factor. Let us assume that in investigating the influence of three types of feed on weight increase, we have tested them by using 27 animals. It is an inept method to allot the 27 animals randomly the treatments A_1, A_2, and A_3. It is better to assign by considering states of the animals which seem to be related to weight increase. In this method, for example the initial period weight, age, and other factors are measured, sequential categorization is performed according to these and a 3^2-type indicative factor created, and this is assigned to orthogonal array L_{27} together with A.

As the data in this case, as was shown in the previous section, we analyze

$$y = 10 \log \frac{\text{body weight before experiment}}{\text{initial body weight}} \text{ db} \tag{28.10}$$

It is therefore not necessary to use the analysis of covariance of this chapter. The same applies in the case of an animal experiment to determine whether or not a febrifacient substance is contained in an injection medicine, and we select animals with various body temperatures (the method of using rabbits with body temperature in the standard state is totally meaningless inasmuch as the injection medicine is often used for humans with fever). One need only categorize, having measured the body temperature, and proceed as in the case of body weight above.

Let us assume that there is no means of regulating temperature, and that the temperature at the time of measurement influences the measured values. Products are produced under various experimental conditions, and we measure. There is a method in which the room temperature is obtained by supplementary measurement, and its factorial effect on the characteristic value is investigated. This is termed analysis of covariance. In analysis of covariance we hypothesize first that, for example, a linear equation holds between the supplementary factor x and the experimental value y, which enables us, by estimating its linear coefficients from the data, *to correct y to a value much as though x had had a constant value* (usually the mean value), and by which one analyzes the factorial effects toward this. Such a case, where the value of the factor which is the randomized experimental condition is obtained by supplementary measurement for each experimental unit beforehand, and where one finds the factorial effects by data analysis of a modified case that can be treated as though these environmental factors had only one level, is termed an experiment with a *supplementary factor*. Section 28.3 shows how to perform data analysis in such a case.

In the case of a supplementary factor, *if factorial analysis is performed for the supplementary factor x, there is no reason why there should be a significant source.* This is because x must have been randomly distributed with respect to the various combinations of the experiment. On the other hand, in cases such as those of *intermediate experimental values* or *the analysis of survey data,* usually the value which has been measured as the supplement possesses factorial effect. Such a case is referred to as one where there is a *supplementary experimental value* (or supplementary survey value).

For example, when attempting to survey the clarity of tone of telephones which are being used at work sites, let us assume that we have divided the targets of the survey into shopping centers, plants, and private residences. When measuring the clarity by selecting telephones at several locations from each category, it will be assumed that the intensity of noise at each location at the time of measurement has been measured. Since the amount of noise clearly differs from category to category, there is a strong relationship between the category and the supplementary measured value; and thus in this case one may not use ordinary analysis of covariance without modification.

When it can be assumed that the supplementary measured value has a relationship with the source, three steps are necessary:

(1) The relationship between the assigned source and the supplementary experimental value is analyzed.
(2) The relationship between the supplementary measured value and the final characteristic is analyzed.
(3) The influence of the assigned source and the final characteristic value is divided into two parts, the part which influences the supplementary measured value and which, as a result, influences the final characteristic value, and the part which influences the final characteristic value without passing by way of the supplementary measured value.

With respect to the experiment of Section 28.1 of this chapter, let us assume that

$$x = 10 \log \frac{\text{number of threadworms at the time of harvest}}{\text{number of threadworms at the beginning}} \tag{28.11}$$

is the supplementary measured value and y is the quantity of yield of the oats. Since the effect of the insectide appears in the supplementary measured value x in this case, x is a supplementary measured value which is influenced by the source. But now, if we ask, Why are we concerned with insecticidal effect? it is because we wish to increase the yield thereby.

In effect, therefore, we wish to investigate the relationship between the harvested quantity and the insecticidal effect. Now, although it might be supposed that it suffices to analyze directly the relationship of the source to the insecticide and to the amount harvested, since certain types of insecticides contain substances that decompose within the soil and release nitrogen, it is conceivable that such insecticides might be influencing the amount harvested not only through their insecticidal effect but also as substances which impart a fertilizing effect, albeit slightly. If this were the case, we would wish to separate the effect of such insecticides on the harvest into two parts, the part which proceeds by way of the insecticidal effect and the part which does not.

Since something like the foregoing can always be said with regard to intermediate experimental values, it is necessary to analyze data with an attitude which differs from that toward the random case discussed at the beginning. In fact, nearly all discussions of analysis of covariance up to now have been in connection with random cases; there has been no adequate study of the other type of case, which is technically even more important.

28.3 Experiment on Purification of Penicillin

28.3.1 Objective

In producing penicillin, the penicillin in the culture solution which has been obtained by tank cultivation is purified. Let us suppose that the problem is, for example, how to decide on A, the quantity of sulfuric acid that is used for purification, or B, the quantity of the solvent. One should take A and B at a certain number of levels and perform an experiment by two-way layout, but often the potency of the penicillin in the solution that has been removed from the tank differs from batch to batch even if the steps up to then have been performed in a specific controlled state. Therefore, if one measures only the yield (%) from the purification by performing an experiment on A and B by the two-way layout method, it will be an experiment of poor precision unless fairly numerous replications are used, since the potency of the penicillin in the solution removed from the tanks shows dispersion. Although in such a case one might perform an experiment by two-way layout of A and B only for the solution in one batch, often reproducibility is lacking because only a small amount of material is available, as when the facility for purification has been constructed on a scale that is just right for one batch. In such a case, it will probably be judged that one should obtain as supplementary measurement the potency of the penicillin in the culture solution within the tank beforehand, that one should conduct an experiment by two-way layout with respect to A and B, and that one should consider the value of the potency of the penicillin in the data analysis.

Thus, when A and B are each at four levels, we perform an experiment of purification with all of the $4 \times 4 = 16$ runs of experiment in a random sequence, but before conducting each purification experiment we measure the potency x of the penicillin in the solution that has been removed from the tank. The data of Table 28.6 are those obtained when both potency x and yield of purification y have been measured.

In order to determine the quantity of sulfuric acid A and the quantity of solvent B that will benefit us by raising the yield from purification the most, each was taken at four levels,

$$A_1=1.8,\ A_2=2.0,\ A_3=2.2,\ A_4=2.4\ [\%]$$

$$B_1=2.0,\ B_2=2.4,\ B_3=2.8,\ B_4=3.2\ [\%]$$

and the combinations were experimented on once each in random order.

Since the potency of the solution removed from the tank influenced the yield y in this case, the potency of the solution before purifying, x, was also determined beforehand as a supplementary measurement.

Since in order to remove the part due to x from the total variation of the experimental value y we use the method of calculating the covaria-

Table 28.6 Data of Experiment (Upper Level Is Value of Potency x, Lower Level Is Value of Yield y)

	B_1	B_2	B_3	B_4
A_1	1 290 49.9	1 400 64.4	1 430 62.0	1 320 57.1
A_2	1 300 53.3	1 200 60.3	1 340 64.5	1 430 63.4
A_3	1 350 54.4	1 450 63.0	1 320 65.3	1 200 59.4
A_4	1 350 53.2	1 230 58.6	1 430 64.7	1 230 57.0

tion by hypothesizing a suitable functional relationship (including a certain number of unknown parameters) between x and y, this is termed the *analysis of covariation method.* Usually, in this case, the relationship with x (regression function) is held down to only about a linear approximation. Sometimes a variable transformation is performed on either or both of x and y, and linear or quadratic approximation is carried out for the new variable. Even if the relationship between x and y is a rather complicated function, a good deal of the variation due to x is eliminated even by linear regression; and most of it is eliminated if quadratic approximation is used. In a case where the function increases or decreases monotonically, most of the variation due to x becomes eliminated by linear approximation only, and if x is one of the major sources for the experimental values the precision of the experiment improves considerably. Nor is it always necessary that supplementary factor x be a quantitative variable; it can, for example, be a scale which has been appropriately applied to something qualitative that can be ranked in order, or it can be data of categorized values or groupings such as rank or class 1, 2, etc. This is because usually the experimental values change monotonically with respect to such ranks.

28.3.2 Analysis of Variance

Table 28.7 is obtained by subtracting the working mean as shown by Equations (28.12) and (28.13) and eliminating the decimal point.

$$\frac{1}{10}(x-1300) \tag{28.12}$$

$$10(y-60) \tag{28.13}$$

The variation (sum of squares of deviations) is calculated for x and y, while, at the same time, covariation (sum of products) of x and y is

Table 28.7 Processing Data

	B_1	B_2	B_3	B_4	Total
A_1	−1 −101	10 44	13 20	2 −29	24 −66
A_2	0 −67	−10 3	4 45	13 34	7 15
A_3	5 −56	15 30	2 53	−10 −6	12 21
A_4	5 −68	−7 −14	13 47	−7 −30	4 −65
Total	9 −292	8 63	32 165	−2 −31	47 −95

calculated. If the symbols CF = correction factor, sources A and B, error e, and total variation T are used as previously and if the variation with respect to x is expressed by the symbol (xx) and the covariation with respect to xy by the symbol (xy), we have

$$CF \begin{cases} (xx) = \dfrac{(\text{sum of } x)^2}{\text{total number of experiments}} = \dfrac{47^2}{16} \fallingdotseq 138 & (28.14) \\[2ex] (xy) = \dfrac{(\text{sum of } x)(\text{sum of } y)}{\text{total number of experiments}} = \dfrac{47 \times (-95)}{16} \fallingdotseq -279 & (28.15) \\[2ex] (yy) = \dfrac{(\text{sum of } y)^2}{\text{total number of experiments}} = \dfrac{(-95)^2}{16} \fallingdotseq 564 & (28.16) \end{cases}$$

$$S_T \begin{cases} (xx) = (-1)^2 + 10^2 + 13^2 + \cdots + (-7)^2 - CF(xx) = 1067 & (28.17) \\ (xy) = (-1) \times (-101) + 10 \times 44 + 13 \times 20 + \cdots + (-7)(-30) & \\ \qquad - CF(xy) = 2529 & (28.18) \\ (yy) = (-101)^2 + 44^2 + 20^2 + \cdots + (-30)^2 - CF(yy) = 35\,303 & (28.19) \end{cases}$$

$$S_A \begin{cases} (xx) = \dfrac{1}{4}[(\text{sum of } A_1 \text{ of } x)^2 + \cdots + (\text{sum of } A_4 \text{ of } x)] - CF(xx) & \\[1ex] \quad = \dfrac{1}{4}(24^2 + 7^2 + 12^2 + 4^2) - 138 \fallingdotseq 58 & (28.20) \\[1ex] (xy) = \dfrac{1}{4}[(\text{sum of } A_1 \text{ of } x)(\text{sum of } A_1 \text{ of } y) + \cdots + (\text{sum of } A_4 \text{ of } x) & \\ \quad \times (\text{sum of } A_4 \text{ of } y)] - CF(xy) & (28.21) \\ \quad = \dfrac{1}{4}[(24)(-66) + 7 \times 15 + 12 \times 21 + 4(-65)] - (-279) = -93 & \\[1ex] (yy) = \dfrac{1}{4}[(\text{sum of } A_1 \text{ of } y)^2 + \cdots + (\text{sum of } A_4 \text{ of } y)^2] - CF(yy) & \\[1ex] \quad = \dfrac{1}{4}[(-66)^2 + 15^2 + 21^2 + (-65)^2] - 564 \fallingdotseq 1748 & (28.22) \end{cases}$$

Similarly, we obtain

$$S_B\begin{cases}(xx)=\frac{1}{4}[9^2+8^2+32^2+(-2)^2]-CF(xx)=293-138\fallingdotseq 155 & (28.23)\\ (xy)=\frac{1}{4}[9(-292)+8\times 63+32\times 165+(-2)(-31)]-CF(xy) & \\ \quad =805-(-279)=1084 & (28.24)\\ (yy)=\frac{1}{4}[(-292)^2+63^2+165^2+(-31)^2]-CF(yy)=29\,355-564=28\,791 & \\ & (28.25)\end{cases}$$

$$S_e\begin{cases}(xx)=1067-58-155=854 & (28.26)\\ (xy)=2\,229-(-93)-1\,084=1538 & (28.27)\\ (yy)=25\,303-1\,748-28\,791=4\,764 & (28.28)\end{cases}$$

The table of *variation and covariation,* Table 28.8, is obtained from these.

The influence $\hat{b}$ on yield y when the potency of the solution x changes by a unit quantity is given by

$$\hat{b}=\frac{S_e(xy)}{S_e(xx)}=\frac{1\,538}{854}\times\frac{1}{100}=0.0180 \tag{28.29}$$

By using this coefficient, S_A, S_B, and S_e under experimental conditions where the potency of the solution is equal can be found as follows.

To obtain S_A, S_B, and S_e from which bias due to x has been eliminated, one calculates in the following sequence.

$$S_e=S_e(yy)-\frac{S_e^2(xy)}{S_e(xx)}=4\,764-\frac{1\,538^2}{854}=4\,764-2\,770=1\,994 \tag{28.30}$$

The number of degrees of freedom becomes 1 less because of the regression term, and it is $9-1=8$.

$$S_A=S_A(yy)+S_e(yy)-\frac{[S_A(xy)+S_e(xy)]^2}{S_A(xx)+S_e(xx)}-S_e$$

$$=1748+4764-\frac{(-93+1\,538)^2}{58+854}-1\,994$$

$$=6\,512-2\,287-1994=2\,231 \tag{28.31}$$

Table 28.8 Table of Variation and Covariation

Source	f	S (xx)	S (xy)	S (yy)
A	3	58	−93	1 748
B	3	155	1 084	28 791
e	9	854	1 538	4 764
Total	15	1 067	2 529	35 303

$$S_B = S_B(yy) + S_e(yy) - \frac{[S_B(xy) + S_e(xy)]^2}{S_B(xx) + S_e(yy)} - S_e$$

$$= 28\,791 + 4\,764 - \frac{(1\,084 + 1\,538)^2}{155 + 854} - 1\,994$$

$$= 33\,555 - 6\,841 - 1\,994 = 24\,747 \qquad (28.32)$$

And the residual variation including variation to x becomes

$$S_x = S_T(yy) - S_A - S_B - S_e = 35\,303 - 2\,228 - 24\,747 - 1\,994 = 6\,334 \qquad (28.33)$$

These results can be organized as in Tables 28.9.

"Regress (uncorrected)" designates the mixed effect of the effect due to potency x and the effect of A and B. However, the part of this which is unmistakably the effect of potency x can be found by the following equation. This is $S(x)$.

$$S(x) = \frac{S_e^2(xy)}{S_e(xx)} = \frac{1\,538^2}{854} = 2\,770 \qquad (28.34)$$

Therefore, the difference between this and 6331, 3561, is variation that cannot be used in explaining the cause. This cannot be helped since x and A and B are sources that are not orthogonal.

Note: The method of testing and estimation of regression in analysis of covariance when it is not assumed that x is randomly allocated is still incomplete, together with its theory. A complete study of this problem, or in other words analysis which is conducted by recovering interfactor information, still does not exist. For example, if x is a normal distribution

Tables 28.9 Analysis of Variance Tables

Source	f	S			
		(xx)	(xy)	(yy)	$S(yy) - S(xy)^2/S(xx)$
A	3	58	−93	1 748	
B	3	155	1 084	28 791	
e	9	854	1 538	4 764	1 994
T	15	1 067	2 529	35 303	
$e+A$		913	1 445	6 512	4 222
$e+B$		1 009	2 622	33 555	26 741

Source	f	S	V
A	3	2 231	743
B	3	24 747	8 249**
Regress (uncorrected)	1	6 331	
e	8	1 994	249
Total	15	35 303	
(x)	(1)	(2 770)	2 770**

and $S_e(xx)$ was at most two degrees of freedom, the expected value of the estimated error of regression coefficient β becomes infinity. The expected value of S_x (uncorrected) in the case of the two-way layout above becomes as follows. Assuming $\mu = \Sigma_{ij}\, x_{ij}/ab$, when given x, we have

$$\begin{aligned} E[S\text{ (regress (uncorrected)}] &= \sigma_e^2 + \beta^2[S_A(xx)+S_B(xx)+S_e(xx)] \\ &+2b\beta \sum_i a_i(\bar{x}_i-\mu)+2a\beta \sum_j b_j(\bar{x}_j-\mu)-b^2 \sum_i a_i(\bar{x}_i-\mu)^2/[S_A(xx) \\ &+S_e(xx)]-a^2 \sum_j b_j^2(\bar{x}_j-\mu)^2/[S_B(xx)+S_e(xx)] \end{aligned}$$

Therefore, it is not unusual for the following to be better as an estimate of the regression coefficient β.

$$\hat{\beta}=\frac{S_e(xy)+\frac{1}{F_A}S_A(xy)+\frac{1}{F_B}S_B(xy)}{S_e(xx)+\frac{1}{F_A}S_A(xx)+\frac{1}{F_B}S_B(xx)} \tag{28.35}$$

F_A and F_B here are the variance ratios of souces A and B. Something like this also holds in more general cases.

28.3.3 Estimations

Next, since A is not significant, it is not always necessary to find the curve for A, but if the number of replications is about 4, one can consider an error of the second kind if the F value is 3 or more. Let us therefore consider the problem of finding the curves for A and B. As explained earlier, we return x and y to the original units and we have

$$\hat{b}=\frac{S_e(xy)}{S_e(xx)}\pm\sqrt{F\times V_e\times\frac{1}{S_e(xx)}} \tag{28.36}$$

$$=\frac{1\,538}{85\,400}\pm\sqrt{5.32\times 2.49\times\frac{1}{85\,400}}=0.0180\pm 0.0125 \tag{28.37}$$

Therefore, the value of $\bar{A}_1$ when the value of x has been taken at its mean value is

$$\bar{A}_1=60+\frac{-6.6}{4}-\hat{b}\left(\frac{240}{4}-\frac{470}{16}\right)=57.8\pm 1.9 \tag{28.38}$$

Similarly,

$$\bar{A}_2=60.6\pm 1.9$$

$$\bar{A}_3=60.5\pm 1.9$$

$$\bar{A}_4=58.7\pm 1.9$$

For $\bar{A}_1$, the confidence limits were found by

$$\pm\sqrt{F\times V_e\left[\frac{1}{4}+\frac{1}{S_e(xx)}\left(\frac{240}{4}-\frac{470}{16}\right)^2\right]}=\pm 1.9 \tag{28.39}$$

It is the same for the others.

$$\bar{B}_1 = 52.8 \pm 1.8$$
$$\bar{B}_2 = 61.8 \pm 1.8$$
$$\bar{B}_3 = 63.2 \pm 1.8$$
$$\bar{B}_4 = 59.8 \pm 1.8 \qquad (28.40)$$

When these are represented graphically, Figure 28.1 and Figure 28.2 are obtained.

Next, let us assume that we have selected A_2 as the best quantity of sulfuric acid A and B_3 as the quantity of solvent B. The estimated value $\hat{\mu}$ of the process average μ for A_2B_3 is given by the following equation. For an arbitrary x, it is

$$\hat{\mu} = \hat{\mu}_y - b(\mu_x - x) \qquad (28.41)$$

Here, $\hat{\mu}_y$ is calculated by calculating the process average μ_y of A_2B_3 with regard to y in the same way as when there is no x. Thus, we have

$$\hat{\mu}_y = \bar{A}_2(y) + \bar{B}_3(y) - \bar{T}(y)$$
$$= 60 + \frac{1.5}{4} + \frac{16.5}{4} - \frac{-9.5}{16} = 65.1 \qquad (28.42)$$

$$\mu_x = 1300 + \frac{70}{4} + \frac{320}{4} - \frac{470}{16} = 1368 \qquad (28.43)$$

Therefore, the realized value y of the yield when a raw material of potency x has been purified is

$$y = 65.1 + 0.0180(x - 1368) \pm \sqrt{F \times V_e\left[\frac{7}{16} + \frac{(x-1368)^2}{85\,400} + 1\right]} \qquad (28.44)$$

For example, for $x = 1400$, we have

$$\hat{\mu} = 65.1 + 0.0180(1400 - 1368) \pm \sqrt{5.32 \times 2.49 \times \left[\frac{7}{16} + \frac{(-32)^2}{85\,400} + 1\right]}$$
$$= 65.7 \pm 4.4 \qquad (28.45)$$

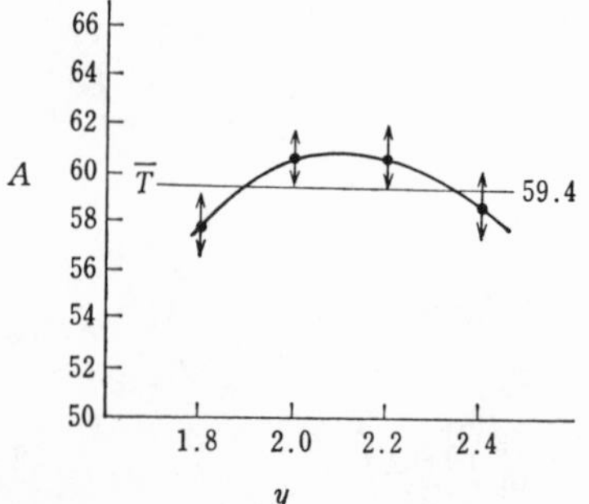

FIGURE 28.1 Relationship Between *A* and *y*

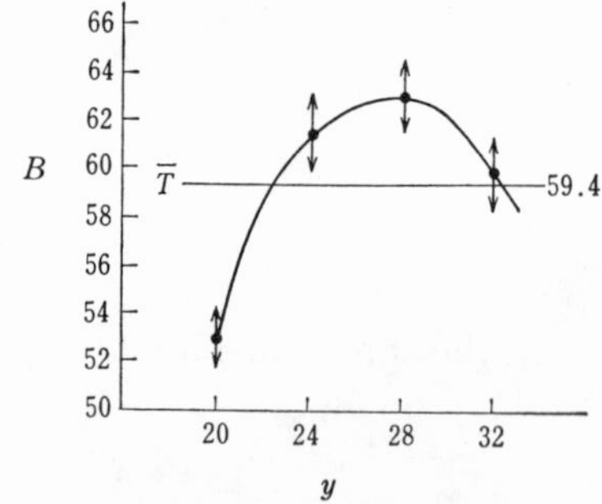

FIGURE 28.2 Relationship Between *B* and *y*

28.4 Efficiency of Analysis of Covariance

Next, the amount that the efficiency of the experiment has risen since the supplementary measurement has been used can be found as follows.

It is only natural that no matter how ingeniously we have performed stratification, supplementary measurement, etc., it is not possible to do away with the effect of environmental conditions. Thus, the objective of such methods is none other than to *change the distribution of error.*

In other words, assuming that there are environmental conditions, measurement error, etc., for a total of 100, if it is possible to distribute most of this to sources that are not very important (effects such as those of strata, blocks, and regression), it means that the quantity of error which becomes distributed to detection of the effects of more important control sources is decreased to only a part, such as 30 or 40, so that the power of test regarding factorial effects is heightened. This is comparable to halving the error variation by using a measuring instrument with no measurement error whatsoever when measurement error takes up one-half of the error variation.

In the case of improvement of the measuring instrument, however, the power of test of the sources was doubled by halving the total error variation by substantial technique; but in the case, for example, of stratification or supplementary measurement, we do not decrease the total error by using methods such as the randomized block method, Latin square, confounding method, incomplete block design, and analysis of covariance, but only raise the power of test of sources as much as possible by reducing the distribution of error that hinders the estimation of important factorial effects. Therefore, since we are not decreasing the error itself but merely changing the distribution of the error, no matter whether we use the randomized block method or the confounding method or the incomplete block design or analysis of covariance, *the total variation, having also added the correction factor, should not differ even the least bit from that of a case where such an assignment is not used.* This should be borne in mind since it is a most fundamental problem of design of experiments.

Let us discuss the efficiency of analysis of covariance based on the theory above. It is assumed that x is random, its population mean is μ, and the variance is τ^2. The analysis of variance table by two-way layout, not using analysis of covariance, is shown in Table 28.10.

When the same data are treated by analysis of covariance, the expected values of the mean square become as given in Table 28.11. (Proof will be omitted because of lack of space.)

What is clear from the analysis of variance table is that even when the supplementary measured value x is completely random, there is a tendency to increase the effective number of replications of sources A and B. The ratio obtained by dividing the effective number of replications by the

Table 28.10 When Supplementary Measured Value x Is Not Considered (The Case of Two-Way Layout)

Source	f	$E(V)$
m	1	$\sigma^2+\beta^2\tau^2+abm^2$
A	$a-1$	$\sigma^2+\beta^2\tau^2+b\sigma_A{}^2$
B	$b-1$	$\sigma^2+\beta^2\tau^2+a\sigma_B{}^2$
e	$(a-1)(b-1)$	$\sigma^2+\beta^2\tau^2$
Total	ab	$\sigma^2+\beta^2\tau^2+m^2+\frac{a-1}{a}\sigma_A{}^2+\frac{b-1}{b}\sigma_B{}^2$

number of replications when there is no supplementary measurement is termed the efficiency factor. Unlike the case of the incomplete block design, the efficiency factor in analysis of covariance is a stochastic variable, and its mean value m_λ is, in the case of source A,

$$m_\lambda=1-\frac{1}{ab-a-2}$$

This means that only in the case of

$$\frac{\sigma^2}{\sigma^2+\beta^2\tau^2}<m_\lambda \tag{28.46}$$

does there exist a value according to ordinary analysis of covariance. For example, if $a=3$ and $b=2$, m_λ becomes 0, so that ordinary analysis of covariance is meaningless. For a calculation method by which there is never such a loss, one should resort to experimental analysis of covariance, although this will not be discussed here.

Increase of efficiency in the case of this problem is

$$\sigma^2+\beta^2\tau^2=\frac{S_e(yy)}{(a-1)(b-1)}=\frac{4764}{9}=529$$

$$\hat{\sigma}^2=\frac{S_e}{(a-1)(b-1)-1}=\frac{1994}{8}=249$$

Table 28.11 The Case of Analysis of Covariance

Source	f	$E(V)$
m	1	$\sigma^2+\beta^2\tau^2+abm^2$
A	$a-1$	$\sigma^2+\left[1-\frac{1}{(ab-b-2)}\right]b\sigma_A{}^2$
B	$b-1$	$\sigma^2+\left[1-\frac{1}{(ab-a-2)}\right]a\sigma_B{}^2$
Regress (uncorrected)	1	$\sigma^2+(ab-1)\beta^2\tau^2+\frac{b(a-1)}{ab-b-2}\sigma_A{}^2+\frac{a(b-1)}{ab-a-2}\sigma_B{}^2$
e	$(a-1)(b-1)-1$	σ^2
Total	ab	$\sigma^2+\beta^2\tau^2+m^2+\frac{a-1}{a}\sigma_A{}^2+\frac{b-1}{b}\sigma_B{}^2$

and since

$$a=b=4$$

this means that for each of A and B, the efficiency of test has risen by analysis of covariance by

$$\left[\frac{529}{249}\times\left(1-\frac{1}{16-4-2}\right)-1.00\right]\times 100=91[\%] \tag{28.47}$$

Of course, the labor of calculation has increased considerably and no doubt the calculation error will also become large, so that if the increase of efficiency is 10% or 20%, probably analysis of covariance is useless, but if it has increased as much as 90%, it would be close to having conducted the experiment twice as often. Perhaps one should not object to the labor of calculating if one considers the alternative of performing the experiment twice as often.

28.5 Calculation Method In a General Case

28.5.1 The Case When There Is One Supplementary Variable and a Linear Equation Is Assumed

Analysis of Covariance in General Is Performed in the Following Sequence

(1) *Analysis of Variance*

(*i*) Variations $S(xx)$ and $S(yy)$ are calculated for x and y as a matter of form by the ordinary formula and, at the same time, covariation $S(xy)$ is calculated by the same procedure (Table 28.12).

(*ii*) The variation for each source is calculated.

$$S_e=S_e(yy)-\frac{S_e^2(xy)}{S_e(xx)} \tag{28.48}$$

$$S_A=S_A(yy)+S_e(yy)-\frac{[S_A(xy)+S_e(xy)]^2}{S_A(xx)+S_e(xx)}-S_e \tag{28.49}$$

Table 28.12 Table of Variations

Source	f	S (xx)	(xy)	(yy)
A	$a-1$	$S_A(xx)$	$S_A(xy)$	$S_A(yy)$
B	$b-1$	$S_B(xx)$	$S_B(xy)$	$S_B(yy)$
$\vdots$	$\vdots$	$\vdots$	$\vdots$	$\vdots$
e	ν	$S_e(xx)$	$S_e(xy)$	$S_e(yy)$
Total	$n-1$			

Table 28.13 Analysis of Variance Table

Source	f	S	V	F
A	$a-1$	S_A		
B	$b-1$	S_B		
$\vdots$	$\vdots$	$\vdots$		
res	1	S_{res}		
e	$\nu-1$	S_e		
Total	$n-1$	$S_T(yy)$		

$$S_B = S_B(yy) + S_e(yy) - \frac{[S_B(xy)+S_e(xy)]^2}{S_B(xx)+S_e(xx)} - S_e \tag{28.50}$$

$$\vdots$$

$$S_{\text{res}} = S_T(yy) - (S_e + S_A + S_B + \cdots) \tag{28.51}$$

The analysis of variance table becomes as given by Table 28.13. It is the same in a case where A is separated into linear and quadratic components.

Regression variation $S_{\text{regression}}$ is given by

$$S_e^2(xy)/S_e(xx) \tag{28.52}$$

(2) *Estimations*

(*i*) The estimated value $\hat{\mu}_y$ of the population mean μ_y for y is the same as when x is absent.

(*ii*) μ_x is found by applying, as a matter of form, the formula of (*i*).

(*iii*) b is found by the following equation, provided that $\dfrac{S_e(xx)}{\nu}$ is not very small compared with σ_x^2.

$$\hat{b} = \frac{S_e(xy)}{S_e(xx)} \tag{28.53}$$

$$\sigma^2(\hat{b}) = \frac{\sigma_e^2}{S_e(xx)} \tag{28.54}$$

$$\text{Confidence limits of } \hat{b} = \hat{b} \pm \sqrt{F \times V_e \times \frac{1}{S_e(xx)}} \tag{28.55}$$

Sometimes sources that are not significant are pooled with V_e.

(*iv*) The estimated value $\hat{\mu}$ of the population mean μ when x assumes the value x_0 is

$$\hat{\mu} = \hat{\mu}_y + b(x_0 - \mu_x) \tag{28.56}$$

$$\sigma^2(\hat{\mu}) = \left[\frac{1}{n_e} + \frac{(x_0-\mu_x)^2}{S_e(xx)}\right]\sigma_e^2 \tag{28.57}$$

n_e is the effective number of replications of $\hat{\mu}_y$ when x is absent.

28.5.2 The Case When There Is One Supplementary Variable and a Quadratic Equation Is Assumed

Only the formulas will be given. It is suggested that the reader use these formulas and try to solve the example of penicillin. We wish to find the values of x and x^2. Note that in this case, when the working mean a has been subtracted, the quadratic equation becomes

$$y = m + b_1(x-a) + b_2(x-a)^2 \tag{28.58}$$

Therefore, when for the relationship of x and y, one can expect

$$y = m + b_1 x + b_2 x^2 \tag{28.59}$$

the square of x must be used as is for the value of x^2. Since the procedure of calculation is the same, we will consider only the case of x^2 here. Note that in many cases it is best to take a as a value close to the mean of x.

We find the variations as shown in Table 28.14. Note that it suffices with the sources exactly as in the general case.

The estimated values of b_1 and b_2 are given by the following equations.

$$\hat{b}_1 = \frac{S_e(x^2x^2)\,S_e(xy) - S_e(xx^2)S_e(x^2y)}{S_e(xx)S_e(x^2x^2) - S_e^{\,2}(xx^2)} \tag{28.60}$$

$$\hat{b}_2 = \frac{S_e(xx)S_e(x^2y) - S_e(xx^2)S_e(xy)}{S_e(xx)S_e(x^2x^2) - S_e^{\,2}(xx^2)} \tag{28.61}$$

$$\sigma^2(\hat{b}_1) = \frac{S_e(x^2x^2)}{\Delta}\sigma_e^{\,2} \tag{28.62}$$

$$\sigma^2(\hat{b}_2) = \frac{S_e(xx)}{\Delta}\sigma_e^{\,2} \tag{28.63}$$

$$\mathrm{cov}(\hat{b}_1\hat{b}_2) = \frac{-S_e(xx^2)}{\Delta}\sigma_e^{\,2} \tag{28.64}$$

where

$$\Delta = S_e(xx)S_e(x^2x^2) - S_e^{\,2}(xx^2) \tag{28.65}$$

Table 28.14 Variations, Covariations

Source	f	(xx)	(xx^2)	(x^2x^2)	(xy)	(x^2y)	(yy)
A	$a-1$	$S_A(xx)$	$S_A(xx^2)$	$S_A(x^2x^2)$	$S_A(xy)$	$S_A(x^2y)$	$S_A(yy)$
B	$b-1$	$S_B(xx)$	$S_B(xx^2)$	$S_B(x^2x^2)$	$S_B(xy)$	$S_B(x^2y)$	$S_B(yy)$
$A\times B$	$(a-1)\times(b-1)$	$S_{A\times B}(xx)$	$S_{A\times B}(xx^2)$	$S_{A\times B}(x^2x^2)$	$S_{A\times B}(xy)$	$S_{A\times B}(x^2y)$	$S_{A\times B}(yy)$
⋮	⋮	⋮	⋮	⋮	⋮	⋮	⋮
e	ν	$S_e(xx)$	$S_e(xx^2)$	$S_e(x^2x^2)$	$S_e(xy)$	$S_e(x^2y)$	$S_e(yy)$
Total	$n-1$	$S_T(xx)$	$S_T(xx^2)$	$S_T(x^2x^2)$	$S_T(xy)$	$S_T(x^2y)$	$S_T(yy)$

With these, it becomes possible to estimate and test the confidence limits of b_i and the population mean for arbitrary x.

Analysis of variance is performed as follows. For example, if $S_A + e(x^2x^2)$ is an abbreviated symbol which expresses $S_A(x^2x^2) + S_e(x^2x^2)$, and so forth, we have

$$S_e = S_e(yy) - \hat{b}_1 S_e(xy) - \hat{b}_2 S_e(x^2y) \tag{28.66}$$

$$S_A = S_{A+e}(yy) - \hat{b}_1' S_{A+e}(xy) - \hat{b}_2' S_{A+e}(x^2y) - S_e \tag{28.67}$$

Here, $\hat{b}$ and $\hat{b}_2'$ are values which are obtained by replacing e with A in the equations of $\hat{b}_1$ and $\hat{b}_2$. We thus have

$$\hat{b}_1' = \frac{S_{A+e}(x^2x^2)S_{A+e}(xy) - S_{A+e}(xx^2)S_{A+e}(x^2y)}{S_{A+e}(xx)S_{A+e}(x^2x^2) - S_{A+e}{}^2(xx^2)} \tag{28.68}$$

It is the same for B, $A \times B$, . . . , and it is the same for the others except that the number of degrees of freedom of error decreases by 2.

28.5.3 When There Are Two Supplementary Variables

When there are two supplementary variables, if we call them x and z, we proceed exactly the same as before, making the substitution

$$x = x$$

$$z = x^2$$

in Equation (28.59). When the variables are even more numerous, it is necessary to calculate the variation and covariation for the respective variations including y as the variation. The labor of calculation increases in such a case, and it becomes enormous. Therefore, it is better to use the methods of Chapter 15 and Chapter 31.

Exercises (28)

(1) The atmospheric temperature x (°C) and the number of persons y admitted to a certain pool (unit: 100 persons), classed by the week as A_1 (first week), A_2 (second week), and A_3 (third week) and by the day as B_1 (Monday), B_2 (Tuesday), . . . , B_7 (Sunday), were as follows (upper row, x; lower row, y).

	B_1	B_2	B_3	B_4	B_5	B_6	B_7
A_1	26	28	29	30	30	28	28
	46	56	51	57	48	48	70
A_2	28	25	26	28	27	30	34
	36	31	27	37	32	50	77
A_3	34	26	27	31	33	32	30
	56	27	33	41	55	62	59

(a) Perform an analysis of covariance with x as a supplementary variable.

(b) Predict the number of persons admitted classed by day of week and by atmospheric temperature in the first week by using the significant sources. Vary the atmospheric temperature as the three levels 26, 30, and 34°C.

(2) The data of two replications were as follows (unit: m) when, for the braking distance of two types of vehicle, A_1 and A_2, the road conditions B were taken at four levels:

B_1 = dry concrete (friction coefficient 0.85)
B_2 = concrete during rainy weather (″ 0.60)
B_3 = pebble road (″ 0.55)
B_4 = solidified snow road (″ 0.15).

The speed was 40 km/h.

	B_1		B_2		B_3		B_4	
A_1	10.0	10.2	15.2	15.3	20.2	20.4	40.4	41.2
A_2	10.4	10.3	15.2	15.4	20.6	20.8	41.6	42.0

Perform an analysis of variance with the friction coefficient as the supplementary variable. Resolve the main effect of B into the linear effect of the friction coefficient and the residual effect. Interpret the effect of B having eliminated the effect of the friction coefficient.

29 Formula for Effective Number of Replications n_e and the Confidence Limits of Split-Unit Design

The formula for finding the effective number of replications n_e of the estimated value of the process average in the case of the orthogonal form will be explained.

29.1 Population Variance of Linear Estimated Values (Effective Number of Replications n_e)

It is a well-known fact that the unit number D in the confidence limits of a linear equation with constant coefficients in n observed values,

$$\hat{\mu}=c_1y_1+c_2y_2+\cdots+c_ny_n\pm\sqrt{F\times V_e\times\frac{1}{n_e}}$$

is given by

$$D=c_1^2+c_2^2+\cdots+c_n^2=\frac{1}{n_e}$$

But when numerous sources are involved it becomes rather difficult to find the value of $c_1^2+\cdots+c_n^2$ by finding the coefficients $c_1, c_2, \ldots,$ c_n of the equation of $\hat{\mu}$.

For example, even in the case of the simplest two-way layout, when interactions are not considered, the estimate of the population mean μ for A_iB_j is given by

Table 29.1 Coefficients of Linear Equation (29.1)

	B_1	B_2	·········	B_j	·········	B_b
A_1	$\left(-\frac{1}{ab}\right)$	$\left(-\frac{1}{ab}\right)$	·········	$\left(\frac{1}{a}-\frac{1}{ab}\right)$	·········	$\left(-\frac{1}{ab}\right)$
A_2	$\left(-\frac{1}{ab}\right)$	$\left(-\frac{1}{ab}\right)$	·········	$\left(\frac{1}{a}-\frac{1}{ab}\right)$	·········	$\left(-\frac{1}{ab}\right)$
⋮	⋮	⋮		⋮		⋮
A_i	$\left(\frac{1}{b}-\frac{1}{ab}\right)$	$\left(\frac{1}{b}-\frac{1}{ab}\right)$	·········	$\left(\frac{1}{a}+\frac{1}{b}-\frac{1}{ab}\right)$	·········	$\left(\frac{1}{b}-\frac{1}{ab}\right)$
⋮	⋮	⋮		⋮		⋮
A_a	$\left(-\frac{1}{ab}\right)$	$\left(-\frac{1}{ab}\right)$	·········	$\left(\frac{1}{a}-\frac{1}{ab}\right)$	·········	$\left(-\frac{1}{ab}\right)$

$$\hat{\mu}=\bar{A}_i+\bar{B}_j-\bar{T}$$

$$=\frac{A_i}{b}+\frac{B_j}{a}-\frac{T}{ab} \tag{29.1}$$

but its unit number is found only by proceeding as follows. The weights for the respective experimental values y_{ij} of the estimated value of Equation (29.1) are

A_iB_j	$\frac{1}{b}+\frac{1}{a}-\frac{1}{ab}$
Row of A_i	$\frac{1}{b}-\frac{1}{ab}$ (except A_iB_j)
Column of B_j	$\frac{1}{a}-\frac{1}{ab}$ (except A_iB_j)
Others	$-\frac{1}{ab}$

When written in detail, it is as given in Table 29.1.

In other words, $\hat{\mu}$ is given by a linear equation with weights (coefficients) as shown by the table above for the individual experimental values. For its unit number, therefore, one need only calculate the sum of squares of the coefficients. But since we have

1 of	$\left(\frac{1}{a}+\frac{1}{b}-\frac{1}{ab}\right)$
$(b-1)$ of	$\left(\frac{1}{b}-\frac{1}{ab}\right)$
$(a-1)$ of	$\left(\frac{1}{a}-\frac{1}{ab}\right)$
$(a-1)(b-1)$ of	$\left(-\frac{1}{ab}\right)$

the sum of squares of the coefficients becomes as follows.

$$\frac{1}{n_e}=\left(\frac{1}{a}+\frac{1}{b}-\frac{1}{ab}\right)^2\times 1+\left(\frac{1}{b}-\frac{1}{ab}\right)^2\times(b-1)+\left(\frac{1}{a}-\frac{1}{ab}\right)^2\times(a-1)$$

$$+\left(-\frac{1}{ab}\right)^2\times(a-1)(b-1)$$

$$=\left(\frac{1}{ab}\right)^2[(b+a-1)^2+(a-1)^2(b-1)+(b-1)^2(a-1)+(a-1)(b-1)]$$

$$=\left(\frac{1}{ab}\right)^2(a+b-1)[a+b-1+(a-1)(b-1)]$$

$$=\frac{a+b-1}{ab}$$

Thus,

$$\mu=\hat{\mu}\pm\sqrt{F\times V_e\times\frac{(a+b-1)}{ab}} \tag{29.2}$$

One can imagine how troublesome these calculations are for other, more complicated, assignments.

It was in 1952 that I became aware that when the experiment is of the *orthogonal form,* it is possible to find the unit number of $\hat{\mu}$ simply as a function only of the number of degrees of freedom of the sources that have not been disregarded. The reciprocal of $c_1^2+\cdots+c_n^2$ is termed the *effective number of relications* n_e, *or* NR, and the formula for this is called the *formula for effective number of replications, or Taguchi's formula.* In the case of the orthogonal form, this formula is

$$n_e=\frac{\text{size of experiment}}{\begin{array}{c}\text{sum of the numbers of degrees of freedom of}\\ \text{the sources that have not been disregarded}\\ \text{in estimating the population mean }\mu\end{array}} \tag{29.3}$$

For example, we will assume that in an experiment by orthogonal array $L_{32}(2^{31})$, only the six main effects A, B, C, E, F, and G and three two-factor interactions, $A\times B$, $A\times C$, and $E\times F$, are significant; considering only these significant sources, if we estimate the population mean μ for a combination of arbitrary levels of the six factors A, B, C, E, F, and G, *viz.*, $A_aB_bC_cE_eF_fG_g$, $\hat{\mu}$ is given by the following equation.

$$\hat{\mu}=\overline{m+a_a+b_b+I_{ab}+c_c+e_e+f_f+g_g+I_{ac}+I_{ef}}$$

$$=\overline{m+a_a+b_b+I_{ab}}+\overline{m+a_a+c_c+I_{ac}}$$

$$+\overline{m+e_e+f_f+I_{ef}}-\widehat{m+a_a}+\widehat{m+g_g}-2\hat{m}$$

$$=\overline{A_aB_b}+\overline{A_aC_c}+\overline{E_eF_f}-\bar{A}_a+\bar{G}_g-2\bar{T} \tag{29.4}$$

Here, $\overline{A_aB_b}$ and so forth represent the means of the experimental values obtained by A_aB_b and so forth, and $\bar{T}$ is the mean of the whole. Essentially, the value of n_e is given by

$$n_e = \frac{32}{1(m)+1(A)+1(B)+1(C)+1(E)+1(F)+1(G)+1(A\times B)+1(A\times C)+1(E\times F)}$$

$$= \frac{32}{10} = 3.2 \qquad (29.5)$$

from Formula (29.3).

Once n_e is known, the 95% confidence limits of $\hat{\mu}$ are given by

$$\hat{\mu} \pm \sqrt{F \times V_e \times \frac{1}{3.2}} \qquad (29.6)$$

29.2 When There Are Dummy Levels

With respect to Formula (29.3) for the effective number of replications n_e for the orthogonal form,

$$n_e = \frac{\text{total number of experiments (total number of degrees of freedom of experiment)}}{\text{total of the independent unknowns that have not been disregarded (or sum of the numbers of degrees of freedom of the sources that were not disregarded in estimating the process average)}}$$

the question here is how the denominator changes when there are factors with dummies. It does not happen very often that dummy levels are required for a multi-way layout, but this is a convenient technique which often becomes necessary in experiments using orthogonal arrays. For example, given a $2^4 \times 3^5$-type experiment where A, B, C, and D are all at two levels and E, F, G, H, and I are all at three levels and one is to experiment so as to find $E \times B$ and $E \times F$, can there be any ingenious

Table 29.2 Analysis of Variance Table

Source	f	
A	1	**
B	1	**
$A\times B$	1	*
C	1	**
D	1	**
E	2	**
$A\times E$	2	*
F	2	**
G	2	
H	2	**
I	2	
e_1	9	$\}$ $e=e_1+e_2$, $27r-18$
e_2	$27(r-1)$	
Total	$27r-1$	

array other than the 3^9 form obtained by inserting dummy levels into the two-level factors and using orthogonal array L_{27}? It immediately comes to the author's mind that an experiment which is smaller than a 108-run mixed-form orthogonal array is inconceivable. Generally speaking, this is far too large. But if dummies are inserted into all of the two-level factors and they become three-level factors in terms of format, we have a 3^9-type experiment, so that it becomes possible to use a $L_{27}(3^{13})$-type orthogonal array and 27 runs suffice. Therefore, the *dummy level technique* is one of the convenient techniques which are actually used the most frequently.

Then, what should we do with the formula for n_e when there are dummy levels? It becomes as follows.

$$n_e = \frac{\text{total number of experiments}}{\begin{array}{c}\text{In the equation of the total number}\\ \text{of independent unknowns which has been}\\ \text{formulated on the assumption that there}\\ \text{are no dummy levels, when there are dummy}\\ \text{levels one substitutes for the number of}\\ \text{levels of that factor the value obtained by}\\ \text{dividing that number by the multiplicity}\\ \text{of that level}\end{array}} \tag{29.7}$$

For example, given the case of $2^4 \times 3^5$, let us assume that we have performed an experiment with A, B, C, and D all at two levels and E, F, G, H, and I all at three levels by inserting dummy levels into A_1, B_1, C_2, and D_2 and by using orthogonal array $L_{27}(3^{13})$ to find $A \times B$ and $A \times E$. Let us assume that we have taken r test pieces for each of the 27 runs and that Table 29.2 has been obtained as a result of analysis of variance.

It will be assumed that e_1 is not significant relative to e_2, and that as a result of having conducted factorial testing by pooling e_1 and e_2, only G and I have been found not significant and thus can be disregarded. In this instance, several cases of number of replications appear for various combinations of levels of $A, B, \ldots, H$.

For n_e of a combination which does not include dummies, for example $A_2B_2C_1D_1E_1F_1H_1$, we can use the formula for the case where there is no dummy as it stands, and n_e can be found as

$$n_e = \frac{27r}{1 + (a-1) + (b-1) + (a-1)(b-1) + (c-1) + (d-1) + (e-1) + (a-1)(e-1) + (f-1) + (h-1)}$$

$$= \frac{27r}{1+2+2+4+2+2+2+4+2+2} = \frac{27}{23}r \tag{29.8}$$

Here, $a, b, c, \ldots, d$ and so forth are the level numbers of $A, B, C, D, \ldots$ (the level numbers counting the dummy levels as distinct levels).

Since, according to the rule given above, in calculating n_e for $A_1B_1C_2D_1E_1F_1H_1$, inasmuch as A_1, B_1, and C_2 are dummy levels we need only substitute

$$\frac{a}{2}, \quad \frac{b}{2}, \quad \frac{c}{2}$$

in place of a, b, and c in the formula of Equation (29.8). Thus it becomes

$$n_e = \frac{27r}{1+\left(\frac{3}{2}-1\right)+\left(\frac{3}{2}-1\right)+\left(\frac{3}{2}-1\right)\left(\frac{3}{2}-1\right)+\left(\frac{3}{2}-1\right)+(3-1)+ +(3-1)+\left(\frac{3}{2}-1\right)(3-1)+(3-1)+(3-1)}$$

$$= \frac{27r}{11.75} \tag{29.9}$$

and it can be seen that the number of replications has become appreciably greater than with levels that are not dummies. Thus, if one sets up a formula for n_e much as though the dummies were distinct levels, it becomes possible to find immediately the n_e of the process average of a combination in which dummy levels are included, and by extension, its confidence limits as well.

Let us give one more example. It will be assumed that the following experiment has been carried out when A, B, C, and D are at two levels, E, F, and G are at three levels, and H, I, J, and K are at four levels.

A: In order to experiment on A_1 three times as much as A_2 although the materials of A_2 were meager, to obtain four levels 1, 2, 3, and 4, two dummy levels were inserted into A_1 and the levels were arranged as A_1, A_2, A_1, and A_1.

B: One dummy level each was inserted into B_1 and B_2 and the four levels 1, 2, 3, and 4 were taken to be B_1, B_1, B_2, and B_2.

C': For C and D, four combinations were created as follows: levels 1, 2, 3, 4 = C_1D_1, C_1D_2, C_2D_1, and C_2D_2.

As to E, F, and G, four levels were created as follows using a dummy level for the second level in each case.

1	2	3	4
E_1	E_2	E_2	E_3
F_1	F_2	F_2	F_3
G_1	G_2	G_2	G_3

Since H, I, J, and K were all at four levels, the experiment was finally so contrived that it could be regarded as a 4^{10}-form experiment, and, so as to be able to find $A \times B$, $A \times E$, and $B \times E$, it was carried out by using orthogonal array $L_{64}(4^{21})$. Four test pieces were produced from each experiment. Let us assume that as a result of testing, the main effects of A, B, C, D, E, F, H, I, and J and $A \times B$ and $A \times E$ turned out to be significant and could not be disregarded. Also, C constitutes brands of a certain auxiliary material, and although there is a relation of superiority-inferiority between the two brands, it will be assumed that both had to be used because they were on hand. In other words, the two brands, C_1 and C_2, were used as stratification factors; since they were selected in order to raise the logicality and precision of estimation of the other factorial effects, even if they emerge as significant, estimating their effect level by level serves no purpose, so that usually they are disregarded in estimating the process average. Even though one says process average, to be accurate only the relative value of the process average is useful in determining action; since therefore the confidence limits with respect to their variation are also not important, usually the main effect of C is disregarded.

Therefore, in this instance, the sources that are not disregarded are only A, B, D, E, F, H, I, and J, and $A \times B$ and $A \times E$. For example, for n_e for the process average of $A_1B_2D_1E_2F_2H_1I_2J_3$, we take the formula for n_e for the case where there is no dummy level,

$$n_e = \frac{64 \times 4}{\underset{m}{1} + \underset{A}{(4-1)} + \underset{B}{(4-1)} + \underset{A \times B}{(4-1)(4-1)} + \underset{D}{(4-1)} + \underset{E}{(4-1)} + \underset{F}{(4-1)} + \underset{H}{(4-1)} + \underset{I}{(4-1)} + \underset{J}{(4-1)} + \underset{A \times E}{(4-1)(4-1)}} \tag{29.10}$$

and in place of the level number 4 we just substitute

$\frac{4}{3}$ in the case of A_1

$\frac{4}{2}$ in the cases of B_2 and D_1

$\frac{4}{2}$ in the cases of E_2 and F_2

Therefore, the n_e that we seek becomes

$$\begin{aligned} n_e &= \frac{64 \times 4}{\underset{m}{1} + \underset{A}{\left(\frac{4}{3}-1\right)} + \underset{B}{\left(\frac{4}{2}-1\right)} + \underset{A \times B}{\left(\frac{4}{3}-1\right)\left(\frac{4}{2}-1\right)} + \underset{D}{\left(\frac{4}{2}-1\right)} + \underset{E}{\left(\frac{4}{2}-1\right)} + \underset{F}{\left(\frac{4}{2}-1\right)} + \underset{H}{(4-1)} + \underset{I}{(4-1)} + \underset{J}{(4-1)} + \underset{A \times E}{\left(\frac{4}{3}-1\right)\left(\frac{4}{2}-1\right)}} \\ &= \frac{256}{1 + \frac{1}{3} + 1 + \frac{1}{3} \times 1 + 1 + 1 + 1 + 3 + 3 + 3 + \frac{1}{3} \times 1} = \frac{256}{16} \end{aligned} \tag{29.11}$$

Calculation of n_e is unrelated to whether or not e_1 is significant relative to the replication error e_2. If e_1 is significant relative to e_2, it merely means that it is necessary to use e_1 as the estimate of error and, accompanying this, that the number of degrees of freedom changes.

29.3 Formula for Obtaining Effective Number of Replications n_e in an Experiment in Split-Unit Form, and How to Obtain the Confidence Limits of the Process Average

It is most important in assigning an experiment to manage logically the two procedures,

(1) Selecting an assignment that makes it possible to find the necessary factorial effects
(2) Assigning factors by thoroughly considering the difficulty of changing the levels of the factors and designing the experiment so as to be able to perform it easily and yet be able to obtain adequate information.

Now, to take care of point (2) means to perform a so-called split-unit-form experiment; and often factors are divided into several classes in accordance with the difficulty of changing their levels.

As was explained in Chapter 9, when factors up to third order are used and furthermore several test pieces are taken for the respective combinations, the error terms become:

Error of first-order sources, e_1
Error of second-order sources, e_2
Error of third-order sources, e_3
Error of replications, e_4

Of course, if e_3 is not significant relative to e_4, usually e_3 and e_4 are pooled and the third-order sources are tested by this, but e_3 is not necessarily negligibly small compared with e_4. A similar state of affairs develops regarding e_2 and e_3, and e_1 and e_2. If all the errors are of about the same magnitude and all can be pooled, it is possible to use the ordinary formula as it stands since there is in effect only one error term.

But since one cannot necessarily pool a certain number of errors, one would wish for a formula for estimation of the process average in such a case and a simple method of finding the n_e of its estimate. A formula for the confidence limits is also important. Since the calculation equation for the formula of estimation is exactly the same no matter how many stages of error there are, this will be omitted here. The formula for n_e or, equivalently, the *population variance of the estimated value,* is given by the following equation.

$$\sigma^2(\hat{\mu}) = \frac{\begin{array}{c}\text{sum of the numbers of degrees of freedom of the}\\ \text{sources that have not been disregarded in}\\ \text{testing the first-order error variance } e_1\\ \text{(including the 1 degree of freedom of the general mean)}\end{array}}{\text{total number of degrees of freedom}} v_1{}^2$$

$$+\frac{\begin{array}{c}\text{sum of the numbers of degrees of freedom of the}\\ \text{sources that have not been disregarded in}\\ \text{testing the second-order error variance } e_2\\ \text{(not including the 1 degree of freedom of the}\\ \text{general mean)}\end{array}}{\text{total number of degrees of freedom}}\, v_2^2$$

$$+\frac{\begin{array}{c}\text{sum of the numbers of degrees of freedom of the}\\ \text{sources that have not been disregarded in}\\ \text{testing the third-order error variance } e_3\\ \text{(not including the 1 degree of freedom of the}\\ \text{general mean)}\end{array}}{\text{total number of degrees of freedom}}\, v_3^2$$

$$=\frac{v_1^2}{(n_e)_1}+\frac{v_2^2}{(n_e)_2}+\frac{v_3^2}{(n_e)_3} \tag{29.12}$$

Here, v_1^2, v_2^2, and v_3^2 are the means of V_{e_1}, V_{e_2}, and V_{e_3} with the numbers of degrees of freedom of e_1, e_2, and e_3 as ν_1, ν_2, and ν_3, or in other words they are the true values of the variances.

For example, let us assume that an experiment for finding the main effects of eighteen factors each at three levels, $A, B, C, D, E, F, G, H, I, J, K, L, M, N, O, P, Q$, and R, and $A \times D$, $A \times J$, and $E \times K$, has been assigned by using orthogonal array $L_{81}(3^{40})$. It will be assumed that A, B, and C have been selected as first-order factors, D, E, F, G, H, and I as second-order factors, and J, K, L, M, N, O, P, Q, and R as third-order factors, and that four test pieces have been produced in each experiment.

Let us consider a case where the results of testing are as given in Table 29.3.

The population variance of the estimated value $\hat{\mu}$ of the process average for, as an example, $A_2B_1C_3D_2E_2G_3I_2J_1K_1M_3O_2P_2$ is given by the following equation.

$$\sigma^2(\hat{\mu})=\frac{\overset{A}{2}+\overset{B}{2}+\overset{C}{2}+\overset{m}{1}}{324}(\sigma_4^2+4\sigma_3^2+12\sigma_2^2+36\sigma_1^2)$$

$$+\frac{\overset{D}{2}+\overset{A\times D}{4}+\overset{E}{2}+\overset{G}{2}}{324}(\sigma_4^2+4\sigma_3^2+12\sigma_2^2)$$

$$+\frac{\overset{I}{2}+\overset{J}{2}+\overset{A\times J}{4}+\overset{K}{2}+\overset{E\times K}{4}+\overset{M}{2}+\overset{O}{2}+\overset{P}{2}}{324}(\sigma_4^2+4\sigma_3^2)$$

$$=\frac{7}{324}v_1^2+\frac{10}{324}v_2^2+\frac{20}{324}v_3^2 \tag{29.13}$$

The confidence limits of μ are given approximately by the following equation.

$$\hat{\mu}\pm\sqrt{F_2^1\times\frac{7}{324}V_{e_1}+F_4^1\frac{10}{324}V_{e_2}+F_{26}^1\times\frac{20}{324}\times V_{e_3}} \tag{29.14}$$

Table 29.3 Analysis of Variance Table

Source		f	$E(V)$
1st-order source	A	2**	$\sigma_4^2+4\sigma_3^2+12\sigma_2^2+36\sigma_1^2+108\sigma_A^2$
	B	2**	〃 $+108\sigma_B^2$
	C	2*	〃 $+108\sigma_C^2$
1st order error	e_1	2*	$\sigma_4^2+4\sigma_3^2+12\sigma_2^2+36\sigma_1^2(\equiv v_1^2)$
2nd order source	D	2**	$\sigma_4^2+4\sigma_3^2+12\sigma_2^2+108\sigma_D^2$
	$A\times D$	4**	〃 $+36\sigma_{A\times D}^2$
	E	2**	〃 $+108\sigma_E^2$
	F	2	〃 $+108\sigma_F^2$
	G	2**	〃 $+108\sigma_G^2$
	H	2	〃 $+108\sigma_H^2$
2nd order error	e_2	4**	$\sigma_4^2+4\sigma_3^2+12\sigma_2^2(\equiv v_2^2)$
3rd-order source	I	2**	$\sigma_4^2+4\sigma_3^2+108\sigma_I^2$
	J	2**	〃 $+108\sigma_J^2$
	$A\times J$	4**	〃 $+36\sigma_{A\times J}^2$
	K	2**	$\sigma_4^2+4\sigma_3^2+108\sigma_K^2$
	$E\times K$	4**	〃 $+36\sigma_{E\times K}^2$
	L	2	〃 $+108\sigma_L^2$
	M	2*	〃 $+108\sigma_M^2$
	N	2	〃 $+108\sigma_N^2$
	O	2**	〃 $+108\sigma_O^2$
	P	2**	〃 $+108\sigma_P^2$
	Q	2	〃 $+108\sigma_Q^2$
	R	2	〃 $+108\sigma_R^2$
3rd-order error	e_3	26**	$\sigma_4^2+4\sigma_3^2(\equiv v_3^2)$
4th-order error	e_4	243	$\sigma_4^2(\equiv v_4^2)$
Total		323	

Thus, in general,

$$\mu \doteqdot \hat{\mu} \pm \sqrt{F_{\nu_1}^1 \frac{1}{(n_e)_1} V_e + F_{\nu_2}^1 \frac{1}{(n_e)_2} V_{e_2} + F_{\nu_3}^1 \frac{1}{(n_e)_3} V_{e_3}} \tag{29.15}$$

is the equation which is on the safe side and which is also the most convenient.

It is possible to construct a formula in an exactly similar manner even for a more general case. One may also pool sources that have not been found significant with corresponding errors. When this is done, only the number of degrees of freedom of these respective errors changes.

It is also exactly the same when there are dummy levels in a number of factors; one would find the formula for when there are no dummy levels, then for the number of levels of the respective factors in this formula one would merely substitute the value obtained by dividing this number by the multiplicity of the dummy. For example, when A_2, B_1, D_2, E_2, and J_1 are dummy levels in the example above, the population variance of $\hat{\mu}$, $\sigma^2(\hat{\mu})$, is found as

$$\sigma^2(\hat{\mu}) = \frac{\overset{A}{\left(\frac{3}{2}-1\right)} + \overset{B}{\left(\frac{3}{2}-1\right)} + \overset{C}{2} + \overset{m}{1}}{324} v_1^2$$

$$+ \frac{\overset{D}{\left(\frac{3}{2}-1\right)} + \overset{A \times D}{\left(\frac{3}{2}-1\right)\left(\frac{3}{2}-1\right)} + \overset{E}{\left(\frac{3}{2}-1\right)} + 2}{324} v_2^2$$

$$+ \frac{2 + \overset{J}{\left(\frac{3}{2}-1\right)} + \overset{A \times J}{\left(\frac{3}{2}-1\right)\left(\frac{3}{2}-1\right)} + \overset{K}{2} + \overset{E \times K}{\left(\frac{3}{2}-1\right)2} + \overset{M}{2} + \overset{O}{2} + \overset{P}{2}}{324} v_3^2$$

$$= \frac{4}{324} v_1^2 + \frac{3\frac{1}{4}}{324} v_2^2 + \frac{11\frac{3}{4}}{324} v_3^2 \qquad (29.16)$$

Therefore, the confidence limits become as follows when sources that are not significant have been pooled with the errors.

$$\mu = \hat{\mu} \pm \sqrt{F_4^1 \frac{4}{324} V_{e_1} + F_{13}^1 \frac{13}{1\,296} V_{e_2} + F_{39}^1 \frac{47}{1\,296} V_{e_3}} \qquad (29.17)$$

Calculations such as the foregoing can be carried out exactly similarly for assignments by orthogonal-form arrays and multi-way orthogonal arrays. It is simple to find n_e and the confidence limits for them analogously to the example given above.

29.4 Proof of Formula for n_e*

Let us express the main effect a_i of factor A at a levels by *orthogonal components* b_j of one degree of freedom. With the coefficient multiplied by the orthogonal component as ϕ_{ij}, we express this as

$$a_i = \sum_{j=1}^{a-1} \varphi_{ij} b_j \qquad (i = 1, 2, \cdots, a) \qquad (29.18)$$

ϕ_{ij} satisfies the orthogonality conditions

$$\left.\begin{aligned} &\sum_{i=1}^{a} \varphi_{ij}^2 = W_j \\ &\sum_{i=1}^{a} \varphi_{ij}\varphi_{ik} = 0 \qquad (k \neq j) \end{aligned}\right\} \qquad (29.19)$$

ϕ_{ij} such as this can be found by setting $x = x_j\,(j = 1, 2, \ldots, a)$ in $\phi_i(x)$ in the expansion of an orthogonal polynomial, or one can simply use coefficients such as the following.

* Method according to a letter from Dr. Genzaburō Masuyama to the author, 1953.

	A_1	A_2	$A_3 \cdots\cdots A_{a-1}$	A_a	W (sum of squares)
φ_{i1}	$a-1$	-1	$-1 \cdots\cdots -1$	-1	$a(a-1)$
φ_{i2}	0	$a-2$	$-1 \cdots\cdots -1$	-1	$(a-1)(a-2)$
φ_{i3}	0	0	$a-3 \cdots\cdots -1$	-1	$(a-2)(a-3)$
$\vdots$	$\vdots$	$\vdots$	$\vdots \quad\quad \vdots$	$\vdots$	$\vdots$
$\varphi_{i,a-1}$	0	0	$0 \cdots\cdots 1$	-1	2

The least-squares estimate of b_j is

$$\hat{b}_j = \frac{\sum_{i=1}^{a} \varphi_{ij} A_i}{r W_j} \tag{29.20}$$

and its variance is

$$\mathrm{Var}(\hat{b}_j) = \frac{\sigma^2}{r W_j} \tag{29.21}$$

Of course, $\hat{b}_j$ are mutually independent. Therefore, we have

$$\mathrm{Var}(\hat{a}_i) = \frac{\sigma^2}{r} \sum_j \frac{\varphi_{ij}^2}{W_j} \tag{29.22}$$

Now, since it does not matter what is the sequence of the levels A_1, A_2, . . . , A_a, the values of the right side must be the same for all $\hat{a}_i$. The coefficient of σ^2/r will be expressed as K. On the other hand,

$$\begin{aligned} \sum_i \mathrm{Var}(\hat{a}_i) &= \frac{\sigma^2}{r} \sum_i \sum_j \frac{\varphi_{ij}^2}{W_j} \\ &= \frac{\sigma^2}{r} \sum_j \frac{W_j}{W_j} \\ &= \frac{\sigma^2}{r}(a-1) \end{aligned} \tag{29.23}$$

Therefore,

$$K = \sum_j \frac{\varphi_{ij}^2}{W_j} = \frac{a-1}{a} \tag{29.24}$$

Similarly, two-factor interactions, too, can be expanded in orthogonal functions.

$$\hat{I}_{ij} = \sum_{kl} \varphi_{ijkl} \hat{b}_{kl} \quad (i=1, 2, \cdots, a-1, j=1, 2, \cdots, b-1) \tag{29.25}$$

The variance of this is, with the number of replications of A_iB_j as r,

$$\mathrm{Var}(\hat{I}_{ij}) = \sum_{kl} \frac{\varphi_{ijkl}^2}{\sum_{ij} \varphi_{ijkl}^2} \frac{\sigma^2}{r} \tag{29.26}$$

The number of units of b_{kl} is $(a-1)(b-1)$ and from symmetry we find, similarly,

$$\sum_{kl} \frac{\varphi_{ijkl}^2}{\sum_{ij} \varphi_{ijkl}^2} = \frac{(a-1)(b-1)}{ab} \tag{29.27}$$

This method also holds for higher-order interactions in general. Also, since the respective contrasts of main effects and interactions are mutually independent, one has, for example,

$$\mathrm{Var}(\hat{a}_1+\hat{I}_{11})=\left[\frac{a-1}{abr}+\frac{(a-1)(b-1)}{abr}\right]\sigma^2$$

$$=\frac{(a-1)+(a-1)(b-1)}{abr}\sigma^2 \tag{29.28}$$

From the foregoing we obtain the following theorem:

[THEOREM]

When the sources are mutually orthogonal, the effective number of replications n_e of the population mean estimate $\hat{\mu}$ which is obtained by adding some or all of these sources is given by

$$n_e=\frac{\text{total number of experiments}}{\begin{array}{c}\text{total number of independent unknowns}\\ \text{that have not been disregarded in}\\ \text{its estimation}\end{array}}$$

$$=\frac{\text{total number of degrees of freedom}}{\begin{array}{c}\text{sum of the numbers of degrees of free-}\\ \text{dom of the sources considered for}\\ \text{its estimation}\end{array}} \tag{29.29}$$

The reader is encouraged to try a separate proof of the theorem above by calculating directly using an expansion by an orthogonal polynomial. The proof consists merely of a direct calculation of the variance of $f(x)$ in orthogonal polynomial form,

$$f(x)=\sum_{i=0}^{a-1}c_i\varphi_i(x)$$

Even by this method, one page is sufficient for proof of the theorem given above.

29.5 Proof of Formula for Confidence Limits by Split-Unit Design (Generalization of the Behrens-Fisher Problem)

29.5.1 Problem

It will be proved that the formula for the confidence limits in the case of the split-unit design, discussed in Chapter 9 and Section 29.3, is one that is on the safe side and is also one of sufficiently good precision. This is also a generalization of the *Behrens-Fisher problem* concerning the testing and estimation of the difference between two mean values where there is no guarantee of homoscedasticity. Here, we will introduce a generalization of the proof by P. B. Patanaik.

[LEMMA CONCERNING CONCAVE FUNCTIONS]

When, for a function $F(x)$ that takes real number values in the interval $[a, b]$,

$$F\left(\frac{x_1+x_2}{2}\right) \geq \frac{1}{2}[F(x_1)+F(x_2)] \tag{29.30}$$

always holds for two arbitrary points x_1 and x_2 in the interval $[a, b]$, we say that $F(x)$ is a concave function in the interval $[a, b]$.

If $F(x)$ is a concave function, then with $\lambda_1, \lambda_2, \ldots, \lambda_k$ as arbitrary positive numbers,

$$F\left(\frac{\lambda_1 x_1+\lambda_2 x_2+\cdots+\lambda_k x_k}{\lambda_1+\lambda_2+\cdots+\lambda_k}\right) \geq \frac{\lambda_1}{\lambda_1+\cdots+\lambda_k} F(x_1)+\cdots+\frac{\lambda_k}{\lambda_1+\cdots+\lambda_k} F(x_k) \tag{29.31}$$

holds for arbitrary $x_1, x_2, \ldots, x_k$ in $[a, b]$. This can be derived simply from Equation (29.30), and since it is proved in many books the proof will be omitted.

29.5.2 General Formula and Its Proof

The following general theorem will be proved.

"$\hat{\mu}_1, \hat{\mu}_2, \ldots, \hat{\mu}_k$ is distributed according to a normal distribution of population mean $\mu_1, \mu_2, \ldots, \mu_k$ and population variance $\frac{\sigma_1^2}{n_1}$, $\frac{\sigma_2^2}{n_2}, \ldots, \frac{\sigma_k^2}{n_k}$. Also, it is assumed that variances $V_1, V_2, \ldots, V_k$ are unbiased estimates of $\sigma_1^2, \sigma_2^2, \ldots, \sigma_k^2$ of numbers of degrees of freedom $\nu_1, \nu_2, \ldots, \nu_k$, and that $\hat{\mu}_1, \ldots, \hat{\mu}_k$ and $V_1, \ldots, V_k$ are mutually independent.

"Then the confidence limits of $\mu_1 + \mu_2 + \cdots + \mu_k$,

$$\hat{\mu}_1+\hat{\mu}_2+\cdots+\hat{\mu}_k \pm \sqrt{F_{\nu_1}^{1} \times \frac{V_1}{n_1}+F_{\nu_2}^{1} \times \frac{V_2}{n_2}+\cdots+F_{\nu_k}^{1} \times \frac{V_k}{n_k}} \tag{29.32}$$

are confidence limits on the safe side. F is the F-table value of the level of significance α."

[PROOF]

$$\chi_i^2=\frac{\nu_i V_i}{\sigma_i^2} \qquad (i=1, 2, \cdots, k) \tag{29.33}$$

assumes a χ^2 distribution of number of degrees of freedom ν_i. If its probability density function is written as $f(\chi_i^2)$, we have

$$f(\chi_i^2)=\frac{1}{2}\frac{1}{\Gamma\left(\frac{\nu_i}{2}\right)}\left(\frac{\chi_i^2}{2}\right)^{\frac{\nu_i}{2}-1}e^{-\frac{\chi_i^2}{2}} \qquad (i=1, 2, \cdots, k) \tag{29.34}$$

Next, we consider the distribution of the variable

$$\chi_0^2=\frac{(\hat{\mu}_1+\cdots+\hat{\mu}_k-\mu_1-\cdots-\mu_k)^2}{\frac{\sigma_1^2}{n_1}+\frac{\sigma_2^2}{n_2}+\cdots+\frac{\sigma_k^2}{n_k}} \tag{29.35}$$

χ_0^2 follows a χ^2 distribution of one degree of freedom. Its probability density function is expressed as $f(\chi_0^2)$.

$$f(\chi_0^2)=\frac{1}{2}\frac{1}{\Gamma\left(\frac{1}{2}\right)}\left(\frac{\chi_0^2}{2}\right)^{-\frac{1}{2}}e^{-\frac{\chi_0^2}{2}} \tag{29.36}$$

We therefore need only calculate the following expression for the probability P of being within the confidence limits of Equation (29.32).

$$P=\int_0^\infty\cdots\int_0^\infty f(\chi_1^2)\cdots f(\chi_k^2)d\chi_1^2\cdots d\chi_k^2\int_0^R f(\chi_0^2)d\chi_0^2 \tag{29.37}$$

Here, the limit of integration R is to be chosen so that the following inequality is satisfied:

$$(\hat{\mu}_1+\cdots+\hat{\mu}_k-\mu_1-\cdots-\mu_k)^2\leq F_{\nu_1}^1\times\frac{V_1}{n_1}+\cdots+F_{\nu_k}^1\times\frac{V_k}{n_k}$$

This inequality is, if the variables are expressed as $\chi_0^2, \chi_1^2, \ldots, \chi_k^2$,

$$\frac{(\hat{\mu}_1+\cdots+\hat{\mu}_k-\mu_1-\cdots-\mu_k)^2}{\frac{\sigma_1^2}{n_1}+\cdots+\frac{\sigma_k^2}{n_k}}\leq\frac{\left(F_{\nu_1}^1\times\frac{\sigma_1^2}{n_1}\times\frac{\chi_1^2}{\nu_1}+\cdots+F_{\nu_k}^1\times\frac{\sigma_k^2}{n_k}\times\frac{\chi_k^2}{\nu_k}\right)}{\frac{\sigma_1^2}{n_1}+\cdots+\frac{\sigma_k^2}{n_k}} \tag{29.38}$$

Thus we have

$$\chi_0^2\leq\frac{\frac{\sigma_1^2}{n_1}\times F_{\nu_1}^1\times\frac{\chi_1^2}{\nu_1}+\cdots+\frac{\sigma_k^2}{n_k}\times F_{\nu_k}^1\times\frac{\chi_k^2}{\nu_k}}{\frac{\sigma_1^2}{n_1}+\cdots+\frac{\sigma_k^2}{n_k}} \tag{29.39}$$

If we put

$$\lambda_i=\frac{\sigma_i^2}{n_i} \qquad (i=1, 2, \cdots, k)$$

Equation (29.39) can be written as follows.

$$\chi_0^2\leq\frac{\lambda_1\times F_{\nu_1}^1\times\frac{\chi_1^2}{\nu_1}+\cdots+\lambda_k\times F_{\nu_k}^1\times\frac{\chi_k^2}{\nu_k}}{\lambda_1+\cdots+\lambda_k} \tag{29.40}$$

If, further, we put

$$x_i=F_{\nu_i}^1\times\frac{\chi_i^2}{\nu_i} \qquad (i=1, 2, \cdots, k) \tag{29.41}$$

the upper limit R of the integration in Equation (29.37) is given by the following equation.

$$R=\frac{\lambda_1x_1+\cdots+\lambda_kx_k}{\lambda_1+\cdots+\lambda_k} \tag{29.42}$$

It will be proved that the cumulative distribution function $F(x)$ of the χ^2 distribution of one degree of freedom,

$$F(x)=\frac{1}{2}\int_0^x\frac{1}{\Gamma\left(\frac{1}{2}\right)}(\chi^2)^{-\frac{1}{2}}e^{-\frac{\chi^2}{2}}d\chi^2 \tag{29.43}$$

is a concave function in interval [0, ∞]. We need only prove the following inequality:

$$F\left(\frac{\lambda_1x_1+\cdots+\lambda_kx_k}{\lambda_1+\cdots+\lambda_k}\right)\geq\frac{\lambda_1F(x_1)}{\lambda_1+\cdots+\lambda_k}+\cdots+\frac{\lambda_kF(x_k)}{\lambda_1+\cdots+\lambda_k} \tag{29.44}$$

Since $x_1, x_2, \ldots, x_k$ are positive and $\lambda_1, \lambda_2, \ldots, \lambda_k$ are also positive, if we put

$$\bar{x}=\frac{\lambda_1x_1+\cdots+\lambda_kx_k}{\lambda_1+\cdots+\lambda_k} \tag{29.45}$$

then $\bar{x}$ too is positive. We now redefine $\lambda_ix_i/(\lambda_1+\lambda_2+\cdots+\lambda_k)$ as x_i; since $F(x)$ is twice continuously differentiable, there exists a real number θ_i such that

$$\begin{aligned}F(x_i)&=F(\bar{x}+x_i-\bar{x})\\&=F(\bar{x})+\frac{F'(\bar{x})}{1!}(x_i-\bar{x})+\frac{F''[\bar{x}+\theta_i(x_i-\bar{x})]}{2!}(x_i-\bar{x})^2\quad(0<\theta_i<1)\end{aligned} \tag{29.46}$$

by the mean value theorem of differential calculus. Therefore, the right side of Equation (29.44) becomes

$$\begin{aligned}&F(\bar{x})+F'(\bar{x})\sum_{i=1}^k(x_i-\bar{x})+\sum_{i=1}^k\frac{F''[\bar{x}+\theta_i(x_i-\bar{x})]}{2!}(x_i-\bar{x})^2\\&=F(\bar{x})+0+\sum_{i=1}^k\frac{F''[\bar{x}+\theta_i(x_i-\bar{x})]}{2!}(x_i-\bar{x})^2\end{aligned} \tag{29.47}$$

The third term of the right side of the equation above is always negative. This is because

$$F'(x)=\frac{1}{2}\frac{1}{\Gamma\left(\frac{1}{2}\right)}x^{-\frac{1}{2}}e^{-\frac{x}{2}}$$

$$F''(x)=\frac{1}{2}\frac{1}{\Gamma\left(\frac{1}{2}\right)}\left(\frac{-1}{2}\right)e^{-\frac{x}{2}}\left(x^{-\frac{3}{2}}+x^{-\frac{1}{2}}\right)<0\quad(x>0) \tag{29.48}$$

This means that the right side of the equation is smaller than the $F(\bar{x})$ of the left side.

We substitute this result into Equation (29.37).

$$\begin{aligned}P&=\int_0^\infty\cdots\int_0^\infty f(\chi_1^2)\cdots f(\chi_k^2)\,d\chi_1^2\cdots d\chi_k^2\int_0^R f(\chi_0^2)d\chi_0^2\\&\geq\int_0^\infty\cdots\int_0^\infty f(\chi_1^2)\cdots f(\chi_k^2)d\chi_1^2\cdots d\chi_k^2\sum_{i=1}^k\frac{\lambda_i}{\lambda_1+\cdots+\lambda_k}\int_0^{x_i}f(\chi_0^2)d\chi_0^2\end{aligned}$$

$$=\sum_{i=1}^{k}\frac{\lambda_i}{\lambda_1+\cdots+\lambda_k}\int_0^\infty\cdots\int_0^\infty f(\chi_1^2)\cdots f(\chi_k^2)\,d\chi_1^2\cdots d\chi_k^2\int_0^{F_{\nu_i}^1\times\frac{\chi_i^2}{\nu_i}} f(\chi_0^2)d\chi_0^2 \tag{29.49}$$

Since the last integral of each term involves only χ_i^2, the integrals of the χ^2 distributions other than the i-th are 1 and the equation above becomes as follows.

$$P\geq\sum_{i=1}^{k}\frac{\lambda_i}{\lambda_1+\cdots+\lambda_k}\int_0^\infty f(\chi_i^2)\,d\chi_i^2\int_0^{F_{\nu_i}^1\times\frac{\chi_i^2}{\nu_i}} f(\chi_0^2)d\chi_0^2 \tag{29.50}$$

The integration

$$P_i=\int_0^\infty\int_0^{F_{\nu_i}^1\times\frac{\chi_i^2}{r_i}} f(\chi_i^2)f(\chi_0^2)d\chi_i^2d\chi_0^2 \tag{29.51}$$

when the variance of ν_i degrees of freedom is V_i and the variance of one degree of freedom is V_0, is carried out with respect to the variables

$$\left.\begin{aligned}\chi_i^2&=\frac{\nu_i V_i}{\sigma_i^2}\\ \chi_0^2&=\frac{n_i(\hat{\mu}_i-\mu_i)^2}{\sigma_i^2}\end{aligned}\right\} \tag{29.52}$$

in the domain

$$\left.\begin{aligned}&0\leq\chi_i^2<\infty\\ &\chi_0^2=\frac{n_i(\hat{\mu}_i-\mu_i)^2}{\sigma_i^2}\leq F_{\nu_i}^1\times\frac{\chi_i^2}{\nu_i}\end{aligned}\right\} \tag{29.53}$$

The range of integration given in Equation (29.53) becomes as follows if the χ_i^2 are transformed according to Equation (29.52).

$$0\leq V_i<\infty$$

$$\frac{n_i(\hat{\mu}_i-\mu_i)^2}{\sigma_i^2}\leq F_{\nu_i}^1\times\frac{1}{\nu_i}\times\frac{\nu_i V_i}{\sigma_i^2}$$

Finally, this becomes an integration in the domain

$$\left.\begin{aligned}&0\leq V_i<\infty\\ &(\hat{\mu}_i-\mu_i)^2\leq\frac{F_{\nu_i}^1\times V_i}{n_i}\end{aligned}\right\} \tag{29.54}$$

This is clearly the probability that μ_i falls within the confidence limits

$$\mu_i=\hat{\mu}_i\pm\sqrt{F_{\nu_i}^1\times\frac{V_i}{n_i}} \tag{29.55}$$

Since $F_{\nu_i}^1$ is the F-table value for level of significance α, the P_i in Equation (29.51) all become the reliability $(1-\alpha)$:

$$P_i=1-\alpha \tag{29.56}$$

By substituting this into Equation (29.50), we obtain

$$P \geq \sum_{i=1}^{k} \frac{\lambda_i}{\lambda_1 + \cdots + \lambda_k} \times (1-\alpha) \tag{29.57}$$

and from this we obtain

$$P \geq 1-\alpha \tag{29.58}$$

This completes the proof.

Therefore, the formula of Chapter 9, which is a special case of the theorem stated above, and Equations (29.15) and (29.17) are *confidence limits on the safe side. Formulas using the effective degrees of freedom of Satterthwaite are used* in many Japanese books, *but proof as to what becomes of the precise reliability does not exist by Satterthwaite's formula.* Nor has there appeared any report of statistical data showing that Satterthwaite's formula fits reality well. Proof of Satterthwaite's formula is nearly impossible since, for example, the effective number of degrees of freedom becomes zero. Not only do the confidence limits of the general theorem given here become confidence limits on the safe side in the case of a normal distribution, but there exists a study by Masaharu Tsuchiya* for the case $k=2$ as to how close to the correct reliability the reliability comes. The correct levels of significance are as given in Table 29.4 according to that author.

29.5.3 Applications of General Theorem

The general theorem of Equation (29.32) can be used for the testing of significant difference between two mean values from two normal populations when it is unclear whether the population variances are equal, which has been a problem since long ago. What is involved is this: With the two mean values of populations n_1 and n_2 as $\bar{x}_1$ and $\bar{x}_2$, one examines whether or not it is significant by whether the following variance ratio is at least 1 or not.

$$F_0 = \frac{(\bar{x}_1 - \bar{x}_2)^2}{F^1_{n_1-1} \times V_1/n_1 + F^1_{n_2-1} \times V_2/n_2} \tag{29.59}$$

It is judged as being significant if F_0 is at least 1 and as being not significant if F_0 is less than 1.

The reader is referred to Chapter 9 for an actual example of finding the confidence limits when the population mean μ has been estimated for combinations of levels of various sources when, in an experiment by the split-unit design, the first-order error variance, second-order error variance, etc., cannot be pooled.

* Masaharu Tsuchiya: "The Robustness of Taguchi's Formula For the Behrens-Fisher Problem," Aoyama Gakuin University," Dissertation in candidacy for Master of Arts, 1971.

Table 29.4 Values of Correct Reliability Toward Confidence Limits of 95% Reliability

σ_2^2/σ_1^2	$n_1=n_2=5$	$n_1=n_2=10$	$n_1=10,\ n_2=5$
10^0	0.97594	0.96370	0.96821
10^1	0.96251	0.95531	0.95463
10^2	0.95192	0.95069	0.95055
10^3	0.95020	0.95007	0.95006
10^4	0.95002	0.95001	0.95001
10^5	0.95000	0.95000	0.95000
10^6	0.95000	0.95000	0.95000

The general theorem can also be used when one wishes to find the confidence limits of the total of weights when, in a case where variances among k types of weights have been obtained, one has used $n_1, n_2, \ldots, n_k$ for each of them.

30 How to Treat Missing Values

Here we will discuss the data analysis method for cases such as when one has lost a part of the data as a result of carelessness or accident, or when patients drop out of drug effectiveness tests, or when it is necessary to discontinue after half of the experiments by an orthogonal array have been finished.

30.1 What Is a Missing Value?

When it unfortunately happens that one has lost test pieces as a result of carelessness or has lost a part of the data in the course of an experiment, it is best to redo only the experiment for that combination. If it is difficult to redo it, it is also possible to use an improved version of the method developed by Yates, which will be discussed in this chapter. One such instance occurs when, in conducting an experiment by the combinations of $L_{27}(3^{13})$, although the schedule was to make measurements on five test pieces for each experiment number, it was impossible to obtain five pieces for several of them and groups of two and four resulted. Another example is when test pieces of the experiments for a certain number of combinations have been accidentally rendered useless. However, if, for example, a product turns into jelly as one is attempting to produce something by a certain experiment number and it is impossible to measure the values, or if it is impossible to assemble a product according to a certain combination in an assembly experiment and therefore a product is not obtained, so it becomes impossible to get complete data, these are by no means missing values. In such situations it has been demonstrated that a good product cannot be obtained by such a combination or that assembling itself is impossible, so it will never work out well if such a combination is used; and thus one has obtained extremely valuable information.

If, in designing an experiment, it is known that the product will turn to jelly and nothing will result if it is experimented on by a single combination of levels of certain factors together with all levels of a certain number of other factors, or if there are combinations such that it is known even without acquiring data that assembling is impossible, it means that the way in which one has taken the levels of the factors is inept. But if it is a case where, although one had expected the product probably to be good before the experiment, it has turned into jelly when the experiment was tried, *this information is very important since one has acquired information regarding something that the experimenter had not known.* In this case, it is *absolutely not a missing value.* In the author's experience, nearly every inquiry to the effect: I have trouble since there are many missing values; what should I do? has turned out not to be a case of missing values.

Merely that one has not been able to obtain continuous values as in other cases does not constitute a case of missing values; thus, it is necessary to analyze the case under consideration appropriately. The reader is referred to the section on fixed marginal enumerative values in Chapter 3 regarding data analysis in this case.

30.2 When the Number of Replications Is Unequal

We will consider the following example. For an example involving fixed marginal enumerative values, please refer to Section 8.4.

[EXAMPLE]

When an experiment was carried out by a two-way layout of A at three levels and B at four levels, the number of test pieces was *unequal* for the respective A_iB_j, and the data were as given in Table 30.1. How should we analyze these data?

When there is at least one measured value under each of the experimental conditions, we find the mean value for each condition. In the present case, Table 30.2 is obtained.

Table 30.1 Data with Missing Values in Replications

	B_1	B_2	B_3	B_4
A_1	68	69 81 76	79 84 92	56 62
A_2	60 62 55	64 70	76 81	49 45
A_3	48 51	46 51 49	58	40 40

Calculations by the ordinary two-way layout method are performed on the basis of this.

$$S_{T_1}=68.0^2+75.3^2+\cdots+40.0^2-\frac{735.0^2}{12}=2\,108.53 \qquad (f=11) \tag{30.1}$$

$$S_A=\frac{287.3^2+251.5^2+196.2^2}{4}-\frac{735.0^2}{12}=1053.24 \qquad (f=2) \tag{30.2}$$

$$S_B=\frac{176.5^2+191.0^2+221.5^2+146.0^2}{3}-\frac{735.0^2}{12}=985.08 \qquad (f=3) \tag{30.3}$$

$$S_{e_1}=S_{T_1}-S_A-S_B=70.21 \qquad (f=6) \tag{30.4}$$

Next, we find the variation between replications, S_{e_2}.

$$S_{e_2}=\frac{1}{\bar{r}}S_{e_2}' \tag{30.5}$$

$\bar{r}$ here is the *harmonic mean of the numbers of replications* (reciprocal of mean of reciprocals) under the respective experiment conditions, and in this instance it is obtained by the following equation.

$$\frac{1}{\bar{r}}=\frac{1}{12}\left(\frac{1}{1}+\frac{1}{3}+\frac{1}{3}+\frac{1}{2}+\frac{1}{3}+\frac{1}{2}+\frac{1}{2}+\frac{1}{2}+\frac{1}{2}+\frac{1}{3}+\frac{1}{1}+\frac{1}{2}\right)=0.52778 \tag{30.6}$$

The number of replications in this case varies: 1, 3, 3, . . . , 2; if it is treated as though it were constant, it is equivalent to a number of replications equal to the harmonic mean $\bar{r}=1.89$. S_{e_2}' is the sum of the variations between replications for the respective experimental conditions.

Table 30.2 Table of Mean Values

	B_1	B_2	B_3	B_4	Total
A_1	68.0	75.3	85.0	59.0	287.3
A_2	59.0	67.0	78.5	47.0	251.5
A_3	49.5	48.7	58.0	40.0	196.2
Total	176.5	191.0	221.5	146.0	735.0

$$S_{e_2}' = \left[\left(69^2+81^2+76^2-\frac{226^2}{3}\right)+\left(79^2+84^2+92^2-\frac{255^2}{3}\right)+\cdots+\left(40^2+40^2-\frac{80^2}{2}\right)\right]$$
$$=258.33 \qquad (f=14) \tag{30.7}$$

For the number of degrees of freedom, one might find the numbers of degrees of freedom of replication for the respective experimental conditions and obtain a total for them, or one might also find it as 14 by subtracting the number of experimental conditions, 12, from the total of the number of data, 26. We therefore have

$$S_{e_2} = 0.52778 \times 258.33 = 136.34 \qquad (f=14) \tag{30.8}$$

The analysis of variance table, Table 30.3, is obtained from this.

Instead of dividing S_{e_2}' by the harmonic mean $\bar{r}$, as shown above, it is also possible to multiply S_A, S_B, and S_{e_1} by $\bar{r}$. In this case, everything in the column of expected values $E(V)$ in the analysis of variance table becomes multiplied by $\bar{r}$.

We next estimate the factorial effects.

$$\bar{A}_1 = \frac{287.3}{4} = 71.8 \pm 3.4 \tag{30.9}$$

$$\bar{A}_2 = \frac{251.5}{4} = 62.9 \quad ''$$

$$\bar{A}_3 = \frac{196.2}{4} = 49.0 \quad ''$$

The confidence limits were found as

$$\pm\sqrt{\frac{4.35 \times 10.33}{4}} = \pm 3.4 \tag{30.10}$$

Although, fundamentally, $\bar{A}_1$, $\bar{A}_2$, and $\bar{A}_3$ are means of $4\bar{r}$ items, it suffices to use Equation (30.10) since the error variance itself is $1/\bar{r}$ (one-$\bar{r}$-th).

Similarly, we have

$$\bar{B}_1 = 58.8 \pm 3.9$$

$$\bar{B}_2 = 63.7 \quad ''$$

$$\bar{B}_3 = 73.8 \quad ''$$

$$\bar{B}_4 = 48.7 \quad ''$$

Table 30.3 Analysis of Variance Table

Source	f	S	V	$E(V)$	S'	ρ[%]
A	2	1 053.24	526.62	$\sigma_2^2/\bar{r}+\sigma_1^2+4\sigma_A^2$	1 032.58	46.0
B	3	985.08	328.36	$\sigma_2^2/\bar{r}+\sigma_1^2+3\sigma_B^2$	954.09	42.5
e_1	6	70.21	11.70	$\sigma_2^2/\bar{r}+\sigma_1^2$		
e_2	14	136.34	9.74	$\sigma_2^2/\bar{r}$		
(e)	(20)	(206.55)	(10.33)		258.20	11.5
T	25	2 244.87			2 244.87	100.0

The conditions by which one obtains the maximum value are A_1B_3, and the prediction of the actually realized value when confirmatory trials are run under these conditions is

$$\hat{\mu}\pm\sqrt{F\times V_e\left(\frac{1}{n_e}+\bar{r}\right)}=\bar{A}_1+\bar{B}_3-\bar{T}\pm\sqrt{4.35\times 10.32\left(\frac{6}{12}+1.89\right)}$$

$$=71.8+73.8-\frac{735.0}{12}\pm 10.4$$

$$=84.4\pm 10.4 \qquad (30.11)$$

30.3 The General Case (1): Fisher-Yates Method

30.3.1 When Degrees of Freedom of Error Remain

In the previous section, we treated the case where the number of replications for the combinations of levels of several factors is unequal; what should one do when there is not even one for certain combinations? This problem will be discussed in this section. For the case where there is one missing value in the two-way layout method, please refer to Section 1.1.6. A somewhat more general case will be treated in this chapter. For example, we have conducted an experiment of the 3^8 type by L_{27}, to include a determination of $A\times B$, but in the two experiments No. 5 and No. 8 the raw materials were mistakenly treated and became unusable. The other 25 experiments were carried out and measured values were obtained for three test pieces each. What numerical values should we insert into No. 5 and No. 8 to analyze the experiment?

In general, when an orthogonal array is used, one finds the error variation from columns to which errors correspond (in the present example, three columns correspond to error since of the 13 columns, eight columns have been used for the main effects and two columns have been used for $A\times B$), with x inserted into No. 5 and y inserted into No. 8. Error variation S_e is a quadratic function of x and y since x and y appear in the data. If this is written as

$$S_e=ax^2+2hxy+by^2+2gx+2fy+c \qquad (30.12)$$

then to obtain the values of x and y which cause S_e to be minimum, we need only differentiate S_e with respect to x and y and use the two equations simultaneously.

$$\left.\begin{aligned}\frac{1}{2}\frac{\partial S_e}{\partial x}&=ax+hy+g=0\\ \frac{1}{2}\frac{\partial S_e}{\partial y}&=hx+by+f=0\end{aligned}\right\} \qquad (30.13)$$

Now, if only the data of No. 5 are missing, S_e becomes a quadratic function of x and we solve

$$\frac{1}{2}\frac{dS_e}{dx} = ax + (hy + g) = 0 \tag{30.14}$$

This has the same form as the first of Equations (30.13). Therefore, the simultaneous linear equations, Equations (30.13), are obtained if one constructs a formula for estimation of the missing value considering only x, then substitutes y for the experimental value of No. 8 in this formula, thereby obtaining an estimation formula for the missing value related only with y, then substitutes x for the value of No. 5 therein. Since this means that it is unnecessary to form a long equation such as Equation (30.12), calculations become far easier.

	B_1	B_2	$B_3 \cdots\cdots B_b$
A_1	x	y_{12}	$y_{13} \cdots\cdots y_{1b}$
A_2	y_{21}	y_{22}	$y \cdots\cdots\cdots y_{2b}$
A_3	z	y_{32}	$y_{33} \cdots\cdots y_{3b}$
$\vdots$	$\vdots$	$\vdots$	$\vdots \qquad \vdots$
A_a	y_{a1}	y_{a2}	$y_{a3} \cdots\cdots y_{ab}$

For example, let us assume that in a two-way layout of the $a \times b$ form, there are no data at the three locations x, y, and z in the table above.

For estimation by the method of least squares of the missing values when they are missing only for A_1B_1, we differentiate

$$\begin{aligned} S_e &= x^2 + y_{12}^2 + \cdots + y_{ab}^2 - CF - S_A - S_B \\ &= x^2 + y_{12}^2 + \cdots + y_{ab}^2 - \frac{1}{b}[(x + A_1')^2 + \cdots + A_a^2] \\ &\quad - \frac{1}{a}[(x + B_1')^2 + \cdots + B_b^2] + \frac{(x + T')^2}{ab} \end{aligned}$$

with respect to x, and if we put this equal to 0 we obtain

$$\frac{d}{dx}S_e = 2x - \frac{2}{b}(x + A_1') - \frac{2}{a}(x + B_1') + \frac{2}{ab}(x + T') = 0$$

$$\left(1 - \frac{1}{b} - \frac{1}{a} + \frac{1}{ab}\right)x = \frac{A_1'}{b} + \frac{B_1'}{a} - \frac{T'}{ab}$$

$$x = \frac{aA_1' + bB_1' - T'}{(a-1)(b-1)} \tag{30.15}$$

If the values of y and z are inserted into the totals A_1', B_1', and T' which do not contain x in this equation, we have

$$\begin{aligned} (a-1)(b-1)x - bz + y + z &= aA_1' + bB_1'' - T''' \\ (a-1)(b-1)x + y - (b-1)z &= aA_1' + bB_1'' - T''' \end{aligned} \tag{30.16}$$

Similarly, from equations related with y and z, we have

$$(a-1)(b-1)y+x+z=aA_2'+bB_3'-T''' \tag{30.17}$$

$$(a-1)(b-1)z-(b-1)x+y=aA_3'+bB_1''-T''' \tag{30.18}$$

Therefore, we need only solve Equations (30.16), (30.17), and (30.18) simultaneously.

Therefore, to get a formula for estimation of missing values, it suffices if one has constructed a formula for the case where there is only one missing value.

30.3.2 When $A_iB_jC_k$ Are Missing Values in a Three-Way Layout

$$x=\frac{-aA_i'-bB_j'-cC_k'+ab(A_iB_j)'+ac(A_iC_k)'+bc(B_jC_k)'+T'}{(a-1)(b-1)(c-1)} \tag{30.19}$$

30.3.3 The Case of an Orthogonal Array

When, in the case of an orthogonal array, there is a column that corresponds to error, one finds the error variation from this column only with unknowns x and y inserted, and one finds it by partially differentiating with respect to x and y. Refer to Section 30.4 *when there are no degrees of freedom of error or when the number of degrees of freedom of the error is very small.*

30.3.4 The Case of Split-Unit Form

When an experiment in split-unit form has been carried out, it is necessary to proceed as follows. For simplicity, we will explain various cases of missing values in which we have used L_{27} and it is assumed that the first-order factors are A, B, and C, that the second-order factors are D, E, F, and G, and that $A \times B$ exists.

(1) When, among the nine combinations of columns 1, 2, 3, and 4 to which first-order factors correspond, *one combination consists entirely of missing values,* as for example when $A_1B_1C_2$ are all missing, S_{e_1} is found from column 4 assuming the correspondences $A \rightarrow 1$, $B \rightarrow 2$, and $C \rightarrow 3$. We *estimate* $A_1B_1C_2$ *so as to minimize* S_{e_1}. Next, estimates for the three experiment numbers in $A_1B_1C_2$ are performed by using method (2).

(2) When several among the 27 runs are missing, but all three experiments in any one combination among the nine combinations of first-order factors are not missing, we estimate so as to minimize S_{e_2}. In other words, we need only cause S_{e_2} to be minimum except in the case where all of one or two blocks among the nine blocks are missing. When all of a certain block is missing, or in other words as in case (1), one puts two of the three values as x and y, and with the third as *(value estimated by*

(1) $- x - y$*)*, one finds the x and y that minimize S_{e_2}. It is the same when all of one or two blocks are missing, and in addition parts of some blocks are missing.

When missing values are very numerous, calculation is troublesome by the method of this section; please therefore refer to the method of the next section.

30.4 The General Case (2): Sequential Approximation Method

30.4.1 A General Account

This method can be used even when sources are numerous and the *number of degrees of freedom of error is zero.* The basis of this method of thinking is: fundamentally, the method of the previous section—that of minimizing error variation—is *one whose precision is the highest only when none of the factorial effects can be disregarded.* Clearly, however, when there are some among the sources which can be disregarded, it is better to minimize the error variation including such sources as well. In other words, one would wish to add these small factorial effects to the error and to estimate missing values so as to minimize the total. We experimented only because it was not clear which factorial effects could be disregarded, but since data have already been obtained, it is possible to estimate analogously therefrom. The method of this section should therefore be as follows.

(1) Analysis of variance is performed by substituting a suitable value, such as the mean of all the measured values, for the missing values.
(2) One considers only non-negligible factorial effects and estimates the process average for the factor combination of the missing values. As sources which are not negligible one takes for example several of higher rank whose variance is large.
(3) Analysis of variance is performed once more by using these process average values for the missing values.
(4) If necessary, the procedure is repeated.

From (2) on, we may also proceed as follows. All sources that were not large in (1) are pooled with error, and missing values are found so as to minimize the error variation which is thus obtained. The method explained above is more realistic when missing values are numerous.

30.4.2 An Example with Two-Way Layout

The method shown here can be used especially effectively when orthogonal arrays are used, but since calculations become long in large experi-

Table 30.4 Data

	B_1	B_2	B_3	B_4
A_1	−51	−74	−89	−108
A_2	−9	1	−32	−64
A_3	61	40	13	−28
A_4	129	100	60	23

Table 30.5 0th-Order Approximation

	B_1	B_2	B_3	B_4	Total
A_1	−51	−74	−89	**−12**	−226
A_2	−9	1	**−12**	−64	−84
A_3	**−12**	40	**−12**	−28	−12
A_4	129	**−12**	**−12**	23	128
Total	57	−45	−125	−81	−194

mental examples, let us try applying this method by assuming that, in the example of Section 1.3 Volume 1, the underlined parts in Table 30.4 are missing. We subtract the working mean 70.0 and multiply by 10.

(1) We find the *mean of all except the missing values* and insert this where the values are missing. In this case, it is −6. It does not necessarily have to be the mean of the whole; it is better to insert the mean of several values in the neighborhood of the missing value, but since this is impossible in the case of an orthogonal array, we have shown a case where the mean of all values has been inserted.

(2) Analysis of variance of Table 30.5 is performed. We calculate by assuming that the data are all available. For example,

$$S_{A_l}=\frac{[-3(-226)-1(-84)+1(-12)+3\times 128]^2}{80}=16074 \qquad (f=1) \qquad (30.20)$$

The analysis of variance table then becomes as given by Table 30.6.

The only significant effect is A_l. If not even a single source is significant, one estimates one or two of the larger effects. In the equation of the linear effect of A,

$$y=\bar{y}+a_1(A-45)=-8.69+\frac{1134}{4\times 10\times 10}(A-45) \qquad (30.21)$$

we substitute $A_1=30$, $A_2=40$, $A_3=50$, and $A_4=60$, and we substitute the first approximate values of the missing values, −55, −26, 2, and 30, in the locations of the missing values (Table 30.7).

We perform analysis of variance a second time. Table 30.8 is obtained.

Table 30.6 Analysis of Variance Table

Source		f	S	V
A	l	1	16074*	
	q	1	0	
	c	1	238	
B	l	1	3050	
	q	1	1332	
	c	1	130	
$A_l \times B_l$		1	4537	
e		8	12881	1610
T		15	38242	

Table 30.7 Data with First Approximations Inserted

	B_1	B_2	B_3	B_4	Total
A_1	−51	−74	−89	**−55**	−269
A_2	−9	1	**−26**	−64	−98
A_3	**2**	40	**2**	−28	16
A_4	129	**30**	**30**	23	212
Total	71	−3	−83	−124	−139

Table 30.8 Analysis of Variance Table the Second Time

Source		f	S	V
A	l	1	30381	30381**
	q	1	3	3
	c	1	419	419
B	l	1	4606	4606**
	q	1	115	115
	c	1	21	21
$A_l \times B_l$		1	2030	2030
e		8	5179	647
T		15	42754	

Table 30.9 Second Approximate Values of Missing Values

	B_1	B_2	B_3	B_4	Total
A_1	−51	−77	−89	**−92**	−306
A_2	−9	1	**−6**	−64	−108
A_3	**36**	40	**2**	−28	50
A_4	129	**58**	**41**	23	251
Total	105	25	−82	−161	−113

Based on Table 30.8, we estimate considering A_l and B_l.

$$y=\frac{-139}{16}+\frac{1557}{400}(A-45)-\frac{665}{2000}(B-225) \tag{30.22}$$

By substituting in the A and B values of $A_1B_4, A_2B_3, A_3B_1, A_3B_3, A_4B_2$, and A_4B_3, we obtain the second approximate values of the missing values. For example, A_1B_4 is

$$\begin{aligned} y&=\frac{-139}{16}+\frac{1557}{400}(30-45)-\frac{665}{2000}(300-225) \\ &=-92 \end{aligned} \tag{30.23}$$

When the other two are calculated similarly, Table 30.9 is obtained.

If analysis of variance is performed one more, Table 30.10 is obtained.

Therefore, the third approximate values of the missing values are as given by Table 30.11. Since V_e has become small again, it is best to approximate about one more time. Usually, however, one may stop with the third time.

Data analysis of Table 30.11 is carried out. An analysis of variance table such as Table 30.10 is drawn up and, after distinguishing between large and small effects, this is organized into an adjusted analysis of

Table 30.10 Analysis of Variance Table the Third Time

Source		f	S	V
A	l	1	41816	41816**
	q	1	1	1
	c	1	86	86
B	l	1	10238	10238**
	q	1	0	0
	c	1	38	38
$A_l \times B_l$		1	958	958
e		8	1900	237
T		15	55037	

Table 30.11 Third Approximate Values of Missing Values

	B_1	B_2	B_3	B_4
A_1	−51	−74	−89	**−96**
A_2	−9	1	**−40**	−46
A_3	**54**	40	**3**	−28
A_4	129	**77**	**46**	23

variance table. When so doing, we *subtract 6 from the number of degrees of freedom of the whole and the number of degrees of freedom of error*. Other than this, one may analyze in the same way as when there are no missing values.

30.4.3 The Case of Medical Treatment Effect Data

When surveying the effects of medical treatment methods and the like, one conducts a test during a certain period of time, such as five years, with A_1 as the control and A_2 as the tested medicine; and it will be assumed that five years have passed since the end of the test. When, of 30 persons given A_1 and 18 persons given A_2, patients who have survived for five years number ten persons and nine persons, respectively, and the remaining persons have only survived two years, three years, or four years after the treatment, it is necessary to analyze data with many missing values (they are missing at the present stage although they will become known in the future). In such a case, the method of the previous section is applied, but for 0, 1 data, estimation of the missing values must be done by the omega method.

The data given in Table 30.12 constitute observed data of the results of treatment for several years for cancer of the prostate by the eight combinations of

A: four types of treatment method
  A_1, A_2, A_3, A_4
B: symptoms
  $B_1 =$ Term I or II, $B_2 =$ Term III

To the respective combinations $A_1B_1, \ldots, A_4B_2$, patients R_1–R_{65} can be seen to correspond as 3 persons, 7 persons, 13 persons, 18 persons, 3 persons, 7 persons, 7 persons, and 7 persons, respectively. When, at the time-points

$K_1 = 1$ year later, $K_2 = 2$ years later,
$K_3 = 3$ years later, $K_4 = 4$ years later

the patient was alive, it was indicated as 1, and when the patient was dead, it was indicated as 0. Thus,

0 = died, 1 = still alive
( ) = unknown (there is no data since enough time has not passed; this is a missing value)

Distribution of the degrees of freedom is as given by Table 30.13.

The missing values are filled in as follows.

When they are in the same group (same level of A, B and same group of K levels), as in the case of R_2K_2, R_2K_3, and R_3K_3, if there are data on other

Table 30.12 Data as to Living or Dead After Treatment for Cancer

A	B	R	K_1	K_2	K_3	K_4
1	1	1	1	1	1	( )
1	1	2	1	( )	( )	( )
1	1	3	1	1	( )	( )
1	2	4	1	1	1	1
1	2	5	1	1	0	0
1	2	6	1	1	1	1
1	2	7	1	1	1	1
1	2	8	1	0	0	0
1	2	9	1	1	( )	( )
1	2	10	1	1	( )	( )
2	1	11	1	1	1	1
2	1	12	1	1	1	1
2	1	13	1	1	1	1
2	1	14	1	1	1	1
2	1	15	1	1	1	( )
2	1	16	1	1	( )	( )
2	1	17	1	1	1	1
2	1	18	1	1	1	1
2	1	19	1	1	0	0
2	1	20	1	1	0	0
2	1	21	1	1	1	1
2	1	22	1	1	0	0
2	1	23	1	1	1	1
2	2	24	1	1	0	0
2	2	25	1	1	1	1
2	2	26	1	0	0	0
2	2	27	1	1	1	1
2	2	28	1	1	1	1
2	2	29	0	0	0	0
2	2	30	1	1	1	1
2	2	31	1	1	0	0
2	2	32	1	1	1	1
2	2	33	1	( )	( )	( )
2	2	34	1	1	1	1
2	2	35	1	1	0	0
2	2	36	1	1	1	0
2	2	37	1	1	1	1
2	2	38	1	0	0	0
2	2	39	1	1	1	1
2	2	40	1	1	1	1
2	2	41	1	1	0	0
3	1	42	1	1	1	( )
3	1	43	1	1	1	( )
3	1	44	1	1	( )	( )
3	2	45	1	1	0	0
3	2	46	1	0	0	0
3	2	47	1	1	1	1
3	2	48	1	0	0	0
3	2	49	1	1	1	1
3	2	50	1	1	0	0
3	2	51	1	( )	( )	( )
4	1	52	1	1	1	0
4	1	53	1	1	1	1
4	1	54	1	1	0	0
4	1	55	1	0	0	0
4	1	56	1	1	1	1
4	1	57	1	0	0	0
4	1	58	1	1	1	0
4	2	59	1	1	1	1
4	2	60	0	0	0	0
4	2	61	0	0	0	0
4	2	62	1	0	0	0
4	2	63	1	0	0	0
4	2	64	0	0	0	0
4	2	65	1	0	0	0

Table 30.13
Distribution of Number of Degrees of Freedom

Source	f
A	3
B	1
$A \times B$	3
$e_1(R)$	57
K	3
$A \times K$	9
$B \times K$	3
$A \times B \times K$	7
e_2	150
T	236

persons in the group the mean for the other persons is inserted. For R_2K_2, R_2K_3, and R_3K_3, for example, 1 is inserted. For K_3 and K_4 of R_9 and R_{10}, we insert $\frac{6}{10} = 0.6$. For the K_3 of R_{16} we insert $\frac{9}{12} = 0.75$, and for the K_4 of R_{15} and K_4 of R_{16} we insert $\frac{8}{11} \doteqdot 0.7$. When this is done, there remain only K_4 of R_1, R_2, and R_3 and K_4 of R_{42}, R_{43}, and R_{44}. As the values of these, the mean of all of K_4, $\frac{25}{53} \doteqdot 0.5$, may be used for the 0-th order approximation. Or one may insert the mean of the whole as the 0-th approximation. One then performs an analysis of variance. Calculations are left to the reader.

The number of degrees of freedom of the whole has been decreased only by the total number of missing values, 23, and this is balanced by subtracting degrees of freedom 2 from $A \times B \times K$ and 21 from e_2. The process average is estimated by the omega method, using the significant sources, and the first approximations are found. Thereafter, it becomes exactly the same as in the previous section.

30.5 The Case of Fixed Marginal Enumerative Value

Even in the case of fixed marginal enumerative values, when they are analyzed by minute accumulating analysis (Chapter 32) it is exactly the same as in Section 30.4.3. Here, however, we will explain as an example a case of accumulating analysis.

By the 2^5 type, in order to be able to obtain the main effects of A, B, C, D, and E and $A \times B$, assignment to L_8 (2^7) was performed as follows. Three test pieces each were taken from the respective batch-finished products of eight runs, and the data categorized into three classes by qualitative analysis of a certain impurity as

− No trace
+ Trace found
# More than a trace

were as follows.

The data of the first line were obtained when three test pieces were taken from products obtained by method No. 1 under the conditions $A_1B_1C_1D_1E_1$ and qualitative analysis was performed regarding the quantity of the impurity. This means that there was no case of no trace, that there was one run where a trace was found, and that there were two runs where there was more than a trace.

In No. 2, there were two replications; in No. 3, one replication; and in No. 6, zero replications. The data are given in Table 30.14.

After taking the cumulative frequency, in the cases of experiments where the replications gave different results we find the mean percent for each class. In the case of No. 6, where there are no data at all, the unknowns x and y are inserted into Class I and Class II. The results are as given in the Mean section of Table 30.14.

When there is error between experiments, the unknowns x and y can be estimated from it. In the present case, we find the x and y that minimize the variation of column 7. The totals of the first level and second level of column 7 are as given in the table below.

	I	II	III
$(7)_1$	x+1.00	y+2.00	4.00
$(7)_2$	1.50	2.67	4.00

The x and y that minimize

$$S_{e_1}=\frac{(x+1.00-1.50)^2}{4}\times W_1+\frac{(y+2.00-2.67)^2}{4}\times W_2 \tag{30.24}$$

Table 30.14 Assignment and Data

No.	A 1	B 2	$A\times B$ 3	C 4	D 5	E 6	e 7	Data −	Data +	Data #	Total	Mean I	Mean II	Mean III
1	1	1	1	1	1	1	1	0	1	2	3	0.00	0.33	1.00
2	1	1	1	2	2	2	2	1	1	0	2	0.50	1.00	1.00
3	1	2	2	1	1	2	2	1	0	0	1	1.00	1.00	1.00
4	1	2	2	2	2	1	1	0	2	1	3	0.00	0.67	1.00
5	2	1	2	1	2	1	2	0	0	3	3	0.00	0.00	1.00
6	2	1	2	2	1	2	1	—	—	—	0	x	y	1.00
7	2	2	1	1	2	2	1	3	0	0	3	1.00	1.00	1.00
8	2	2	1	2	1	1	2	0	2	1	3	0.00	0.67	1.00

are

$$x = 0.50, \quad y = 0.67$$

What we do is substitute this into Equation (30.24) and find the variation.

However, the method explained above, known as the Fisher-Yates method, cannot be used if there is no column of inter-experiment error, and furthermore, if missing values are numerous, calculation is troublesome. In spite of the fact that it is advisable to minimize even factorial effects that are not fundamentally significant by including them in error variation, often the method cannot be used for such reasons as that consideration therefore is absent. The sequential approximation method will be explained here. The means of Class I and Class II are substituted for x and y. This is termed 0-th order approximation.

$$x(0) = 2.5/7 = 0.36 \tag{30.25}$$

$$y(0) = 4.67/7 = 0.67 \tag{30.26}$$

These are inserted in the locations of x and y, and analysis of variance is performed. The totals of Class I and Class II then become 2.86 and 5.34.

$$W_1 = \frac{8^2}{2.86 \times 5.14} = 4.354$$

$$W_2 = \frac{8^2}{5.34 \times 2.66} = 4.506$$

$$CF = \frac{2.86^2 \times W_1 + 5.34^2 \times W_2}{8} = 20.51 \tag{30.27}$$

$$S_A = \frac{(1.50 - 1.36)^2 \times 4.354 + (3 - 2.34)^2 \times 4.506}{8} = 0.26 \qquad (f = 2) \tag{30.28}$$

$$\vdots$$

$$S_{e_1} = \frac{(1.36 - 1.50)^2 \times 4.354 + (2.67 - 2.67)^2 \times 4.506}{8}$$

$$= 0.01 \qquad (f = 2) \tag{30.29}$$

$$S_{e_2} = \frac{1}{\bar{r}} S_e{}' = 0.45238 \times 11.189 = 5.06 \qquad (f = 22) \tag{30.30}$$

Here,

$$\frac{1}{\bar{r}} = \frac{1}{7}\left(\frac{1}{3} + \frac{1}{2} + \frac{1}{1} + \frac{1}{3} + \cdots + \frac{1}{3}\right) = 0.45238 \tag{30.31}$$

$$S_{e_1}{}' = \left[\left(0 - \frac{0^2}{3}\right) + \left(1 - \frac{1^2}{2}\right) + \cdots + \left(0 - \frac{0^2}{3}\right)\right] \times W_1$$

$$+ \left[\left(1 - \frac{1^2}{3}\right) + \left(2 - \frac{2^2}{2}\right) + \cdots + \left(2 - \frac{2^2}{3}\right)\right] \times W_2$$

$$= (5 - 4.5) \times 4.354 + (11 - 9) \times 4.506 = 11.189 \tag{30.32}$$

Therefore, the analysis of variance table is as given in Table 30.15.

Table 30.15 Analysis of Variance Table (The First Time)

Source	f	S	V
m	2	20.51	10.26**
A	1.71	0.26 ○	0.15
B	1.71	1.72	1.01**
$A\times B$	1.71	0.26 ○	0.15
C	1.71	0.97	0.57
D	1.71	0.01 ○	0.01
E	1.71	6.70	3.92**
e_1	1.71	0.01 ○	0.01
e_2	22	5.06 ○	0.23
○ marks pooled (e)	(33.97)	(5.60)	(0.165)
T	36.00	35.50	

The number of degrees of freedom between experiments, including the general mean, has decreased to 14 from 16 and we have taken it as

	I	II
$\bar{B}_1$	0.215	0.500
$\bar{E}_2$	0.715	0.918
$\bar{T}$	0.358	0.668

$$2\times\frac{12}{14}=1.71 \tag{30.33}$$

when equally distributed over all of the sources excluding the general mean. The number of degrees of freedom of the second-order error variance, 22, was obtained by doubling the total of 11, obtained from 1 for No. 2, 0 for No. 3, 0 for No. 6, and 2 for the others.

Only B and E are large. Since No. 6 is B_1E_2, we estimate the effects of B and E, and we estimate the first approximate values of x and y for No. 6, $x(1)$ and $y(1)$, by the process average.

We use decibel values, and from Appendix Table 5 we have

$$\text{Decibel value of Class I} = -5.62+4.00-(-2.54)=0.92 \text{ [db]}$$

$$\text{Decibel value of Class II} = 0.00+10.49-3.04=7.45 \text{ [db]}$$

$$x(1)=0.55 \tag{30.34}$$

$$y(1)=0.85 \tag{30.35}$$

These are substituted for the x and y of Table 30.14. Analysis of variance is again performed on these data.

$$W_1=\frac{8^2}{3.05\times4.95}=4.239$$

$$W_2=\frac{8^2}{5.52\times2.48}=4.675$$

$$CF=\frac{3.05^2\times4.239+5.52^2\times4.675}{8}=22.74 \tag{30.36}$$

$$S_A=\frac{(1.50-1.55)^2\times4.239+(3.00-2.52)^2\times4.675}{8}=0.14 \tag{30.37}$$

$$\vdots$$

$$S_{e_1}=\frac{(1.55-1.50)^2\times4.239+(2.85-2.67)^2\times4.675}{8}=0.02 \tag{30.38}$$

$$S_{e_2}=0.45238[(5-4.5)\times4.239+(11-9)\times4.675] \tag{30.39}$$

$$=5.19 \tag{30.40}$$

The analysis of variance table, Table 30.16, is obtained. The number of degrees of freedom of first-order sources is 14 in all; the number of degrees of freedom of the main effects of general mean m, B, and E is taken as 2, and eight degrees of freedom are equally distributed among the others, $2\times\frac{8}{10}=1.6$ each.

Although B and C are rather large, they are not significant. Whether to regard E alone as significant, or to re-estimate the missing values of No. 6 assuming that the effects of B and C, too, are significant inasmuch as B and C are rather large although they are not significant, does not matter. This is because, fundamentally, it is forcing matters to decide on large sources by the F table of level of significance 5% when estimating missing values.

Table 30.16 Analysis of Variance Table (The Second Time)

Source	f	S	V
m	2	22.74	11.37 **
A	1.6	0.14 ○	0.09
B	2	1.26	0.63 △
$A\times B$	1.6	0.14 ○	0.09
C	1.6	0.91	0.57 △
D	1.6	0.02 ○	0.01
E	2	7.71	3.86 **
e_1	1.6	0.02 ○	0.01
e_2	22	5.19	0.24
○ marks pooled (e)	(28.4)	(5.51)	(0.194)
T	36.0	38.12	

Here, let us find the $\hat{\mu}$ of the conditions of No. 6, $B_1C_1E_2$, assuming that m, B, C, and E are large, and let us estimate the second approximate values of the missing values.

	Mean value I	Mean value II	Decibel value I	Decibel value II
$\bar{B}_1$	0.262	0.545	−4.50	0.78
$\bar{C}_2$	0.262	0.798	−4.50	5.97
$\bar{E}_2$	0.762	0.962	5.05	14.03
$\bar{T}$	0.381	0.690	−2.09	3.48

For the estimated values of the process averages, therefore, we find the rates of occurrence as

$$-4.50-4.50+5.05-2(-2.09)=0.23 \tag{30.41}$$

$$0.78+5.97+14.03-2(3.48)=13.82 \tag{30.42}$$

from Appendix Table 5 and obtain

$$x(2)=0.51,\ y(2)=0.96 \tag{30.43}$$

When these are substituted for the x and y in Table 30.14 and analysis of variance is performed the third time, the table becomes as presented in Table 30.17.

E is still the only significant main effect. Therefore the analysis finishes with Table 30.17. It is best to pool the effects of B and C with error and to regard the error variance as having 32 degrees of freedom. However, we do not test with it but use it merely for the confidence limits and contribution ratios. As a fundamental rule, once the error variance for testing has been decided on, even if there are sources which are not significant, they are merely pooled with the error variation and one does not test with the pooled error variance.

The best countermeasure against missing values is to redo the experiment to the extent possible and not produce missing values.

Table 30.17 Analysis of Variance Table (The Third Time)

Source	f	S	V
m	2	23.83	11.92**
A	1.5	0.08	0.05
B	2	1.18	0.59
$A\times B$	1.5	0.08	0.05
C	2	1.09	0.54
D	1.5	0.05	0.03
E	2	7.97	3.98**
e_1	1.5	0.05	0.03
e_2	22.0	5.30	0.24
(e)	(28.0)	(5.56)	(0.199)
T	36.0	39.89	

31 Sequential Categorization Method

In this chapter, an analysis method for data that have not been designed for orthogonalization, randomization, etc., will be explained. Mainly, the *sequential categorization method,* which is being widely used as an analysis method when there are numerous factors and the relationships are complicated, as in the cases of recorded data and survey data at work sites and prediction formulae, will be explained. Please refer to Chapter 15 in Volume 1 regarding calculations by the ordinary method of least squares. The method described here is useful when the number of sources is especially great and when *functions and vector variables* exist among the sources, or in other words for *data analysis of functionals.*

31.1 A General Explanation

31.1.1 Objective

Analysis of survey data becomes necessary when, for example, one wishes to determine the effects of various conditions on the target characteristic from observed data where the states of a process have been recorded, or one wishes to probe causes from observed data of an epidemiological nature. The method of experimental regression analysis, discussed in Chapter 15 of Volume 1, is useful when the variables that have effect are fairly clear, but when this is not the case the method discussed in this chapter should be useful. What poses the greatest problem when analyzing observed data is when, in analyzing a certain factor or several factors by an appropriate method, *the effect has not been orthogonalized or randomized* with regard to other factors, and *there is no basis for certainty as to the validity of the conclusions.* For example, if there are similar tendencies in the temperature changes at a certain plant and the changes of product quality of a certain material, it might happen that results which are thought to be the effect of temperature might actually be the effect of some such material.

If one wishes to analyze simultaneously various characteristic values of the material as well, the number of variables increases by leaps and bounds; even if a simple approximation by a quadratic equation is possible for the effects of these factors and if it is assumed that most interactions can be omitted, it is still usually necessary to estimate regression equations that include several tens or several hundreds of unknowns by the method of least squares; and still further, if one wishes to find unknowns among them that may safely be omitted in testing, it would probably be necessary to divert several days or several months to calculations even with the use of the newest computers.

Since it is the only way to achieve such an analysis method which is nearly accurate, we have usually begun by omitting most unknowns by resorting to experience and hunch, leaving only a few unknowns, and proceeded by applying the method of least squares. But since there is no guarantee, at least theoretically, of the validity of having omitted the other unknowns by such a method, as stated before, it has often happened that the conclusions are not correct.

31.1.2 Correlation and Regression, Especially Causality

The popularization of superficial calculation techniques, borrowing their names from statistics, has displayed the erroneous interpretations explained to researchers who have been plagued by difficulties such as those mentioned above, causing them to experience the feeling that linear equations suffice in nearly every case as regression equations (observation equations), and it has come about that statistical terms such as correlation and partial correlation are often used erroneously in data analysis.

According to statistical theory, *if both variable x and variable y are stochastic variables and if both assume a normal distribution, the relationship between the two (correlation) is linear, and they cannot possess any other relationship.* This fact holds even if there are more variables x. Of course, the theorem above is correct. But what is most important in the statement is that the relationship equation is linear only when both x and y are stochastic variables that follow the normal distribution law.

For example, the body height x and weight y of a human being each possess a specific distribution, and since the relationship between the two is not a causal relationship (cause and effect), it is of course a correlation; and if both follow normal distributions it is theoretically guaranteed that their correlation is linear. Similarly, in regard to the correlation between the hardness and tensile strength of the various items within a given lot, if each follows a normal distribution it can be theoretically guaranteed that the relationship is linear.

However, most of the data we wish to analyze at plants and research laboratories describe the relationship between various conditions or intermediate characteristics of processes and the quality or yield of the final product, and this is a *causal relationship;* since, moreover, the various conditions and intermediate characteristic values of processes do not vary randomly and are usually not stochastic variables, finding the relationship between these and the quality and yield of the final products is not a calculation of correlation but a probing of cause and effect, and it is a pursuit of a general functional relationship.

We see from time to time studies that confuse causal relationships with correlations; and in most cases, the difficult distribution of the correlation coefficient (according to Fisher) is not used in testing so-called correlation, but testing of regression relationship or regression coefficient (in this case x is not a stochastic variable, and therefore there is no guarantee anywhere that the relationship between x and y is linear!) is being used (this means the t or F test is being used), so that the beginning and end are inconsistent and it is contradictory.

What we understand from the discussion above is that since, generally speaking, the factors (variables) are not stochastic variables, in cases of data analysis for the obtaining of some hint that will be useful in probing causal relationships, or in other words cases where statistical treatment is applied to a process, one should usually hypothesize higher-order relationships since it is impossible to affirm definitely that the effect of the factors on product quality and productivity is linear.

31.1.3 Factors and Target Characteristics

The greatest problem encountered when analyzing data of the past is that, generally speaking, balance is lacking among the factors and moreover there is no guarantee of randomization. When certain factorial effects have been estimated by the method of least squares, other sources might be confounded with them, so it is necessary to correct for this. But even if other factorial effects, to the extent known, are corrected for, this does not guarantee that no unknown sources are confounded; therefore *one must be resigned to the fact that sometimes the conclusions found can be somewhat mistaken.*

It is when factors have been expressed quantitatively or have been ordered in sequence that the analysis method explained below can be used comfortably. For example, the hardness of a material, the percentage of a component, the temperature of a certain process, magnetic property, pressure, time, size measurement, dispersion, and date are all quantitative variables; and, for example, the degree of surface flaws, glossiness, appearance, and color-tone can often be ordered. Therefore, all such factors become subject to analysis.

Furthermore, for example, a certain number of stands of machinery, a certain number of furnaces, a distinction by steps of a process can also be treated by this analysis, but a clear picture cannot be obtained if the numbers produced by different machines differ greatly. Also, when performing an analysis such as this, one should take at least two seasons, such as winter and autumn or summer and winter, and investigate whether the factorial effects differ between them. If a certain factorial effect does not depend on the season, it means that the best conditions are always the same in regard to that source, and thus one can probably assume that an appreciable guarantee has been obtained regarding the logic of the analysis.

If possible, therefore, it is better to acquire data from each of the four seasons, even if one is to obtain data of one month each.

Please refer to Section 31.3 when the number of target characteristics is far fewer than the number of factors. Also, please refer to Section 31.3 when functions such as curves of temperature rise and fall and vector quantities exist among the factors.

31.1.4 Size of Data

Next, what the size of data should be is an important problem; when one lot is produced every day and the various conditions of the production steps, quality of the finished product, yield, etc., are recorded each time, if one obtains data from each season, as explained above, approximately one year's worth becomes necessary. In this case, the data size becomes greater than 300. Usually about 200 is considered best; if it is too much trouble to select 200 from among 300, one may decide to use all of the 300 data.

The number of factors may be as great as one wishes. However, when there is vector information, or continuously measured quantities such as temperature curves plotted by automatic apparatus or paper beating curves, among the factors, please refer to Section 31.3.

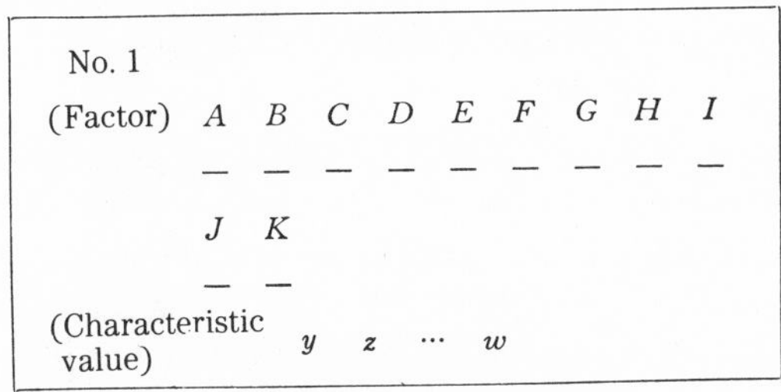

FIGURE 31.1 Recording Onto Card

31.1.5 Transcribing Onto Cards

The size of the data which constitute the object of analysis will be expressed as N. To simplify explanation, we will assume $N = 200$. We bring over 200 hole-sort cards or suitable computer cards and we write or punch in the data, card by card, as shown in Figure 31.1. If the number of factors is 30, all 30 are entered. As for the characteristic values, if these number 20, all 20 are entered. When seasons have been selected as a factor, as mentioned in the general comment, it is best to place the factor that corresponds to season at A, at the beginning. If we have chosen the four seasons, this would be for example, A_1: spring, A_2: summer, A_3: autumn, and A_4: winter. Since the seasons change continuously, one may for example add several cards at the beginning of summer to A_1 in order to arrange that the number of cards will be the same for each of the seasons.

In other words, we line up the 200 cards in order of time and categorize them in sets of 50, as the first 50 sheets and the next 50 sheets, etc. It would be advisable for beginners to use data of size $N = 81$, 108, 162, or 128.

31.1.6 Categorization

Next, if factor B is a continuous quantity such as temperature, we first take the 50 sheets of A_1 and arrange them from lower to higher values of temperature B. In many cases factors are categorized into, say, three classes, and actually it is unnecessary to arrange them in sequence as above. This is because we are to regard about $\frac{1}{3}$ of the cards whose value of B is small as the class B_1 and the next $\frac{1}{3}$ as cards of B_2, so that it is necessary to dependably arrange only the cards on the border between B_1 and B_2 and to perform categorization of B_1 and B_2 without mistakes.

Once the 50 cards of A_1 have been arranged in the order of magnitude of the B value, the 17 cards whose value is smallest are termed the class B_1 (of A_1). The next 16 cards are termed class B_2 (of A_1), and the 17 cards of the greatest values are termed the class B_3 (of A_1). Categorization is carried out here in such a manner that, with respect to the number of cards,

(1) When it can be evenly divided by 3, the number of cards in each class B_1, B_2, and B_3 is to be the same.
(2) If 1 is left over after dividing by 3, the number of cards of class B_2 at the center is to be one card more than in the other classes.
(3) If 2 is left over after dividing by 3, one card each is put in B_1, which has the smallest values, and B_3, which has the largest.

A categorizing method such as that described above will be called the trisecting method henceforth. Similarly, the 50 cards of A_2 are also divided by the trisecting method into the three classes B_1, B_2, and B_3 of A_2. What is important here is that the cards are arranged in order of magnitude only with regard to the 50 cards of A_2, and that there is no relationship with the B values of the other A values. It could happen that the value of B of class B_1 of A_1 is far greater than the value of B of the class B_1 of A_2, and is about the same as the value of B of the class B_2 of A_2. To get the picture here, we perform an analysis of variance of the value of B by the categorization of A and B, as depicted by Table 31.1.

Then, if A is significant compared with e, it means that the value of B is not balanced depending on the levels of A (see Table 31.9). In such a case, the mean value of B should be calculated for each level of A.

The 50 cards of A_3 and A_4 are similarly arranged in the order of the B values, and classes B_1, B_2, and B_3 are created by the trisecting method. Next, we take the 17 cards of class A_1B_1 and arrange them in order of the value of C, and again by the trisecting method we create the three classes C_1 (6 cards), C_2 (5 cards), and C_3 (6 cards). We similarly take the 16 cards of class A_1B_2 and categorize them by the trisecting method into the three classes consisting of five cards in C_1, with the smaller values of C, six cards in C_2, which are the next, and five cards in C_3 which are on the larger side. Similarly for the remaining A_iB_j classes, too, we arrange them in order according to the value of C and create the three classes C_1, C_2, and C_3.

Table 31.1
Analysis of Effect of *A* on the Value of *B*

Source	f
A	3
B	2
$A \times B$	6
e	188
T	199

Factor D is treated similarly. The six cards of $A_1B_1C_1$ are arranged in the order of the value of D and we categorize them in three classes: two cards of D_1 (of $A_1B_1C_1$), two cards of D_2 (of $A_1B_1C_1$), and two cards of D_3 (of $A_1B_1C_1$). Therefore, in this instance, the respective categories of cards become as indicated in the parentheses:

			D_1	D_2	D_3
$A_1(50)$	$B_1(17)$	$C_1(6)$	(2)	(2)	(2)
		$C_2(5)$	(2)	(1)	(2)
		$C_3(6)$	(2)	(2)	(2)
	$B_2(16)$	$C_1(5)$	(2)	(1)	(2)
		$C_2(6)$	(2)	(2)	(2)
		$C_3(5)$	(2)	(1)	(2)
	$B_3(17)$	$C_1(6)$	(2)	(2)	(2)
		$C_2(5)$	(2)	(1)	(2)
		$C_3(6)$	(2)	(2)	(2)

We have shown the results only for the 50 cards of A_1 because of space limitations, but is is exactly the same for A_2, A_3, and A_4.

31.1.7 Analysis of Variance

Once categorization is finished, we move on to data analysis, and unbalance of the number of cards is handled as follows. Usually, when categorizing such as this is performed, the number of cards in the respective classes $A_iB_jC_kD_l$ is either r or $(r - 1)$, the latter being one card fewer than the former. One might randomly discard one card from the classes of r cards and calculate with an equal number of replications. Moreover, this is probably easier when the number of data is over 100. If the number of data is insufficient, one uses the method of augmenting the classes of $(r - 1)$ cards with one card containing mean values of the factors as well as the characteristic values. Thus, we draw up a table such as Table 31.2.

It is allowable to pool e_1 and e_2 to obtain a single error, e. For persons who have a problem with the time and labor required for numerical calculations, we recommend the following abbreviated calculation method.

(1) *Sources are considered only up to two-factor interactions.*
(2) *Calculation by class mark is performed.*

To carry out (2), for example when one is to analyze characteristic value y, the minimum and maximum values of y are estimated and the range is divided into about 8–20 classes. For example, if the minimum is about 50 and the maximum is 140, we divide the range into 10 classes, as follows.

Table 31.2
Distribution of Degrees of Freedom

Source	f
A	3
B	2
C	2
D	2
$A \times B$	6
$A \times C$	6
$A \times D$	6
$B \times C$	4
$B \times D$	4
$C \times D$	4
e_1	68
e_2	92
T	199

class 1 = 59 or less, class 2 = 60–69,
class 3 = 70–79, . . . , class 10 = 140 or more

Or, if the minimum is about 3 and the maximum is between 7 and 8, the classes would be

class 1 = 3 or less, class 2 = 3.1–3.5,
class 3 = 3.6–4.0, . . . , class 9 = 7.1 or more

Or, if the minimum is of the order of magnitude of 10^5 and the maximum is of the order of magnitude of 10^{12}, since we know that $\sqrt{10} = 3.16$, we would choose

class 1 = 3.16×10^5 or less class 2 = $3.17 \times 10^5 - 10^6$,
class 3 = $1.01 \times 10^6 - 3.16 \times 10^6$, . . . , class 15 = 1.01×10^{12} or more

Once the values have been thus grouped, we write the upper limits of the values in the respective classes on a large sheet of paper and spread this on the desktop; and, inspecting the numerals on the cards, we proceed to distribute the cards into the classes to which these values belong. Of course, since we wish to construct a two-way table, we do this only with respect to cards that belong to A_1B_1. Separately, we give the class mark 0 to the class which is approximately at the center, and class marks such as -1 for the one just below and -2 for the one below that are written in ahead of time on the paper mentioned above. Since there were 10 classes in the first example, if we assign 0 to the 5th class, the class marks become

The 1st class = -4, the 2nd class = -3, . . . ,
the 5th class = 0, the 6th class = 1, . . . ,
the 10th class = 5

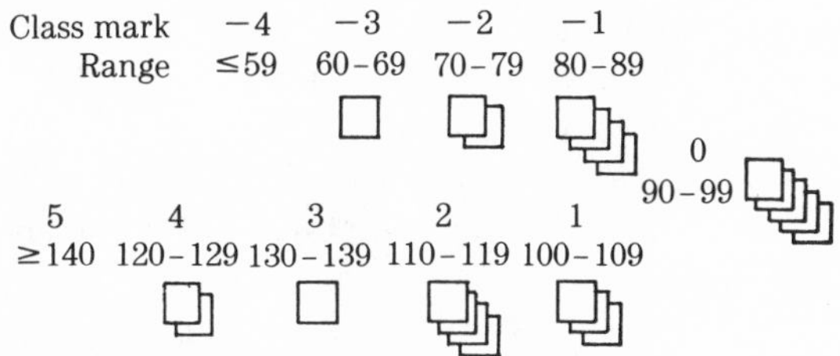

FIGURE 31.2 Categorization of Cards

We will assume that there were 22 A_1B_1 cards and that the cards were distributed as depicted in Figure 31.2.

Then, inasmuch as the total values are represented by class marks, that of class -3 and that of class $+3$ become cancelled, and therefore these two cards are stacked at the location of class 0. We also cancel two cards at -2 and two cards at $+2$ and stack these at the location of class 0. Two cards at -1 and one cards at $+2$ are cancelled; then, of the remainder, two cards at -1 and two cards at $+1$ are cancelled and these are stacked upon the class of 0. Then, only the plus classes remain, and since there are two cards at 4, one card at 2, and one card at 1, the total value becomes 11 by mental calculation. Thus, by using class marks, it becomes possible to perform the calculation by mental arithmetic, and probably even a child could do it without mistake.

The analysis of variance and estimations are also carried out by class mark. Lastly, if estimates of raw values are necessary, the multiplier for width of class would, in this case, be 10, and one need only add the central value of class 0, 95.

31.1.8 Correction Calculation

Correction is not always necessary if factors A, B, C, . . . , etc., have been obtained in the sequence of process steps. For example, when B is a certain characteristic value of the raw material and C is a certain characteristic value of an intermediate product, C is influenced by B and there is a causal relationship between B and C; thus, no matter how B influences C and, as a result, influences y, all of this can be regarded as the effect of B, and thus correction is not always necessary. When, in this way, A, B, C, . . . , have been written in the order of the steps and there is a necessary causal relationship among A, B, C, . . . , then if the effect of A influencing B, C, . . . , and, as a result, influencing y is all regarded as the effect of A on y, no correction whatsoever is necessary to the analysis of variance above and the regression equation.

S_A indicates the *simple regression relationship* of A, and S_B indicates the *partial regression relationship* (simple regression relationship for C

onward) when A is fixed. This is the same even in the case of ordinary correlation, and simple correlation includes the effects of all variables which have been related with A, as with S_A here.

It is to be hoped that the importance of this method, which is neither simple correlation nor partial correlation but employs *Schmidt's orthogonal expansion,* will be recognized.

If relationships do not necessarily exist among A, B, C, . . . , generally speaking it is better to correct the data as follows. For example, when A and C turn out to be significant, first we estimate the effect of C as by the accompanying table.

	Mean value of C	Mean value of y
C_1	$\bar{C}_1$	$\bar{C}_1(y)$
C_2	$\bar{C}_2$	$\bar{C}_2(y)$
C_3	$\bar{C}_3$	$\bar{C}_3(y)$

Further, we will assume that in the analysis of variance of A and B toward C, both B and A are significant but $A \times B$ has been found not significant.

We find the mean value of C for each of $A_1, A_2, A_3, B_1, B_2, B_3$, then we substitute these values into the relationship curve of C and y and obtain the difference from the mean value of y. These differences are corrected to the total value of y for each of A_1, A_2, A_3 or B_1, B_2, B_3, and we find the effect of A and B having eliminated the effect of C. If, when this is done, B turns out to be significant, it means that B and C become mutually cancelled and it had not been significant before, and therefore it is necessary to estimate the effect of B and perform correction on the effect of A.

If the number of factors is great, we perform analysis as explained above on the first four, for example, among them and estimate the factorial effects that have become significant; and with respect to what is obtained by subtracting these from the data and performing correction, we analyze the next four factorial effects. If this is repeated thenceforth, it is possible to pursue the relationship with y no matter how many factors there are.

However, please see Section 31.3 when the factors are very numerous.

31.2 An Example of Application (Characteristics of Carbon for Rubber)

31.2.1 Problem

The data of Table 31.3 were obtained by borrowing a part of a study on the relationship between the characteristics of carbon for rubber and the

Table 31.3 Characteristics of Carbon and Characteristics of Rubber (Class Marks Are Parenthesized)

No.	Iodine adsorption quant.	pH	Ash	Volatile components	Tensile strength	
					E_1 (20 min)	E_2 (40 min)
1	114.9	7.4	0.15	1.23	248.8(6)	237.5(4)
2	82.6	7.5	0.14	1.82	223.6(1)	215.4(0)
3	90.3	7.1	0.14	3.87	210.0(−1)	202.6(−3)
4	61.6	6.1	0.37	1.59	204.4(−3)	207.1(−2)
5	61.5	5.8	0.19	1.59	231.2(3)	202.5(−3)
6	95.6	4.7	0.37	1.67	219.5(0)	213.9(−1)
7	48.4	6.1	0.39	1.48	204.6(−3)	165.0(−10)
8	76.9	7.0	0.37	1.99	216.5(0)	210.1(−1)
9	92.7	3.6	6.67	3.02	196.9(−4)	180.3(−7)
10	84.5	7.6	0.11	0.99	232.2(3)	228.9(2)
11	93.4	7.5	0.08	1.41	234.7(3)	224.6(1)
12	118.9	8.0	0.12	1.51	221.5(1)	218.1(0)
13	120.8	7.8	0.21	2.77	242.9(5)	223.5(1)
14	120.8	7.8	0.21	2.77	238.1(4)	225.7(2)
15	120.8	7.8	0.21	2.77	229.2(2)	222.5(1)
16	108.0	7.8	0.18	1.61	210.1(−1)	179.8(−8)
17	59.8	7.6	0.30	1.66	201.4(−3)	199.9(−4)
18	104.1	7.5	0.27	2.88	239.0(4)	237.5(4)
19	54.2	8.4	0.03	1.40	216.0(0)	212.9(−1)
20	80.6	8.8	0.05	1.40	216.8(0)	196.4(−5)
21	82.1	8.7	0.04	1.53	227.9(2)	213.9(−1)
22	110.5	8.6	0.06	1.51	222.3(1)	203.3(−3)
23	121.4	8.1	0.31	2.57	239.1(4)	230.5(3)
24	121.4	8.1	0.31	2.57	241.3(5)	229.5(2)
25	112.2	8.4	0.71	1.49	191.9(−5)	173.6(−9)
26	121.4	8.1	0.31	2.57	240.5(5)	218.5(0)
27	121.4	8.1	0.31	2.57	243.4(5)	237.3(4)
28	145.4	5.6	0.08	1.77	232.3(3)	219.9(0)
29	153.3	5.6	0.11	2.00	240.3(5)	225.0(2)
30	174.3	6.2	0.05	4.34	241.0(5)	227.0(2)
31	146.1	3.8	0.18	1.78	244.7(5)	211.6(−1)
32	164.7	6.7	0.17	2.25	233.8(3)	222.0(1)
33	136.7	5.7	0.13	2.35	246.5(6)	235.6(4)
34	127.1	5.3	0.21	2.09	213.4(−1)	208.5(−2)
35	137.8	4.4	0.66	2.83	221.0(1)	218.0(0)
36	126.6	5.1	0.77	3.48	211.7(−1)	200.5(−3)
37	138.8	7.5	0.06	1.18	241.7(5)	229.7(2)
38	167.3	7.2	0.08	2.06	237.1(4)	224.2(1)
39	153.2	7.4	0.01	3.08	270.2(11)	262.0(9)
40	146.9	7.5	0.09	2.32	243.3(5)	227.9(2)
41	120.8	7.8	0.21	2.77	237.9(4)	228.5(2)
42	178.4	7.6	0.11	3.72	211.7(−1)	203.5(−3)
43	122.8	7.9	0.13	1.14	244.6(5)	234.1(3)
44	170.1	7.3	0.24	2.03	241.6(5)	237.0(4)
45	153.1	8.0	0.80	2.05	258.9(8)	223.5(1)
46	165.1	8.2	0.12	1.80	215.6(0)	216.9(0)
47	121.4	8.1	0.31	2.57	242.7(5)	232.9(3)
48	143.7	8.5	0.10	6.59	222.2(1)	217.8(0)
49	176.2	8.1	0.57	1.93	240.5(5)	236.3(4)
50	121.4	8.1	0.31	2.57	256.2(8)	228.9(2)
51	159.5	8.0	0.35	2.99	251.8(7)	252.1(7)
52	160.2	8.6	0.77	2.22	257.1(8)	241.9(5)
53	167.8	8.2	1.15	2.62	243.8(5)	233.9(3)
54	171.8	9.1	1.84	6.48	210.5(−1)	206.3(−2)
55	189.2	6.2	0.04	2.05	227.2(2)	225.2(2)

(continued on next page)

Table 31.3 (Continued)

No.	Iodine adsorption quant.	pH	Ash	Volatile components	Tensile strength	
					E_1 (20 min)	E_2 (40 min)
56	192.1	2.9	0.21	2.31	211.8(−1)	221.9(1)
57	234.6	6.0	0.19	2.81	237.5(4)	238.8(3)
58	229.1	2.3	0.26	2.81	231.6(3)	221.8(1)
59	204.1	6.3	0.24	2.90	245.0(6)	222.6(1)
60	220.7	5.6	0.45	4.92	235.3(4)	232.7(3)
61	218.8	5.1	0.46	2.50	191.4(−4)	204.0(−3)
62	178.7	1.9	1.12	3.93	223.7(1)	227.9(2)
63	190.2	4.3	1.04	4.28	235.9(4)	232.8(3)
64	180.3	6.5	0.14	1.33	221.6(1)	223.4(1)
65	215.1	7.1	0.11	2.68	212.3(−1)	213.1(−1)
66	226.8	8.1	0.12	3.27	226.8(2)	223.2(1)
67	227.0	6.8	0.17	2.53	198.8(−4)	215.8(0)
68	194.7	7.5	0.18	2.74	247.9(6)	235.6(4)
69	217.0	7.6	0.16	4.26	227.2(2)	223.2(1)
70	193.4	6.6	0.19	1.58	235.1(4)	231.7(3)
71	179.3	7.9	0.20	2.38	214.9(−1)	212.5(−1)
72	198.0	6.5	0.30	3.92	271.1(11)	256.1(8)
73	210.7	8.2	0.07	4.12	227.8(2)	219.8(0)
74	219.7	8.2	0.07	4.12	210.7(−1)	218.3(0)
75	219.7	8.2	0.07	4.12	224.7(1)	217.0(0)
76	210.0	8.2	0.24	2.70	212.9(−1)	209.9(−2)
77	234.0	9.1	0.26	2.95	234.0(3)	224.1(1)
78	213.5	8.3	0.15	3.56	210.2(−1)	215.1(0)
79	191.1	8.4	0.31	2.16	212.0(−1)	210.6(−1)
80	191.1	8.4	0.31	2.16	228.8(2)	222.6(1)
81	243.0	9.1	0.26	2.95	247.6(6)	230.1(3)

quality of the rubber made with it, which study was carried out over a number of years by the Nippon Carbon Kyōkai (Japan Carbon Association). The original data were voluminous, but since the information is used here only to show the calculation method, we give the results of investigation of the relationship between four characteristics of carbon,

A = quantity of iodine adsorbed, B = pH, C = ash,
D = volatile components

and the tensile strength of the rubber, y. For the respective carbon batches, we decided to use data taken with the vulcanization time at two levels,

$E_1 = 20$ minutes, $E_2 = 40$ minutes

Since the number of batches of carbon was 81 and for these there were two levels of vulcanization time, E_1 and E_2, the size of the data was $81 \times 2 = 162$.

The data constitute the results after sequential categorization by A, B, C, and D. In other words, they are already categorized first into the three classes A_1, A_2, and A_3 by the value of A, and the data in each of them divided into three classes according to the value of B, and so forth. Only factor E is a usual control factor.

31.2.2 Analysis of Variance

Since the data had been entered onto cards from the beginning, the time and labor of transcribing was saved. Also, to facilitate calculations, all the data were divided into classes and calculation was performed by class marks. Data which were dispersed from 165.0 to 270.2 were divided into 22 classes of class width 5; and −10 was assigned to the class of the lowest values and +11 to the class of the highest. These classes are indicated within parentheses.

For analysis of variance, in many cases it suffices to include up to two-factor interactions, and one constructs a two-way array, but since E was a second-order control factor in this case, it was treated specially and supplementary tables such as Tables 31.4 were constructed.

Table 31.4 Supplementary Tables

E_1	B_1	B_2	B_3	Total
A_1	−1	18	17	34
A_2	26	46	38	110
A_3	19	20	10	49
Total	44	84	65	193

E_2	B_1	B_2	B_3	Total
A_1	−23	−1	−10	−34
A_2	3	21	22	46
A_3	13	16	2	31
Total	−7	36	14	43

E_1+E_2	B_1	B_2	B_3	Total
A_1	−24	17	7	0
A_2	29	67	60	156
A_3	32	36	12	80
Total	37	120	79	236

E_1	C_1	C_2	C_3	Total
A_1	15	21	−2	34
A_2	39	42	29	110
A_3	9	18	22	49
Total	63	81	49	193

E_2	C_1	C_2	C_3	Total
A_1	−3	0	−31	−34
A_2	19	18	9	46
A_3	7	9	15	31
Total	23	27	−7	43

E_1+E_2	C_1	C_2	C_3	Total
A_1	12	21	−33	0
A_2	58	60	38	156
A_3	16	27	37	80
Total	86	108	42	236

E_1	D_1	D_2	D_3	Total
A_1	3	17	14	34
A_2	35	40	35	110
A_3	2	14	33	49
Total	40	71	82	193

E_2	D_1	D_2	D_3	Total
A_1	−26	−7	−1	−34
A_2	13	18	15	46
A_3	1	8	22	31
Total	−12	19	36	43

E_1+E_2	D_1	D_2	D_3	Total
A_1	−23	10	13	0
A_2	48	58	50	156
A_3	3	22	55	80
Total	28	90	118	236

E_1	C_1	C_2	C_3	Total
B_1	24	27	−7	44
B_2	29	23	32	84
B_3	10	31	24	65
Total	63	81	49	193

E_2	C_1	C_2	C_3	Total
B_1	11	3	−21	−7
B_2	16	10	10	36
B_3	−4	14	4	14
Total	23	27	−7	43

E_1+E_2	C_1	C_2	C_3	Total
B_1	35	30	−28	37
B_2	45	33	42	120
B_3	6	45	28	79
Total	86	108	42	236

E_1	D_1	D_2	D_3	Total
B_1	8	19	17	44
B_2	23	21	40	84
B_3	9	31	25	65
Total	40	71	82	193

E_2	D_1	D_2	D_3	Total
B_1	−11	3	1	−7
B_2	6	8	22	36
B_3	−7	8	13	14
Total	−12	19	36	43

E_1+E_2	D_1	D_2	D_3	Total
B_1	−3	22	18	37
B_2	29	29	62	120
B_3	2	39	38	79
Total	28	90	118	236

(continued on next page)

Table 31.4 (Continued)

E_1	D_1	D_2	D_3	Total
C_1	22	15	26	63
C_2	16	41	24	81
C_3	2	15	32	49
Total	40	71	82	82

E_2	D_1	D_2	D_3	Total
C_1	10	2	11	23
C_2	0	13	14	27
C_3	−22	4	11	−7
Total	−12	19	36	43

E_1+E_2	D_1	D_2	D_3	Total
C_1	32	17	37	86
C_2	16	54	38	108
C_3	−20	19	43	42
Total	28	90	118	236

$$CF=\frac{236^2}{162}=343.8$$

$$S_A=\frac{1}{54}(0^2+156^2+80^2)-CF=225.4$$

$$S_B=\frac{1}{54}(37^2+120^2+79^2)-CF=63.8$$

$$S_C=41.8$$

$$S_D=78.6$$

$$S_E=138.9$$

$$S_{A\times B}=\frac{1}{18}[(-24)^2+17^2+7^2+29^2+67^2+60^2+32^2+36^2+12^2]-CF-S_A-S_B$$

$$=50.8$$

$$S_{A\times C}=79.9$$

$$S_{A\times D}=45.8$$

$$S_{B\times C}=141.3$$

$$S_{B\times D}=31.1$$

$$S_{C\times D}=86.2$$

$$S_{e_1}=\text{sum of squares of sum of two}-CF-(S_A+\cdots+S_{C\times D})$$

$$=1\,122.2-844.7=277.5 \qquad (31.1)$$

$$S_{A\times E}=102.6$$

$$S_{B\times E}=0.1$$

$$S_{C\times E}=2.8$$

$$S_{D\times E}=0.4$$

$$S_{e_2}=\text{total variation}-(\text{sum of source variations})$$

$$=1805.2-(1122.2+S_E+S_{A\times E}+S_{B\times E}+S_{C\times E}+S_{D\times E})=1805.2-1367.0$$

$$=438.2 \qquad (31.2)$$

We calculated the contribution ratios in the total variation S_{T_1} among lots of carbon. Although in principle it is necessary to subtract measurement error, etc., from S_{T_1}, this was omitted here.

Table 31.5 Analysis of Variance Table

Source	f	S	V	S'	ρ(%)
A	2	225.4	112.7**	213.5	19.1
B	2	63.8	31.9**	51.9	4.6
C	2	41.8	20.9**	29.9	2.7
D	2	78.6	39.3**	66.7	6.0
$A\times B$	4	50.8	12.7		
$A\times C$	4	79.9	20.0**	56.5	5.0
$A\times D$	4	45.8	12.0		
$B\times C$	4	141.3	35.3**	117.4	10.5
$B\times D$	4	31.1	7.8		
$C\times D$	4	86.2	21.6**	62.3	5.5
e_1	48	277.5	5.78	(res)	(46.6)
(T_1)	80	1 122.2			100.0
E	1	138.9	138.9**	132.9	
$E\times A$	2	102.6	51.3**	90.7	
$E\times B$	2	0.1	0.0		
$E\times C$	2	2.8	1.4		
$E\times D$	2	0.4	0.2		
e_2	72	438.2	6.09		
(e_1+e_2)	(120)	(715.7)	(5.97)		
T	161	1 805.2			

31.2.3 Estimations

Main effect A can be estimated as follows. From Table 31.4, by using the central value of the classes, 217.5, and class width 5, we have

$$\left.\begin{aligned} \bar{A}_1 &= 217.5+\frac{0}{54}\times 5 = 217.5\pm 3.3 \\ \bar{A}_2 &= 217.5+\frac{156}{54}\times 5 = 231.9\pm 3.3 \\ \bar{A}_3 &= 217.5+\frac{80}{54}\times 5 = 224.9\pm 3.3 \end{aligned}\right\} \tag{31.3}$$

The confidence limits were found as follows.

$$\pm\sqrt{\frac{3.92\times 5.97}{54}}\times 5 = \pm 3.3 \tag{31.4}$$

It is necessary to obtain a two-way array of A and C as in Table 31.6 since, actually, $A \times C$ is significant.

When Table 31.6 is drawn as graphs, they are as shown in Figures 31.3. Here, the mean of the values of A, A_1, A_2, and A_3, was obtained from the analysis of the value of A by categorization of A in (1) of Section 31.2.4, and C_1, C_2, and C_3 were obtained from the mean value of C by categorization of A, B, and C in (3) of that section. The broken-line curve of curve C_1 is a correction of the effect of B.

Table 31.6 Estimation of Two-Way Array of A and C

	C_1	C_2	C_3	
A_1	220.8	223.3	208.3	
A_2	233.6	234.2	228.1	±5.7
A_3	221.9	225.0	227.8	

31.2.4 Interrelationship Among Factors

The question is whether or not Equation (31.3) or the graphs of Figures 31.3 can be trusted. For this, it becomes necessary to study mutual interrelations among factors. Of course, if interrelations among A, B, C, and D necessarily exist, it is certainly unnecessary to perform the study explained below. One need only draw the significant sources as graphs directly, as explained in Section 31.1.8.

We study the correlations among A, B, C, and D, and if we wish to correct the factorial effects that are significant, we perform the following analyses.

(1) Analysis of the value of A by categorization of A
(2) Analysis of the value of B by categorization of A and B
(3) Analysis of the value of C by categorization of A, B, and C
(4) Analysis of the value of D by categorization of A, B, C, and D

(1) *Analysis of the value of A by Categorization of A* With A_1, A_2, and A_3 as three levels, let us investigate to what extent the variation of the value of A is expressed by A_1, A_2, and A_3.

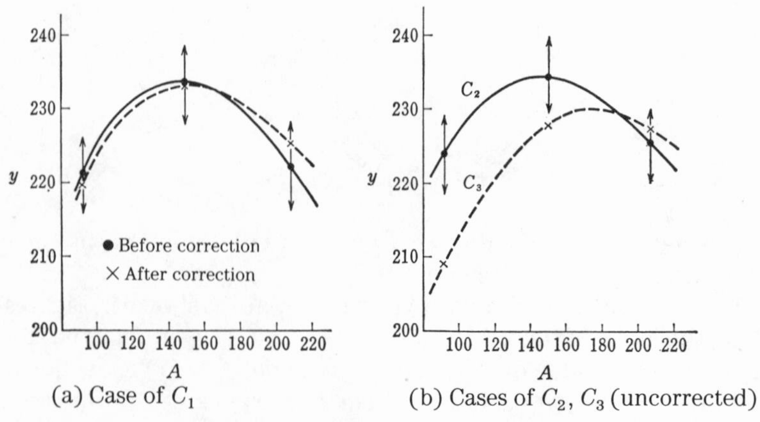

FIGURE 31.3 Relationship Between *A* and *y* by Level of *C*

Supplementary Table

	Sum of A	Mean
A_1	2 580.8	95.6
A_2	4 050.5	150.0
A_3	5 621.9	208.2
Total	12 253.2	

$$CF(A)=\frac{12\,253.2^2}{81}=185\,359.48$$

$$S_A(A)=\frac{1}{27}(2580.8^2+4050.5^2+5621.9^2)-CF\fallingdotseq 171\,328 \tag{31.5}$$

Table 31.7 Analysis of Variance $A \rightarrow A$

Source	f	S	V	ρ[%]
A	2	171 328	85 664	83.6
e	78	32 656	418.7	16.4
T	80	203 984		100.0

$$\rho_{A\rightarrow A}=\frac{171\,328-2(418.7)}{203\,984}=83.6[\%]$$

This indicates that the fact that A was categorized into three classes has captured 83.6% of the variation of A. If the calculation had been done with A at 81 levels, it can be estimated that the contribution ratio of A to the target characteristic y would have become about

$$\rho_A(y)=\frac{19.1}{0.836}\fallingdotseq 22.8\ [\%]$$

However, perhaps the effect of B and the effect of C have entered into the effect of A. For this, analysis of two-way layout of A and B toward the value of B becomes necessary.

(2) *Analysis of Value of B by Categorization of A and B* We construct a two-way Array of A and B toward the value of B.

Table 31.8 Two-Way Array of A and B Toward the Value of B

k=9	A_1	A_2	A_3	Total
B_1	55.3	48.4	40.6	144.3
B_2	69.4	68.2	64.6	202.2
B_3	75.3	74.9	76.1	226.3
Total	200.0	191.5	181.3	572.8

$$CF=\frac{572.8^2}{81}=4\,050.62$$

$$S_A=\frac{1}{27}(200.0^2+191.5^2+181.3^2)-CF=6.49$$

$$S_B=\frac{1}{27}(144.3^2+202.2^2+226.3^2)-CF=131.50$$

$$S_{A\times B}=\frac{1}{9}(55.3^2+69.4^2+\cdots+76.1^2)-CF-S_A-S_B=7.00 \quad (31.6)$$

$$S_T=\text{sum of squares of respective values}-\text{CF}=195.74 \quad (31.7)$$

Table 31.9 Analysis of Variance Table
AB → B

Source	f	S	V	ρ[%]
A	2	6.49	3.24**	2.5
B	2	131.50	65.75**	66.4
$A\times B$	4	7.00	1.75 }	31.1
e	72	50.73	0.704 }	
T	80	195.74		100.0

Depending on the level of A, there is some slight difference in the mean of the distribution of the value of B (A is significant at 1%), but there is no significant difference regarding dispersion ($A\times B$ is not significant). Since everything which is related with A is considered as the effect of A in the sequential categorization method (this is the case with the usual simple correlation and simple regression, as well), but since the contribution ratio of A to B is only 2.5%, it is clear that in the analysis of variance table for tensile strength, the effect of B is confounded only to the degree of about $4.6\%\times(2.5\div66.4)=0.2\%$ in the contribution ratio of A of 19.1%. This means that there is no need to correct the effect of A. However, if the effect of A is to be corrected, we proceed as follows. First, we draw a graph of the effect of B. From the mean of the values of B, B_1, B_2, and B_3 and the mean of the values of y, the graph becomes as shown by Figure 31.4.

	Mean of B	Mean of y
B_1	5.3	220.9±3.3
B_2	7.5	228.6±3.3
B_3	8.4	224.8±3.3
T		224.8

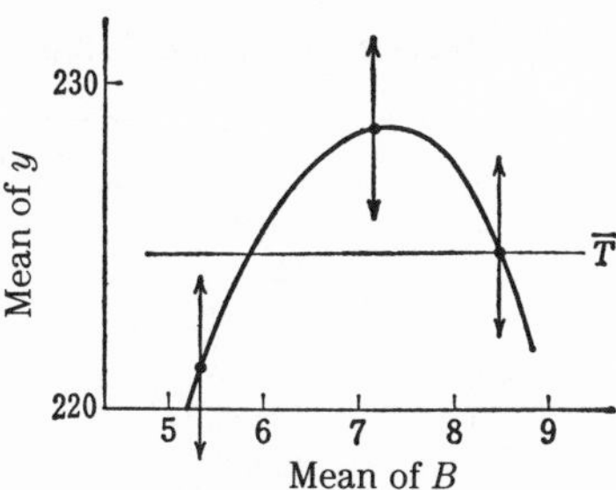

FIGURE 31 Relationship Between *B* and *y*

Since the relationship between B and y can be discerned from Figure 31.4, we substitute the nine values of B for the case of A_1C_1 on this curve and find the mean of the difference between the value of y and $\overline{T}$. It is the same with regard to A_2 and A_3. This difference is the correction quantity.

	Correction quantity
A_1	−1.2
A_2	−0.3
A_3	+3.1

(31.8)

The three correction quantities in Equation (31.8) are corrected to the mean values of A_1, A_2, and A_3 of Figure 31.3. Therefore, the graph of A after correction has been shown only for the case of C_1 in Figure 31.3, but there is change from the • symbol to the × symbol. It can be seen that there is no great difference even with correction.

(3) *Analysis of the Value of C by Categorization of A, B, and C*

Since all except C are not significant in the analysis of variance table,

Tables 31.10 Supplementary Tables

C_1	A_1	A_2	A_3	Total
B_1	0.43	0.24	0.44	1.11
B_2	0.31	0.15	0.37	0.83
B_3	0.12	0.53	0.21	0.86
Total	0.86	0.92	1.02	2.80

C_2	A_1	A_2	A_3	Total
B_1	0.93	0.48	0.95	2.36
B_2	0.63	0.41	0.51	1.55
B_3	0.68	1.23	0.65	2.56
Total	2.24	2.12	2.11	6.47

C_3	A_1	A_2	A_3	Total
B_1	7.43	3.64	2.62	13.69
B_2	4.75	2.17	0.69	7.61
B_3	1.33	3.76	0.88	5.97
Total	13.51	9.57	4.19	27.27

$C_1+C_2+C_3$	A_1	A_2	A_3	Total
B_1	8.79	4.36	4.01	17.16
B_2	5.69	2.73	1.57	9.99
B_3	2.13	5.52	1.74	9.39
Total	16.61	12.61	7.32	36.54

Table 31.11 Analysis of Variance Table, $A, B, C \rightarrow C$

Source	f	S	ρ[%]
A	2	16 081	
B	2	12 840	
C	2	128 999**	17.1
$A \times B$	4	30 335	
$A \times C$	4	32 586	
$B \times C$	4	24 610	
e	62	427 626	
	80	673 076	100.0

Table 31.11, it is clear that the values of A and B correspond randomly to C. We can therefore expect that

$$\rho_C(y) = \frac{2.7}{0.171} = 15.8[\%]$$

but we need not correct the effect of C from the effect of A and B on y.

(4) *Analysis of Value of D by Categorization of A, B, C, and D* We analyze the value of D by A, B, C, and D.

Since A greatly influences the distribution of the value of D, it is better to correct the effect of D from the effect of A. We draw the graph of D versus y, then, classed by level of A_1, A_2, and A_3, 97 each of the values of D

Tables 31.12 Supplementary Tables

	A_1	A_2	A_3	Total
B_1	18.26	22.89	28.51	69.66
B_2	18.37	20.35	24.69	63.41
B_3	17.61	29.77	28.84	76.22
Total	54.24	73.01	82.04	209.29

	A_1	A_2	A_3	Total
C_1	15.16	25.39	26.81	67.86
C_2	19.83	22.68	29.37	71.86
C_3	19.27	24.94	25.86	70.07
Total	54.24	73.01	82.04	209.29

	D_1	D_2	D_3	Total
A_1	14.07	17.78	22.39	54.24
A_2	16.23	21.70	35.08	73.01
A_3	21.78	26.17	34.09	82.04
Total	52.08	65.65	91.56	209.29

	C_1	C_2	C_3	Total
B_1	22.20	21.86	25.60	69.66
B_2	17.51	26.65	19.25	63.41
B_3	27.65	23.35	25.22	76.22
Total	69.36	71.86	76.07	209.29

	D_1	D_2	D_3	Total
B_1	17.30	21.62	30.74	69.66
B_2	15.45	20.50	27.46	63.41
B_3	19.33	23.53	33.36	76.22
Total	52.08	65.65	91.56	209.29

	D_1	D_2	D_3	Total
C_1	15.87	20.37	31.12	69.36
C_2	19.94	23.11	28.81	71.86
C_3	16.27	22.17	31.63	76.07
Total	52.08	65.65	91.56	209.29

Table 31.13 Analysis of Variance Table, *A, B, C, D* → *D*

Source	f	S
A	2	14.90**
B	2	3.04*
C	2	0.38
D	2	29.80**
$A \times B$	4	3.46*
$A \times C$	4	2.26
$A \times D$	4	3.60
$B \times C$	4	6.84**
$B \times D$	4	0.26
$C \times D$	4	1.67
e	48	28.02
T	80	94.22

are substituted into this graph; we then find the value of y and then its difference from $\overline{T}$, and average this. What is obtained by changing the sign is the correction quantity.

31.3 When the Factors (Descriptive Variables) Include Functions and Vector Variables

31.3.1 When There Are Factors That Are Vector Quantities or Continuously Measured Values

In certain cases, the value of a factor is not a scalar quantity but is given by a vector; when automation is used, as recently, it happens in no few instances that the factor is a continuously measured value. In such a case, if we attempt to express the components of such vectors or continuously measured values at several time-points, for example by A, B, C, . . . , and strive to find the relationship between these and y, correction calculation becomes very difficult. Of course, it is also possible to express the continuously measured values by factors that express a number of different aspects such as averaged values or the degree of dispersion, but sometimes such a device does not work well when the vector components differ rather considerably, as in the cases of the temperature at various locations in a tunnel furnace, the voltage-time cycle in the Alumite formation process, and the beating curve in the production of a special paper.

When, in such a case, the characteristic values we wish to analyze number three, such as x, y, and z, it is advisable to perform the following analysis, reversing the causal relationship. We regard x, y, and z as though they were factors, and we categorize the cards by season A and x,

y, and z; then, when the vector is of k dimensions, the positions where these respective components exist are designated $K_1, K_2, \ldots, K_k$.

For example, we will assume that the temperature has been measured at 20 locations from the inlet to the outlet of a tunnel kiln, *viz.*, at the points $K_1, K_2, \ldots, K_{20}$. We assume that from both last month A_1 and this month A_2, 54 batches of temperature curves (the values of the temperature at 20 locations) and the characteristics x (compression strength), y (temperature) and z (rate of water absorption) of tiles which have been produced under these conditions have been measured. The 54 data each of A_1 and A_2 are categorized as:

$x_1 = 18$ whose compression strength is small
$x_2 = 18$ whose compression strength is medium
$x_3 = 18$ whose compression strength is large

and we categorize within x_1, x_2, and x_3 as follows by the temperature coefficient (thermal expansion coefficient) y.

$y_1 = 6$ whose temperature coefficient is small
$y_2 = 6$ whose temperature coefficient is medium
$y_3 = 6$ whose temperature coefficient is large

Six batches within $x_1y_1, \ldots, x_3y_3$ are sequentially categorized by small, medium, and large rate of water absorption. We then obtain data as shown schematically in Table 31.14. The data constitute the values of the temperature at locations $K_1, K_2, \ldots, K_{20}$ in the tunnel kiln with respect to the batches which have been sequentially categorized by A, x, y, and z. Since there are two batches in each category, in effect these are data of two replications.

The total number of data of Table 31.14 is $2 \times 3 \times 3 \times 3 \times 2 \times 20 = 2160$. An analysis of variance such as shown in Table 31.15 is carried out from this. If main effect x is large, it is indicated that there was a difference in the mean temperature of the tunnel kiln between when the compression strength was small and when it was large. If $x \times K$ is signifi-

Table 31.14 Data of Temperatures at 20 Locations (Curves)

	K_1	K_2	$\cdots$	K_{20}
$A_1x_1y_1z_1$	=	=	$\cdots$	=
$A_1x_1y_1z_2$	=	=	$\cdots$	=
$A_1x_1y_1z_3$	=	=	$\cdots$	=
$A_1x_1y_2z_1$	=	=	$\cdots$	=
$\vdots$				
$A_2x_3y_3z_3$	=	=	$\cdots$	=

Table 31.15 Analysis of Temperature

Source		f
First-order sources	A	1
	x	2
	y	2
	z	2
	$A \times x$	2
	⋮	⋮
	$y \times z$	4
First-order error e_1		27
Replication error e_2		54
Second-order sources	K	19
	$A \times K$	38
	$x \times K$	38
	⋮	⋮
	$y \times z \times K$	76
Second-order error e_3		1 539
T		2 159

cant, it is indicated that the shape of the temperature curve (effect of K) differed between when the compression strength was small and when it was large. For example, even if the mean temperature is the same (main effect x is not significant) for x_1, x_2, and x_3, if the compression strength increases more when the temperature is raised rapidly, then slowly lowered, rather than slowly raised and then quenched, the interaction of x and K becomes significant. Thus, the relationship between the shape of the curve of ascending and descending temperature and the strength becomes clear by analysis of the interaction $x \times K$. Interactions with K, such as $y \times K$ and $z \times K$, constitute the main objective of analysis such as this.

If analysis is difficult because the data are too voluminous, one does not lessen the number of levels of K; but it is best to pick up only those of x_1, x_3, y_1, y_3, z_1, and z_3. The size of the data becomes small: $16 \times 2 \times 20 = 640$.

31.3.2 When the Factors Are Too Numerous

When the number of factors is too large, 20 or 50 or 100, we reverse the sequence of analysis. Thus, we categorize by the seasons and the characteristic values x, y, and z, say, and by a four-way layout we perform analysis for each of the sources. If, even without conducting analysis of variance, we can obtain some idea as to which factors seem to have effect merely by calculating the total values for the respective levels of x, y, and z, we start over with only such factors as seem to have effect, then add the seasons and analyze in the usual order.

31.3.3 When There Are Discontinuous Factors

Let us assume that, for example, in analyzing medical treatment effect or the like, we wish to examine the relationship between sex, age, symptoms, etc., and the cure rate. Since the sexes are only two, male and female, this is a *qualitative factor;* but sometimes one creates levels such as, when there are 60 men and 40 women, $A_1 = 33$ men (random sample from among 60 persons), $A_2 = 27$ men and 7 women (the remaining men and 7 randomly chosen from among 40 women) and $A_3 = 33$ women, and, by putting the ratio of men to women on the x-axis one creates the three continuous levels 100:0, 27:7, and 0:100.

If the number of cards for men and women is sufficient, in this case one might analyze by equalizing the number of cards by discarding 20 cards from among 60 cards for the men. The best method is to discard after having categorized by age, etc.

31.4 Number of Levels of Categorization

Shichirō Ishikawa investigated what is the best number of classes of categorization, based on experience, and showed that the power of detection is the highest with three or four levels. Calculations have been performed regarding the theory of this, but it will be omitted here since this would mean going into depth concerning a special problem. The accuracy of the expected values of variance of an analysis of variance table, too, has been calculated, but it will be omitted.

Exercise (31)

(1) Survey data on the magnitude of noise of a certain electrical auto part (unit: phon; working mean has been subtracted),

A: bearing clearance (unit: μm)
B: deflection of rotor (unit: 0.01 mm)
C: unbalance of rotor (unit: g · cm)

were as follows (Nippondensō Company, Ltd., Yutaka Ōsuga, 1966). Perform data analysis by the sequential categorization method and the method of regression analysis, and compare the two.

No	A	B	C	y
1	25	46	8	9
2	22	22	0	5
3	20	10	8	4
4	23	11	12	4
5	17	26	11	7
6	25	30	2	5
7	21	39	4	7
8	20	31	0	6
9	21	20	10	10
10	20	18	3	5
11	17	11	8	6
12	19	16	7	6
13	16	18	5	6
14	16	24	10	6
15	18	19	7	6
16	17	11	8	4
17	17	18	11	5
18	21	25	12	8
19	18	13	2	5
20	21	24	4	6
21	21	15	7	7
22	22	17	5	6
23	22	10	5	4
24	24	30	11	9
25	21	13	5	5
26	23	5	10	8
27	22	32	11	8
28	23	26	6	7
29	21	17	10	4
30	19	16	13	6
31	22	10	3	5
32	19	17	6	4
33	22	5	11	4
34	20	10	3	6
35	23	12	10	7
36	27	28	4	10
37	20	22	5	5
38	20	25	15	6
39	21	20	8	5
40	10	13	10	6
41	9	16	11	4
42	9	14	11	4
43	7	24	10	6
44	10	19	7	4
45	11	17	3	4
46	9	21	6	6
47	10	13	7	6
48	8	27	7	4
49	9	17	7	4
50	8	17	9	4
51	7	16	4	4
52	10	32	12	8
53	10	10	2	4
54	9	15	5	4

32 Minute Accumulating Analysis

32.1 Objective

The method of attack of data analysis to be discussed here takes the viewpoint that any characteristic value can be expressed by the two classes of categorization "exists" and "does not exist" at each point of the coordinate, and makes it clear that even continuous values, which have been playing a principal role in data analysis, are nothing but analyses for variations of one-lower order. Thus, by taking the standpoint that fundamentally, according to the analysis method to be discussed here, each individual characteristic value is nothing else than an estimate of the distribution function no matter what type of characteristic value it is, not only does it become possible to analyze continuous values, discrete values, categorizing values, etc., all in a unified manner, but also significant-difference testing of the distribution form becomes performed simultaneously, so that not only differences of mean values but differences of distribution due to dispersion, etc., become analyzed simultaneously. And as to estimation, it is not estimation of the mean value; rather, the distribution function itself is estimated.

Table 32.1 Data of Life Test

Cycle	1	2	3	4	5	6	7	8	9	10	11	12	13	14	15	16	17	18	19	20	≥21
A_1	0	0	0	0	0	0	0	0	0	0	0	0	0	0	1	3	0	1	0	2	3
A_2	0	0	0	0	0	0	0	0	0	0	0	1	2	0	1	1	1	2	1	0	1

32.2 A Simple Example of Length of Life Test (Data and Analysis of Variance)

Since it would probably be difficult to grasp if the data analysis described above were explained generally, here we will explain it by an example which is easy to understand.

The data given in Table 32.1 were obtained when ten samples each were taken of two types of magnet wire,

A_1 = polyethylene-covered

A_2 = silicone-covered

and a length-of-life test was carried out. In performing the life test, testing was conducted at a certain high temperature and it was determined whether or not dielectric breakdown had taken place at regular intervals of a specific duration. (This is termed the minimum unit or the cycle, or the like; it does not matter whether the minimum unit is one second or two minutes or five hours or one day or one year; if we are discussing the length of life of light bulbs, probably one hour or one day will be used as the minimum unit.) Therefore, essentially the data constituted an investigation of which minimum unit (cycle) the end of the life took place in, with regard to the ten test pieces of A_1 and of A_2. Usually, in such life tests, it is adequate if they are performed to a certain target value or until about one-half break down; in the present case, testing was carried out to 20 cycles, which was the target value.

These data are expressed as in Table 32.2. Thus, in each cycle, if the individual test piece is alive, it is expressed as 1, and if it is dead, it is expressed as 0.

If these data are added vertically for A_1 and for A_2, two numbers-surviving are obtained, while if they are added horizontally one obtains the length of life of the individual test pieces (the life up to 20 cycles; those still alive are counted as 20). It has been the tendency up to now in mathematical statistics to lay stress on the length of life per test piece, while among practical investigators stress has been laid on the *survival rate curve,* but here, in effect, these are both analyzed simultaneously.

The levels of the *minimum units* are to be expressed as $\omega_1, \omega_2, \ldots, \omega_{20}$. Although there are only 20 levels in this case, if an electric light bulb

Table 32.2 Data of Life Test Represented by 0, 1

	Test piece \ Cycle	1	2	3	4	5	6	7	8	9	10	11	12	13	14	15	16	17	18	19	20	Total
	1	1	1	1	1	1	1	1	1	1	1	1	1	1	1	0	0	0	0	0	0	14
	2	1	1	1	1	1	1	1	1	1	1	1	1	1	1	1	0	0	0	0	0	15
	3	1	1	1	1	1	1	1	1	1	1	1	1	1	1	1	0	0	0	0	0	15
	4	1	1	1	1	1	1	1	1	1	1	1	1	1	1	1	0	0	0	0	0	15
	5	1	1	1	1	1	1	1	1	1	1	1	1	1	1	1	1	1	0	0	0	17
A_1	6	1	1	1	1	1	1	1	1	1	1	1	1	1	1	1	1	1	1	1	0	19
	7	1	1	1	1	1	1	1	1	1	1	1	1	1	1	1	1	1	1	1	0	19
	8	1	1	1	1	1	1	1	1	1	1	1	1	1	1	1	1	1	1	1	1	20
	9	1	1	1	1	1	1	1	1	1	1	1	1	1	1	1	1	1	1	1	1	20
	10	1	1	1	1	1	1	1	1	1	1	1	1	1	1	1	1	1	1	1	1	20
	Total	10	10	10	10	10	10	10	10	10	10	10	10	10	10	9	6	6	5	5	3	174
	1	1	1	1	1	1	1	1	1	1	1	1	0	0	0	0	0	0	0	0	0	11
	2	1	1	1	1	1	1	1	1	1	1	1	1	0	0	0	0	0	0	0	0	12
	3	1	1	1	1	1	1	1	1	1	1	1	1	0	0	0	0	0	0	0	0	12
	4	1	1	1	1	1	1	1	1	1	1	1	1	1	1	0	0	0	0	0	0	14
	5	1	1	1	1	1	1	1	1	1	1	1	1	1	1	1	0	0	0	0	0	15
A_2	6	1	1	1	1	1	1	1	1	1	1	1	1	1	1	1	1	0	0	0	0	16
	7	1	1	1	1	1	1	1	1	1	1	1	1	1	1	1	1	1	0	0	0	17
	8	1	1	1	1	1	1	1	1	1	1	1	1	1	1	1	1	1	0	0	0	17
	9	1	1	1	1	1	1	1	1	1	1	1	1	1	1	1	1	1	1	0	0	18
	10	1	1	1	1	1	1	1	1	1	1	1	1	1	1	1	1	1	1	1	1	20
	Total	10	10	10	10	10	10	10	10	10	10	10	9	7	7	6	5	4	2	1	1	152
Combined total		20	20	20	20	20	20	20	20	20	20	20	19	17	17	15	11	10	7	6	4	326

life test were conducted for 100 days in one-day units, the number of levels of ω would become 101, including defectives at hour zero. Henceforth, the factor that corresponds to the levels of the minimum unit (minimum width of measured values) will always be expressed by ω and will be termed the ω-order factor. The level number of ω should be at most 50. Moreover, since all of the data of ω_1–ω_{10} are 1, we can dispense with them and calculate only with ω_{11}–ω_{20}, and probably this would have been better. It does not matter if ω is not equispaced.

The 10 test pieces of A_1 will be designated as $R_1', R_2', \ldots, R_{10}'$ and those of A_2 will be called $R_1'', R_2'', \ldots, R_{10}''$. The analysis of variance can then be performed as follows. The numbers are rounded to one decimal place.

$$S_m = \frac{(\text{total})^2}{400} = \frac{326^2}{400} = 265.7 \qquad (f=1) \tag{32.1}$$

$$S_A = \frac{(A_1 - A_2)^2}{400} = \frac{(174-152)^2}{400} = 1.2 \qquad (f=1) \tag{32.2}$$

$$S_\omega=\frac{1}{20}(20^2+20^2+\cdots+6^2+4^2)-CF=\frac{5\,886}{20}-265.7$$
$$=28.6 \qquad (f=19) \tag{32.3}$$

$$S_{A\times\omega}=\frac{1}{20}[(10-10)^2+(10-10)^2+\cdots+(5-1)^2+(3-1)^2]-S_A=\frac{62}{20}-1.2$$
$$=1.9 \qquad (f=19) \tag{32.4}$$

$$S_R'=\frac{1}{20}(14^2+15^2+\cdots+20^2)-\frac{174^2}{200}=\frac{3\,082}{20}-151.4=2.7 \quad (f=9) \tag{32.5}$$

$$S_R''=\frac{1}{20}(11^2+12^2+\cdots+20^2)-\frac{150^2}{200}=3.9 \qquad (f=9) \tag{32.6}$$

The total variation S_T including also S_m is

$$S_T=1^2+1^2+\cdots+1^2=326 \qquad (f=400) \tag{32.7}$$

As to main effect A, since inter-test piece differences have been mixed in, main effect A is tested by:

First-Order Error Variance

$$S_{e_1}=S_R=S_R'+S_R''=2.7+3.9=6.6 \qquad (f=18) \tag{32.8}$$

Sources related with ω, such as the main effect of ω and $\omega\times A$, are tested by the second-order error variance.

Second-Order Error Variance

$$S_{e_2}=S_T-(S_m+S_A+S_{e_1}+S_\omega+S_{A\times\omega})$$
$$=326.0-(265.7+1.2+6.6+28.6+1.9)$$
$$=22.0 \qquad (f=400-1-18-19-19) \tag{32.9}$$

Although in this instance main effect A does not have a significant variation by the 5% level of significance, because the variance ratio is

Table 32.3 Analysis of Variance Table

Source		f	S	V	F
O-order source	m	1	265.7	265.7	**
First-order source	A	1	1.2	1.2	3.2△
First-order error e_1	R'	9	2.7	0.30	
	R''	9	3.9	0.43	
	(e_1)	(18)	(6.6)	(0.37)	3.5**
Second(ω)-order error	ω	19	28.6	1.51	21.8**
	$A\times\omega$	19	1.9	0.10	1.44
Second(ω)-order source	e_2	342	22.0	0.0693	
Total	T	400	326.0		

greater than 3, we decide to perform estimation considering the four sources m, A, e_1, and ω. In terms of testing, the main effects of m and ω should always be regarded as significant, and it is unnecessary to test each time. Analysis of ω source-related values has not been performed up to now, but this is for no other reason than that we have regarded e_1 as always significant by the ω-order error. If e_1 is not significant by the ω-order error e_2, the main effect of A is tested by the result of pooling e_1 and e_2. Also, when the main effect of A is not significant, it means that $A \times \omega$ is testing whether or not the shape of the distribution differs between A_1 and A_2, and in this case, it means that it is not significant. This fact shows the advantage of having used the present method: Up to now, hypotheses about $A \times \omega$—nearly meaningless hypotheses such as that both follow a normal distribution and the variances are equal—had to be used as premises.

32.3 A Simple Example of Life Test (Estimation)

Next, we take up the problem of estimation. Of course, in the course of estimation, the survival rate (distribution function) among the 0-20 cycles actually experimented with becomes estimated.

We estimate by using the facts that the general mean m and the type of product A are significant with respect to the first-order error variance and that the first-order error variance e_1 and main effect ω are significant by the second-order error variance (ω-order error). Since we have 0, 1 data, we calculate by using omega conversion. Estimation in the case of arithmetic additivity becomes as follows for arbitrary $A_i\omega_j$.

$$\hat{\mu}_{ij}=\bar{A}_i+\bar{\omega}_j-\bar{T} \qquad (i=1,\ 2;\ j=1,\ 2,\ \cdots,\ 20) \tag{32.10}$$

Therefore, we need only find the decibel values from Appendix Table 5 and evaluate the expression above in decibels.

$$\left.\begin{aligned} \bar{A}_1&=174/200=0.870=8.26\,[\text{db}] \\ \bar{A}_2&=152/200=0.760=5.01\,[\text{db}] \end{aligned}\right\} \tag{32.11}$$

$$\bar{T}=326/400=0.815=6.44\,[\text{db}] \tag{32.12}$$

Similarly, for $\omega_1, \omega_2, \ldots, \omega_{20}$ the decibel values become as follows.

$$\left.\begin{aligned} \bar{\omega}_1&=\bar{\omega}_2=\cdots=\bar{\omega}_{11}=1.000=\infty\,[\text{db}] \\ \bar{\omega}_{12}&=19/20=0.95=12.79\ [\text{db}] \\ \bar{\omega}_{13}&=17/20=0.85=7.53\ [\text{db}] \\ &\cdots \\ \bar{\omega}_{20}&=4/20=0.20=-6.02\ [\text{db}] \end{aligned}\right\} \tag{32.13}$$

From this, the decibel values of, for example, $\hat{\mu}_{11}$ and $\hat{\mu}_{2,20}$ are:

$$\text{decibel value of } \hat{\mu}_{11} = \text{decibel value of } \overline{A}_1 + \text{decibel value of } \overline{\omega}_1 - \text{decibel value of } \overline{T}$$

$$= 8.26 + \infty - 6.44 = \infty \text{ [db]} \quad (32.14)$$

$$\hat{\mu}_{11} = 100.0 [\%] \quad (32.15)$$

$$\text{decibel value of } \hat{\mu}_{2,20} = \text{decibel value of } \overline{A}_2 + \text{decibel value of } \overline{\omega}_{20} - \text{decibel value of } \overline{T}$$

$$= 5.01 + (-6.02) - 6.44 = -7.45 \text{ [db]} \quad (32.16)$$

$$\hat{\mu}_{2,20} = 15.2 [\%] \quad (32.17)$$

The confidence limits of Equations (32.15) and (32.17) are found as follows.

$$(\text{decibel value of } \hat{\mu}) \pm \varepsilon \text{ (decibel value)} \quad (32.18)$$

Here,

$$\varepsilon = \frac{4.343}{\overline{T}(1-\overline{T})} \sqrt{F_{18}^{1} \times V_1 \times \frac{1}{n_1} + F_{361}^{1} \times V_2 \times \frac{1}{n_2}} \text{ [db]} \quad (32.19)$$

For proof, see the Note to this chapter. In this case,

$$\overline{T} = 0.815$$

$$F_{18}^{1} = 4.41, \quad F_{361}^{1} = 3.87$$

$$V_1 = 0.37 \qquad V_2 = 0.066$$

$$n_1 = \frac{\text{total number of experiments}}{\text{sum of degrees of freedom of sources significant for } e_1} = \frac{400}{1(m) + 1(A)} = 200$$

$$n_2 = \frac{400}{\text{sum of degrees of freedom of sources significant for } e_2} = \frac{400}{19 \text{ (main effect of } \omega)} = 21.1$$

Therefore,

$$\varepsilon = \frac{4.343}{0.815 \times 0.185} \sqrt{4.41 \times \frac{0.37}{200} + 3.87 \times \frac{0.066}{21.1}} = 4.10 \text{ [db]} \quad (32.20)$$

From this, the confidence limits become

$$\pm 4.10 \text{ db}$$

For example, for $\hat{\mu}$ of $A_2\omega_{20}$, we have

$$-7.45 \pm 4.10 = -11.55 \sim -3.35 \text{ [db]} \quad (32.21)$$

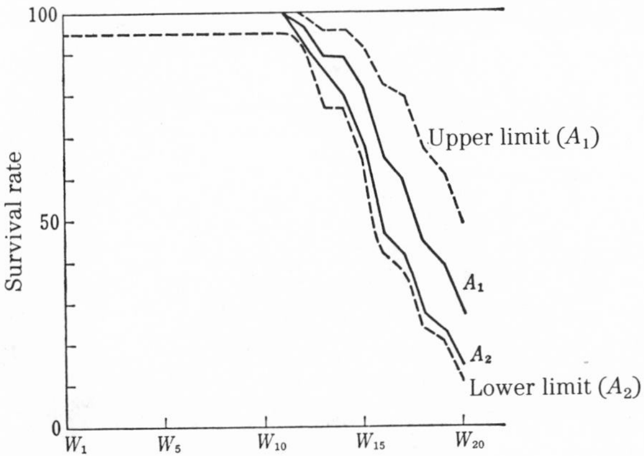

FIGURE 32.1 Survival Rate Curves and Confidence Limits (Confidence Limits of A_2 Are Omitted)

$$\text{lower limit of } \hat{\mu} = 6.5\%$$

$$\text{upper limit of } \hat{\mu} = 31.6\% \tag{32.22}$$

For $\hat{\mu} = 0$ and $\hat{\mu} = 1$, in decibel units, we assume that the limits are:

$$\text{When } \hat{\mu} = 0, \left\{-\infty, \text{ decibel value of } \frac{1}{2}\left(\frac{1}{n_1} + \frac{1}{n_2}\right) + \varepsilon\right\} \tag{32.23}$$

$$\text{When } \hat{\mu} = 1, \left\{\text{decibel value of } -\left[\frac{1}{2}\left(\frac{1}{n_1} + \frac{1}{n_2}\right) + \varepsilon\right], +\infty\right\} \tag{32.24}$$

For example, for $\hat{\mu}$ of $A_1\omega_1$, we use Equation (32.24).

$$\begin{aligned} &\text{decibel value of } \frac{1}{2}\left(\frac{1}{n_1} + \frac{1}{n_2}\right) + \varepsilon \\ &= \text{decibel value of } \tfrac{1}{2}\left(\tfrac{1}{200} + \tfrac{1}{21.1}\right) + 4.10 \\ &= \text{decibel value of } 0.0262 + 4.10 \\ &= -15.82 + 4.10 = -11.72 \text{ [db]} \end{aligned} \tag{32.25}$$

Therefore the lower confidence limit becomes 11.72 in decibel value. From this, the lower limit is 93.7%. When these are illustrated graphically, Figure 32.1 is obtained.

32.4 Life Test of Coned Disc Spring (Analysis of Variance)

The following factors were selected in order to raise the durability of coned disc springs.

A = shape, 3 levels

B = hole ratio, 2 levels

C = coining, 2 levels

D = stress σ_t, 90 65 40

E = stress σ_c, 200 175 140

F = shot peening, 3 levels

G = outer diameter cutting, 3 levels

B and C were combined and, so as to be able to obtain all of the main effects and the interactions $D \times E$ and $D \times F$, data were assigned to orthogonal array L_{27} as shown in Table 32.4. These are data obtained when three coned disc springs were produced under each of the respective 27 types of conditions and an endurance test of 1.1 million times was carried out. We take

ω_1 = 100,000th time, ω_2 = 200,000th time, . . . ,

ω_{10} = 1 millionth time, ω_{11} = 1.1 millionth time

and, at each level, if the sample was alive it was given the value 1 and if it was no longer good it was given the value 0. Therefore, if the sample is still alive at the 1.1 millionth time, ω_{11} is 1.

The data of Table 32.4 are converted into data of 0, 1 for each of the coned disc springs, as shown by Table 32.5; we construct a two-way array of A, B, . . . , and ω and find the variations.

$$S_m = \frac{476^2}{27 \times 3 \times 11} = 254.29 \tag{32.26}$$

$$S_T = 476 \qquad (f = 891) \tag{32.27}$$

$$S_{T_1} = \frac{1}{33}(3^2 + 20^2 + 15^2 + \cdots + 33^2) - S_m = 120.31 \qquad (f = 26) \tag{32.28}$$

$$S_{T_2} = \frac{1}{11}(1^2 + 1^2 + 1^2 + 4^2 + \cdots + 11^2) - S_m = 134.79 \qquad (f = 80) \tag{32.29}$$

Here, S_{T_1} is the inter-experiment variation of the 27 runs and S_{T_2} is the variation among the 81 coned disc springs.

$$S_A = \frac{120^2 + 169^2 + 187^2}{297} - 254.29 = 8.10 \tag{32.30}$$

$$S_{(BC)} = \frac{155^2 + 190^2 + 131^2}{297} - 254.29 = 5.92$$

$$S_B = \frac{(155 - 131)^2}{594} = 0.97$$

$$S_C = \frac{(155 - 190)^2}{594} = 2.06$$

$$S_D = \frac{95^2 + 157^2 + 224^2}{297} - 254.29 = 28.03$$

Table 32.4 Assignment and Data

Col. no. / No.	D 1	E 2	$D\times E$ 3	4	A 5	(BC) 6	e 7	F 8	$D\times F$ 9	10	G 11	e 12	e 13	Data ω_1	ω_2	ω_3	ω_4	ω_5	ω_6	ω_7	ω_8	ω_9	ω_{10}	ω_{11}	Total
1	1	1	1	1	1	1	1	1	1	1	1	1	1	3	0	0	0	0	0	0	0	0	0	0	3
2	1	1	1	1	2	2	2	2	2	2	2	2	2	3	3	3	3	2	1	1	1	1	1	1	20
3	1	1	1	1	3	3	3	3	3	3	3	3	3	3	3	1	1	1	1	1	1	1	1	1	15
4	1	2	2	2	1	1	1	2	2	2	3	3	3	3	3	2	0	0	0	0	0	0	0	0	8
5	1	2	2	2	2	2	2	3	3	3	1	1	1	3	3	3	3	3	2	2	2	2	2	2	27
6	1	2	2	2	3	3	3	1	1	1	2	2	2	3	0	0	0	0	0	0	0	0	0	0	3
7	1	3	3	3	1	1	1	3	3	3	2	2	2	3	1	1	0	0	0	0	0	0	0	0	5
8	1	3	3	3	2	2	2	1	1	1	3	3	3	3	1	0	0	0	0	0	0	0	0	0	4
9	1	3	3	3	3	3	3	2	2	2	1	1	1	3	3	3	1	0	0	0	0	0	0	0	10
10	2	1	2	3	1	2	3	1	2	3	1	2	3	3	1	0	0	0	0	0	0	0	0	0	4
11	2	1	2	3	2	3	1	2	3	1	2	3	1	3	3	3	3	3	3	3	3	3	3	3	33
12	2	1	2	3	3	1	2	3	1	2	3	1	2	3	3	3	3	3	3	2	2	2	2	2	28
13	2	2	3	1	1	2	3	2	3	1	3	1	2	3	3	3	3	3	3	3	3	3	3	3	33
14	2	2	3	1	2	3	1	3	1	2	1	2	3	3	3	0	0	0	0	0	0	0	0	0	6
15	2	2	3	1	3	1	2	1	2	3	2	3	1	3	2	0	0	0	0	0	0	0	0	0	5
16	2	3	1	2	1	2	3	3	1	2	2	3	1	3	3	2	1	0	0	0	0	0	0	0	9
17	2	3	1	2	2	3	1	1	2	3	3	1	2	3	3	0	0	0	0	0	0	0	0	0	6
18	2	3	1	2	3	1	2	2	3	1	1	2	3	3	3	3	3	3	3	3	3	3	3	3	33
19	3	1	3	2	1	3	2	1	3	2	1	3	2	3	3	3	2	0	0	0	0	0	0	0	11
20	3	1	3	2	2	1	3	2	1	3	2	1	3	3	3	3	3	3	3	3	3	3	3	3	33
21	3	1	3	2	3	2	1	3	2	1	3	2	1	3	3	3	3	3	3	3	3	3	3	3	33
22	3	2	1	3	1	3	2	2	1	3	3	2	1	3	3	3	3	3	3	3	3	3	3	3	33
23	3	2	1	3	2	1	3	3	2	1	1	3	2	3	3	3	3	3	3	3	3	3	3	3	33
24	3	2	1	3	3	2	1	1	3	2	2	1	3	3	3	3	3	3	2	2	2	2	2	2	27
25	3	3	2	1	1	3	2	3	2	1	2	1	3	3	3	3	3	1	1	0	0	0	0	0	14
26	3	3	2	1	2	1	3	1	3	2	3	2	1	3	3	1	0	0	0	0	0	0	0	0	7
27	3	3	2	1	3	2	1	2	1	3	1	3	2	3	3	3	3	3	3	3	3	3	3	3	33
													Total	81	68	52	44	37	34	32	32	32	32	32	476

Table 32.5 Data Converted to 0, 1

No.		ω_1	ω_2	ω_3	ω_4	ω_5	ω_6	ω_7	ω_8	ω_9	ω_{10}	ω_{11}	Total
1	1	1	0	0	0	0	0	0	0	0	0	0	1
	2	1	0	0	0	0	0	0	0	0	0	0	1
	3	1	0	0	0	0	0	0	0	0	0	0	1
2	1	1	1	1	1	0	0	0	0	0	0	0	4
	2	1	1	1	1	1	0	0	0	0	0	0	5
	3	1	1	1	1	1	1	1	1	1	1	1	11
3	1	1	1	0	0	0	0	0	0	0	0	0	2
	2	1	1	0	0	0	0	0	0	0	0	0	2
	3	1	1	1	1	1	1	1	1	1	1	1	11
⋮	⋮						⋮						⋮
27	1	1	1	1	1	1	1	1	1	1	1	1	11
	2	1	1	1	1	1	1	1	1	1	1	1	11
	3	1	1	1	1	1	1	1	1	1	1	1	11
Total		81	68	52	44	37	34	32	32	32	32	32	476

$$S_E=\frac{180^2+175^2+121^2}{297}-254.29=7.21$$

$$S_{D\times E}=\frac{38^2+38^2+\cdots+54^2}{99}-254.29-S_D-S_E=5.50 \qquad (f=4)$$

$$S_F=\frac{70^2+236^2+170^2}{297}-254.29=47.04 \qquad (f=2)$$

$$S_{D\times F}=\frac{10^2+38^2+47^2+\cdots+80^2}{99}-254.29-S_D-S_F=12.60 \qquad (f=4)$$

$$S_G=\frac{160^2+149^2+167^2}{297}-254.29=0.55$$

$$S_{e_1}=S_{T_1}-(S_A+S_{(BC)}+\cdots+S_G)=5.36 \qquad (f=6) \tag{32.31}$$

Table 32.6 Analysis of Variance Table

Source	f	S	V	Source	f	S	V
m	1	254.29	254.29	ω	10	35.46	3.55
A	2	8.10	4.05	$A\times\omega$	20	2.26	0.11 ○
(BC)	2	5.93	2.96	$(BC)\times\omega$	20	2.47	0.12 ○
(B)	(1)	0.97	0.97 ○	$(B)\times\omega$	(10)	1.29	0.13 ○
(C)	(1)	2.06	2.06	$(C)\times\omega$	(10)	0.59	0.06 ○
D	2	28.03	14.02	$D\times\omega$	20	4.00	0.20
E	2	7.21	3.60	$E\times\omega$	20	2.59	0.13 ○
$D\times E$	4	5.50	1.38 ○	$D\times E\times\omega$	40	2.03	0.05 ○
F	2	47.04	23.52	$F\times\omega$	20	5.80	0.29
$D\times F$	4	12.60	3.15	$D\times F\times\omega$	40	11.23	0.28
G	2	0.55	0.28 ○	$G\times\omega$	20	1.10	0.06 ○
e_1	6	5.36	0.89** ○	e_3	600	19.98	0.033
e_2	54	14.48	0.27**	T	891	476.00	

Table 32.7 Adjusted Analysis of Variance Table

Source	f	S	V	S'	ρ[%]
m	1	254.29	254.29	253.34	53.22
A	2	8.10	4.05	6.20	1.30
C	1	4.96	4.96	4.01	0.84
D	2	28.03	14.02	26.13	5.49
E	2	7.21	3.60	5.31	1.12
F	2	47.04	23.52	45.14	9.48
$D\times F$	4	12.60	3.15	8.80	1.85
e_1	13	12.37	0.95	18.58	3.90
e_2	54	14.48	0.263	18.155	3.82
ω	10	35.46	3.546	35.037	7.36
$D\times\omega$	20	4.00	0.200	3.154	0.66
$F\times\omega$	20	5.80	0.290	4.954	1.04
$D\times F\times\omega$	40	11.23	0.281	9.538	2.00
e_3	720	30.43	0.0423	37.652	7.92
T	891	476.00		476.000	100.00

$$S_{e_2}=S_{T_2}-S_{T_1}=134.79-120.31=14.48 \qquad (f=54) \tag{32.32}$$

$$S_\omega=\frac{81^2+68^2+\cdots+32^2}{27}-254.29=35.46 \qquad (f=10) \tag{32.33}$$

$$S_{A\times\omega}=\frac{27^2+20^2+\cdots+14^2}{27}-S_m-S_A-S_\omega=2.26 \qquad (f=20) \tag{32.34}$$

$$\vdots$$

$$S_{G\times\omega}=\frac{27^2+22^2+\cdots+12^2}{27}-S_m-S_G-S_\omega=1.10 \qquad (f=20)$$

$$S_{e_3}=S_T-(S_m+S_{T_2}+S_\omega+S_{A\times\omega}+\cdots+S_{G\times\omega})=19.98 \qquad (f=600) \tag{32.35}$$

We therefore obtain the analysis of variance table, Table 32.6.

In Table 32.6, e_2 is significant by e_3, e_1 is significant by e_2, and for e_1, the ○ marks are pooled and a newly pooled e_1 is thus created. Many sources are significant as ω-related sources, but small sources are included in error and the sources are narrowed down to only important ones; and ◎ marks are pooled with e_3. Then, the adjusted analysis of variance table is Table 32.7.

32.5 Life Test of Coned Disc Spring (Estimation)

From the analysis of variance table, we estimate the main effects of m, A, C, and E and the three-way array of D, F, and ω.

$$\bar{T}=\frac{476}{891}=0.534\pm0.071 \quad \left(\pm\sqrt{4.67\times\frac{0.95}{891}}=\pm0.071\right) \tag{32.36}$$

$$\left.\begin{aligned}\bar{A}_1&=\frac{120}{297}=0.404\pm0.122\\ \bar{A}_2&=\frac{169}{297}=0.569\pm0.122\\ \bar{A}_3&=\frac{187}{297}=0.630\pm0.122\end{aligned}\right\} \quad (32.37)$$

$$\bar{C}_1=\frac{286}{594}=0.481\pm0.086$$

$$\bar{C}_2=\frac{190}{297}=0.640\pm0.122$$

$$\bar{E}_1=\frac{180}{297}=0.606\pm0.122$$

$$\bar{E}_2=\frac{175}{297}=0.589\pm0.122$$

$$\bar{E}_3=\frac{121}{297}=0.407\pm0.122$$

The confidence limits of the two-way array of DF are, by using V_{e_1},

$$\pm\sqrt{4.67\times\frac{0.95}{99}}=\pm0.212 \quad (32.38)$$

As to interaction with ω, the table above is divided by 9 and revised to percent values; then, for example for $D_1F_1\omega_2$, we have

$$\frac{1}{9}\pm\sqrt{3.85\times\frac{0.0423}{9}}=0.111\pm0.135 \quad (32.39)$$

The optimum conditions are $A_3C_2E_1D_2F_2$ or $A_3C_2E_1D_3F_2$. Since in this case the survival rate is 100% for D_3F_2 and D_2F_2, it is unnecessary to find the process average, but one should conduct adequate testing by confirmatory trials.

Table 32.8 Three-Way Array of $DF\omega$ (Shown as Totals: Divide by 9 to Obtain Mean)

	ω_1	ω_2	ω_3	ω_4	ω_5	ω_6	ω_7	ω_8	ω_9	ω_{10}	ω_{11}	Total	Percent
D_1F_1	9	1	0	0	0	0	0	0	0	0	0	10	0.101
D_1F_2	9	9	8	4	2	1	1	1	1	1	1	38	0.384
D_1F_3	9	7	5	4	4	3	3	3	3	3	3	47	0.474
D_2F_1	9	6	0	0	0	0	0	0	0	0	0	15	0.152
D_2F_2	9	9	9	9	9	9	9	9	9	9	9	99	1.000
D_2F_3	9	9	5	4	3	3	2	2	2	2	2	43	0.434
D_3F_1	9	9	7	5	3	2	2	2	2	2	2	45	0.455
D_3F_2	9	9	9	9	9	9	9	9	9	9	9	99	1.000
D_3F_3	9	9	9	9	7	7	6	6	6	6	6	80	0.808
Total	81	68	52	44	37	34	32	32	32	32	32	476	0.534
Percent	1.000	0.840	0.642	0.543	0.457	0.420	0.395	0.395	0.395	0.395	0.395		

To show the calculation method, we will estimate the process average for $A_2C_2D_1E_3F_2$, which constituted the onsite conditions at the time. We calculate

$$\hat{\mu}=\overline{D_1F_2\omega_j}+\bar{A}_2+\bar{C}_2+\bar{E}_3-3\bar{T} \text{ [db]} \quad (32.40)$$

by decibel values.

$\bar{A}_2 = 0.569$ 1.21 (decibel value)

$\bar{C}_2 = 0.640$ 2.50 (decibel value)

$\bar{E}_3 = 0.407$ −1.63 (decibel value)

$\bar{T} = 0.534$ 0.59 (decibel value)

Since the decibel values of $D_1F_2\omega_1$, $D_1F_2\omega_2$, . . . , $D_1F_2\omega_{11}$ are

$$\infty,\ \infty,\ 9.04,\ -0.98,\ -5.44,\ -9.04,\ \cdots,\ -9.04$$

we add to these,

$$\bar{A}_2+\bar{C}_2+\bar{E}_3-3\times\bar{T}=1.21+2.50-1.63-3\times 0.59=0.31$$

and obtain

$$\infty,\ \infty,\ 9.35,\ -0.67,\ -5.13,\ -8.73,\ \cdots,\ -8.73 \text{ [db]} \quad (32.41)$$

Therefore, the survival rates become as given by Table 32.9.

The upper and lower confidence limits were drawn up by adding the following decibel values to, or subtracting them from, the decibel values (32.41).

$$\pm\varepsilon=\mp\left[\frac{4.343}{\bar{T}(1-\bar{T})}\sqrt{F\times\frac{V_1}{n_1}+F\times\frac{V_3}{n_3}}\ \right]$$

$$=\mp\left(\frac{4.343}{0.534\times 0.466}\sqrt{4.67\times 0.95\times\frac{13.5}{891}+3.85\times 0.0423\times\frac{90}{891}}\right)$$

$$=\mp 5.05 \quad (32.42)$$

Where ω_1 and ω_2 were 100.0, 96.0% was found by assuming that

$$1-\frac{1}{2}\times \text{decibel value of } \frac{22.5}{891} - 5.05 = 13.90 \text{ db} \quad (32.43)$$

n_1 and n_2 were

$$\frac{1}{n_1}=\frac{1(m)+2(A)+0.5(\text{dummy of } C_2)+2(E)+8(D_1F_2)}{891}=\frac{13.5}{891} \quad (32.44)$$

Table 32.9 Estimation of Process Average (%)

	ω_1	ω_2	ω_3	ω_4	ω_5	ω_6	ω_7	ω_8	ω_9	ω_{10}	ω_{11}
	100.0	100.0	89.6	46.1	23.5	11.8	11.8	11.8	11.8	11.8	11.8
Lwr. Conf. Lmt.	96.0	96.0	72.9	21.1	8.7	4.0	4.0	4.0	4.0	4.0	4.0
Upp. Conf. Lmt.	100.0	100.0	96.5	73.3	49.5	30.0	30.0	30.0	30.0	30.0	30.0

$$\frac{1}{n_2} = \frac{10(\omega) + 20(D \times \omega) + 20(F \times \omega) + 40(D \times F \times \omega)}{891} = \frac{90}{891} \qquad (32.45)$$

Exercises (32)

(1) In producing fluorescent lamps, experiments were carried out by assigning the information to orthogonal array L_8 in order to find the main effects of five two-level factors, A, B, C, D, and E, and interaction $A \times B$. The data shown here are those of brightness (target value 40 watts has been subtracted) and of a length of life test under certain simulated conditions. The number of replications was 2 in every case; the length of life test data are of 20 days' span, and it was investigated every 2 days whether the end of life had arrived. If the figure 4 is given it means that the bulb was alive on the fourth day but was dead on the sixth day, and a ◎ symbol indicates that it was still alive on the 20th day.

	A 1	B 2	A×B 3	C 4	D 5	E 6	e 7	Brightness		Life	
1	1	1	1	1	1	1	1	0	5	14	◎
2	1	1	1	2	2	2	2	3	−4	20	◎
3	1	2	2	1	1	2	2	0	9	8	10
4	1	2	2	2	2	1	1	2	1	18	◎
5	2	1	2	1	2	1	2	2	2	◎	◎
6	2	1	2	2	1	2	1	−4	3	13	◎
7	2	2	1	1	2	2	1	−6	−1	17	◎
8	2	2	1	2	1	1	2	2	5	12	14

(a) With the levels of ω of brightness and life, respectively, as $\omega_1 = -6, \ldots, \omega_{16} = 9$; and $\omega_1 = 0$, $\omega_2 = 2$ days, $\ldots, \omega_{11} = 20$ days, perform an analysis of variance by minute accumulating analysis.

(b) Estimate the process average under the best conditions using significant sources.

(2) Analyze the data of operability in Section 7.2.2 of Volume 1 by minute accumulating analysis, with the data still as easy, ordinary, and difficult.

Chapter 32 Note

Note 32.1 Proof of Formula for Confidence Limits, Equation (32.18)

If the confidence limits as raw values are expressed as $\pm\varepsilon$, we will consider the problem of what happens to them on the omega scale. It can be assumed, to begin with, that the confidence limits have become ε by addition of the effects of small causes ε_i.

$$\varepsilon = \varepsilon_1 + \varepsilon_2 + \cdots + \varepsilon_n$$

We then have

$$\hat{\mu} \pm \varepsilon = \hat{\mu} \pm (\varepsilon_1 + \varepsilon_2 + \cdots + \varepsilon_n) \tag{* 32.1}$$

$$= \hat{\mu} \pm [(\bar{T} + \varepsilon_1 - \bar{T}) + \cdots + (\bar{T} + \varepsilon_n - \bar{T})] \tag{* 32.2}$$

The value of omega for $\hat{\mu}$ is expressed as $\mathbf{\Omega}$. Then, if the value of omega for $\hat{\mu} \pm \varepsilon$ is expressed as $\mathbf{\Omega}\pm$, we have

$$\Omega_{\pm} = \Omega \times \left[\frac{\left(\frac{1}{\bar{T} + \varepsilon_1} - 1\right)}{\left(\frac{1}{\bar{T}} - 1\right)} \times \cdots \times \frac{\left(\frac{1}{\bar{T} + \varepsilon_n} - 1\right)}{\left(\frac{1}{\bar{T}} - 1\right)} \right]^{\pm 1} \tag{* 32.3}$$

$$= \Omega \exp\left\{ \pm \left[\sum_{i=1}^{n} \log \frac{\left(\frac{1}{\bar{T} + \varepsilon_i} - 1\right)}{\left(\frac{1}{\bar{T}} - 1\right)} \right] \right\} \tag{* 32.4}$$

Since ε_i is small, we substitute

$$\frac{1}{\bar{T} + \varepsilon_i} - 1 \fallingdotseq \frac{1}{\bar{T}}\left(1 - \frac{\varepsilon_i}{\bar{T}}\right) - 1 = \left(\frac{1}{\bar{T}} - 1\right) - \frac{1}{\bar{T}^2}\varepsilon_i \tag{* 32.5}$$

into Equation (* 32.4) and obtain

$$\Omega_{\pm} = \Omega \exp\left\{ \pm \log\left[1 - \frac{\varepsilon_i}{\bar{T}(1 - \bar{T})}\right] \right\} \tag{* 32.6}$$

Here, we substitute in

$$\log\left[1 - \frac{\varepsilon_i}{\bar{T}(1 - \bar{T})}\right] \fallingdotseq -\frac{\varepsilon_i}{\bar{T}(1 - \bar{T})} \tag{* 32.7}$$

and we obtain

$$\Omega_{\pm} = \Omega \exp\left[\mp \frac{1}{\bar{T}(1 - \bar{T})} \sum \varepsilon_i \right]$$

$$= \Omega\, 10^{\mp \frac{0.4343}{\bar{T}(1 - \bar{T})}\varepsilon} \tag{* 32.8}$$

This completes the proof.

33 How to Decide on the Frequency of Experimentation

In this chapter we will discuss the method of deciding on the scale and frequency of experimentation from an economic viewpoint, and we will show its conclusions.

33.1 A Simple Example

33.1.1 A Simple Problem

Let us consider an experiment to decide, when there is a factor A at two levels, whether A_1 or A_2 is to be used. Such a case occurs, for example, when deciding whether the temperature of a certain production step is to be 150°C or 160°C, or when deciding whether an operation should be carried out by method A_1 or method A_2. Cases where there are more than two levels will be discussed later. We assume that n replications of experiment will be performed for A_1 and A_2.

It will be assumed that the cost difference of A_1 and A_2 is negligibly small. The discussion below will probably apply similarly even when there is a difference of cost.

We perform an experiment of n replications for A_1 and A_2 and find the mean values for A_1 and A_2, $\overline{A}_1$ and $\overline{A}_2$. It will be assumed that the method of

$$\begin{cases} \text{using } A_1 & \text{if } \overline{A}_1 \text{ is greater than } \overline{A}_2 \\ \text{using } A_2 & \text{if } \overline{A}_2 \text{ is greater than } \overline{A}_1 \\ \text{deciding to use } A_1 \text{ or } A_2 \text{ by lot} & \text{if } \overline{A}_1 \text{ and } \overline{A}_2 \text{ are exactly equal} \end{cases}$$

is being used. If there is a cost difference between A_1 and A_2, it is exactly the same except that we decide to use A_1 or A_2 according to whether the difference in gain, after having subtracted the cost, is greater or less than zero. If the difference is equal, one just decides by lot. Such a way of deciding is the most commonly used; *in choosing either of A_1 or A_2, since it is necessary to use one, it is best to use the one that turns out to be better even if only slightly.* Thus, in an actual situation, whether or not there is a significant difference between A_1 and A_2, one should use the one that has been estimated to provide the greater gain, even if only by a little. However, when using a determination method such as described above, the design must be carefully worked out since unless the decision is made by performing an experiment of exactly the right number of replications, the better method might be handicapped by error, as of environmental conditions, and there is a chance that it will come out looking worse. How many times one should experiment will be discussed from here on. A decision method such as the foregoing can be considered with any kind of assignment.

The error variance per unit experiment will be expressed as σ^2. To explain briefly what is to be done from here on, we wish to find out, when performing an experiment on percent yield with regard to A_1 and A_2, how many replications of the experiment it would be most economical to use in deciding between A_1 and A_2. If there is clearly no difference whatsoever between A_1 and A_2, there would be a loss by having performed the experiment since there is no economic difference whichever is used. Again, when there is a very great difference between A_1 and A_2, it is still desirable to use fewer replications of the experiment since the mean value of the better one emerges as almost certainly greater than that of the worse one even by one replication of the experiment. Thus, the reason why we require an experiment of a certain number of replications is that there is a possibility that the unknown difference between A_1 and A_2 harbors a difference which renders rather questionable the seeming certainty of judgment by the data. As will become clear later, in ordinary experiments the greatest number of replications is required when the true difference between A_1 and A_2 is about $0.1\sigma - 0.4\sigma$. *Of course, since the true difference between A_1 and A_2 is unknown, we have almost no information prior to the experiment regarding the amount of this* *difference. Therefore, when deciding on the number of replications of the experiment, one must also take into account the fact that it might possess a true difference which is close to the worst.*

In order to decide logically on the number of replications, rough estimates must be made for two quantities, a and b.

First, the value of b will be defined. The value of b is the amount by which the gain will rise if the yield becomes higher by σ%. Calculation of this quantity is rather difficult. This is because, first, the estimated value of the standard error σ is not clear. If the estimated value of σ can be found from past experiments, it is possible to use it. It can be seen that in this case, even if the estimate were about one-half or double, there would be no great difference in the conclusions. In certain cases, one performs an experiment of several replications, then estimates σ from this and decides on the number of replications by which the experiment is to be supplemented. Another difficult point is that in the calculation of how much gain there will be when the yield rises by σ%, it is necessary to estimate for about how many years one is to use what has been decided on from these experimental results. For example, if an erroneous judgment has been made when the decision from these experimental results is to be adhered to for as long as 100 years, the loss thereby becomes enormous by integration and a corresponding accuracy of experiment is required. In actual production plants, however, it is rare that standard operations which have been decided on at a certain time are to be adhered to without question for such a long time; usually new raw materials or apparatus, development of energy resources, etc., do not permit such stagnation of technology. It is usually sufficient to assume that operation standards are to be used for about ten years at the longest, and actually one may probably assume that it will be for only several years at most. Re-experimenting may be carried out three years later, or, if there are developments in techniques and raw materials, it may be necessary to decide newly on the temperature (or whatever) at that point. In any case, it is impossible to estimate the value of b unless estimation of σ is done. However, as long as it is an ordinary experiment, the number of replications does not depart greatly from the approximately standard value even if the values of b and a differ to a certain degree, as will become clear later, so that even if one does not necessarily estimate the values of a and b by individual experiments, it is probably possible to find the approximately reasonable number of replications. Therefore, probably serious error will not arise if an experimenter who finds such calculations to be difficult experiments according to the standard values.

The process averages of the yields of A_1 and A_2 will be expressed as m_1 and m_2. Also, the value obtained by dividing the unknown difference, $m_1 - m_2$, by σ will be denoted Δ.

$$\Delta = \frac{m_1 - m_2}{\sigma} \tag{33.1}$$

This means that if Δ is positive, A_1 is better than A_2, and if it is negative, A_2 is better than A_1. The value of Δ is of course unknown.

Next, we will explain the other quantity, a. The value of a signifies the loss per replication of the experiment. Often, not only the expense required for the experiment but actually *loss due to delay of judgment* accounts for a large part of this. Therefore, accurate calculation of a, too, is rather difficult. As long as the operational standard in the present state is not the best, loss due to using the worse operational standard while taking time in experimenting becomes equal to something of the order of the foregoing b value divided by the number of batches.

33.1.2 Risk Function and Methods for Its Determination

The *expected loss* when deciding on the use of either A_1 or A_2 by an experiment of n runs of replication is expressed as follows.

$$I_n(\Delta) = an + b|\Delta| \times (\text{probability of erroneously using the worse one}) \qquad (33.2)$$

The expression in parentheses means the probability of erroneously using A_2 when A_1 is better than A_2 by $\Delta\sigma$, and it is also the probability of erroneously using A_1 when A_2 is better than A_1 by $\Delta\sigma$. Since the same conclusion can be arrived at by using the expected gain instead of the expected loss, we need only calculate the expected loss. For, since the expected gain is

$$\begin{aligned} & b|\Delta| \times (\text{probability of using the better}) - an \\ & \quad = b|\Delta| \times [1 - (\text{probability of erroneously using the worse})] - an \\ & \quad = b|\Delta| - [an + b|\Delta| \times (\text{probability of erroneously using the worse})] \end{aligned}$$

to minimize the expected loss $I_n(\Delta)$ means the same as to maximize the expected gain. For this reason, we consider the expected loss. Henceforth, expected loss will be expressed by the term "risk."

Since the expected loss function, or in other words the *risk function*, of Equation (33.2) is a symmetric function of Δ, generality will not be lost if one considers only the case of positive Δ. We will therefore consider only the case where Δ is always positive, or in other words where A_1 is better than A_2 by $\Delta\sigma$. Then, $I_n(\Delta)$ becomes

$$\begin{aligned} I_n &= an + b\Delta \times (\text{probability of erroneously using } A_2) \\ &= an + b\Delta \times (\text{probability that } \bar{A}_1 \text{ is smaller than } \bar{A}_2) \qquad (33.3) \end{aligned}$$

Here, it will be assumed that both $\bar{A}_1$ and $\bar{A}_2$ are normally distributed. This assumption is sufficiently useful since even if the individual experimental values are not normal, the arithmetic mean assumes a distribution that is close to normal. A similar study can be carried out even without assuming normal distribution, but for practical purposes it is believed that the conclusion for the normal case suffices.

Further, instead of $I_n(\Delta)$ in Equation (33.3), we consider what is obtained by dividing both sides by a. In a problem of maximum and minimum, the conclusion does not change if one considers the quantity obtained by dividing by a specific constant. If we put

$$\lambda=\frac{a}{b} \tag{33.4}$$

and if we redefine $I_n(\Delta)/a$ as $I_n(\Delta)$, Equation (33.3) becomes as follows.

$$I_n(\Delta)=n+\frac{\Delta}{\lambda}P_r\{\bar{A}_1\leq\bar{A}_2\}=n+\frac{\Delta}{\lambda}P_r\{\bar{A}_1-\bar{A}_2\leq 0\} \tag{33.5}$$

But since $x=\bar{A}_1-\bar{A}_2$ distributes according to the normal distribution of mean m_1-m_2 and variance $(2/n)\sigma^2$, by assumption, we have

$$I_n(\Delta)=n+\frac{\Delta}{\lambda}\int_{-\infty}^{0}\frac{\sqrt{n}}{\sqrt{2\pi}\sqrt{2\sigma^2}}e^{-\frac{n}{2(2\sigma^2)}[x-(m_1-m_2)]^2}dx$$

If we assume that $t=\dfrac{\sqrt{n}(x-m_1+m_2)}{\sqrt{2}\sigma}$, we have

$$I_n(\Delta)=n+\frac{\Delta}{\lambda}\int_{-\infty}^{-\frac{\sqrt{n}}{\sqrt{2}}\Delta}\frac{1}{\sqrt{2\pi}}e^{-\frac{t^2}{2}}dt \tag{33.6}$$

Figures 33.1, 33.2, and 33.3 show results of calculation of the values of the risk function $I_n(\Delta)$ of Equation (33.6) for various n and Δ for the cases of $\lambda=0.01$, 0.001, and 0.0001.

One respect in which actual problems differ greatly from ordinary mathematics is in the fact that often *we treat the maximum and minimum of equations that contain unknown quantities.* For example, when causing Equation (33.6) to become minimum, if Δ is known, the n that causes it to become minimum becomes immediately known; but Δ is not known, or even if it is known, we have only *incomplete knowledge.*

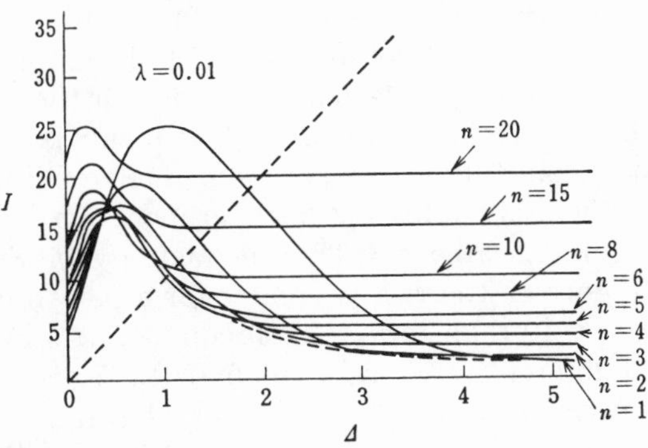

FIGURE 33.1 Risk Function ($\lambda=0.01$)

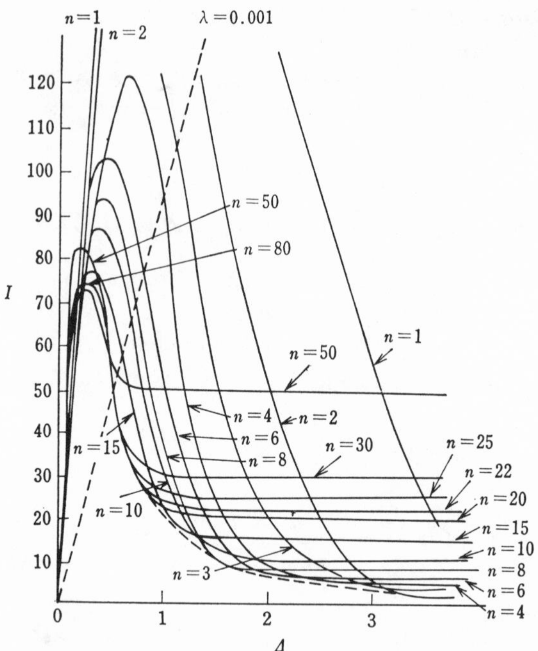

FIGURE 33.2 Risk Function ($\lambda = 0.001$)

Most of the daily judgments which we make harbor difficult problems in having to deal with the maximum and minimum of functions which contain parameters of which we have only incomplete knowledge. Even though there are only questionable estimates regarding the economic condition tomorrow or half a year later, we must take action on them now. Since we are trying to find the difference between A_1 and A_2 by experiment, it is probably better to state that we have no knowledge whatsoever regarding the difference between A_1 and A_2. How should we go about logically deciding on the number of replications of experiment in such circumstances? First, let us begin with the graph of $I_n(\Delta)$.

What is clear from the graph is that, for example in the case of $\lambda = 0.01$, the expected loss, or in other words the risk, is always greater in the case of a person who is conducting an experiment of ten replications ($n = 10$) than for a person who is conducting an experiment of $n = 5$, whether the difference Δ is 0 or 0.1 or 0.5 or 3. Thus, no matter what the value of the unknown difference Δ, the risk is greater for a person conducting an experiment of $n = 10$ than for a person conducting an experiment of $n = 5$. To put it differently, when λ is about 0.01 it is disadvantageous to perform an experiment of $n = 10$ no matter what the value of the unknown difference.

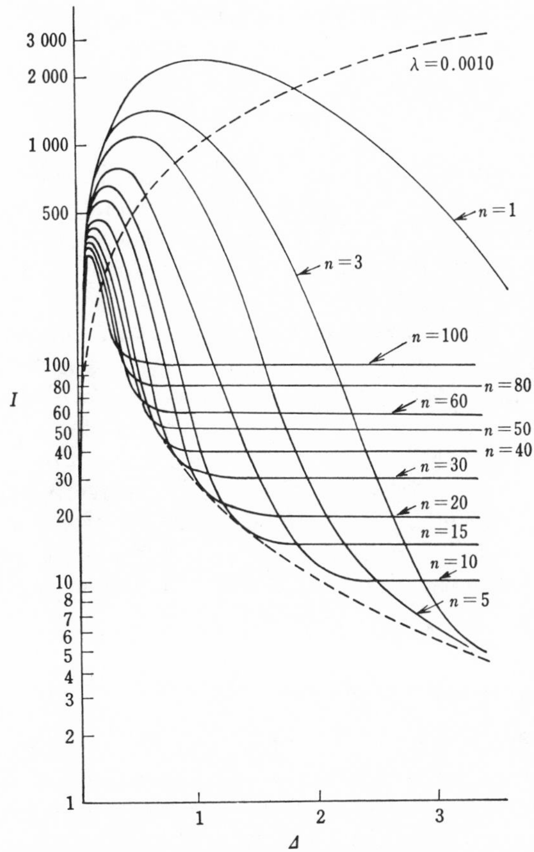

FIGURE 33.3 Risk Function ($\lambda = 0.0001$)

But in the case of a person who is deciding which level to use by an experiment of $n = 1$, if Δ is within the range

$$0.3 \leq \Delta \leq 3.2$$

it is more to his disadvantage than for a person experimenting with $n = 5$. However, in other cases, of smaller or greater difference Δ than this, the risk is smaller than for a person using $n = 5$, as is obvious from the graph. This means that except for instances where the difference is very great (at least 3.2σ) or where there is almost no difference (at most 0.3σ), if one decides whether to use A_1 or A_2 by an experiment of about $n = 1$, the danger of suffering loss is greater than that for a person experimenting by $n = 5$ since judgment error is appreciably greater. When the difference is about σ, the risk is at least double that for a person experimenting by $n = 5$; and thus one may probably regard an experiment of $n = 1$ as not very favorable. When it is an experiment of $n = 2$, if $0.3 < \Delta < 2.3$, the risk is greater than with $n = 5$, but even in the worst case there is only about a 50% increase compared with a person experimenting by $n = 5$, so that it is much safer than in the case of $n = 1$.

When there is a large difference, an experiment of $n = 5$ possesses a risk which is as much as 2.5 times that of an experiment of $n = 2$. By an experiment of $n = 3$, compared with an experiment of $n = 5$, the risk is slightly greater in the range of $0.3 \leq \Delta \leq 1.8$, but when Δ is outside this range there is an appreciably greater gain than by an experiment of $n = 5$.

The broken curves represent the loss function of a person who is able to select the correct level with 99% accuracy by experience or hunch. When $\lambda = 0.001$, *loss becomes approximately the same between a person who performs an experiment of* $n = 8$ *and a person who can tell the better level with* 99% *accuracy*. This means that when there are 100 two-level factors, a person who is able to select correctly the better level except in one case and a person who decides to experiment even if cost and time are involved are approximately the same in terms of loss and gain. This shows succinctly the impotance of actual proof by experimentation.

33.1.3 Dodge-Romig's Solution

When one possesses some knowledge regarding the unknown difference Δ in conducting an experiment such as this, say, that the difference between A_1 and A_2 is at least 0.1σ and is about 1.5σ, it means that an experiment of $n = 4$ or $n = 5$ is clearly more favorable than one of $n = 3$. When one possesses fairly accurate information regarding the degree of difference, it is possible to determine the number of replications n by which the risk is least for this estimated difference. Since such a method of deciding on the number of replications n possesses special qualities

Table 33.1 Dodge-Romig's Solution

$\lambda = \frac{a}{b}$	Δ							
	0.4	0.6	0.8	1.0	1.5	2.0	3.0	4.0
0.05	1	2	2	2	2	2	1	1
0.01	2	4	4	4	3	2	1	1
0.005	9	9	7	5	3	2	1	1
0.002	17	13	10	7	4	3	2	1
0.001	24	16	12	9	5	3	2	1
0.0005	31	20	14	10	6	4	2	2
0.0001	48	28	18	13	7	4	3	2
0.00005	55	32	20	14	7	5	3	2
0.00001	75	49	25	17	9	6	3	2

similar to those of Dodge-Romig's extraction testing table, the author terms this the *Dodge-Romig solution.* Table 33.1 gives Dodge-Romig's solutions for various λ and Δ values. Since clearly Dodge-Romig's solution is *dangerous unless the estimated value is fairly accurate,* it is best used only by persons who are quite confident.

33.1.4 Minimax Solution

Another method of solution takes exactly the opposite standpoint from that of the Dodge-Romig solution above, that is, it finds the most pessimistic solution, and it is named the *minimax solution.* This method takes the attitude of minimizing the risk when the difference Δ is, more than unknown, the worst difference, in other words the difference which requires the greatest number of replications of experiment. The idea is to minimize one's own loss in the worst possible case since the worst might happen. From Figure 33.1, the difference for which the greatest number of replications is desirable is clearly about $\Delta = 0.5$; and the solution with the smallest risk in the case of $\Delta = 0.5$ is $n = 5$ or so. Similarly, in the case of $\lambda = 0.001$, it occurs at around $\Delta = 0.2$, and the minimax solution becomes $n =$ about 24. However, by this solution, unless the difference is the worst, as here, the risk becomes greater than by an experiment of fewer replications than this. *The risk in an experiment with a number of replications which is greater than the minimax solution is greater than that in an experiment using the minimax solution, whatever the difference* Δ.

Stated differently, the minimax solution gives us the upper limit of the number of replications of experiment.

Although some mathematical equations now come on stage, since the minimax solution has not been explained much we will show here how to find the solution in this case. (Persons who dislike analytical equations would do well to skip this part and proceed; the most important part of this chapter will be presented immediately afterward.)

To find the minimax solution one need only calculate, in regard to the risk function $I_n(\Delta)$ of Equation (33.6), the n such that

$$\min_{n} \max_{\Delta}\left(n+\frac{\Delta}{\lambda}\int_{-\infty}^{-\sqrt{\frac{n}{2}}\Delta}\frac{1}{\sqrt{2\pi}}e^{-\frac{t^2}{2}}dt\right) \tag{33.7}$$

For this, it is first necessary to find the value of Δ which will maximize

$$I_n(\Delta)=n+\frac{\Delta}{\lambda}\int_{-\infty}^{-\sqrt{\frac{n}{2}}\Delta}\frac{1}{\sqrt{2\pi}}e^{-\frac{t^2}{2}}dt$$

The value of Δ which maximizes $I_n(\Delta)$ can be found by differentiating Equation (33.6) with respect to Δ.

$$\frac{\partial I_n(\Delta)}{\partial \Delta}=\frac{1}{\lambda}\int_{-\infty}^{-\sqrt{\frac{n}{2}}\Delta}\frac{1}{\sqrt{2\pi}}e^{-\frac{t^2}{2}}\,dt+\frac{\Delta}{\lambda}\frac{1}{\sqrt{2\pi}}e^{-\frac{1}{2}\left(\sqrt{\frac{n}{2}}\Delta\right)^2}\left(-\sqrt{\frac{n}{2}}\right)$$

We set this expression equal to zero and transpose the second term, and if we divide both sides by $\frac{1}{\lambda}-\frac{1}{\sqrt{2\pi}}e^{-\frac{1}{2}\left(\sqrt{\frac{n}{2}}\Delta\right)^2}$ we obtain

$$\sqrt{\frac{n}{2}}\Delta=\frac{\int_{-\infty}^{-\sqrt{\frac{n}{2}}\Delta}e^{-\frac{t^2}{2}}\,dt}{e^{-\frac{1}{2}\left(\sqrt{\frac{n}{2}}\Delta\right)^2}}$$

If we put $\sqrt{\frac{n}{2}}\Delta=A$, we obtain

$$A=\frac{\int_{-\infty}^{-A}e^{-\frac{t^2}{2}}\,dt}{e^{-\frac{A^2}{2}}} \tag{33.8}$$

Thus, A is the solution of Equation (33.8). Since, fortunately, there exists a numerical table where the right side of Equation (33.8) has been calculated as a function of A, it is possible to find from the table the value of A such that the value of the right side becomes equal to A. The answer is

$$A\fallingdotseq 0.7518$$

Therefore,

$$\Delta=\sqrt{\frac{2}{n}}\times 0.7518 \tag{33.9}$$

If this value is substituted into $I_n(\Delta)$ of equation (33.6) we obtain

$$I_n(\Delta)=n+\frac{\sqrt{2}\,(0.7518)}{\lambda\sqrt{n}}\int_{-\infty}^{-0.7518}\frac{1}{\sqrt{2\pi}}e^{-\frac{t^2}{2}}\,dt=n+\frac{0.2404}{\lambda\sqrt{n}} \tag{33.10}$$

To find the value of n which causes Equation (33.10) to be minimum, we differentiate again, with respect to n, and we get

$$\frac{\partial}{\partial n}\left(n+\frac{0.2404}{\lambda\sqrt{n}}\right)=1+\frac{0.2404}{\lambda}\left(-\frac{1}{2}\right)n^{-\frac{3}{2}}=0$$

From this we have

$$n=\left(\frac{0.1202}{\lambda}\right)^{\frac{2}{3}} \tag{33.11}$$

And the value of I for the worst difference Δ is

$$I=0.7307\left(\frac{1}{\lambda}\right)^{\frac{2}{3}} \tag{33.12}$$

Table 33.2 Recommended Solution and Minimax Solution

Continuous value	Fraction defective	$\lambda = \frac{a}{b}$	Recommended no. of replications	Minimax solution	Worst difference Δ
Light	×	0.03 0.01	1 (1–3) 3 (2–5)	3 5	0.67 0.46
Ordinary	×	0.003 0.001 0.0003	5 (4–8) 8 (6–10) 12 (8–18)	12 24 54	0.31 0.22 0.14
Important	Light	0.0001 0.000 03	20 (18–30) 40 (30–50)	113 252	0.10 0.07
×	Ordinary	0.000 01 0.000 003 0.000 001	70 (50–100) 140 (100–200) 200 (150–300)	525 1 171 2 436	0.05 0.03 0.02
×	Important	0.000 000 3 0.000 000 1	300 (200–400) 400 (300–500)	5 435 11 300	0.01 0.01

This is the answer which is sought. A similar calculation can be performed when A_1 is the temperature used up to now and we know the arithmetic mean of the yield n_1 in that case, and we wish to find how many replications of experiment should be used to determine whether or not to change to a new temperature. If we put $n_1 = \infty$ in that equation, it becomes the case of judgment as to whether below or above the standard.

The n values for various λ values are given in the minimax solution section in Table 33.2.

These solutions are the maximum values of the envelopes of the graphs of Figures 33.1–33.3, and they match the n value of the curves which contact them. The values of Δ corresponding to the minimax solutions, too, are given in Table 33.2. It can be seen from this that the worst difference is in the case where the difference is fairly small.

33.1.5 Recommended Solution

In the discussions so far, we have shown

(1) Dodge-Romig's solution — the most optimistic
(2) Minimax solution — the most pessimistic

Although these give us the lower limit and upper limit of the number of replications required for our experiments, depending on the special nature of each, it might be said that neither informs us of the most desirable number of replications. Because of this, a third solution is created. The third solution requires even more detailed investigation, and the author terms it the *Bayes-Wald solution*. We will not touch on this here, but will directly read off from the graphs the conclusion as to the most desirable number of replications.

Clearly, in the case of $\lambda = 0.01$, $n = 3$ or so constitutes the most suitable number of experiments, in view of the foregoing considerations. If the case of $\lambda = 0.001$ is similarly considered it can probably be said that $n =$ about 8 is best. Similarly, it would be $n =$ about 20 in the case of $\lambda = 0.0001$ and $n =$ about 70 in the case of $\lambda = 0.000\,01$. These values are given in Table 33.2 in the column of recommended number of replications.

Thus, in practice, one need only estimate what is about the best λ value for one's own experiment, and roughly read off the desirable number of replications for this from Table 33.2. If it is difficult to estimate the value of λ, it would probably be safe to assume that *an experiment of about 8–10 replications is favorable.*

The number of replications in the case of discrete values such as fraction defective differs according to the fraction defective which is estimated, but it is probably safe to assume that it does not exceed the range of $\lambda = 0.000\,03 - 0.000\,000\,1$ throughout all values. When an experimental value of fraction defective of several percent can be expected, it means that a number of test-constructed items equal to at least about 100 is favorable both for A_1 and for A_2. Since all of this has been organized in Section 9.1 of Volume 1, the reader is referred thereto.

33.2 A Worked Example and the Conclusions

33.2.1 A Worked Example

[EXERCISE]

When deciding by experimentation whether to direct that the temperature affecting the yield of a certain process be A_1: 150°C or A_2: 155°C, it can be estimated that if an experiment by two-way layout is performed the standard deviation of error will be about 1.3%. If the yield differs by 1%, a difference in profit of 5000 yen per day can be expected.

The cost of an experiment of one replication is about 10,000 yen. For the time being, we will not consider loss incurred by delay of judgment. It will be assumed that the temperature in this experiment by which the mean value of the yield is higher will be used, and that the product is to be made with this value of temperature, without change, for about three years. What should the number of replications of the experiment be?

[SOLUTION]

The loss when A_2 has been mistakenly used when the yield by A_1 is greater than by A_2 by σ% is, since there will be about 1000 batches in three years assuming one batch in a day, more or less,

$$b = 1.3 \times 5000 \times 1000 = 6.5 \text{ million yen}$$

Also, since a is 10,000 yen,

$$\lambda=\frac{a}{b}=\frac{1}{650}=0.0015 \tag{33.13}$$

Since a is 15,000 yen when A_2 is used, as now, we have

$$\lambda=\frac{1.5}{650}\fallingdotseq 0.0023 \tag{33.14}$$

When the proportion in which the better one is being used is about 50%, $\lambda = 0.0020$.

Whatever the case, a more or less logical number of replications is 6–8 from Table 33.2, and if one uses an orthogonal array, this would be L_{16}. Even if there is a difference of cost between A_1 and A_2, unless the cost difference is at least several thousand yen per day, there is no change in the conclusion given above.

Also, if the three years above is changed to one year, five years, and 20 years, the number of replications n becomes 4 times, 8 times, and 15 times, so orthogonal arrays L_8, L_{16}, and L_{32} are just right.

33.2.2 Conclusions

Generally speaking, as was mentioned in Section 9.1 of Volume 1, the most suitable sizes of experiment would be one run each in the case of an especially difficult experiment ($\lambda \geq 0.03$), by L_8, L_9, or L_{18} for a fairly difficult experiment ($0.03 \geq \lambda \geq 0.003$), by L_{16} or L_{27} for ordinary cases ($0.003 \geq \lambda \geq 0.0003$), and by L_{32}, L_{64}, L_{54}, L_{81}, etc., for an easy experiment ($0.0003 \geq \lambda \geq 0.000\,03$). Whether the number of factors to be assigned to such orthogonal arrays is one or several or several tens, there is no difference in these conclusions. Therefore, when there is only one control factor such as a production condition, it is wasteful to use L_{16} when λ is only about 0.001 but L_{16} should nevertheless be used as the orthogonal array. Thus, it becomes desirable to make use of a larger number of factors. Conversely, when there are 30 factors in an experiment which is comparatively difficult, it is indicated that experiments by L_8 should be duplicated several times and all of the factorial effects should be found. Just because there are 30 factors, one should not use L_{32} or L_{27}.

When factors whose replication is difficult and factors whose replication is easy coexist within a single experiment, please refer to the split-unit design of Chapter 9.

33.3 Optimum Number of Replications

Generally speaking, when control factors are assigned to an orthogonal array, a number of test pieces are taken under the respective experimental conditions and the temperature, for example, is varied over r levels and the characteristic is studied. This is termed the *number of replications* r in the broad sense. When the inter-experiment error variance is expressed as σ_1^2 and the inter-replication (second-order) error variance is expressed as σ_2^2, if the number of replications is r, σ^2 in Section 33.1 is

$$\sigma^2 = \sigma_1^2 + \frac{\sigma_2^2}{r} \tag{33.15}$$

This means that σ^2 becomes determined only after the optimum number of replications r is decided on, and the theory of Section 33.1 holds. If the orthogonal array is L_N and the number of replications is r, the total number of data becomes Nr. We will designate the cost increase when the number of replications has been increased by 1 as C_2 and the cost when the experiment itself is conducted once more from the beginning as C_1 here. The optimum number of replications r is given by the following equation.

$$r = \sqrt{\frac{C_1 \times \sigma_2^2}{C_2 \times \sigma_1^2}} \tag{33.16}$$

This formula can be derived as a method of minimizing the equation of cost when using r replications and r runs of repeats by holding the precision of comparison of $\bar{A}_1$ and $\bar{A}_2$ at a specific value. Thus, we need only solve the following equation.

$$\frac{1}{n}\left(\sigma_1^2 + \frac{\sigma_2^2}{r}\right) = \text{const} \tag{33.17}$$

$$nC_1 + nrC_2 = \min \tag{33.18}$$

The n in Equation (33.17) is substituted into (33.18), then we differentiate with respect to r and put it equal to zero.

$$\frac{d}{dr}\left[\frac{1}{\text{const}}\left(\sigma_1^2 + \frac{\sigma_2^2}{r}\right)(C_1 + rC_2)\right] = \frac{1}{\text{const}}\left(-\frac{\sigma_2^2 C_1}{r^2} + C_2\sigma_1^2\right) = 0 \tag{33.19}$$

Equation (33.16) is obtained from this.

In the tile experiment in Chapter 6, if the cost increase when 101 tiles are studied instead of 100 is 10 yen, the cost when the experiment is re-run from mixing of the raw materials is 10,000 yen, and the percentage of tiles which do not become grade-one products is assumed to be 20%, we have

$$\sigma_2^2 = 0.2 \times 0.8 = 0.16 \tag{33.20}$$

Estimation of inter-experiment error variance σ_1^2 is the most difficult. We will assume that the variation in fraction defective between batches of pulverization and mixing (this is inter-experiment dispersion) is $\pm 10\%$

when a very large number of tiles is being produced daily. Since this may be regarded as being approximately $\pm 2\sigma_1$, we have

$$\sigma_1^2 = (5\%)^2 = (0.05)^2 = 0.0025 \tag{33.21}$$

Therefore the optimum number of replications r is

$$r = \sqrt{\frac{10\,000 \times 0.16}{10 \times 0.0025}} \fallingdotseq 250 \tag{33.22}$$

It is therefore indicated that one should test by producing about 250 tiles in each experiment of orthogonal array L_8. When 100 tiles are studied on the outer side and 100 on the inner side by orthogonal array L_8, the sum would be 200 tiles, and this approximately matches the optimum solution of Equation (33.22).

But in the case of the compression strength of tiles, not only does the cost per sheet become several hundred yen but the dispersion of the data also becomes small, and it suffices when r is about 3–5. In practice, it is important to decide on a more or less logical number by referring to the table of the optimum number of replications in Section 9.1.

34 Randomly Combined Design and Almost Orthogonal Array

As methods of finding all of the unknowns from a number of data which is less than the number of unknowns, there are the experimental regression analysis of Chapter 15 and the randomly combined design and almost orthogonal array of this chapter. The randomly combined design has not been used recently, but the method of Chapter 15 and the almost orthogonal array are being used. The two new methods are introduced in this chapter, but they cannot be recommended for beginners.

34.1 Experiment on Foamed PVC Electrical Wire*

34.1.1 Objective

It is an interesting process to improve the insulating property of PVC resin by producing fine foam within it, in order to lessen the loss of messages and to conserve material. This experiment is one excellent example of a classical analysis of a completely new problem. However, assignment by the randomly combined design, which is shown here, cannot be recommended for beginners.

* Performed by the randomly combined design, by the Electric Wires Division, Line Section, The Electrical Communication Laboratory.

34.1.2 Factors and Levels

Factors such as the following were selected for an experiment of test production of foamed PVC electrical wires.

Mixing Factors

A: quantity of stabilizing agent added
$A_1 = 5, A_2 = 7, A_3 = 3$ (parts)
B: quantity of addition of plasticizer added
$B_1 = 50, B_2 = 40, B_3 = 30$ (parts)
C: type of supplemental plasticizer
$C_1 =$ n-DOP, $C_2 =$ n-DOP, $C_3 =$ DOA (brand)
D: quantity of base plasticizer, DOP : C
The case of n-DOP
$D_1 = 100:0, D_2 = 70:30, D_3 = 50:50$
The case of DOA
$D_1 = 90:10, D_2 = 80:20, D_3 = 70:30$
E: quantity of foaming agent added
$E_1 = 5.5, E_2 = 4.0, E_3 = 7.0$ (parts)
F: type of foaming assistant
$F_1 =$ salicylic acid, $F_2 =$ BK, $F_3 =$ adipic acid
G: quantity of assistant added
$G_1 = 0.5, G_2 = 1.0, G_3 = 1.5$ (parts)

Processing Factors

H: head temperature
$H_1 = 170, H_2 = 180, H_3 = 175$ (°C)
J: take-up speed
$J_1 = 7.5, J_2 = 12.5, J_3 = 10.0$ (m/min)
L: meshes of screen
$L_1 = 100, L_2 = 120, L_3 = 60$ (mesh)
M: shape of breaker plate
$M_1 = 1, M_2 = 0.5, M_3 = 2.0$ (dimension ratio of cylinder side and head side)
T: cylinder temperature
$T_1 = 145, T_2 = 155$ (°C)

The reason why the specific content of the levels of D differs between n-DOP and DOA is that for a given weight, a lesser amount of DOA is needed than of n-DOP. Since n-DOP possesses about the same characteristics as DOP, fundamentally in the case of n-DOP it should be $D_1 = 100:15$, $D_2 = 100:32.5$, and $D_3 = 100:50$. Since two types among the three types of foaming assistants were used for the first time ever, the existence of $F \times G$ was believed evident. T was used as a second-order factor since it was simple to change its level during the experiment.

34.1.3 Assignment and Distribution of Degrees of Freedom

It was decided that the factors related with mixing, A, B, C, D, E, F, and G, would be taken together in one class and would be designated Group I, and, in consideration of the interactions $E \times F$, $E \times G$, and $F \times G$, it was decided to assign them orthogonal array L_{27}. A dummy was inserted into the two-level factor C: $C_1 = C_2$.

Next, regarding the five factors from steps such as kneading and extrusion, H, J, L, M, and T, since H, J, L, and M were first-order factors, they constituted Group II; factor T was a second-order factor and it was decided that every combination with L_{27} would be experimented on. It was hoped that the interactions $J \times H$, $J \times L$, and $J \times M$ among H, J, L, and M would also be found, for the most part.

According to the usual assignment, except for $F \times G$ where the existence of interaction is clear-cut, interactions among E, F, and G would be omitted; also, H, J, L, and M too would be incorporated into Group I and assignment would be to the ordinary L_{27}, and in terms of the result, it is suspected that this would have been better. However, since there was the desire to use the randomly combined design for the first time in this case, the assignment turned out to be as follows.

Assignment to orthogonal array $L_{27}(3^{13})$:

Group I

A	B	C	D	E	F	G
(9)	(10)	(12)	(13)	(5)	(1)	(2)

Since $E \times F$ appeared in columns (6) and (7), $E \times G$ appeared in columns (8) and (11), and $F \times G$ appeared in columns (3) and (4), and A, B, C, and D were assigned to the remaining four columns, avoiding the previously mentioned columns. For the error term, therefore, only one degree of freedom could be obtained from the dummy portion of C.

Group II

H	J	L	M
(3)	(5)	(1)	(2)

$J \times H$ appears in columns (9) and (13), $J \times L$ appears in columns (6) and (7), and $J \times M$ appears in columns (8) and (11). Therefore, the number of degrees of freedom of error becomes 6 from the remaining columns, *viz.*, (4), (10), and (12).

Then, by using $L_{27}(3^{13})$ for each group, the data were assigned as explained above, and then the experiment numbers of the two groups were caused to correspond randomly. The results are as given in Table 34.1.

As a result of this, experiments No. 1 – 3 are performed by $A_1, B_1, C_1, D_1, E_1, F_1, G_1, H_1, J_3, L_1$, and M_1 and experiments No. 27 – 20 are performed by $A_1, B_3, C_3, D_2, E_3, F_3, G_3, H_3, J_2, L_3$, and M_1. Thus, considering the six interactions, the eleven factors have been incorporated into the experiments of 27 runs.

Table 34.1 Assignment Array

Class	Group I							Group II			
Factor	*A*	*B*	*C*	*D*	*E*	*F*	*G*	*H*	*J*	*L*	*M*
Experiment No. / Column	(9)	(10)	(12)	(13)	(5)	(1)	(2)	(3)	(5)	(1)	(2)
1—3	1	1	1	1	1	1	1	1	3	1	1
2—4	2	2	2	2	2	1	1	2	1	1	2
3—15	3	3	3	3	3	1	1	3	3	2	2
4—10	2	2	3	3	1	1	2	2	1	2	1
5—7	3	3	1	1	2	1	2	3	1	1	3
6—22	1	1	2	2	3	1	2	1	1	3	2
7—5	3	3	2	2	1	1	3	2	2	1	2
8—2	1	1	3	3	2	1	3	1	2	1	1
9—19	2	2	1	1	3	1	3	3	1	3	1
10—9	2	3	2	3	1	2	1	3	3	1	3
11—17	3	1	3	1	2	2	1	1	2	2	3
12—27	1	2	1	2	3	2	1	2	3	3	3
13—26	3	1	1	2	1	2	2	2	2	3	3
14—24	1	2	2	3	2	2	2	1	3	3	2
15—8	2	3	3	1	3	2	2	3	2	1	3
16—12	1	2	3	1	1	2	3	2	3	2	1
17—23	2	3	1	2	2	2	3	1	2	3	2
18—21	3	1	2	3	3	2	3	3	3	3	1
19—14	3	2	3	2	1	3	1	3	2	2	2
20—1	1	3	1	3	2	3	1	1	1	1	1
21—11	2	1	2	1	3	3	1	2	2	2	1
22—25	1	3	2	1	1	3	2	2	1	3	3
23—18	2	1	3	2	2	3	2	1	3	2	3
24—13	3	2	1	3	3	3	2	3	1	2	2
25—16	2	1	1	3	1	3	3	1	1	2	3
26—6	3	2	2	1	2	3	3	2	3	1	2
27—20	1	3	3	2	3	3	3	3	2	3	1

Correspondence between the numerals indicating the levels and the actual concrete levels was randomized for each factor, and the entire experiment of 27 runs was carried out by newly deciding on a random order.

Then, T_1 and T_2 were combined with these 27 experiments so that it would be possible to find the interactions of the temperature factor T of two levels, which was a second-order factor, with all the other factors. Of course, it was not that the 54 combinations were randomized, but T_1 and T_2 were experimented on consecutively in two experiments of 27 runs. Therefore, the assignment of the 27 runs became first-order units and T_1 and T_2 formed second-order units, and it was a split-unit design. Moreover, samples were taken at the beginning of the experiments, K_1, and at the end, K_2. Therefore, K_1 and K_2 became third-order units; moreover, since three samples apiece were taken, the final data appeared as fourth-order units and essentially this was a four-step split-unit design. The total number of data was $27 \times 2 \times 2 \times 3 = 324$. Assignment of the numbers of degrees of freedom was as follows.

(1) *Degrees of freedom of first-order sources* For the respective groups, the distributions of degrees of freedom were as presented in Table 34.2.

(2) *Degrees of freedom of second-order sources* The number of second-order units was 54. Therefore, the number of degrees of freedom of the variation among second-order units was 53. This is actually assigned to both the first-order sources and the second-order sources. The number of degrees of freedom of the first-order sources was 26, as investigated above. Therefore, the total of the degrees of freedom of the second-order sources was $53 - 26 = 27$. Clearly, this is also [number of second-order units (54) − number of first-order units (27)]. Let us consider

Table 34.2 Number of Degrees of Freedom of First-Order Sources

Group I

Source	f
A	2
B	2
C	1
D	2
E	2
F	2
G	2
$E \times F$	4
$E \times G$	4
$F \times G$	4
e_1	1
Total	26

Group II

Source	f
H	2
J	2
L	2
M	2
$H \times J$	4
$L \times J$	4
$M \times J$	4
e_1	6
Total	26

how this is assigned second-order sources. Since T_1 and T_2 were combined with first-order units and second-order units were thereby formed, the main effect of T appeared first of all as a second-order source, and interactions between T and all first-order sources were obtained. In other words,

$$\text{Second-order sources} = T + (\text{first-order sources}) \times T$$

Regarding the number of degrees of freedom, too,

$$27 = 1 + 26 \times 1$$

and this relationship is satisfied.

Therefore the numbers of degrees of freedom of the second-order sources become as given by Tables 34.3.

Here, the interaction between the first-order error variance e_1 and T clearly forms the entire second-order error variance e_2; but actually three-factor interactions are usually pooled with error, and if we decide to pool them we obtain 13 and 18, respectively, as the numbers of degrees of freedom of e_2 for Group I and Group II.

(3) *Degrees of freedom of third-order sources* The number of third-order units is 108. Therefore, the number of degrees of freedom of the variation among third-order units is 107. Actually, this is assigned to the first-order, second-order, and third-order sources. The sum of the degrees of freedom of the first-order sources and second-order sources is 53, as found above. Therefore, the sum of the degrees of freedom of third-order sources is $107 - 53 = 54$. Clearly, this is also [number of third-order units (108) − number of second-order units (54)]. In exactly the same way as above, these are distributed to the main effect of K and

Table 34.3 Number of Degrees of Freedom of Second-Order Sources

Group I

Source	f	
T	1	
$A \times T$	2	
$B \times T$	2	
$C \times T$	1	
$D \times T$	2	
$E \times T$	2	
$F \times T$	2	
$G \times T$	2	
$E \times F \times T$	4	13
$E \times G \times T$	4	
$F \times G \times T$	4	
$e_1 \times T \equiv e_2$	1	
Total	27	

Group II

Source	f	
T	1	
$H \times T$	2	
$J \times T$	2	
$L \times T$	2	
$M \times T$	2	
$H \times J \times T$	4	18
$L \times J \times T$	4	
$M \times J \times T$	4	
$e_1 \times T \equiv e_2$	6	
Total	27	

interactions between K and the first-order sources and second-order sources. Thus,

$$\text{Third-order sources} = K + (\text{first-order sources}) \times K + (\text{second-order sources}) \times K$$

Regarding the number of degrees of freedom, too, we have

$$54 = 1 + 26 \times 1 + 27 \times 1$$

and this relationship is satisfied.

Therefore, the numbers of degrees of freedom of the third-order sources become as given by Tables 34.4.

Here, $e_1 \times K$ and $e_2 \times K$ (a part of this last is actually four-factor interaction) alone form the third-order error variance e_3, but actually, three-factor interactions are usually pooled with this. Then 39 and 44, respectively, are obtained as the numbers of degrees of freedom of e_3.

Now, in the present case, in view of technical considerations regarding factor K, it is believed that two-factor interactions related with K do not have much meaning. Since K is a factor that indicates sequence—the beginning and end of the experiment, as explained before—it can be

Table 34.4 Number of Degrees of Freedom of Third-Order Sources

Group I

Source	f	
K	1	
$A \times K$	2	
$B \times K$	2	
$C \times K$	1	
$D \times K$	2	
$E \times K$	2	
$F \times K$	2	
$G \times K$	2	
$E \times F \times K$	4	13
$E \times G \times K$	4	
$F \times G \times K$	4	
$e_1 \times K \equiv e_3$	1	
$T \times K$	1	
$A \times T \times K$	2	26
$B \times T \times K$	2	
$C \times T \times K$	1	
$D \times T \times K$	2	
$E \times T \times K$	2	
$F \times T \times K$	2	
$G \times T \times K$	2	
$e_2 \times K \equiv e_3$	13	
Total	54	

Group II

Source	f	
K	1	
$H \times K$	2	
$J \times K$	2	
$L \times K$	2	
$M \times K$	2	
$H \times J \times K$	4	18
$L \times J \times K$	4	
$M \times J \times K$	4	
$e_1 \times K \equiv e_3$	6	
$T \times K$	1	
$H \times T \times K$	2	26
$J \times T \times K$	2	
$L \times T \times K$	2	
$M \times T \times K$	2	
$e_2 \times K \equiv e_3$	18	
Total	54	

Table 34.5 Number of Degrees of Freedom of Third-Order Sources

Group I

Source	f
K	1
$T \times K$	1
e_3	52
Total	54

Group II

Source	f
K	1
$T \times K$	1
e_3	52
Total	54

expected that interactions with K do not exist, and even if they do exist, they are not very useful. It was therefore decided to disregard all two-factor interactions except $T \times K$ (in other words, interactions between first- and second-order sources and K) from the start and to incorporate them in error.

Although there is no special significance to the fact that $T \times K$ was left remaining, only $T \times K$ can be found simply, and furthermore it facilitates calculation of e_3 if left in. In this case, therefore, the numbers of degrees of freedom of the third-order factors became simple, as given by Tables 34.5.

Of course, if K is a factor with more meaning, usually $A \times K, \ldots, G \times K$ and $H \times K, \ldots, M \times K$ are also analyzed, even if only as a matter of form.

(4) *Degrees of freedom of fourth-order sources* The number of fourth-order units, or in other words the total number of data, is 324. Therefore the number of degrees of freedom of the whole is 323. But, now, the sum total of the number of degrees of freedom of the first-order, second-order, and third-order sources is 107, as found above. Therefore, the number of degrees of freedom of the fourth-order sources is $323 - 107 = 216$. But factors have not been assigned to fourth-order units. In other words, there is no meaning to the sequence of the three samples each of the data which are included in the third-order units. Therefore, the fourth-order source which possesses 216 degrees of freedom is, in the last analysis, formed only by the fourth-order error variance e_4. Since this is also the dispersion of the three samples of the data among the third-order units, each posesses two degrees of freedom, and since 108 classes of these exist (by just the number of the third-order units), it may also be considered that the number of degrees of freedom becomes

$$(3-1) \times 108 = 216$$

Therefore, the fourth-order source becomes as follows.

Group I

Source	f
e_4	216

Group II

Source	f
e_4	216

34.1.4 Analysis the First Time for Group I

In this experiment, measurement and analysis were carried out on the following 15 characteristics, and a comprehensive judgment was arrived at; moreover, interrelations among characteristic values were also analyzed, but we have decided to show here only the method of data analysis for the *apparent specific gravity*.

a. Apparent specific gravity
b. Size of gas bubble particles
c. Uniformity of gas bubble particles
d. Beauty of surface finish
e. Change of color of surface finish
f. 50% modulus
g. Tensile strength

Table 34.6 Experimental Results (Raw Data)

No.	T_1						T_2					
	K_1			K_2			K_1			K_2		
1	1.04	1.04	1.05	1.04	1.10	1.09	0.96	0.98	0.97	1.00	1.02	1.05
2	1.09	1.09	1.07	1.10	1.07	1.10	0.86	0.86	0.87	1.19	1.18	1.17
3	0.78	0.79	0.76	0.78	0.78	0.78	0.65	0.66	0.67	0.68	0.69	0.69
4	0.81	0.80	0.82	0.84	0.83	0.85	0.71	0.73	0.73	0.77	0.77	0.76
5	0.67	0.66	0.66	0.72	0.70	0.73	0.63	0.65	0.66	0.66	0.64	0.62
6	0.89	0.88	0.91	0.90	0.87	0.88	0.74	0.74	0.71	0.73	0.73	0.74
7	0.61	0.62	0.63	0.71	0.70	0.73	0.66	0.67	0.67	0.69	0.68	0.68
8	1.18	1.17	1.16	1.22	1.21	1.21	1.15	1.15	1.16	1.19	1.16	1.18
9	0.85	0.84	0.82	0.87	0.87	0.88	0.65	0.67	0.64	0.67	0.67	0.68
10	0.67	0.69	0.69	0.76	0.77	0.75	0.75	0.71	0.71	0.71	0.68	0.67
11	1.26	1.24	1.22	1.24	1.25	1.26	1.13	1.15	1.17	1.14	1.14	1.13
12	0.89	0.87	0.89	0.84	0.85	0.85	0.69	0.71	0.70	0.74	0.72	0.72
13	1.13	1.14	1.13	1.15	1.14	1.15	1.22	1.15	1.14	1.18	1.14	1.16
14	0.96	0.95	0.97	1.06	1.07	1.05	0.93	0.91	0.92	0.92	0.88	0.94
15	0.79	0.81	0.79	0.81	0.81	0.80	0.78	0.78	0.79	0.81	0.81	0.81
16	0.85	0.87	0.88	0.86	0.86	0.85	0.80	0.81	0.80	0.79	0.83	0.81
17	0.83	0.84	0.84	0.87	0.86	0.85	0.77	0.75	0.77	0.83	0.82	0.82
18	0.85	0.87	0.88	0.88	0.89	0.87	0.81	0.85	0.79	0.84	0.85	0.91
19	1.13	1.15	1.17	1.17	1.13	1.16	1.05	1.07	1.04	1.07	1.08	1.08
20	0.97	0.98	0.97	1.02	1.07	1.03	0.78	0.81	0.86	0.87	0.90	0.90
21	1.00	1.00	1.02	1.04	1.05	1.04	1.15	1.16	1.18	1.13	1.15	1.15
22	0.66	0.65	0.64	0.77	0.77	0.77	0.60	0.61	0.60	0.60	0.60	0.61
23	1.25	1.23	1.24	1.25	1.25	1.24	1.15	1.18	1.15	1.11	1.16	1.13
24	0.84	0.85	0.87	0.93	0.89	0.89	0.76	0.73	0.75	0.73	0.74	0.73
25	1.04	1.03	1.02	1.05	1.04	1.02	0.86	0.87	0.85	0.85	0.85	0.84
26	0.72	0.71	0.72	0.68	0.67	0.70	0.67	0.67	0.66	0.63	0.61	0.64
27	0.82	0.79	0.81	0.82	0.84	0.82	0.73	0.73	0.76	0.73	0.73	0.72

Table 34.7 Processing Data and Totals

No.	T_1								Total
	K_1			Subtotal	K_2			Subtotal	
1	24	24	25	73	24	30	29	83	156
2	29	29	27	85	30	27	30	87	172
3	−2	−1	−4	−7	−2	−2	−2	−6	−13
4	1	0	2	3	4	3	5	12	15
5	−13	−14	−14	−41	−8	−10	−7	−25	−66
6	9	8	11	28	10	7	8	25	53
7	−19	−18	−17	−54	−9	−10	−7	−26	−80
8	38	37	36	111	42	41	41	124	235
9	5	4	2	11	7	7	8	22	33
10	−13	−11	−11	−35	−4	−3	−5	−12	−47
11	46	44	42	132	44	45	46	135	267
12	9	7	9	25	4	5	5	14	39
13	33	34	33	100	35	34	35	104	204
14	16	15	17	48	26	27	25	78	126
15	−1	1	−1	−1	1	1	0	2	1
16	5	7	8	20	6	6	5	17	37
17	3	4	4	11	7	6	5	18	29
18	5	7	8	20	8	9	7	24	44
19	33	35	37	105	37	33	36	106	211
20	17	18	17	52	22	27	23	72	124
21	20	20	22	62	24	25	24	73	135
22	−14	−15	−16	−45	−3	−3	−3	−9	−54
23	45	43	44	132	45	45	44	134	266
24	4	5	7	16	13	9	9	31	47
35	24	23	22	69	25	24	22	71	140
26	−8	−9	−8	−25	−12	−13	−10	−35	−60
27	2	−1	1	2	2	4	2	8	10
Total				897				1 127	2 024
	4th-order units			3rd order	4th-order units			3rd order	2nd order

h. Elongation
i. Abrasion test
j. Bite-in quantity
k. Brittleness characteristics
l. Insulation resistance
m. Breakdown voltage
n. Permittivity
o. tan δ

First, the 324 raw data are given in Table 34.6. The experiment numbers are indicated by *those of Group I.*

Table 34.7 (Continued)

T_2								Total	Grand total
K_1			Subtotal	K_2			Subtotal		
16	18	17	51	20	22	25	67	118	274
6	6	7	19	39	38	37	114	133	305
−15	−14	−13	−42	−12	−11	−11	−34	−76	−89
−9	−7	−7	−23	−3	−3	−4	−10	−33	−18
−17	−15	−14	−46	−14	−16	−18	−48	−94	−160
−6	−6	−9	−21	−7	−7	−6	−20	−41	12
−14	−13	−13	−40	−11	−12	−12	−35	−75	−155
35	35	36	106	39	36	38	113	219	454
−15	−13	−16	−44	−13	−13	−12	−38	−82	−49
−5	−9	−9	−23	−9	−12	−13	−34	−57	−104
33	35	37	105	34	34	33	101	206	473
−11	−9	−10	−30	−6	−8	−8	−22	−52	−13
42	35	34	111	38	34	36	108	219	423
13	11	12	36	12	8	14	34	70	196
−2	−2	−1	−5	1	0	1	2	−3	−2
0	1	0	1	−1	3	1	3	4	41
−3	−5	−3	−11	3	2	2	7	−4	25
1	5	−1	5	4	5	11	20	25	69
25	27	24	76	27	28	28	83	159	370
−2	1	6	5	7	10	10	27	32	156
35	36	38	109	33	35	35	103	212	347
−20	−19	−20	−59	−20	−20	−19	−59	−118	−172
35	38	35	108	31	36	33	100	208	474
−4	−7	−5	−16	−7	−6	−7	−20	−36	11
6	7	5	18	5	5	4	14	32	172
−13	−13	−14	−40	−17	−19	−16	−52	−92	−152
−7	−7	−4	−18	−7	−7	−8	−22	−40	−30
			332				502	834	2 858
4th-order units			3rd order	4th-order units			3rd order	2nd order	1st order

We use the device of subtracting 0.8 from these raw data and multiplying by 100, and then take the necessary sums for each experiment number and construct Table 34.7. The subtotal columns show the sum of the three data at the left, and essentially this is the sum of third-order units. The total columns give the sum of the two subtotals, and essentially this is the sum of second-order units. Furthermore, the grand total columns show the sum of the two totals, and in effect this is the sum of first-order units. For each of these three types of sum, the 27 are added vertically and the totals are arrived at beforehand. Analysis of variance is performed as follows from Group I, wherein large sources are believed to exist.

Table 34.8 Supplementary Tables (Group I, the First Time)

$r=54$	T_1	T_2	Total
A_1	726	192	918
A_2	744	406	1 150
A_3	554	236	790
B_1	1 500	1 198	2 698
B_2	620	71	691
B_3	−96	−435	−531

$r=54$	T_1	T_2	Total
E_1	582	249	831
E_2	1 093	678	1 771
E_3	349	−93	256
F_1	505	69	574
F_2	700	408	1 108
F_3	819	357	1 176
G_1	1 044	675	1 719
G_2	592	172	764
G_3	388	−13	375
Total	2 024	834	2 858

$r=81$	T_1	T_2	Total
K_1	897	332	1 229
K_2	1 127	502	1 629
Total	2 024	834	2 858

$r=54$	T_1	T_2	Total
C_1	706	133	839
C_2	289	57	346
C_3	1 029	944	1 673
D_1	449	151	600
D_2	904	507	1 411
D_3	671	176	847

$r=36$	F_2	F_2	F_3	Total
E_1	101	360	370	831
E_2	599	694	478	1 771
E_3	−126	54	328	256
Total	574	1 108	1 176	2 858

$r=36$	G_1	G_2	G_3	Total
E_1	540	233	58	831
E_2	934	510	327	1 771
E_3	245	21	−10	256
Total	1 719	764	375	2 358

$r=36$	G_1	G_2	G_3	Total
F_1	490	−166	250	574
F_2	356	617	135	1 108
F_3	873	313	−10	1 176
Total	1 719	764	375	2 858

(1) *Preparation of Supplementary Tables* (Table 34.8)

(2) *Calculation of Variation*

$$CF=\frac{2858^2}{324}=\frac{8168164}{324}=25210 \qquad (f=1) \tag{34.1}$$

Next, the total variation is found by squaring the 324 individual data which appear in Table 34.7

$$S_{\text{Total}}=24^2+24^2+25^2+29^2+\cdots+(-16)^2+(-7)^2+(-7)^2+(-8)^2-CF$$
$$=134962-25210=109752 \qquad (f=323) \tag{34.2}$$

Then, the total variation among third-order units, S_{T_3}, is found by squaring the 108 data which appear in the subtotal columns of Table 34.7.

$$S_{T_3}=\frac{1}{3}[73^2+85^2+\cdots+2^2+83^2+87^2+\cdots+8^2+51^2+19^2+\cdots$$
$$+(-18)^2+67^2+114^2+\cdots+(-22)^2]-CF$$
$$=\frac{403\,418}{3}-CF=134\,473-25\,210=109\,263 \qquad (f=107) \qquad (34.3)$$

Next, we find the total variation among second-order units, S_{T_2}, by squaring the 54 data which appear in the total columns of Table 34.7.

$$S_{T_2}=\frac{1}{6}[156^2+172^2+\cdots+10^2+118^2+133^2+\cdots+(-40)^2]-CF$$
$$=\frac{790\,196}{6}-CF=131\,699-25\,210=106\,489 \qquad (f=53) \qquad (34.4)$$

We further find the total variation among first-order units, S_{T_1}, from the 27 data of the grand total column of Table 34.7.

$$S_{T_1}=\frac{1}{12}[274^2+305^2+\cdots+(-30)^2]-CF$$
$$=\frac{1\,483\,376}{12}-CF=123\,615-25\,210=98\,405 \qquad (f=26) \qquad (34.5)$$

Next, we find the variation of first-order sources. By using Table 34.8, we obtain

$$S_A=\frac{1}{108}(918^2+1\,150^2+790^2)-CF=617$$
$$S_B=\frac{1}{108}[2\,698^2+691^2+(-531)^2]-CF=49\,222 \qquad (34.6)$$

Paying heed to the dummy $C_1 \equiv C_2$, we have

$$S_C=\frac{[(839+346)-1\,673\times 2]^2}{216\times 1^2+108\times(-2)^2}=\frac{(-2\,161)^2}{648}=7\,207$$
$$S_D=\frac{1}{108}(600^2+1\,411^2+847^2)-CF=3\,200$$
$$S_E=\frac{1}{108}(831^2+1771^2+256^2)-CF=10\,832$$
$$S_F=\frac{1}{108}(574^2+1\,108^2+1\,176^2)-CF=2\,013$$
$$S_G=\frac{1}{108}(1\,719^2+764^2+375^2)-CF=8\,857$$
$$S_{E\times F}=\frac{1}{36}(101^2+360^2+\cdots+328^2)-CF-S_E-S_F=2\,834$$
$$S_{E\times G}=\frac{1}{36}[540^2+233^2+\cdots+(-10)^2]-CF-S_E-S_G=912$$
$$S_{F\times G}=\frac{1}{36}[490^2+(-166)^2+\cdots+(-10)^2]-CF-S_F-S_G=11\,587$$

From the dummy of C, we have

$$S_{e_1}=\frac{(839-346)^2}{216}=1\,125 \tag{34.7}$$

Next, as regards second-order sources, we have

$$S_T=\frac{(2\,024-834)^2}{324}=4\,371 \tag{34.8}$$

$$S_{A\times T}=\frac{1}{54}(726^2+744^2+\cdots+236^2)-CF-S_A-S_T=264$$

$$S_{B\times T}=238$$

$$S_{C\times T}=\frac{[(706+289)+644\times 2-(133+57)-1\,029\times 2]^2}{108\times 1^2+54\times 2^2+108\times(-1)^2+54\times(-2)^2}=2$$

$$S_{D\times T}=180$$

$$S_{E\times T}=59$$

$$S_{F\times T}=155$$

$$S_{G\times T}=12$$

$$S_{e_2}=S_{T_2}-S_{T_1}-(S_T+S_{A\times T}+\cdots+S_{G\times T})=2\,713 \tag{34.9}$$

As for third-order sources, we have

$$S_K=\frac{(1\,229-1\,629)^2}{324}=494 \tag{34.10}$$

$$S_{T\times K}=\frac{(897+502-332-1\,127)^2}{324}=11$$

$$S_{e_3}=S_{T_3}-S_{T_2}-S_K-S_{T\times K}=2\,269 \tag{34.11}$$

Since the fourth-order source is only error, we have

$$S_{e_4}=S_{\text{Total}}-S_{T_3}=489 \tag{34.12}$$

The analysis of variance table, Table 34.9, is obtained from these results.

(3) *Analysis of Variance Table*

Since each of the errors e_1, e_2, and e_3 is significant for the error which is next higher in order, none can be pooled. It is therefore necessary to perform testing for each unit. However, since the first-order error variance possesses only one degree of freedom, testing was done by creating an error of 15 degrees of freedom by pooling A, D, F, $E\times F$, and $E\times G$, whose variances were comparatively small. The pooled error was as shown in the following table.

Source	f	S	V
e_1	15	10 700	713

Table 34.9 Analysis of Variance Table (Group I, the First Time)

Source	f	S	V	
A	2	617	308	
B	2	49 222	24 611	**
C	1	7 207	7 207	**
D	2	3 200	1 600	
E	2	10 832	5 416	**
F	2	2 013	1 006	
G	2	8 857	4 428	*
$E \times F$	4	2 834	708	
$E \times G$	4	912	228	
$F \times G$	4	11 587	2 897	*
e_1	1	1 124	1 124	**
T_1	26	98 405		
T	1	4 371	4 371	**
$A \times T$	2	264	132	
$B \times T$	2	328	164	
$C \times T$	1	2	2	
$D \times T$	2	180	90	
$E \times T$	2	59	30	
$F \times T$	2	155	78	
$G \times T$	2	12	6	
e_2	13	2 713	209	**
T_2	53	106 489		
K	1	494	494	**
$T \times K$	1	11	11	
e_3	52	2 269	44	**
T_3	107	109 263		
e_4	216	489	2.26	
Total	323	109 752		

As a result, the significant sources are the first-order sources, B, C, E, G, and $F \times G$, the second-order source, T, and third-order source, K.

Now, since this is randomly combined design, we correct the data with regard to sources which have emerged as significant and move on to analysis of Group II, but first let us take a bit of time for discussion of the method of correction. Correction means to erase the effect of a source which has emerged as significant; for example, regarding main effect A, if the mean value of A_1 is higher than the mean of the whole, we decrease data that contain A_1 by that much. As a basic rule, this correction should be performed on the individual data. But in the present case, only main effects T and K are significant as second-order and third-order sources. Since T and K have been combined completely orthogonally with the assignment of 27 runs that have been designed by the randomly combined design, it is unnecessary to correct them even if they emerge as significant (their effects are preserved when analyzing Group II). Therefore, only the effects of first-order sources must be corrected; and even if correction is performed on the individual data, the 12 data which belong to the same first-order units receive the same amount of correction.

Then, since the fourth-order error which constitutes the fourth-order source is the dispersion among the three data which belong to the same third-order units, it is clear that its value does not change even if corrected. Actually, both S_{Total} and S_{T_3} change and their difference remains constant.

Let us next consider what becomes of the third-order sources. Earlier, we investigated the distribution of the number of degrees of freedom; with regard to third-order sources, at the time, ultimately only K and $T \times K$ and e_3 remained for both Groups I and II, and they had exactly the same form. This is because all interactions with K except $T \times K$ were disregarded and were incorporated into e_3. Of course, if $A \times K$ or $H \times K$ were considered the make-up of Group I and Group II would differ, but since we have decided not to calculate them individually, the make-up of the third-order sources has the same form for both, as a whole. In this case, therefore, even if correction is performed, the variation of the third-order sources does not change as a whole.

As is clear from the discussion above, in analysis after correction it is necessary to calculate first-order and second-order sources but it is unnecessary to calculate third-order and fourth-order sources. Then, instead of correcting the individual data, it is sufficient if we correct with second-order units as the units. For facility in calculation, it is more convenient if the correction quantity is made to belong to the same unit as the original data, but if correction quantities are obtained for the individual data, it is troublesome since the rounding error becomes great; but it becomes small if second-order units are used as the units, and it is convenient since it is possible to stay with the same unit as that of the original data.

34.1.5 Analysis the First Time for Group II

(1) *Correction of Original Data.* To embark on analysis of Group II, let us correct the data. It is adequate if correction is performed with second-order units as the units, and the sources that should be corrected are B, C, E, and G, and $F \times G$. Since $F \times G$ is included, the correction quantity changes at each combination of the levels of F and G; and since main effect F is not significant, care is required in calculating the correction quantities. For example, considering F_1G_1, since the effects which

are to be corrected are main effect G and interaction $F \times G$, we find the information by reversing the sign of

$$\overline{F_1G_1} - \overline{F}_1 = \frac{490}{6} - \frac{574}{18} = 50 \tag{34.13}$$

The correction quantities are

$$\begin{aligned} &F_1G_1:\ -50, \quad F_1G_2:\ 60, \quad F_1G_3:\ -10 \\ &F_2G_1:\ 2, \quad F_2G_2:\ -41, \quad F_2G_3:\ 39 \\ &F_3G_1:\ -80, \quad F_3G_2:\ 13, \quad F_3G_3:\ 67 \end{aligned} \tag{34.14}$$

Similarly, for B, C, and E also, for example for the data of $\overline{B}_1$, we find the corrected quantities by assuming that

Table 34.10 Correction Quantities and Data After Correction

No.	Correction quantity	Data after correction		
		T_1	T_2	Total
1	−120	36	−2	34
2	−61	111	72	183
3	31	18	−45	−27
4	42	57	9	66
5	116	50	22	72
6	22	75	−19	56
7	99	19	24	43
8	−193	42	26	68
9	64	97	−18	79
10	111	64	54	118
11	−181	86	25	111
12	76	115	24	139
13	−111	93	108	201
14	−52	74	18	92
15	40	41	37	78
16	21	58	25	83
17	95	124	91	215
18	1	45	26	71
19	−98	113	61	174
20	−24	100	8	108
21	−118	17	94	111
22	122	68	4	72
23	−170	96	38	134
24	87	134	51	185
25	−3	137	29	166
26	56	−4	−36	−40
27	148	158	108	266
Total	0	2 024	834	2 858

$$\bar{T}-\bar{B}_1=\frac{2858}{54}-\frac{2698}{18}=-97 \tag{34.15}$$

And

$$B_1:\ -97,\ B_2:\ 15,\quad B_3:\ 82$$

$$C_1=C_2:\ 20,\qquad C_3:\ -40 \tag{34.16}$$

$$E_1:\ 7,\quad E_2:\ -46,\ E_3:\ 39 \tag{34.17}$$

Then, since the assignment of No. 7 in Table 34.10 is $B_3C_2E_1F_1G_3$, its corrected quantity is found as

$$\underset{B_3}{82}+\underset{C_2}{20}+\underset{E_1}{7}+\underset{F_1G_3}{(-10)}=99 \tag{34.18}$$

(2) *Preparation of Supplementary Tables.* The values of the data after correction (T_1, T_2, and total) are transcribed onto cards on which the data have been entered (actually, only the correction quantities are first found, and the correction is done on the cards; they are written in directly and the sums are obtained on the cards); these are sorted according to the levels of the factors of Group II, and supplementary tables such as Tables 34.11 are constructed.

(3) *Analysis of Variance.*

$$CF=25210 \tag{34.19}$$

$$S_{T_2}=\frac{1}{6}(36^2+111^2+\cdots+108^2)-CF=18789 \tag{34.20}$$

$$S_{T_1}=\frac{1}{12}(34^2+183^2+\cdots+266^2)-CF=10704 \tag{34.21}$$

Table 34.11 Supplementary Tables (Group II, the First Time)

$k=54$	T_1	T_2	Total
H_1	770	214	984
H_2	534	324	858
H_3	720	296	1 016
J_1	829	158	987
J_2	693	574	1 267
J_3	502	102	604
L_1	459	205	664
L_2	716	287	1 003
L_3	849	342	1 191
M_1	610	276	886
M_2	664	217	881
M_3	750	341	1 091
Total	2 024	834	2 858

$k=36$	J_1	J_2	J_3	Total
H_1	330	394	260	984
H_2	321	355	182	858
H_3	336	518	162	1 016
L_1	363	189	112	664
L_2	417	396	190	1 003
L_3	207	682	302	1 191
M_1	253	445	188	886
M_2	424	432	25	881
M_3	310	390	391	1 091
Total	987	1 267	604	2 858

Table 34.12 Analysis of Variance Table (Group II, the First Time)

Source	f	S	V	
H	2	130	65	
J	2	2 052	1 026	*
L	2	1 321	660	
M	2	266	133	
$H\times J$	4	424	106	
$L\times J$	4	3 251	813	
$M\times J$	4	2 069	517	
e_1	6	1 191	198	
T_1	26	10 704		
T	1	4 371	4 371	**
$H\times T$	2	564	282	*
$J\times T$	2	1 410	705	**
$L\times T$	2	311	156	
$M\times T$	2	61	30	
e_2	18	1 368	76	
T_2	53	18 789		
K	1	494	494	**
$T\times K$	1	11	11	
e_3	52	2 269	44	

$$S_H=\frac{1}{108}(984^2+858^2+1016^2)-CF=130$$

$$S_J=\frac{1}{108}(987^2+1267^2+604^2)-CF=2052$$

$$S_L=\frac{1}{108}(664^2+1003^2+1191^2)-CF=1321$$

$$S_M=\frac{1}{108}(886^2+881^2+1091^2)-CF=266$$

$$S_{H\times J}=\frac{1}{36}(330^2+394^2+\cdots+162^2)-CF-S_H-S_J=424$$

$$S_{L\times J}=3251$$

$$S_{M\times J}=2069$$

$$S_{e_1}=S_{T_1}-(S_H+\cdots+S_{M\times J})=1191$$

$$S_{H\times T}=\frac{1}{54}(770^2+534^2+\cdots+296^2)-CF-S_H-S_J=564$$

$$S_{J\times T}=1410$$

$$S_{L\times T}=311$$

$$S_{M\times T}=61$$

$$S_{e_2}=S_{T_2}-S_{T_1}-(S_T+S_{H\times J}+\cdots+S_{M\times T})=1368$$

(4) *Analysis of Variance Table.* From this, the analysis of variance table becomes as given by Table 34.12. For the third-order sources, the values at the time of Group I were used, and e_4 was omitted since it could not be pooled anyhow.

Both e_1 and e_2 became far smaller than in the case of the analysis of Group I. As a result, in the case of Group II, first-order source J, second-

Table 34.13 Correction Quantities

No.	Correction quantity		Data after correction		
	T_1	T_2	T_1	T_2	Total
1	10	29	166	147	313
2	−7	3	165	136	301
3	18	21	5	−55	−50
4	−7	3	8	−30	−22
5	−18	14	−84	−80	−164
6	−26	22	27	−19	8
7	8	−43	−72	−118	−190
8	−11	−24	224	195	419
9	−18	14	15	−68	−53
10	18	21	−29	−36	−65
11	−11	−24	256	182	438
12	29	10	68	−42	26
13	8	−43	212	176	388
14	10	29	136	99	235
15	−3	−32	−2	−35	−37
16	29	10	66	14	80
17	−11	−24	18	−28	−10
18	18	21	62	46	108
19	−3	−32	208	127	335
20	−26	22	98	54	152
21	8	−43	143	169	312
22	−7	3	−61	−115	−176
23	10	29	276	237	513
24	−18	14	29	−22	7
25	−26	22	114	54	168
26	29	10	−31	−82	−113
27	−3	−32	7	−72	−65
Total	0	0	2 025	834	2 858

order sources T, $H \times T$, and $J \times T$, and third-order source K are significant. Therefore, we correct the original data by J and $H \times T$ and $J \times T$ and analyze again for Group I.

34.1.6 Analysis the Second Time for Group I

(1) *Correction of Data.* Correction should of course be performed with second-order units as the units, but since interactions with T are significant this time, the correction quantities will differ between T_1 and T_2.

Table 34.14 Analysis of Variance Table (Group I, the Second Time)

Source	f	S	V
A	2	583	292
B	2	49 906	24 953 **
C	1	6 019	6 019 **
D	2	2 308	1 154
E	2	10 832	5 416 **
F	2	2 124	1 062
G	2	9 869	4 934 **
$E \times F$	4	1 164	291
$E \times G$	4	907	227
$F \times G$	4	9 238	2 310 *
e_1	1	766	766 **
T'''	26	93 716	
T	1	4 371	4 371 **
$A \times T$	2	122	61
$B \times T$	2	278	139
$C \times T$	1	108	108
$D \times T$	2	185	92
$E \times T$	2	53	26
$F \times T$	2	38	19
$G \times T$	2	80	40
e_2	13	876	67
T''	53	99 827	
K	1	494	494
$T \times K$	1	11	11
e_3	52	2 269	44

Table 34.15 Analysis of Variance Chart Group I

Source	f	(1) S	(1) V		(2) S	(2) V		(3) S	(3) V	
A	2	617	308		583	292		687	344	
B	2	49 222	24 611	**	49 906	24 953	**	49 893	24 946	**
C	1	7 207	7 207	**	6 019	6 019	**	5 174	5 174	**
D	2	3 200	1 600		2 308	1 154		2 008	1 008	
E	2	10 832	5 416	**	10 832	5 416	**	10 832	5 416	**
F	2	2 013	1 006		2 124	1 062		1 932	966	
G	2	8 857	4 428	*	9 869	4 934	**	10 171	5 086	**
$E\times F$	4	2 834	708		1 164	291		801	200	
$E\times G$	4	912	228		907	227		1 167	292	
$F\times G$	4	11 578	2 897	*	9 238	2 310	*	8 253	2 063	*
e_1	1	1 124	1 124	**	766	766	**	721	721	**
(e_1)	(15)	(10 700)	(713)		(7 852)	(523)		(7 316)	(488)	
T_1	26	98 405			93 716			91 639		
T	1	4 371	4 371	**	4 371	4 371	**			
$A\times T$	2	264	132		122	61				
$B\times T$	2	328	164		278	139				
$C\times T$	1	2	2		108	108				
$D\times T$	2	180	90		185	92				
$E\times T$	2	59	30		53	26				
$F\times T$	2	155	78		38	19				
$G\times T$	2	12	6		80	40				
e_2	13	2 713	209	**	876	67				
T_2	53	106 489			99 827					
K	1	494	494	**						
$T\times K$	1	11	11							
e_3	52	2 269	44							
T_3	107	109 263								
e_4	216	489	2.26							
Total	323	109 752								

In regard to the combination of J and T, since the effects to be corrected are main effect J and interaction $J\times T$, we have for J_1T_1, for example,

Group II

Source	f	(1) S	(1) V		(2) S	(2) V		(3) S	(3) V	
H	2	130	65		286	143		388	194	
J	2	2 052	1 026	*	3 999	2 000	*	5 154	2 577	**
L	2	1 321	660		1 101	550		937	468	
M	2	266	133		378	189		453	226	
$H\times J$	4	424	106		428	107		473	118	
$L\times J$	4	3 251	813		2 487	622		2 238	560	
$M\times J$	4	2 069	517		1 669	417		1 525	381	
e_1	6	1 191	198		1 151	192		1 227	204	
(e_1)	(24)							(7 241)	(302)	**
T_1	26	10 704			10 704			12 395		
T	1	4 371	4 371	**						
$H\times T$	2	564	282	*						
$J\times T$	2	1 410	705	**						
$L\times T$	2	311	156							
$M\times T$	2	61	30							
e_2	18	1 368	76							
T_2	53	18 789								

$$\overline{T}_1-\overline{J_1T_1} \tag{34.22}$$

But for the combination of H and T, since the effect to be corrected is only the interaction $H\times T$, for example for H_1T_1 we have

$$\overline{T}-\overline{H}_1-\overline{T}_1+\overline{H_1T_1} \tag{34.23}$$

Therefore, by adding these, we get the correction quantities given in Table 34.13.

Supplementary tables were constructed and analysis the second time performed for Group I, and Table 34.14 was obtained.

34.1.7 Recapitulation

In regard to the significant sources of Group I, the original data were corrected as the previous time, and we perform analysis the second time for Group II. Tables 34.15 were obtained when analysis of variance was performed the third time by iterating the same procedure.

Tables 34.16 Supplementary Tables (Group I)

r=108	Total
A_1	1 022
A_2	1 101
A_3	735
B_1	2 643
B_2	850
B_3	−635
Total	2 858

r=108	Total
C_1	845
C_2	450
C_3	1 563
D_1	600
D_2	1 252
D_3	1 006
Total	2 858

r=36	F_1	F_2	F_3	Total
E_1	101	409	321	831
E_2	550	639	582	1 771
E_3	−71	103	224	256
Total	580	1 151	1 127	2 858

r=36	G_1	G_2	G_3	Total
E_1	589	184	58	831
E_2	885	614	272	1 771
E_3	294	−28	−10	256
F_1	594	−160	146	580
F_2	405	562	184	1 151
F_3	769	368	−10	1 127
Total	1 768	770	320	2 858

34.1.8 Estimation

The supplementary tables for the third time were as given by Tables 34.16 and 34.17.

Tables 34.17 Supplementary Tables (Group II)

k=54	T_1	T_2	Total
H_1	791	235	1 026
H_2	498	288	786
H_3	735	311	1 046
J_1	785	114	899
J_2	812	693	1 505
J_3	427	27	454
Total	2 024	834	2 858

k=54	J_1	J_2	J_3	Total
H_1	358	450	218	1 026
H_2	237	449	100	786
H_3	304	606	136	1 046

Sources that have become significant by analysis of variance the third time are estimated by using Tables 34.16 and 34.17.

(1) *Group I.* We need only draw the graphs of B, C, and E, the graphs of G for each level F, and the graphs of T and—although it is not very meaningful—of K.

First, as to B, C, and E and G per F, we have, from Tables 34.16,

$$B_1: 0.8+\frac{1}{100}\,\frac{2643}{108}=1.045\pm0.045 \tag{34.24}$$

$$B_2: 0.8+\frac{1}{100}\,\frac{850}{108}=0.879 \quad ''$$

$$B_3: 0.8+\frac{1}{100}\,\frac{-635}{108}=0.741\pm0.045$$

$$C_1: 0.8+\frac{1}{100}\,\frac{845+450}{108\times2}=0.860\pm0.032 \tag{34.25}$$

$$C_3: 0.8+\frac{1}{100}\,\frac{1563}{108}=0.945\pm0.045$$

$$E_1: 0.8+\frac{1}{100}\,\frac{831}{108}=0.877 \quad ''$$

$$E_2: 0.8+\frac{1}{100}\,\frac{1771}{108}=0.964 \quad ''$$

$$E_3: 0.8+\frac{1}{100}\,\frac{256}{108}=0.824 \quad ''$$

$$F_1\begin{cases} G_1: 0.8+\dfrac{1}{100}\,\dfrac{594}{36}=0.965\pm0.078 \\ G_2: 0.8+\dfrac{1}{100}\,\dfrac{-160}{36}=0.756 \quad '' \\ G_3: 0.8+\dfrac{1}{100}\,\dfrac{146}{36}=0.841 \quad '' \end{cases} \tag{34.26}$$

$$F_2\begin{cases} G_1: 0.8+\dfrac{1}{100}\,\dfrac{405}{36}=0.912 \quad '' \\ G_2: 0.8+\dfrac{1}{100}\,\dfrac{562}{36}=0.956 \quad '' \\ G_3: 0.8+\dfrac{1}{100}\,\dfrac{184}{36}=0.851 \quad '' \end{cases}$$

$$F_3\begin{cases} G_1: 0.8+\dfrac{1}{100}\,\dfrac{769}{36}=1.014\pm0.078 \\ G_2: 0.8+\dfrac{1}{100}\,\dfrac{368}{36}=0.902 \quad '' \\ G_3: 0.8+\dfrac{1}{100}\,\dfrac{-10}{36}=0.797 \quad '' \end{cases}$$

The formula for the confidence limits, using $V_e = 488$ in the case of main effect B, is

$$\pm\sqrt{4.54\times\frac{488}{108}}\times\frac{1}{100}=\pm0.045 \tag{34.27}$$

For the mean of C_1 the effective number of replications 108 becomes 216, and for the combinations of F and G it becomes 36.

Next, T and K are found from Group II whose error variance is small.

(2) *Group II.* We draw the graphs of H and J for the levels of T and the graph of K. From Tables 34.15 we have

$$T_1\begin{cases} H_1: 0.8+\frac{1}{100}\frac{791}{54}=0.947\pm0.037 \\ H_2: 0.8+\frac{1}{100}\frac{498}{54}=0.892 \quad '' \\ H_3: 0.8+\frac{1}{100}\frac{735}{54}=0.936 \quad '' \end{cases} \tag{34.28}$$

$$T_2\begin{cases} H_1: 0.8+\frac{1}{100}\frac{235}{54}=0.844 \quad '' \\ H_2: 0.8+\frac{1}{100}\frac{288}{54}=0.853 \quad '' \\ H_3: 0.8+\frac{1}{100}\frac{311}{54}=0.858 \quad '' \end{cases}$$

$$T_1\begin{cases} J_1: 0.8+\frac{1}{100}\frac{785}{54}=0.945 \quad '' \\ J_2: 0.8+\frac{1}{100}\frac{812}{54}=0.950 \quad '' \\ J_3: 0.8+\frac{1}{100}\frac{427}{54}=0.872 \quad '' \end{cases} \tag{34.29}$$

$$T_2\begin{cases} J_1: 0.8+\frac{1}{100}\frac{114}{54}=0.821 \quad '' \\ J_2: 0.8+\frac{1}{100}\frac{693}{54}=0.928 \quad '' \\ J_3: 0.8+\frac{1}{100}\frac{27}{54}=0.805 \quad '' \end{cases}$$

In order to obtain the 95% confidence limits with regard to e_1, we pool the sources other than J from Tables 34.15 and we have

$$S_{e_1}=7\,241,\quad f=24$$

Also, since e_2 is not significant relative to e_3, we pool $L\times T, M\times T, e_2$, and $T\times K$ with e_3 and obtain

$$S_{e_3}=4\,020,\quad f=75$$

$$F_{24}^{1}=4.26 \qquad F_{75}^{1}=3.94$$

$$(n_e)_1=\frac{324}{2+1}=108, \qquad (n_e)_3=\frac{324}{1+2+2}=64.8$$

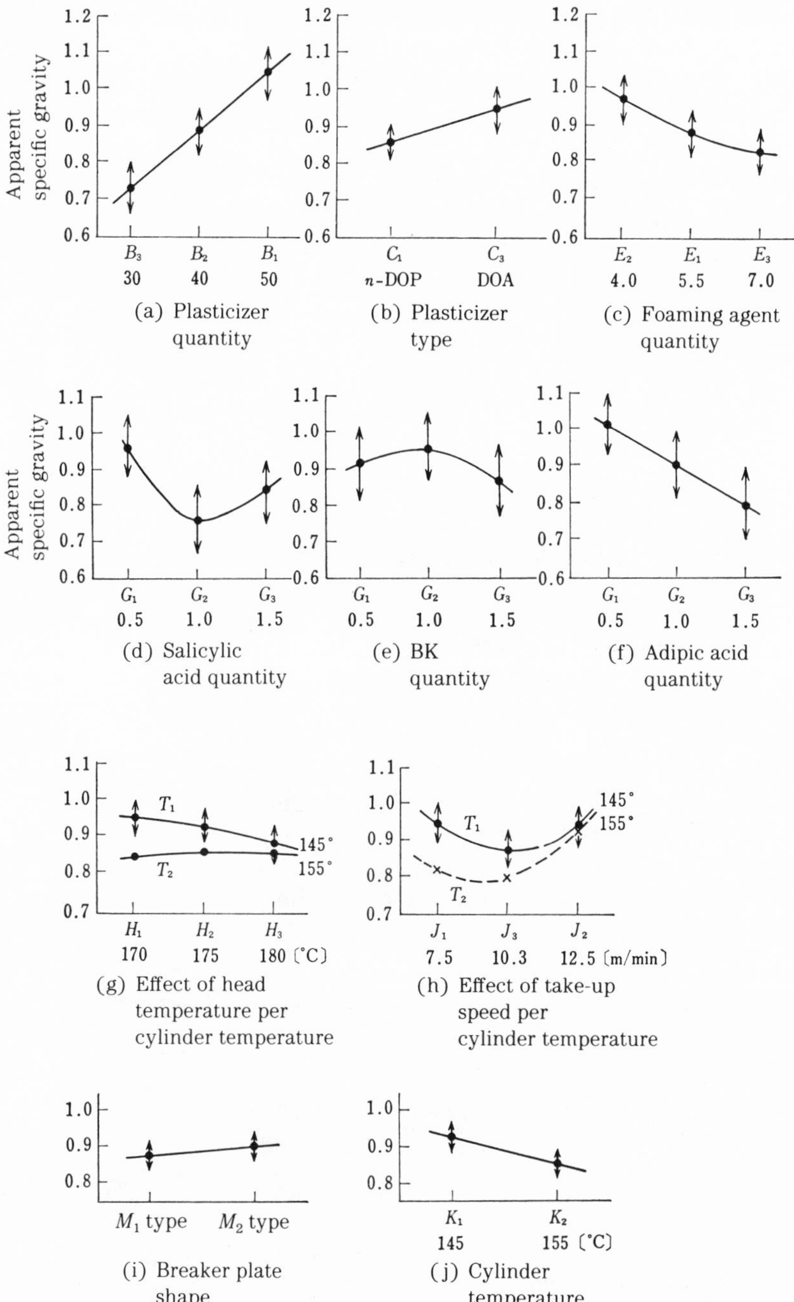

FIGURE 34.1 Graphs of Significant Sources in Relation to Apparent Specific Gravity

Therefore, by using the formula for the confidence limits in the split-unit design, we have

$$\pm\sqrt{4.26\times\frac{7241}{24}\times\frac{1}{108}+3.94\times\frac{4020}{75}\times\frac{1}{64.8}}\times\frac{1}{100}=\pm0.036 \tag{34.30}$$

The main effects of third-order source K are

$$K_1:\ 0.8+\frac{1}{100}\ \frac{2024}{162}=0.925\pm0.010 \tag{34.31}$$

$$K_2:\ 0.8+\frac{1}{100}\ \frac{1629}{162}=0.901 \qquad ''$$

$$\pm\sqrt{4.02\times\frac{2280}{53}\times\frac{1}{162}}\times\frac{1}{100}=\pm0.010 \tag{34.32}$$

When the results above are drawn as graphs, we obtain Figures 34.1.

34.1.9 Estimation of Process Average

If the best conditions are read off from the graphs, assuming that the characteristic values are better the smaller they are and the degree of foaming is better the greater it is, we obtain

$$B_3C_1E_3F_1G_2H_1J_3T_2$$

Since K is not a control factor, it is necessary to estimate separately with K_1 and K_2. Since it is the same thing, we decide to estimate in the case of K_2. From Groups I and II, we find *the combined process average having estimated the factorial effects* of each.

From Tables 34.16 and 34.17, and from Table 34.8 with regard to K, the process average $\hat{\mu}$ is

$$\hat{\mu}=\text{working mean}+\frac{1}{100}\left(\frac{F_1G_2}{36}-\frac{F_1}{108}+\frac{B_3}{108}+\frac{C_1}{216}+\frac{E_3}{108}+\frac{J_3T_2}{54}+\frac{H_1T_2}{54}\right.$$
$$\left.-\frac{H_1}{108}-\frac{T_2}{162}+\frac{K_2}{162}-3\times\frac{T}{324}\right)$$
$$=0.8+\frac{1}{100}\left(\frac{-160}{36}-\frac{580}{108}+\frac{-635}{108}+\frac{1295}{216}+\frac{256}{108}+\frac{27}{54}+\frac{235}{54}\right.$$
$$\left.-\frac{1026}{108}-\frac{834}{162}+\frac{1629}{162}-3\times\frac{2858}{324}\right)=0.465 \tag{34.33}$$

The range of the measured values of the apparent specific gravity, x, under the best conditions is

$$x=\hat{\mu}\pm\sqrt{F\times\frac{V_{e_1}}{(n_e)_1}+F\times\frac{V_{e_2}}{(n_e)_2}+F\times\frac{V_{e_3}}{(n_e)_3}+F\times\frac{V_{e_4}}{(n_e)_4}} \tag{34.34}$$

$$=0.465\pm\sqrt{4.54\times\frac{40.5}{324}\times 0.04877+4.22\times\frac{32\times 0.01264}{324}+4.02\times\frac{55\times 0.00421}{324}}$$

$$\overline{+3.89\times\frac{216\times 0.000226}{324}}=0.465\pm 0.191 \qquad (34.35)$$

This formula was obtained as follows.

In the confidence limits, since there is a correlation between the error variances of the two groups, we use the error variance which is greater to be on the safe side. It does not matter if the smaller is used.

First-Order Error Variance

$$V_{e_1}=\frac{0.7316}{15}=0.04877\ (f=15) \qquad \text{Group I} \qquad (34.36)$$

Second-Order Error Variance

$$V_{e_2}=\frac{0.1740}{22}=0.01264\ (f=22) \qquad \text{Group II} \qquad (34.37)$$

Third-Order Error Variance

$$V_{e_3}=\frac{0.2280}{53}=0.00421\ (f=53) \qquad \text{In common} \qquad (34.38)$$

Fourth-Order Error Variance

$$V_{e_4}=0.000226\ (f=216) \qquad \text{In common} \qquad (34.39)$$

As regards the effective number of replications of the process average, for the first-order error variance we use the sum of the numbers of degrees of freedom of m, B, C, E, G, $F\times G$, and J, and we have

$$\frac{1}{(n_e)_1}=\frac{1+2+\left(\frac{3}{2}-1\right)+2+2+4+2}{324}=\frac{13.5}{324} \qquad (34.40)$$

Regarding the second-order error variance, since T, $H\times T$, and $J\times T$ are significant, we have

$$\frac{1}{(n_e)_2}=\frac{1+2+2}{324}=\frac{5}{324} \qquad (34.41)$$

As for the third-order error variance, since there is only K, we have

$$\frac{1}{(n_e)_3}=\frac{1}{324} \qquad (34.42)$$

Next, the variance of the value x which has been measured by taking just one sample from among K_2 of the products produced under the best conditions, is

$$\mathrm{Var}(x) = \sigma_4^2 + \sigma_3^2 + \sigma_2^2 + \sigma_1^2$$
$$= \frac{1}{12}(\sigma_4^2 + 3\sigma_3^2 + 6\sigma_2^2 + 12\sigma_1^2) + \frac{1}{12}(\sigma_4^2 + 3\sigma_3^2 + 6\sigma_2^2)$$
$$+ \frac{1}{6}(\sigma_4^2 + 3\sigma_3^2) + \frac{2}{3}\sigma_4^2 \tag{34.43}$$

Therefore, the effective numbers of replications for the first-order, second-order, third-order, and fourth-order error variances are

$$\frac{1}{(n_e)_1} = \frac{1}{12}, \quad \frac{1}{(n_e)_2} = \frac{1}{12} \tag{34.44}$$

$$\frac{1}{(n_e)_3} = \frac{1}{6}, \quad \frac{1}{(n_e)_4} = \frac{2}{3} \tag{34.45}$$

Taking these together with Equations (34.40), (34.41), and (34.42), the effective numbers of replications of Equation (34.44) become

$$\frac{1}{(n_e)_1} = \frac{13.5}{324} + \frac{1}{12} = \frac{40.5}{324} \tag{34.46}$$

$$\frac{1}{(n_e)_2} = \frac{5}{324} + \frac{1}{12} = \frac{32}{324} \tag{34.47}$$

$$\frac{1}{(n_e)_3} = \frac{1}{324} + \frac{1}{6} = \frac{55}{324} \tag{34.48}$$

$$\frac{1}{(n_e)_4} = \frac{2}{3} = \frac{216}{324} \tag{34.49}$$

34.2 Interpretation

We have explained an example where the sources were divided into two classes and assigned to Group I and Group II of orthogonal array L_{27}. Let us assume here that the factorial effects of the class of Group II, compared with the class of Group I which includes the large sources, are, when totaled, about twice as great as the maximum factorial effects of the latter. If the largest source of the class of the greater sources is called A, the following equations hold for the expected values of variance. It will be assumed that there is no replication here.

$$E(V_A) = \sigma^2 + \sigma_\mu^2 + 9\sigma_A^2 \tag{34.50}$$
$$E(V_e) = \sigma^2 + \sigma_\mu^2$$

σ_μ^2 here is such that, when the population means under the 27 conditions of Group II are expressed as $\mu_1, \mu_2, \ldots, \mu_{27}$,

$$\sigma_\mu^2 = \frac{1}{26}\left[(\mu_1^2 + \mu_2^2 + \cdots + \mu_{27}^2) - \frac{(\mu_1 + \mu_2 + \cdots + \mu_{27})^2}{27}\right] = \frac{36}{26}\sigma_A^2 \tag{34.51}$$

It may therefore be assumed that the variance ratio is significant since it follows a distribution centered about

$$F = \frac{\sigma_\mu^2 + 9\sigma_A^2}{\sigma_\mu^2} = \frac{\frac{36}{26} + 9}{\frac{36}{26}} = 7.5 \tag{34.52}$$

assuming that σ_μ^2 is far greater than the error variance. Even if the factorial effects of Group II are randomized, effects of Group I which are greater than the greatest in Group II will be found more or less dependably. Therefore, if we estimate the significant effects of Group I and erase them from the data, since there is nothing among them which is of greater effect than the greatest in Group II, the greatest factorial effects in Group II and the effects of those subordinate thereto will be found dependably.

However, since calculation is difficult for such a correction method, recently the almost orthogonal array of the next section has been used almost exclusively. It is therefore advised that the method of the next section be used.

34.3 Almost Orthogonal Array

Table 34.18 is the almost orthogonal array L_{27}' which was constructed by Masao Ina.

We will assume that all of A, B, C, D, E, F, G, H, I, J, K, L, and M are at three levels and that N and O are at two levels. We will assume that we wish all of the main effects and the interaction $A \times B$. The interaction of column 1 and column 11 appears in columns 12 and 13. We will assume that the factors have been assigned as shown by Table 34.19.

We will assume that r replicated data have been obtained by each experiment of L_{27}'. Essentially, we will be constructing two sets of analysis of variance tables. The sources of columns 1–13 will be termed the sources of Class a and those of columns 1, 11, 12, and 13–22 will be termed the sources of Class b.

Probably in many cases it is adequate to perform these two sets of analyses of variance separately. But if one wishes to perform an analysis of variance which is somewhat more precise, assuming that large factorial effects are more numerous in Class a, one estimates those which have emerged as significant among the sources from C on and erases their effects from the data. For example, assuming that D is significant, three differences,

$$\bar{T} - \bar{D}_1,\ \bar{T} - \bar{D}_2,\ \bar{T} - \bar{D}_3$$

Table 34.18 Almost Orthogonal Array L_{27}'

No. \ Col.	1	2	3	4	5	6	7	8	9	10	11	12	13	14	15	16	17	18	19	20	21	22
1	1	1	1	1	1	1	1	1	1	1	1	1	1	3	3	3	2	2	2	1	1	1
2	1	1	1	1	2	2	2	2	2	2	2	2	2	3	3	3	3	3	3	2	2	2
3	1	1	1	1	3	3	3	3	3	3	3	3	3	1	1	1	2	2	2	2	2	2
4	1	2	2	2	1	1	1	2	2	2	3	3	3	2	2	2	3	3	3	1	1	1
5	1	2	2	2	2	2	2	3	3	3	1	1	1	1	1	1	3	3	3	3	3	3
6	1	2	2	2	3	3	3	1	1	1	2	2	2	1	1	1	1	1	1	1	1	1
7	1	3	3	3	1	1	1	3	3	3	2	2	2	2	2	2	2	2	2	3	3	3
8	1	3	3	3	2	2	2	1	1	1	3	3	3	3	3	3	1	1	1	3	3	3
9	1	3	3	3	3	3	3	2	2	2	1	1	1	2	2	2	1	1	1	2	2	2
10	2	1	2	3	1	2	3	1	2	3	1	2	3	1	2	3	3	1	2	3	1	2
11	2	1	2	3	2	3	1	2	3	1	2	3	1	3	1	2	3	1	2	2	3	1
12	2	1	2	3	3	1	2	3	1	2	3	1	2	3	1	2	1	2	3	3	1	2
13	2	2	3	1	1	2	3	2	3	1	3	1	2	1	2	3	2	3	1	2	3	1
14	2	2	3	1	2	3	1	3	1	2	1	2	3	2	3	1	1	2	3	2	3	1
15	2	2	3	1	3	1	2	1	2	3	2	3	1	1	2	3	1	2	3	1	2	3
16	2	3	1	2	1	2	3	3	1	2	2	3	1	2	3	1	2	3	1	3	1	2
17	2	3	1	2	2	3	1	1	2	3	3	1	2	2	3	1	3	1	2	1	2	3
18	2	3	1	2	3	1	2	2	3	1	1	2	3	3	1	2	2	3	1	1	2	3
19	3	1	3	2	1	3	2	1	3	2	1	3	2	2	1	3	1	3	2	2	1	3
20	3	1	3	2	2	1	3	2	1	3	2	1	3	3	2	1	3	2	1	2	1	3
21	3	1	3	2	3	2	1	3	2	1	3	2	1	2	1	3	3	2	1	1	3	2
22	3	2	1	3	1	3	2	2	1	3	3	2	1	3	2	1	1	3	2	3	2	1
23	3	2	1	3	2	1	3	3	2	1	1	3	2	3	2	1	2	1	3	1	3	2
24	3	2	1	3	3	2	1	1	3	2	2	1	3	1	3	2	1	3	2	1	3	2
25	3	3	2	1	1	3	2	3	2	1	2	1	3	2	1	3	2	1	3	3	2	1
26	3	3	2	1	2	1	3	1	3	2	3	2	1	1	3	2	2	1	3	2	1	3
27	3	3	2	1	3	2	1	2	1	3	1	3	2	1	3	2	3	2	1	3	2	1

Note: Columns 1, 11, 12, and 13 are orthogonal mutually as well as with columns 2–10 and with every one of the columns 14–22. Columns 2–10 are mutually orthogonal but are not orthogonal with columns 14–22. However, even in the worst case, the levels of columns 14–22 confound in the proportions of 2, 2, and 5. Columns 14–22 are mutually orthogonal.

are to be added to all of the data of D_1, D_2, and D_3. Correction of A, B, and $A \times B$ is unnecessary. Or, to word it differently, the process average for No. 1–No. 27 is estimated by the significant sources among $C, D, \ldots, K$, and we find the difference between this value $\hat{\mu}_i$ and $\overline{T}$. The values obtained by adding the difference,

$$\overline{T} - \hat{\mu}_i \qquad (i=1,\ 2,\ \cdots,\ 27)$$

Table 34.19 Assignment by L_{27}'

Source	A	C	D	E	F	G	H	I	J	K	B	$A \times B$	$A \times B$	L	M	N	O	e	e	e	e	e
Column	1	2	3	4	5	6	7	8	9	10	11	12	13	14	15	16	17	18	19	20	21	22

Tables 34.20

Analysis of Variance Table for Class *a*

Source	f
A	2
B	2
$A \times B$	4
C	2
D	2
E	2
F	2
G	2
H	2
I	2
J	2
K	2
e_1	0
e_2	$27(r-1)$
T	$27r-1$

Analysis of Variance Table for Class *b*

Source	f
A	2
B	2
$A \times B$	4
L	2
M	2
N	1
O	1
e_1	12
e_2	$27(r-1)$
T	$27r-1$

to the data of No. 1 – No. 27 are now regarded as the data and we perform an analysis of the factorial effects of Class *b*. The procedure for this is exactly the same as in the case of the randomly combined design.

35 Compounding Technique

An assignment method that can be used when certain types of technical information exist regarding the relationship between the target characteristics and the sources will be discussed. In particular, Section 35.1 is important for promoting the efficiency of an assignment in many cases.

35.1 When Sources Differ According to the Characteristic Values

When conducting a length of life test of a copier, for example, one usually wishes to study the reliability of the machine operation and the reliability of the copy quality, together. Usually the *sources for the reliability of the machine and the sources for the copy quality are different*. In such a case, it is possible to assign the sources related with the former and the sources related with the latter doubled on the same orthogonal array. We analyze by assuming that the sources of the orthogonal array are related with the machine when data-analyzing the machine characteristics, and by regarding them as sources related with the copy quality when data-analyzing the copy quality. Thus, it becomes possible to perform two types of study with one orthogonal array.

Table 35.1 Factors and Characteristic Values
(O marks: could be related)

	A	B	C	D	E	F	G	H	I	J	K	L	M	N	O	P	Q	R	S	T	U	V	W
x	O	O	O	O	O	O	O	O	O	O													
y	O	O				O	O	O		O	O	O									O		
z	O	O											O	O	O	O	O	O	O	O	O		O
u	O				O									O		O	O	O	O	O	O	O	O

Researchers are prone to focus on the one characteristic value that is posing a problem. It is astonishing how much this hinders the efficiency of experimental research. One should simultaneously consider all characteristic values related with function and product quality, and should consider factors that are believed to influence these characteristic values. Then one constructs a chart such as Table 35.1 and does the assignment. Let us assume that there are the characteristic values x, y, z, and u and that the factors number 23, *viz.*, $A, B, \ldots, W$. The O marks are factors that have influence or that might have influence, and those with no mark are factors that clearly have no influence. To simplify the explanation, we will assume that all of the factors are at two levels. The method is exactly the same even if the level number is large. Let us assign to orthogonal array L_{16} so as to be able to find all of the main effects with the O marks and the interactions $A \times B$ and $A \times E$ for x, $A \times F$ for y, $A \times M$ for z, and $A \times E$ for u.

Even though 42 main effects and four interactions are being sought, they can be assigned neatly to orthogonal array L_{16}, as shown by Table 35.2, and there are still plenty of extra columns. It is exactly the same when x, y, z, and u are not four characteristic values but are groups of characteristic values.

An analysis of variance table is constructed separately for each of x, y, z, and u. For example, for x and z, they become as shown by Tables 35.3. It is assumed that the number of replications is two for x and three for z.

Table 35.2 Assignment by Compounding Technique

x	A	B	$A\times B$	C	D	E	$A\times E$	F	I	G	H	J	e	e	e				
y	A	B	U	K	L	e	e	F	$A\times F$	G	H	J	e	e	e				
z	A	B	U	R	S	M	$A\times M$	T	W	N	O	P	Q	e	e				
u	A	e	U	R	S	E	$A\times E$	T	W	N	V	P	Q	e	e				
No. \ Col.	1	2	3	4	5	6	7	8	9	10	11	12	13	14	15	x	y	z	u
1	1	1	1	1	1	1	1	1	1	1	1	1	1	1	1				
2	1	1	1	1	1	1	1	2	2	2	2	2	2	2	2				
⋮	⋮	⋮	⋮	⋮	⋮	⋮	⋮	⋮	⋮	⋮	⋮	⋮	⋮	⋮	⋮				
16	2	2	1	2	1	1	2	2	1	1	2	1	2	2	1				

Tables 35.3 Analysis of Variance Tables for x and z

Analysis of Variance Table for x

Source	f
A	1
B	1
$A \times B$	1
C	1
D	1
E	1
$A \times E$	1
F	1
I	1
G	1
H	1
J	1
e_1	3
e_2	16
T	31

Analysis of Variance Table for z

Source	f
A	1
B	1
U	1
R	1
S	1
M	1
$A \times M$	1
T	1
W	1
N	1
O	1
P	1
Q	1
e_1	2
e_2	32
T	47

35.2 The Case of One Factor and Many Levels

Suppose the question is, are there any among 1000 types of mold which are effective against cancer, and if there are, which are they? Technically speaking, something such as the following can be considered. Usually, *most of the 1000 types are harmless and also valueless against cancer*, and the substances that are effective or that are negative in their effects number at most several types if they exist at all. Let us assume that we wish to generate cancer artificially in rabbits and to investigate which types are effective. We dose 100 rabbits with ten types each of mold and watch their conditions. If it seems that the cancer of five of the rabbits is, say, improving, we regard the 50 types of molds that were used for those five rabbits as being potentially curative, and we generate cancer in 50 new rabbits and investigate. Then we select several types among the molds which appear to be good, and this time, so as also to find out about the dosing method, quantity of dose, combined use, and other factors, we experiment by using, for example, 32 rabbits.

A method such as this is commonly used in experiments of one factor and many levels; although occasionally it could happen that effects of certain types and certain other types become mutually cancelled, since it can be proved that the level of significance of such a happening is very small, would it not be best to use a method such as this more courageously? Details of the calculation of this level of significance will be omitted.

35.3 When the Qualitative Tendency Is Determinable

For example, regarding three types of parts, A, B, and C, we will assume that in every case those of the first level have little eccentricity and those of the second level have considerable eccentricity. We form the following two-level factor (ABC):

$$(ABC)_1 = A_1B_1C_1 \qquad (ABC)_2 = A_2B_2C_2$$

If, when assembled products are produced by assigning these to an orthogonal array, it is found that their effect on the magnitude of noise is not significant, we may assume that there is no effect by any of the three factors. This is because whichever the part, it cannot happen that the second level is better than the first level. Thus, suppose that there is no difference with the second level as compared with the first level, or if there is the noise becomes better, and it turns out that there is no significant difference when they are assigned compounded. That shows in one sweep that none of the three has any effect. If the difference is great, it is necessary to separate the effect of one of them, but regarding this, the reader is referred to the direct sum design of Chapter 13 of Volume 1.

36 Incomplete Design

A general account of incomplete design will be presented in this chapter. However, since it is convenient to use the junction technique except in special cases such as sports and games, it is wise to avoid using the method of this chapter as much as possible. Beginners are advised to read starting with the example of Section 36.3. It might be interesting to use the cyclic design of Section 36.2 in connection with games. This is because it is possible to rank all of the participants by a number of games which is one round more than the number in games of the tournament. Theoretically, by the junction technique it is possible to rank them all by the same number as in the tournament.

36.1 Balanced Incomplete Design (BIB, League Battle System)

36.1.1 Balanced Incomplete Block Design

An assignment that renders the precision of comparison among all treatments equal under a given block size is, generally speaking, termed the *balanced incomplete block design.* There are many problems concerning the actual usefulness of such an assignment, and it would seem better for an experimenter to avoid it as much as possible. But since it might be unavoidable in certain circumstances, we will introduce here several assignments that might conceivably be used.

We will assume that in an experiment of the 2^{12} type, we have assigned the necessary sources to $L_{16}(2^{15})$. If we desire 12 main effects and three two-factor interactions, 15 columns are filled, so that there is no margin for introduction of block factors. How should we proceed to cause the block size to be four when we wish to compare 16 modes of treatment T_1, T_2, . . . , T_{16}?

Of course, even if there are no extra columns, since in the partial confounding method a part of the sources are confounded with the blocks, if we confound such confoundings with the same frequency as all of the sources, it should be possible to create an assignment with uniform confounding. For example, in the case of this problem, let us try using type 5 of assignment by L_{16} (Figure 36.1).

First, if block R is caused to correspond to the three columns 1, 2, and 3, the sources corresponding to columns 1, 2, and 3 confound with R. Next, if the three columns 4, 8, and 12 and R are confounded, the sources corresponding to these columns confound with R. If we proceed thereafter as in Table 36.1, drawing up an assignment which has been confounded with R five times, the original objective will be achieved since all of the sources are confounded with R with the same frequency.

It can be seen that all treatments show their faces just once each in the same blocks as the other treatments. An assignment such as this is termed a *balanced incomplete block design*.

The analysis of variance table, Table 36.2, is a case of analysis by intra-block information, as will be explained later. *Since the effective number of replications is not very large with the incomplete blocks compared with when there are no blocks, the problem of recovering inter-block information is not as important as in the case of analysis of covariance.* Therefore, except for special cases, as for example those where block effect is not significant, it is probably unnecessary to recover inter-block information.

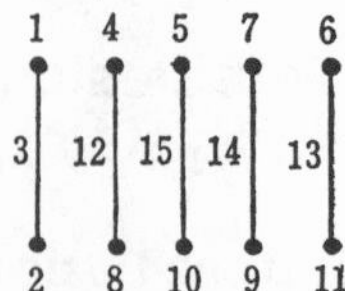

FIGURE 36.1
Assignment Type 5

Table 36.1 Assignment Whereby 16 Types of Treatment Have Been Confounded with Block Factors of Size 4
[Numerical values are experiment numbers (treatment numbers)]

R_1	R_2	R_3	R_4	R_5	R_6	R_7	R_8	R_9	R_{10}	R_{11}	R_{12}	R_{13}	R_{14}	R_{15}	R_{16}	R_{17}	R_{18}	R_{19}	R_{20}
1	5	9	13	1	2	3	4	1	2	3	4	1	2	3	4	1	2	3	4
2	6	10	14	5	6	7	8	6	5	8	7	7	8	5	6	8	7	6	5
3	7	11	15	9	10	11	12	11	12	9	10	12	11	10	9	10	9	12	11
4	8	12	16	13	14	15	16	16	15	14	13	14	13	16	15	15	16	13	14

Since the calculation methods for the orthogonal arrays can be used straightforwardly as the method of analysis in the foregoing case, an explanation is probably unnecessary.

Although it is possible to construct an incomplete block design by using an orthogonal array when the number of treatments is 16, 32, 9, or 27, this is usually not the case. Of course, when the factors are numerous, probably the case of other numbers of treatments is irrelevant since an orthogonal array will be used, but if there is only one factor, sometimes an assignment by BIB becomes necessary. Several assignments of balanced incomplete blocks are shown in Tables 36.3.

Several assignments which might be useful have been shown in Tables 36.3, but often balanced incomplete assignments are not practical since the size of the experiment usually becomes enormous. At work sites in plants, L_{16}, L_{27}, and L_{32} are assignments of a size which can barely be used; to replicate them, or experiments of a size over 50, is rather difficult.

For example, in the case of the 5×3 type, if it is necessary to have the block size be 3, when it is possible to disregard interactions it is possible to use $L_9(3^4)$ and to assign three levels among the five levels and a three-level factor and a three-level block. The remaining two levels among the five levels are handled by partial supplementing according to Chapter 11 of Volume 1. The size of the experiment is then $9 + 6 = 15$, and is far smaller than the 105 of Assignment (16).

Table 36.2 Number of Degrees of Freedom and Effective Number of Replications

Source	f	n_e
Main effect	12	32
Two-factor interaction	3	16
Block (unadjusted)	19	
e	45	
Total	79	

Tables 36.3 Balanced Incomplete Assignment Arrays

v: number of treatments
k: block size
b: levels of block factor
r: number of replications of treatment
λ: frequency with which two treatments meet in the same block
E: (effective number of replications of treatment)/(number of replications of treatment)

Assignment (1) $v = 6, k = 2, b = 15, r = 5, \lambda = 1, E = 0.60$

R_1	R_2	R_3	R_4	R_5	R_6	R_7	R_8	R_9	R_{10}	R_{11}	R_{12}	R_{13}	R_{14}	R_{15}
1	1	1	1	1	2	2	2	2	3	3	3	4	4	5
2	3	4	5	6	3	4	5	6	4	5	6	5	6	6

Note: Every combination of two is performed by each block. The assignment for the case of $k = 2$ is omitted since regardless of the number of treatments, it is necessary to perform every combination in order to cause the block size to become 2.

Assignment (2) $v = 6, k = 3, b = 10, r = 5, \lambda = 2, E = 0.80$

R_1	R_2	R_3	R_4	R_5	R_6	R_7	R_8	R_9	R_{10}
1	1	1	1	1	2	2	2	3	4
2	2	3	3	4	3	3	4	5	5
5	6	4	6	5	4	5	6	6	6

Assignment (3) $v = 7, k = 2, b = 21, r = 6, \lambda = 1, E = 0.58$
(Omitted since block size is 2)

Assignment (4) $v = 7, k = 3, b = 7, r = 3, \lambda = 1, E = 0.78$

	R_1	R_2	R_3	R_4	R_5	R_6	R_7
V_1	1	2	3	4	5	6	7
V_2	2	3	4	5	6	7	1
V_3	4	5	6	7	1	2	3

Assignment (5) $v = 7, k = 4, b = 7, r = 4, \lambda = 2, E = 0.87$

	R_1	R_2	R_3	R_4	R_5	R_6	R_7
V_1	1	2	3	4	5	6	7
V_2	7	1	2	3	4	5	6
V_3	6	7	1	2	3	4	5
V_4	4	5	6	7	1	2	3

Assignment (6) $v = 8, k = 2, b = 28, r = 7, \lambda = 1, E = 0.58$
(Every combination of two)

Assignment (7) $v = 8, k = 4, b = 14, r = 7, \lambda = 3, E = 0.86$

R_1	R_2	R_3	R_4	R_5	R_6	R_7	R_8	R_9	R_{10}	R_{11}	R_{12}	R_{13}	R_{14}
1	5	1	3	1	3	1	2	1	2	1	2	1	2
2	6	2	4	2	4	3	4	3	4	4	3	4	3
3	7	5	7	7	5	5	6	6	5	5	6	6	5
4	8	6	8	8	6	7	8	8	7	8	7	7	8

(From $L_8(2^7)$)

Assignment (8) $v=9, k=2, b=36, r=8, \lambda=1, E=0.56$
(Every combination of two)

Assignment (9) $v=9, k=3, b=12, r=4, \lambda=1, E=0.75$

R_1	R_2	R_3	R_4	R_5	R_6	R_7	R_8	R_9	R_{10}	R_{11}	R_{12}
1	4	7	1	2	3	1	2	3	1	2	3
2	5	8	4	5	6	6	4	5	5	6	4
3	6	9	7	8	9	8	9	7	9	7	8

(From the respective columns of $L_9(3^4)$)

Assignment (10) $v=9, k=4, b=18, r=8, \lambda=3, E=0.83$

R_1	R_2	R_3	R_4	R_5	R_6	R_7	R_8	R_9	R_{10}	R_{11}	R_{12}	R_{13}	R_{14}	R_{15}	R_{16}	R_{17}	R_{18}
1	1	1	1	1	1	1	1	2	2	2	2	2	3	3	3	4	4
2	2	2	3	3	4	4	5	3	3	4	5	6	4	5	6	5	6
3	5	7	5	6	6	8	7	8	4	5	6	7	5	8	7	7	7
4	6	8	7	9	8	9	9	9	7	9	8	9	6	9	8	8	9

Assignment (11) $v=10, k=2, b=45, r=9, \lambda=1, E=0.55$
(Combinations of two)

Assignment (12) $v=10, k=3, b=30, r=9, \lambda=2, E=0.74$

R_1	R_2	R_3	R_4	R_5	R_6	R_7	R_8	R_9	R_{10}	R_{11}	R_{12}	R_{13}	R_{14}	R_{15}
1	1	1	1	1	1	1	1	1	2	2	2	2	2	2
2	2	3	4	5	6	7	8	9	3	4	5	5	6	7
3	4	5	6	7	8	9	10	10	6	10	8	9	7	9

R_{16}	R_{17}	R_{18}	R_{19}	R_{20}	R_{21}	R_{22}	R_{23}	R_{24}	R_{25}	R_{26}	R_{27}	R_{28}	R_{29}	R_{30}
2	3	3	3	3	3	3	4	4	4	4	5	5	6	6
8	4	4	5	7	8	9	5	5	6	7	6	7	7	8
10	7	8	6	10	9	10	9	10	9	8	10	8	10	9

Assignment (13) $v=10, k=4, b=15, r=6, \lambda=2, E=0.83$

R_1	R_2	R_3	R_4	R_5	R_6	R_7	R_8	R_9	R_{10}	R_{11}	R_{12}	R_{13}	R_{14}	R_{15}
1	1	1	1	1	1	2	2	2	2	3	3	3	4	4
2	2	3	4	5	6	3	4	5	7	5	6	4	5	6
3	5	7	9	7	8	6	7	8	8	9	7	5	6	8
4	6	8	10	9	10	9	10	10	9	10	10	8	7	9

Again, the five-level factor is divided into two classes, as follows.

$$A'\begin{cases}A_1\\A_2\\A_3\end{cases}\qquad A''\begin{cases}A_3\\A_4\\A_5\end{cases}$$

We assign A' and B at three levels to L_9, and it is possible to introduce a three-level block factor therein. If, further, we replicate L_9 with A'', B, and a three-level block, it becomes possible to estimate one-half of A', A'', and B and $A' \times B$ and $A'' \times B$ in 18 runs, excluding block effect. Therefore, even though this is a partial confounding design, it is possible to estimate all effects if the experiment is performed 36 times. Moreover, it is almost sufficient to know the calculation methods for an orthogonal array. Such a method of dividing the main effects, etc., of certain factors into a number of classes and of so arranging that it is possible to compare inter-class differences by means of levels in common is nothing other than a special case of the method termed the pseudo-factor design in Chapter 10 of Volume 1.

36.1.2 Analysis Method for Balanced Incomplete Blocks

There are the following three calculation methods for assignments by balanced incomplete blocks.

(1) Calculation by intra-block information
(2) Method of using inter-block information by the weight of the ratio of inter-block variance to intra-block variance
(3) Sequential correction method

Only calculation by method (1) will be shown.

(a) The total per level of each treatment is drawn up. The unit number is r.
(b) To perform estimation for treatment at the levels T_i, since it is necessary to correct the block effect where T_i is contained, we find the total S_i of treatment among the blocks containing T_i. The unit number of S_i is (block size − 1) times T_i. Since the block size is k, we compute

$$W_i=(k-1)T_i-S_i \tag{36.1}$$

Since clearly block effects are contained in the same number in $(k-1)T_i$ and S_i, there is no block effect in their difference W_i. If the main effect of treatment T_i is expressed as t_i, t_i is contained in $(k-1)T_i$ $(k-1)r$ times, and since $t_1, t_2, \ldots, t_{i-1}, t_{i+1}, \ldots, t_v$ are contained a total of λ times in S_i, if we solve the relationship equation,

$$t_1+t_2+\cdots+t_i+\cdots+t_v=0$$

with respect to t_i, in effect it means that t_i is contained $-\lambda$ times. Now, for the balanced incomplete design the following equations hold, as should be evident upon a bit of reflection.

$$\left.\begin{aligned} vr&=kb \\ \lambda&=\frac{(k-1)r}{v-1} \end{aligned}\right\} \tag{36.2}$$

If the relationship above is used, it means that t_i is contained $[(k-1)r+\lambda)]$ times, or in other words $\dfrac{(k-1)rv}{v-1}$ times, in W_i. Thus, the expected value of W_i is

$$E(W_i)=\frac{(k-1)rv}{v-1}t_i \tag{36.3}$$

Also, since the variance of W_i is

$$\sigma^2(W_i)=[(k-1)^2r+(k-1)r]\sigma^2=(k-1)rk\sigma^2 \tag{36.4}$$

we have

$$E(W_i^2)=(k-1)r\left[k\sigma^2+(k-1)r\left(\frac{v}{v-1}\right)^2t_i^2\right] \tag{36.5}$$

Therefore, the expected value of the sum of squares of W_i is

$$E(\sum_i W_i^2)=(k-1)r\left[kv\sigma^2+(k-1)r\left(\frac{v}{v-1}\right)^2\Sigma t_i^2\right]$$

$$=(k-1)r\left[kv\sigma^2+(k-1)r\frac{v^2}{v-1}\sigma_T^2\right] \tag{36.6}$$

Hence, if we calculate the variation of the sum of treatments S_T by

$$S_T=\frac{v-1}{kvr(k-1)}\sum_{i=1}^{v}W_i^2 \tag{36.7}$$

the expected value of the mean square after dividing S_T by the number of degrees of freedom $v-1$ is

$$E\left(\frac{S_T}{v-1}\right)=\sigma^2+\frac{k-1}{k}\,\frac{v}{v-1}r\sigma_T^2 \tag{36.8}$$

If the block size is equal to v, the coefficient of σ_T^2 should become r. Since k is smaller than v,

$$E_C\equiv\frac{k-1}{k}\,\frac{v}{v-1}\leq 1 \tag{36.9}$$

is smaller than 1, and it is termed the *efficiency factor*. For the *effective number of replications for treatment T*, n_e, we need only multiply the number of replications when there are no blocks, r, by the aforementioned efficiency factor E_C.

$$n_e=E_C\times r \tag{36.10}$$

The inter-block uncorrected variation S_R in Table 36.4 is found by calculating with the treatment effect still confounded.

$$S_R=\frac{1}{k}(R_1^2+R_2^2+\cdots+R_b^2)-\text{CF} \tag{36.11}$$

Table 36.4 Analysis of Variance Table

Source	f
Inter-treatment	$v-1$
Inter-block (uncorrected)	$b-1$
Error	$vr-v-b+1$
Total	$vr-1$

The residual sum of squares S_e is found by subtracting S_T and S_R from the total variation.

$$S_e = \text{total variation} - S_R - S_T \tag{36.12}$$

To resolve the inter-treatment effect into the main effects and interactions of various factors, one need only proceed as follows. We will explain by assuming $v = 12$ for simplicity, with Table 36.5 of the $2^2 \times 3$ type.

Main effect A is found from the following quantity. Since the correction factor is 0, with K as an appropriate constant, we need only arrange that the expected value of the mean square of

$$S_A = K \times (A_1^2 + A_2^2)$$

becomes

$$E(S_A) = \sigma^2 + n_A \sigma_A^2 \tag{36.13}$$

n_A is the effective number of replications of main factor A, and in the case of a balanced incomplete block it is given by

$$n_A = (\text{effective number of replications of } T) \times \frac{v}{\text{level of } A}$$

$$= (\text{number of replications of main effect } A \text{ when there is no block}) \times \text{efficiency factor}$$

$$= r\frac{v}{a}E_C \tag{36.14}$$

Since in this case $v = 12$ and $a = 2$, we have $n_A = 6rE_C$.

The coefficient K can be calculated as follows. The method of calculating by returning the variances of the respective W_i to the original values is troublesome since they are not generally independent. We use the fact that the effective number of replications is given by formula (36.14). For

Table 36.5

A	B	C	T	A	B	C	T	A	B	C	T	A	B	C	T	A	B	C	T	A	B	C	T
1	1	1	1	1	1	3	3	1	2	2	5	2	1	1	7	2	1	3	9	2	2	2	11
1	1	2	2	1	2	1	4	1	2	3	6	2	1	2	8	2	2	1	10	2	2	3	12

example, since t_6 is the population mean for $A_1B_2C_3$, we have the correspondence

$$t_6 = a_1 + b_2 + c_3 + I_{12} + I_{23} + I_{13} + I_{123}$$

Therefore, if we draw up the totals for A_1 and A_2 much as though it were a three-way layout of A, B, and C, only six of a_1 or a_2 remain and we have

$$E(A_1) = \frac{(k-1)rv}{v-1} \times 6 \times a_1$$

$$E(A_2) = \frac{(k-1)rv}{v-1} \times 6 \times a_2$$

Therefore, the expected value of the part of main effect A of $A_1^2 + A_2^2$ becomes

$$\left[\frac{(k-1)rv}{v-1}\right]^2 (6^2 a_1^2 + 6^2 a_2^2) = \left[\frac{(k-1)rv}{v-1}\right]^2 \left(\frac{v}{a}\right)^2 \sigma_A^2 \qquad (36.15)$$

and for the coefficient of σ_A^2 to agree with the value of n_A,

$$K\left[\frac{(k-1)rv}{v-1}\right]^2 \left(\frac{v}{a}\right)^2 = \frac{rv}{a}\,\frac{k-1}{k}\,\frac{v}{v-1}$$

must hold. Therefore,

$$K = \frac{a(v-1)}{v^2 k(k-1)r} \qquad (36.16)$$

Since $a = 2$ and $v = 12$ in this case, it becomes

$$K = \frac{11}{72k(k-1)r} \qquad (36.17)$$

Therefore,

$$S_A = \frac{a(v-1)}{v^2 k(k-1)r}(A_1^2 + A_2^2) \qquad (36.18)$$

Since in the calculation of K the part which changes by level is only $(v/a)^2$, we have

$$\left.\begin{aligned} S_B &= \frac{b(v-1)}{v^2 k(k-1)r}(B_1^2 + B_2^2) \\ S_C &= \frac{c(v-1)}{v^2 k(k-1)r}(C_1^2 + C_2^2 + C_3^2) \end{aligned}\right\} \qquad (36.19)$$

$$\left.\begin{aligned} S_{A\times B} &= \frac{ab(v-1)}{v^2 k(k-1)r}(A_1B_1^2 + A_1B_2^2 + A_2B_1^2 + A_2B_2^2) - S_A - S_B \\ S_{A\times C} &= \frac{ac(v-1)}{v^2 k(k-1)r}(A_1C_1^2 + A_1C_2^2 + A_1C_3^2 + A_2C_1^2 + A_2C_2^2 + A_2C_3^2) - S_A - S_C \\ S_{B\times C} &= \frac{bc(v-1)}{v^2 k(k-1)r}(B_1C_1^2 + B_1C_2^2 + B_1C_3^2 + B_2C_1^2 + B_2C_2^2 + B_2C_3^2) - S_B - S_C \end{aligned}\right\} \qquad (36.20)$$

$$S_{A\times B\times C}=\frac{abc(v-1)}{v^2k(k-1)r}(W_1^2+W_2^2+\cdots+W_v^2)-S_A-S_B-S_C$$

$$-S_{A\times B}-S_{A\times C}-S_{B\times C} \qquad (36.21)$$

To draw the curve for A, it is easier to draw the main effect directly. Its value is given, of course, by

$$\hat{a}_1=\frac{v-1}{(k-1)rv}\frac{a}{v}A_1 \qquad (36.22)$$

The confidence limits are given by

$$\pm\sqrt{\frac{F\times V_e}{n_e}} \qquad (n_e=rv/a) \qquad (36.23)$$

n_A is the effective number of replications of A. For combinations such as A_iB_j and so forth, the n value becomes

$$n_{AB}=r\frac{v}{ab}E_C \qquad (36.24)$$

It is the same for the others, too.

To find the adjusted block effect, it is exactly the same if we exchange the positions of block factor R and treatment factor T. A part of the confounded treatment effect is included in the unadjusted block factor. Thus, with R as the random factor, we have

$$E\left(\frac{S_R}{b-1}\right)=\sigma^2+k\sigma_R^2+\frac{r(v-k)}{(v-1)(vr-k)}\sigma_T^2 \qquad (36.25)$$

Thus, it is possible to estimate the effect of T from between blocks, as well. The error variance when so doing is, although intra-block error is σ^2,

$$\sigma^2+k\sigma_R^2 \qquad (36.26)$$

Finding main effect T and the variations by using this constitutes the recovering of inter-block information. The method of analysis for this is not difficult, but it will not be discussed in this book.

36.2 Cyclic Design

With the block size as 2, to compare treatments at v levels, $T_1, T_2, \ldots, T_v$, there is a method according to Table 36.6. $R_1, R_2, \ldots, R_v$ are the levels of the blocks.

Table 36.6 Cyclic Design

	R_1	R_2	R_3	$\cdots$	R_v
V_1	T_1	T_2	T_3	$\cdots$	T_v
V_2	T_2	T_3	T_4	$\cdots$	T_1

Table 36.7 Analysis of Variance Table

Source	f
V	1
T	$v-1$
R (uncorrected)	$v-1$
e	0
T	$2v-1$

Table 36.8 Assignment by Symmetry Method

R_1	R_2	$R_3 \cdots R_{v-1}$
T_1	T_1	$T_1 \cdots T_1$
T_2	T_3	$T_4 \cdots T_v$

The observation equations, with the data of R_iV_j as y_{ij}, are

$$\left.\begin{aligned} &m+v_1+r_1+t_1=y_{11} \\ &m+v_2+r_1+t_2=y_{12} \\ &m+v_1+r_2+t_2=y_{21} \\ &\vdots \qquad\qquad\quad \vdots \\ &m+v_2+r_v+t_1=y_{v2} \end{aligned}\right\} \tag{36.27}$$

where

$$v_1+v_2=0$$

$$r_1+r_2+\cdots+r_v=0$$

$$t_1+t_2+\cdots+t_v=0$$

It is possible to solve these equations since the number of unknowns is $2v$ and the number of data is also $2v$. Usually, a computer is necessary for the calculation. The analysis of variance table is as given in Table 36.7. The number of degrees of freedom of error is 0.

When, in the assignment of Table 36.6, the block size is 3, we take, cyclically, T_1, T_2, . . . , T_v for R_1; T_2, T_3, . . . , T_1 for R_2; and T_4, T_5, . . . , T_3 for R_3. These are assignments which also include the Youden square.

Pseudo-factor design is the same as the symmetry method, but Table 36.8 is used instead of Table 36.6. With that factor believed to be important among T_1, T_2, . . . , T_v as T_1, it is a method by which one compares T_1 with T_2, T_1 with T_3, . . . , T_1 with T_v. Please refer to Section 10.8 of Volume 1 for the analysis method.

36.3 Assignment by Incomplete Blocks (Example of Sensory Test)

36.3.1 Objective and Data

In comparing the tone quality of four types of record discs, A_1, A_2, A_3, and A_4, a total of 12 pairwise comparisons, *viz.*, A_1 is listened to first, then A_2;

Table 36.9 Coordinate Variable ω and Data

	ω_1	ω_2	ω_3	ω_4	Total
Bad	0	0	0	0	0
Somewhat bad	1	0	0	0	1
No difference	1	1	0	0	2
Somewhat good	1	1	1	0	3
Good	1	1	1	1	4

A_2 is listened to first, then A_1; . . . ; A_4 is listened to, then A_3, are carried out in random sequence. Since the 12 pairs of experiments were performed with four persons, D_1, D_2, D_3, and D_4, the total number of blocks is 48 since each person corresponds to 12 blocks. The data are categorized into five classes, *viz.*, compared with the former (record heard earlier), the latter was bad, somewhat bad, not different, somewhat good, and good.

If one disc among A_1, A_2, A_3, and A_4 or a competing product had been used as the standard and comparisons with it were ranked into five classes, this would have been a simple two-way layout experiment of A and D. In this sense, this example is one of an unskillful assignment for practical purposes, but the calculation will be explained as an example of BIB.

Coordinate variable ω is defined as in Table 36.9 in order to perform a calculation by the minute accumulating analysis of Chapter 32.

A:	type of record	4 levels
B:	sequence of experiment	B_1 = earlier, B_2 = later
D:	testing team members	4 levels
R:	blocks	12 levels per person

and with these, the data were as given by Table 36.10.

An experiment of 12 blocks was carried out by each person; the interblock number of degrees of freedom per person was 11 and for all of the four persons it was 44. In regard to the difference A among the records, its main effect and $A \times B$, $A \times D$, and $A \times B \times D$ are the effects we wish to find. Since this is minute accumulating analysis, further interactions of these with ω become a problem. Distribution of the number of degrees of freedom is therefore as given by Table 36.11.

36.3.2 Calculations of Variations

Correction for incomplete type is unnecessary for the effects of ω, D, and sequence of experiment B. The supplementary tables related to these are given by Tables 36.12.

Table 36.10 Data Readjusted for Minute Accumulation

R	A	B	D_1					D_2					D_3					D_4					Total				
			ω_1	ω_2	ω_3	ω_4	Total	ω_1	ω_2	ω_3	ω_4	Total	ω_1	ω_2	ω_3	ω_4	Total	ω_1	ω_2	ω_3	ω_4	Total	ω_1	ω_2	ω_3	ω_4	Total
1	1	1	1	1	0	0	2	1	1	0	0	2	1	1	0	0	2	1	1	0	0	2	4	4	0	0	8
	2	2	1	1	1	1	4	1	1	1	1	4	1	1	1	1	4	1	1	1	1	4	4	4	4	4	16
2	2	1	1	1	0	0	2	1	1	0	0	2	1	1	0	0	2	1	1	0	0	2	4	4	0	0	8
	1	2	1	0	0	0	1	0	0	0	0	0	0	0	0	0	0	0	0	0	0	0	1	0	0	0	1
3	1	1	1	1	0	0	2	1	1	0	0	2	1	1	0	0	2	1	1	0	0	2	4	4	0	0	8
	3	2	1	1	1	0	3	1	1	1	0	3	1	1	1	0	3	1	1	0	0	2	4	4	3	0	11
4	3	1	1	1	0	0	2	1	1	0	0	2	1	1	0	0	2	1	1	0	0	2	4	4	0	0	8
	1	2	1	0	0	0	1	0	0	0	0	0	1	1	1	0	3	1	0	0	0	1	3	1	1	0	5
5	1	1	1	1	0	0	2	1	1	0	0	2	1	1	0	0	2	1	1	0	0	2	4	4	0	0	8
	4	2	1	1	0	0	2	1	0	0	0	1	1	1	1	0	3	1	0	0	0	1	4	2	1	0	7
6	4	1	1	1	0	0	2	1	1	0	0	2	1	1	0	0	2	1	1	0	0	2	4	4	0	0	8
	1	2	0	0	0	0	0	1	0	0	0	1	1	0	0	0	1	1	1	1	0	3	3	1	1	0	5
7	2	1	1	1	0	0	2	1	1	0	0	2	1	1	0	0	2	1	1	0	0	2	4	4	0	0	8
	3	2	1	0	0	0	1	1	0	0	0	1	1	1	0	0	2	1	0	0	0	1	4	1	0	0	5
8	3	1	1	1	0	0	2	1	1	0	0	2	1	1	0	0	2	1	1	0	0	2	4	4	0	0	8
	2	2	1	1	1	1	4	1	1	1	0	3	1	1	0	0	2	1	1	1	1	4	4	4	3	2	13
9	2	1	1	1	0	0	2	1	1	0	0	2	1	1	0	0	2	1	1	0	0	2	4	4	0	0	8
	4	2	1	0	0	0	1	0	0	0	0	0	1	0	0	0	1	1	0	0	0	1	3	0	0	0	3
10	4	1	1	1	0	0	2	1	1	0	0	2	1	1	0	0	2	1	1	0	0	2	4	4	0	0	8
	2	2	1	1	1	1	4	1	1	1	1	4	1	1	0	0	2	1	1	1	0	3	4	4	3	2	13
11	3	1	1	1	0	0	2	1	1	0	0	2	1	1	0	0	2	1	1	0	0	2	4	4	0	0	8
	4	2	1	0	0	0	1	1	1	1	0	3	1	1	1	0	3	1	0	0	0	1	4	2	2	0	8
12	4	1	1	1	0	0	2	1	1	0	0	2	1	1	0	0	2	1	1	0	0	2	4	4	0	0	8
	3	2	1	1	0	0	2	1	1	1	0	3	1	1	1	0	3	1	1	1	0	3	4	4	3	0	11
			23	18	4	3	48	21	18	6	2	47	23	21	6	1	51	23	18	5	2	48	90	75	21	8	194

Table 36.11 Distribution of Degrees of Freedom

Source	f
D	3
Block (uncorrected)	44
A	3
B	1
$A\times B$	3
$A\times D$	9
$B\times D$	3
$A\times B\times D$	9
e_1	20
ω	3
$D\times\omega$	9
Block $\times\,\omega$	132
$A\times\omega$	9
$B\times\omega$	3
$A\times B\times\omega$	9
$A\times D\times\omega$	27
$B\times D\times\omega$	9
$A\times B\times D\times\omega$	27
e_2	60
T	383

From Tables 36.12 we have

$$S_\omega=\frac{90^2+75^2+21^2+8^2}{96}-\frac{194^2}{384}=148.23-98.01$$

$$=50.22 \tag{36.28}$$

$$S_D=\frac{48^2+47^2+51^2+48^2}{96}-CF=0.09 \tag{36.29}$$

Tables 36.12 Supplementary Tables Related with ω, D, and B

Three-Way Array of B, D, and ω

$r=12$	ω_1	ω_2	ω_3	ω_4	Total
B_1 D_1	12	12	0	0	24
D_2	12	12	0	0	24
D_3	12	12	0	0	24
D_4	12	12	0	0	24
Total	48	48	0	0	96
B_2 D_1	11	6	4	3	24
D_2	9	6	6	2	23
D_3	11	9	6	1	27
D_4	11	6	5	2	24
Total	42	27	21	8	98

Two-Way Array of D and ω

$r=24$	ω_1	ω_2	ω_3	ω_4	Total
D_1	23	18	4	3	48
D_2	21	18	6	2	47
D_3	23	21	6	1	51
D_4	23	18	5	2	48
Total	90	75	21	8	194

$$S_B=\frac{(96-98)^2}{384}=0.01 \tag{36.30}$$

$$S_{\omega\times D}=\frac{23^2+21^2+\cdots+2^2}{24}-CF-S_\omega-S_D=0.51 \tag{36.31}$$

$$S_{B\times D}=0.09 \tag{36.32}$$

$$S_{\omega\times B}=10.22 \tag{36.33}$$

$$S_{B\times D\times\omega}=0.51 \tag{36.34}$$

Next, in order to find the variations related with A, which is our objective, we find the statistics after having corrected the block effect, the W of Equation (36.1) for the level combinations of D and B and ω. For example, A and S of D_1B_1 become

Block		A_1				S			
		ω_1	ω_2	ω_3	ω_4	ω_1	ω_2	ω_3	ω_4
1	A_1B_1	1	1	0	0	1	1	1	1
3	A_1B_1	1	1	0	0	1	1	1	0
5	A_1B_1	1	1	0	0	1	1	0	0
Total		3	3	0	0	3	3	2	1

Table 36.13 comprises the supplementary tables for these values. However, T_i of Equation (36.1) becomes A_i in this case. Therefore, $A-S$ becomes 0, 0, -2, and -1.

A-related supplementary tables, Tables 36.14, are constructed from this. From Equation (36.7), we have

$$S_A=\frac{v-1}{kvr(k-1)}(A_1^2+A_2^2+A_3^2+A_4^2)$$

$$=\frac{4-1}{2\times4\times96\times(2-1)}[(-23)^2+33^2+1^2+(-11)^2]=\frac{1740}{256}$$

$$=6.80 \qquad (f=3) \tag{36.35}$$

$$S_{A\times B}=\frac{1}{256}\left[3^2+(-3)^2+(-5)^2+1^2-\frac{(-4)^2}{4}\right]$$

$$=0.16 \qquad (f=3) \tag{36.36}$$

$$S_{A\times\omega}=\frac{1}{64}[(-5)^2+(-8)^2+\cdots+(-2)^2]-S_A=0.95 \qquad (f=9) \tag{36.37}$$

$$S_{A\times B\times\omega}=\frac{1}{64}\left[5^2+12^2+\cdots+(-2)^2-\frac{3^2+(-3)^2+(-5)^2+1^2}{4}\right.$$

$$\left.-\frac{12^2+42^2+(-42)^2+(-16)^2}{4}+\frac{(-4)^2}{16}\right]$$

$$=0.56 \qquad (f=9) \tag{36.38}$$

Table 36.13 *A*-Related Calculation Tables

	A				*S*				*W*=*A*−*S*			
	ω_1	ω_2	ω_3	ω_4	ω_1	ω_2	ω_3	ω_4	ω_1	ω_2	ω_3	ω_4
$D_1B_1A_1$	3	3	0	0	3	3	2	1	0	0	−2	−1
〃 A_2	3	3	0	0	3	0	0	0	0	3	0	0
〃 A_3	3	3	0	0	3	1	1	1	0	2	−1	−1
〃 A_4	3	3	0	0	2	2	1	1	1	1	−1	−1
$D_1B_2A_1$	2	0	0	0	3	3	0	0	−1	−3	0	0
〃 A_2	3	3	3	3	3	3	0	0	0	0	3	3
〃 A_3	3	2	1	0	3	3	0	0	0	−1	1	0
〃 A_4	3	1	0	0	3	3	0	0	0	−2	0	0
$D_2B_1A_1$	3	3	0	0	3	2	2	1	0	1	−2	−1
〃 A_2	3	3	0	0	1	0	0	0	2	3	0	0
〃 A_3	3	3	0	0	2	2	2	0	1	1	−2	0
〃 A_4	3	3	0	0	3	2	2	1	0	1	−2	−1
$D_2B_2A_1$	1	0	0	0	3	3	0	0	−2	−3	0	0
〃 A_2	3	3	3	2	3	3	0	0	0	0	3	2
〃 A_3	3	2	2	0	3	3	0	0	0	−1	2	0
〃 A_4	2	1	1	0	3	3	0	0	−1	−2	1	0
$D_3B_1A_1$	3	3	0	0	3	3	3	1	0	0	−3	−1
〃 A_2	3	3	0	0	2	1	0	0	1	2	0	0
〃 A_3	3	3	0	0	3	3	2	0	0	0	−2	0
〃 A_4	3	3	0	0	3	2	1	0	0	1	−1	0
$D_3B_2A_1$	2	1	1	0	3	3	0	0	−1	−2	1	0
〃 A_2	3	3	1	1	3	3	0	0	0	0	1	1
〃 A_3	3	3	2	0	3	3	0	0	0	0	2	0
〃 A_4	3	2	2	0	3	3	0	0	0	−1	2	0
$D_4B_1A_1$	3	3	0	0	3	2	1	1	0	1	−1	−1
〃 A_2	3	3	0	0	2	0	0	0	1	3	0	0
〃 A_3	3	3	0	0	3	1	1	1	0	2	−1	−1
〃 A_4	3	3	0	0	3	3	3	0	0	0	−3	0
$D_4B_2A_1$	2	1	1	0	3	3	0	0	−1	−2	1	0
〃 A_2	3	3	3	2	3	3	0	0	0	0	3	2
〃 A_3	3	2	1	0	3	3	0	0	0	−1	1	0
〃 A_4	3	0	0	0	3	3	0	0	0	−3	0	0

$$S_{A\times D}=\frac{1}{64}[(-7)^2+9^2+\cdots+(-6)^2]-S_A=0.82 \qquad (f=9) \tag{36.39}$$

$$S_{A\times D\times\omega}=\frac{1}{16}[(-1)^2+(-3)^2+\cdots+0^2]-S_A-S_{A\times D}-S_{A\times\omega}$$

$$=1.18 \qquad (f=27) \tag{36.40}$$

Table 36.14 *A*-Related Supplementary Table *W* Values

$r=16$		B_1+B_2				Total	B_1-B_2				Total
		ω_1	ω_2	ω_3	ω_4		ω_1	ω_2	ω_3	ω_4	
D_1	A_1	−1	−3	−2	−1	7	1	3	−2	−1	1
	A_2	0	3	3	3	9	0	3	−3	−3	−3
	A_3	0	1	0	−1	0	0	3	−2	−1	0
	A_4	1	−1	−1	−1	−2	1	3	−1	−1	2
Total		0	0	0	0	0	2	12	−8	−6	0
D_2	A_1	−2	−2	−2	−1	−7	2	4	−2	−1	3
	A_2	2	3	3	2	10	2	3	−3	−2	0
	A_3	1	0	0	0	1	1	2	−4	0	−1
	A_4	−1	−1	−1	−1	−4	1	3	3	−1	0
Total		0	0	0	0	0	6	12	−12	−4	2
D_3	A_1	−1	−2	−2	−1	−6	1	2	−4	−1	−2
	A_2	1	2	1	1	5	1	2	−1	−1	1
	A_3	0	0	0	0	0	0	0	−4	0	−4
	A_4	0	0	1	0	1	0	2	−3	0	−1
Total		0	0	0	0	0	2	6	−12	−2	−6
D_4	A_1	−1	−1	0	−1	−3	1	3	−2	−1	1
	A_2	1	3	3	2	9	1	3	−3	−2	−1
	A_3	0	1	0	−1	0	0	3	−2	−1	0
	A_4	0	−3	−3	0	−6	0	3	−3	0	0
Total		0	0	0	0	0	2	12	−10	−4	0
Total	A_1	−5	−8	−6	−4	−23	5	12	−10	−4	3
	A_2	4	11	10	8	33	4	11	−10	−8	−3
	A_3	1	2	0	−2	1	1	8	−12	−2	−5
	A_4	0	−5	−4	−2	−11	2	11	−10	−2	1
Total		0	0	0	0	0	12	42	−42	−16	−4

$$S_{A\times B\times D}=\frac{1}{64}\left[1^2+(-3)^2+0^2+\cdots+0^2-\frac{0^2+2^2+(-6)^2+0^2}{4}\right.$$

$$\left.-\frac{3^2+(-3)^2+(-5)^2+1^2}{4}+\frac{(-4)^2}{16}\right]=0.44 \qquad (f=9) \tag{36.41}$$

$$S_{A\times B\times D\times\omega}=\frac{1}{16}\left[1^2+3^2+(-2)^2+(-1)^2+\cdots+0^2-\frac{1^2+(-3)^2+\cdots+0^2}{4}\right.$$

$$-\frac{2^2+12^2+\cdots+(-4)^2}{4}-\frac{5^2+4^2+\cdots+2^2}{4}+\frac{6^2+2^2+(-6)^2+0^2}{16}$$

$$\left.+\frac{12^2+42^2+\cdots+(-16)^2}{16}+\frac{3^2+(-3)^2+(-5)^2+1^2}{16}-\frac{(-4)^2}{64}\right]$$

$$=0.595 \qquad (f=27) \tag{36.42}$$

Next, as to block effect (uncorrected) R, in Table 36.10 we draw up the sum of B_1 and B_2 for each block and calculate as follows.

$$S_R=\frac{1}{8}[(2+4)^2+(2+1)^2+(2+3)^2+\cdots+(2+3)^2]-\frac{194^2}{384}-S_D$$

$$=10.40 \qquad (f=44) \tag{36.43}$$

$$S_{R\times\omega}=\frac{1}{2}[(1+1)^2+(1+1)^2+(0+1)^2+(0+1)^2+\cdots+(0+0)^2]$$

$$-CF-S_R-S_\omega-S_{D\times\omega}=3.86 \qquad (f=132) \tag{36.44}$$

Inter-experiment variation S_{T_1} and total variation S_T are

$$S_{T_1}=\frac{2^2+4^2+2^2+\cdots+3^2}{4}-CF=20.99 \qquad (f=95) \tag{36.45}$$

$$S_T=194-\frac{194^2}{384}=95.99 \qquad (f=383) \tag{36.46}$$

36.3.3 Analysis of Variance Table

The analysis of variance table, Table 36.15, is obtained from the foregoing calculations.

Table 36.15 Analysis of Variance Table

Source	f	S	V
D	3	0.09	0.030
R (block)	44	10.40	0.236
A	3	6.80	2.267
B	1	0.01	0.010
$A\times B$	3	0.16	0.053
$A\times D$	9	0.82	0.091
$B\times D$	3	0.09	0.030
$A\times B\times D$	9	0.44	0.048
e_1	20	2.18	0.109
ω	3	50.22	16.740
$D\times\omega$	9	0.51	0.057
$R\times\omega$	132	3.86	0.029
$A\times\omega$	9	0.95	0.106
$B\times\omega$	3	10.22	3.407
$A\times B\times\omega$	9	0.56	0.062
$A\times D\times\omega$	27	1.18	0.044
$B\times D\times\omega$	9	0.51	0.057
$A\times B\times D\times\omega$	27	0.60	0.120
e_2	60	6.39	0.106
T	383	95.99	

Table 36.16 Adjusted Analysis of Variance Table

Source	f	S	V	F_0	ρ[%]
$R(D)$	47	10.49	0.223	4.00**	8.2
A	3	6.80	2.267	40.63**	6.9
ω	3	50.22	16.740	300.00**	52.1
$B\times\omega$	3	10.22	3.407	61.06**	10.5
e	327	18.26	0.0558		22.3
T	383	95.99			100.0

That only main effect A is significant and all of the interactions with A are not significant indicates, in view of Table 36.15, that the superiority-inferiority of A_1, A_2, A_3, and A_4 is constant, regardless of the testing team members D and experiment sequence B. The readjusted analysis of variance table, leaving only the large sources, is as given by Table 36.16.

36.3.4 Conclusions

In this instance, two important sources are only those related with A. Since only main effect A is significant, it is indicated that only the mean values differ among A_1, A_2, A_3, and A_4. This means that it is sufficient if we compare the mean values of grading, with the marks:

Bad = −2
Somewhat bad = −1
No difference = 0
Somewhat good = 1
Good = 2

$$\bar{A}_1=\frac{-23}{64}=-0.36\pm0.12$$

$$\bar{A}_2=\frac{33}{64}=0.52\pm0.12$$

$$\bar{A}_3=\frac{1}{64}=0.02\pm0.12$$

$$\bar{A}_4=\frac{-11}{64}=-0.17\pm0.12$$

The confidence limits were found as follows.

$$\pm\sqrt{3.87\times\frac{0.0558}{256}\times4}=\pm0.12$$

Analysis of first-order sources in minute accumulating analysis matches completely with the case when analysis of continuous values has been performed straightforwardly with −2, −1, 0, 1, and 2 as the marks. On the other hand, with regard to ω-related sources, essentially we are examining whether or not the distribution form differs between A_1 and A_2 for the sources A, B, . . . , etc.

To test first-order sources such as A, B, . . . , calculations of compared parts have been derived by assuming, for example, that the variances are equal for A_1, A_2, . . . , or that the distribution forms are equal, but most truly, something must be the matter somewhere. Usually, it cannot be that it is known from the outset whether or not such hypotheses hold true. Of course, as shown by the calculations above, $B \times \omega$ is significant as interaction with ω. *This means that the distribution forms are not equal in the respective experiments.* One may probably say that calculations under an assumption such as that it is a linear structure or that the distribution is normal are valueless. A model is not something such that one should calculate under the hypothesis that it is correct, but it is a type of theory which says that it is thus and so if the structure is of a certain kind, and *the correctness of the mathematical model itself must constitute the object of the evaluation by calculation.*

No matter what the form of the distribution, in order to examine to what extent a model of synthesis of factorial effects as a certain type of expression of observed data is correct, one need only find out to what degree the individual data are able to express the situation as a sum of such sources as a sum of residual squares of the estimated values and observed values by the model in question. I am saying that therefore, no matter what the distribution form, it is allowable to test first-order source relations as above (this means performing them as in the ordinary calculation method). I am saying that it is stupid to pretentiously write out mathematical models of such calculation methods.

The reason for examining ω-source relations here is not in order to examine the correctness or incorrectness of first-order source relations by examining whether or not there is a difference in distribution form. I am saying that factorial effects should not be viewed only from the mean. For example, let us assume that in regard to liking, it is as presented by Table 36.17 for A_1 and A_2.

In this case, with A_2, people who completely detest it and people who completely like it are divided in the ratio 7:3, but with A_1, everyone regards it as ordinary. As long as A_1 and A_2 are studied by the mean marks,

Table 36.17 Effects of A_1 and A_2

	Dislike	Dislike somewhat	Neither	Like somewhat	Like	Total
A_1	0	0	100	0	0	100
A_2	70	0	0	0	30	100

A_2 is worse than A_1. If one considers a new product and wishes to attract some of the people to one's own merchandise, as long as it is not a first-class society, it is better to produce products like A_2 and to attract 30% of the people to the product of one's own company. People who think that comparison is good enough by only the mean mark are just like those who commit the error of regarding persons who are no good in ordinary school subjects but who are wonderful in music and pictorial arts, or persons who are excellent in physics but who cannot memorize anything, as being useless persons, below the mean mark. This means that articles with characteristics should not be viewed according to mean marks even in sensory tests. It means that interactions with ω, or in other words influences of the distribution form, are important. In practical science, what is important is not the mathematical correctness or incorrectness of what one is doing, but thoroughly understanding its effect on society.

37 Various Problems Related to Distributions

We will treat in this chapter statistics where the distribution of characteristic values has been assumed to follow a normal distribution. Even if the assumption of normal distribution does not hold, it is probably allowable to use the results of this chapter without modification if they are taken as approximate conclusions. Actually, the mean value $\hat{\mu}$ and the value of the variance V_e are themselves important, and the method of this chapter is an antiquated one which is not important.

37.1 Tolerance Limits (Appendix Table 9)

37.1.1 Definition

Sometimes we wish to estimate the ranges of various characteristic values that manufactured products possess. In such a case, the statistical concept of tolerance limits becomes useful.

For example, if it can be said with a specific reliability $1 - \alpha$ that the characteristic value of most (for example 99%) of the product has a value between two numbers that have been obtained by calculation from the data, T_1 and T_2, these two numbers are termed the *tolerance limits of reliability* $1 - \alpha$.

For example, that the tolerance limits of the length of life of electric light bulbs at reliability 95% for 99% of a certain lot of light bulbs are

$T_1 = 1200$ hours
$T_2 = 1600$ hours

means that when one states that at least 99% of the light bulbs in this lot possess a life of 1200–1600 hours, the apparent accuracy of this assertion is 95%.

Since tolerance limits such as these give both the upper and lower limits of the characteristic value, they are termed *two-sided tolerance limits.* In certain cases we wish to find only one of the limits. Such a limit is termed a *one-sided tolerance limit.* A one-sided tolerance limit declares that the characteristic value of most of the products is at least T_1, or at most T_2, with a specific reliability $1 - \alpha$.

The former is termed the *lower tolerance limit* and the latter is termed the *upper tolerance limit.* The lower tolerance limit is equal to a two-sided tolerance limit where $+\infty$ has been used as T_2, and the upper tolerance limit is the same where $-\infty$ has been used as T_1.

For example, in the case of the electric light bulbs above, that the lower tolerance limit for 99% of the lot is

$$T_1 = 1250 \text{ hours}$$

at 95% reliability means that we may suppose that at least 99% of the light bulbs in this lot possess a life of at least 1250 hours. Or, that the upper tolerance limit of the percentage of water absorption for 95% of certain manufactured products at 95% reliability is 5.3% means that at least 95% of these products have a percentage of water absorption of at most 5.3% with a reliability of 95%. Appendix Table 9 (1) is a numerical table for the finding of two-sided tolerance limits, and Appendix Table 9 (2) is the same for the finding of one-sided tolerance limits.

The formula for two-sided tolerance limits such that at least 99% of the characteristic values or data values of the product are included at 95% reliability is as follows:

$$T = \hat{\mu} \pm k\sqrt{V_e} \qquad (37.1)$$

where

$\hat{\mu}$ = Estimate of process average (population mean); its effective number of replications is n_e
V_e = Value of error variance; its number of degrees of freedom is f_2
k = The read-off value given in Appendix Table 9 (1) for n_e and f_2

37.1.2 Calculation Example of One-Sided Tolerance Limit

When ten samples were taken from a certain wire and their tensile strength was measured, the values were as follows.

89, 92, 91, 86, 85, 88, 89, 90, 92, 87 (unit: kg/mm^2)

Based on this, at least what tensile strength value is possessed by at least 99% of the measured values within this lot? The reliability is to be 95%.

We find the mean value $\hat{\mu}$ and the error variance V_e.

$$\hat{\mu}=\frac{1}{10}(89+92+\cdots+87)=88.9 \qquad (n_e=10)$$

$$V_e=\frac{1}{9}(89^2+92^2+\cdots+87^2-CF)=5.88 \qquad (f_2=9)$$

Since this problem involves a one-sided tolerance limit, from Appendix Table 9 (2), we find, for $n_e=10$ and $f_2=9$, $k=3.96$.

$$88.9-3.96\sqrt{5.88}=88.9-9.6=79.3 \tag{37.2}$$

This means that if we bring over many many samples from this wire and measure the tensile strength each time, probably all of the values and at least 99% of the values will be at least 79.3.

37.1.3 Calculation Example of Two-Sided Tolerance Limits

The following data were obtained for the gain and loss of the number of oscillations of three types of hairsprings, A_1, A_2, and A_3, depending on temperature B, $B_1=-10$, $B_2=10$, $B_3=30$, and $B_4=50$°C. Here, the gain and loss in 24 hours in a steady state are given in seconds. Let us try to find the two-sided tolerance limits of the temperature coefficient of A_2, whose temperature coefficient is the smallest.

	B_1	B_2	B_3	B_4	Total	L
A_1	−15	20	62	100	167	387
A_2	−62	−68	−71	−75	−276	−42
A_3	−21	−7	6	19	−3	133
Total	−98	−55	−3	44	−112	478

Here, L gives the values of the linear effect, obtained by multiplying the values of B_1, B_2, B_3, and B_4 by -3, -1, 1, and 3. For example,

$$-3\times(-15)-1\times 20+1\times 62+3\times 100=387$$

How much the mean value of A_1, A_2, and A_3 differs from 0 is given by

$$S_A=\frac{167^2+(-276)^2+(-3)^2}{4}=26018.5 \qquad (f=3) \tag{37.3}$$

How much the linear coefficients of A_1, A_2, and A_3 for the temperature differ from zero is given by

$$S_{A(B_l)}=\frac{387^2+(-42)^2+133^2}{20}=8461.1 \qquad (f=3) \tag{37.4}$$

The total variation S_T is given by

$$S_T=(-15)^2+20^2+\cdots+19^2=34490.0 \qquad (f=12) \tag{37.5}$$

$$S_e=S_T-S_A-S_{A(B_l)}=10.4 \qquad (f_2=6) \tag{37.6}$$

Therefore, the error variance V_e is

$$V_e=\frac{S_e}{f_2}=\frac{10.4}{6}=1.73$$

And the temperature coefficient of A_2 is

$$\hat{b}=\frac{-3(-62)-(-68)+(-71)+3(-75)}{r(\lambda S)h}=\frac{-42}{1\times 10\times 20}=-0.21 \tag{37.7}$$

The effective number of replications n_e of this is

$$n_e=rSh^2=1\times 5\times 20^2=2000 \tag{37.8}$$

Therefore, from Appendix Table 9 (1), the coefficient k for the two-sided tolerance limits for $n_e=2000$, $f_2=6$, and $P=0.99$ is $k=4.94$. Therefore, the formula for the tolerance limits becomes

$$\hat{b}\pm k\sqrt{\mathrm{Var}(\hat{b})}=\hat{b}\pm k\sqrt{V_e/n_e}=-0.21\pm 4.94\times\sqrt{\frac{1.73}{2000}}=-0.21\pm 0.15 \tag{37.9}$$

This means that the temperature coefficients of at least 99% of the hairsprings are dispersed within the range -0.36–-0.06.

37.2 Tolerance Limits of Fraction Defective (Appendix Table 10)

37.2.1 Sequence of Calculation

It will be assumed that $\hat{\mu}$ and V_e are the same as in the preceding section. The procedure for using the nomogram of Appendix Table 10 is as follows. It will be assumed that the number of degrees of freedom f_2 of V_e is at least 4.

Procedure (I)

The following deviation value k is found.

(1) When the upper specification U has been given,

$$k=(U-\hat{\mu})/\sqrt{V_e} \tag{37.10}$$

(2) When the lower specification limit L has been given,

$$k=(\hat{\mu}-L)/\sqrt{V_e} \tag{37.11}$$

(3) When the upper and lower specification limits $U=M+\Delta$, $L=M-\Delta$ have been given,

$$k_1=(U-\hat{\mu})/\sqrt{V_e} \tag{37.12}$$

$$k_2=(\hat{\mu}-L)/\sqrt{V_e} \tag{37.13}$$

Procedure (II)

The following value of y is found.

$$y=\frac{1}{n_e}+\frac{k^2}{2f_2} \tag{37.14}$$

Procedure (III)

The value of y is taken on scale y of the nomogram; depending on whether (1) one wishes to find what the one-sided tolerance limit is, or in other words that the fraction defective is at most what percent, or (2) one wishes to find the two-sided tolerance limits, or in other words the limits on both sides such that the fraction defective is at least what percent and at most what percent, one finds the desired reliability, in each case on the reliability scale, then connects it with y by a straight line and finds the intersection A with scale A.

Procedure (IV)

One connects intersection A and point k on the k scale; in the case of a *one-sided tolerance limit,* if the intersection with the p_1 scale (or p_0 scale) is read, the fraction defective is at most p_1% (at least p_0%) with the chosen reliability. In the case of *two-sided tolerance limits,* if one reads the intersections with both the p_0 and p_1 scales, it means that the fraction defective is at least p_0% and at most p_1% with that reliability.

37.2.2 Worked Example (1)

The following values were obtained when the tensile strength of a certain wire was measured by producing ten samples under certain conditions when the specification that the tensile strength was to be at least 80 kg/mm^2:

89, 92, 91, 86, 85, 88, 89, 90, 92, 87 (unit: kg/mm^2)

At least what percent and at most what percent can we say will not satisfy the specification when products are produced under these conditions?

$$\hat{\mu}=\frac{1}{10}(89+92+\cdots+87)=88.9$$

$$S_e=89^2+92^2+\cdots+87^2-CF=52.9$$

$$V_e=\frac{52.9}{9}=5.88$$

Procedure (I)

We find k. Since this is case (2), we have

$$k=\frac{\hat{\mu}-L}{\sqrt{V_e}}=\frac{88.9-80}{\sqrt{5.88}}=3.67 \qquad (37.15)$$

Procedure (II)

We find y.

$$y=\frac{1}{10}+\frac{3.67^2}{2\times 9}=0.848 \qquad (37.16)$$

Procedure (III)

Since we are dealing with two-sided tolerance limits, we use two sides, 0.95, and we have

$$p_0=0.0 \qquad p_1=2.5 \qquad (37.17)$$

This means that the tolerance limits of fraction defective are 0.0%–2.5%.

37.2.3 Worked Example (2): The Case of Two-Way Layout

The specification for the insulation resistance of a certain product is at least 2×10^4 MΩ. In test-constructing this product, the following were the results obtained by experimenting by two-way layout, varying the raw material A as six types and additive B as three percentage levels. Since insulation resistance assumes a distribution which is close to the logarithmic normal distribution, the common logarithms were taken. When common logarithms are taken, the specification becomes at least

	B_1 (0%)	B_2 (0.5%)	B_3 (1%)	Total
A_1	4.4	4.7	4.9	14.0
A_2	3.7	3.8	3.9	11.4
A_3	4.2	4.3	4.6	13.1
A_4	3.5	3.8	3.7	11.0
A_5	5.3	5.5	5.3	16.1
A_6	4.6	4.7	4.8	14.1
Total	25.7	26.8	27.2	79.7

4.30. As a result of analysis, A_5B_3 was designated as good production conditions. State at most what percent the fraction defective would be if produced from here on by A_5B_3.

We perform an analysis of variance and find the error variance.

$$V_e=0.013 \qquad (f_2=10)$$

$$\hat{\mu}=\bar{A}_5+\bar{B}_3-\bar{T}=\frac{16.1}{3}+\frac{27.2}{6}-\frac{79.7}{18}=5.47$$

$$n_e=\frac{18}{1+5+2}=2.25$$

Procedure (I)

$$k=\frac{5.47-4.30}{\sqrt{0.013}}=10.3 \tag{37.18}$$

Procedure (II)

$$y=\frac{1}{2.25}+\frac{10.3^2}{2\times 10}=5.74 \tag{37.19}$$

Procedure (III)

Since $k=$ at least 10, from the nomogram, both p_0 and p_1 become 0.00%.

37.3 Explanation of Formula for Tolerance Limits

The *formula for tolerance limits in the case of a normal population* was obtained by Wald and Wolfowitz in 1946.* Although it was proved by them for the case of $f=n-1$, since it is exactly the same for the general case we will explain their derivation of the formula using arbitrary f and n.

Assuming that $\hat{\mu}$ and $u=\sqrt{S_e/f}$ distribute independently, we consider the following random variable A.

$$A=A(\hat{\mu},u,\lambda)=\frac{1}{\sqrt{2\pi}\,\sigma}\int_{\hat{\mu}-\lambda u}^{\hat{\mu}+\lambda u} e^{-\frac{1}{2\sigma^2}(t-\mu)^2}\,dt \tag{37.20}$$

Since $\hat{\mu}$ and u are statistics, A is a random variable and it expresses the proportion covering $N(\mu, \sigma^2)$. Our purpose is to find λ such as to obtain, for given P,

$$P_r\{A>P\}=1-\alpha$$

We may put $\mu=0$ and $\sigma=1$ without losing generality. Then Equation (37.20) becomes

$$A=A(\hat{\mu},u,\lambda)=\frac{1}{\sqrt{2\pi}}\int_{\hat{\mu}-\lambda u}^{\hat{\mu}+\lambda u} e^{-\frac{t^2}{2}}\,dt \tag{37.21}$$

* Wald, A., and Wolfowitz, J.: "Tolerance Limits for a Normal Distribution," A.M.S. Vol. 7, 1946.

A is a monotonic increasing function of u for given $\hat{\mu}$. For an arbitrary λ, we write $P_r(A > P)$ as $f(P, \lambda)$. If the conditional probability when given $\hat{\mu}$ is $f(P, \lambda|\hat{\mu})$, we have

$$f(P, \lambda)=\frac{\sqrt{n}}{\sqrt{2\pi}}\int_{-\infty}^{\infty} f(P, \lambda|\hat{\mu})e^{-\frac{n}{2}(\hat{\mu})^2}\,d\hat{\mu}$$

This is, of course, the mean with respect to $\hat{\mu}$ of $f(P, \lambda|\hat{\mu})$. If the solution of

$$A(\hat{\mu}, u, \lambda)=P$$

with respect to u is written as $g(\hat{\mu}, P, \lambda)$, and if this is substituted into Equation (37.21), $g(\hat{\mu}, P, \lambda)$ is the root of

$$\frac{1}{\sqrt{2\pi}}\int_{\hat{\mu}-\lambda g(\hat{\mu},P,\lambda)}^{\hat{\mu}+\lambda g(\hat{\mu},P,\lambda)} e^{-\frac{z^2}{2}}\,dz=P$$

For this equation to hold always, $\lambda g(\hat{\mu}, P, \lambda)$ must not contain λ. We therefore put

$$\lambda g(\hat{\mu}, P, \lambda)=g(\hat{\mu}, P)$$

Therefore, we have

$$\frac{1}{\sqrt{2\pi}}\int_{\hat{\mu}-g(\hat{\mu},P)}^{\hat{\mu}+g(\hat{\mu},P)} e^{-\frac{t^2}{2}}\,dt=P \tag{37.22}$$

If the equation above is used, as long as $\hat{\mu}$ and P are given, we may find the value of g. Since $A(\hat{\mu}, \mu, \lambda) > P$ is the same as $u > g(\hat{\mu}, P)/\lambda$, we have

$$f(P, \lambda|\hat{\mu})=P_r\{u>g(\hat{\mu}, P)/\lambda\}$$

Since, on the other hand, fu^2 assumes a χ^2 distribution of f degrees of freedom, we have

$$P_r\{u>g(\hat{\mu}, P)/\lambda\}=P_r\left\{\chi^2>\frac{fg^2(\hat{\mu}, P)}{\lambda^2}\right\} \tag{37.23}$$

Next, we will prove that the difference between $f(P, \lambda)$ and $f\left(P, \lambda\middle|\frac{1}{\sqrt{n}}\right)$ is $O\left(\frac{1}{n^2}\right)$. We perform a Taylor expansion with respect to $\hat{\mu}$ and find its mean value, but since the odd-number-power expected value of $\hat{\mu}$ is 0, we have

$$f(P, \lambda|\hat{\mu})=f(P, \lambda|0)+\frac{(\hat{\mu})^2}{2!}\frac{\partial^2}{\partial\hat{\mu}^2}f(P, \lambda|\hat{\mu})\Big|_{\hat{\mu}=0}$$

$$+\frac{(\hat{\mu})^4}{4!}\frac{\partial^4}{\partial\hat{\mu}^4}f(P, \lambda|\hat{\mu})\Big|_{\hat{\mu}=\xi}\quad(0\leq\xi\leq\hat{\mu})$$

Taking the expected value of the right side, we have

$$f(P, \lambda)=f(P, \lambda|0)+\frac{1}{2n}\frac{\partial^2}{\partial\hat{\mu}^2}f(P, \lambda|\hat{\mu})\Big|_{\hat{\mu}=0}+O\left(\frac{1}{n^2}\right) \tag{37.24}$$

On the other hand,

$$f\left(P, \lambda \middle| \frac{1}{\sqrt{n}}\right) = f(P, \lambda|0) + \frac{1}{2n}\frac{\partial^2}{\partial\hat{\mu}^2} f(P, \lambda|\hat{\mu})\bigg|_{\hat{\mu}=0} + O\left(\frac{1}{n^2}\right) \tag{37.25}$$

From Equations (37.24) and (37.25), we have

$$f(P, \lambda) - f\left(P, \lambda \middle| \frac{1}{\sqrt{n}}\right) = O\left(\frac{1}{n^2}\right) \tag{37.26}$$

This means that we may use $f\left(P, \lambda \middle| \frac{1}{\sqrt{n}}\right)$ instead of $f(P, \lambda)$ with sufficiently high precision. Thus we may substitute $\frac{1}{\sqrt{n}}$ in place of $\hat{\mu}$ of $f(P, \lambda|\hat{\mu})$. If the $1 - \alpha$ point of the χ^2 distribution is expressed as $\chi^2_{1-\alpha}$, then by solving Equation (37.23) with respect to λ we get

$$\lambda = \sqrt{\frac{f}{\chi^2_{1-\alpha}}}\, g(\hat{\mu}, P) \tag{37.27}$$

If the λ of equation (37.27) is used, the probability that A is greater than P becomes $1 - \alpha$. On the other hand, the value of $g(\hat{\mu}, P)$ is obtained for the given $\hat{\mu}$ from Equation (37.22), but since the right side of Equation (37.22) is $f(P, \lambda|\hat{\mu})$, we substitute in $\frac{1}{\sqrt{n}}$ instead of $\hat{\mu}$ and solve

$$\int_{\frac{1}{\sqrt{n}}-g}^{\frac{1}{\sqrt{n}}+g} \frac{1}{\sqrt{2\pi}} e^{-\frac{t^2}{2}}\, dt = P$$

If g is found from the equation above, then if λ is replaced by k in order to agree with this text, k is obtained as

$$k = g\sqrt{\frac{f}{\chi^2_{1-\alpha}}} \tag{37.28}$$

from Equation (37.27).

The *formula for one-sided tolerance limit* was found by assuming that $\hat{\mu} \pm k\sqrt{S_e/f}$ approximately distributes according to the normal distribution $N\left(\mu \pm k\sigma, \frac{1}{n} + \frac{k^2}{2f}\right)$. This becomes a problem of determining k so that

$$P_r\{\hat{\mu} + ku \geq \mu + K_{1-P}\sigma\} = \frac{1}{\sqrt{2\pi}\,\Delta\sigma}\int_{\mu+K_{1-P}\sigma}^{\infty} e^{-\frac{1}{2\Delta^2\sigma^2}(t-\mu-k\sigma)^2}\, dt$$

$$= \int_{(K_{1-P}-k)/\Delta}^{\infty} \frac{1}{\sqrt{2\pi}} e^{-\frac{t^2}{2}}\, dt \tag{37.29}$$

does not become smaller than α, assuming, for an arbitrary P, that the standard normal $(1-P)$ point is K_{1-P} and $\Delta^2 = \dfrac{1}{n} + \dfrac{k^2}{2f}$. For this, we need merely put

$$\frac{K_{1-P}-k}{\Delta}=K_\alpha$$

If the value of Δ is substituted in, we have

$$K_{1-P}-k=K_\alpha\sqrt{\frac{1}{n}+\frac{k^2}{2f}}$$

If both sides are squared and we collect equal powers of k, we have

$$\left(1-\frac{K_\alpha{}^2}{2f}\right)k^2-2K_{1-P}k+K_{1-P}{}^2-\frac{K_\alpha{}^2}{n}=0 \tag{37.30}$$

We need merely solve this quadratic equation and find k.

37.4 Explanation of Formula for Tolerance Limits of Fraction Defective

It will be assumed that the effective number of replications of the estimate $\hat{\mu}$ of the population mean μ is n_e and that the number of degrees of freedom of the error variance S_e is f. $\hat{\mu}$ and S_e are in general independent, and if $f \geq 4$,

$$\hat{\mu}+k\sqrt{\frac{S_e}{f}}\equiv\hat{\mu}+k\hat{\sigma} \tag{37.31}$$

is known to assume the normal distribution with

$$\left.\begin{aligned}&\text{Population mean} = \mu + k\sigma\\ &\text{Population variance} = \left(\frac{1}{n}+\frac{k^2}{2f}\right)\sigma^2\end{aligned}\right\} \tag{37.32}$$

approximately. When a lower limit L has been given, we consider the confidence limits of statistic

$$X=\frac{\hat{\mu}-L}{\sqrt{S_e/f}}$$

We consider only the lower limit.

$$P_r\{X\geq K_L\}=P_r\left\{\frac{\hat{\mu}-L}{\hat{\sigma}}\geq K_L\right\}=P_r\{\hat{\mu}-K_L\hat{\sigma}\geq L\}$$

In order to cause this value to be at most $\dfrac{\alpha}{2}$, K_L must satisfy the following inequality. Since the population mean of $\hat{\mu} - K_L\hat{\sigma}$ is $\mu - K_L\sigma$ and the population variance is $\left(\frac{1}{n} + \frac{K_L{}^2}{2f}\right)\sigma^2$, we have

$$\int_L^{\infty} \frac{1}{\sqrt{2\pi}\,\sigma\sqrt{\frac{1}{n}+\frac{K_L{}^2}{2f}}} e^{-\frac{1}{2\left(\frac{1}{n}+\frac{K_L{}^2}{2f}\right)\sigma^2}(t-\mu+K_L\sigma)^2} dt \geq 1-\frac{\alpha}{2}$$

Thus,

$$\int_{\frac{L-\mu+K_L\sigma}{\sqrt{\frac{1}{n}+\frac{K_L{}^2}{2f}}\sigma}}^{\infty} e^{-\frac{t^2}{2}} dt \geq 1-\frac{\alpha}{2}$$

So

$$\frac{L-\mu+K_L\sigma}{\sqrt{\frac{1}{n}+\frac{K_L{}^2}{2f}}\sigma} = -K_{\alpha/2} \tag{37.33}$$

We need merely solve the following quadratic equation with respect to K_L:

$$\left(1-\frac{K_{\alpha/2}{}^2}{2f}\right)K_L{}^2 - 2\frac{\mu-L}{\sigma}K_L + \left(\frac{\mu-L}{\sigma}\right)^2 - \frac{K_{\alpha/2}{}^2}{n} = 0 \tag{37.34}$$

We therefore form the confidence limits by using the root of the equation above, K_L, but actually K_L which is the root of Equation (37.34) is a function of modulus K,

$$K = \frac{\mu-L}{\sigma}$$

Since k, the estimate of K, is

$$k = \frac{\hat{\mu}-L}{\hat{\sigma}}$$

the lower limit of confidence, K_{P_0} can be obtained by assuming, for Equation (37.33), that

$$\left.\begin{aligned} K_{P_0} &= \frac{\mu-L}{\sigma} \\ k &= K_L \end{aligned}\right\}$$

Thus,

$$K_{P_0} = k + K_{\alpha/2}\sqrt{\frac{1}{n}+\frac{k^2}{2f}}$$

Therefore, if we find the P_0 for the normal deviation of the left side, P_0 gives the lower limit of confidence of the fraction defective. Similarly, from

$$K_{p_1} = k - K_{\alpha/2}\sqrt{\frac{1}{n}+\frac{k^2}{2f}}$$

the upper limit of confidence p_1 can be obtained.

In the case of a one-sided limit, for example in the case above, we have

$$K_{p_1}=k-K_\alpha\sqrt{\frac{1}{n}+\frac{k^2}{2f}}$$

The solution of Equation (37.34) for K_L is

$$K_L=\frac{k+\sqrt{\left(\frac{1}{n}+\frac{k^2}{2f}\right)K_{\alpha/2}{}^2-\frac{K_{\alpha/2}{}^4}{2fn}}}{1-K_{\alpha/2}{}^2/2f} \tag{37.35}$$

With this as a function of k, we need only plot it in the K_L and k plane.

For finding the coefficient k of two-sided confidence limits, an even better approximation method has been devised.

37.5 Tolerance Limits of Fraction Defective by Number of Defectives Data (Neyman's Method)

The estimation method of Neyman is one that uses the F table, and it is performed by a procedure such as the following. It will be assumed that there have been x defectives in size n (in general, effective number of replications n_e). We put

$$f_1=2(n-x+1)$$

$$f_2=2x$$

and we find the lower limit of confidence P_l by the following equation using the F value of level of significance $\frac{\alpha}{2}$.

$$P_l=\frac{f_2}{f_1F_{f_2}^{f_1}(\alpha/2)+f_2} \tag{37.36}$$

For the upper limit of confidence P_u, we put

$$f_1'=2(x+1)$$

$$f_2'=2(n-x)$$

and we estimate by the following equation by using the F value of level of significance $\frac{\alpha}{2}$.

$$P_u=\frac{f_1'F_{f_2'}^{f_1'}(\alpha/2)}{f_1'F_{f_2'}^{f_1'}(\alpha/2)+f_2'} \tag{37.37}$$

38 How to Construct an Orthogonal Array and How to Apply It

It will be shown how to construct an orthogonal array. Please refer to Section 40.5 regarding the application of orthogonal arrays to circuit design.

38.1 Method of Construction Using Orthogonal Latin Squares

When there exists a *complete orthogonal system* of Latin squares of dimensions $n \times n$, denoted by $L_1, L_2, \ldots, L_{n-1}$, it is simple to construct an orthogonal array of size n^r ($r = 2, 3, \ldots$) and number of columns $(n^r - 1)/n - 1$. An array such as this will be termed a *higher-order Latin square*. For example, since there exists a 3×3 Latin square orthogonal system,

L_1				L_2		
1	2	3		1	3	2
2	3	1		2	1	3
3	1	2		3	2	1

it is possible to form, for example, $L_9(3^4)$, $L_{27}(3^{13})$, $L_{81}(3^{40})$, and $L_{243}(3^{121})$. The method of this is as follows: When constructing $L_{27}(3^{13})$, say, we expand the Graeco-Latin square $L_9(3^4)$ which has been constructed by using L_1 and L_2,

No. \ Col.	1	2	3	4
1	1	1	1	1
2	1	2	2	2
3	1	3	3	3
4	2	1	2	3
5	2	2	3	1
6	2	3	1	2
7	3	1	3	2
8	3	2	1	3
9	3	3	2	1

to three times the size, then we add the fifth column wherein 1, 2, and 3 have merely been written vertically in a line, as shown by Table 38.1.

Now, the numerals of column 1 and column 5 correspond in the nine combinations 11, 12, 13, 21, 22, 23, 31, 32, and 33; these nine combinations are just columns 1 and 2 of $L_9(3^4)$, and we let the numerals of column 3 and column 4 in $L_9(3^4)$ correspond to columns 6 and 7. To put it differently, since Nos. 1, 2, and 3 of column 1 and column 5 have assumed the

Table 38.1 L_{27} Constructed by Using Orthogonal Latin Squares

No. \ Column	1	2	3	4	5	1 and 5 / 6 and 7		2 and 5 / 8 and 9		3×5		4×5	
1	1	1	1	1	1	1	1	1	1				
2	1	1	1	1	2	2	2	2	2				
3	1	1	1	1	3	3	3	3	3				
4	1	2	2	2	1	1	1	2	3				
5	1	2	2	2	2	2	2	3	1				
6	1	2	2	2	3	3	3	1	2				
7	1	3	3	3	1	1	1	3	2				
8	1	3	3	3	2	2	2	1	3				
9	1	3	3	3	3	3	3	2	1				
10	2	1	2	3	1	2	3	1	1				
11	2	1	2	3	2	3	1	2	2				
12	2	1	2	3	3	1	2	3	3				
13	2	2	3	1	1	⋮	⋮	⋮	⋮				
14	2	2	3	1	2	⋮	⋮	⋮	⋮				
15	2	2	3	1	3	⋮	⋮	⋮	⋮				
16	2	3	1	2	1								
17	2	3	1	2	2								
18	2	3	1	2	3								
19	3	1	3	2	1								
20	3	1	3	2	2								
21	3	1	3	2	3								
22	3	2	1	3	1								
23	3	2	1	3	2								
24	3	2	1	3	3								
25	3	3	2	1	1								
26	3	3	2	1	2								
27	3	3	2	1	3								

combinations 11, 12, and 13, we insert in the locations of columns 6 and 7 the combinations 11, 22, and 33, which correspond to 11, 12, and 13, from $L_9(3^4)$. Thus, since in the cases of No. 1, No. 4, and No. 7, column 1 and column 5 are always 1, we need only write into columns 6 and 7 the value of columns 3 and 4 which corresponds to 11 in L_9, namely 11. In this way, columns 6 and 7 become completely filled.

Next, we regard column 2 and column 5 as columns 1 and 2 in L_9, and we write into columns 8 and 9 the numerals which correspond to columns 3 and 4. Similarly, columns 10 and 11 are found from column 5 and column 3, and columns 12 and 13 are found from column 5 and column 4. In this way, we obtain $L_{27}(3^{13})$.

To construct $L_{81}(3^{40})$, we expand $L_{27}(3^{13})$ to three times the size, and we construct an array wherein 1, 2, and 3 have been written in vertically as column 14. From column 1 and column 14, we create columns 15 and 16 by using $L_9(3^4)$. The remainder is constructed in the same way. It is suggested that the reader try to construct $L_{64}(4^{21})$ by using the following three 4 × 4 orthogonal Latin squares.

L_1				L_2				L_3			
1	2	3	4	1	2	3	4	1	2	3	4
2	1	4	3	3	4	1	2	4	3	2	1
3	4	1	2	4	3	2	1	2	1	4	3
4	3	2	1	2	1	4	3	3	4	1	2

What we do is expand $L_{16}(4^5)$ to four times the size and add 1, 2, 3, and 4 replicated 16 times as column 6. Next, we substitute L_1, L_2, and L_3 into the interaction of column 1 and column 6. If, similarly, we substitute L_1, L_2, and L_3, respectively, into the interactions of columns 2, 3, 4, 5, and column 6, we obtain four-level columns in the number of $5 + 1 + 3 \times 5 = 21$ columns, and thus obtain $L_{64}(4^{21})$.

Speaking generally, if orthogonal array $L_n(k^r)$ exists and if $L_{k^2}(k^{k+1})$ exists, it is possible to create $[r(k - 1) + 1]$ columns of at least k levels by an orthogonal array of size nk. For example, we construct $L_{24}(2^{23})$ from $L_{12}(2^{11})$ and $L_4(2^3)$, and $L_{54}(6 \times 3^{24})$ from $L_{18}(6 \times 3^6)$ and $L_9(3^4)$.

38.2 Masuyama's Method

In constructing an orthogonal array for a k-level system, Bose and Bush* showed a method of using a matrix whose elements are k-tuples of 1, 2, . . . , k, taken cyclicly:

$$e_0 = (1, 2, 3, \cdots, k),\ e_1 = (2, 3, \cdots, k, 1),\ e_{k-1} = (k, 1, 2, \cdots, k-1)$$

* Reference 4.

Thus, if in

$$A=\begin{pmatrix} x_{11} & \cdots & x_{1r} \\ \vdots & \ddots & \vdots \\ x_{s1} & \cdots & x_{sr} \end{pmatrix}$$

x_{ij} are any of $e_0, \ldots, e_{k-1}$, and if

0 correspondence	$e_i \sim e_i$	$(i=0, 1, \ldots, k-1)$
1 correspondence	$e_i \sim e_{i+1}$	$(i=0, 1, \ldots, k-1; e_k=e_0)$
⋮	⋮	
$(k-1)$ correspondence	$e_i \sim e_{k-1+i}$	$(i=0, \ldots, k-1; e_{k-1+i}=e_{i-1})$

appear the same number of times no matter which two columns are selected, A is regarded as $L_{sk}(k^r)$. This is only natural since it is no more than another way of stating that no matter which two columns are caused to correspond, the numerals 1, 2, . . . , k of the other column, relative to 1, 2, . . . , k, correspond the same number of times. If e_i is abbreviated as i, the fact above can be stated in terms of a matrix whose components are residues mod k.

Masuyama considered the application of the algebraic theory of numbers as a method of analytically expressing the operation mod k, and he showed a method of constructing $L_{2k^2}(2k \times k^{2k})$ when k is a prime number by using the cyclotomic field.*† We will introduce Masuyama's method of constructing $L_{50}(10 \times 5^{10})$, which is one such case.

Let us consider a 10×10 matrix all of whose elements are powers of one of the 5th roots of 1, $\omega = e^{2\pi i/5}$. This matrix will be expressed as $\mathbf{T}$. We are to consider $\mathbf{T}$ and its conjugate transposed matrix, $\mathbf{T}^*$; now, if A is the orthogonal array being sought, then with $\mathbf{E}$ as the unit matrix,

$$\boldsymbol{TT}^*=10\boldsymbol{E} \tag{38.1}$$

must necessarily hold, so that the columns other than the diagonals must be quartic equations of ω and, moreover, their value must be zero.

Masuyama demonstrated a method of forming a matrix such as $\mathbf{T}$ by using 5×5 matrices $\mathbf{A}$, $\mathbf{C}$, and $\mathbf{D}$. If the 10×10 matrix

$$\mathbf{T}=\begin{pmatrix} \mathbf{A} & \mathbf{C} \\ \mathbf{C}^* & \mathbf{D} \end{pmatrix} \quad \text{or} \quad \begin{pmatrix} \mathbf{A} & \mathbf{C} \\ \mathbf{C}' & \mathbf{D} \end{pmatrix} \qquad (\mathbf{C}' \text{ is a transposed matrix}) \tag{38.2}$$

is that which is desired, then

$$\boldsymbol{TT}^*=\begin{pmatrix} \boldsymbol{A} & \boldsymbol{C} \\ \boldsymbol{C}^* & \boldsymbol{D} \end{pmatrix}\begin{pmatrix} \boldsymbol{A}^*\boldsymbol{C} \\ \boldsymbol{C}^*\boldsymbol{D}^* \end{pmatrix}=\begin{pmatrix} \boldsymbol{AA}^*+\boldsymbol{CC}^* & \boldsymbol{AC}+\boldsymbol{CD}^* \\ \boldsymbol{C}^*\boldsymbol{A}^*+\boldsymbol{DC}^* & \boldsymbol{C}^*\boldsymbol{C}+\boldsymbol{DD}^* \end{pmatrix}=\begin{pmatrix} 10\boldsymbol{E} & 0 \\ 0 & 10\boldsymbol{E} \end{pmatrix} \tag{38.3}$$

* Reference 26.

† A cyclotomic field is sufficient since, according to class field theory, an Abelian entity upon a rational number field is only a cyclotomic field.

and therefore we need only choose **A**, **B**, and **C** as follows.

$$AA^* = CC^* = DD^* = 5E \tag{38.4}$$

$$AC + CD^* = C^*A^* + DC^* = 0 \tag{38.5}$$

Since the algebraic norm of $(1-\omega)$ is $(1-\omega)(1-\omega^2)(1-\omega^3)(1-\omega^4) = 5$, we divide this into two conjugate parts and designate one as μ. For example, $\mu = (1-\omega)(1-\omega^2)$. Then, it is claimed that **A** and **C** become determined as follows if it can be so arranged that the elements of the first column of the cyclic matrix $\mathbf{B} = (b_{ij})$, whose elements are the powers of ω, b_{00}, b_{10}, b_{20}, b_{30}, b_{40}, are such that

$$b_{00} + b_{10}\omega^k + b_{20}\omega^{2k} + b_{30}\omega^{3k} + b_{40}\omega^{4k} \quad (k=0, 1, 2, 3, 4) \tag{38.6}$$

will all become associate numbers of μ.

$$A = (\omega^{ij})(i, j = 0, 1, 2, 3, 4), \quad C = AB/\mu, \quad D = -C^*B/\mu \tag{38.7}$$

According to Masuyama, b_{i0} are in general the roots of indefinite equations, and

$$b_{00} = \omega^3, \quad b_{10} = \omega^4, \quad b_{20} = \omega^3, \quad b_{30} = 0, \quad b_{40} = 1$$

constitute one of the solutions; and **B**, **AB**, and **C * B** (in this case **C ′ B**) can be found immediately as follows. He has also shown a fairly general method of determining b_{i0} by using the triangular numbers, 0, 1, 3, 6, 10, . . . , as the powers of ω.

We obtain

$$B = \begin{bmatrix} \omega^3 & 1 & 1 & \omega^3 & \omega^4 \\ \omega^4 & \omega^3 & 1 & 1 & \omega^3 \\ \omega^3 & \omega^4 & \omega^3 & 1 & 1 \\ 1 & \omega^3 & \omega^4 & \omega^3 & 1 \\ 1 & 1 & \omega^3 & \omega^4 & \omega^3 \end{bmatrix} \tag{38.8}$$

Therefore,

$$AB = (1-\omega)(1-\omega^2)C = (1-\omega)(1-\omega^2) \begin{bmatrix} 1 & 1 & 1 & 1 & 1 \\ 1 & \omega & \omega^2 & \omega^3 & \omega^4 \\ \omega^3 & 1 & \omega^2 & \omega^4 & \omega \\ \omega^4 & \omega^2 & 1 & \omega^3 & \omega \\ \omega^3 & \omega^2 & \omega & 1 & \omega^4 \end{bmatrix} = \mu C \tag{38.9}$$

Further, if we use the transposed matrix of **C**, **C′**, and **B** and obtain

$$C'B = -(1-\omega)(1-\omega^2)D = -\mu D \tag{38.10}$$

then

$$T = \begin{pmatrix} A & C \\ C' & D \end{pmatrix} \tag{38.11}$$

becomes the isometric matrix which we are seeking. $L_{50}(2 \times 5^{11})$ of Table 38.2 is obtained by substituting **A**, **C**, **C′**, and **D**. Interaction between column 1 and column 2 can be obtained without sacrificing other columns. $L_{50}(2 \times 5^{11})$ is also $L_{50}(10 \times 5^{10})$.

Table 38.2 $L_{50}(2 \times 5^{11})$

No. \ Col.	1	2	3	4	5	6	7	8	9	10	11	12
1	1	1	1	1	1	1	1	1	1	1	1	1
2	1	1	2	2	2	2	2	2	2	2	2	2
3	1	1	3	3	3	3	3	3	3	3	3	3
4	1	1	4	4	4	4	4	4	4	4	4	4
5	1	1	5	5	5	5	5	5	5	5	5	5
6	1	2	1	2	3	4	5	1	2	3	4	5
7	1	2	2	3	4	5	1	2	3	4	5	1
8	1	2	3	4	5	1	2	3	4	5	1	2
9	1	2	4	5	1	2	3	4	5	1	2	3
10	1	2	5	1	2	3	4	5	1	2	3	4
11	1	3	1	3	5	2	4	4	1	3	5	2
12	1	3	2	4	1	3	5	5	2	4	1	3
13	1	3	3	5	2	4	1	1	3	5	2	4
14	1	3	4	1	3	5	2	2	4	1	3	5
15	1	3	5	2	4	1	3	3	5	2	4	1
16	1	4	1	4	2	5	3	5	3	1	4	2
17	1	4	2	5	3	1	4	1	4	2	5	3
18	1	4	3	1	4	2	5	2	5	3	1	4
19	1	4	4	2	5	3	1	3	1	4	2	5
20	1	4	5	3	1	4	2	4	2	5	3	1
21	1	5	1	5	4	3	2	4	3	2	1	5
22	1	5	2	1	5	4	3	5	4	3	2	1
23	1	5	3	2	1	5	4	1	5	4	3	2
24	1	5	4	3	2	1	5	2	1	5	4	3
25	1	5	5	4	3	2	1	3	2	1	5	4
26	2	1	1	1	4	5	4	3	2	5	2	3
27	2	1	2	2	5	1	5	4	3	1	3	4
28	2	1	3	3	1	2	1	5	4	2	4	5
29	2	1	4	4	2	3	2	1	5	3	5	1
30	2	1	5	5	3	4	3	2	1	4	1	2
31	2	2	1	2	1	3	3	2	4	5	5	4
32	2	2	2	3	2	4	4	3	5	1	1	5
33	2	2	3	4	3	5	5	4	1	2	2	1
34	2	2	4	5	4	1	1	5	2	3	3	2
35	2	2	5	1	5	2	2	1	3	4	4	3
36	2	3	1	3	3	1	2	5	5	4	2	4
37	2	3	2	4	4	2	3	1	1	5	3	5
38	2	3	3	5	5	3	4	2	2	1	4	1
39	2	3	4	1	1	4	5	3	3	2	5	2
40	2	3	5	2	2	5	1	4	4	3	1	3
41	2	4	1	4	5	4	1	2	5	2	3	3
42	2	4	2	5	1	5	2	3	1	3	4	4
43	2	4	3	1	2	1	3	4	2	4	5	5
44	2	4	4	2	3	2	4	5	3	5	1	1
45	2	4	5	3	4	3	5	1	4	1	2	2
46	2	5	1	5	2	2	5	3	4	4	3	1
47	2	5	2	1	3	3	1	4	5	5	4	2
48	2	5	3	2	4	4	2	5	1	1	5	3
49	2	5	4	3	5	5	3	1	2	2	1	4
50	2	5	5	4	1	1	4	2	3	3	2	5
	Group 1	Group 2	Group 3									

$1^{(2)}$ ●———● $2^{(5)}$

38.3 Orthogonal Array and Non-Orthogonal Array

An orthogonal array possesses the following property. For example, with regard to the 3^8 type, let us assume that we have performed an experiment by using $L_{27}(3^{13})$ to find the main effects of A, B, C, D, E, F, G, and H at three levels and two two-factor interactions, $A \times B$ and $A \times C$. Assuming that it has become known as a result of experimentation that some of the sources can be disregarded, the population variance of the estimate of the population mean μ for any desired level combination of the respective factors is given by

$$\sigma^2(\hat{\mu}) = \frac{1}{n_e}\sigma^2 \tag{38.12}$$

$$n_e = \frac{\text{total number of experiments}}{\text{sum of numbers of degrees of freedom of sources which are not disregarded}} \tag{38.13}$$

There is the very notable property that n_e has exactly the *same value* whether for the combination by which the experiment was performed or for a combination by which it was not performed.* Even when an array other than orthogonal array $L_{27}(3^{13})$ has been used, generally it is possible to estimate the value of the population mean μ for an arbitrary combination of levels, but usually the value of the effective number of replications n_e differs for different combinations of levels, and for the worst among them it is always smaller than the value of n_e of Equation (38.12).† Worded differently, it can be said that as long as one is comparing combinations for which the precision of estimation is the worst, the orthogonal array is the best. An orthogonal array is said to be *the optimum in the minimax sense.*

Moreover, that estimations and testing for main effects and interactions are best in the minimax sense for every source by use of an orthogonal array can be easily seen. For example, for main effect A, the effective number of replications becomes 9, and the confidence limits are given by

$$\pm \sqrt{F_f^1 \times V_e/n_e} \tag{38.14}$$

pooling disregardable sources with error. If L_{81} is used, n_e increases from 9 to 27, and it can be expected that the confidence limits above will become narrowed to approximately $1/\sqrt{3}$.

But now, regarding the main effects of the various factors, the precision is adequate even with 9 replications and it is unnecessary to raise the number to as much as $n_e = 27$. The problem is: There are 50 factors and

* Genichi Taguchi: Jikken Keikakuhō Text (Textbook of design of experiments method), internal reference material of Denki Tsūshin Kenkyūjo (Electrical Communication Laboratory), 1953.

† Moriguti, S.: Optimality of orthogonal designs, *Rep. Stat. Appl. Res. Un. Jap. Sci. Eng.* 3, 1954.

we also wish to find several interactions, but since n_e = about 9 is sufficient, is there not a method of experimenting without using a large orthogonal array such as $L_{243}(3^{121})$? To present it differently, it is known that results of precision which is greater than necessary can be obtained by using a large orthogonal array to which all of these sources can be assigned, but when less precision is permissible, could there not exist an array where it is possible to toss in a greater number of sources simultaneously—an array which is not an orthogonal array? This is a question which anyone comes to face, naturally, once he understands orthogonal arrays.

Especially when there are several characteristics or several tens of characteristics in question and there are as many as 100 factors which might have an effect on some one of them, it is not unusual that the sources which are effective for individual characteristic values number several, or several more than ten, or so. But since it is not clear which factor is effective for which characteristic value, for the experiment one would still wish to take up as many as 100 simultaneously. If all of the factors are at three levels, as long as an orthogonal array is used it is necessary to use $L_{243}(3^{121})$. To make up for this, the effective number of replications of the main effects then becomes $n_e = 81$, so that even very small effects appear significant. This would actually be regarded as *an excess of experimentation,* and naturally it would be more advantageous to perform an experiment of less precision and thereby to arrive at conclusions sooner.

Even though one says lower the precision, if it is lowered in any old fashion, the experiment becomes distant from a logical assignment. One can consider the following two methods of maintaining things at a suitable precision, while making the greatest effort not to permit the precision of the whole to drop.

(1) *Randomly combined design*
(2) *Almost orthogonal array*

Please refer to Chapter 34 regarding the content of these.

39 Method of Quantification

A general discussion of how to quantify data of good, medium, and bad, and of how to perform changes of variables, will be given in this chapter.

39.1 Purpose of Quantification

39.1.1 Introduction

We often wish to assign marks of evaluation such as grade-one products, grade-two products, and grade-three products or good, medium, and bad, and we often wonder what mark to give. Actually, what quantification should be performed cannot be determined until there is a purpose. When an oxidized film (Alumite) is produced by applying chromic acid treatment to an aluminum alloy, the method of measuring the electrostatic capacity of the film instead of its thickness and studying how many microfarads there were can be regarded as giving a different expression to the same quantity, using electrical characteristics rather than thickness.

In other words, no matter what change of variable, $z = f(y)$, is performed to place a certain characteristic value z in a one-to-one relationship with y, *there is absolutely no superiority or inferiority mathema-*

tically, since z becomes determined if y is determined and y is determined if z is determined. Since the electrostatic capacity becomes greater the thinner the oxidized film, the relationship between them is one to one, and therefore instead of thickness, which is difficult to measure, the electrostatic capacity may be measured. Therefore, as long as they are numbers that have been converted anyhow, even if the electrostatic capacity becomes substantially close to infinitely great since there are pinholes or the film is too thin, and it goes off-scale, there is no reason whatsoever why one should not convert this value z further, as follows.

$$w = \tan^{-1} z \tag{39.1}$$

When this is done, by virtue of the inverse tangent transformation the value of z from 0 to ∞ falls into the finite range of 0 to $\pi/2$. And there does not exist anywhere any purely mathematical reason why it would be bad to use w instead of z for data analysis.

It is common sense to take the logarithm when the data are in the interval $(0, \infty)$ and to perform conversion by $\tanh^{-1} x$ when the data are in $(0, 1)$. Therefore, for example, temperature should be expressed as absolute temperature, then one should take its logarithm and perform decibel calculation. Please refer to the omega method of Note 3.5 of Chapter 3 of Volume 1 concerning $\tanh^{-1} x$.

To tell the truth, even *MKS units,* which we have so plausibly set up, are used merely for convenience. That is why we express the apparent intensity of the light of a star not only by candlepower, which is a unit of brightness, but also by magnitude. The method of taking the logarithm of a certain value and converting to decibel units is a practice commonly resorted to. Why do we go to the trouble of performing such changes of variables? Is it only for convenience? Mathematically speaking, it is clear that there is no absolute relationship of superiority, one way or the other, between measures between which we can switch by one-to-one conversion. Therefore, *the problem of quantification is not a problem of mathematics but one of substantial science.*

This chapter is one answer to this problem. Before embarking on it, we will present one or two ways of thinking about methods of quantification that are used in other areas, although the purpose differs.

39.1.2 Problems of Quantification in Other Areas

In the world of law, it is usual to parole a convict while his term still remains uncompleted, considering for example his background and performance. In so doing, one would not wish to release a person who would commit a repetition of his offense, while one would like to release a person who would not commit further crimes, insofar as this is possible. Although it is not completely certain whether or not a particular criminal would continue to commit crimes, it is conceivable that the percentage of commission of new crimes will depend on sources such as the following.

A: sex
B: age
C: whether single or married
D: whether there are children, and if so how many
E: living standard of the household
F: amount of education
G: attitude during serving of term
H: environment of life
I: type of crime and reasons for commission of crime

For example, generally speaking, it is believed that the person would tend to commit new crimes less when married than when single, when the number of children is greater, and when the living standard is higher. If, then, many hundreds of paroles have been granted so far, if there are persons among the parolees who have committed new crimes and persons who have not, and if, for each of them, information has been obtained regarding each of the items A–I, it is possible to give a critical value L of parole or no parole to the marks for the respective classes of each item and the total mark so as to minimize the following W.

$$W = \text{proportion of persons committing crime among the paroled persons} \times W_1 + \text{proportion of persons not committing crime among persons who are not paroled} \times W_2 \qquad (39.2)$$

W_1 and W_2 here are weights; they are determined according to which mistake is regarded as important, and usually stress is laid on the former. To put it differently, the superiority or inferiority of one method of marking, φ_1, to another method of marking, φ_2, is determined by which method causes W to become smaller when the respective methods of marking have been tried on the data acquired so far. Of course, when so doing, there will also be two ways that the critical value of specification L appears. In the present case, the method of quantification is determined according to what is the quantification which makes the most correct judgment. One should use *experimental regression analysis* (Chapter 15 of Volume 1) for the actual calculation method.

It is said that in demobilizing military persons from theaters of war during the Second World War, the Americans considered for each military person, for example,

A: whether a professional military person or not
B: number of years of service
C: age
D: condition of household
E: scholastic history
F: meritorious achievements

and then gave marks for each level of each item and demobilized beginning with persons with greater composite marks, also considering the desire of the person in question. Since there is no error of judgment in this case, the problem of quantification here differs from the preceding one. Of course, since this was America, quantification was done by also considering the opinions of civilian representatives. In this case, quantification was performed so as to fit the opinions of civilians the most.

39.2 Quantification in the Case of Experimental Data

39.2.1 Quantification and Efficiency

As is clear from the two examples given above, there must always be a purpose in the problem of quantification. Then, what purpose should there be in the case of experimental data? The author would wish to regard the matter as follows, emulating the method of Fisher.

The purpose of quantification in the case of research is to maximize the efficiency of studying science and technology. To state it differently, the quantification is best when the contribution ratios of the factorial effects are the greatest.

Then, how should we render this concrete? For example, when the experimental data constituted first-grade products, second-grade products, and third-grade products, we were able to detect the respective factorial effects by accumulating analysis; and it would be possible to give specific marks, a, b, and c, to these respectively, and to analyze by continuous values. If appropriate quantification is carried out, the power of detection of sources would improve considerably since sometimes this would mean greater accuracy. But if inept quantification is performed, of course it would be inferior to accumulating analysis. And in fact, no matter what the data, it is possible to lose significant sources by inept quantification.

But when all of the effects of the sources are truly small, it is not possible to cause the factorial effects to become significant no matter what quantification is performed. One can examine the superiority or inferiority of quantification by inspecting the following ratio of factorial effect to error effect, Q. Q is* the SN ratio, as it is termed in communications.

* Although it becomes the same thing as Q above, the idea of minimizing

$$Q' = \frac{\text{error variation}}{\text{total variation}}$$

was suggested by Fisher. In Japan, too, it has been used for data of agricultural experiments by Genzaburō Masuyama. This method is also an important extension of the method of least squares (since the total variation is constant, to render Q' minimum is to render S_e minimum). The author feels that it is easier to understand when it is explained that this agrees with rendering signal/noise, or the so-called SN ratio of communication theory, maximum; and this is the interpretation which has existed longer.

$$Q = \frac{\text{sum of factorial variations}}{\text{total error variation}} \tag{39.3}$$

The author uses the following definition since it is undesirable for interactions to exist.

$$Q = \frac{\text{sum of main effect variations}}{\text{total error variation} + \text{total interaction variation}} \tag{39.4}$$

Therefore, in deciding on superiority or inferiority when there are a number of methods of quantification or of change of variables, $M_1, M_2, \ldots, M_m$, one should resolve the sum of squares for each of these methods and calculate $Q_i\,(i = 1, 2, \ldots, m)$, and that method for which this quantity becomes the greatest would be the best. In principle, the problem has thus been solved, but actually it is a great deal of work to find the value of Q each time for the data which have been found by various methods of quantification. Thus, a method of finding the solution directly, going at once to the heart of the matter, would be desirable.

Formulae of quantification to maximize the value of Q will now be explained.

39.2.2 When the Number of Classes Is Two

When the number of classes is two, as for example non-defectives and defectives or grade-one products and grade-two products, it suffices to assign 0 to non-defectives and 1 to defectives. No matter what logical values a and b are given non-defectives and defectives, both ease and the value of Q agree, the choice is $a = 0$, $b = 1$. This means that it suffices to use $a = 0$ and $b = 1$, by which calculation is simple.

39.2.3 When the Number of Classes Is Three

When the data are separated into three groups, such as grade one-products, grade-two products, and grade-three products or good, somewhat defective, and defective, we may without losing generality designate grade-one products by 0, grade-two products by 1, and grade-three products by x, and we need only decide on x in such a way that the value of the SN ratio Q becomes maximum.

If the sum of the variations of main effects is called ΣS_{F_i} and the error variation, which also includes interaction variations, is now denoted by S_e, we must find the x which will maximize

$$Q = \frac{\sum S_{F_i}}{S_e} \tag{39.5}$$

But since S_{F_i} and S_e are quadratic expressions in x, we may put

$$\sum S_{F_i} = Ax^2 - 2Bx + C \tag{39.6}$$

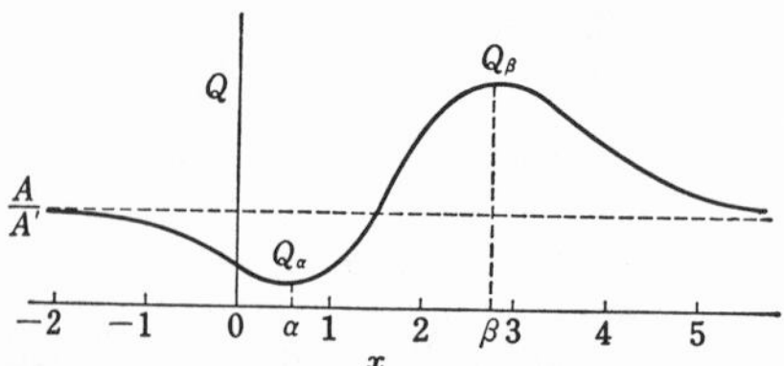

FIGURE 39.1 Relationship Between Quantified Value x and Variance Ratio Q

$$S_e = A'x^2 - 2B'x + C' \tag{39.7}$$

We then have

$$Q = \frac{Ax^2 - 2Bx + C}{A'x^2 - 2B'x + C'} \tag{39.8}$$

In Equation (39.8), both the numerator and the denominator are positive-definite quadratic expressions (quadratic equations which do not become negative) in x, and Q is as depicted by Figure 39.1 as a function of x.

It is clear from Figure 39.1 that the variance ratio Q has the asymptotic value A/A' at $x \to \pm\infty$, and that, in general, it possesses two extreme values, minimal value Q_α and maximal value Q_β, at $x = \alpha$ and $x = \beta$. Of these, $x = \beta$, which gives the maximal value, is the solution which is sought.

To find $x = \beta$, we differentiate Equation (39.8) with respect to x and set it equal to 0, then we find the two roots, α and β. The equation is

$$(AB' - A'B)x^2 - (AC' - A'C)x + (BC' - B'C) = 0 \tag{39.9}$$

Of the two roots of Equation (39.9), that by which Q becomes large is the solution we are seeking. If this solution is $\beta = 4.0$, grade-one products 0, grade-two products 1, and grade-three products 4 constitutes the quantification which we desire. It is indicated that in this case, to improve grade-three products to grade-two products requires *three times as much factorial effect* (three times as much substantial quantity of work) as improving grade-two products to grade-one products. Please refer to Section 39.3 for a calculation example.

When finding the solution, or after having found the solution, various cases develop. For example, when $AB' - A'B = 0$, $x = +\infty$ becomes the solution, and in this case, in effect grade-one products and grade-two products are taken together and are regarded as 0 and grade-three products are regarded as 1. In the case of the optimal solution, $\beta \leq 1.0$, it is indicated that sources which are effective toward dispersion are numerous, so that it is better to stop quantifying and to use accumulating analysis or minute accumulating analysis. Even if quantification has been performed well, calculations of accumulating analysis should also be performed separately. The reader is referred to Section 39.3.

39.2.4 When the Number of Classes Is Four or More

When the number of classes is four, when we form equations for the finding of x and y to maximize Q by putting class three and class four as x and y, these are simultaneous equations of a higher degree and it becomes rather troublesome to solve them. Therefore, a different method becomes desirable when the number of classes is four or more.*

The method to be explained here might be termed the sequential approximation method; we regard class three and class four tentatively as the same class and give this class the mark x. We give class one 0, class two 1, and classes three and four x, and, using the formula of Section 39.2.3, we perform a quantification which is appropriate to the combined group. Let us call the result x_0.

Next, we assume that x_0 is the mean of the marks which fundamentally should be given class three and class four, x_1 and x_2. Thus, if the number of items in class three is n_1 and the number of items in class four is n_2, we have

$$x_0 = \frac{n_1 x_1 + n_2 x_2}{n_1 + n_2} \tag{39.10}$$

Since x_0 is known, we solve the equation above for x_2 and obtain

$$x_2 = \frac{1}{n_2}[(n_1+n_2)x_0 - n_1 x_1] = (p_1+1)x_0 - p_1 x_1 \qquad \left(p_1 = \frac{n_1}{n_2}\right) \tag{39.11}$$

Next, we calculate the ratio Q by using x_1 and x_2 as a function of x_1 as given by the equation above. Since x_2 too is a linear function of x_1, Q becomes the ratio of quadratic equations in x_1. Since in this way we can find x_1 by again using the formula of Section 39.2.3, we find out about x_2, too. Or, one may also perform sequential approximation by the method of Chapter 15 of Volume 1 by giving the ranges of x_1 and x_2.

39.2.5 Quantification When Off-Scale and When a Life Test Is Discontinued Midway

In this case, the value which has gone off scale is taken as the unknown x. If it is a case of a length-of-life test, the life of that whose life did not end

*According to Genzaburō Masuyama, even using the method in which x is assigned to class 3 and y is assigned to class 4 and then x and y are found directly, it is possible to get a cubic equation for Q_m if Q is taken as the maximum value and Q_m is taken as a parameter; and it is possible to calculate x and y by using its roots.

after a specific period of time is taken as x. *We choose the value of x so that the accuracy becomes the best.* Therefore, if range of measurement of the measuring instrument is wider, it might be that the value when it has become off-scale is different from the value of the true life if the life test were conducted to the end. Therefore, if this value turns out to be contrary to common sense, it is better to resort to accumulating analysis.

Finding the value of x is, strictly, in order to improve the precision of detection of various sources that are influencing a characteristic value such as the life. If we find the sum of the factorial variations and the error variation with x still inserted into the off-scale values, the ratio will be the ratio of quadratic expressions in x, so that it is posssible to use the method of Section 39.2.3.

39.3 An Example of a Case When the Number of Classes of Categorization Is Three

We will quantify the data of operability of electric welding, which was discussed in Section 7.2 of Volume 1. In that experiment, operability was categorized into the three classes of easy, ordinary, and difficult. Supposing that easy is 0, ordinary is 1, and difficult is x, we quantify x. In this experiment, assignment was to orthogonal array L_{16} in order to find nine main effects and four interactions, but since with good quantification interactions too are small, here let us try quantifying so as to cause

$$Q = \frac{\text{sum of main effect variations}}{\text{total variation} - \text{sum of main effect variations}} \tag{39.12}$$

to be maximum.

The supplementary table for the main effects of A–I when easy is 0, ordinary is 1, and difficult is x is given by Table 39.1.

Therefore, the variations as continuous values after the quantification become as follows.

Table 39.1 Supplementary Table

	Total		Total		Total
A_1	$3+2x$	D_1	5	G_1	7
A_2	$6+x$	D_2	$4+3x$	G_2	$2+3x$
B_1	$4+2x$	E_1	$5+x$	H_1	$4+2x$
B_2	$5+x$	E_2	$4+2x$	H_2	$5+x$
C_1	$4+x$	F_1	$5+3x$	I_1	$5+2x$
C_2	$5+2x$	F_2	4	I_2	$4+x$
Total	$9+3x$	Total	$9+3x$	Total	$9+3x$

$$S_T=0^2\times4+1^2\times9+x^2\times3-\frac{(9+3x)^2}{16}=\frac{63-54x+39x^2}{16} \tag{39.13}$$

$$S_A=\frac{(3+2x-6-x)^2}{16}=\frac{9-6x+x^2}{16} \tag{39.14}$$

$$16S_B=(4+2x-5-x)^2=1-2x+x^2$$

$$16S_C=1+2x+x^2$$

$$16S_D=1-6x+9x^2$$

$$16S_E=1-2x+x^2$$

$$16S_F=1+6x+9x^2$$

$$16S_G=25-30x+9x^2$$

$$16S_H=1-2x+x^2$$

$$16S_I=1+2x+x^2$$

Therefore, the sum of the variations of the main effects is

$$S_A+S_B+\cdots+S_I=\frac{41-38x+33x^2}{16} \tag{39.15}$$

From this, the error variation S_e including the interaction variations is

$$S_e=S_T-(S_A+S_B+\cdots+S_I)=\frac{63-54x+39x^2-41+38x-33x^2}{16}$$

$$=\frac{22-16x+6x^2}{16} \tag{39.16}$$

Therefore, the SN ratio Q is

$$Q=\frac{S_A+S_B+\cdots+S_I}{S_e}=\frac{33x^2-38x+41}{6x^2-16x+22} \tag{39.17}$$

The x which causes the SN ratio Q to be maximum is the greater of the two roots of the following equations.

$$\begin{vmatrix}33 & 6\\19 & 8\end{vmatrix}x^2-\begin{vmatrix}33 & 6\\41 & 22\end{vmatrix}x+\begin{vmatrix}19 & 8\\41 & 22\end{vmatrix}=0 \tag{39.18}$$

$$150x^2-480x+90=0$$

$$5x^2-16x+3=0 \tag{39.19}$$

From this, the two roots are 0.2 and 3, and the solution being sought is

$$x=3 \tag{39.20}$$

Thus, in regard to operability, this means that *if easy is expressed as 0 and ordinary as 1, 3 is the optimum quantification for difficult.* This indicates that improving operability from difficult to ordinary requires

Table 39.2 Assignment and Quantified Data

	A 1	G 2	$A\times G$ 3	H 4	$A\times H$ 5	$G\times H$ 6	B 7	D 8	E 9	F 10	I 11	e 12	e 13	$A\times C$ 14	C 15	Operability			Quantification	Minute accumulating analysis			
																Easy	Ordinary	Difficult		ω_1	ω_2	ω_3	Total
1	1	1	1	1	1	1	1	1	1	1	1	1	1	1	1	0	1	0	1	1	0	0	1
2	1	1	1	1	1	1	1	2	2	2	2	2	2	2	2	0	1	0	1	1	0	0	1
3	1	1	1	2	2	2	2	1	1	1	1	2	2	2	2	0	1	0	1	1	0	0	1
4	1	1	1	2	2	2	2	2	2	2	2	1	1	1	1	1	0	0	0	0	0	0	0
5	1	2	2	1	1	2	2	1	1	2	2	1	1	2	2	1	0	0	0	0	0	0	0
6	1	2	2	1	1	2	2	2	2	1	1	2	2	1	1	0	0	1	3	1	1	1	3
7	1	2	2	2	2	1	1	1	1	2	2	2	2	1	1	1	0	0	0	0	0	0	0
8	1	2	2	2	2	1	1	2	2	1	1	1	1	2	2	0	0	1	3	1	1	1	3
9	2	1	2	1	2	1	2	1	2	1	2	1	2	1	2	0	1	0	1	1	0	0	1
10	2	1	2	1	2	1	2	2	1	2	1	2	1	2	1	0	1	0	1	1	0	0	1
11	2	1	2	2	1	2	1	1	2	1	2	2	1	2	1	0	1	0	1	1	0	0	1
12	2	1	2	2	1	2	1	2	1	2	1	1	2	1	2	0	1	0	1	1	0	0	1
13	2	2	1	1	2	2	1	1	2	2	1	1	2	2	1	1	0	0	0	0	0	0	0
14	2	2	1	1	2	2	1	2	1	1	2	2	1	1	2	0	0	1	3	1	1	1	3
15	2	2	1	2	1	1	2	1	2	2	1	2	1	1	2	0	1	0	1	1	0	0	1
16	2	2	1	2	1	1	2	2	1	1	2	1	2	2	1	0	1	0	1	1	0	0	1
Total																4	9	3	18	12	3	3	18

about twice as much source effect as improving operability from ordinary to easy.

Let us analyze the data as continuous values by using the quantified data. The data after quantification are as given by Table 39.2. We analyze by minute accumulating analysis. This is because first-order source relations analyzed by minute accumulating analysis give exactly the same results as analysis of continuous values for quantified data.

$$S_m=\frac{18^2}{48}=6.75 \qquad (f=1) \tag{39.21}$$

$$S_{T_1}=\frac{1}{3}(1^2+1^2+\cdots+3^2+1^2+1^2)-S_m=5.25 \qquad (f=1) \tag{39.22}$$

$$S_A=\frac{(9-9)^2}{48}=0.00 \qquad (f=1) \tag{39.23}$$

$$S_B=\frac{(10-8)^2}{48}=0.08 \qquad (f=1) \tag{39.24}$$

It is the same from S_C on.

$$S_\omega=\frac{1}{16}(12^2+3^2+3^2)-6.25=3.88 \qquad (f=2) \tag{39.25}$$

$$S_{A\times\omega}=\frac{1}{8}(5^2+2^2+2^2+7^2+1^2+1^2)-CF-S_A-S_\omega$$

$$=0.38 \qquad (f=2) \tag{39.26}$$

Table 39.3 Minute Accumulation Analysis of Variance Table

Source	f	S	V	ρ[%]
A	1	0.00	0.00○	
B	1	0.08	0.08○	
C	1	0.33	0.33	
D	1	1.33	1.33**	23.8
E	1	0.08	0.08○	
F	1	2.08	2.08**	38.1
G	1	0.33	0.33	
H	1	0.08	0.08○	
I	1	0.33	0.33	
$A\times C$	1	0.08	0.08○	
$A\times G$	1	0.08	0.08○	
$A\times H$	1	0.00	0.00○	
$G\times H$	1	0.00	0.00○	
e_1	2	0.42	0.21○	
○ marks pooled (e_1)	(10)	(0.82)	0.082**	
ω	2	3.38	1.94**	100.0
$A\times\omega$	2	0.38	0.19**	
$B\times\omega$	2	0.04	0.02○	
$C\times\omega$	2	0.04	0.02○	
$D\times\omega$	2	0.04	0.02○	
$E\times\omega$	2	0.04	0.02○	
$F\times\omega$	2	0.04	0.02○	
$G\times\omega$	2	1.04	0.52**	
$H\times\omega$	2	0.04	0.02○	
$I\times\omega$	2	0.04	0.02○	
$A\times C\times\omega$	2	0.04	0.02○	
$A\times G\times\omega$	2	0.04	0.02○	
$A\times H\times\omega$	2	0.38	0.19**	
$G\times H\times\omega$	2	0.38	0.19**	
e_2	4	0.08	0.02○	
○ marks pooled (e_2)	(22)	(0.44)	(0.020)	
T	47	11.25		

$$S_{B\times\omega}=\frac{1}{8}(6^2+2^2+2^2+6^2+1^2+1^2)-CF-S_B-S_\omega$$

$$=0.04 \qquad (f=2) \tag{39.27}$$

It is exactly the same with regard to the other sources. The total variation is

$$S_T=18-\frac{18^2}{48}=11.25 \qquad (f=47) \tag{39.28}$$

Table 39.3 is obtained when these are organized. Of the sources, the variations from A to e_1 are exactly the same as (one-third of) the variations of continuous values.

Table 39.4 Effect of G

	Easy	Ordinary	Difficult	Total	Mean value
G_1	1	7	0	8	0.875
G_2	3	2	3	8	1.375

When inter-experiment variation is regarded as 100, the contribution ratios of D and F are 23.8% and 38.1%, and they have great effect, but G, which was significant by accumulating analysis, has become not significant. The effect of G is such that, as seen in Table 39.4, either dispersion of operability has effect or two large main effects, in this case D and F, produced that effect on their interaction column. This means that for an effect such as this, quantification leads to a great error.

The author therefore recommends that even when quantifying, one always perform accumulating analysis (or minute accumulating analysis). Also, even if continuous values have been given originally, he recommends that calculation also be conducted by minute accumulating analysis or, after categorizing, by accumulating analysis. This is because often, whether by quantification or by continuous values, it does not constitute a solution of the problem merely by analyzing continuous values.

Exercise (39)

(1) In the life test data of Exercise (32), quantify the data marked with ◎. Put the ◎ mark as x and find x which causes (main effect variations)/(variations of the remainder) to be maximum. Perform an analysis of variance using the quantified data, and compare with minute accumulating analysis when not quantified.

40 Mathematical Principles

An explanation of most of the methods of traditional mathematical statistics which are relevant to this work will be presented in this chapter. However, testing and confidence limits are left for reference elsewhere, as their importance here is small.

40.1 Proof of Formula for Randomized Experiments

When samples of size n, $x_1, x_2, \ldots, x_n$ have been randomly selected from among $2n$ numerical values $X_1, X_2, \ldots, X_{2n}$, the mean value of these, $\bar{x}$, and its variance $\mathrm{Var}(\bar{x})$ are as follows.

The ways of taking n things from $X_1, X_2, \ldots, X_{2n}$ total $\binom{2n}{n}$, and one of them is selected. Thus, if we obtain the sum for each of the $\binom{2n}{n}$ combinations, the respective X_i will in effect exist therein to the extent of $\binom{2n}{n} \times n \times \frac{1}{2n}$ each, in every case, so that we have

$$E(\bar{x})=\frac{1}{\binom{2n}{n}}\times\frac{1}{n}\times\binom{2n}{n}\times n\times\frac{1}{2n}(X_1+X_2+\cdots+X_{2n})$$

$$=\bar{X} \tag{40.1}$$

$$\text{Var}(\bar{x})=E[(\bar{x}-\bar{X})^2]=E\left\{\left[\frac{(x_1-\bar{X})+\cdots+(x_n-\bar{X})}{n}\right]^2\right\}$$

$$=\frac{1}{n^2}E\{(x_1-\bar{X})^2+\cdots+(x_n-\bar{X})^2+\sum_{i=1}^{n}(x_i-\bar{X})[(x_1-\bar{X})+\cdots$$

$$+(x_{i-1}-\bar{X})+(x_{i+1}-\bar{X})+\cdots+(x_n-\bar{X})]\} \tag{40.2}$$

If we obtain the sum of $(x_1-\bar{X})^2, \ldots, (x_n-\bar{X})^2$ for every combination of taking n from $2n$, $\binom{2n}{n}$, $(X_1-\bar{X})^2, \ldots, (X_{2n}-\bar{X})^2$ appear with equal frequency by virtue of symmetry; and moreover, since its total frequency is $\binom{2n}{n}\times n$, the frequency with which $(X_i-\bar{X})^2$ appears is $1/2n$ thereof, or

$$\frac{1}{2n}\times\binom{2n}{n}\times n=\frac{1}{2}\binom{2n}{n} \tag{40.3}$$

times. Also, the number of the whole in the sum of the product terms is $n(n-1)$, and since the frequency with which $(X_i-\bar{X})(X_j-\bar{X})(i\neq j)$ appears should be equal for all, the frequency with which each term appears would be

$$\frac{1}{2n(2n-1)}\times\binom{2n}{n}\times n(n-1)=\frac{n-1}{2(2n-1)}\binom{2n}{n} \tag{40.4}$$

But now,

$$(X_i-\bar{X})[(X_1-\bar{X})+\cdots+(X_{i-1}-\bar{X})+(X_{i+1}-\bar{X})+\cdots+(X_{2n}-\bar{X})]$$

$$=-(X_i-\bar{X})^2 \tag{40.5}$$

Thus, the value of Equation (40.2) becomes

$$\text{Var}(\bar{x})=\frac{1}{n^2}\frac{1}{\binom{2n}{n}}\left\{\frac{1}{2}\binom{2n}{n}[(X_1-\bar{X})^2+\cdots+(X_{2n}-\bar{X})^2]\right.$$

$$\left.+\frac{n-1}{2(2n-1)}\binom{2n}{n}[-(X_1-\bar{X})^2-\cdots-(X_{2n}-\bar{X})^2]\right\}$$

$$=\frac{1}{n^2}\left\{\left[\frac{1}{2}-\frac{n-1}{2(2n-1)}\right](2n-1)\sigma_X{}^2\right\}=\frac{1}{2n}\sigma_X{}^2 \tag{40.6}$$

Similarly, we have

$$E(\bar{y})=\bar{Y} \tag{40.7}$$

$$\text{Var}(\bar{y})=\frac{1}{2n}\sigma_Y{}^2 \tag{40.8}$$

Next, we consider the expected value of $(\overline{x} - \overline{X})(\overline{y} - \overline{Y})$. In

$$E[(\bar{x}-\bar{X})(\bar{y}-\bar{Y})]=E\left\{\frac{[(x_1-\bar{X})+\cdots+(x_n-\bar{X})][(y_1-\bar{Y})+\cdots+(y_n-\bar{Y})]}{n^2}\right\}$$

$$=\frac{1}{n^2}E[\sum_{ij}(x_i-\bar{X})(y_j-\bar{Y})] \tag{40.9}$$

we consider the frequency with which $(X_i - \overline{X})(Y_j - \overline{Y})$ appears in the n^2 terms within the [ ]. It does not happen that $i=j$, and if $i \neq j$, the frequency with which this combination appears should be equal for all. This means that the mean frequency with which the respective product terms appear is

$$\frac{1}{\binom{2n}{n}}\times\frac{n^2}{2n(2n-1)}\times\binom{2n}{n}=\frac{n}{2(2n-1)} \tag{40.10}$$

Therefore, with ρ as the correlation coefficient, we have

$$E[(\bar{x}-\bar{X})(\bar{y}-\bar{Y})]=\frac{1}{n^2}\times\frac{n}{2(2n-1)}[\sum_{i\neq j}(X_i-\bar{X})(Y_j-\bar{Y})]$$

$$=\frac{1}{2n(2n-1)}\sum_{i=1}^{2n}(X_i-\bar{X})[(Y_1-\bar{Y})+\cdots+(Y_{i-1}-\bar{Y})$$

$$+(Y_{i+1}-\bar{Y})+\cdots+(Y_{2n}-\bar{Y})]$$

$$=\frac{-1}{2n(2n-1)}\sum_{i=1}^{2n}(X_i-\bar{X})(Y_i-\bar{Y})=-\frac{1}{2n}\rho\sigma_X\sigma_Y \tag{40.11}$$

Therefore,

$$E(\bar{x}-\bar{y})=\bar{X}-\bar{Y} \tag{40.12}$$

$$\mathrm{Var}(\bar{x}-\bar{y})=E[(\bar{x}-\bar{X})-(\bar{y}-\bar{Y})]^2$$

$$=E[(\bar{x}-\bar{X})^2-2(\bar{x}-\bar{X})(\bar{y}-\bar{Y})+(\bar{y}-\bar{Y})^2] \tag{40.13}$$

By substituting from Equations (40.6), (40.8), and (40.11), we get

$$\mathrm{Var}(\bar{x}-\bar{y})=\frac{1}{2n}\sigma_X{}^2-2\times\frac{-1}{2n}\rho\sigma_X\sigma_Y+\frac{1}{2n}\sigma_Y{}^2$$

$$=\frac{1}{2n}(\sigma_X{}^2+2\rho\sigma_X\sigma_Y+\sigma_Y{}^2) \tag{40.14}$$

This ends the proof.

In an actual problem, values such as of $\sigma_X{}^2$, $\sigma_Y{}^2$, and ρ are unknown. However, the following inequality holds.

$$\mathrm{Var}(\bar{x}-\bar{y})\leq\frac{2}{n}\times\frac{\sigma_X{}^2+\sigma_Y{}^2}{2} \tag{40.15}$$

This is because

$$\frac{1}{2n}(\sigma_X^2+2\rho\sigma_X\sigma_Y+\sigma_Y^2)\leq\frac{1}{2n}(\sigma_X^2+2\sigma_X\sigma_Y+\sigma_Y^2)=\frac{1}{2n}(\sigma_X+\sigma_Y)^2$$

$$\leq\frac{1}{2n}[(\sigma_X+\sigma_Y)^2+(\sigma_X-\sigma_Y)^2]=\frac{2}{n}\times\frac{\sigma_X^2+\sigma_Y^2}{2}$$

Thus, the more σ_X^2 and σ_Y^2 differ, the more does the quantity in Equation (40.14) become smaller than the right side of (40.15). This indicates that even if the variances of A_1 and A_2 are unequal, we are on the safe side when we analyze by using the means of the variances of A_1 and A_2. The means of these variances can be found by dividing the error variance S_e by the number of degrees of freedom. Therefore, *even if the distributions are not homoscedastic,*

$$F_0=\frac{(A_1-A_2)^2/2n}{V_e} \tag{40.16}$$

$$\bar{A}_1-\bar{A}_2\pm\sqrt{F\times V_e\times\frac{2}{n}} \tag{40.17}$$

and this means that *it does not matter if one performs the F test or creates the confidence limits of the difference.* It also proves that *the method of testing beforehand whether or not the variance is equal for A_1 and A_2, before testing the difference of the measured values, is useless.*

40.2 *F* Test

40.2.1 Probability and Statistics

The *F* test was derived by Fisher. It was not by any means derived as a variance ratio but was derived as a likelihood ratio test. An explanation of the shortest way of deriving the *F* test from the likelihood ratio test is the purpose of this section. Since the author considers the *F* value as the ratio of variance, he does not take the standpoint of this section. Variance ratio is a concept that holds more broadly. Please refer to Chapter 18 concerning this. In this section, however, an explanation will be presented from Fisher's standpoint.

The probability P_r that a phenomenon which occurs with probability p occurs r times when tried independently n times is determined by the theory of probability as a binomial distribution term,

$$P_r=\binom{n}{r}p^r(1-p)^{n-r}\quad(r=0,\ 1,\ \cdots,\ n) \tag{40.18}$$

According to the theory of probability, if certain hypotheses are used regarding the frequency of occurrence of phenomena, one makes a prediction regarding the result of the trials, such as that it will occur in the proportion r times in n trials or that the probability that it will occur at least r times is so much. If this is shown schematically, we ask,

Premise regarding phenomenon	→	With what degree of ease does what phenomenon occur?

But in statistics, we ask: What is the true proportion in which the phenomenon will occur, based on observed data that a certain phenomenon has occurred? Here, given that the phenomenon has occurred r times as a result of n trials, we consider, About what may we assume will be the proportion p in which this phenomenon occurs? In other words, we probe the cause from the result.

Data that a certain kind of phenomenon has occurred	→	What is the probability that the phenomenon will occur?

If we use a medical example, "What will be the symptoms when the disease is A?" is the standpoint of the theory of probability, and "These were the symptoms. What is the disease?" is the standpoint of the theory of statistics. Thus, the theory of probability and the theory of statistics have completely opposite standpoints; whereas the former is deductive, the latter is inductive.

40.2.2 Likelihood

It can be said that the foundation of the theory of statistics was laid by R. A. Fisher of Great Britain. The most important idea thereof is the concept of *likelihood.* For example, let us assume that in 100 independent observations, the frequency of occurrence of a certain phenomenon was 30 times. If the probability of occurrence of the phenomenon is p, the probability that it will occur 30 times in 100 trials is, from probability theory,

$$L_p = \binom{100}{30} p^{30}(1-p)^{70} \tag{40.19}$$

Fisher regarded the quantity above as a function of p and denoted it by L_p; he named it the *likelihood function,* and gave us the following important concept.

[FISHER'S THEORY REGARDING LIKELIHOOD]

After having obtained the data that the phenomenon has occurred 30 times in 100 independent trials, the likelihood that the probability of occurrence of this phenomenon is p is proportional to L_p.

The value of L_p does not give the probability that p is p, but it indicates the likelihood that it is p. For example, let us assume that high fever is always suffered when one has disease A or disease B. With disease C, high fever is suffered only by about 30% of people. With disease D, let us assume that high fever does not occur. In this case, if we find out only that a certain patient has high fever, the likelihood becomes as follows.

The likelihood of disease A $= C \times 1$
The likelihood of disease B $= C \times 1$
The likelihood of disease C $= C \times 0.3$
The likelihood of disease D $= C \times 0.0$

C here is a constant. What to do with respect to its value is completely unknown for the time being. However, if there is no additional information (for example, the information that disease A is common but disease B is rare), the likelihood of the disease being A and the likelihood of its being B are the same; the likelihood of its being C is 0.3 times that of its being A and that of its being B, and the likelihood of its being D is 0.

In the case of Equation (40.19), we can consider the possibility that the value of p as the cause is every value from 0 to 1, but the likelihood that p is 0.9 is only $L_{0.9}/L_{0.3}$ times as much as the likelihood that it is $p = 0.3$. If actually calculated,

$$\frac{L_{0.9}}{L_{0.3}} = \frac{\binom{100}{30}(0.9)^{30}(0.1)^{70}}{\binom{100}{30}(0.3)^{30}(0.7)^{70}} = 1.435 \times 10^{-45} \tag{40.20}$$

This means that the likelihood that it is $p = 0.9$ is 1.435×10^{-45} times as much as the likelihood that it is $p = 0.3$.

Likelihood function L_p is a function of p, but it does not indicate the probability that the true value is a value between p and $p + dp$. It should be clearly understood that it is not a distribution function of the true value p.

The p which renders L_p maximum is the value of p such that the likelihood that it is so is maximum. This is termed the maximum likelihood estimate of p. One need only differentiate the L_p of Equation (40.19) with respect to p and put it equal to zero.

$$\frac{d}{d_p}L_p = \binom{100}{30}[30p^{30-1}(1-p)^{70} + p^{30} \times 70(1-p)^{69}(-1)] = 0$$

We obtain the maximum likelihood estimate of p, $\hat{p}$.

$$\hat{p} = 0.3 \tag{40.21}$$

For the case where the event has occurred r times in n trials, too, by exactly the same calculation, the maximum likelihood estimate is

$$\hat{p} = \frac{r}{n} \tag{40.22}$$

It is a real-life problem that the likelihood changes as the investigation is pushed ahead. For example, assuming that a murder has been committed, although at the beginning five persons, A, B, C, D, and E, all had about the same degree of likelihood of being the criminal, if it is discovered as a result of investigation that A had an alibi, the likelihood of A having been the criminal instantly becomes zero. If a machine breaks down, even if the likelihood of causation by A is twice as much as by B and that by C is three times as much as by B, once investigated it becomes determined that the cause is just one among A, B, and C. Although likelihood *should be used only for mutual comparisons* in a condition of insufficient knowledge, *there was a great problem point in that Fisher attempted to use it in calculating statistically.* The author feels that it is more desirable to use testing and confidence limits, deciding on the likely range and not-likely range only by the likelihood ratio, even when taking the viewpoint of so-called mathematical statistics. Here, however, he will introduce only the traditional method of the *F* test. We will explain by a simple example.

40.2.3 *F* Test as a Likelihood Ratio Test

We will explain by the example of Section 21.1. Let us assume that we wish to investigate whether or not the law of equipartition of energy for the random electric current which derives from the thermal motions of electrons in the circuit,

$$y = k\frac{T}{C} \tag{40.23}$$

holds or does not hold. Here, T is the absolute temperature, C is the capacity of the circuit, k is Boltzmann's constant, 1.38×10^{-16} erg/°K, and y is the mean square of the current. Let us assume that the absolute temperature T is varied over the a levels, $T_1, T_2, \ldots, T_a$ and the circuit capacity is varied over b-levels $C_1, C_2, \ldots, C_b$, and that we have measured the power of the random current y under the combinations ab. This constitutes experimental data by a two-way layout.

The measured value of y for T_iC_j is expressed as y_{ij}'. If formula (40.23) holds,

$$y_{ij}'=k\frac{T_i}{C_j} \quad (i=1, 2, \cdots, a, j=1, 2, \cdots, b)$$

should hold within the range of measurement error.

When there is such an expectation of a formula, it is best to perform a conversion such as the following on the data y_{ij}' so that one's own expectation becomes cancelled.

$$y_{ij}=y_{ij}'\times\frac{C_j}{T_i}-k \tag{40.24}$$

Table 40.1 Data by Two-Way Layout

T \ C	C_1	$C_2 \cdots\cdots C_b$
T_1	y_{11}'	$y_{12}' \cdots\cdots y_{1b}'$
T_2	y_{21}'	$y_{22}' \cdots\cdots y_{2b}'$
$\vdots$	$\vdots$	$\vdots \quad \ddots \quad \vdots$
T_a	y_{a1}'	$y_{a2}' \cdots\cdots y_{ab}'$

We revise the data of Table 40.1 to y_{ij}. If y' is not proportional to T, $y = y'/T$ becomes a function of T. Thus, for y_{ij} we consider a structure such as

$$y_{ij}=m+t_i+c_j+\varepsilon_{ij} \tag{40.25}$$

Here, $t_1, \ldots, t_a$ show the degrees to which the current has departed from proportionality to T at $T_1, T_2, \ldots, T_a$, and without losing generality we may assume the condition

$$t_1+t_2+\cdots+t_a=0 \tag{40.26}$$

If y' is exactly proportional to T, $t_1 = t_2 = \cdots = t_a = 0$ should obtain.

Similarly, c_j is a modulus which indicates the degree to which y' has departed from inverse proportionality to C; if $c_1 = c_2 = \cdots = c_b = 0$, it means that y does not contain the effect of C, so it is indicated that y' is inversely proportional to C. ϵ_{ij} is termed experimental error, but in a case such as this it may be regarded as measurement error. m is termed the general mean, and if m is zero it is indicated that the proportionality constant k is exactly Boltzmann's constant.

Thus, this problem is one of testing three hypotheses based on ab measured values, *viz.*,

$$\left.\begin{array}{l} H_0\colon m = 0 \text{ (the proportionality constant is Boltzmann's constant)} \\ H_0'\colon t_1 = t_2 = \cdots t_a = 0 \text{ (}y\text{ is proportional to }T\text{)} \\ H_0''\colon c_1 = c_2 = \cdots = c_b = 0 \text{ (}y\text{ is inversely proportional to }C\text{)} \end{array}\right\} \tag{40.27}$$

Regarding this testing problem, we will determine what test to perform assuming that the distribution of the measurement errors follows a normal distribution of mean 0 and variance σ^2. Since y_{ij} follows the normal distribution $N(m + t_i + c_j, \sigma^2)$, its probability density $f(y_{ij})$ is

$$f(y_{ij})=\frac{1}{\sqrt{2\pi}\,\sigma}e^{-1/2\sigma^2(y_{ij}-m-t_i-c_j)^2} \tag{40.28}$$

The population mean of y_{ij} becomes zero only when m, a_i, and b_j are all zero, but since we wish to test whether or not m, a_i, and b_j are zero, it might not be zero; and therefore, we are considering a density function such as Equation (40.28). To get the likelihood for the whole of the data of size ab, we need only multiply together Equation (40.28) for each i and j.

$$L(m, t, c, \sigma^2)=\left(\frac{1}{\sqrt{2\pi}\,\sigma}\right)^{ab} e^{-\frac{1}{2\sigma^2}\sum_{ij}(y_{ij}-m-t_i-c_j)^2} \tag{40.29}$$

First, let us derive the formula for testing of m. We use all of m, t_i, c_j, and σ^2, and we calculate the maximum likelihood estimate so that the L of the equation above becomes maximum. If we partially differentiate the logarithm of Equation (40.29) by with respect to m, t_i, c_j, and σ^2 and put the results equal to zero, we have,

$$\left.\begin{array}{ll} \text{For } m & \hat{m}=\frac{1}{ab}(\sum_{ij} y_{ij}) \\ \text{For } t_i & \hat{t}_i=\frac{1}{b}\sum_{j} y_{ij}-\frac{1}{ab}(\sum_{ij} y_{ij}) \\ \text{For } c_j & \hat{c}_j=\frac{1}{a}\sum_{i} y_{ij}-\frac{1}{ab}(\sum_{ij} y_{ij}) \\ \text{For } \sigma^2 & \hat{\sigma}^2=\frac{1}{ab}\sum_{ij}(y_{ij}-\frac{1}{b}\sum_{j} y_{ij}-\frac{1}{a}\sum_{i} y_{ij}+\frac{1}{ab}\sum_{ij} y_{ij})^2 \end{array}\right\} \tag{40.30}$$

Since the notation is cumbersome in the equations above, we introduce symbols such as the following. G is the total of all data, T_i is the total of the experimental values y at temperature T_i, and similarly with regard to C_j:

$$G=\sum_{ij} y_{ij}, \quad T_i=\sum_{j} y_{ij}, \quad C_j=\sum_{i} y_{ij}$$

When these are substituted, Equations (40.30) become

$$\left.\begin{array}{l} \hat{m}=G/ab \\ \hat{t}_i=T_i/b-G/ab \\ \hat{c}_j=C_j/a-G/ab \\ \hat{\sigma}^2=\sum_{ij}(y_{ij}-T_i/b-C_j/a+G/ab)^2/ab \end{array}\right\} \tag{40.31}$$

When Equations (40.31) are substituted into (40.29), the maximum likelihood L_0 becomes

$$L_0=\left(\frac{1}{2\pi}\times\frac{1}{\hat{\sigma}^2}\right)^{ab/2} e^{-ab/2} \tag{40.32}$$

To test hypothesis H_0: $m=0$, we need merely investigate to what degree the likelihood L becomes unlikely when $m=0$. If the likelihood is close to L_0 even when $m=0$, hypothesis H_0 is accepted; and if the likelihood is small, hypothesis H_0 is rejected. For this, we need only find the likelihood L for $m=0$ and compare this with L_0.

We put $m = 0$ and find the maximum likelihood estimates of t_i, c_j, and σ^2, *viz.*, $\hat{t}_i$, $\hat{c}_j$, and $\hat{\sigma}^2$.

$$\left.\begin{aligned} \hat{t}_i &= T_i/b - G/ab \\ \hat{c}_j &= C_j/a - G/ab \\ \hat{\sigma}^2 &= \sum_{ij}(y_{ij} - \hat{t}_i - \hat{c}_j)^2 \end{aligned}\right\} \tag{40.33}$$

When these are substituted into Equation (40.29), we find the maximum likelihood for $m = 0$.

$$L(0, \hat{t}, \hat{c}, \hat{\sigma}^2) = \left[\frac{1}{2\pi} \times \frac{1}{\sum_{ij}(y_{ij} - \hat{t}_i - \hat{c}_j)^2}\right]^{ab/2} e^{-ab/2} \tag{40.34}$$

We divide Equation (40.34) by (40.32) and obtain the likelihood ratio λ.

$$\lambda = \left[\frac{\sum(y_{ij} - \hat{m} - \hat{t}_i - \hat{c}_j)^2}{\sum(y_{ij} - \hat{t}_i - \hat{c}_j)^2}\right]^{ab/2} \tag{40.35}$$

Now, we have

$$\begin{aligned} \sum(y_{ij} - \hat{t}_i - \hat{c}_j)^2 &= \sum(y_{ij} - \hat{m} - \hat{t}_i - \hat{c}_j + \hat{m})^2 \\ &= \sum(y_{ij} - \hat{m} - \hat{t}_i - \hat{c}_j)^2 + 2\hat{m}\sum(y_{ij} - \hat{m} - \hat{t}_i - \hat{c}_j) + \sum(\hat{m})^2 \\ &= \sum(y_{ij} - \hat{m} - \hat{t}_i - \hat{c}_j)^2 + \frac{G^2}{ab} \end{aligned}$$

If we set

$$S_e = \sum_{ij}(y_{ij} - \hat{m} - \hat{t}_i - \hat{c}_j)^2$$

$$S_m = G^2/ab$$

the λ of Equation (40.35) becomes

$$\lambda = \left(\frac{S_e}{S_e + S_m}\right)^{ab/2} = \left(1 + \frac{S_m}{S_e}\right)^{-ab/2} \tag{40.36}$$

The magnitude of λ corresponds 1-to-1 to the magnitude of S_m/S_e. Although there is a problem in considering the distribution of λ, if it is assumed that this is considered, to cause the point of level of significance α of λ to be λ_α corresponds 1-to-1 with the point of level of significance α of the statistic S_m/S_e. But now, under the hypothesis $m = 0$ it can be proved that S_m/σ^2 assumes a χ^2 distribution of one degree of freedom, and that S_e/σ^2 assumes a χ^2 distribution of $(a - 1)(b - 1)$ degrees of freedom. Please refer to the next section concerning distributions. Since, moreover, S_m and S_e distribute independently,

$$F = \frac{S_m}{S_e/(a-1)(b-1)} \tag{40.37}$$

assumes an F distribution of number of degrees of freedom of the numerator 1 and number of degrees of freedom of the denominator $(a-1)(b-1)$. Since the critical values of F of levels of significance 5% and 1% are given in the F distribution table, it is possible to perform the F test by comparing these values with the actual statistics.

Similarly, for zero hypotheses H_0' and H_0'', the statistics

$$F_T = \frac{S_T/(a-1)}{S_e/(a-1)(b-1)} \quad \text{[F distribution of number of degrees of freedom $(a-1)$ and $(a-1)(b-1)$]} \tag{40.38}$$

$$F_C = \frac{S_C/(b-1)}{S_e/(a-1)(b-1)} \quad \text{[F distribution of number of degrees of freedom $(b-1)$ and $(a-1)(b-1)$]} \tag{40.39}$$

correspond if

$$S_T = \frac{T_1^2 + \cdots + T_a^2}{b} - \frac{G^2}{ab}$$

$$S_C = \frac{C_1^2 + \cdots + C_b^2}{a} - \frac{G^2}{ab}$$

Calculations specifically of, for example, S_T, S_C, S_m, and S_e are performed as follows.

$$S_m = \frac{G^2}{ab}$$

$$S_T = \frac{T_1^2 + \cdots + T_a^2}{b} - S_m$$

$$S_C = \frac{C_1^2 + \cdots + C_b^2}{a} - S_m$$

$$S_e = y_{11}^2 + y_{12}^2 + \cdots + y_{ab}^2 - (S_m + S_T + S_C)$$

These calculations can be organized as in Table 40.2. Here, the variance $V = S/f$ and the variance ratio $F = V/V_e$. Please refer to Chapter 18 and Chapter 19 regarding the meaning of the analysis of variance table and how to use it.

Table 40.2 Analysis of Variance Table

Source	f	S	V	F
m (prop. const.)	1	S_m	V_m	F_m
T (absol. temp.)	$a-1$	S_T	V_T	F_T
C (capacity)	$b-1$	S_C	V_C	F_C
e (error)	$(a-1)(b-1)$	S_e	V_e	
Total	ab	S_{Total}		

40.3 Distribution of Statistics

Some statistics discussed in mathematical statistics will be explained. Reference 32) gives details regarding various distributions.

The density function $f(y)$ of a normal distribution of mean value m and variance σ^2 is

$$f(y)=\frac{1}{\sqrt{2\pi}\,\sigma}e^{-\frac{1}{2\sigma^2}(y-m)^2} \tag{40.40}$$

According to the central limit theorem, the distribution of measured values assumes a normal distribution when the dispersion of the data has resulted from the influences of countless small causes.

The mean value $\bar{y}$ of n measured values which have the same normal distribution, $y_1, y_2, \ldots, y_n$, assumes a normal distribution of mean m and variance σ^2/n. Also, the quotient obtained by dividing error variation S_e by σ^2 assumes a χ^2 distribution of $n-1$ degrees of freedom. If its density function is expressed as $f(\chi^2)$, we have

$$f(\chi^2)=\frac{1}{2\Gamma\left(\frac{n-1}{2}\right)}\left(\frac{\chi^2}{2}\right)^{\frac{n-1}{2}-1}e^{-\frac{\chi^2}{2}} \qquad (\chi^2\geq 0) \tag{40.41}$$

The quantity obtained by dividing the value of χ^2 by $n-1$ degrees of freedom, or in other words the quantity which is obtained by dividing the variance V_e by σ^2, assumes an F distribution of number of degrees of freedom of the numerator $n-1$ and number of degrees of freedom of the denominator ∞.

Also, the statistic t,

$$t=\frac{\bar{y}-m}{\sqrt{V_e}}\times\sqrt{n} \tag{40.42}$$

is a statistic which originated in mathematical statistics, and it is termed Student's t distribution. Its density function $f(t)$ is given by

$$f(t)=\frac{1}{\sqrt{n-1}\,B\left(\frac{1}{2},\frac{n-1}{2}\right)}\left(1+\frac{t^2}{n-1}\right)^{-\frac{n}{2}} \qquad (-\infty<t<+\infty) \tag{40.43}$$

This has a distribution which is symmetric about the origin. The distribution of t^2 is equal to an F distribution of number of degrees of freedom of the numerator 1 and number of degrees of freedom of the denominator $n-1$.

Therefore, nearly all distributions which are used in mathematical statistics are F distributions themselves or special cases thereof. If the probability density of an F distribution of number of degrees of freedom of numerator f_1 and number of degrees of freedom of denominator f_2 is expressed as $f(F)$, we have

$$f(F)=\frac{\Gamma\left(\frac{f_1+f_2}{2}\right)f_1^{\frac{f_1}{2}}f_2^{\frac{f_2}{2}}}{\Gamma\left(\frac{f_1}{2}\right)\Gamma\left(\frac{f_2}{2}\right)}\frac{F^{\frac{f_1}{2}-1}}{(f_1F+f_2)^{(f_1+f_2)/2}} \qquad (F>0) \tag{40.44}$$

And the mean and variance of these are

$$E(F)=\frac{f_2}{f_2-2} \qquad (f_2>2) \tag{40.45}$$

$$\text{Var}(F)=\frac{2f_2^2(f_1+f_2-2)}{f_1(f_2-2)^2(f_2-4)} \qquad (f_2>4) \tag{40.46}$$

If the variance of a source of number of degrees of freedom f_1 is expressed as V_A and if the error variance is expressed as V_e, then with the eccentricity (same as SN ratio) as λ,

$$E(V_A)=\sigma^2(1+\lambda) \qquad (\lambda=r\sigma_A^2/\sigma^2) \tag{40.47}$$

$$E(V_e)=\sigma^2 \tag{40.48}$$

hold. If $\lambda=0$, the variance ratio

$$F=\frac{V_A}{V_e} \tag{40.49}$$

assumes an F distribution of number of degrees of freedom of numerator f_1 and number of degrees of freedom of denominator f_2. If $\lambda>0$, the variance ratio F of Equation (40.49) assumes a noncentral F distribution. Its distribution density $f(F)$ is given by the following equation.

$$f(F)=\sum_{i=0}^{\infty}\left[\frac{\Gamma\left(\frac{2i+f_1+f_2}{2}\right)\left(\frac{f_1}{f_2}\right)^{\frac{1}{2}(2i+f_1)}e^{-\frac{f_1\lambda}{2}}}{\Gamma\left(\frac{f_2}{2}\right)\Gamma\left(\frac{2i+f_1}{2}\right)i!}\frac{F^{(2i+f_1-2)/2}}{\left(1+\frac{f_1}{f_2}F\right)^{(2i+f_1+f_2)/2}}\right] \tag{40.50}$$

In Appendix Table 9, the confidence limits of the SN ratio λ have been calculated by using Equation (40.50). However, there is often a cause of error with regard to the actual value of λ, a cause which it has in common with the V_e of the denominator. If the degree of non-centrality of V_A is expressed as λ and the degree of non-centrality of the denominator is expressed as λ_1, the variance ratio F distributes according to the following double non-central F distribution density.

$$f(F)=\sum_{j=0}^{\infty}\sum_{i=0}^{\infty}\frac{\left(\frac{f_1\lambda}{2}\right)^i\left(\frac{f_2\lambda_1}{2}\right)^j e^{-\frac{1}{2}(f_1\lambda+f_2\lambda_1)}}{i!j!B\left(\frac{f_1}{2}+i,\ \frac{f_2}{2}+j\right)}\left(\frac{F}{1+F}\right)^{\frac{f_1}{2}+i-1}\left(\frac{1}{1+F}\right)^{\frac{f_2}{2}+j+1} \tag{40.51}$$

Here, $B(f_1, f_2)$ is a beta function. The SN ratio λ behaves according to the distribution of Equation (40.51) even if the error part is random and it follows a normal distribution. This is a difficult distribution, but by using

Patnike's approximation for the denominator and numerator, it can be approximated by an F distribution of the following number of degrees of freedom of numerator f_1' and number of degrees of freedom of denominator f_2'.

Number of Degrees of Freedom of Numerator:

$$f_1' = f_1 F_1^2/(2F_1 - 1),$$

Number of Degrees of Freedom of Denominator:

$$f_2' = f_2 F_2^2/(2F_2 - 1)$$

Here, F_1 and F_2 are the ratios of the variances of the numerator and the denominator to the error variance.

40.4 Communication Engineering and Design of Experiments

40.4.1 Theory of Communication and Design of Experiments

It is said that the comprehensive theoretical study of information began with the thesis by *C. Shannon* in 1948 [reference 40]. However, the author feels that, separately from Shannon's thesis, Fisher's reference 5 also constitutes an initial, important study for acquisition of information.

Regarding "matter" and "energy," many studies have been carried out on rationalization of their production, transportation, and storage.

It is said that from here on this will be a world where information is at the center. In regard to the production, transportation, and storage of information, Shannon's theory of communication is nothing other than a discussion of the efficiencies of "communication," which is transportation of information, and "memory," which is storage of information. Of course, without being information theory as such, there is also the theory of systems, which discusses the efficiency of the system as a whole by including the value of information as one element within the system; the science of information, which attempts to consider all problems related with information; and the science of information processing, which pursues only the treatment of information itself; and thus the range has expanded even more. However, these are centered about the transportation, storage, processing, and use of information, and they do not touch very much upon the production of information itself.

In regard to the production of information, there is information for consumer-type entertainment, such as information related with the performing arts, and that which is related with production and development, such as information obtained by calculation, experimentation, and surveys. Consumer-type information does not have direct influence on rationalization of the society as a whole.

Here, we will present a discussion dealing with information that is obtained from calculation, experimentation, and surveys, which is related with rationalization.

The theory of communication can be defined as "the whole of a universal technique for heightening the transmission (transportation and storage) efficiency of information." The method of coding, how to introduce a degree of redundancy, the method of design of filters, the concept of channel capacity, etc., which are used for this purpose can be said to be countermeasures for problem points which arise in common in the transmission of information, regardless of the individual instrument or apparatus.

The Design of Experiments method which was developed by Fisher can be defined as the whole of a universal technique for the heightening of the efficiency of acquisition of information by (technical and market) experiments.

In 1962, at the Bell Telephone Laboratories in the United States, the author conducted studies related with the theory of communication and Design of Experiments, and he presented the conclusion that the two are completely the same in regard to mathematical concepts. For example, relationships of correspondence such as are given in Table 40.3 exist between the terms which are used by each.

Also, *the fundamental theorem of the theory of information,* that "there exists coding by which it is possible to transmit given information with the required arbitrary accuracy in the presence of an arbitrary given noise; moreover, it is possible to minimize its loss as much as desired by means of redundancy," corresponds to the following proposition in Design of Experiments: "No matter how many sources exist, and no matter how much error there is, there exists a method of assignment for finding all of these factorial effects with the required accuracy, and moreover, the ratio of the size of experiment, n, and the sum of the

Table 40.3 Relationships of Terminology

Design of Experiments Method	Theory of Communications
Source	Signal
Error	Noise
Resolution of variation	Spectral resolution
Variance ratio	SN ratio
Estimation	Filter
Assignment	Coding
Orthogonal array	Orthogonal circuit
Interaction	Nonlinearity

degrees of freedom of the sources, f, can be brought as close to 1 as one wishes by making f sufficiently large." This corresponds to the fact that the precision rises as much as one wishes if a large orthogonal array is used.

In regard to the production of information in research techniques, if we include not only physics, chemistry, electricity, and machinery but, for example, medicine, agriculture, market surveying and so forth, the diversity of methods and apparatus for acquisition of information is very great. Therefore, compared with the theory of information by which information is considered only in its quantitative aspects, methods to counter various situations have been developed more for Design of Experiments. It is the purpose of this section to discuss the third and fourth problems in Table 40.3, *viz.*, resolution of variation and the variance ratio according to Design of Experiments and spectral resolution and SN ratio according to the theory of communication, in terms of the mathematics the two fields have in common.

40.4.2 Spectral Resolution

As a simple example, we consider the case of the telephone. When we lift the transmitter-receiver of a telephone, 48 volts DC is applied to the telephone circuit. But there is no sound just because direct current passes through. It is necessary for the transmitting person to speak. Once the speaker speaks, longitudinal waves in the air are created correspondingly, causing vibrations in a diaphragm which is arranged to slide within the cylinder immediately behind the mouthpiece. The region behind this diaphragm is packed with carbon powder. The state of contact of the carbon powder changes in conformity with the vibrations. The cylinder filled with the carbon powder constitutes the resistance of the 48-volt direct-current circuit. Therefore, since the state of contact of the carbon powder changes in response to the vibrations of the diaphragm, the contact resistance changes and as a result an oscillating current corresponding to the vibrations of the vocal sounds is produced.

Let us assume that this oscillating current, expressed as $y(t)$, was as depicted by Figure 40.1.

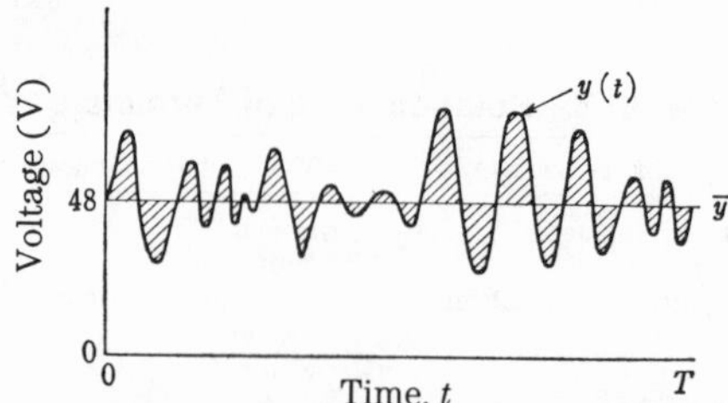

FIGURE 40.1 Vocal Sound Voltage of Telephone Circuit

The above description means that the designer of the telephone pours out his heart's blood in regard to the following two matters.

(1) Up to how high a frequency of sound waves can the diaphragm vibrate accordingly, with motion of the carbon powder permitting the creation of the oscillating current? Up to how many thousand hertz of sound waves is it possible to vary current? Since the diaphragm is designed to slide within the cylinder, moisture inevitably enters. It is essentially difficult to produce carbon powder that flows as many as several thousand times in a second. In the future, it will be desirable to use, for example, something where free electrons within crystals move in response to sound pressure. How many thousand hertz of sound waves it is possible to render into oscillating current is termed the degree of hifi-ness of a communication system.

(2) How to design the system so that the sound waves resulting from speech of a specific magnitude, given a specific direct-current voltage, become an oscillating current whose amplitude is as great as possible. In other words, how to heighten the sensitivity. If the sensitivity becomes ten times as great, a sound which is ten times greater becomes transmitted to the receiver, and this means that the loss in the telephone cables in between may be as much as ten times less. One may, for example, then reduce the thickness of the copper wires of the cables correspondingly.

Only the part of the current hatched with diagonal lines in Figure 40.1 is effective for transmission of vocal sounds, and the following equation gives the magnitude of this considered as an oscillating current, or in other words *the power of the alternating-current part, P.*

$$P=\int_0^T [y(t)-\bar{y}]^2 dt \tag{40.52}$$

Strictly speaking, the equation above defines *the quantity of work performed by the alternating current.* Since time does not have very great significance in this case, assuming that we make comparisons with the time always the same, we will use the term "power" for the quantity of work, too.

$\bar{y}$ in Equation (40.52) corresponds to the direct-current voltage. Equation (40.52) can be changed in form, as follows.

$$P=\int_0^T [y(t)-\bar{y}]^2 dt=\int_0^T y^2(t)dt-\frac{\left[\int_0^T y(t)\,dt\right]^2}{T}$$

$$= \text{sum of squares of } y(t) - \frac{[\text{sum of } y(t)]^2}{T} \tag{40.53}$$

Resolving the total power of the alternating-current part, P, into various frequency components is termed spectral resolution. The frequency components may include a discrete frequency component which is termed a discrete spectrum and a dense frequency component which is termed a continuous spectrum. The theory when both are mixed is discussed in detail in the paper by Paley and Wiener [reference 41].

Here, we will assume that there are only discrete spectra for vocal sounds; we consider a case where there could be every component for noise, and we will express the components of such a discrete spectrum of vocal sounds as $\omega_1, \omega_2, \ldots, \omega_k$. Then, the total power P of the oscillating current becomes resolved into the powers of the components of the discrete spectrum, $P_{\omega_1}, P_{\omega_2}, \ldots, P_{\omega_k}$, and the remaining power of noise, P_N; and the following equality holds.

$$P = P_{\omega_1} + P_{\omega_2} + \cdots + P_{\omega_k} + P_N \tag{40.54}$$

When the result of Equation (40.53) is substituted into this, we obtain the following equality.

$$\text{sum of squares of } y(t) = (\text{power of direct current}) + P_{\omega_1} + P_{\omega_2} + \cdots + P_{\omega_k} + P_N \tag{40.55}$$

In mathematics, this equality is known as the *Parseval equality*.

The components of direct current and ω_i are all found by normalizing the result of multiplying $y(t)$ by the constant 1 or a coefficient such as $\cos(\omega_1 t)$ and integrating. Normalization means to perform division by the sum of squares of the coefficients by which $y(t)$ is multiplied.

For example,

$$\text{power of direct current} = \frac{[\text{sum of } y(t)]^2}{\text{sum of squares of coefficient 1}} = \frac{\left[\int_0^T y(t)dt\right]^2}{\int_0^T 1^2 dt} = \frac{\left[\int_0^T y(t)dt\right]^2}{T} \tag{40.56}$$

Also,

$$P_{\omega_i} = \frac{[\text{sum of } y(t)\cos(\omega_1 t)]^2}{\text{sum of squares of } \cos(\omega_1 t)} = \frac{\left[\int_0^T y(t)\cos(\omega_i t)dt\right]^2}{\int_0^T \cos^2(\omega_i t)dt} \tag{40.57}$$

It is assumed that T is sufficiently larger than the reciprocal of ω_i. Next, it will be proved that analysis of variance in Design of Experiments is exactly the same as the above.

40.4.3 Resolution of Variation (1): Resolution Into Components of One Degree of Freedom

We will express the target variable by y and the cause variable expected to influence it by x; assuming that n pairs of data concerning x and y have been observed, these will be expressed as (x_1, y_1), (x_2, y_2), . . . , (x_n, y_n). We assume that y is a linear regression equation of x, and this is put as

$$y=m+b(x-\bar{x}) \tag{40.58}$$

where

$$\bar{x}=\frac{1}{n}(x_1+x_2+\cdots+x_n)$$

Then, from the method of least squares, the estimates of m and b, m and b, become as follows.

$$m=\frac{1}{n}(y_1+y_2+\cdots+y_n)=\bar{y}$$

$$b=\frac{\sum(x_i-\bar{x})y_i}{\sum(x_i-\bar{x})^2}=\frac{\sum(x_i-\bar{x})(y_i-\bar{y})}{\sum(x_i-\bar{x})^2}$$

In this case, the residual sum of squares, S_e, is

$$\begin{aligned} S_e&=\sum_i\{y_i-[\bar{y}+b(x_i-\bar{x})]\}^2=\sum(y_i-\bar{y})^2-\frac{\sum[(x_i-\bar{x})(y_i-\bar{y})]^2}{\sum(x_i-\bar{x})^2}\\ &=\sum y_i^2-\frac{(\sum y_i)^2}{n}-\frac{[\sum(x_i-\bar{x})y_i]^2}{\sum(x_i-\bar{x})^2} \end{aligned} \tag{40.59}$$

When this is rewritten, we have

$$\sum y_i^2=\frac{(\sum y_i)^2}{n}+\frac{[\sum(x_i-\bar{x})y_i]^2}{\sum(x_i-\bar{x})^2}+S_e \tag{40.60}$$

The left side of the equation above is a unit quadratic form in y_1, y_2, . . . , y_n, while the three terms on the right side are all quadratic forms in y_1, y_2, . . . , y_n. When the matrices of these quadratic forms are represented as $\boldsymbol{M}_m$, $\boldsymbol{M}_x$, and $\boldsymbol{M}_e$, we have

$$\boldsymbol{M}_m=\begin{bmatrix} \frac{1}{n} & \frac{1}{n} \cdots & \frac{1}{n} \\ \frac{1}{n} & \frac{1}{n} \cdots & \frac{1}{n} \\ \vdots & \vdots \ddots & \vdots \\ \frac{1}{n} & \frac{1}{n} \cdots & \frac{1}{n} \end{bmatrix} \tag{40.61}$$

$$\boldsymbol{M}_x=\begin{bmatrix} \frac{(x_1-\bar{x})^2}{S(xx)}, & \frac{(x_1-\bar{x})(x_2-\bar{x})}{S(xx)}, & \cdots, & \frac{(x_1-\bar{x})(x_n-\bar{x})}{S(xx)} \\ \frac{(x_2-\bar{x})(x_1-\bar{x})}{S(xx)}, & \frac{(x_2-\bar{x})^2}{S(xx)}, & \cdots, & \frac{(x_2-\bar{x})(x_n-\bar{x})}{S(xx)} \\ \vdots & \vdots & \ddots & \vdots \\ \frac{(x_n-\bar{x})(x_1-\bar{x})}{S(xx)}, & \frac{(x_n-\bar{x})(x_2-\bar{x})}{S(xx)}, & \cdots, & \frac{(x_n-\bar{x})^2}{S(xx)} \end{bmatrix} \tag{40.62}$$

$$M_e=\begin{bmatrix} \frac{n-1}{n}-\frac{(x_1-\bar{x})^2}{S(xx)}, & -\frac{1}{n}-\frac{(x_1-\bar{x})(x_2-\bar{x})}{S(xx)}, & \cdots, & -\frac{1}{n}-\frac{(x_1-\bar{x})(x_n-\bar{x})}{S(xx)} \\ -\frac{1}{n}-\frac{(x_2-\bar{x})(x_1-\bar{x})}{S(xx)}, & \frac{n-1}{n}-\frac{(x_2-\bar{x})^2}{S(xx)}, & \cdots, & -\frac{1}{n}-\frac{(x_2-\bar{x})(x_n-\bar{x})}{S(xx)} \\ \vdots & \vdots & \ddots & \vdots \\ -\frac{1}{n}-\frac{(x_n-\bar{x})(x_1-\bar{x})}{S(xx)}, & -\frac{1}{n}-\frac{(x_n-\bar{x})(x_2-\bar{x})}{S(xx)}, & \cdots, & \frac{n-1}{n}-\frac{(x_n-\bar{x})^2}{S(xx)} \end{bmatrix} \quad (40.63)$$

In this case, the rank of matrices $\boldsymbol{M}_m$ and $\boldsymbol{M}_x$ is 1 in each case, and the rank of matrix $\boldsymbol{M}_e$ is $n-2$. The three matrices $\boldsymbol{M}_m$, $\boldsymbol{M}_x$, and $\boldsymbol{M}_e$ are mutually orthogonal. With 0 as the zero matrix, we have

$$\boldsymbol{M}_m\cdot\boldsymbol{M}_x=0$$

$$\boldsymbol{M}_m\cdot\boldsymbol{M}_e=0$$

$$\boldsymbol{M}_x\cdot\boldsymbol{M}_e=0$$

Since it will be needed later, we will prove here the following more general theorem.

[THEOREM 1]

For n observed values $y_1, y_2, \ldots, y_n$, the k $(<n)$ normalized linear equations

$$L^{(i)}=C_1^{(i)}y_1+C_2^{(i)}y_2+\cdots+C_n^{(i)}y_n \quad (i=1, 2, \cdots, k) \quad (40.64)$$

are mutually orthogonal; in other words

$$C_1^{(i)}C_1^{(j)}+C_2^{(i)}C_2^{(j)}+\cdots+C_n^{(i)}C_n^{(j)}=\delta_i^j \quad (40.65)$$

Here, δ_i^j is Kronecker's symbol.

$$\delta_i^j=\begin{cases}1 & (i=j)\\ 0 & (i\neq j)\end{cases} \quad (40.66)$$

Then, the quadratic form matrices $\boldsymbol{M}_i$ which are obtained by the squaring of linear equation $L^{(i)}$ are all of rank 1 and they are mutually orthogonal. Also, the rank of matrix $\boldsymbol{M}_e$ which is obtained by subtracting these matrices from the unit matrix $\boldsymbol{E}$,

$$\boldsymbol{M}_e=\boldsymbol{E}-\boldsymbol{M}_1-\boldsymbol{M}_2-\cdots-\boldsymbol{M}_k \quad (40.67)$$

is $n-k$, and $\boldsymbol{M}_e$ is orthogonal to any of the matrices $\boldsymbol{M}_1, \boldsymbol{M}_2, \ldots, \boldsymbol{M}_k$.

[PROOF]

It will be proved first that the value of the determinant of an arbitrary 2×2 submatrix of matrices $\boldsymbol{M}_i$ is 0. An i_1 row, i_2 row, j_1 column, and j_2 column 2×2 submatrix is

$$\begin{vmatrix} C_{j1}^{(i)}C_{i1}^{(i)} & C_{j1}^{(i)}C_{i2}^{(i)} \\ C_{j2}^{(i)}C_{i1}^{(i)} & C_{j2}^{(i)}C_{i2}^{(i)} \end{vmatrix} = C_{j1}^{(i)}C_{j2}^{(i)}C_{i1}^{(i)}C_{i2}^{(i)} - C_{j1}^{(i)}C_{j2}^{(i)}C_{i1}^{(i)}C_{i2}^{(i)} = 0 \tag{40.68}$$

But since from the condition of normalization we have

$$C_1^{(i)2} + C_2^{(i)2} + \cdots + C_n^{(i)2} = 1 \tag{40.69}$$

there is at least one which is not 0 among $C_1^{(i)}, C_2^{(i)}, \ldots, C_n^{(i)}$. If this one is called $C_j^{(i)}$, the (j, j) component of the matrix $\boldsymbol{M}_i$ is not 0. This means that the rank of the matrices $\boldsymbol{M}_i$ is 1.

Next, it will be proved that matrix $\boldsymbol{M}_1$ and matrix $\boldsymbol{M}_2$ are orthogonal.

$$\boldsymbol{M}_1 = \begin{bmatrix} C_1^{(1)}C_1^{(1)}, & C_1^{(1)}C_2^{(1)}, & \cdots\cdots, & C_1^{(1)}C_n^{(1)} \\ C_2^{(1)}C_1^{(1)}, & C_2^{(1)}C_2^{(1)}, & \cdots\cdots, & C_2^{(1)}C_n^{(1)} \\ \vdots & \vdots & \ddots & \vdots \\ C_n^{(1)}C_1^{(1)}, & C_n^{(1)}C_2^{(1)}, & \cdots\cdots, & C_n^{(1)}C_n^{(1)} \end{bmatrix} \tag{40.70}$$

$$\boldsymbol{M}_2 = \begin{bmatrix} C_1^{(2)}C_1^{(2)}, & C_1^{(2)}C_2^{(2)}, & \cdots\cdots, & C_1^{(2)}C_n^{(2)} \\ C_2^{(2)}C_1^{(2)}, & C_2^{(2)}C_2^{(2)}, & \cdots\cdots, & C_2^{(2)}C_n^{(2)} \\ \vdots & \vdots & \ddots & \vdots \\ C_n^{(2)}C_1^{(2)}, & C_n^{(2)}C_2^{(2)}, & \cdots\cdots, & C_n^{(2)}C_n^{(2)} \end{bmatrix} \tag{40.71}$$

When the ij component of the product of matrix $\boldsymbol{M}_1$ and matrix $\boldsymbol{M}_2$ is found, it is 0, as follows.

$$\begin{aligned} & C_i^{(1)}C_1^{(1)} \times C_1^{(2)}C_j^{(2)} + C_i^{(1)}C_2^{(1)} \times C_2^{(2)}C_j^{(2)} + \cdots + C_i^{(1)}C_n^{(1)} \times C_n^{(2)}C_j^{(2)} \\ & = C_i^{(1)}C_j^{(2)}[C_1^{(1)} \times C_1^{(2)} + C_2^{(1)} \times C_2^{(2)} + \cdots + C_n^{(1)} \times C_n^{(2)}] \\ & = 0 \text{ (from orthogonality condition)} \end{aligned} \tag{40.72}$$

Therefore, $\boldsymbol{M}_1$ and $\boldsymbol{M}_2$ are orthogonal. Exactly similarly, we find that $\boldsymbol{M}_1$, $\boldsymbol{M}_2, \ldots, \boldsymbol{M}_k$ are mutually orthogonal. Next, we consider the matrix of the quadratic form of error variation. If we multiply

$$\boldsymbol{M}_e = \boldsymbol{E} - (\boldsymbol{M}_1 + \boldsymbol{M}_2 + \cdots + \boldsymbol{M}_k) \tag{40.73}$$

by matrices $\boldsymbol{M}_j$ $(i = 1, 2, \ldots, k)$ from the left, we have

$$\boldsymbol{M}_j\boldsymbol{M}_e = \boldsymbol{M}_i - \boldsymbol{M}_i^2$$

Now, for

$$M_i^2=\begin{bmatrix} C_1^{(i)}C_1^{(i)}, & C_1^{(i)}C_2^{(i)}, & \cdots\cdots, & C_1^{(i)}C_n^{(i)} \\ C_2^{(i)}C_1^{(i)}, & C_2^{(i)}C_2^{(i)}, & \cdots\cdots, & C_2^{(i)}C_n^{(i)} \\ \vdots & \vdots & \ddots & \vdots \\ C_n^{(i)}C_1^{(i)}, & C_n^{(i)}C_2^{(i)}, & \cdots\cdots, & C_n^{(i)}C_n^{(i)} \end{bmatrix}^2$$

we find the (j, k) component.

$$\begin{aligned} &C_j^{(i)}C_1^{(i)}\times C_1^{(i)}C_k^{(i)}+C_j^{(i)}C_2^{(i)}\times C_2^{(i)}C_k^{(i)}+\cdots+C_j^{(i)}C_n^{(i)}\times C_n^{(i)}\times C_k^{(i)} \\ &=C_j^{(i)}C_k^{(i)}[C_1^{(i)2}+C_2^{(i)2}+\cdots+C_n^{(i)2}]=C_j^{(i)}C_k^{(i)} \end{aligned}$$

This is the same as the (j, k) component of $\boldsymbol{M}_i$. Therefore,

$$\boldsymbol{M}_i^2=\boldsymbol{M}_i \tag{40.74}$$

Therefore,

$$\boldsymbol{M}_i\cdot\boldsymbol{M}_e=0$$

It is evident that $\boldsymbol{M}_i$ and $\boldsymbol{M}_e$ are orthogonal.

That $\boldsymbol{M}_1, \boldsymbol{M}_2, \ldots, \boldsymbol{M}_k$ and $\boldsymbol{M}_e$ are mutually orthogonal and, moreover, their sum is a unit matrix means that the rank of $\boldsymbol{M}_e$ is $n-k$.

$\boldsymbol{M}_m$ of Equation (40.61) and $\boldsymbol{M}_x$ of Equation (40.62) are both matrices of the square of a normalized linear equation in $y_1, y_2, \ldots, y_n$.

$$\boldsymbol{M}_m \equiv \text{matrix of } \left(\frac{y_1+\cdots+y_n}{\sqrt{n}}\right)^2 \tag{40.75}$$

$$\boldsymbol{M}_x = \text{matrix of } \left[\frac{\sum(x_i-\bar{x})y_i}{\sqrt{\sum(x_i-\bar{x})^2}}\right]^2 \tag{40.76}$$

Normalization has been carried out with regard to both $\boldsymbol{M}_m$ and $\boldsymbol{M}_x$, as follows.

$$\left(\frac{1}{\sqrt{n}}\right)^2+\left(\frac{1}{\sqrt{n}}\right)^2+\cdots+\left(\frac{1}{\sqrt{n}}\right)^2=1 \tag{40.77}$$

$$\left[\frac{(x_1-\bar{x})}{\sqrt{S(xx)}}\right]^2+\left[\frac{(x_2-\bar{x})}{\sqrt{S(xx)}}\right]^2+\cdots+\left[\frac{(x_n-\bar{x})}{\sqrt{S(xx)}}\right]^2=1 \tag{40.78}$$

Also, the inner product of both is 0, as follows.

$$\frac{1}{\sqrt{n}}\times\frac{(x_1-\bar{x})}{\sqrt{S(xx)}}+\frac{1}{\sqrt{n}}\times\frac{(x_2-\bar{x})}{\sqrt{S(xx)}}+\cdots+\frac{1}{\sqrt{n}}\times\frac{(x_n-\bar{x})}{\sqrt{S(xx)}}=0 \tag{40.79}$$

From this, matrices $\boldsymbol{M}_m$ and $\boldsymbol{M}_x$ are mutually orthogonal, and moreover the rank is 1. Therefore, the rank of the residual sum of squares S_e is $n-2$, and it is orthogonal to either $\boldsymbol{M}_m$ or $\boldsymbol{M}_x$.

As a case of an even more general regression function, assuming that y is a function of x, we consider the case of a linear equation with no unknown constant in $\varphi_1(x), \varphi_2(x), \ldots, \varphi_k(x)$.

$$y = a_1\varphi_1(x) + a_2\varphi_2(x) + \cdots + a_k\varphi_k(x) \quad (40.80)$$

Calculation becomes complicated if Equation (40.80) is used as it stands. To avoid this, we perform Schmidt's orthogonal expansion. We put

$$y_i = b_1\varphi_1(x_i) \quad (40.81)$$

and we estimate b_1.

$$b_1 = \frac{(\varphi_1 y)}{(\varphi_1\varphi_1)} \quad (40.82)$$

where

$$(\varphi_1 y) = \sum_{i=1}^{n} \varphi_1(x_i) y_i \quad (40.83)$$

$$(\varphi_1\varphi_1) = \sum_{i=1}^{n} \varphi_1(x_i)\varphi_1(x_i) \quad (40.84)$$

We then fit $\varphi_2(x)$ to the difference from y_i, which cannot be completely expressed by $\varphi_1(x)$.

$$y_i^{(1)} = y_i - b_1\varphi_1(x_i) \quad (40.85)$$

However, instead of $\varphi_2(x)$ itself we use the function $\varphi_2^{(1)}$ of the remainder when $\varphi_2(x)$ has been expressed by $\varphi_1(x)$.

$$\varphi_2^{(1)}(x_i) = \varphi_2(x_i) - b_{21}\varphi_1(x_i) \quad (40.86)$$

Here,

$$b_{21} = \frac{(\varphi_1\varphi_2)}{(\varphi_1\varphi_1)}$$

In this case, $\varphi_1(x_i)$ and $\varphi_2^{(1)}(x_i)$ are orthogonal. For,

$$\sum \varphi_1(x_i)\varphi_2^{(1)}(x_i) = \sum \varphi_1(x_i)\left[\varphi_2(x_i) - \frac{(\varphi_1\varphi_2)}{(\varphi_1\varphi_1)}\varphi_1(x_i)\right]$$

$$= (\varphi_1\varphi_2) - \frac{(\varphi_1\varphi_2)}{(\varphi_1\varphi_1)}(\varphi_1\,\varphi_1) = 0$$

Therefore,

$$y_i = b_1\varphi_1(x_i) + b_2\varphi_2^{(1)}(x_i) \quad (40.87)$$

Here, b_2 is

$$b_2 = \frac{(\varphi_2^{(1)} y)}{(\varphi_2^{(1)}\varphi_2^{(1)})}$$

Next, if we put

$$\varphi_3^{(1)} = \varphi_3 - b_{31}\varphi_1 - b_{32}\varphi_2^{(1)}$$

we have

$$b_{31} = \frac{(\varphi_1\varphi_3)}{(\varphi_1\varphi_1)}$$

$$b_{32} = \frac{(\varphi_2^{(1)}\varphi_3)}{(\varphi_2^{(1)}\varphi_2^{(1)})}$$

By using this, we get

$$b_3 = \frac{(\varphi_3^{(1)} y)}{(\varphi_3^{(1)} \varphi_3^{(1)})}$$

From this, we obtain the expansion to the third term.

$$y = b_1\varphi_1 + b_2\varphi_2^{(1)} + b_3\varphi_3^{(1)} \tag{40.88}$$

By proceeding similarly, we obtain an orthogonal expansion to term k.

In this way, even if y is a linear function of arbitrary functions of x, $\varphi_1(x), \varphi_2(x), \ldots, \varphi_k(x)$, it is possible to resolve the unit matrix into the sum of a matrix of rank 1 for $\varphi_1(x), \varphi_2^{(1)}(x), \varphi_3^{(1)}(x), \ldots, \varphi_k^{(1)}(x)$ and a matrix for the residual sum of squares of rank $n-k$.

In analysis of variance, the quadratic equation for each matrix is termed variation, and it is written as $S_1, S_2, \ldots, S_k$ and S_e; and the rank of the matrix of these quadratic equations is termed the number of degrees of freedom. Therefore, the total sum of squares S of n observed values, $y_1, y_2, \ldots, y_n$, has been resolved as:

$$\text{Variation } S = S_1 + S_2 + \cdots + S_k + S_e \tag{40.89}$$

Number of degrees of freedom

$$n = 1 + 1 + \cdots + 1 + (n-k) \tag{40.90}$$

40.4.4 Resolution of Variation (2): Resolution Into Degrees of Freedom in General

In Design of Experiments, the total sum of squares of n experimental values $y_1, y_2, \ldots, y_n$,

$$S_T = y_1^2 + y_2^2 + \cdots + y_n^2 \tag{40.91}$$

is not necessarily resolved into components of one degree of freedom.

For example, when two factors, A and B, are at five levels and four levels, respectively, let us assume that, having experimented on every one of the 20 combinations, we have obtained the data in the two-way array form of Table 40.4.

Table 40.4 Data of Two-Way Layout

	B_1	B_2	B_3	B_4	Total
A_1	y_1	y_2	y_3	y_4	A_1
A_2	y_5	y_6	y_7	y_8	A_2
A_3	y_9	y_{10}	y_{11}	y_{12}	A_3
A_4	y_{13}	y_{14}	y_{15}	y_{16}	A_4
A_5	y_{17}	y_{18}	y_{19}	y_{20}	A_5
Total	B_1	B_2	B_3	B_4	T

In this case, the total sum of squares S_T is resolved as follows.

$$S_T = S_m + S_A + S_B + S_e \tag{40.92}$$

where

$$S_T = y_1^2 + y_2^2 + \cdots + y_{20}^2$$

$$S_m = \frac{(y_1 + y_2 + \cdots + y_{20})^2}{20}$$

$$S_A = \frac{A_1^2 + A_2^2 + \cdots + A_5^2}{4} - S_m = \frac{(y_1 + \cdots + y_4)^2 + \cdots + (y_{17} + \cdots + y_{20})^2}{4} - S_m$$

$$S_B = \frac{B_1^2 + B_2^2 + \cdots + B_4^2}{5} - S_m = \frac{(y_1 + y_5 + \cdots + y_{17})^2 + \cdots + (y_4 + y_8 + \cdots + y_{20})^2}{5} - S_m$$

$$S_e = S - S_m - S_A - S_B$$

The terms on the right side of Equation (40.92) are quadratic forms in $y_1, y_2, \ldots, y_{20}$ for the variations of m, A, B, and e of rank *1*, *4*, *3*, and 12, respectively, and their matrices are as follows.

$$\boldsymbol{M}_m = \begin{bmatrix} \frac{1}{20} & \frac{1}{20} \cdots & \frac{1}{20} \\ \frac{1}{20} & \frac{1}{20} \cdots & \frac{1}{20} \\ \vdots & \vdots \ddots & \vdots \\ \frac{1}{20} & \frac{1}{20} \cdots & \frac{1}{20} \end{bmatrix}$$

$$\boldsymbol{M}_A = \begin{bmatrix} \frac{4}{20} & \frac{4}{20} \cdots & -\frac{1}{20} \\ \frac{4}{20} & \frac{4}{20} \cdots & -\frac{1}{20} \\ \vdots & \vdots \ddots & \vdots \\ -\frac{1}{20} & -\frac{1}{20} \cdots & \frac{4}{20} \end{bmatrix}$$

$$\boldsymbol{M}_B = \begin{bmatrix} \frac{3}{20} & -\frac{1}{20} \cdots & -\frac{1}{20} \\ -\frac{1}{20} & \frac{3}{20} \cdots & -\frac{1}{20} \\ \vdots & \vdots \ddots & \vdots \\ -\frac{1}{20} & -\frac{1}{20} \cdots & \frac{3}{20} \end{bmatrix}$$

$$\boldsymbol{S}_e = \begin{bmatrix} \frac{12}{20} & -\frac{4}{20} & \cdots & \frac{1}{20} \\ -\frac{4}{20} & \frac{12}{20} & \cdots & \frac{1}{20} \\ \vdots & \vdots & \ddots & \vdots \\ \frac{1}{20} & \frac{1}{20} & \cdots & \frac{12}{20} \end{bmatrix}$$

Matrices such as $\boldsymbol{M}_A$ and $\boldsymbol{M}_B$ can be resolved into orthogonal components of one degree of freedom. Such methods of resolution are infinite in number, and the following is one example. Only resolution $\boldsymbol{M}_A$ will be explained.

As the four normal linear equations among $A_1, A_2, \ldots, A_5$, we consider the following.

$$L_1=\frac{4A_1-A_2-A_3-A_4-A_5}{\sqrt{80}}$$

$$L_2=\frac{3A_2-A_3-A_4-A_5}{\sqrt{48}}$$

$$L_3=\frac{2A_3-A_4-A_5}{\sqrt{24}}$$

$$L_4=\frac{A_4-A_5}{\sqrt{8}}$$

L_1, L_2, L_3, and L_4 are mutually orthogonal linear equations. For example, with regard to L_1 and L_2, the sum of products of the coefficients of $y_1, y_2, \ldots, y_{20}$ is

$$\left[\left(\frac{4}{\sqrt{80}}\right)\times 0+\left(-\frac{1}{\sqrt{80}}\right)\left(\frac{3}{\sqrt{48}}\right)+\left(\frac{-1}{\sqrt{80}}\right)\left(\frac{-1}{\sqrt{48}}\right)+\left(\frac{-1}{\sqrt{80}}\right)\left(\frac{-1}{\sqrt{48}}\right)+\left(\frac{-1}{\sqrt{80}}\right)\left(\frac{-1}{\sqrt{48}}\right)\right]\times 4$$

$$=\frac{(-3+1+1+1)\times 4}{\sqrt{80}\times\sqrt{48}}=0$$

Also, since we have normalized,

$$\left[\left(\frac{4}{\sqrt{80}}\right)^2+\left(\frac{-1}{\sqrt{80}}\right)^2+\left(\frac{-1}{\sqrt{80}}\right)^2+\left(\frac{-1}{\sqrt{80}}\right)^2+\left(\frac{-1}{\sqrt{80}}\right)^2\right]\times 4=\frac{(16+1+1+1+1)\times 4}{80}=1$$

$$\left[0^2+\left(\frac{3}{\sqrt{48}}\right)^2+\left(\frac{-1}{\sqrt{48}}\right)^2+\left(\frac{-1}{\sqrt{48}}\right)^2+\left(\frac{-1}{\sqrt{48}}\right)^2\right]\times 4=\frac{(0+9+1+1+1)\times 4}{48}=1$$

Thus, it is evident from Theorem 1 that the quadratic form matrices which are obtained by squaring L_1 and L_2 are mutually orthogonal.

Similarly, the matrices corresponding to L_1, L_2, L_3, and L_4 are mutually orthogonal. Next, for B we consider the following three linear equations.

$$L_5=\frac{3B_1-B_2-B_3-B_4}{\sqrt{60}}$$

$$L_6=\frac{2B_2-B_3-B_4}{\sqrt{30}}$$

$$L_7=\frac{B_3-B_4}{\sqrt{10}}$$

That L_5, L_6, and L_7 are normalized mutually orthogonal linear equations can be proved as in the case of A. Next, it will be proved that L_1–L_4 and L_5–L_7 are mutually orthogonal. For this, we prove the following general theorem.

[THEOREM 2]

When A at k levels and B at l levels have been experimented on an equal number of times for all of the level combinations, an arbitrary linear equation in A, L_A, the sum of whose coefficients is zero, and an arbitrary linear equation in B, L_B, the sum of whose coefficients is zero, are mutually orthogonal.

[PROOF]

The number of replications will be expressed as r. We write the two linear equations as

$$L_A = a_1A_1 + a_2A_2 + \cdots + a_kA_k$$

$$L_B = b_1B_1 + b_2B_2 + \cdots + b_lB_l$$

where

$$a_1 + a_2 + \cdots + a_k = 0$$

$$b_1 + b_2 + \cdots + b_l = 0$$

It will be assumed that A_iB_j represents the sum of r data under conditions A_iB_j.

Then, the sum of the products of the coefficients of the data of L_A and L_B becomes zero, as follows. By substituting in

$$A_1 = A_1B_1 + A_1B_2 + \cdots + A_1B_l$$

$$A_2 = A_2B_1 + A_2B_2 + \cdots + A_2B_l$$

$$\cdots \qquad \cdots$$

$$A_k = A_kB_1 + A_kB_2 + \cdots + A_kB_l$$

$$B_1 = A_1B_1 + A_2B_1 + \cdots + A_kB_1$$

$$B_2 = A_1B_2 + A_2B_2 + \cdots + A_kB_2$$

$$\cdots \qquad \cdots$$

$$B_l = A_1B_l + A_2B_l + \cdots + A_kB_l$$

we have

$$[(a_1b_1 + a_1b_2 + \cdots + a_1b_l) + (a_2b_1 + a_2b_2 + \cdots + a_2b_l) + \cdots + (a_kb_1 + a_kb_2 + \cdots + a_kb_l)]r$$

$$= 0$$

It has been proved in Theorem 2 that $L_1, L_2, \ldots, L_7$ are mutually orthogonal normalized linear equations. By using Theorem 1, it has been proved that S_m, S_A, S_B, and S_e are mutually orthogonal quadratic equations.

Resolution of the number of degrees of freedom, in general, is the same as where, in spectral resolution in communication, the frequencies are divided into k mutually exclusive bands, $\Omega_1, \Omega_2, \ldots, \Omega_k$, and the total power P is resolved into the sum of the powers of the respective bands.

$$P = P(\Omega_1) + P(\Omega_2) + \cdots + P(\Omega_k) + P(N) \tag{40.93}$$

$P(N)$ is the total power of the remaining frequency regions, and it is the power of the noise. It can therefore be said that the number of degrees of freedom in analysis of variance is a concept that corresponds to the number of frequency components or the band width.

40.4.5 Variance Ratio and SN Ratio

In the world of communication engineering, signal-to-noise ratio has been used in comparing the merits of communication methods. This ratio is abbreviated "SN ratio." When the total power of the output is divided into the power of the signals and the power of the noise, the ratio of these is the SN ratio.

$$\text{SN ratio in communication } \eta = \frac{\text{power of signals}}{\text{power of noise}} \tag{40.94}$$

It is possible to obtain this ratio if one knows the observed value of the total power of the output and either the observed value of the power of the signals or that of the noise. One can define a good communication system as one whose SN ratio is great.

Now, the object of resolution of variation has been to divide the total sum of squares of the data into the variation attributable to the sources and that attributable to the error. Therefore, as with the SN ratio in communication systems, it should be possible to find the SN ratio, even when the data are of a finite number.

Let us consider the SN ratio of Equation (40.58). In this case, the total quantity of variation of the data, S_T, is

$$S_T = y_1^2 + y_2^2 + \cdots + y_n^2 - \frac{(y_1 + \cdots + y_n)^2}{n}$$

and the variation due to x therein, S_x, is

$$S_x = \frac{[\sum(x_i - \bar{x})y_i]^2}{\sum(x_i - \bar{x})^2}$$

If we use the same equation of definition as in the case of the SN ratio in communication, the SN ratio η will be given by

$$\eta = \frac{S_x}{S_e}$$

In statistics, the correlation coefficient r between y and x is defined by the following equation.

$$r = \frac{V_{xy}}{\sqrt{V_x V_y}} \tag{40.95}$$

Here,

$$\left.\begin{aligned} V_x &= \frac{1}{n-1}\sum(x_i-\bar{x})^2 \\ V_y &= \frac{1}{n-1}\sum(y_i-\bar{y})^2 \\ V_{xy} &= \frac{1}{n-1}\sum(x_i-\bar{x})(y_i-\bar{y}) \end{aligned}\right\} \tag{40.96}$$

But from resolution of variation, we have

$$\sum(y_i-\bar{y})^2=\frac{[\sum(x_i-\bar{x})(y_i-\bar{y})]^2}{\sum(x_i-\bar{x})^2}+S_e \tag{40.97}$$

If we divide both sides of Equation 40.97 by $\Sigma(y_1-\bar{y})^2$, we have

$$1=\frac{[\sum(x_i-\bar{x})(y_i-\bar{y})]^2}{\sum(x_i-\bar{x})^2\sum(y_i-\bar{y})^2}+\frac{S_e}{\sum(y_i-\bar{y})^2}$$

$$1=\frac{\left[\dfrac{\sum(x_i-\bar{x})(y_i-\bar{y})}{n-1}\right]^2}{\left[\dfrac{1}{n-1}\sum(x_i-\bar{x})^2\right]\left[\dfrac{1}{n-1}\sum(y_i-\bar{y})^2\right]}+\frac{S_e}{\sum(y_i-\bar{y})^2}$$

Substituting from Equations (40.95) and (40.96), we verify:

$$1=r^2+(1-r^2)$$

The square of the correlation coefficient r is that proportion of the total variation which is taken up by the magnitude of the influence of x, and it is termed the contribution ratio.

$$\rho=r^2 \tag{40.98}$$

If this symbol is used, we obtain

$$\eta=\frac{S_x}{S_e}=\frac{S_x/\sum(y_i-\bar{y})^2}{S_e/\sum(y_i-\bar{y})^2}=\frac{\rho}{1-\rho} \tag{40.99}$$

Actually, there are only $n-2$ error variances in the error variation, and one of them is in S_m, the other in S_x. The error variance which is within S_x should be subtracted from S_x. Moreover, when we remember that the SN ratio is used for comparisons of various methods of measurement, testing, analysis, and prediction, it is better to arrange matters so that it is possible to compare even data whose $x_1, x_2, \ldots, x_n$ values differ. Now, the expected value of S_x, $E(S_x)$, is

$$E(S_x)=\sigma^2+b^2\sum(x_i-\bar{x})^2$$

The part of the variation due to the dispersion of x corresponds to the power of signals in communication, and this part constitutes the square of the coefficient b of the unit variation of x multiplied by the total variation of x. It is better to divide b^2 by σ^2, patterning after the practice in the case of communication when finding the SN ratio for a unit electric field strength.

Therefore, as the SN ratio, the following equation is better than Equation (40.99).

$$\eta=\frac{b^2}{\sigma^2}=\frac{\frac{1}{S(xx)}(S_x-V_e)}{V_e}=\frac{1}{S(xx)}\left(\frac{S_x}{V_e}-1\right) \tag{40.100}$$

But, now, since S_x/V_e is the variance ratio F_0, the equation above can also be written

$$\eta=\frac{1}{S(xx)}(F_0-1)$$

Here $S(xx)$ is the variation of x.

The magnitude of the variance ratio F_0 constitutes statistical testing, in itself. However, the author does not feel that a qualitative conclusion such as is obtained by statistical testing is particularly important. This is because he feels that one should regard the SN ratio, defined by Equation (40.100), or the contribution ratio ρ_x of cause x to the total variation, as being more important than the variance ratio F_0.

40.5 Application of Orthogonal Array to Circuit Design

An orthogonal array is a method of estimating with fairly good precision factorial effects which are smaller than the effects of environmental conditions, under the assumption that the factorial effects have the monotonic property. When there is complete additivity of the factorial effects, the F ratio of the factorial effects and the error variance which is an effect of the environmental conditions increases in proportion to replication. There is a complete resemblance between this and the method of delivering electric waves to every part on the earth even when there is great noise, such as atmospherics, merely by varying weak electric fields.

The following study was carried out as to whether or not it is possible to use orthogonal arrays for communication methods. This was related to the use of orthogonal arrays as follows, in regard to the method of transmission of information. We consider orthogonal array L_8. Messages by seven persons are entered as + or − into the seven columns of L_8. The conversations of the seven persons appear in channel No. 1 all as −. In channel No. 2, they become mixed as −, −, −, +, +, +, +, according to No. 2 of the orthogonal array. Similarly, for channels No. 3–8, the messages of the seven persons are mixed from instant to instant according to the symbols of the orthogonal array (− if 1 and + if 2), and they are sent to the receiving end through the eight channels. At the receiving end the information of No. 1, . . . , No. 4 is multiplied by − and that of No. 5, . . . , No. 8 is multiplied by +, and the message by the communicator corresponding to column 1 is separated from the others. If the losses of the eight channels are the same, and if the speeds of the current passing through the eight channels (or electric waves in the case of wireless) are equal, then if messages are transmitted in this way, it is possible to improve the SN ratio appreciably in respect to the problem of nonlinearity of the circuits.

Or, if it is possible to narrow the message circuits to six circuits (at present it is four circuits) by the four pairs or in other words eight channels, there arises the possibility of designing a communication system such that even if one arbitrary channel of the eight is severed, the SN ratio hardly changes at all, even without switching. A method such as this gives us a way of introducing redundancy with about the same efficiency as with codification where self-correction is possible. Communication systems and Design of Experiments can be said to be partners in nearly the same theories, as in regard to orthogonal arrays and coding relations. At the present stage, design of experiments possesses more numerous methods than the theory of communication. The method of Chapter 23 can be used for calculation of the SN ratio in communication by pulses.

References

References

We present here only a part of the literature in this field, in order to give historical references and to supplement inadequate parts of the text.

1) Hotelling, H. 1944. Some improvements in weighing and other experimental techniques. *Ann. Math. Statistics* 15:297–306.

2) Plackett, R. L., and J. P. Burman. 1946. The design of optimum multifactorial experiments. *Biometrika* 33:305–325.

3) Brownlee, K. A. 1949. *Industrial experimentation.* Brooklyn: Chemical Publishing Co.

4) Bose, R. C., and K. A. Bush. 1952. Orthogonal arrays of strength two and three. *Ann. Math. Statistics* 23:508–524.

5) Fisher, R. A. 1951. *Design of Experiments.* Edinburgh: Oliver & Boyd.

   This is the original work on design of experiments. The manner of thinking of this field is central to it; but assignment techniques, etc., are only of the early period.

6) Cochran, W. G., and G. M. Cox. 1957. *Experimental designs.* 2nd ed. New York: John Wiley & Sons.

A detailed explanation of assignment of the incomplete type. This is an easy-to-read book.

7) Kempthorne, O. 1952. *The design and analysis of experiments.* New York: John Wiley & Sons.

This is a major work with in-depth discussion of derivations of formulas.

8) Kitagawa, Toshio, and Michio Mitsutome. 1953. *Jikken Keikaku Yōin Haichihyō* (Assignment arrays of sources in Design of Experiments). Tokyo: Baifūkan.

Although well presented as a treatment of source arrays, not everything is included; the interpretations are centered about theoretical concepts.

9) Davis, O. L. 1954. *Design and analysis of industrial experiments.* Edinburgh: Oliver & Boyd.

Easy to understand but somewhat redundantly long; to be useful, it should probably be read through to the end.

10) Masuyama, Genzaburō. 1955. *Kōjō Gijutsusha no tame no Jikken Keikakuhō no Hanashi, Keikaku-hen* (Design of Experiments explained for the plant technician, volume on design). Tokyo: Japanese Standards Association.

Rather than discussion of assignment, theories concerning the formation of assignments and a general discussion of analyses are given.

11) Shimada, Seizō. 1955. *Suikeigaku Nyūmon* (Introduction to stochastics). Tokyo: Denki Shoin.

The whole of stochastics is organized into a small volume, and it is a very easy-to-read book, but the treatment of each topic is on an introductory level.

12) Masuyama, Genzaburō. 1956. *Jikken Keikakuhō* (Design of Experiments). Iwanami Zensho.

Fundamental principles and the method of constructing various assignment types are discussed. From the theoretical aspect, this is the best-organized book.

13) Taguchi, Genichi. 1956. *Kōsa no Kimekata* (How to determine tolerances). Tokyo: Japanese Standards Association.

This is not for the general reader since it is centered on assembled products. See Chapter 16 of this book.

14) Kayano, Ken. 1956. Kōjō Jikken ni okeru Saigensei ni tsuite (On

reproducibility in plant experiments). *Hinshitsu Kanri* (Statistical Quality Control).

Problems of randomization, stratification, and models are explained so as to be easily understood by technicians.

15) Itō, Fumirō *et al.* 1955. Mōbō ni okeru Senjō Kōtei to Kaado Kōtei no Sōsa ni kansuru Jikken (An experiment on the operations of the washing step and carding step in wool spinning). *The 5th Quality Control Symposium.*

An example of $L_{16}(2^{15}) \otimes L_4(2^3)$.

16) Tsurumaki, Ryōsuke, and Suzuki, Akira. 1956. Datsurō Sōchi no Kōjō Jikken ni tsuite (On a plant experiment on dewaxing apparatus). *The 6th Quality Control Symposium.*

An example of $L_{16}(2^{15})$.

17) Kakehashi, Shigeo. 1956. Kami Parupu Kōgyō ni okeru Jikken Keikakuhō no Tekiyō (Applications of Design of Experiments in the paper and pulp industry). *The 6th Quality Control Symposium.*

This is an example of the randomly combined design.

18) Taguchi, Genichi. 1956. *Jikken Keikakuhō Tekisuto* (A text on Design of Experiments). 2nd rev. ed. Internal Data of Denki Tsūshin Kenkyūjo (Electrical Communication Laboratory).

The parent of the original edition of this book.

19) Kitagawa, Toshio, and Genzaburō Masuyama, compilers. 1952. *Shinpen Tōkei Sūchihyō* (Statistical numerical tables, new compilation).

Various types of distribution functions and the calculation equation for errors of the second kind in the F test are explained in detail.

20) Wilks, S. S. 1943. *Mathematical statistics.* Princeton: Princeton Univ. Press.

This is good as a summary of statistics. It is especially detailed on the normal regression theory.

21) Doob, J. L. 1953. *Stochastic processes.* New York: John Wiley and Sons.

This is an easily read work which is the most organized regarding statistical processes, but it is a large volume.

22) Feller, W. 1950. *An introduction to probability theory and its applications,* vol. 1. New York: John Wiley and Sons.

Whereas 21) is based on measure theory, this book is based on idealized experiments, like physics. Its content is similar to that of

Yasuji Fushimi's *Kakuritsuron* (Theory of probability), Tokyo: Kawade Shobō, but there are new interpretations, especially concerning statistical treatments.

23) Rice, S. O. 1944-1945. Mathematical analysis of random noise. *Bell System Tech. J.* 23:282-332; 24:46-156.
The zero point of time series and the theory of extreme values are treated.

24) Ishikawa, Kaoru. 1952. *Kōjō ni okeru Sanpuring* (Sampling at plants). Tokyo: Maruzen.
How to actually perform sampling is carefully explained.

25) Kawada, Tatsuo. 1952. *Ōyō Sūgaku Gairon II* (An outline of applied mathematics, II). Iwanami Zensho.
The problem of time series is treated from the standpoint of the transformation theory; the content is diverse and it is an interesting volume. The problem of the best approximation in numerical calculations is also well presented.

26) Masuyama, Genzaburō. On difference sets for constructing orthogonal arrays of index two and of strength two. 1957. *Rep. Statist. Appl. Res. Un. Jap. Sci. Eng.* 5:27-34.
This is a revolutionary work, having introduced for the first time the use of the theory of algebraic numbers for construction of orthogonal arrays. Will be generally used from now on.

27) Hatamura, Sasaki, and Okuno. 1954. *Tanbo Jikken no Sekkei to Bunseki* (Design and analysis of agricultural field experiments). Nōgiken Hōkoku (Nōgyō Gijutsu Kenkyūsho Hōkoku: Bulletin of the National Institute of Agricultural Sciences), Ser. A, No. 3.
How departures from normal distribution and homoscedasticity influence the F ratio in analysis of variance is studied thoroughly.

28) Shinohara, Takeo. 1959. *Denji Keidenki no Sekkei* (Design of electromagnetic relays). Kenkyū Jitsuyōka Hokoku (Research and development reports), vol. 10, no. 9.
Many formulas which are necessary for the designing of relays have been obtained efficiently and accurately by the extensive use of Design of Experiments together with more specialized techniques.

29) Takeuchi, Hiroshi. 1960. On a special class of regression problem and its applications; random combined fractional factorial designs, *Rep. Statist. Appl. Res. Un. Japan. Sci. Eng.* 8, No. 1.
Calculation of $E(V)$ by the randomly combined design is performed in detail.

30) Taguchi, Genichi, *et al.* 1972. *Hinshitsu no Hyōka* (Evaluation of product quality). Tokyo: Japanese Standards Association.

Evaluation of product quality in the development of new products is illustrated by various actual examples.

31) Taguchi, Genichi. 1962. *Shinpan Tōkei Kaiseki* (Statistical analysis, new edition). Tokyo: Maruzen.

Methods of statistical analysis for office, business, home economics, and social science are explained by means of actual examples.

32) Takeuchi, Hiroshi. 1975. *Kakuritsu Bunpu to Tōkei Kaiseki* (Probability Distributions and Statistical Analysis). Tokyo: Japanese Standards Association.

An interpretation of distribution from the aspect of mathematical statistics is explained in detail.

33) Taguchi, Jōya. 1973. *Kōkateki na Shijō Kaihatsu* (Effective market development). Tokyo: Nikkan Kōgyō Shinbun-Sha.

The market development method for introduction of new products and the design of experiments for the effects of advertising and propaganda are introduced.

34) Taguchi, Genichi, and Yukiko Yokoyama. 1975. *Business Data no Kaiseki* (Analysis of business data). Tokyo: Maruzen.

For actual data analysis in areas of office and business, new methods of experimental regression analysis and statistical analysis are introduced in detail.

35) Taguchi, Genichi. 1961. The bias on an estimate due to random error, *Bull. Inst. Internat.*

The beta coefficient (discount coefficient) method is discussed, assuming a normal distribution. However, it is better to use the method of Chapter 19, where there is no need for a distribution form.

36) Taguchi, Genichi. 1969. *Hinshitsu no Kanri* (Quality control). Tokyo: Japanese Standards Association.

The beta coefficient method is introduced in Chapter 11.

37) Masuyama, Genzaburō. 1964. *Shōsūrei no Matomekata* (How to organize small number examples) (I), (II).

Examples of experiments with few sources and methods for their analysis are discussed in detail from the standpoint of mathematical statistics.

38) *Dainamikku na Tokusei to SN-hi* (Dynamic characteristics and SN ratio) 1976. *SN-hi Bunkakai* (Committee on SN ratio) Tokyo: Japanese Standards Association.

The discussion of maneuverability in Chapter 24 was based on the data here.

39) Masuyama, Genzaburō, *et al.* 1974. *Jikken Keikakuhō, sono Hatten to Saikin no Wadai* (The Design of Experiments method, its development and recent topics of discussion.) *Tōdai Shuppankai* (Tokyo University Publication Society).
Recent topics in the Design of Experiments method are collected.

40) C. E. Shannon, 1948. A mathematical theory of communication. *Bell System Tech. J.* 27:379–423; 623–656.
Mathematical treatment of the quantity of information was systematically studied here for the first time, and this is a pioneering paper in the science of information.

41) Paley, Raymond E. A. C., and N. Wiener. 1934. *Fourier transforms in the complex domain.* New York: American Math. Soc.

APPENDIX TABLES
Numerical Tables for Design of Experiments

The numerical tables used in the text are assembled here.

APPENDIX TABLE 1	Table of squares (square roots are also obtained from this table)
APPENDIX TABLE 2	*F* Table
APPENDIX TABLE 3	Orthogonal polynomials for equispaced levels
APPENDIX TABLE 4	Table of common logarithms
APPENDIX TABLE 5	Omega (logit) conversion table
APPENDIX TABLE 6	Decibel tables
APPENDIX TABLE 7	Formulas related to SN ratio
APPENDIX TABLE 8	Confidence limits of SN ratio
APPENDIX TABLE 9	Coefficients for tolerance limits (1) Coefficients for two-sided tolerance limits (2) Coefficients for one-sided tolerance limits
APPENDIX TABLE 10	Nomograms for finding confidence limits of fraction defective by measured data
APPENDIX TABLE 11	Orthogonal arrays and linear graphs

Explanation of Appendix Tables 1–11

1. How to Use Appendix Table 1 (Table of Squares)

Worked Example 1. Find the square of 2.57.

The square of 257 is found in Appendix Table 1 as 66,049; the decimal point is entered four decimal places from the end in order to obtain a number with four decimal places, by mental calculation, and the answer is found as 6.6049.

Worked Example 2. Find the square roots of 81.52 and 0.0764.

The first digit of the square root of 81.52 is found to be 9 by mental calculation; from Appendix Table 1, we find that square of a number whose first digit is 9 which is closest to 8152 . . . . When n is 903, n^2 is closest to 815,200; positioning of the decimal point is performed by mental calculation, and the answer is 9.03.

$$\sqrt{81.52} = 9.03$$

$\sqrt{0.0764}$ should be 0.2 . . . ; the number for which the square of 2 . . . is closest to 76,400 is found from the table of squares, Appendix Table 1, as 276. Therefore, we have

$$\sqrt{0.0764} = 0.276$$

2. How to Use Appendix Table 2 (F Table)

Worked Example 1. Find the 5% and 1% values in the F table where the number of degrees of freedom of the numerator is $f_1 = 1$ and the number of degrees of freedom of the denominator is 10.

From Appendix Table 2, we obtain the 5% value as 4.96 and the 1% value is 10.04.

Worked Example 2. Find the 5% value and 1% in the F table where the number of degrees of freedom of the numerator is 1 and the number of degrees of freedom of the denominator is 2500.

From Appendix Table 2, we obtain 3.84 and 6.65. Midway between 1000 degrees of freedom and ∞ degrees of freedom is 2000 degrees of freedom.

3. How to Use Appendix Table 3 (Orthogonal Polynomials for Equispaced Levels)

Worked Example 1. $A_1, A_2, \ldots, A_5$ are the sums of r data, classed by level, of a five-level factor A of equal level spacing h. Find the equations of estimation of the coefficients of the linear, quadratic, and cubic terms of A, the equation for the confidence limits, and the calculating equations for the variations.

The coefficients of the linear, quadratic, and cubic terms of A are expressed as b_1, b_2, and b_3. From $k = 5$ levels in Appendix Table 3, we have

$$\hat{b}_1 = \frac{-2A_1 - A_2 + A_4 + 2A_5}{r \times 10 \times h} \pm \sqrt{\frac{F \times V_e}{r \times 10 \times h^2}}$$

$$\hat{b}_2 = \frac{2A_1 - A_2 - 2A_3 - A_4 + 2A_5}{r \times 14 \times h^2} \pm \sqrt{\frac{F \times V_e}{r \times 14 \times h^4}}$$

$$\hat{b}_3 = \frac{-A_1 + 2A_2 - 2A_4 + A_5}{r \times 12 \times h^3} \pm \sqrt{\frac{F \times V_e}{r \times \frac{72}{5} \times h^6}}$$

$$S_{\hat{b}_1} = \frac{(-2A_1 - A_2 + A_4 + 2A_5)^2}{r \times 10}$$

$$S_{\hat{b}_2} = \frac{(2A_1 - A_2 - 2A_3 - A_4 + 2A_5)^2}{r \times 14}$$

$$S_{\hat{b}_3} = \frac{(-A_1 + 2A_2 - 2A_4 + A_5)^2}{r \times 10}$$

$r \cdot \lambda S \cdot h^i$ enters the denominator of the estimated value of the i-th order coefficient; $r \cdot S \cdot h^{2i}$ enters the denominator within the square root sign of the confidence limits; and $r \cdot \lambda^2 S$ enters the denominator of the variation.

For example, let us assume that factor A is the five levels related with temperature of -10, 0, 10, 20, and 30 (°C) and that the error variance of six degrees of freedom, V_e, has been found to be 3.2. It will be assumed that the sums of the data for A_1, A_2, A_3, A_4 and A_5, with $r=3$, are

$$A_1=280, A_2=252, A_3=225, A_4=196, A_5=166$$

Then, from the equations given above, $F=5.99$, we have

$$\hat{b}_1=\frac{-2\times 280-252+196+2\times 166}{3\times 10\times 10}\pm\sqrt{\frac{5.99\times 3.2}{3\times 10\times 10^2}}=-0.947\pm 0.080$$

$$\hat{b}_2=\frac{2\times 280-252-2\times 225-196+2\times 195}{3\times 14\times 10^2}\pm\sqrt{\frac{5.99\times 3.2}{3\times 14\times 10^4}}=0.0124\pm 0.0068$$

$$\hat{b}_3=\frac{-280+2\times 252-2\times 196+166}{3\times 12\times 10^3}\pm\sqrt{\frac{5.99\times 3.2}{3\times\frac{72}{5}\times 10^6}}=-0.00006\pm 0.00067$$

$$S_{A_l}=\frac{(-2\times 280-252+195+2\times 166)^2}{3\times 10}=2688.53$$

$$S_{A_q}=\frac{(2\times 280-252-2\times 225-195+2\times 166)^2}{3\times 14}=64.38$$

$$S_{A_c}=\frac{(-280+2\times 252-2\times 195+166)^2}{3\times 10}=0.13$$

Also, we have

$$E(S_{A_l})=\sigma^2+r\cdot s\cdot h^2b_1^2=\sigma^2+3\times 10\times 10^2b_1^2$$
$$=\sigma^2+3000\,b_1^2$$
$$E(S_{A_q})=\sigma^2+3\times 14\times 10^4\,b_2^2$$
$$=\sigma^2+420000\,b_2^2$$
$$E(S_{A_c})=\sigma^2+3\times\frac{72}{5}\times 10^6\,b_3^2$$
$$=\sigma^2+43200000\,b_3^2$$

In actual practice, b_3 is not significant; therefore the cubic term is not estimated.

4. How to Use Appendix Table 4 (Table of Common Logarithms)

Worked Example 1. Find the common logarithms of 5.82, 68,040, and 0.001 529 3.

(1) For the logarithm of 5.82, the value corresponding to 582 is found from Appendix Table 4 as 7649. Since this value is the decimal-place part, or in other words the mantissa, 0.7649 is the value being sought in this case.

(2) The characteristic of the logarithm for 68,040 is 4. What remains, therefore, is to find the logarithm value for 6.804, or in other words the mantissa. For the logarithm of 6.800, 0.8325, we read off the value, 3, which corresponds to 4 in the top row of the proportional part section at the right side; we then add this to the last place of 0.8325 and find the answer as 4.8328.

(3) For the common logarithm of 0.001 529 3, first the characteristic is −3, and to this we add the logarithm value of 1.5293. The logarithm of 1.52 is 0.1818; we take the remaining digits 93 as 9, and if we add the proportional part for 9, 26 (or the proportional part for 9.3, 27) to 0.1818, it becomes 0.1844 (or 0.1845). Therefore, we have

$$\log 0.0015293 = -3 + 0.1844 = -2.8156 \text{ (or } -2.8155)$$

Worked Example 2. Find the antilogarithms of 0.5424, 8.6326, and −4.5255.

The table entry which is closest to 0.5425 on the low side is 0.5416, which has the antilogarithm 3.48. From this, however, there is a difference in logarithm value of 0.5425 − 0.5416 = 0.0009. We add the numerical value in the top row of 7, corresponding to the difference 9 in the proportional part table, and 3.487 is the antilogarithm which we wish to find.

As to the antilogarithm of 8.6326, first we find the antilogarithm of 0.6326 as 4.291, and we then multiply this by 10^8 and the answer is found as 429,100,000.

The antilogarithm of −4.5255 is the antilogarithm of −5 + 0.4745. The antilogarithm of 0.4745, 2.982, is multiplied by 10^{-5}, and the answer is found as 0.00002982.

5. How to Use Appendix Table 5 (Omega Conversion Table)

Worked Example 1. Find the omega value of fraction defective 80.25% to the second decimal place.

From Appendix Table 5, we have

$$6.075+0.027\times0.5=6.0885\doteqdot6.09 \text{ [db]}$$

Worked Example 2. Find the reliability p for decibel value 34.32. In this case, we use the logarithm table, Appendix Table 4. In the definition formula,

$$-10\log\left(\frac{1}{p}-1\right)=34.32$$

if we let $x = 1/p - 1$, we have

$$\log x = -3.432 = -4 + 0.568$$

Therefore, the antilogarithm for the mantissa of x, 0.568, is 3.698; therefore, x is 0.000 369 8, and from this we obtain

$$p = \frac{1}{1+x} = \frac{1}{1.0003698} = 0.999630$$

Therefore, the reliability is 99.9630%.

6. How to Use Appendix Table 6 (Decibel Tables)

For the SN ratio η, these tables show the values of 10 log η, or in other words decibel values which correspond to $1.00 \leqq \eta \leqq 10.00$ (in steps of 0.01). As a general rule, one may round off the second decimal place and stop at the first decimal place. When more decimal places are desired, find the value by multiplying the value given in Appendix Table 4 (Table of Common Logarithms) by 10.

Worked Example 1. Find the decibel values of 1.75, 17.5, and 0.0175.

From Appendix Table 6, the decibel value of 1.75 is 2.43 db. For the decibel value of 17.5, since a factor of 10 is 10 db, we add this to the decibel value for 1.75, thus,

$$2.43 + 10 = 12.43 \text{ [db]} \tag{1}$$

Similarly, for the decibel value of 0.0175, since one-one hundredth is −20 db, we have

$$2.43 + (-20) = -17.57 \text{ [db]} \tag{2}$$

Worked Example 2. Find the decibel values of 0.1054 and of 0.000 105 4.

One-tenth is −10 db. To find the decibel value of 1.054, we perform interpolation using Appendix Table 6 and we find 0.23 db. Therefore, the result is

$$0.23 + (-10) = -9.77 \text{ [db]} \tag{3}$$

For the decibel value of 0.000 105 4, we add the decibel value of one-ten thousandth (1/10,000), −40 db, to the decibel value of 1.054, 0.23 db, and we find

$$0.23 + (-40) = -39.77 \text{ [db]} \tag{4}$$

Worked Example 3. When two measuring instruments, A_1 and A_2, are compared, A_2 gives us a gain of 13.5 db. What fraction of the error variance of A_1 is that of A_2?

Since the η value corresponding to 3.5 db is 2.24 from Appendix Table 6, the SN ratio for 13.5 db is 22.4 (multiplying by 10); and therefore the answer is 1/22.4.

Worked Example 4. Chemical analysis of A_2 shows it to be inferior to A_1 by -25.67 db. If method A_2 is used, how many times greater will the analysis error be?

We write $-25.67 = (-30) + 4.33$. Appendix Table 6 gives the η value 2.71 for 4.33 db. Because of the -30 db, the SN ratio is 1/1000 of this, or 0.002 71; therefore, the error variance is its reciprocal, or 369 times, as great and the error is $\sqrt{369} = 19.2$ times as great.

7. How to Use Appendix Table 7 (Formulas Related with SN Ratio)

Refer to Section 22.1.

8. How to Use Appendix Table 8 (Confidence Limits of SN Ratio)

The SN ratio, discussed in Chapter 22 and Chapter 24, was given by the following equation when the number of degrees of freedom of the numerator is 1.

$$\eta = \frac{\frac{1}{r}(V_\beta - V_e)}{V_e} \tag{1}$$

In the confidence limits for the decibel value of the SN ratio η, $-\log r$ becomes a constant and they are unrelated to r. Therefore, the confidence limits for the decibel value are a function of the number of degrees of freedom of the error variance V_e, f_2, and the following variance ratio:

$$F_0 = \frac{V_\beta}{V_e} \tag{2}$$

Often, in actual cases, effects of error factors that are common to the error variances are included; thus, usually the correct confidence limits are narrower than the confidence limits shown in these tables. (The calculations for the tables were done by Takumi Nakamura.)

Worked Example 1. Confidence Limits of SN Ratio

The values of Table 1 were obtained when the alcohol content of soy sauce was varied as the four levels (in percent):

$$M_1 = x \text{ (unknown)}, M_2 = x + 0.3, M_3 = x + 0.6, M_4 = x + 0.9$$

and two persons, R_1 and R_2, read them off from the gas chromatographs. The analysis of variance table of this is as given in Table 2. Since r_0, the number of replications, is 2, h_0 is the spacing 0.3%, and the S of the four equispaced levels is 5, we have $r = r_0 \times S \times h_0^2 = 2 \times 5 \times 0.3^2 = 0.9$

Table 1 Data of Read Values

	M_1	M_2	M_3	M_4
R_1	0.46	0.96	1.48	1.79
R_2	0.48	0.92	1.36	1.56

Table 2 Analysis of Variance Table

Source	f	S	V	F_0	$E(V)$
β	1	1.6769	1.6769	158.2	$\sigma^2+0.9\,\beta^2$
e	6	0.0633	0.0106		σ^2
T	7	1.7402			

Therefore, we have

$$\eta=\frac{\frac{1}{0.9}(V_\beta-V_e)}{V_e}=\frac{\frac{1}{0.9}(1.6769-0.0106)}{0.0106}$$

$$=174.70=22.4\ [\text{db}] \tag{3}$$

Since the variance ratio F_0 is 158.2, we perform interpolation calculation between $F_0=60$ of $f_2=6$ and $F_0=\infty$ by using Appendix Table 8. If this is troublesome, one may also use the nearer value. Reciprocal interpolation is used. Since the value is -8.2 if $F_0=60$ and -6.9 if $F_0=\infty$, the difference is -1.3. We therefore have

$$-6.9+(-1.3)\times\frac{60}{158.2}=-7.4 \tag{4}$$

Similarly, the upper limit is

$$3.8+0.3\times\frac{60}{158.2}=3.9 \tag{5}$$

Thus, we have

$$\eta=22.4^{-7.4}_{+3.9}\ [\text{db}] \tag{6}$$

Worked Example 2. Comparison of SN Ratios (Testing)

In order to compare the SN ratios of two types of portable hardness meters, A_1 = stand type and A_2 = manually operated type, standard hardness pieces at five levels, $M_1=20$, $M_2=30$, $M_3=40$, $M_4=50$, and $M_5=60$, were measured once each by three persons, R_1, R_2, and R_3; and the read-off values and analysis of variance table were as given by Table 3 and Table 4.

Therefore, the SN ratios of A_1 and A_2 are

$$\eta(A_1)=\frac{\frac{1}{3000}(2837.269-0.0158)}{0.0158}=59.9=17.8\ [\text{db}] \tag{7}$$

$$\eta(A_2)=\frac{\frac{1}{3000}(2820.760-0.0459)}{0.0459}=20.5=13.1\ [\text{db}] \tag{8}$$

Table 3 Data of Read Values

The Case of A_1

	M_1	M_2	M_3	M_4	M_5
R_1	21.0	30.6	40.5	50.25	59.8
R_2	21.2	30.5	40.4	50.2	59.9
R_3	21.1	30.6	40.4	50.4	59.9

The Case of A_2

	M_1	M_2	M_3	M_4	M_5
R_1	20.4	29.9	40.0	49.8	59.2
R_2	20.5	30.0	39.9	49.6	59.3
R_3	21.0	30.3	40.3	49.7	59.4

Table 4 Analysis of Variance Table

The Case of A_1

Source	f	S	V	F_0	$E(V)$
β	1	2 837.269	2 837.269	179 574.0	$\sigma^2+3\,000\beta^2$
e	13	0.206	0.0158		σ^2
T	14	2 837.475			

The Case of A_2

Source	f	S	V	F_0	$E(V)$
β	1	2 820.760	2 820.760	61 454.5	$\sigma^2+3\,000\beta^2$
e	13	0.597	0.0459		
T	14	2 821.357			

Since the variance ratio is large, we assume that $F_0 \doteqdot \infty$, and we interpolate calculation between $f_2 = 10$ and $f_2 = 15$; we then have

$$\text{SN ratio of } A_1 = 17.8^{-4.2}_{+2.8} \text{ (db)}$$

$$\text{SN ratio of } A_2 = 13.1^{-4.2}_{+2.8} \text{ (db)}$$

Since the SN ratio of A_1 is greater, we find the square root of the sum of squares of the minus side -4.2 for A_1 and the plus side 2.8 for A_2, and we compare this with the difference between A_1 and A_2.

$$\sqrt{(-4.2)^2+2.8^2} = 5.0 \text{ (db)}$$

Since the difference between A_1 and A_2 is 4.7 db, it means that there is no significant difference. Actually, since the difference between R's has entered both errors in common, it is very possible that there is a significant difference when this is considered, but this will be omitted since calculation becomes troublesome. Testing should be kept to about reference degree.

9. How to Use Appendix Table 9 (Coefficients for Tolerance Limits)

Refer to Section 37.1.

10. How to Use Appendix Table 10 (Nomograms for Finding Confidence Limits of Fraction Defective by Measured Data)

Refer to Section 37.2.

11. How to Use Appendix Table 11 (Orthogonal Arrays and Linear Graphs)

Refer to the opening text for the Appendix (page 1126) for explanations.

Appendix Table 1 Table of Squares

n	n^2	n	n^2	n	n^2	n	n^2	n	n^2
1	1	51	2601	101	10201	151	22801	201	40401
2	4	52	2704	102	10404	152	23104	202	40804
3	9	53	2809	103	10609	153	23409	203	41209
4	16	54	2916	104	10816	154	23716	204	41616
5	25	55	3025	105	11025	155	24025	205	42025
6	36	56	3136	106	11236	156	24336	206	42436
7	49	57	3249	107	11449	157	24649	207	42849
8	64	58	3364	108	11664	158	24964	208	43264
9	81	59	3481	109	11881	159	25281	209	43681
10	100	60	3600	110	12100	160	25600	210	44100
11	121	61	3721	111	12321	161	25921	211	44521
12	144	62	3844	112	12544	162	26244	212	44944
13	169	63	3969	113	12769	163	26569	213	45369
14	196	64	4096	114	12996	164	26896	214	45796
15	225	65	4225	115	13225	165	27225	215	46225
16	256	66	4356	116	13456	166	27556	216	46656
17	289	67	4489	117	13689	167	27889	217	47089
18	324	68	4624	118	13924	168	28224	218	47524
19	361	69	4761	119	14161	169	28561	219	47961
20	400	70	4900	120	14400	170	28900	220	48400
21	441	71	5041	121	14641	171	29241	221	48841
22	484	72	5184	122	14884	172	29584	222	49284
23	529	73	5329	123	15129	173	29929	223	49729
24	576	74	5476	124	15376	174	30276	224	50176
25	625	75	5625	125	15625	175	30625	225	50625
26	676	76	5776	126	15876	176	30976	226	51076
27	729	77	5929	127	16129	177	31329	227	51529
28	784	78	6084	128	16384	178	31684	228	51984
29	841	79	6241	129	16641	179	32041	229	52441
30	900	80	6400	130	16900	180	32400	230	52900
31	961	81	6561	131	17161	181	32761	231	53361
32	1024	82	6724	132	17424	182	33124	232	53824
33	1089	83	6889	133	17689	183	33489	233	54289
34	1156	84	7056	134	17956	184	33856	234	54756
35	1225	85	7225	135	18225	185	34225	235	55225
36	1296	86	7396	136	18496	186	34596	236	55696
37	1369	87	7569	137	18769	187	34969	237	56169
38	1444	88	7744	138	19044	188	35344	238	56644
39	1521	89	7921	139	19321	189	35721	239	57121
40	1600	90	8100	140	19600	190	36100	240	57600
41	1681	91	8281	141	19881	191	36481	241	58081
42	1764	92	8464	142	20164	192	36864	242	58564
43	1849	93	8649	143	20449	193	37249	243	59049
44	1936	94	8836	144	20736	194	37636	244	59536
45	2025	95	9025	145	21025	195	38025	245	60025
46	2116	96	9216	146	21316	196	38416	246	60516
47	2209	97	9409	147	21609	197	38809	247	61009
48	2304	98	9604	148	21904	198	39204	248	61504
49	2401	99	9801	149	22201	199	39601	249	62001
50	2500	100	10000	150	22500	200	40000	250	62500

Appendix Table 1 (Continued)

n	n^2	n	n^2	n	n^2	n	n^2	n	n^2
251	63001	301	90601	351	123201	401	160801	451	203401
252	63504	302	91204	352	123904	402	161604	452	204304
253	64009	303	91809	353	124609	403	162409	453	205209
254	64516	304	92416	354	125316	404	163216	454	206116
255	65025	305	93025	355	126025	405	164025	455	207025
256	65536	306	93636	356	126736	406	164836	456	207936
257	66049	307	94249	357	127449	407	165649	457	208849
258	66564	308	94864	358	128164	408	166464	458	209764
259	67081	309	95481	359	128881	409	167281	459	210681
260	67600	310	96100	360	129600	410	168100	460	216100
261	68121	311	96721	361	130321	411	168921	461	212521
262	68644	312	97344	362	131044	412	169744	462	213444
263	69169	313	97969	363	131769	413	170569	463	214369
264	69696	314	98596	364	132496	414	171396	464	215296
265	70225	315	99225	365	133225	415	172225	465	216225
266	70756	316	99856	366	133956	416	173056	466	217156
267	71289	317	100489	367	134689	417	173889	467	218089
268	71824	318	101124	368	135424	418	174724	468	219024
269	72361	319	101761	369	136161	419	175561	469	219961
270	72900	320	102400	370	136900	420	176400	470	220900
271	73441	321	103041	371	137641	421	177241	471	221841
272	73984	322	103684	372	138384	422	178084	472	222784
273	74529	323	104329	373	139129	423	178929	473	223729
274	25076	324	104976	374	139876	424	179776	474	224676
275	75625	325	105625	375	140625	425	180625	475	225625
276	76176	326	106276	376	141376	426	181476	476	226576
277	76729	327	106929	377	142129	427	182329	477	227529
278	77284	328	107584	378	142884	428	183184	478	228484
279	77841	329	108241	379	143641	429	184041	479	229441
280	78400	330	108900	380	144400	430	184900	480	230400
281	78961	331	109561	381	145161	431	185761	481	231361
282	79524	332	110224	382	145924	432	186624	482	232324
283	80089	333	110889	383	146689	433	187489	483	233289
284	80656	334	111556	384	147456	434	188356	484	234256
285	81225	335	112225	385	148225	435	189225	485	235225
286	81796	336	112896	386	148996	436	190096	486	236196
287	82369	337	113569	387	149769	437	190969	487	237169
288	82944	338	114244	388	150544	438	191844	488	238144
289	83521	339	114921	389	151321	439	192721	489	239121
290	84100	340	115600	390	152100	440	193600	490	240100
291	84681	341	116281	391	152881	441	194481	491	241081
292	85264	342	116964	392	153664	442	195364	492	242064
293	85849	343	117649	393	154449	443	196249	493	243049
294	86436	344	118336	394	155236	444	197136	494	244036
295	87025	345	119025	395	156025	445	198025	495	245025
296	87616	346	119716	396	156816	446	198916	496	246016
297	88209	347	120409	397	157609	447	199809	497	247009
298	88804	348	121104	398	158404	448	200704	498	248004
299	89401	349	121801	399	159201	449	201601	499	249001
300	90000	350	122500	400	160000	450	202500	500	250000

Appendix Table 1 (Continued)

n	n^2	n	n^2	n	n^2	n	n^2	n	n^2
501	251001	551	303601	601	361201	651	423801	701	491401
502	252004	552	304704	602	362404	652	425104	702	492804
503	253009	553	305809	603	363609	653	426409	703	494209
504	254016	554	306916	604	364816	654	427716	704	495616
505	255025	555	308025	605	366025	655	429025	705	497025
506	256036	556	309136	606	367236	656	430336	706	498436
507	257049	557	310249	607	368449	657	431649	707	499849
508	258064	558	311364	608	369664	658	432964	708	501264
509	259081	559	312481	609	370881	659	434281	709	502681
510	260100	560	313600	610	372100	660	435600	710	504100
511	261121	561	314721	611	373321	661	436921	711	505521
512	262144	562	315844	612	374544	662	438244	712	506944
513	263169	563	316969	613	375769	663	439569	713	508369
514	264196	564	318096	614	376996	664	440896	714	509796
515	265225	565	319225	615	378225	665	442225	715	511225
516	266256	566	320356	616	379456	666	443556	716	512656
517	267289	567	321489	617	380689	667	444889	717	514089
518	268324	568	322624	618	381924	668	446224	718	515524
519	269361	569	323761	619	383161	669	447561	719	516961
520	270400	570	324900	620	384400	670	448900	720	518400
521	271441	571	326041	621	385641	671	450241	721	519841
522	272484	572	327184	622	386884	672	451584	722	521284
523	273529	573	328329	623	388129	673	452929	723	522729
524	274576	574	329476	624	389376	674	454276	724	524176
525	275625	575	330625	625	390625	675	455625	725	525625
526	276676	576	331776	626	391876	676	456976	726	527076
527	277729	577	332929	627	393129	677	458329	727	528529
528	278784	578	334084	628	394384	678	459684	728	529984
529	279841	579	335241	629	395641	679	461041	729	531441
530	280900	580	336400	630	396900	680	462400	730	532900
531	281961	581	337561	631	398161	681	463761	731	534361
532	283024	582	338724	632	399424	682	465124	732	535824
533	284089	583	339889	633	400689	683	466489	733	537289
534	285156	584	341056	634	401956	684	467856	734	538756
535	286225	585	342225	635	403225	685	469225	735	540225
536	287296	586	343396	636	404496	686	470596	736	541696
537	288369	587	344569	637	405769	687	471969	737	543169
538	289444	588	345744	638	407044	688	473344	738	544644
539	290521	589	346921	639	408321	689	474721	739	546121
540	291600	590	348100	640	409600	690	476100	740	547600
541	292681	591	349281	641	410881	691	477481	741	549081
542	293764	592	350464	642	412164	692	478864	742	550564
543	294849	593	351649	643	413449	693	480249	743	552049
544	295936	594	352836	644	414736	694	481636	744	553536
545	297025	595	354025	645	416025	695	483025	745	555025
546	298116	596	355216	646	417316	696	484416	746	556516
547	299209	597	356409	647	418609	697	485809	747	558009
548	300301	598	357604	648	419904	698	487204	748	559504
549	301401	599	358801	649	421201	699	488601	749	561001
550	302500	600	360000	650	422500	700	490000	750	562500

Appendix Table 1 (Continued)

n	n^2	n	n^2	n	n^2	n	n^2	n	n^2
751	564001	801	641601	851	724201	901	811801	951	904401
752	565504	802	643204	852	725904	902	813604	952	906304
753	567009	803	644809	853	727609	903	815409	953	908209
754	568516	804	646416	854	729316	904	817216	954	910116
755	570025	805	648025	855	731025	905	819025	955	912025
756	571536	806	649636	856	732736	906	820836	956	913936
757	573049	807	651249	857	734449	907	822649	957	915849
758	574564	808	652864	858	736164	908	824464	958	917764
759	576081	809	654481	859	737881	909	826281	959	919681
760	577600	810	656100	860	739600	910	828100	960	921600
761	579121	811	657721	861	741321	911	829921	961	923524
762	580644	812	659344	862	743044	912	831744	962	925444
763	582169	813	660969	863	744769	913	833569	963	927369
764	583696	814	662596	864	746496	914	835396	964	929296
765	585225	815	664225	865	748225	915	837225	965	931225
766	586756	816	665856	866	749956	916	839056	966	933156
767	588289	817	667489	867	751689	917	840889	967	935089
768	589824	818	669124	868	753424	918	842724	968	937024
769	591361	819	670761	869	755161	919	844561	969	938961
770	592900	820	672400	870	756900	920	846400	970	940900
771	594441	821	674041	871	758641	921	848241	971	942841
772	595984	822	675684	872	760384	922	850084	972	944784
773	597529	823	677329	873	762129	923	851929	973	946729
774	599076	824	678976	874	763876	924	853776	974	948676
775	600625	825	680625	875	765625	925	855625	975	950625
776	602176	826	682276	876	767376	926	857476	976	952576
777	603729	827	683929	877	769129	927	859329	977	954529
778	605284	828	685584	878	770884	928	861184	978	956484
779	606841	829	687241	879	772641	929	863041	979	958441
780	608400	830	688900	880	774400	930	864900	930	960400
781	609961	831	690561	881	776161	931	866761	981	962361
782	611524	832	692224	882	777924	932	868624	982	964324
783	613089	833	693889	883	779689	933	870489	983	966289
784	614656	834	695556	884	781456	934	872356	984	968256
785	616225	835	697225	885	783225	935	874225	985	970225
786	617796	836	698896	886	784996	936	876096	986	972196
787	619369	837	700569	887	786769	937	877969	987	974169
788	620944	838	702244	888	788544	938	879844	988	976144
789	622521	839	703921	889	790321	939	881721	989	978121
790	624100	840	705600	890	792100	940	883600	990	980100
791	625681	841	707281	891	793881	941	885481	991	982081
792	627264	842	708964	892	795664	942	887364	992	984064
793	628849	843	710649	893	797449	943	889249	993	986049
794	631436	844	712336	894	799236	944	891136	994	988036
795	632025	845	714025	895	801025	945	893025	995	990025
796	633616	846	715716	896	802816	946	894916	996	992016
797	635209	847	717409	897	804609	947	896809	997	994009
798	636804	848	719104	898	806404	948	898704	998	996004
799	638401	849	720801	899	808201	949	900601	999	998001
800	640000	850	722500	900	810000	950	902500	1000	1000000

Appendix Table 2 *F* Table

Number of degrees of freedom of denominator, f_2	Number of degrees of freedom of numerator, f_1										
	1	2	3	4	5	6	7	8	9	10	11
1	161 **4052**	200 **4999**	126 **5403**	225 **5625**	230 **5764**	234 **5859**	237 **5928**	239 **5981**	241 **6022**	242 **6056**	243 **6082**
2	18.51 **98.49**	19.00 **99.01**	19.16 **99.17**	19.25 **99.25**	19.30 **99.30**	19.33 **99.33**	19.36 **99.34**	19.37 **99.36**	19.38 **99.38**	19.39 **99.40**	19.40 **99.41**
3	10.13 **34.12**	9.55 **30.81**	9.28 **29.46**	9.12 **28.71**	9.01 **28.24**	8.94 **27.91**	8.88 **27.67**	8.84 **27.49**	8.81 **27.34**	8.78 **27.23**	8.76 **27.13**
4	7.71 **21.20**	6.94 **18.00**	6.59 **16.69**	6.39 **15.98**	6.26 **15.52**	6.16 **15.21**	6.09 **14.98**	6.04 **14.80**	6.00 **14.66**	5.96 **14.54**	5.93 **14.45**
5	6.61 **16.26**	5.79 **13.27**	5.41 **12.06**	5.19 **11.39**	5.05 **10.97**	4.95 **10.67**	4.88 **10.45**	4.82 **10.27**	4.78 **10.15**	4.74 **10.05**	4.70 **9.96**
6	5.99 **13.74**	5.14 **10.92**	4.76 **9.78**	4.53 **9.15**	4.39 **8.75**	4.28 **8.47**	4.21 **8.26**	4.15 **8.10**	4.10 **7.98**	4.06 **7.87**	4.03 **7.79**
7	5.59 **12.25**	4.74 **9.55**	4.35 **8.45**	4.12 **7.85**	3.97 **7.46**	3.87 **7.19**	3.79 **7.00**	3.73 **6.84**	3.68 **6.71**	3.63 **6.62**	3.60 **6.54**
8	5.32 **11.26**	4.46 **8.65**	4.07 **7.59**	3.84 **7.01**	3.69 **6.63**	3.58 **6.37**	3.50 **6.19**	3.44 **6.03**	3.39 **5.91**	3.34 **5.82**	3.31 **5.74**
9	5.12 **10.56**	4.26 **8.02**	3.86 **6.99**	3.63 **6.42**	3.48 **6.06**	3.37 **5.80**	3.29 **5.62**	3.23 **5.47**	3.18 **5.35**	3.13 **5.26**	3.10 **5.18**
10	4.96 **10.04**	4.10 **7.56**	3.71 **6.55**	3.48 **5.99**	3.33 **5.64**	3.22 **5.39**	3.14 **5.21**	3.07 **5.06**	3.02 **4.95**	2.97 **4.85**	2.94 **4.78**
11	4.84 **9.65**	3.98 **7.20**	3.59 **6.22**	3.36 **5.67**	3.20 **5.32**	3.09 **5.07**	3.01 **4.88**	2.95 **4.74**	2.90 **4.63**	2.86 **4.54**	2.82 **4.46**
12	4.75 **9.33**	3.88 **6.93**	3.49 **5.95**	3.26 **5.41**	3.11 **5.06**	3.00 **4.82**	2.92 **4.65**	2.85 **4.50**	2.80 **4.39**	2.76 **4.30**	2.72 **4.22**
13	4.67 **9.07**	3.80 **6.70**	3.41 **5.74**	3.18 **5.20**	3.02 **4.86**	2.92 **4.62**	2.84 **4.44**	2.77 **4.30**	2.72 **4.19**	2.67 **4.10**	2.63 **4.02**
14	4.60 **8.86**	3.74 **6.51**	3.34 **5.56**	3.11 **5.03**	2.96 **4.69**	2.85 **4.46**	2.77 **4.28**	2.70 **4.14**	2.65 **4.03**	2.60 **3.94**	2.56 **3.86**
15	4.54 **8.68**	3.68 **6.36**	3.29 **5.42**	3.06 **4.89**	2.90 **4.56**	2.79 **4.32**	2.70 **4.14**	2.64 **4.00**	2.59 **3.89**	2.55 **3.80**	2.51 **3.73**
16	4.49 **8.53**	3.63 **6.23**	3.24 **5.29**	3.01 **4.77**	2.85 **4.44**	2.74 **4.20**	2.66 **4.03**	2.59 **3.89**	2.54 **3.78**	2.49 **3.69**	2.45 **3.61**
17	4.45 **8.40**	3.59 **6.11**	3.20 **5.18**	2.96 **4.67**	2.81 **4.34**	2.70 **4.10**	2.62 **3.93**	2.55 **3.79**	2.50 **3.68**	2.45 **3.59**	2.41 **3.52**
18	4.41 **8.28**	3.55 **6.01**	3.16 **5.09**	2.93 **4.58**	2.77 **4.25**	2.66 **4.01**	2.58 **3.85**	2.51 **3.71**	2.46 **3.60**	2.41 **3.51**	2.37 **3.44**
19	4.38 **8.18**	3.52 **5.93**	3.13 **5.01**	2.90 **4.50**	2.74 **4.17**	2.63 **3.94**	2.55 **3.77**	2.48 **3.63**	2.43 **3.52**	2.38 **3.43**	2.34 **3.36**
20	4.35 **8.10**	3.49 **5.85**	3.10 **4.94**	2.87 **4.43**	2.71 **4.10**	2.60 **3.87**	2.52 **3.71**	2.45 **3.56**	2.40 **3.45**	2.35 **3.37**	2.31 **3.30**
21	4.32 **8.02**	3.47 **5.78**	3.07 **4.87**	2.84 **4.37**	2.68 **4.04**	2.57 **3.81**	2.49 **3.65**	2.42 **3.51**	2.37 **3.40**	2.32 **3.31**	2.28 **3.24**
22	4.30 **7.94**	3.44 **5.72**	3.05 **4.82**	2.82 **4.31**	2.66 **3.99**	2.55 **3.76**	2.47 **3.59**	2.40 **3.45**	2.35 **3.35**	2.30 **3.26**	2.26 **3.18**
23	4.28 **7.88**	3.42 **5.66**	3.03 **4.76**	2.80 **4.26**	2.64 **3.94**	2.53 **3.71**	2.45 **3.54**	2.38 **3.41**	2.32 **3.30**	2.28 **3.21**	2.24 **3.14**
24	4.26 **7.82**	3.40 **5.61**	3.01 **4.72**	2.78 **4.22**	2.62 **3.90**	2.51 **3.67**	2.43 **3.50**	2.36 **3.36**	2.30 **3.25**	2.26 **3.17**	2.22 **3.09**
25	4.24 **7.77**	3.38 **5.57**	2.99 **4.68**	2.76 **4.18**	2.60 **3.86**	2.49 **3.63**	2.41 **3.46**	2.34 **3.32**	2.28 **3.21**	2.24 **3.13**	2.20 **3.05**
26	4.22 **7.72**	3.37 **5.53**	2.89 **4.64**	2.74 **4.14**	2.59 **3.82**	2.47 **3.59**	2.39 **3.42**	2.32 **3.29**	2.27 **3.17**	2.22 **3.09**	2.18 **3.02**
	1	2	3	4	5	6	7	8	9	10	11

Appendix Table 2 (Continued)

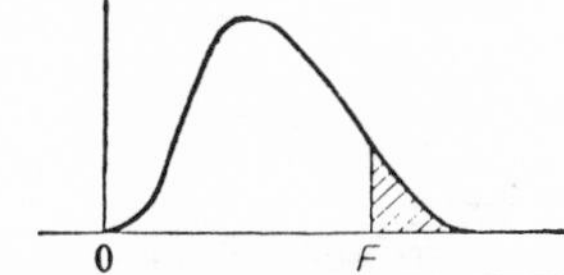

Lighter figures are 5% point of F distribution.
Darker figures are 1% point of F distribution.

12	14	16	20	24	30	40	50	75	100	200	500	∞
244 **6106**	245 **6142**	246 **6169**	248 **6208**	249 **6234**	250 **6258**	251 **6286**	252 **6302**	253 **6323**	253 **6334**	254 **6352**	254 **6361**	254 **6366**
19.41 **99.42**	19.42 **99.43**	19.43 **99.44**	19.44 **99.45**	19.45 **99.46**	19.46 **99.47**	19.47 **99.48**	19.47 **99.48**	19.48 **99.49**	19.49 **99.49**	19.49 **99.49**	19.50 **99.50**	19.50 **99.50**
8.74 **27.05**	8.71 **26.92**	8.69 **26.83**	8.66 **26.69**	8.64 **26.60**	8.62 **26.50**	8.60 **26.41**	8.58 **26.30**	8.57 **26.27**	8.56 **26.23**	8.54 **26.18**	8.54 **26.14**	8.53 **26.12**
5.91 **14.37**	5.87 **14.24**	5.84 **14.15**	5.80 **14.02**	5.77 **13.93**	5.74 **13.83**	5.71 **13.74**	5.70 **13.69**	5.68 **13.61**	5.66 **13.57**	5.65 **13.52**	5.64 **13.48**	5.63 **13.46**
4.68 **9.89**	4.64 **9.77**	4.60 **9.68**	4.56 **9.55**	4.53 **9.47**	4.50 **9.38**	4.46 **9.29**	4.44 **9.24**	4.42 **9.17**	4.40 **9.13**	4.38 **9.07**	4.37 **9.04**	4.36 **9.02**
4.00 **7.72**	3.96 **7.60**	3.92 **7.52**	3.87 **7.39**	3.84 **7.31**	3.81 **7.23**	3.77 **7.14**	3.75 **7.09**	3.72 **7.02**	3.71 **6.99**	3.69 **6.94**	3.68 **6.90**	3.67 **6.88**
3.57 **6.47**	3.52 **6.35**	3.49 **6.27**	3.44 **6.15**	3.41 **6.07**	3.38 **5.98**	3.34 **5.90**	3.32 **5.85**	3.29 **5.78**	3.28 **5.75**	3.25 **5.70**	3.24 **5.67**	3.23 **5.65**
3.28 **5.67**	3.23 **5.56**	3.20 **5.48**	3.15 **5.36**	3.12 **5.28**	3.08 **5.20**	3.05 **5.11**	3.03 **5.06**	3.00 **5.00**	2.98 **4.96**	2.96 **4.91**	2.94 **4.88**	2.93 **4.86**
3.07 **5.11**	3.02 **5.00**	2.98 **4.92**	2.93 **4.80**	2.90 **4.73**	2.86 **4.64**	2.82 **4.56**	2.80 **4.51**	2.77 **4.45**	2.76 **4.41**	2.73 **4.36**	2.72 **4.33**	2.71 **4.31**
2.91 **4.71**	2.86 **4.60**	2.82 **4.52**	2.77 **4.41**	2.74 **4.33**	2.70 **4.25**	2.67 **4.17**	2.64 **4.12**	2.61 **4.05**	2.59 **4.01**	2.56 **3.96**	2.55 **3.93**	2.54 **3.91**
2.79 **4.40**	2.74 **4.29**	2.70 **4.21**	2.65 **4.10**	2.61 **4.02**	2.57 **3.94**	2.53 **3.86**	2.50 **3.80**	2.47 **3.74**	2.45 **3.70**	2.42 **3.66**	2.41 **3.62**	2.40 **3.60**
2.69 **4.16**	2.64 **4.05**	2.60 **3.98**	2.54 **3.86**	2.50 **3.78**	2.46 **3.70**	2.42 **3.61**	2.40 **3.56**	2.36 **3.49**	2.35 **3.46**	2.32 **3.41**	2.31 **3.38**	2.30 **3.36**
2.60 **3.96**	2.55 **3.85**	2.51 **3.78**	2.46 **3.67**	2.42 **3.59**	2.38 **3.51**	2.34 **3.42**	2.32 **3.37**	2.28 **3.30**	2.26 **3.27**	2.24 **3.21**	2.22 **3.18**	2.21 **3.16**
2.53 **3.80**	2.48 **3.70**	2.44 **3.62**	2.39 **3.51**	2.35 **3.43**	2.31 **3.34**	2.27 **3.26**	2.24 **3.21**	2.21 **3.14**	2.19 **3.11**	2.16 **3.06**	2.14 **3.02**	2.13 **3.00**
2.48 **3.67**	2.43 **3.56**	2.39 **3.48**	2.33 **3.36**	2.29 **3.29**	2.25 **3.20**	2.21 **3.12**	2.18 **3.07**	2.15 **3.00**	2.12 **2.97**	2.10 **2.92**	2.08 **2.89**	2.07 **2.87**
2.42 **3.55**	2.37 **3.45**	2.33 **3.37**	2.28 **3.26**	2.24 **3.18**	2.20 **3.10**	2.16 **3.01**	2.13 **2.96**	2.09 **2.89**	2.07 **2.86**	2.04 **2.80**	2.02 **2.77**	2.01 **2.75**
2.38 **3.45**	2.33 **3.35**	2.29 **3.27**	2.23 **3.16**	2.19 **3.08**	2.15 **3.00**	2.11 **2.92**	2.08 **2.86**	2.04 **2.79**	2.02 **2.76**	1.99 **2.70**	1.97 **2.67**	1.96 **2.65**
2.34 **3.37**	2.29 **3.27**	2.25 **3.19**	2.19 **3.07**	2.15 **3.00**	2.11 **2.91**	2.07 **2.83**	2.04 **2.78**	2.00 **2.71**	1.98 **2.68**	1.95 **2.62**	1.93 **2.59**	1.92 **2.57**
2.31 **3.30**	2.26 **3.19**	2.21 **3.12**	2.15 **3.00**	2.11 **2.92**	2.07 **2.84**	2.02 **2.76**	2.00 **2.70**	1.96 **2.63**	1.94 **2.60**	1.91 **2.54**	1.90 **2.51**	1.88 **2.49**
2.28 **3.23**	2.23 **3.13**	2.18 **3.05**	2.12 **2.94**	2.08 **2.86**	2.04 **2.77**	1.99 **2.69**	1.96 **2.63**	1.92 **2.56**	1.90 **2.53**	1.87 **2.47**	1.85 **2.44**	1.84 **2.42**
2.25 **3.17**	2.20 **3.07**	2.15 **2.99**	2.09 **2.88**	2.05 **2.80**	2.00 **2.72**	1.96 **2.63**	1.93 **2.58**	1.89 **2.51**	1.87 **2.47**	1.84 **2.42**	1.82 **2.38**	1.81 **2.36**
2.23 **3.12**	2.18 **3.02**	2.13 **2.94**	2.07 **2.83**	2.03 **2.75**	1.98 **2.67**	1.93 **2.58**	1.91 **2.53**	1.87 **2.46**	1.84 **2.42**	1.81 **2.37**	1.80 **2.33**	1.78 **2.31**
2.20 **3.07**	2.14 **2.97**	2.10 **2.89**	2.04 **2.78**	2.00 **2.70**	1.96 **2.62**	1.91 **2.53**	1.88 **2.48**	1.84 **2.41**	1.82 **2.37**	1.79 **2.32**	1.77 **2.28**	1.76 **2.26**
2.18 **3.03**	2.13 **2.93**	2.09 **2.85**	2.02 **2.74**	1.98 **2.66**	1.94 **2.58**	1.89 **2.49**	1.86 **2.44**	1.82 **2.36**	1.80 **2.33**	1.76 **2.27**	1.74 **2.23**	1.73 **2.21**
2.16 **2.99**	2.11 **2.89**	2.06 **2.81**	2.00 **2.70**	1.96 **2.62**	1.92 **2.54**	1.87 **2.45**	1.84 **2.40**	1.80 **2.32**	1.77 **2.29**	1.74 **2.23**	1.72 **2.19**	1.71 **2.17**
2.15 **2.96**	2.10 **2.86**	2.05 **2.77**	1.99 **2.66**	1.95 **2.58**	1.90 **2.50**	1.85 **2.41**	1.82 **2.36**	1.78 **2.28**	1.76 **2.25**	1.72 **2.19**	1.70 **2.15**	1.69 **2.13**
12	14	16	20	24	30	40	50	75	100	200	500	∞

Appendix Table 2 (Continued)

Number of degrees of freedom of denominator, f_2	Number of degrees of freedom of numerator, f_1										
	1	2	3	4	5	6	7	8	9	10	11
27	4.21 **7.68**	3.35 **5.49**	2.96 **4.60**	2.73 **4.11**	2.57 **3.79**	2.46 **3.56**	2.37 **3.39**	2.30 **3.26**	2.25 **3.14**	2.20 **3.06**	2.16 **2.98**
28	4.20 **7.64**	3.34 **5.45**	2.95 **4.57**	2.71 **4.07**	2.56 **3.76**	2.44 **3.53**	2.36 **3.36**	2.29 **3.23**	2.24 **3.11**	2.19 **3.03**	2.15 **2.95**
29	4.18 **7.60**	3.33 **5.52**	2.93 **4.54**	2.70 **4.04**	2.54 **3.73**	2.43 **3.50**	2.35 **3.33**	2.28 **3.20**	2.22 **3.03**	2.18 **3.00**	2.14 **2.92**
30	4.17 **7.56**	3.32 **5.39**	2.92 **4.51**	2.69 **4.02**	2.53 **3.70**	2.42 **3.47**	2.34 **3.30**	2.27 **3.17**	2.21 **3.06**	2.16 **2.98**	2.12 **2.90**
32	4.15 **7.50**	3.30 **5.34**	2.90 **4.46**	2.67 **3.97**	2.51 **3.66**	2.40 **3.42**	2.32 **3.25**	2.25 **3.12**	2.19 **3.01**	2.14 **2.94**	2.10 **2.86**
34	4.13 **7.44**	3.28 **5.29**	2.88 **4.42**	2.65 **3.93**	2.49 **3.61**	2.38 **3.38**	2.30 **3.21**	2.23 **3.08**	2.17 **2.97**	2.12 **2.89**	2.08 **2.82**
36	4.11 **7.39**	3.26 **5.25**	2.86 **4.38**	2.63 **3.89**	2.48 **3.58**	2.36 **3.35**	2.28 **3.18**	2.21 **3.04**	2.15 **2.94**	2.10 **2.86**	2.06 **2.78**
38	4.10 **7.35**	3.25 **5.21**	2.85 **4.34**	2.62 **3.86**	2.46 **3.54**	2.35 **3.32**	2.26 **3.15**	2.19 **3.02**	2.14 **2.91**	2.09 **2.82**	2.05 **2.75**
40	4.08 **7.31**	3.23 **5.18**	2.84 **4.31**	2.61 **3.83**	2.45 **3.51**	2.34 **3.29**	2.25 **3.12**	2.18 **2.99**	2.12 **2.88**	2.07 **2.80**	2.04 **2.73**
42	4.07 **7.27**	3.22 **5.15**	2.83 **4.29**	2.59 **3.80**	2.44 **3.49**	2.32 **3.26**	2.24 **3.10**	2.17 **2.96**	2.11 **2.86**	2.06 **2.77**	2.02 **2.70**
44	4.06 **7.24**	3.21 **5.12**	2.82 **4.26**	2.58 **3.78**	2.43 **3.46**	2.31 **3.24**	2.23 **3.07**	2.16 **2.94**	2.10 **2.84**	2.05 **2.75**	2.01 **2.68**
46	4.05 **7.21**	3.20 **5.10**	2.81 **4.24**	2.57 **3.76**	2.42 **3.44**	2.30 **3.22**	2.22 **3.05**	2.15 **2.92**	2.09 **2.82**	2.04 **2.73**	2.00 **2.66**
48	4.04 **7.19**	3.19 **5.08**	2.80 **4.22**	2.56 **3.74**	2.41 **3.42**	2.30 **3.20**	2.21 **3.04**	2.14 **2.90**	2.08 **2.80**	2.03 **2.71**	1.99 **2.64**
50	4.03 **7.17**	3.18 **5.06**	2.79 **4.20**	2.56 **3.72**	2.40 **3.41**	2.29 **3.18**	2.20 **3.02**	2.13 **2.88**	2.07 **2.78**	2.02 **2.72**	1.98 **2.62**
55	4.02 **7.12**	3.17 **5.01**	2.78 **4.16**	2.54 **3.68**	2.38 **3.37**	2.27 **3.15**	2.18 **2.98**	2.11 **2.85**	2.05 **2.75**	2.00 **2.66**	1.97 **2.59**
60	4.00 **7.08**	3.15 **4.98**	2.76 **4.13**	2.52 **3.65**	2.37 **3.34**	2.25 **3.12**	2.17 **2.95**	2.10 **2.82**	2.04 **2.72**	1.99 **2.63**	1.95 **2.56**
65	3.99 **7.04**	3.14 **4.95**	2.75 **4.10**	2.51 **3.62**	2.36 **3.31**	2.24 **3.09**	2.15 **2.93**	2.08 **2.79**	2.02 **2.70**	1.98 **2.61**	1.94 **2.54**
70	3.98 **7.01**	3.13 **4.92**	2.74 **4.08**	2.50 **3.60**	2.35 **3.29**	2.32 **3.07**	2.14 **2.91**	2.07 **2.77**	2.01 **2.67**	1.97 **2.59**	1.93 **2.51**
80	3.96 **6.96**	3.11 **4.88**	2.72 **4.04**	2.48 **3.56**	2.33 **3.25**	2.21 **3.04**	2.12 **2.87**	2.05 **2.74**	1.99 **2.64**	1.95 **2.55**	1.91 **2.48**
100	3.94 **6.90**	3.09 **4.82**	2.70 **3.98**	2.46 **3.51**	2.30 **3.20**	2.19 **2.99**	2.10 **2.82**	2.03 **2.69**	1.97 **2.59**	1.92 **2.51**	1.88 **2.43**
125	3.92 **6.84**	3.07 **4.78**	2.68 **3.94**	2.44 **3.47**	2.29 **3.17**	2.17 **2.95**	2.08 **2.79**	2.01 **2.65**	1.95 **2.56**	1.90 **2.47**	1.86 **2.40**
150	3.91 **6.81**	3.06 **4.75**	2.67 **3.91**	2.43 **3.44**	2.27 **3.13**	2.16 **2.92**	2.07 **2.76**	2.00 **2.62**	1.94 **2.53**	1.89 **2.44**	1.85 **2.37**
200	3.89 **6.76**	3.04 **4.71**	2.65 **3.88**	2.41 **3.41**	2.26 **3.11**	2.14 **2.90**	2.05 **2.73**	1.98 **2.60**	1.92 **2.50**	1.87 **2.41**	1.83 **2.34**
400	3.86 **6.70**	3.02 **4.66**	2.62 **3.83**	2.39 **3.36**	2.23 **3.06**	2.12 **2.85**	2.03 **2.69**	1.96 **2.55**	1.90 **2.46**	1.85 **2.37**	1.81 **2.29**
1000	3.85 **6.66**	3.00 **4.62**	2.61 **3.80**	2.38 **3.34**	2.22 **3.04**	2.10 **2.82**	2.02 **2.66**	1.95 **2.53**	1.89 **2.43**	1.84 **2.34**	1.80 **2.26**
∞	3.84 **6.64**	2.99 **4.60**	2.60 **3.78**	2.37 **3.32**	2.21 **3.02**	2.09 **2.80**	2.01 **2.64**	1.94 **2.51**	1.88 **2.41**	1.83 **2.32**	1.79 **2.24**
	1	2	3	4	5	6	7	8	9	10	11

Appendix Table 2 (Continued)

Lighter figures are 5% point of F distribution.
Darker figures are 1% point of F distribution.

12	14	16	20	24	30	40	50	75	100	200	500	∞
2.13 **2.93**	2.08 **2.83**	2.03 **2.74**	1.97 **2.63**	1.93 **2.55**	1.88 **2.47**	1.84 **2.38**	1.80 **2.33**	1.76 **2.25**	1.74 **2.21**	1.71 **2.16**	1.68 **2.12**	1.67 **2.10**
2.12 **2.90**	2.06 **2.80**	2.02 **2.71**	1.96 **2.60**	1.91 **2.52**	1.87 **2.44**	1.81 **2.35**	1.78 **2.30**	1.75 **2.22**	1.72 **2.18**	1.69 **2.13**	1.67 **2.09**	1.65 **2.06**
2.10 **2.87**	2.05 **2.77**	2.00 **2.68**	1.94 **2.57**	1.90 **2.49**	1.85 **2.41**	1.80 **2.32**	1.77 **2.27**	1.73 **2.19**	1.71 **2.15**	1.68 **2.10**	1.65 **2.06**	1.64 **2.03**
2.09 **2.84**	2.04 **2.74**	1.99 **2.66**	1.93 **2.55**	1.89 **2.47**	1.84 **2.38**	1.79 **2.29**	1.76 **2.24**	1.72 **2.16**	1.69 **2.13**	1.66 **2.07**	1.64 **2.03**	1.62 **2.01**
2.07 **2.80**	2.02 **2.70**	1.97 **2.62**	1.91 **2.51**	1.86 **2.42**	1.82 **2.34**	1.76 **2.25**	1.74 **2.20**	1.69 **2.12**	1.67 **2.08**	1.64 **2.02**	1.61 **1.98**	1.59 **1.96**
2.05 **2.76**	2.00 **2.66**	1.95 **2.58**	1.89 **2.47**	1.84 **2.38**	1.80 **2.30**	1.74 **2.21**	1.71 **2.15**	1.67 **2.08**	1.64 **2.04**	1.61 **1.98**	1.59 **1.94**	1.57 **1.91**
2.03 **2.72**	1.98 **2.62**	1.93 **2.54**	1.87 **2.43**	1.82 **2.35**	1.78 **2.26**	1.72 **2.17**	1.69 **2.12**	1.65 **2.04**	1.62 **2.00**	1.59 **1.94**	1.56 **1.90**	1.55 **1.87**
2.02 **2.69**	1.96 **2.59**	1.92 **2.51**	1.85 **2.40**	1.80 **2.32**	1.76 **2.22**	1.71 **2.14**	1.67 **2.08**	1.63 **2.00**	1.60 **1.97**	1.57 **1.90**	1.54 **1.86**	1.53 **1.84**
2.00 **2.66**	1.95 **2.56**	1.90 **2.49**	1.84 **2.37**	1.79 **2.29**	1.74 **2.20**	1.69 **2.11**	1.66 **2.05**	1.61 **1.97**	1.59 **1.94**	1.55 **1.88**	1.53 **1.84**	1.51 **1.81**
1.99 **2.64**	1.94 **2.54**	1.89 **2.46**	1.82 **2.35**	1.78 **2.26**	1.73 **2.17**	1.68 **2.08**	1.64 **2.02**	1.60 **1.94**	1.57 **1.91**	1.54 **1.85**	1.15 **1.80**	1.49 **1.78**
1.98 **2.62**	1.92 **2.52**	1.88 **2.44**	1.81 **2.32**	1.76 **2.24**	1.72 **2.15**	1.66 **2.06**	1.63 **2.00**	1.58 **1.92**	1.56 **1.88**	1.52 **1.82**	1.50 **1.78**	1.48 **1.75**
1.97 **2.60**	1.91 **2.50**	1.87 **2.42**	1.80 **2.30**	1.75 **2.22**	1.71 **2.13**	1.65 **2.04**	1.62 **1.98**	1.57 **1.90**	1.54 **1.86**	1.51 **1.80**	1.48 **1.76**	1.46 **1.72**
1.96 **2.58**	1.90 **2.48**	1.86 **2.40**	1.79 **2.28**	1.74 **2.20**	1.70 **2.11**	1.64 **2.02**	1.61 **1.96**	1.56 **1.88**	1.53 **1.84**	1.50 **1.78**	1.47 **1.73**	1.45 **1.70**
1.95 **2.56**	1.90 **2.46**	1.85 **2.39**	1.78 **2.26**	1.74 **2.18**	1.69 **2.10**	1.63 **2.00**	1.60 **1.94**	1.55 **1.86**	1.52 **1.82**	1.48 **1.76**	1.46 **1.71**	1.44 **1.68**
1.93 **2.53**	1.88 **2.43**	1.83 **2.35**	1.76 **2.23**	1.72 **2.15**	1.67 **2.06**	1.61 **1.96**	1.58 **1.90**	1.52 **1.82**	1.50 **1.78**	1.46 **1.71**	1.43 **1.66**	1.41 **1.64**
1.92 **2.50**	1.86 **2.40**	1.81 **2.32**	1.75 **2.20**	1.70 **2.12**	1.65 **2.03**	1.59 **1.93**	1.56 **1.87**	1.50 **1.79**	1.48 **1.74**	1.44 **1.68**	1.41 **1.63**	1.39 **1.60**
1.90 **2.47**	1.85 **2.37**	1.80 **2.30**	1.73 **2.18**	1.68 **2.09**	1.63 **2.00**	1.57 **1.90**	1.54 **1.84**	1.49 **1.76**	1.46 **1.71**	1.42 **1.64**	1.39 **1.60**	1.37 **1.56**
1.89 **2.45**	1.84 **2.35**	1.79 **2.28**	1.72 **2.15**	1.67 **2.07**	1.62 **1.98**	1.56 **1.88**	1.53 **1.82**	1.47 **1.74**	1.45 **1.69**	1.40 **1.62**	1.37 **1.56**	1.35 **1.53**
1.88 **2.41**	1.82 **2.32**	1.77 **2.24**	1.70 **2.11**	1.65 **2.03**	1.60 **1.94**	1.54 **1.84**	1.51 **1.78**	1.45 **1.70**	1.42 **1.65**	1.38 **1.57**	1.35 **1.52**	1.32 **1.49**
1.85 **2.36**	1.79 **2.26**	1.75 **2.19**	1.68 **2.06**	1.63 **1.98**	1.57 **1.89**	1.51 **1.79**	1.48 **1.73**	1.42 **1.64**	1.39 **1.59**	1.34 **1.51**	1.30 **1.46**	1.28 **1.43**
1.83 **2.33**	1.77 **2.23**	1.72 **2.15**	1.65 **2.03**	1.60 **1.94**	1.55 **1.85**	1.49 **1.75**	1.45 **1.68**	1.39 **1.59**	1.36 **1.54**	1.31 **1.46**	1.27 **1.40**	1.25 **1.37**
1.82 **2.30**	1.76 **2.20**	1.71 **2.12**	1.64 **2.00**	1.59 **1.91**	1.54 **1.83**	1.47 **1.72**	1.44 **1.66**	1.37 **1.56**	1.34 **1.51**	1.29 **1.43**	1.25 **1.37**	1.22 **1.33**
1.80 **2.28**	1.74 **2.17**	1.69 **2.09**	1.62 **1.97**	1.57 **1.88**	1.52 **1.79**	1.45 **1.69**	1.42 **1.62**	1.35 **1.53**	1.32 **1.48**	1.26 **1.39**	1.22 **1.33**	1.19 **1.28**
1.78 **2.23**	1.72 **2.12**	1.67 **2.04**	1.60 **1.92**	1.54 **1.84**	1.49 **1.74**	1.42 **1.64**	1.38 **1.57**	1.32 **1.47**	1.28 **1.42**	1.22 **1.32**	1.16 **1.24**	1.13 **1.19**
1.76 **2.20**	1.70 **2.09**	1.65 **2.01**	1.58 **1.89**	1.53 **1.81**	1.47 **1.71**	1.41 **1.61**	1.36 **1.54**	1.30 **1.44**	1.26 **1.38**	1.19 **1.28**	1.13 **1.19**	1.08 **1.11**
1.75 **2.18**	1.69 **2.07**	1.64 **1.99**	1.57 **1.87**	1.52 **1.79**	1.46 **1.69**	1.40 **1.59**	1.35 **1.52**	1.28 **1.41**	1.24 **7.36**	1.17 **1.25**	1.11 **1.15**	1.00 **1.00**
12	14	16	20	24	30	40	50	75	100	200	500	∞

Appendix Table 3 Orthogonal Polynomial for Equispaced Levels

b_1: 1st-order
b_2: 2nd-order
⋮

$$\hat{b}_i=\frac{W_1A_1+\cdots\cdots+W_aA_a}{r\cdot\lambda S\cdot h^i}$$

$$\mathrm{Var}(b_i)=\frac{\sigma^2}{r\cdot S\cdot h^{2i}}$$

$$S_{\hat{b}i}=\frac{(W_1A_1+\cdots\cdots+W_aA_a)^2}{r\cdot\lambda^2 S}$$

Number of levels / Coefficient	$k=2$	$k=3$		$k=4$			$k=5$			
	b_1	b_1	b_2	b_1	b_2	b_3	b_1	b_2	b_3	b_4
W_1	−1	−1	1	−3	1	−1	−2	2	−1	1
W_2	1	0	−2	−1	−1	3	−1	−1	2	−4
W_3		1	1	1	−1	−3	0	−2	0	6
W_4				3	1	1	1	−1	−2	−4
W_5							2	2	1	1
$\lambda^2 S$	2	2	6	20	4	20	10	14	10	70
λS	1	2	2	10	4	6	10	14	12	24
S	$\frac{1}{2}$	2	$\frac{2}{3}$	5	4	$\frac{9}{5}$	10	14	$\frac{72}{5}$	$\frac{288}{35}$
λ	2	1	3	2	1	$\frac{10}{3}$	1	1	$\frac{5}{6}$	$\frac{35}{12}$

Appendix Table 3 (Continued)

	$k=6$					$k=7$				
	b_1	b_2	b_3	b_4	b_5	b_1	b_2	b_3	b_4	b_5
W_1	−5	5	−5	1	−1	−3	5	−1	3	−1
W_2	−3	−1	7	−3	5	−2	0	1	−7	4
W_3	−1	−4	4	2	−10	−1	−3	1	1	−5
W_4	1	−4	−4	2	10	0	−4	0	6	0
W_5	3	−1	−7	−3	−5	1	−3	−1	1	5
W_6	5	5	5	1	1	2	0	−1	−7	−4
W_7						3	5	1	3	1
$\lambda^2 S$	70	84	180	28	252	28	84	6	154	84
λS	35	56	108	48	120	28	84	36	264	240
S	$\frac{35}{2}$	$\frac{112}{3}$	$\frac{324}{5}$	$\frac{576}{7}$	$\frac{400}{7}$	28	84	216	$\frac{3168}{7}$	$\frac{4800}{7}$
λ	2	$\frac{3}{2}$	$\frac{5}{3}$	$\frac{7}{12}$	$\frac{21}{10}$	1	1	$\frac{1}{6}$	$\frac{7}{12}$	$\frac{7}{20}$

	$k=8$					$k=9$				
	b_1	b_2	b_3	b_4	b_5	b_1	b_2	b_3	b_4	b_5
W_1	−7	7	−7	7	−7	−4	28	−14	14	−4
W_2	−5	1	5	−13	23	−3	7	7	−21	11
W_3	−3	−3	7	−3	−17	−2	−8	13	−11	−4
W_4	−1	−5	3	9	−15	−1	−17	9	9	−9
W_5	1	−5	−3	9	15	0	−20	0	18	0
W_6	3	−3	−7	−3	17	1	−17	−9	9	9
W_7	5	1	−5	−13	−23	2	−8	−13	−11	4
W_8	7	7	7	7	7	3	7	−7	−21	−11
W_9						4	28	14	14	4
$\lambda^2 S$	168	168	264	616	2184	60	2772	990	2002	468
λS	84	168	396	1056	3102	60	924	1188	3432	3120
S	42	168	594	$\frac{12672}{7}$	$\frac{31200}{7}$	60	308	$\frac{7128}{5}$	$\frac{41184}{7}$	20800
λ	2	1	$\frac{2}{3}$	$\frac{12}{7}$	$\frac{10}{7}$	1	3	$\frac{5}{6}$	$\frac{7}{12}$	$\frac{3}{20}$

Appendix Table 3 (Continued)

	$k=10$					$k=11$				
	b_1	b_2	b_3	b_4	b_5	b_1	b_2	b_3	b_4	b_5
W_1	−9	6	−42	18	−6	−5	15	−30	6	−3
W_2	−7	2	14	−22	14	−4	6	6	−6	6
W_3	−5	−1	35	−17	−1	−3	−1	22	−6	1
W_4	−3	−3	31	3	−11	−2	−6	23	−1	−4
W_5	−1	−4	12	18	−6	−1	−9	14	4	−4
W_6	1	−4	−12	18	6	0	−10	0	6	0
W_7	3	−3	−31	3	11	1	−9	−14	4	4
W_8	5	−1	−35	−17	1	2	−6	−23	−1	4
W_9	7	2	−14	−22	−14	3	−1	−22	−6	−1
W_{10}	9	6	42	18	6	4	6	−6	−6	−6
W_{11}						5	15	30	6	3
$\lambda^2 S$	330	132	8580	2860	780	110	858	4290	286	156
λS	165	264	5148	6864	7800	110	858	5148	3432	6240
S	$\frac{165}{2}$	528	$\frac{15444}{5}$	$\frac{82368}{5}$	78000	110	858	$\frac{30888}{5}$	41184	249600
λ	2	$\frac{1}{2}$	$\frac{5}{3}$	$\frac{5}{12}$	$\frac{1}{10}$	1	1	$\frac{5}{6}$	$\frac{1}{12}$	$\frac{1}{40}$

	$k=12$					$k=13$				
	b_1	b_2	b_3	b_4	b_5	b_1	b_2	b_3	b_4	b_5
W_1	−11	55	−33	33	−33	−6	22	−11	99	−22
W_2	−9	25	3	−27	57	−5	11	0	−66	33
W_3	−7	1	21	−33	21	−4	2	6	−96	18
W_4	−5	−17	25	−13	−29	−3	−5	8	−54	−11
W_5	−3	−29	19	12	−44	−2	−10	7	11	−26
W_6	−1	−35	7	28	−20	−1	−13	4	64	−20
W_7	1	−35	−7	28	20	0	−14	0	84	0
W_8	3	−29	−19	12	44	1	−13	−4	64	20
W_9	5	−17	−25	−13	29	2	−10	−7	11	26
W_{10}	7	1	−21	−33	−21	3	−5	−8	−54	11
W_{11}	9	25	−3	−27	57	4	2	−6	−96	−18
W_{12}	11	55	33	33	33	5	11	0	−66	−33
W_{13}						6	22	11	99	22
$\lambda^2 S$	572	12012	5148	8008	15912	182	2002	572	68068	6188
λS	286	4004	7722	27456	106080	182	2002	3432	116688	106080
S	143	$\frac{4004}{3}$	11583	$\frac{658944}{7}$	707200	182	2002	20592	$\frac{1400256}{7}$	$\frac{12729600}{7}$
λ	2	3	$\frac{2}{3}$	$\frac{7}{24}$	$\frac{3}{20}$	1	1	$\frac{1}{6}$	$\frac{7}{12}$	$\frac{7}{120}$

Appendix Table 3 (Continued)

	$k=14$					$k=15$				
	b_1	b_2	b_3	b_4	b_5	b_1	b_2	b_3	b_4	b_5
W_1	−13	13	−143	143	−143	−7	91	−91	1001	−1001
W_2	−11	7	−11	−77	187	−6	52	−13	−429	1144
W_3	−9	2	66	−132	132	−5	19	35	−869	979
W_4	−7	−2	98	−92	−28	−4	−8	58	−704	44
W_5	−5	−5	95	−13	−139	−3	−29	61	−249	−751
W_6	−3	−7	67	63	−145	−2	−44	49	251	−1000
W_7	−1	−8	24	108	−60	−1	−53	27	621	−675
W_8	1	−8	−24	108	60	0	−56	0	756	0
W_9	3	−7	−67	63	145	1	−53	−27	621	675
W_{10}	5	−5	−95	−13	139	2	−44	−49	251	1000
W_{11}	7	−2	−98	−92	28	3	−29	−61	−249	751
W_{12}	9	2	−66	−132	−132	4	−8	−58	−704	−44
W_{13}	11	7	11	−77	−187	5	19	−35	−869	−979
W_{14}	13	13	143	143	143	6	52	13	−429	−1144
W_{15}						7	91	91	1001	1001
$\lambda^2 S$	910	728	97240	136136	235144	280	37128	39780	6466460	10581480
λS	455	1456	58344	233376	1007760	280	12376	47736	2217072	10077600
S	$\frac{455}{2}$	2912	$\frac{175032}{5}$	$\frac{2800512}{7}$	$\frac{30232800}{7}$	280	$\frac{12376}{3}$	$\frac{286416}{5}$	$\frac{26604864}{35}$	$\frac{201552000}{21}$
λ	2	$\frac{1}{2}$	$\frac{5}{3}$	$\frac{7}{12}$	$\frac{7}{30}$	1	3	$\frac{5}{6}$	$\frac{35}{12}$	$\frac{21}{20}$

Appendix Table 3 (Continued)

	$k=16$					$k=17$				
	b_1	b_2	b_3	b_4	b_5	b_1	b_2	b_3	b_4	b_5
W_1	−15	35	−455	273	−143	−8	40	−28	52	−104
W_2	−13	21	−91	−91	143	−7	25	−7	−13	19
W_3	−11	9	143	−221	143	−6	12	7	−39	104
W_4	−9	−1	267	−201	33	−5	1	15	−39	39
W_5	−7	−9	301	−101	−77	−4	−8	18	−24	−36
W_6	−5	−15	265	23	−131	−3	−15	17	−3	−83
W_7	−3	−19	179	129	−115	−2	−20	13	17	−88
W_8	−1	−21	63	189	−45	−1	−23	7	31	−55
W_9	1	−21	−63	189	45	0	−24	0	36	0
W_{10}	3	−19	−179	129	115	1	−23	−7	31	55
W_{11}	5	−15	−265	23	131	2	−20	−13	17	88
W_{12}	7	−9	−301	−101	77	3	−15	−17	−3	83
W_{13}	9	−1	−267	−201	−33	4	−8	−18	−24	36
W_{14}	11	9	−143	−221	−143	5	1	−15	−39	−39
W_{15}	13	21	91	−91	−143	6	12	−7	−39	−104
W_{16}	15	35	455	273	143	7	25	7	−13	−91
W_{17}						8	40	28	52	104
$\lambda^2 S$	1360	5712	1007760	470288	201552	408	7752	3876	16796	100776
λS	680	5712	302328	806208	2015520	408	7752	23256	201552	2015520
S	340	5712	$\frac{906984}{10}$	$\frac{9674492}{7}$	20155200	408	7752	139536	2418624	40310400
λ	2	1	$\frac{10}{3}$	$\frac{7}{12}$	$\frac{1}{10}$	1	1	$\frac{6}{1}$	$\frac{1}{12}$	$\frac{1}{20}$

Appendix Table 3 (Continued)

	$k=18$					$k=19$		
	b_1	b_2	b_3	b_4	b_5	b_1	b_2	b_3
W_1	−17	68	−68	68	−884	−9	51	−204
W_2	−15	44	−20	−12	676	−8	34	− 68
W_3	−13	23	13	−47	871	−7	19	28
W_4	−11	5	33	−51	429	−6	6	89
W_5	− 9	−10	42	−36	−156	−5	− 5	120
W_6	− 7	−22	42	−12	−588	−4	−14	126
W_7	− 5	−31	35	13	−733	−3	−21	112
W_8	− 3	−37	23	33	−583	−2	−26	83
W_9	− 1	−40	8	44	−220	−1	−29	44
W_{10}	1	−40	− 8	44	220	0	−30	0
W_{11}	3	−37	−23	33	583	1	−29	− 44
W_{12}	5	−31	−35	13	733	2	−26	− 83
W_{13}	[illegible]	−22	−42	−12	588	3	−21	−112
W_{14}	9	−10	−42	−36	156	4	−14	−126
W_{15}	11	5	−33	−51	−429	5	− 5	−120
W_{16}	13	23	−13	−47	−871	6	6	− 89
W_{17}	15	44	20	−12	−676	7	19	− 28
W_{18}	17	68	68	68	884	8	34	68
W_{19}						9	51	204
$\lambda^2 S$	1938	23256	23256	28424	6953544	570	13566	213180
λS	969	15504	69768	341088	23178480	570	13566	255816
S	$\frac{696}{2}$	10336	209304	4093056	77261600	570	13566	$\frac{1534896}{5}$
λ	2	$\frac{3}{2}$	$\frac{1}{3}$	$\frac{1}{12}$	$\frac{3}{10}$	1	1	$\frac{5}{6}$

Appendix Table 3 (Continued)

	$k=19$		$k=20$				
	b_4	b_5	b_1	b_2	b_3	b_4	b_5
W_1	612	−102	−19	57	−969	1938	−1938
W_2	− 68	68	−17	39	−357	− 102	1122
W_3	−388	98	−15	23	85	−1122	1802
W_4	−453	58	−13	9	377	−1402	1222
W_5	−354	− 3	−11	− 3	539	−1187	187
W_6	−168	− 54	− 9	−13	591	− 687	− 771
W_7	42	− 79	− 7	−21	553	− 77	−1351
W_8	227	− 74	− 5	−27	445	503	−1441
W_9	352	− 44	− 3	−31	287	948	−1076
W_{10}	396	0	− 1	−33	99	1188	− 396
W_{11}	352	44	1	−33	− 99	1188	396
W_{12}	227	74	3	−31	−287	948	1076
W_{13}	42	79	5	−27	−445	503	1441
W_{14}	−168	54	7	−21	−553	− 77	1351
W_{15}	−354	3	9	13	−591	− 687	771
W_{16}	−453	− 58	11	− 3	−539	−1187	− 187
W_{17}	−388	− 98	13	9	−377	−1402	−1222
W_{18}	− 68	− 68	15	23	− 85	−1122	−1802
W_{19}	612	102	17	39	357	− 102	−1122
W_{20}			19	57	969	1938	1938
$\lambda^2 S$	2288132	89148	2660	17556	4903140	22881320	31201800
$S\lambda$	3922512	3565920	1330	17556	1470942	15690048	89148000
S	$\frac{47070144}{7}$	142636800	665	17556	$\frac{4412826}{10}$	$\frac{376561152}{35}$	$\frac{1782960000}{7}$
λ	$\frac{7}{12}$	$\frac{1}{40}$	2	1	$\frac{10}{3}$	$\frac{35}{24}$	$\frac{7}{20}$

Appendix Table 4 Table of Common Logarithms

	0	1	2	3	4	5	6	7	8	9	1	2	3	4	5	6	7	8	9
10	0000	0043	0086	0128	0170						4	9	13	17	21	26	30	34	38
						0212	0253	0294	0334	0374	4	8	12	16	20	24	28	32	36
11	0414	0453	0492	0531	0569						4	8	12	15	19	23	27	31	35
						0607	0645	0682	0719	0755	4	7	11	15	19	22	26	30	33
12	0792	0828	0864	0899	0934						3	7	11	14	18	21	25	28	32
						0969	1004	1038	1072	1106	3	7	10	14	17	20	24	27	31
13	1139	1173	1206	1239	1271						3	7	10	13	16	20	23	26	30
						1303	1335	1367	1399	1430	3	6	10	13	16	19	22	25	29
14	1461	1492	1523	1553	1584						3	6	9	12	15	19	22	25	28
						1614	1644	1673	1703	1732	3	6	9	12	15	18	21	24	27
15	1761	1790	1818	1847	1875						3	6	9	11	14	17	20	23	26
						1903	1931	1959	1987	2014	3	6	8	11	14	17	19	22	25
16	2041	2068	2095	2122	2148						3	5	8	11	14	16	19	22	24
						2175	2201	2227	2253	2279	3	5	8	10	13	16	18	21	23
17	2304	2330	2355	2380	2405						3	5	8	10	13	15	18	20	23
						2430	2455	2480	2504	2529	2	5	7	10	12	15	17	20	22
18	2553	2577	2601	2625	2648						2	5	7	9	12	14	16	19	21
						2672	2695	2718	2742	2765	2	5	7	9	11	14	16	18	21
19	2788	2810	2833	2856	2878						2	4	7	9	11	13	16	18	20
						2900	2923	2945	2967	2989	2	4	6	8	11	13	15	17	19
20	3010	3032	3054	3075	3096	3118	3139	3160	3181	3201	2	4	6	8	11	13	15	17	19
21	3222	3243	3263	3284	3304	3324	3345	3365	3385	3404	2	4	6	8	10	12	14	16	18
22	3424	3444	3464	3483	3502	3522	3541	3560	3579	3598	2	4	6	8	10	12	14	15	17
23	3617	3636	3655	3674	3692	3711	3729	3747	3766	3784	2	4	6	7	9	11	13	15	17
24	3802	3820	3838	3856	3874	3892	3909	3927	3945	3962	2	4	5	7	9	11	12	14	16
25	3979	3997	4014	4031	4048	4065	4082	4099	4116	4133	2	3	5	7	9	10	12	14	15
26	4150	4166	4183	4200	4216	4232	4249	4265	4281	4298	2	3	5	7	8	10	11	13	15
27	4314	4330	4346	4362	4378	4393	4409	4425	4440	4456	2	3	5	6	8	9	11	13	14
28	4472	4487	4502	4518	4533	4548	4564	4579	4594	4609	2	3	5	6	8	9	11	12	14
29	4624	4639	4654	4669	4683	4698	4713	4728	4742	4757	1	3	4	6	7	9	10	12	13
30	4771	4786	4800	4814	4829	4843	4857	4871	4886	4900	1	3	4	6	7	9	10	11	13
31	4914	4928	4942	4955	4969	4983	4997	5011	5924	5038	1	3	4	6	7	8	10	11	12
32	5051	5065	5079	5092	5105	5119	5132	5145	5159	5172	1	3	4	5	7	8	9	11	12
33	5185	5198	5211	5224	5237	5250	5263	5276	5289	5302	1	3	4	5	6	8	9	10	12
34	5315	5328	5340	5353	5366	5378	5391	5403	5416	5428	1	3	4	5	6	8	9	10	11
35	5441	5453	5465	5478	5490	5502	5514	5527	5539	5551	1	2	4	5	6	7	9	10	11
36	5563	5575	5587	5599	5611	5623	5635	5647	5658	5670	1	2	4	5	6	7	8	10	11
37	5682	5694	5705	5717	5729	5740	5752	5763	5775	5786	1	2	3	5	6	7	8	9	10
38	5798	5809	5821	5832	5843	5855	5866	5877	5888	5899	1	2	3	5	6	7	8	9	10
39	5911	5922	5933	5944	5955	5966	5977	5988	5999	6010	1	2	3	4	5	7	8	9	10
40	6021	6031	6042	6053	6064	6075	6085	6096	6107	6117	1	2	3	4	5	6	8	9	10
41	6128	6138	6149	6160	6170	6180	6191	6201	6212	6222	1	2	3	4	5	6	7	8	9
42	6232	6243	6253	6263	6274	6284	6294	6304	6314	6325	1	2	3	4	5	6	7	8	9
43	6335	6345	6355	6365	6375	6385	6395	6405	6415	6425	1	2	3	4	5	6	7	8	9
44	6435	6444	6454	6464	6474	6484	6493	6503	6513	6522	1	2	3	4	5	6	7	8	9
45	6532	6542	6551	6561	6571	6850	6590	6599	6609	6618	1	2	3	4	5	6	7	8	9
46	6628	6637	6646	6656	6665	6675	6684	6693	6702	6712	1	2	3	4	5	6	7	7	8
47	6721	6730	6739	6749	6758	6767	6776	6785	6794	6803	1	2	3	4	5	5	6	7	8
48	6812	6821	6830	6839	6848	6857	6866	6875	6884	6893	1	2	3	4	4	5	6	7	8
49	6902	6911	6920	6928	6937	6946	6955	6964	6972	6981	1	2	3	4	4	5	6	7	8

How to obtain logarithms:

i) When the antilogarithm is a number of two digits, this number is found in the column at the left side, and one need only read off the value in the 0 column.

ii) When the antilogarithm is a three-digit number, the first two digits are read off from the column on the left side, then the third digit corresponds to one of the columns 1 – 9 and one need merely read off where the row and column cross.

iii) When the antilogarithm is a four-digit number, do as in ii) for the first three digits; the fourth digit is found in 1 – 9 at the right of the top row, and the numeral where the column crosses the row is added to the logarithm for the first three digits.

Appendix Table 4 (Continued)

	0	1	2	3	4	5	6	7	8	9	1	2	3	4	5	6	7	8	9
50	6990	6998	7007	7016	7024	7033	7042	7050	7059	7067	1	2	3	3	4	5	6	7	8
51	7076	7084	7093	7101	7110	7118	7126	7135	7143	7152	1	2	3	3	4	5	6	7	8
52	7160	7168	7177	7185	7193	7202	7210	7218	7226	7235	1	2	2	3	4	5	6	7	7
53	7243	7251	7259	7267	7275	7284	7292	7300	7308	7316	1	2	2	3	4	5	6	6	7
54	7324	7332	7340	7348	7356	7364	7372	7380	7388	7396	1	2	2	3	4	5	6	6	7
55	7404	7412	7419	7427	7435	7443	7451	7459	7466	7474	1	2	2	3	4	5	5	6	7
56	7482	7490	7497	7505	7513	7520	7528	7536	7543	7551	1	2	2	3	4	5	5	6	7
57	7559	7566	7574	7582	7589	7597	7604	7612	7619	7627	1	2	2	3	4	5	5	6	7
58	7634	7642	7649	7657	7664	7672	7679	7686	7694	7701	1	1	2	3	4	4	5	6	7
59	7709	7716	7723	7731	7738	7745	7752	7760	7767	7774	1	1	2	3	4	4	5	6	7
60	7782	7789	7796	7803	7810	7818	7825	7832	7839	7846	1	1	2	3	4	4	5	6	6
61	7853	7860	7868	7875	7882	7889	7896	7903	7910	7917	1	1	2	3	4	4	5	6	6
62	7924	7931	7938	7945	7952	7959	7966	7973	7980	7987	1	1	2	3	3	4	5	6	6
63	7993	8000	8007	8014	8021	8028	8035	8041	8048	8055	1	1	2	3	3	4	5	5	6
64	8062	8069	8075	8082	8089	8096	8102	8109	8116	8122	1	1	2	3	3	4	5	5	6
65	8129	8136	8142	8149	8156	8162	8169	8176	8182	8189	1	1	2	3	3	4	5	5	6
66	8195	8202	8209	8215	8222	8228	8235	8241	8248	8254	1	1	2	3	3	4	5	5	6
67	8261	8267	8274	8280	8287	8293	8299	8306	8312	8319	1	1	2	3	3	4	5	5	6
68	8325	8331	8338	8344	8351	8357	8363	8370	8376	8382	1	1	2	3	3	4	4	5	6
69	8388	8395	8401	8407	8414	8420	8426	8432	8439	8445	1	1	2	2	3	4	4	5	6
70	8451	8457	8463	8470	8476	8482	8488	8494	8500	8506	1	1	2	2	3	4	4	5	6
71	8513	8519	8525	8531	8537	8543	8549	8555	8561	8567	1	1	2	2	3	4	4	5	5
72	8573	8579	8585	8591	8597	8603	8609	8615	8621	8627	1	1	2	2	3	4	4	5	5
73	8633	8639	8645	8651	8657	8663	8669	8675	8681	8686	1	1	2	2	3	4	4	5	5
74	8692	8698	8704	8710	8716	8722	8727	8733	8739	8745	1	1	2	2	3	4	4	5	5
75	8751	8756	8762	8768	8774	3779	8785	8791	8797	8802	1	1	2	2	3	3	4	5	5
76	8808	8814	8820	8825	8831	8837	8842	8848	8854	8859	1	1	2	2	3	3	4	5	5
77	8865	8871	8876	8882	8887	8893	8899	8904	8910	8915	1	1	2	2	3	3	4	4	5
78	8921	8927	8932	8938	8943	8949	8954	8960	8965	8971	1	1	2	2	3	3	4	4	5
79	8976	8982	8987	8993	8998	9004	9009	9015	9020	9025	1	1	2	2	3	3	4	4	5
80	9031	9036	9042	9047	9053	9058	9063	9069	9074	9079	1	1	2	2	3	3	4	4	5
81	9085	9090	9096	9101	9106	9112	9117	9122	9128	9133	1	1	2	2	3	3	4	4	5
82	9138	9143	9149	9154	9159	9165	9170	9175	9180	9186	1	1	2	2	3	3	4	4	5
83	9191	9196	9201	9206	9212	9217	9222	9227	9232	9238	1	1	2	2	3	3	4	4	5
84	9243	9248	9253	9258	9263	9269	9274	9279	9284	9289	1	1	2	2	3	3	4	4	5
85	9294	9299	9304	9309	9315	9320	9325	9330	9335	9340	1	1	2	2	3	3	4	4	5
86	9345	9350	9355	9360	9365	9370	9375	9380	9385	9390	1	1	2	2	3	3	4	4	5
87	9395	9400	9405	9410	9415	9420	9425	9430	9435	9440	0	1	1	2	2	3	3	4	4
88	9445	9450	9455	9460	9465	9469	9474	9479	9484	9489	0	1	1	2	2	3	3	4	4
89	9494	9499	9504	9509	9513	9518	9523	9528	9533	9538	0	1	1	2	2	3	3	4	4
90	9542	9547	9552	9557	9562	9566	9571	9576	9581	9586	0	1	1	2	2	3	3	4	4
91	9590	9595	9600	9605	9609	9614	9619	9624	9628	9633	0	1	1	2	2	3	3	4	4
92	9638	9643	9647	9652	9657	9661	9666	9671	9675	9680	0	1	1	2	2	3	3	4	4
93	9685	9689	9694	9699	9703	9708	9713	9717	9722	9727	0	1	1	2	2	3	3	4	4
94	9731	9736	9741	9745	9750	9754	9759	9763	9768	9773	0	1	1	2	2	3	3	4	4
95	9777	9782	9786	9791	9795	9800	9805	9809	9814	9818	0	1	1	2	2	3	3	4	4
96	9823	9827	9832	9836	9841	9845	9850	9854	9859	9863	0	1	1	2	2	3	3	4	4
97	9868	9872	9877	9881	9886	9890	9894	9899	9903	9908	0	1	1	2	2	3	3	4	4
98	9912	9917	9921	9926	9930	9934	9939	9943	9948	9952	0	1	1	2	2	3	3	4	4
99	9956	9961	9965	9969	9974	9978	9983	9987	9991	9996	0	1	1	2	2	3	3	3	4

Appendix Table 5 Omega (Logit) Conversion Table

p (%)	db	p (%)	db	p (%)	db	p (%)	db	p (%)	db	p (%)	db
0.0	∞*	5.0	-12.787	10.0	-9.541	15.0	-7.532	20.0	-6.020	25.0	-4.770
0.1	-29.995	5.1	-12.696	10.1	-9.493	15.1	-7.498	20.1	-5.993	25.1	-4.747
0.2	-26.980	5.2	-12.607	10.2	--9.446	15.2	-7.465	20.2	-5.966	25.2	-4.724
0.3	-25.215	5.3	-12.520	10.3	-9.399	15.3	-7.431	20.3	-5.939	25.3	-4.701
0.4	-23.961	5.4	-12.434	10.4	-9.352	15.4	-7.397	20.4	-5.912	25.4	-4.678
0.5	-22.988	5.5	-12.350	10.5	-9.305	15.5	-7.364	20.5	-5.885	25.5	-4.655
0.6	-22.191	5.6	-12.267	10.6	-9.259	15.6	-7.331	20.6	-5.859	25.6	-4.632
0.7	-21.518	5.7	-12.185	10.7	-9.214	15.7	-7.298	20.7	-5.832	25.7	-4.610
0.8	-20.933	5.8	-12.105	10.8	-9.168	15.8	-7.266	20.8	-5.806	25.8	-4.587
0.9	-20.417	5.9	-12.026	10.9	-9.124	15.9	-7.233	20.9	-5.779	25.9	-4.564
1.0	-19.955	6.0	-11.949	11.0	-9.079	16.0	-7.201	21.0	-5.753	26.0	-4.542
1.1	-19.537	6.1	-11.872	11.1	-9.035	16.1	-7.168	21.1	-5.727	26.1	-4.519
1.2	-19.155	6.2	-11.797	11.2	-8.991	16.2	-7.136	21.2	-5.701	26.2	-4.497
1.3	-18.803	6.3	-11.723	11.3	-8.947	16.3	-7.104	21.3	-5.675	26.3	-4.474
1.4	-18.476	6.4	-11.650	11.4	-8.904	16.4	-7.073	21.4	-5.649	26.4	-4.452
1.5	-18.172	6.5	-11.578	11.5	-8.861	16.5	-7.041	21.5	-5.623	26.5	-4.429
1.6	-17.888	6.6	-11.507	11.6	-8.819	16.6	-7.010	21.6	-5.598	26.6	-4.407
1.7	-17.620	6.7	-11.437	11.7	-8.777	16.7	-6.978	21.7	-5.572	26.7	-4.385
1.8	-17.367	6.8	-11.368	11.8	-8.735	16.8	-6.947	21.8	-5.547	26.8	-4.363
1.9	-17.128	6.9	-11.300	11.9	-8.693	16.9	-6.916	21.9	-5.521	26.9	-4.341
2.0	-16.901	7.0	-11.233	12.0	-8.652	17.0	-6.885	22.0	-5.496	27.0	-4.319
2.1	-16.685	7.1	-11.167	12.1	-8.611	17.1	-6.855	22.1	-5.470	27.1	-4.297
2.2	-16.478	7.2	-11.101	12.2	-8.570	17.2	-6.824	22.2	-5.445	27.2	-4.275
2.3	-16.281	7.3	-11.037	12.3	-8.530	17.3	-6.794	22.3	-5.420	27.3	-4.253
2.4	-16.091	7.4	-10.973	12.4	-8.490	17.4	-6.763	22.4	-5.395	27.4	-4.231
2.5	-15.910	7.5	-10.910	12.5	-8.450	17.5	-6.733	22.5	-5.370	27.5	-4.209
2.6	-15.735	7.6	-10.848	12.6	-8.410	17.6	-6.703	22.6	-5.345	27.6	-4.187
2.7	-15.566	7.7	-10.786	12.7	-8.371	17.7	-6.673	22.7	-5.321	27.7	-4.166
2.8	-15.404	7.8	-10.725	12.8	-8.332	17.8	-6.644	22.8	-5.296	27.8	-4.144
2.9	-15.247	7.9	-10.665	12.9	-8.293	17.9	-6.614	22.9	-5.271	27.9	-4.122
3.0	-15.096	8.0	-10.606	13.0	-8.255	18.0	-6.584	23.0	-5.427	28.0	-4.101
3.1	-14.949	8.1	-10.547	13.1	-8.216	18.1	-6.555	23.1	-5.222	28.1	-4.079
3.2	-14.806	8.2	-10.489	13.2	-8.178	18.2	-6.526	23.2	-5.198	28.2	-4.058
3.3	-14.668	8.3	-10.432	13.3	-8.141	18.3	-6.497	23.3	-5.173	28.3	-4.036
3.4	-14.534	8.4	-10.375	13.4	-8.103	18.4	-6.468	23.4	-5.149	28.4	-4.015
3.5	-14.404	8.5	-10.319	13.5	-8.066	18.5	-6.439	23.5	-5.125	28.5	-3.994
3.6	-14.277	8.6	-10.263	13.6	-8.029	18.6	-6.410	23.6	-5.101	28.6	-3.972
3.7	-14.153	8.7	-10.209	13.7	-7.992	18.7	-6.381	23.7	-5.077	28.7	-3.951
3.8	-14.033	8.8	-10.154	13.8	-7.955	18.8	-6.353	23.8	-5.053	28.8	-3.930
3.9	-13.916	8.9	-10.100	13.9	-7.919	18.9	-6.325	23.9	-5.029	28.9	-3.909
4.0	-13.801	9.0	-10.047	14.0	-7.883	19.0	-6.296	24.0	-5.005	29.0	-3.888
4.1	-13.689	9.1	-9.994	14.1	-7.847	19.1	-6.268	24.1	-4.981	29.1	-3.867
4.2	-13.580	9.2	-9.942	14.2	-7.811	19.2	-6.240	24.2	-4.958	29.2	-3.846
4.3	-13.473	9.3	-9.890	14.3	-7.775	19.3	-6.212	24.3	-4.934	29.3	-3.825
4.4	-13.369	9.4	-9.839	14.4	-7.740	19.4	-6.184	24.4	-4.910	29.4	-3.804
4.5	-13.267	9.5	-9.788	14.5	-7.705	19.5	-6.157	24.5	-4.887	29.5	-3.783
4.6	-13.167	9.6	-9.738	14.6	-7.670	19.6	-6.129	24.6	-4.863	29.6	-3.762
4.7	-13.069	9.7	-9.688	14.7	-7.635	19.7	-6.101	24.7	-4.840	29.7	-3.741
4.8	-12.973	9.8	-9.639	14.8	-7.601	19.8	-6.074	24.8	-4.817	29.8	-3.720
4.9	-12.879	9.9	-9.590	14.9	-7.566	19.9	-6.047	24.9	-4.793	29.9	-3.699

* When p or ρ is less than 0.1%, both p and ρ are rewritten as percent, obtaining $\eta = \frac{p}{1-p}$ $\left(\text{or } \frac{\rho}{1-\rho}\right)$, and the numerical value is found by using Appendix Table 4.

Appendix Table 5 (Continued)

p (%)	db	p (%)	db	p (%)	db	p (%)	db	p (%)	db	p (%)	db
30.0	−3.679	35.0	−2.687	40.0	−1.760	45.0	−0.871	50.0	0.000	55.0	0.872
30.1	−3.658	35.1	−2.668	40.1	−1.742	45.1	−0.853	50.1	0.017	55.1	0.889
30.2	−3.637	35.2	−2.649	40.2	−1.724	45.2	−0.835	50.2	0.035	55.2	0.907
30.3	−3.617	35.3	−2.630	40.3	−1.706	45.3	−0.818	50.3	0.052	55.3	0.924
30.4	−3.596	35.4	−2.611	40.4	−1.688	45.4	−0.800	50.4	0.069	55.4	0.942
30.5	−3.576	35.5	−2.592	40.5	−1.670	45.5	−0.783	50.5	0.087	55.5	0.959
30.6	−3.555	35.6	−2.573	40.6	−1.652	45.6	−0.765	50.6	0.104	55.6	0.977
30.7	−3.535	35.7	−2.554	40.7	−1.634	45.7	−0.748	50.7	0.122	55.7	0.995
30.8	−3.515	35.8	−2.536	40.8	−1.616	45.8	−0.730	50.8	0.139	55.8	1.012
30.9	−3.494	35.9	−2.517	40.9	−1.598	45.9	−0.713	50.9	0.156	55.9	1.030
31.0	−3.474	36.0	−2.498	41.0	−1.580	46.0	−0.695	51.0	0.174	56.0	1.047
31.1	−3.454	36.1	−2.479	41.1	−1.562	46.1	−0.678	51.1	0.191	56.1	1.065
31.2	−3.433	36.2	−2.460	41.2	−1.544	46.2	−0.660	51.2	0.209	56.2	1.083
31.3	−3.413	36.3	−2.441	41.3	−1.526	46.3	−0.643	51.3	0.226	56.3	1.100
31.4	−3.393	36.4	−2.423	41.4	−1.508	46.4	−0.625	51.4	0.243	56.4	1.118
31.5	−3.373	36.5	−2.404	41.5	−1.490	46.5	−0.608	51.5	0.261	56.5	1.136
31.6	−3.353	36.6	−2.385	41.6	−1.472	46.6	−0.591	51.6	0.278	56.6	1.153
31.7	−3.333	36.7	−2.366	41.7	−1.454	46.7	−0.573	51.7	0.295	56.7	1.171
31.8	−3.313	36.8	−2.348	41.8	−1.436	46.8	−0.556	51.8	0.313	56.8	1.189
31.9	−3.293	36.9	−2.329	41.9	−1.419	46.9	−0.538	51.9	0.330	56.9	1.206
32.0	−3.273	37.0	−2.310	42.0	−1.401	47.0	−0.521	52.0	0.348	57.0	1.224
32.1	−3.253	37.1	−2.292	42.1	−1.383	47.1	−0.503	52.1	0.365	57.1	1.242
32.2	−3.233	37.2	−2.273	42.2	−1.365	47.2	−0.486	52.2	0.382	57.2	1.260
32.3	−3.213	37.3	−2.255	42.3	−1.347	47.3	−0.468	52.3	0.400	57.3	1.277
32.4	−3.193	37.4	−2.236	42.4	−1.330	47.4	−0.451	52.4	0.417	57.4	1.295
32.5	−3.173	37.5	−2.217	42.5	−1.312	47.5	−0.434	52.5	0.435	57.5	1.313
32.6	−3.153	37.6	−2.199	42.6	−1.294	47.6	−0.416	52.6	0.452	57.6	1.331
32.7	−3.134	37.7	−2.180	42.7	−1.276	47.7	−0.399	52.7	0.469	57.7	1.348
32.8	−3.114	37.8	−2.162	42.8	−1.259	47.8	−0.381	52.8	0.487	57.8	1.366
32.9	−3.094	37.9	−2.144	42.9	−1.241	47.9	−0.364	52.9	0.504	57.9	1.384
33.0	−3.075	38.0	−2.125	43.0	−1.223	48.0	−0.347	53.0	0.522	58.0	1.402
33.1	−3.055	38.1	−2.107	43.1	−1.205	48.1	−0.329	53.1	0.539	58.1	1.420
33.2	−3.035	38.2	−2.088	43.2	−1.188	48.2	−0.312	53.2	0.557	58.2	1.437
33.3	−3.016	38.3	−2.070	43.3	−1.170	48.3	−0.294	53.3	0.574	58.3	1.455
33.4	−2.996	38.4	−2.051	43.4	−1.152	48.4	−0.277	53.4	0.592	58.4	1.473
33.5	−2.977	38.5	−2.033	43.5	−1.135	48.5	−0.260	53.5	0.609	58.5	1.491
33.6	−2.957	38.6	−2.015	43.6	−1.117	48.6	−0.242	53.6	0.626	58.6	1.509
33.7	−2.938	38.7	−1.996	43.7	−1.099	48.7	−0.225	53.7	0.644	58.7	1.527
33.8	−2.918	38.8	−1.978	43.8	−1.082	48.8	−0.208	53.8	0.661	58.8	1.545
33.9	−2.899	38.9	−1.960	43.9	−1.064	48.9	−0.190	53.9	0.679	58.9	1.563
34.0	−2.880	39.0	−1.942	44.0	−1.046	49.0	−0.173	54.0	0.696	59.0	1.581
34.1	−2.860	39.1	−1.923	44.1	−1.029	49.1	−0.155	54.1	0.714	59.1	1.599
34.2	−2.841	39.2	−1.905	44.2	−1.011	49.2	−0.138	54.2	0.731	59.2	1.617
34.3	−2.822	39.3	−1.887	44.3	−0.994	49.3	−0.121	54.3	0.749	59.3	1.635
34.4	−2.802	39.4	−1.869	44.4	−0.976	49.4	−0.103	54.4	0.766	59.4	1.653
34.5	−2.783	39.5	−1.851	44.5	−0.958	49.5	−0.086	54.5	0.784	59.5	1.671
34.6	−2.764	39.6	−1.832	44.6	−0.941	49.6	−0.068	54.6	0.801	59.6	1.689
34.7	−2.745	39.7	−1.814	44.7	−0.923	49.7	−0.051	54.7	0.819	59.7	1.707
34.8	−2.726	39.8	−1.796	44.8	−0.906	49.8	−0.034	54.8	0.836	59.8	1.725
34.9	−2.707	39.9	−1.778	44.9	−0.888	49.9	−0.016	54.9	0.854	59.9	1.743

Appendix Table 5 (Continued)

p (%)	db	p (%)	db	p (%)	db	p (%)	db	p (%)	db	p (%)	db
60.0	1.761	65.0	2.688	70.0	3.680	75.0	4.771	80.0	6.021	85.0	7.533
60.1	1.779	65.1	2.708	70.1	3.700	75.1	4.794	80.1	6.048	85.1	7.567
60.2	1.797	65.2	2.727	70.2	3.721	75.2	4.818	80.2	6.075	85.2	7.602
60.3	1.815	65.3	2.746	70.3	3.742	75.3	4.841	80.3	6.102	85.3	7.636
60.4	1.833	65.4	2.765	70.4	3.763	75.4	4.864	80.4	6.130	85.4	7.671
60.5	1.852	65.5	2.784	70.5	3.784	75.5	4.888	80.5	6.158	85.5	7.706
60.6	1.870	65.6	2.803	70.6	3.805	75.6	4.911	80.6	6.185	85.6	7.741
60.7	1.888	65.7	2.823	70.7	3.826	75.7	4.935	80.7	6.213	85.7	7.776
60.8	1.906	65.8	2.842	70.8	3.847	75.8	4.959	80.8	6.241	85.8	7.812
60.9	1.924	65.9	2.861	70.9	3.868	75.9	4.982	80.9	6.269	85.9	7.848
61.0	1.943	66.0	2.881	71.0	3.889	76.0	5.006	81.0	6.297	86.0	7.884
61.1	1.961	66.1	2.900	71.1	3.910	76.1	5.030	81.1	6.326	86.1	7.920
61.2	1.979	66.2	2.919	71.2	3.931	76.2	5.054	81.2	6.354	86.2	7.956
61.3	1.997	66.3	2.939	71.3	3.952	76.3	5.078	81.3	6.382	86.3	7.993
61.4	2.016	66.4	2.958	71.4	3.973	76.4	5.102	81.4	6.411	86.4	8.030
61.5	2.034	66.5	2.978	71.5	3.995	76.5	5.126	81.5	6.440	86.5	8.067
61.6	2.052	66.6	2.997	71.6	4.016	76.6	5.150	81.6	6.469	86.6	8.104
61.7	2.071	66.7	3.017	71.7	4.037	76.7	5.174	81.7	6.498	86.7	8.142
61.8	2.089	66.8	3.036	71.8	4.059	76.8	5.199	81.8	6.527	86.8	8.179
61.9	2.108	66.9	3.056	71.9	4.080	76.9	5.223	81.9	6.556	86.9	8.217
62.0	2.126	67.0	3.076	72.0	4.102	77.0	5.248	82.0	6.585	87.0	8.256
62.1	2.145	67.1	3.095	72.1	4.123	77.1	5.272	82.1	6.615	87.1	8.294
62.2	2.163	67.2	3.115	72.2	4.145	77.2	5.297	82.2	6.645	87.2	8.333
62.3	2.181	67.3	3.135	72.3	4.167	77.3	5.322	82.3	6.674	87.3	8.372
62.4	2.200	67.4	3.154	72.4	4.188	77.4	5.346	82.4	6.704	87.4	8.411
62.5	2.218	67.5	3.174	72.5	4.210	77.5	5.371	82.5	6.734	87.5	8.451
62.6	2.237	67.6	3.194	72.6	4.232	77.6	5.396	82.6	6.764	87.6	8.491
62.7	2.256	67.7	3.214	72.7	4.254	77.7	5.421	82.7	6.795	87.7	8.531
62.8	2.274	67.8	3.234	72.8	4.276	77.8	5.446	82.8	6.825	87.8	8.571
62.9	2.293	67.9	3.254	72.9	4.298	77.9	5.471	82.9	6.856	87.9	8.612
63.0	2.311	68.0	3.274	73.0	4.320	78.0	5.497	83.0	6.886	88.0	8.653
63.1	2.330	68.1	3.294	73.1	4.342	78.1	5.522	83.1	6.917	88.1	8.694
63.2	2.349	68.2	3.314	73.2	4.364	78.2	5.548	83.2	6.948	88.2	8.736
63.3	2.367	68.3	3.334	73.3	4.386	78.3	5.573	83.3	6.979	88.3	8.778
63.4	2.386	68.4	3.354	73.4	4.408	78.4	5.599	83.4	7.011	88.4	8.820
63.5	2.405	68.5	3.374	73.5	4.430	78.5	5.624	83.5	7.042	88.5	8.862
63.6	2.424	68.6	3.394	73.6	4.453	78.6	5.650	83.6	7.074	88.6	8.905
63.7	2.442	68.7	3.414	73.7	4.475	78.7	5.676	83.7	7.105	88.7	8.948
63.8	2.461	68.8	3.434	73.8	4.498	78.8	5.702	83.8	7.137	88.8	8.992
63.9	2.480	68.9	3.455	73.9	4.520	78.9	5.728	83.9	7.169	88.9	9.036
64.0	2.499	69.0	3.475	74.0	4.543	79.0	5.754	84.0	7.202	89.0	9.080
64.1	2.518	69.1	3.495	74.1	4 565	79.1	5.780	84.1	7.234	89.1	9.125
64.2	2.537	69.2	3.516	74.2	4.588	79.2	5.807	84.2	7.267	89.2	9.169
64.3	2.555	69.3	3.536	74.3	4.611	79.3	5.833	84.3	7.299	89.3	9.215
64.4	2.574	69.4	3.556	74.4	4.633	79.4	5.860	84.4	7.332	89.4	9.260
64.5	2.593	69.5	3.577	74.5	4.656	79.5	5.886	84.5	7.365	89.5	9.306
64.6	2.612	69.6	3.597	74.6	4.679	79.6	5.913	84.6	7.398	89.6	9.353
64.7	2.631	69.7	3.618	74.7	4.702	79.7	5.940	84.7	7.432	89.7	9.400
64.8	2.650	69.8	3.638	74.8	4.725	79.8	5.967	84.8	7.466	89.8	9.447
64.9	2.669	69.9	3.659	74.9	4.748	79.9	5.994	84.9	7.499	89.9	9.494

Appendix Table 5 (Continued)

p (%)	db	p (%)	db	p (%)	db	p (%)	db	p (%)	db	p (%)	db
90.0	9.542	92.0	10.607	94.0	11.950	96.0	13.802	98.0	16.902	100.0	∞*
90.1	9.591	92.1	10.666	94.1	12.027	96.1	13.917	98.1	17.129		
90.2	9.640	92.2	10.726	94.2	12.106	96.2	14.034	98.2	17.368		
90.3	9.689	92.3	10.787	94.3	12.186	96.3	14.154	98.3	17.621		
90.4	9.739	92.4	10.849	94.4	12.268	96.4	14.278	98.4	17.889		
90.5	9.789	92.5	10.911	94.5	12.351	96.5	14.405	98.5	18.173		
90.6	9.840	92.6	10.974	94.6	12.435	96.6	14.535	98.6	18.477		
90.7	9.891	92.7	11.038	94.7	12.521	96.7	14.669	98.7	18.804		
90.8	9.943	92.8	11.102	94.8	12.608	96.8	14.807	98.8	19.156		
90.9	9.995	92.9	11.168	94.9	12.697	96.9	14.950	98.9	19.538		
91.0	10.048	93.0	11.234	95.0	12.788	97.0	15.097	99.0	19.956		
91.1	10.111	93.1	11.301	95.1	12.880	97.1	15.248	99.1	20.418		
91.2	10.155	93.2	11.369	95.2	12.974	97.2	15.405	99.2	20.934		
91.3	10.210	93.3	11.438	95.3	13.070	97.3	15.567	99.3	21.519		
91.4	10.264	93.4	11.508	95.4	13.168	97.4	15.736	99.4	22.192		
91.5	10.320	93.5	11.579	95.5	13.268	97.5	15.911	99.5	22.989		
91.6	10.376	93.6	11.651	95.6	13.370	97.6	16.092	99.6	23.962		
91.7	10.433	93.7	11.724	95.7	13.474	97.7	16.282	99.7	25.216		
91.8	10.490	93.8	11.798	95.8	13.581	97.8	16.479	99.8	26.981		
91.9	10.548	93.9	11.873	95.9	13.690	97.9	16.686	99.9	29.996		

* When p or ρ exceeds 99.9%, both p and ρ are rewritten as percent, obtaining $\eta = \frac{p}{1-p}$ $\left(\text{or } \frac{\rho}{1-\rho}\right)$, and the numerical value is obtained by using Appendix Table 4.

Appendix Table 6 Decibel Table, The Case of $1 \leqq \eta \leqq 10.00$

(10 Times the Logarithm)

η	Decibels	η	Decibels	η	Decibels	η	Decibels	η	Decibels	η	Decibels
1.00	0.00	1.50	1.76	2.00	3.01	2.50	3.98	3.00	4.77	3.50	5.44
01	0.04	51	1.79	01	3.03	51	4.00	01	4.79	51	5.45
02	0.09	52	1.81	02	3.05	52	4.01	02	4.80	52	5.47
03	0.13	53	1.85	03	3.08	53	4.03	03	4.81	53	5.48
04	0.17	54	1.88	04	3.10	54	4.05	04	4.83	54	5.49
1.05	0.21	1.55	1.90	2.05	3.12	2.55	4.06	3.05	4.84	3.55	5.50
06	0.25	56	1.93	06	3.14	56	4.08	06	4.86	56	5.51
07	0.29	57	1.96	07	3.16	57	4.10	07	4.87	57	5.53
08	0.33	58	1.99	08	3.18	58	4.12	08	4.89	58	5.54
09	0.37	59	2.01	09	3.20	59	4.13	09	4.90	59	5.55
1.10	0.41	1.60	2.04	2.10	3.22	2.60	4.15	3.10	4.91	3.60	5.56
11	0.45	61	2.07	11	3.24	61	4.17	11	4.93	61	5.58
12	0.49	62	2.10	12	3.26	62	4.18	12	4.94	62	5.59
13	0.53	63	2.12	13	3.28	63	4.20	13	4.96	63	5.60
14	0.57	64	2.15	14	3.30	64	4.22	14	4.97	64	5.61
1.15	0.61	1.65	2.18	2.15	3.32	2.65	4.23	3.15	4.98	3.65	5.62
16	0.64	66	2.20	16	3.34	66	4.25	16	5.00	66	5.64
17	0.68	67	2.23	17	3.36	67	4.27	17	5.01	67	5.65
18	0.72	68	2.25	18	3.38	68	4.28	18	5.02	68	5.66
19	0.76	69	2.28	19	3.40	69	4.30	19	5.04	69	5.67
1.20	0.79	1.70	2.30	2.20	3.42	2.70	4.31	3.20	5.05	3.70	5.68
21	0.83	71	2.33	21	3.44	71	4.33	21	5.07	71	5.69
22	0.86	72	2.36	22	3.46	72	4.35	22	5.08	72	5.71
23	0.90	73	2.38	23	3.48	73	4.36	23	5.09	73	5.72
24	0.93	74	2.40	24	3.50	74	4.38	24	5.11	74	5.73
1.25	0.97	1.75	2.43	2.25	3.52	2.75	4.39	3.25	5.12	3.75	5.74
26	1.00	76	2.46	26	3.54	76	4.41	26	5.13	76	5.75
27	1.04	77	2.48	27	3.56	77	4.42	27	5.14	77	5.76
28	1.07	78	2.50	28	3.58	78	4.44	28	5.16	78	5.77
29	1.11	89	2.53	29	3.60	79	4.46	29	5.17	79	5.79
1.30	1.14	1.80	2.55	2.30	3.62	2.80	4.47	3.30	5.18	3.80	5.80
31	1.17	81	2.58	31	3.64	81	4.49	31	5.20	81	5.81
32	1.21	82	2.60	32	3.66	82	4.50	32	5.21	82	5.82
33	1.24	83	2.62	33	3.67	83	4.52	33	5.22	83	5.83
34	1.27	84	2.65	34	3.69	84	4.53	34	5.24	84	5.84
1.35	1.30	1.85	2.67	2.35	3.71	2.85	4.55	3.35	5.25	3.85	5.85
36	1.34	86	2.70	36	3.73	86	4.56	36	5.26	86	5.87
37	1.37	87	2.72	37	3.75	87	4.58	37	5.28	87	5.88
38	1.40	88	2.74	38	3.77	88	4.59	38	5.29	88	5.89
39	1.43	89	2.76	39	3.78	89	4.61	39	5.30	89	5.90
1.40	1.46	1.90	2.79	2.40	3.80	2.90	4.62	3.40	5.32	3.90	5.91
41	1.49	91	2.81	41	3.82	91	4.64	41	5.33	91	5.92
42	1.52	92	2.83	42	3.84	92	4.65	42	5.34	92	5.93
43	1.55	93	2.86	43	3.86	93	4.67	43	5.35	93	5.94
44	1.58	94	2.88	44	3.87	94	4.68	44	5.37	94	5.96
1.45	1.61	1.95	2.90	2.45	3.89	2.95	4.70	3.45	5.38	3.95	5.97
46	1.64	96	2.92	46	3.91	96	4.71	46	5.39	96	5.98
47	1.67	97	2.94	47	3.93	97	4.73	47	5.40	97	5.99
48	1.70	98	2.97	48	3.94	98	4.74	48	5.42	98	6.00
49	1.73	99	2.99	49	3.96	99	4.76	49	5.43	99	6.01

Appendix Table 6 (Continued)

η	Decibels	η	Decibels	η	Decibels	η	Decibels	η	Decibels	η	Decibels
4.00	6.02	4.50	6.53	5.00	6.99	5.50	7.40	6.00	7.78	6.50	8.13
01	6.03	51	6.54	01	7.00	51	7.41	01	7.79	51	8.14
02	6.04	52	6.55	02	7.01	52	7.42	02	7.80	52	8.14
03	6.05	53	6.56	03	7.02	53	7.43	03	7.80	53	8.15
04	6.06	54	6.57	04	7.02	54	7.44	04	7.81	54	8.16
4.05	6.07	4.55	6.58	5.05	7.03	5.55	7.44	6.05	7.82	6.55	8.16
06	6.09	56	6.59	06	7.04	56	7.45	06	7.83	56	8.17
07	6.10	57	6.60	07	7.05	57	7.46	07	7.83	57	8.18
08	6.11	58	6.61	08	7.06	58	7.47	08	7.84	58	8.18
09	6.12	59	6.62	09	7.07	59	7.47	09	7.85	59	8.19
4.10	6.13	4.60	6.63	5.10	7.08	5.60	7.48	6.10	7.85	6.60	8.19
11	6.14	61	6.64	11	7.08	61	7.49	11	7.86	61	8.20
12	6.15	62	6.65	12	7.09	62	7.50	12	7.87	62	8.21
13	6.16	63	6.66	13	7.10	63	7.50	13	7.87	63	8.21
14	6.17	64	6.67	14	7.11	64	7.51	14	7.88	64	8.22
4.15	6.18	4.65	6.68	5.15	7.12	5.65	7.52	6.15	7.89	6.65	8.23
16	6.19	66	6.68	16	7.12	66	7.53	16	7.90	66	8.23
17	6.20	67	6.69	17	7.13	67	7.54	17	7.90	67	8.24
18	6.21	68	6.70	18	7.14	68	7.54	18	7.91	68	8.25
19	6.22	69	6.71	19	7.15	69	7.55	19	7.92	69	8.25
4.20	6.23	4.70	6.72	5.20	7.16	5.70	7.56	6.20	7.92	6.70	8.26
21	6.24	71	6.73	21	7.17	71	7.57	21	7.93	71	8.27
22	6.25	72	6.74	22	7.18	72	7.57	22	7.94	72	8.27
23	6.26	73	6.75	23	7.18	73	7.58	23	7.94	73	8.28
24	6.27	74	6.76	24	7.19	74	7.59	24	7.95	74	8.29
4.25	6.28	4.75	6.77	5.25	7.20	5.75	7.60	6.25	7.96	6.75	8.29
26	6.29	76	6.78	26	7.21	76	7.60	26	7.97	76	8.30
27	6.30	77	6.78	27	7.22	77	7.61	27	7.97	77	8.31
28	6.31	78	6.79	28	7.23	78	7.62	28	7.98	78	8.31
29	6.32	79	6.80	29	7.23	79	7.63	29	7.99	79	8.32
4.30	6.33	4.80	6.81	5.30	7.24	5.80	7.63	6.30	7.99	6.80	8.32
31	6.34	81	6.82	31	7.25	81	7.64	31	8.00	81	8.33
32	6.35	82	6.83	32	7.26	82	7.65	32	8.01	82	8.34
33	6.36	83	6.84	33	7.27	83	7.66	33	8.01	83	8.34
34	6.37	84	6.85	34	7.28	84	7.66	34	8.02	84	8.35
4.35	6.38	4.85	6.86	5.35	7.28	5.85	7.67	6.35	8.03	6.85	8.36
36	6.39	86	6.87	36	7.29	86	7.68	36	8.04	86	8.36
37	6.40	87	6.88	37	7.30	87	7.69	37	8.04	87	8.37
38	6.41	88	6.88	38	7.31	88	7.69	38	8.05	88	8.37
39	6.42	89	6.89	39	7.32	89	7.70	39	8.06	89	8.38
4.40	6.43	4.90	6.90	5.40	7.32	5.90	7.71	6.40	8.06	6.90	8.39
41	6.44	91	6.91	41	7.33	91	7.72	41	8.07	91	8.39
42	6.45	92	6.92	42	7.34	92	7.72	42	8.08	92	8.40
43	6.46	93	6.93	43	7.35	93	7.73	43	8.08	93	8.41
44	6.47	94	6.94	44	7.36	94	7.74	44	8.09	94	8.41
4.45	6.48	4.95	6.95	5.45	7.36	5.95	7.75	6.45	8.10	6.95	8.42
46	6.49	96	6.95	46	7.37	96	7.75	46	8.10	96	8.43
47	6.50	97	6.96	47	7.38	97	7.76	47	8.11	97	8.43
48	6.51	98	6.97	48	7.39	98	7.77	48	8.12	98	8.44
49	6.52	99	6.98	49	7.40	99	7.77	49	8.12	99	8.44

Appendix Table 6 (Continued)

η	Decibels	η	Decibels	η	Decibels	η	Decibels	η	Decibels	η	Decibels
7.00	8.45	7.50	8.75	8.00	9.03	8.50	9.29	9.00	9.54	9.50	9.78
01	8.46	51	8.76	01	9.04	51	9.30	01	9.55	51	9.79
02	8.46	52	8.76	02	9.04	52	9.30	02	9.55	52	9.79
03	8.47	53	8.77	03	9.05	53	9.31	03	9.56	53	9.80
04	8.48	54	8.77	04	9.05	54	9.31	04	9.56	54	9.80
7.05	8.48	7.55	8.78	8.05	9.06	8.55	9.32	9.05	9.57	9.55	9.80
06	8.49	56	8.79	06	9.06	56	9.32	06	9.57	56	9.81
07	8.49	57	8.79	07	9.07	57	9.33	07	9.58	57	9.81
08	8.50	58	8.80	08	9.07	58	9.33	08	9.58	58	9.82
09	8.51	59	8.80	09	9.08	59	9.34	09	9.59	59	9.82
7.10	8.51	7.60	8.81	8.10	9.08	8.60	9.34	9.10	9.59	9.60	9.82
11	8.52	61	8.81	11	9.09	61	9.35	11	9.60	61	9.83
12	8.53	62	8.82	12	9.10	62	9.35	12	9.60	62	9.83
13	8.53	63	8.83	13	9.10	63	9.36	13	9.61	63	9.84
14	8.54	64	8.83	14	9.11	64	9.36	14	9.61	64	9.84
7.15	8.54	7.65	8.84	9.15	9.11	8.65	9.37	9.15	9.61	9.65	9.84
16	8.55	66	8.84	16	9.12	66	9.37	16	9.62	66	9.85
17	8.56	67	8.85	17	9.12	67	9.38	17	9.62	67	9.85
18	8.56	68	8.85	18	9.13	68	9.38	18	9.63	68	9.86
19	8.57	69	8.86	19	9.13	69	9.39	19	9.63	69	9.86
7.20	8.57	7.70	8.86	8.20	9.14	8.70	9.39	9.20	9.64	9.70	9.87
21	8.58	71	8.87	21	9.14	71	9.40	21	9.64	71	9.87
22	8.59	72	8.88	22	9.15	72	9.40	22	9.65	72	9.88
23	8.59	73	8.88	23	9.15	73	9.41	23	9.65	73	9.88
24	8.60	74	8.89	24	9.16	74	9.41	24	9.66	74	9.89
7.25	8.60	7.75	8.89	9.25	9.16	8.75	9.42	9.25	9.66	9.75	9.89
26	8.61	76	8.90	26	9.17	76	9.42	26	9.67	76	9.89
27	8.62	77	8.90	27	9.18	77	9.43	27	9.67	77	9.90
28	8.62	78	8.91	28	9.18	78	9.43	28	9.68	78	9.90
29	8.63	79	8.92	29	9.19	79	9.44	29	9.68	89	9.91
7.30	8.63	7.80	8.92	8.30	9.19	8.80	9.44	9.30	9.68	9.80	9.91
31	8.64	81	8.93	31	9.20	81	9.45	31	9.69	81	9.92
32	8.64	82	8.93	32	9.20	82	9.45	32	9.69	82	9.92
33	8.65	83	8.94	33	9.21	83	9.46	33	9.70	83	9.93
34	8.66	84	8.94	34	9.21	84	9.46	34	9.70	84	9.93
7.35	8.66	7.85	8.95	8.35	9.22	8.85	9.47	9.35	9.71	9.85	9.93
36	8.67	86	8.95	36	9.22	86	9.47	36	9.71	86	9.94
37	8.68	87	8.96	37	9.23	87	9.48	37	9.72	87	9.94
38	8.68	88	8.97	38	9.23	88	9.48	38	9.72	88	9.95
39	8.69	89	8.97	39	9.24	89	9.49	39	9.72	89	9.95
7.40	8.69	7.90	8.98	8.40	9.24	8.90	9.49	9.40	9.73	9.90	9.96
41	8.70	91	8.98	41	9.25	91	9.50	41	9.74	91	9.96
42	8.70	92	8.99	42	9.25	92	9.50	42	9.74	92	9.96
43	8.71	93	8.99	43	9.26	93	9.51	43	9.74	93	9.97
44	8.72	94	9.00	44	9.26	94	9.51	44	9.75	94	9.97
7.45	8.72	7.95	9.00	8.45	9.27	8.95	9.52	9.45	9.75	9.95	9.98
46	8.73	96	9.01	46	9.27	96	9.52	46	9.76	96	9.98
47	8.73	97	9.01	47	9.28	97	9.53	47	9.76	97	9.99
48	8.74	98	9.02	48	9.28	98	9.53	48	9.77	98	9.99
49	8.74	99	9.02	49	9.29	99	9.54	49	9.77	99	9.99
7.50	8.75	8.00	9.03	8.50	9.29	9.00	9.54	9.50	9.78	10.00	10.00

Appendix Table 7 Formulas Related to SN Ratio

The number of levels of the signal factor is expressed as k. Assuming that the number of read-off values of the signal factor per level is in all cases equal to r_0 (it does not matter if $r_0 = 1$), the totals of the readings for the respective levels are $y_1, y_2, \ldots, y_k$. We show here how to find S_β and r when the spacing between signal levels is equal and is assumed to have the value h (when the exact value is unknown, its estimated value may be taken as h), or when the greatest level value is regarded as h in the case of equal-ratio spacing (again, when the exact value is unknown, its estimated value may be used as h). If the formula cannot be found in these tables, it is necessary to use the general formula of Section 22.1.3.

(1) *Linear Equation Calibration (in an equispaced case)*

k	r	S_β
2	$\frac{1}{2}r_0h^2$	$\frac{(-y_1+y_2)^2}{2r_0}$
3	$2r_0h^2$	$\frac{(-y_1+y_3)^2}{2r_0}$
4	$5r_0h^2$	$\frac{(-3y_1-y_2+y_3+3y_4)^2}{20r_0}$
5	$10r_0h^2$	$\frac{(-2y_1-y_2+y_4+2y_5)^2}{10r_0}$
6	$\frac{35}{2}r_0h^2$	$\frac{(-5y_1-3y_2-y_3+y_4+3y_5+5y_6)^2}{70r_0}$
7	$28r_0h^2$	$\frac{(-3y_1-2y_2-y_3+y_5+2y_6+3y_7)^2}{28r_0}$
8	$42r_0h^2$	$\frac{(-7y_1-5y_2-3y_3-y_4+y_5+3y_6+5y_7+7y_8)^2}{168r_0}$
9	$60r_0h^2$	$\frac{(-4y_1-3y_2-2y_3-y_4+y_6+2y_7+3y_8+4y_9)^2}{60r_0}$
10	$\frac{165}{2}r_0h^2$	$\frac{(-9y_1-7y_2-5y_3-3y_4-y_5+y_6+3y_7+5y_8+7y_9+9y_{10})^2}{330r_0}$

(2) *Linear Equation Calibration (in the case of equal-ratio spacing)*

If the maximum value is h and the common ratio is α,

$$M_1=h\alpha^{k-1},\ M_2=h\alpha^{k-2},\ \cdots,\ M_{k-1}=h\alpha,\ M_k=h$$

Common ratio α	No. of levels k	r	S_β
$\frac{1}{2}$	$k=2$	$\frac{1}{8}r_0$	$\frac{(-y_1+y_2)^2}{2r_0}$
	$k=3$	$\frac{7}{24}r_0$	$\frac{(-4y_1-y_2+5y_3)^2}{42r_0}$
	$k=4$	$\frac{115}{256}r_0$	$\frac{(-11y_1-7y_2+y_3+17y_4)^2}{460r_0}$
	$k=5$	$\frac{93}{160}r_0$	$\frac{(-26y_1-21y_2-11y_3+9y_4+49y_5)^2}{3\,720r_0}$
$\frac{1}{3}$	$k=2$	$\frac{2}{9}r_0$	$\frac{(-y_1+y_2)^2}{2r_0}$
	$k=3$	$\frac{104}{243}r_0$	$\frac{(-5y_1-2y_2+7y_3)^2}{78r_0}$
	$k=4$	$\frac{140}{243}r_0$	$\frac{(-9y_1-7y_2-y_3+17y_4)^2}{420r_0}$
	$k=5$	$\frac{22\,264}{32\,805}r_0$	$\frac{(-58y_1-53y_2-38y_3+7y_4+142y_5)^2}{27\,830r_0}$
$\frac{1}{4}$	$k=2$	$\frac{9}{32}r_0$	$\frac{(-y_1+y_2)^2}{2r_0}$
	$k=3$	$\frac{63}{128}r_0$	$\frac{(-2y_1-y_2+3y_3)^2}{14r_0}$
	$k=4$	$\frac{10\,251}{16\,384}r_0$	$\frac{(-27y_1-23y_2-7y_3+57y_4)^2}{4\,556r_0}$
$\frac{1}{5}$	$k=2$	$\frac{8}{25}r_0$	$\frac{(-y_1+y_2)^2}{2r_0}$
	$k=3$	$\frac{992}{1\,875}r_0$	$\frac{(-7y_1-4y_2+11y_3)^2}{186r_0}$
	$k=4$	$\frac{5\,096}{15\,625}r_0$	$\frac{(-19y_1-17y_2-7y_3+43y_4)^2}{2\,548r_0}$
$\frac{1}{10}$	$k=2$	$\frac{81}{200}r_0$	$\frac{(-y_1+y_2)^2}{2r_0}$
	$k=3$	$\frac{2\,997}{5\,000}r_0$	$\frac{(-4y_1-3y_2+7y_3)^2}{74r_0}$

(3) *Proportional Equation Calibration (the case of equal spacing)*

$$r=Sr_0h^2,\qquad S_\beta=\frac{(y_1+2y_2+\cdots+ky_k)^2}{Sr_0}$$

Here, $S = 1^2 + 2^2 + \ldots + k^2$, and it has the following values.

k	1	2	3	4	5	6	7	8	9	10
S	1	5	14	30	55	91	140	204	285	385

(4) *Proportional Equation Calibration (the case of equal-ratio spacing)*

$$M_1=h\alpha^{k-1},\ M_2=h\alpha^{k-2},\ \cdots,\ M_{k-1}=h\alpha,\ M_k=h$$

$$r=\alpha^{k-1}Sr_0h^2 \qquad S_\beta=\frac{(y_1+\alpha^{-1}y_2+\cdots+\alpha^{-k+1}y_k)^2}{Sr_0}$$

k	2					3				
Common ratio α	$\frac{1}{2}$	$\frac{1}{3}$	$\frac{1}{4}$	$\frac{1}{5}$	$\frac{1}{10}$	$\frac{1}{2}$	$\frac{1}{3}$	$\frac{1}{4}$	$\frac{1}{5}$	$\frac{1}{10}$
S	5	10	17	26	101	21	91	273	651	10 101

k	4				5			6	
Common ratio α	$\frac{1}{2}$	$\frac{1}{3}$	$\frac{1}{4}$	$\frac{1}{5}$	$\frac{1}{2}$	$\frac{1}{3}$	$\frac{1}{4}$	$\frac{1}{2}$	$\frac{1}{3}$
S	85	820	4 369	16 276	341	7 381	69 905	1 365	66 430

Appendix Table 8 Confidence Limits of SN Ratio (Decibel Values)
(Upper Line: Lower Limit of Confidence; Lower Line: Upper Limit of Confidence)

F_0 \ f_2	5	6	7	8	10	15	20	30	50	100	200	∞
2	−∞ 10.9	−∞ 10.8	−∞ 10.8	−∞ 10.8	−∞ 10.7	−∞ 10.7	−∞ 10.7	−∞ 10.6	−∞ 10.6	−∞ 10.6	−∞ 10.6	−∞ 10.6
3	−∞ 8.8	−∞ 8.7	−∞ 8.7	−∞ 8.6	−∞ 8.6	−∞ 8.5	−∞ 8.4	−∞ 8.4	−∞ 8.4	−∞ 8.4	−∞ 8.4	−∞ 8.3
4	−∞ 7.8	−∞ 7.7	−∞ 7.6	−∞ 7.6	−∞ 7.5	−∞ 7.4	−∞ 7.3	−∞ 7.3	−∞ 7.2	−∞ 7.2	−∞ 7.2	−∞ 7.1
5	−∞ 7.1	−∞ 7.0	−∞ 7.0	−∞ 6.9	−∞ 6.8	−∞ 6.7	−∞ 6.6	−∞ 6.6	−∞ 6.5	−∞ 6.5	−∞ 6.5	−∞ 6.4
6	−∞ 6.7	−∞ 6.6	−∞ 6.5	−∞ 6.4	−∞ 6.3	−∞ 6.2	−23.6 6.1	−18.0 6.0	−15.9 6.0	−14.7 5.9	−14.2 5.9	−13.8 5.9
7	−∞ 6.4	−∞ 6.2	−∞ 6.1	−∞ 6.1	−28.2 6.0	−16.5 5.8	−14.7 5.7	−13.3 5.6	−12.4 5.6	−11.7 5.5	−11.4 5.5	−11.2 5.5
8	−∞ 6.1	−∞ 6.0	−∞ 5.9	−20.8 5.8	−16.4 5.7	−13.4 5.5	−12.3 5.4	−11.4 5.3	−10.7 5.3	−10.2 5.2	−9.9 5.2	−9.7 5.2
9	−∞ 5.9	−25.7 5.8	−18.3 5.7	−16.1 5.6	−14.0 5.4	−11.9 5.3	−11.0 5.2	−10.2 5.1	−9.6 5.0	−9.1 4.9	−8.9 4.9	−8.7 4.9
10	−∞ 5.8	−18.1 5.6	−15.5 5.5	−14.1 5.4	−12.6 5.3	−10.8 5.1	−10.1 5.0	−9.4 4.9	−8.8 4.8	−8.4 4.7	−8.2 4.7	−7.9 4.6
11	−20.0 5.6	−15.8 5.5	−14.0 5.3	−13.0 5.2	−11.7 5.1	−10.1 4.9	−9.4 4.8	−8.7 4.7	−8.2 4.6	−7.8 4.5	−7.6 4.5	−7.3 4.4
12	−17.3 5.5	−14.5 5.3	−13.1 5.2	−12.1 5.1	−11.0 5.0	−9.5 4.8	−8.8 4.6	−8.2 4.5	−7.6 4.4	−7.3 4.4	−7.1 4.3	−6.9 4.3
13	−15.8 5.4	−13.6 5.2	−12.4 5.1	−11.5 5.0	−10.4 4.8	−9.0 4.6	−8.4 4.5	−7.7 4.4	−7.2 4.3	−6.8 4.2	−6.7 4.2	−6.5 4.1
14	−14.8 5.3	−12.9 5.2	−11.8 5.0	−11.0 4.9	−10.0 4.8	−8.6 4.5	−8.0 4.4	−7.4 4.2	−6.9 4.2	−6.5 4.1	−6.3 4.0	−6.1 3.9
15	−14.1 5.2	−12.4 5.1	−11.3 4.9	−10.6 4.8	−9.6 4.7	−8.3 4.4	−7.7 4.3	−7.1 4.2	−6.6 4.0	−6.2 4.0	−6.0 3.9	−5.8 3.8
16	−13.6 5.2	−12.0 5.0	−11.0 4.9	−10.2 4.7	−9.3 4.6	−8.0 4.3	−7.4 4.2	−6.8 4.1	−6.3 3.9	−5.9 3.8	−5.7 3.7	−5.6 3.6
17	−13.1 5.1	−11.6 4.9	−10.6 4.8	−10.0 4.7	−9.0 4.5	−7.8 4.2	−7.2 4.1	−6.6 4.0	−6.1 3.8	−5.7 3.7	−5.5 3.6	−5.3 3.5
18	−12.7 5.1	−11.3 4.9	−10.4 4.7	−9.7 4.6	−8.8 4.4	−7.6 4.2	−7.0 4.0	−6.4 3.9	−5.9 3.8	−5.5 3.6	−5.3 3.4	−5.1 3.4
19	−12.4 5.0	−11.0 4.8	−10.2 4.7	−9.4 4.6	−8.6 4.4	−7.4 4.1	−6.8 4.0	−6.2 3.8	−5.7 3.7	−5.3 3.5	−5.1 3.2	−5.0 3.2
20	−12.1 5.0	−10.8 4.8	−9.9 4.6	−9.3 4.5	−8.4 4.3	−7.2 4.0	−6.6 3.9	−6.0 3.7	−5.5 3.6	−5.1 3.4	−4.9 3.3	−4.8 3.1
30	−10.5 4.7	−9.4 4.5	−8.6 4.3	−8.0 4.2	−7.2 3.9	−5.9 3.6	−5.3 3.4	−4.9 3.2	−4.6 3.0	−4.0 2.8	−3.8 2.6	−3.7 2.3
60	−9.3 4.3	−8.2 4.1	−7.4 4.0	−6.8 3.8	−6.0 3.6	−4.8 3.1	−4.3 2.8	−3.7 2.6	−3.2 2.3	−2.7 2.0	−2.4 1.7	−2.2 1.3
∞	−7.8 4.1	−6.9 3.8	−6.2 3.6	−5.6 3.4	−4.9 3.1	−3.8 2.6	−3.2 2.3	−2.5 2.0	−1.9 1.6	−1.3 1.1	−0.9 0.8	−0.0 0.0

Appendix Table 9 Coefficients for Tolerance Limits
(Reliability 95%, Tolerance Limits $\hat{\mu} \pm k\sqrt{V_e}$; f_2: Number of Degrees of Freedom, n_e: Effective Number of Replications, Proportion $P = 0.99$)
(1) Coefficient k for Two-Sided Tolerance Limits (Refer to Section 37.1)

n_e \ f_2	1	2	3	4	5	6	7	8	9	10	11	12
0.5	59.7	16.5	10.9	8.87	7.82	7.17	6.72	6.40	6.15	5.96	5.80	5.67
1.0	53.0	14.7	9.71	7.89	6.95	6.37	5.98	5.69	5.47	5.30	5.15	5.04
1.5	50.1	13.9	9.18	7.46	6.57	6.02	5.65	5.38	5.17	5.01	4.87	4.76
2.0	48.4	13.4	8.87	7.20	6.35	5.82	5.46	5.20	4.99	4.84	4.71	4.60
3.0	46.4	12.9	8.50	6.90	6.08	5.58	5.23	4.98	4.79	4.64	4.51	4.41
4.0	45.3	12.5	8.29	6.74	5.94	5.44	5.10	4.86	4.67	4.52	4.40	4.30
5.0	44.6	12.3	8.16	6.63	5.84	5.36	5.02	4.78	4.60	4.45	4.33	4.23
6.0	44.5	12.3	8.15	6.62	5.83	5.35	5.01	4.77	4.59	4.45	4.32	4.23
7.0	44.0	12.2	8.06	6.54	5.77	5.29	4.96	4.72	4.54	4.40	4.28	4.18
8.0	43.7	12.1	8.00	6.50	5.72	5.25	4.92	4.69	4.51	4.36	4.25	4.15
9.0	43.4	12.0	7.94	6.45	5.69	5.21	4.89	4.66	4.48	4.33	4.22	4.12
10.0	43.1	11.9	7.90	6.42	5.65	5.18	4.86	4.63	4.45	4.31	4.19	4.10
12.0	42.8	11.9	7.84	6.37	5.61	5.14	4.82	4.59	4.42	4.28	4.16	4.07
14.0	42.6	11.8	7.79	6.33	5.58	5.11	4.80	4.57	4.39	4.25	4.14	4.04
16.0	42.4	11.7	7.76	6.30	5.55	5.09	4.77	4.55	4.37	4.23	4.12	4.02
18.0	42.2	11.7	7.73	6.28	5.54	5.07	4.76	4.53	4.36	4.22	4.11	4.01
20.0	42.1	11.7	7.71	6.26	5.52	5.06	4.75	4.52	4.34	4.21	4.09	4.00
25.0	41.9	11.6	7.67	6.23	5.49	5.04	4.72	4.50	4.32	4.19	4.07	3.98
30.0	41.8	11.6	7.65	6.21	5.48	5.02	4.71	4.48	4.31	4.17	4.06	3.97
40.0	41.6	11.5	7.62	6.19	5.45	5.00	4.69	4.47	4.29	4.16	4.05	3.95
50.0	41.5	11.5	7.60	6.17	5.44	4.99	4.68	4.45	4.28	4.15	4.03	3.94
60.0	41.4	11.5	7.58	6.16	5.43	4.98	4.67	4.44	4.27	4.14	4.03	3.93
80.0	41.3	11.4	7.57	6.15	5.42	4.97	4.66	4.43	4.26	4.13	4.02	3.93
100.0	41.3	11.4	7.56	6.14	5.41	4.96	4.65	4.43	4.26	4.13	4.01	3.92
200.0	41.2	11.4	7.55	6.13	5.40	4.95	4.64	4.42	4.25	4.12	4.01	3.92
500.0	41.1	11.4	7.53	6.12	5.39	4.94	4.63	4.41	4.24	4.11	4.00	3.91
1000.0	41.1	11.4	7.53	6.12	5.39	4.94	4.63	4.41	4.24	4.11	4.00	3.91
∞	41.1	11.4	7.52	6.11	5.38	4.94	4.63	4.41	4.24	4.10	3.99	3.90

Appendix Table 9 (Continued)

13	14	15	16	17	18	19	20	22	24	26	28
5.55	5.46	5.37	5.30	5.24	5.18	5.12	5.08	4.99	4.92	4.86	4.81
4.94	4.86	4.78	4.72	4.66	4.61	4.56	4.52	4.44	4.38	4.32	4.28
4.67	4.59	4.52	4.46	4.40	4.35	4.31	4.27	4.20	4.14	4.09	4.04
4.51	4.43	4.36	4.31	4.25	4.21	4.16	4.12	4.05	4.00	3.95	3.90
4.32	4.25	4.18	4.13	4.07	4.03	3.99	3.95	3.89	3.83	3.78	3.74
4.22	4.15	4.08	4.03	3.98	3.93	3.89	3.86	3.79	3.74	3.69	3.65
4.15	4.08	4.02	3.96	3.91	3.87	3.83	3.80	3.73	3 68	3.63	3.59
4.14	4.07	4.01	3.96	3.91	3.86	3.82	3.79	3.72	3.67	3.63	3.59
4.10	4.03	3.97	3.91	3.86	3.82	3.78	3.75	3.68	3.63	3.59	3.55
4.07	4.00	3.94	3.88	3.83	3.79	3.75	3.72	3.66	3.60	3.56	3.52
4.04	3.97	3.91	3.86	3.81	3.77	3.73	3.69	3.63	3.58	3.54	3.50
4.02	3.95	3.89	3.84	3.79	3.75	3.71	3.67	3.61	3.56	3.52	3.48
3.99	3.92	3.86	3.81	3.76	3.72	3.68	3.65	3.58	3.53	3.49	3.45
3.96	3.90	3.84	3.79	3.74	3.70	3.66	3.62	3.56	3.51	3.47	3.43
3.94	3.88	3.82	3.77	3.72	3.68	3.64	3.61	3.55	3 50	3.45	3.42
3.93	3.87	3.81	3.76	3.71	3.67	3.63	3.60	3.54	3.49	3.44	3.41
3.92	3.86	3.80	3.74	3.70	3.66	3.62	3.59	3.53	3.48	3.43	3.40
3.90	3.84	3.78	3.73	3.68	3.64	3.60	3.57	3.51	3.46	3.42	3.38
3.89	3.83	3.77	3.72	3.67	3.63	3.59	3.56	3.50	3.45	3.41	3.37
3.88	3.81	3.75	3.70	3.65	3.61	3.58	3.54	3.48	3.43	3.39	3.36
3.86	3.80	3.74	3.69	3.64	3.60	3.56	3.53	3.47	3.42	3.38	3.35
3.86	3.79	3.73	3.68	3.64	3.60	3.56	3.53	3.47	3.42	3.38	3.34
3.85	3.78	3.72	3.68	3.63	3.59	3.55	3.52	3.46	3.41	3.37	3.33
3.84	3.78	3.72	3.67	3.62	3.59	3.55	3.52	3.46	3.41	3.37	3.33
3.84	3.77	3.71	3.66	3.62	3.58	3.54	3.51	3.45	3.40	3.36	3.32
3.83	3.77	3.71	3.66	3.61	3.57	3.53	3.50	3.44	3.39	3.35	3.32
3.83	3.77	3.71	3.66	3.61	3.57	3.53	3.50	3.44	3.39	3.35	3.32
3.83	3.76	3.70	3.65	3.61	3.57	3.53	3.50	3.44	3.39	3.35	3.31

Appendix Table 9 (Continued)

n_e \ f_2	30	35	40	45	50	55	60	65	70	75	80	85
0.5	4.77	4.66	4.59	4.53	4.48	4.44	4.41	4.38	4.35	4.33	4.31	4.29
1.0	4.24	4.15	4.08	4.03	3.99	3.95	3.92	3.89	3.87	3.85	3.83	3.81
1.5	4.00	3.92	3.86	3.81	3.77	3.73	3.71	3.68	3.66	3.64	3.62	3.60
2.0	3.87	3.79	3.73	3.68	3.64	3.61	3.58	3.55	3.53	3.51	3.49	3.48
3.0	3.71	3.63	3.57	3.53	3.49	3.46	3.43	3.41	3.38	3.37	3.35	3.34
4.0	3.62	3.54	3.49	3.44	3.41	3.37	3.35	3.32	3.30	3.29	3.27	3.26
5.0	3.56	3.49	3.43	3.39	3.35	3.32	3.30	3.27	3.25	3.23	3.22	3.20
6.0	3.55	3.48	3.43	3.38	3.35	3.31	3.29	3.26	3.25	3.23	3.21	3.20
7.0	3.52	3.44	3.39	3.34	3.31	3.28	3.25	3.23	3.21	3.19	3.18	3.16
8.0	3.49	3.41	3.36	3.32	3.28	3.25	3.23	3.20	3.19	3.17	3.15	3.14
9.0	3.47	3.39	3.34	3.30	3.26	3.23	3.21	3.18	3.16	3.15	3.13	3.12
10.0	3.45	3.37	3.32	3.28	3.24	3.21	3.19	3.17	3.15	3.13	3.11	3.10
12.0	3.42	3.35	3.30	3.25	3.22	3.19	3.17	3.14	3.12	3.11	3.09	3.08
14.0	3.40	3.33	3.28	3.23	3.20	3.17	3.15	3.12	3.10	3.09	3.07	3.06
16.0	3.38	3.31	3.26	3.22	3.19	3.16	3.13	3.11	3.09	3.07	3.06	3.04
18.0	3.37	3.30	3.25	3.21	3.18	3.15	3.12	3.10	3.08	3.06	3.05	3.04
20.0	3.36	3.29	3.24	3.20	3.17	3.14	3.11	3.09	3.07	3.06	3.04	3.03
25.0	3.35	3.28	3.23	3.18	3.15	3.12	3.10	3.07	3.06	3.04	3.02	3.01
30.0	3.34	3.27	3.22	3.18	3.14	3.11	3.09	3.07	3.05	3.03	3.02	3.00
40.0	3.33	3.25	3.20	3.16	3.13	3.10	3.08	3.05	3.04	3.02	3.00	2.99
50.0	3.32	3.24	3.20	3.15	3.12	3.09	3.07	3.04	3.03	3.01	3.00	2.98
60.0	3.31	3.24	3.19	3.15	3.11	3.08	3.06	3.04	3.02	3.00	2.99	2.98
80.0	3.30	3.23	3.18	3.14	3.11	3.08	3.06	3.03	3.01	3.00	2.98	2.97
100.0	3.30	3.23	3.18	3.14	3.10	3.08	3.05	3.03	3.01	3.00	2.98	2.97
200.0	3.29	3.22	3.17	3.13	3.10	3.07	3.05	3.02	3.01	2.99	2.97	2.96
500.0	3.29	3.22	3.17	3.13	3.09	3.06	3.04	3.02	3.00	2.98	2.97	2.96
1000.0	3.29	3.22	3.17	3.13	3.09	3.06	3.04	3.02	3.00	2.98	2.97	2.96
∞	3.28	3.21	3.16	3.12	3.09	3.06	3.04	3.01	3.00	2.98	2.97	2.95

Appendix Table 9 (Continued)

90	95	100	200	300	400	500	600	700	800	900	1000	∞
4.27	4.25	4.24	4.08	4.01	3.97	3.95	3.93	3.91	3.90	3.89	3.88	3.74
3.79	3.78	3.77	3.63	3.57	3.53	3.51	3.49	3.48	3.47	3.46	3.45	3.33
3.59	3.57	3.56	3.43	3.37	3.34	3.32	3.30	3.29	3.28	3.27	3.26	3.14
3.46	3.45	3.44	3.31	3.26	3.22	3.20	3.19	3.18	3.17	3.16	3.15	3.04
3.32	3.31	3.30	3.18	3.12	3.09	3.07	3.06	3.04	3.04	3.03	3.02	2.91
3.24	3.23	3.22	3.10	3.05	3.02	3.00	2.98	2.97	2.96	2.96	2.95	2.84
3.19	3.18	3.17	3.05	3.00	2.97	2.95	2.94	2.92	2.92	2.91	2.90	2.80
3.18	3.17	3.16	3.04	2.99	2.96	2.94	2.93	2.92	2.91	2.91	2.90	2.79
3.15	3.14	3.13	3.01	2.96	2.93	2.91	2.90	2.89	2.88	2.87	2.86	2.76
3.12	3.11	3.10	2.99	2.94	2.91	2.89	2.88	2.86	2.86	2.85	2.84	2.74
3.10	3.09	3.08	2.97	2.92	2.89	2.87	2.86	2.85	2.84	2.83	2.82	2.72
3.09	3.08	3.07	2.95	2.90	2.87	2.85	2.84	2.83	2.82	2.82	2.81	2.71
3.06	3.05	3.04	2.93	2.88	2.85	2.83	2.82	2.81	2.80	2.80	2.79	2.68
3.05	3.03	3.02	2.91	2.86	2.83	2.82	2.80	2.79	2.78	2.78	2.77	2.67
3.03	3.02	3.01	2.90	2.85	2.82	2.80	2.79	2.78	2.77	2.77	2.76	2.66
3.02	3.01	3.00	2.89	2.84	2.81	2.79	2.78	2.77	2.76	2.76	2.75	2.65
3.01	3.00	2.99	2.88	2.83	2.80	2.79	2.77	2.76	2.75	2.75	2.74	2.64
3.00	2.99	2.98	2.87	2.82	2.79	2.77	2.76	2.75	2.74	2.74	2.73	2.63
2.99	2.98	2.97	2.86	2.81	2.78	2.76	2.75	2.74	2.73	2.73	2.72	2.62
2.98	2.97	2.96	2.85	2.80	2.77	2.75	2.74	2.73	2.72	2.72	2.71	2.61
2.97	2.96	2.95	2.84	2.79	2.76	2.75	2.73	2.72	2.71	2.71	2.70	2.60
2.96	2.95	2.94	2.83	2.79	2.76	2.74	2.73	2.72	2.71	2.70	2.70	2.60
2.96	2.95	2.94	2.83	2.78	2.75	2.73	2.72	2.71	2.70	2.70	2.69	2.59
2.95	2.94	2.93	2.82	2.78	2.75	2.73	2.72	2.71	2.70	2.70	2.69	2.59
2.95	2.94	2.93	2.82	2.77	2.74	2.73	2.71	2.70	2.70	2.69	2.68	2.58
2.94	2.93	2.92	2.81	2.77	2.74	2.72	2.71	2.70	2.69	2.69	2.68	2.58
2.94	2.93	2.92	2.81	2.77	2.74	2.72	2.71	2.70	2.69	2.69	2.68	2.58
2.94	2.93	2.92	2.81	2.76	2.74	2.72	2.71	2.69	2.69	2.68	2.67	2.58

Appendix Table 9 (Continued)

(2) Coefficient k for One-Sided Tolerance Limit (Refer to Section 37.1)

n_e \ f_2	4	5	6	7	8	9	10	12	15	20	30	40	80	∞
0.5	7.05	6.38	6.01	5.78	5.61	5.49	5.39	5.26	5.12	4.98	4.88	4.83	4.74	4.65
0.6	6.84	6.19	5.81	5.57	5.42	5.28	5.18	5.05	4.92	4.80	4.67	4.62	4.54	4.45
0.7	6.69	6.04	5.67	5.43	5.28	5.14	5.04	4.91	4.78	4.64	4.53	4.47	4.38	4.29
0.8	6.58	5.92	5.54	5.32	5.15	5.02	4.91	4.78	4.65	4.51	4.40	4.35	4.25	4.17
0.9	6.49	5.82	5.44	5.21	5.04	4.91	4.81	4.67	4.54	4.40	4.30	4.24	4.15	4.07
1.0	6.41	5.75	5.36	5.12	4.96	4.83	4.73	4.59	4.46	4.32	4.21	4.15	4.06	3.97
1.1	6.35	5.69	5.30	5.06	4.90	4.77	4.67	4.53	4.40	4.26	4.16	4.09	4.00	3.89
1.2	6.29	5.63	5.24	5.00	4.84	4.71	4.61	4.47	4.33	4.19	4.10	4.03	3.94	3.83
1.3	6.25	5.58	5.19	4.95	4.79	4.65	4.55	4.41	4.27	4.13	4.04	3.98	3.88	3.77
1.4	6.21	5.53	5.14	4.90	4.74	4.60	4.50	4.36	4.22	4.08	3.99	3.92	3.83	3.72
1.5	6.17	5.49	5.10	4.87	4.70	4.56	4.46	4.31	4.17	4.04	3.93	3.87	3.77	3.67
1.6	6.14	5.46	5.06	4.83	4.66	4.52	4.42	4.27	4.12	4.00	3.88	3.82	3.73	3.63
1.7	6.10	5.43	5.03	4.80	4.63	4.49	4.39	4.23	4.08	3.96	3.84	3.78	3.69	3.59
1.8	6.08	5.41	5.00	4.77	4.60	4.46	4.35	4.20	4.05	3.92	3.80	3.74	3.66	3.55
1.9	6.05	5.40	4.97	4.75	4.57	4.43	4.32	4.17	4.03	3.88	3.77	3.72	3.62	3.52
2.0	6.03	5.39	4.95	4.73	4.55	4.41	4.30	4.15	4.01	3.86	3.75	3.70	3.59	3.49
2.2	6.00	5.34	4.91	4.67	4.50	4.37	4.26	4.11	3.96	3.81	3.70	3.65	3.55	3.44
2.4	5.97	5.30	4.88	4.63	4.46	4.33	4.22	4.07	3.92	3.77	3.66	3.60	3.50	3.39
2.6	5.94	5.26	4.84	4.59	4.42	4.29	4.18	4.03	3.88	3.73	3.62	3.56	3.46	3.35
2.8	5.92	5.23	4.82	4.56	4.39	4.26	4.15	4.00	3.85	3.70	3.58	3.52	3.42	3.31
3.0	5.89	5.20	4.79	4.53	4.37	4.23	4.12	3.97	3.82	3.67	3.54	3.48	3.37	3.27
3.2	5.87	5.18	4.77	4.51	4.34	4.21	4.10	3.95	3.80	3.64	3.51	3.45	3.36	3.24
3.4	5.85	5.16	4.75	4.49	4.32	4.19	4.07	3.93	3.78	3.62	3.48	3.42	3.32	3.21
3.6	5.83	5.14	4.74	4.47	4.30	4.17	4.05	3.91	3.76	3.60	3.45	3.39	3.30	3.18
3.8	5.82	5.13	4.72	4.45	4.29	4.14	4.04	3.89	3.74	3.58	3.43	3.37	3.28	3.15
4.0	5.81	5.12	4.71	4.44	4.27	4.13	4.02	3.87	3.72	3.56	3.41	3.35	3.24	3.12
4.5	5.79	5.09	4.68	4.42	4.25	4.10	3.98	3.84	3.69	3.53	3.38	3.32	3.21	3.09
5.0	5.77	5.07	4.66	4.40	4.23	4.07	3.95	3.81	3.66	3.50	3.36	3.29	3.18	3.06

Appendix Table 9 (Continued)

n_e \ f_2	4	5	6	7	8	9	10	12	15	20	30	40	80	∞
5.5	5.75	5.05	4.64	4.38	4.21	4.05	3.94	3.79	3.64	3.48	3.33	3.26	3.18	3.03
6.0	5.74	5.04	4.63	4.36	4.19	4.03	3.93	3.78	3.62	3.45	3.31	3.24	3.12	3.00
6.5	5.73	5.03	4.62	4.35	4.17	4.01	3.91	3.76	3.60	3.43	3.29	3.22	3.10	2.98
7.0	5.72	5.02	4.60	4.34	4.16	4.00	3.90	3.75	3.59	3.41	3.27	3.20	3.09	2.96
7.5	5.71	5.01	4.58	4.33	4.14	3.99	3.89	3.74	3.58	3.40	3.26	3.19	3.07	2.94
8.0	5.70	5.00	4.57	4.32	4.13	3.98	3.88	3.73	3.57	3.38	3.24	3.17	3.04	2.92
8.5	5.69	4.99	4.56	4.31	4.12	3.98	3.87	3.71	3.55	3.37	3.22	3.15	3.03	2.90
9.0	5.68	4.98	4.55	4.30	4.12	3.97	3.86	3.70	3.53	3.35	3.20	3.14	3.01	2.88
9.5	5.67	4.97	4.54	4.29	4.11	3.96	3.85	3.69	3.52	3.34	3.19	3.13	2.99	2.86
10.0	5.67	4.97	4.54	4.29	4.11	3.96	3.84	3.68	3.51	3.33	3.18	3.12	2.98	2.85
11.0	5.67	4.96	4.53	4.28	4.10	3.95	3.83	3.67	3.50	3.32	3.17	3.10	2.97	2.83
12.0	5.66	4.96	4.53	4.27	4.10	3.94	3.83	3.66	3.49	3.31	3.16	3.09	2.96	2.81
13.0	5.66	4.95	4.52	4.26	4.09	3.93	3.82	3.66	3.48	3.30	3.16	3.08	2.95	2.79
14.0	5.65	4.95	4.52	4.25	4.08	3.93	3.81	3.65	3.48	3.29	3.15	3.07	2.93	2.77
15.0	5.65	4.94	4.51	2.24	4.08	3.92	3.80	3.64	3.47	3.28	3.14	3.06	2.92	2.76
16.0	5.64	4.94	4.51	4.24	4.07	3.91	3.80	3.64	3.46	3.27	3.13	3.05	2.91	2.74
17.0	5.64	4.93	4.50	4.23	4.07	3.91	3.79	3.63	3.45	3.26	3.12	3.04	2.90	2.73
18.0	5.63	4.93	4.50	4.23	4.06	3.90	3.78	3.62	3.45	3.25	3.11	3.02	2.89	2.72
19.0	5.63	4.92	4.49	4.22	4.06	3.90	3.78	3.62	3.44	3.24	3.10	3.01	2.87	2.70
20.0	5.63	4.92	4.49	4.22	4.05	3.89	3.77	3.61	3.43	3.24	3.09	3.00	2.86	2.69
30.0	5.60	4.88	4.46	4.18	4.01	3.85	3.72	3.56	3.40	3.20	3.04	2.96	2.80	2.61
40.0	5.60	4.88	4.46	4.18	4.01	3.85	3.72	3.56	3.37	3.18	3.01	2.93	2.76	2.59
50.0	5.60	4.88	4.46	4.18	4.00	3.85	3.72	3.55	3.36	3.18	3.01	2.93	2.75	2.56
60.0	5.59	4.88	4.45	4.18	4.00	3.84	3.71	3.54	3.36	3.18	3.01	2.93	2.75	2.54
70.0	5.59	4.88	4.45	4.18	4.00	3.84	3.71	3.54	3.36	3.17	2.99	2.90	2.75	2.52
80.0	5.59	4.88	4.45	4.17	3.99	3.84	3.70	3.54	3.35	3.17	2.99	2.90	2.74	2.51
90.0	5.59	4.88	4.45	4.17	3.99	3.83	3.70	3.53	3.35	3.17	2.99	2.90	2.74	2.50
100.0	5.59	4.88	4.45	4.17	3.99	3.83	3.70	3.53	3.35	3.17	2.99	2.90	2.74	2.49
∞	5.57	4.78	4.42	4.13	3.96	3.81	3.68	3.51	3.32	3.12	2.96	2.86	2.67	2.32

Appendix Table 10 Nomograms for Finding Confidence Limits of Fraction Defective by Measured Data

(Refer to Section 37.2)
Sequence of Use of Nomograms

1. Find the estimated value $\hat{\mu}$ of the population mean μ, and its effective number of replications n_e.
2. Find the error variance V_e and its number of degrees of freedom f_2.
3. Find the following k value.

$$k = \frac{\text{upper limit of specification} - \hat{\mu}}{\sqrt{V_e}}$$

$$\text{or } k = \frac{\hat{\mu} - \text{lower limit of specification}}{\sqrt{V_e}}$$

4. Find the following y value.

$$y = \frac{1}{n_e} + \frac{k^2}{2f_2}$$

5. Take the y value on the y scale, then on the reliability scale, take the desired reliability on the appropriate side according to
   i. whether one wishes to find the one-sided confidence limit, or in other words the one-sided limit for the fraction defective being at most a certain percentage, or
   ii. whether one wishes the two-sided confidence limits, or in other words the two-sided limits for the fraction defective being at least and at most certain percentages;

   then connect the reliability with y by a straight line and find the intersection A with scale A.
6. Connect intersection A and point k on the k scale, and for a one-sided confidence limit, if the intersection with the p_1 scale (or p_0 scale) is read off, the fraction defective is at most p_1% (at least p_0%) at the chosen reliability.

In the case of two-sided confidence limits, if one finds the intersections with both the p_0 and p_1 scales, the fraction defective is at least p_0% and at most p_1% at that reliability.

Nomograms

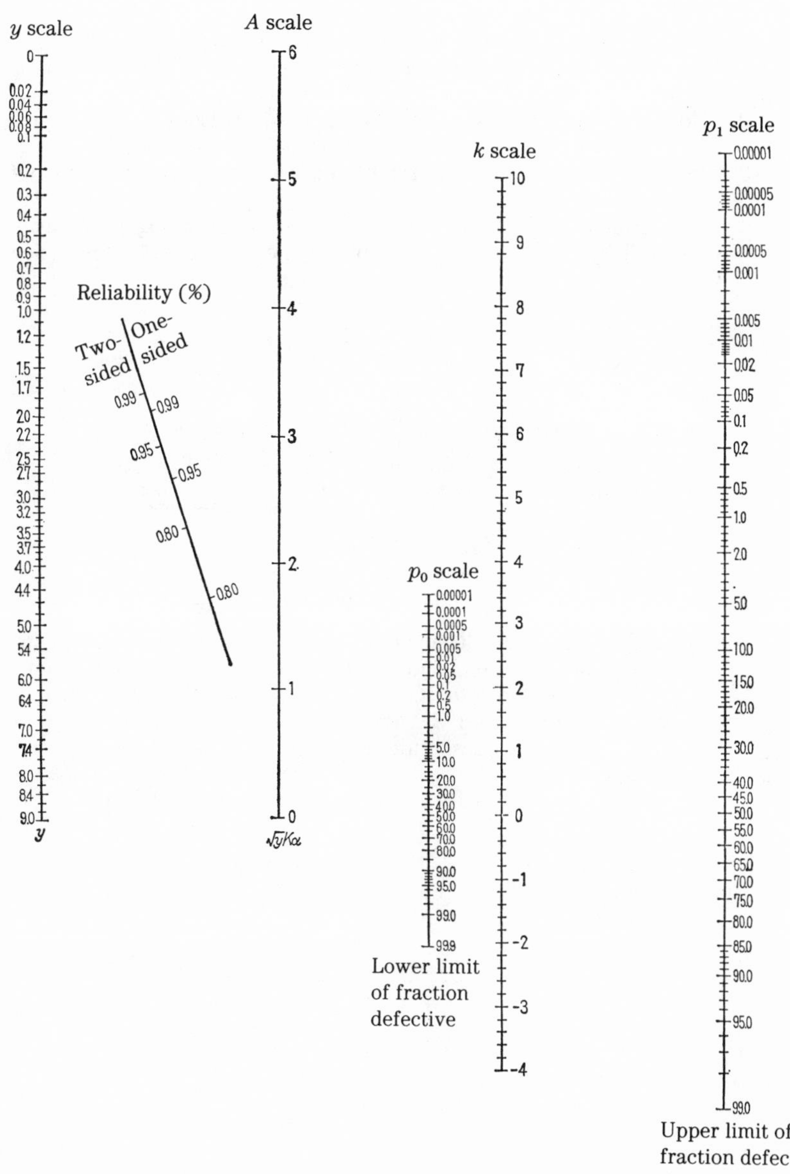

y scale
A scale
k scale
p_0 scale
p_1 scale
Reliability (%)
Two-sided
One-sided
Lower limit of fraction defective
Upper limit of fraction defective

Appendix Table 11 Orthogonal Arrays and Linear Graphs

Explanation

1. Every standard type of orthogonal array and its linear graphs have been collected in order to facilitate assignment of diverse experiments.
2. "No." represents the experiment number, and "Column No." represents the number of the column within the orthogonal array.
3. The table of interactions is for finding two-factor interactions between two columns.
4. The classification by groups of the columns of the orthogonal array is indicated by the following symbols in the assignment types. (See Chapter 9.)

$L_{32}(2^{31})$		$L_{64}(2^{31})$		Other arrays	
Symbol	Group	Symbol	Group	Symbol	Group
○	Gp 1 and Gp 2	○	Gp 1, Gp 2, and Gp 3	○	Group 1
◎	Group 3	◎	Group 4	◎	Group 2
⊙	Group 4	⊙	Group 5	⊙	Group 3
●	Group 5	●	Group 6	●	Group 4

5. $L_{50}(2 \times 5^{11})$ and $L'_{72}(3^{22})$ were taken from the following reference sources. The author extends his appreciation.

$L_{50}(2 \times 5^{11})$ Masuyama, Genzaburō. 1957. On difference sets for constructing orthogonal arrays of index two and strength two, *Rep. Statist. Appl. Res. Un. Jap. Sci. Eng.* 5:27–34.

$L'_{72}(3^{22})$ Ina, Masao. 1957. Aru Kakuritsu Taiohyō (Certain probability correspondence tables), *Hinshitsu Kanri Tokubetsu Kenkyūkai Shiryō* (Data of Special Research Society for Quality Control), Chūbu Sangyō Renmei.

2^n System

$L_4(2^3)$

No. \ Col. no.	1	2	3
1	1	1	1
2	1	2	2
3	2	1	2
4	2	2	1
	Gp 1	Group 2	

Linear Graph of L_4 (1)

1 ○——3——◎ 2

$L_8(2^7)$

No. \ Col. no.	1	2	3	4	5	6	7
1	1	1	1	1	1	1	1
2	1	1	1	2	2	2	2
3	1	2	2	1	1	2	2
4	1	2	2	2	2	1	1
5	2	1	2	1	2	1	2
6	2	1	2	2	1	2	1
7	2	2	1	1	2	2	1
8	2	2	1	2	1	1	2
	Gp 1	Gp 2		Group 3			

Interaction Between Two Columns

Col. \ Col. no.	1	2	3	4	5	6	7
	(1)	3	2	5	4	7	6
		(2)	1	6	7	4	5
			(3)	7	6	5	4
				(4)	1	2	3
					(5)	3	2
						(6)	1
							(7)

Linear Graphs of L_8

(1)

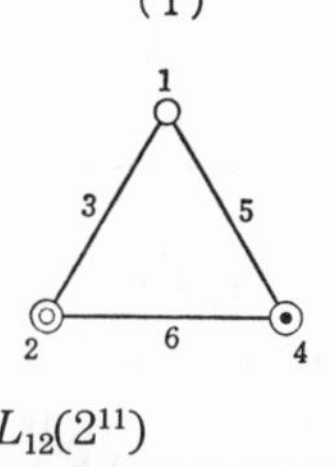

(2)

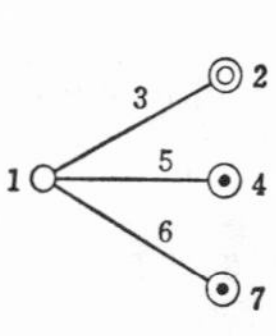

$L_{12}(2^{11})$

No. \ Col. no.	1	2	3	4	5	6	7	8	9	10	11
1	1	1	1	1	1	1	1	1	1	1	1
2	1	1	1	1	1	2	2	2	2	2	2
3	1	1	2	2	2	1	1	1	2	2	2
4	1	2	1	2	2	1	2	2	1	1	2
5	1	2	2	1	2	2	1	2	1	2	1
6	1	2	2	2	1	2	2	1	2	1	1
7	2	1	2	2	1	1	2	2	1	2	1
8	2	1	2	1	2	2	2	1	1	1	2
9	2	1	1	2	2	2	1	2	2	1	1
10	2	2	2	1	1	1	1	2	2	1	2
11	2	2	1	2	1	2	1	1	1	2	2
12	2	2	1	1	2	1	2	1	2	2	1
	Gp 1					Group 2					

(Note) The components of interaction of any two columns become confounded a little with each of the remaining nine columns. The sequential analysis method becomes necessary if one wishes to find the interactions. Therefore, do not use this array for experiments where interactions are necessary.

$L_{16}(2^{15})$

No. \ Col. no.	1	2	3	4	5	6	7	8	9	10	11	12	13	14	15
1	1	1	1	1	1	1	1	1	1	1	1	1	1	1	1
2	1	1	1	1	1	1	1	2	2	2	2	2	2	2	2
3	1	1	1	2	2	2	2	1	1	1	1	2	2	2	2
4	1	1	1	2	2	2	2	2	2	2	2	1	1	1	1
5	1	2	2	1	1	2	2	1	1	2	2	1	1	2	2
6	1	2	2	1	1	2	2	2	2	1	1	2	2	1	1
7	1	2	2	2	2	1	1	1	1	2	2	2	2	1	1
8	1	2	2	2	2	1	1	2	2	1	1	1	1	2	2
9	2	1	2	1	2	1	2	1	2	1	2	1	2	1	2
10	2	1	2	1	2	1	2	2	1	2	1	2	1	2	1
11	2	1	2	2	1	2	1	1	2	1	2	2	1	2	1
12	2	1	2	2	1	2	1	2	1	2	1	1	2	1	2
13	2	2	1	1	2	2	1	1	2	2	1	1	2	2	1
14	2	2	1	1	2	2	1	2	1	1	2	2	1	1	2
15	2	2	1	2	1	1	2	1	2	2	1	2	1	1	2
16	2	2	1	2	1	1	2	2	1	1	2	1	2	2	1
	Gp 1	Gp 2		Group 3				Group 4							

Table of Interactions Between Two Columns

Col. \ Col.	1	2	3	4	5	6	7	8	9	10	11	12	13	14	15
	(1)	3	2	5	4	7	6	9	8	11	10	13	12	15	14
		(2)	1	6	7	4	5	10	11	8	9	14	15	12	13
			(3)	7	6	5	4	11	10	9	8	15	14	13	12
				(4)	1	2	3	12	13	14	15	8	9	10	11
					(5)	3	2	13	12	15	14	9	8	11	10
						(6)	1	14	15	12	13	10	11	8	9
							(7)	15	14	13	12	11	10	9	8
								(8)	1	2	3	4	5	6	7
									(9)	3	2	5	4	7	6
										(10)	1	6	7	4	5
											(11)	7	6	5	4
												(12)	1	2	3
													(13)	3	2
														(14)	1

Linear Graphs of L_{16}

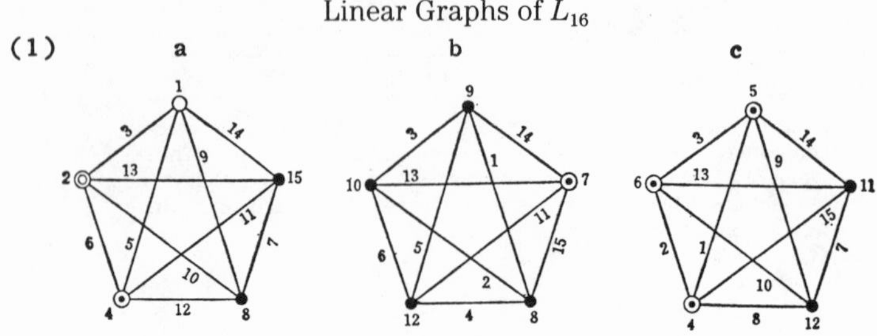

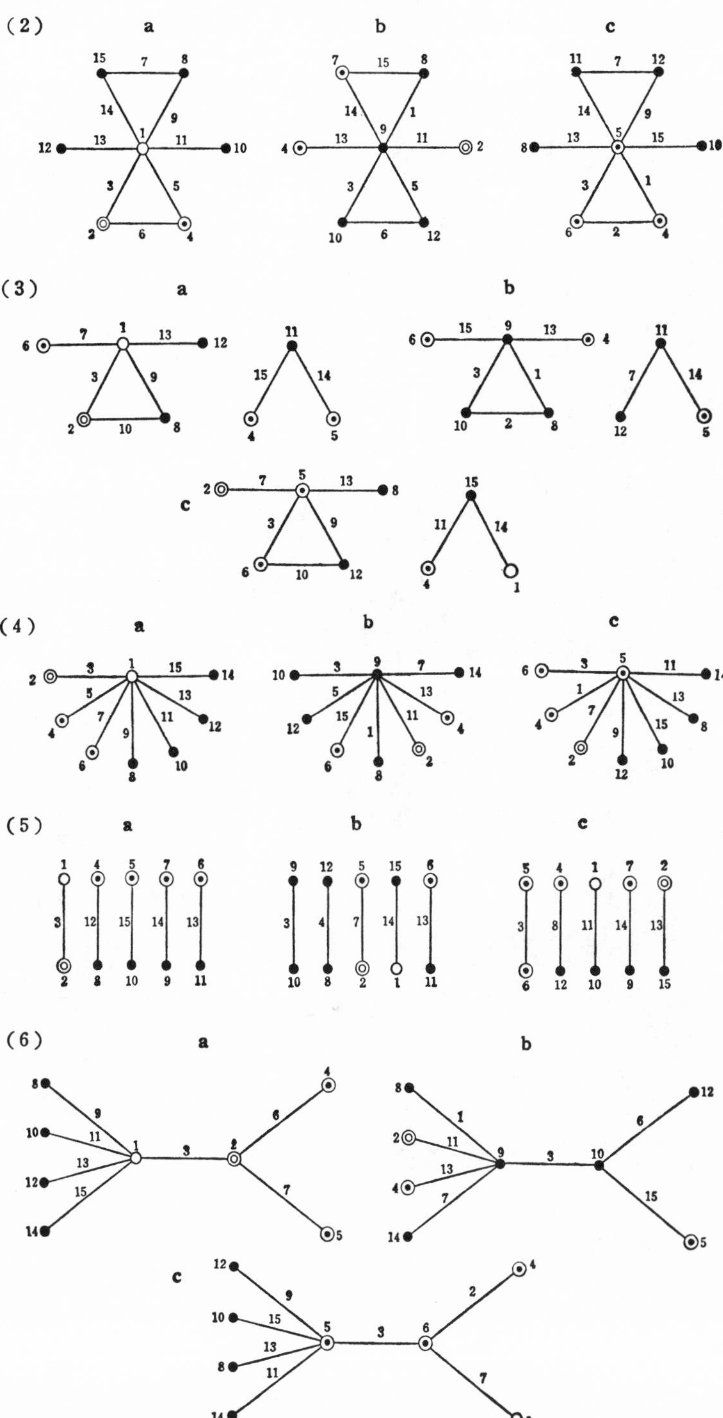
(2) a b c
(3) a b c
(4) a b c
(5) a b c
(6) a b c

$L_{32}(2^{31})$

No. \ Col. no.	1	2	3	4	5	6	7	8	9	10	11	12	13	14	15	16	17	18	19	20	21	22	23	24	25	26	27	28	29	30	31
1	1	1	1	1	1	1	1	1	1	1	1	1	1	1	1	1	1	1	1	1	1	1	1	1	1	1	1	1	1	1	1
2	1	1	1	1	1	1	1	1	1	1	1	1	1	1	1	2	2	2	2	2	2	2	2	2	2	2	2	2	2	2	2
3	1	1	1	1	1	1	1	2	2	2	2	2	2	2	2	1	1	1	1	1	1	1	1	2	2	2	2	2	2	2	2
4	1	1	1	1	1	1	1	2	2	2	2	2	2	2	2	2	2	2	2	2	2	2	2	1	1	1	1	1	1	1	1
5	1	1	1	2	2	2	2	1	1	1	1	2	2	2	2	1	1	1	1	2	2	2	2	1	1	1	1	2	2	2	2
6	1	1	1	2	2	2	2	1	1	1	1	2	2	2	2	2	2	2	2	1	1	1	1	2	2	2	2	1	1	1	1
7	1	1	1	2	2	2	2	2	2	2	2	1	1	1	1	1	1	1	1	2	2	2	2	2	2	2	2	1	1	1	1
8	1	1	1	2	2	2	2	2	2	2	2	1	1	1	1	2	2	2	2	1	1	1	1	1	1	1	1	2	2	2	2
9	1	2	2	1	1	2	2	1	1	2	2	1	1	2	2	1	1	2	2	1	1	2	2	1	1	2	2	1	1	2	2
10	1	2	2	1	1	2	2	1	1	2	2	1	1	2	2	2	2	1	1	2	2	1	1	2	2	1	1	2	2	1	1
11	1	2	2	1	1	2	2	2	2	1	1	2	2	1	1	1	1	2	2	1	1	2	2	2	2	1	1	2	2	1	1
12	1	2	2	1	1	2	2	2	2	1	1	2	2	1	1	2	2	1	1	2	2	1	1	1	1	2	2	1	1	2	2
13	1	2	2	2	2	1	1	1	1	2	2	2	2	1	1	1	1	2	2	2	2	1	1	1	1	2	2	2	2	1	1
14	1	2	2	2	2	1	1	1	1	2	2	2	2	1	1	2	2	1	1	1	1	2	2	2	2	1	1	1	1	2	2
15	1	2	2	2	2	1	1	2	2	1	1	1	1	2	2	1	1	2	2	2	2	1	1	2	2	1	1	1	1	2	2
16	1	2	2	2	2	1	1	2	2	1	1	1	1	2	2	2	2	1	1	1	1	2	2	1	1	2	2	2	2	1	1
17	2	1	2	1	2	1	2	1	2	1	2	1	2	1	2	1	2	1	2	1	2	1	2	1	2	1	2	1	2	1	2
18	2	1	2	1	2	1	2	1	2	1	2	1	2	1	2	2	1	2	1	2	1	2	1	2	1	2	1	2	1	2	1
19	2	1	2	1	2	1	2	2	1	2	1	2	1	2	1	1	2	1	2	1	2	1	2	2	1	2	1	2	1	2	1
20	2	1	2	1	2	1	2	2	1	2	1	2	1	2	1	2	1	2	1	2	1	2	1	1	2	1	2	1	2	1	2
21	2	1	2	2	1	2	1	1	2	1	2	2	1	2	1	1	2	1	2	2	1	2	1	1	2	1	2	2	1	2	1
22	2	1	2	2	1	2	1	1	2	1	2	2	1	2	1	2	1	2	1	1	2	1	2	2	1	2	1	1	2	1	2
23	2	1	2	2	1	2	1	2	1	2	1	1	2	1	2	1	2	1	2	2	1	2	1	2	1	2	1	1	2	1	2
24	2	1	2	2	1	2	1	2	1	2	1	1	2	1	2	2	1	2	1	1	2	1	2	1	2	1	2	2	1	2	1
25	2	2	1	1	2	2	1	1	2	2	1	1	2	2	1	1	2	2	1	1	2	2	1	1	2	2	1	1	2	2	1
26	2	2	1	1	2	2	1	1	2	2	1	1	2	2	1	2	1	1	2	2	1	1	2	2	1	1	2	2	1	1	2
27	2	2	1	1	2	2	1	2	1	1	2	2	1	1	2	1	2	2	1	1	2	2	1	2	1	1	2	2	1	1	2
28	2	2	1	1	2	2	1	2	1	1	2	2	1	1	2	2	1	1	2	2	1	1	2	1	2	2	1	1	2	2	1
29	2	2	1	2	1	1	2	1	2	2	1	2	1	1	2	1	2	2	1	2	1	1	2	1	2	2	1	2	1	1	2
30	2	2	1	2	1	1	2	1	2	2	1	2	1	1	2	2	1	1	2	1	2	2	1	2	1	1	2	1	2	2	1
31	2	2	1	2	1	1	2	2	1	1	2	1	2	2	1	1	2	2	1	2	1	1	2	2	1	1	2	1	2	2	1
32	2	2	1	2	1	1	2	2	1	1	2	1	2	2	1	2	1	1	2	1	2	2	1	1	2	2	1	2	1	1	2
	Group 1	Group 2		Group 3				Group 4								Group 5															

Interactions Between Two Columns

Col. \ Col.	1	2	3	4	5	6	7	8	9	10	11	12	13	14	15	16	17	18	19	20	21	22	23	24	25	26	27	28	29	30	31
	(1)	3	2	5	4	7	6	9	8	11	10	13	12	15	14	17	16	19	18	21	20	23	22	25	24	27	26	29	28	31	30
		(2)	1	6	7	4	5	10	11	8	9	14	15	12	13	18	19	16	17	22	23	20	21	26	27	24	25	30	31	28	29
			(3)	7	6	5	4	11	10	9	8	15	14	13	12	19	18	17	16	23	22	21	20	27	26	25	24	31	30	29	28
				(4)	1	2	3	12	13	14	15	8	9	10	11	20	21	22	23	16	17	18	19	28	29	30	31	24	25	26	27
					(5)	3	2	13	12	15	14	9	8	11	10	21	20	23	22	17	16	19	18	29	28	31	30	25	24	27	26
						(6)	1	14	15	12	13	10	11	8	9	22	23	20	21	18	19	16	17	30	31	28	29	26	27	24	25
							(7)	15	14	13	12	11	10	9	8	23	22	21	20	19	18	17	16	31	30	29	28	27	26	25	24
								(8)	1	2	3	4	5	6	7	24	25	26	27	28	29	30	31	16	17	18	19	20	21	22	23
									(9)	3	2	5	4	7	6	25	24	27	26	29	28	31	30	17	16	19	18	21	20	23	22
										(10)	1	6	7	4	5	26	27	24	25	30	31	28	29	18	19	16	17	22	23	20	21
											(11)	7	6	5	4	27	26	25	24	31	30	29	28	19	18	17	16	23	22	21	20
												(12)	1	2	3	28	29	30	31	24	25	26	27	20	21	22	23	16	17	18	19
													(13)	3	2	29	28	31	30	25	24	27	26	21	20	23	22	17	16	19	18
														(14)	1	30	31	28	29	26	27	24	25	22	23	20	21	18	19	16	17
															(15)	31	30	29	28	27	26	25	24	23	22	21	20	19	18	17	16
																(16)	1	2	3	4	5	6	7	8	9	10	11	12	13	14	15
																	(17)	3	2	5	4	7	6	9	8	11	10	13	12	15	14
																		(18)	1	6	7	4	5	10	11	8	9	14	15	12	13
																			(19)	7	6	5	4	11	10	9	8	15	14	13	12
																				(20)	1	2	3	12	13	14	15	8	9	10	11
																					(21)	3	2	13	12	15	14	9	8	11	10
																						(22)	1	14	15	12	13	10	11	8	9
																							(23)	15	14	13	12	11	10	9	8
																								(24)	1	2	3	4	5	6	7
																									(25)	3	2	5	4	7	6
																										(26)	1	6	7	4	5
																											(27)	7	6	5	4
																												(28)	1	2	3
																													(29)	3	2
																														(30)	1

Linear Graphs of L_{32}

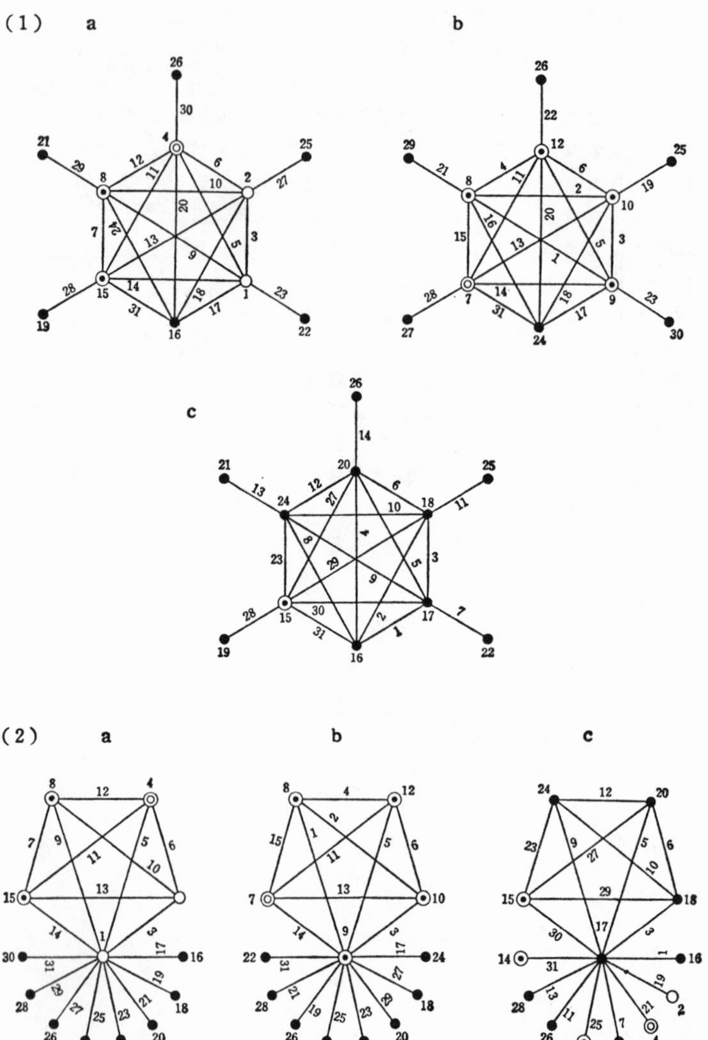

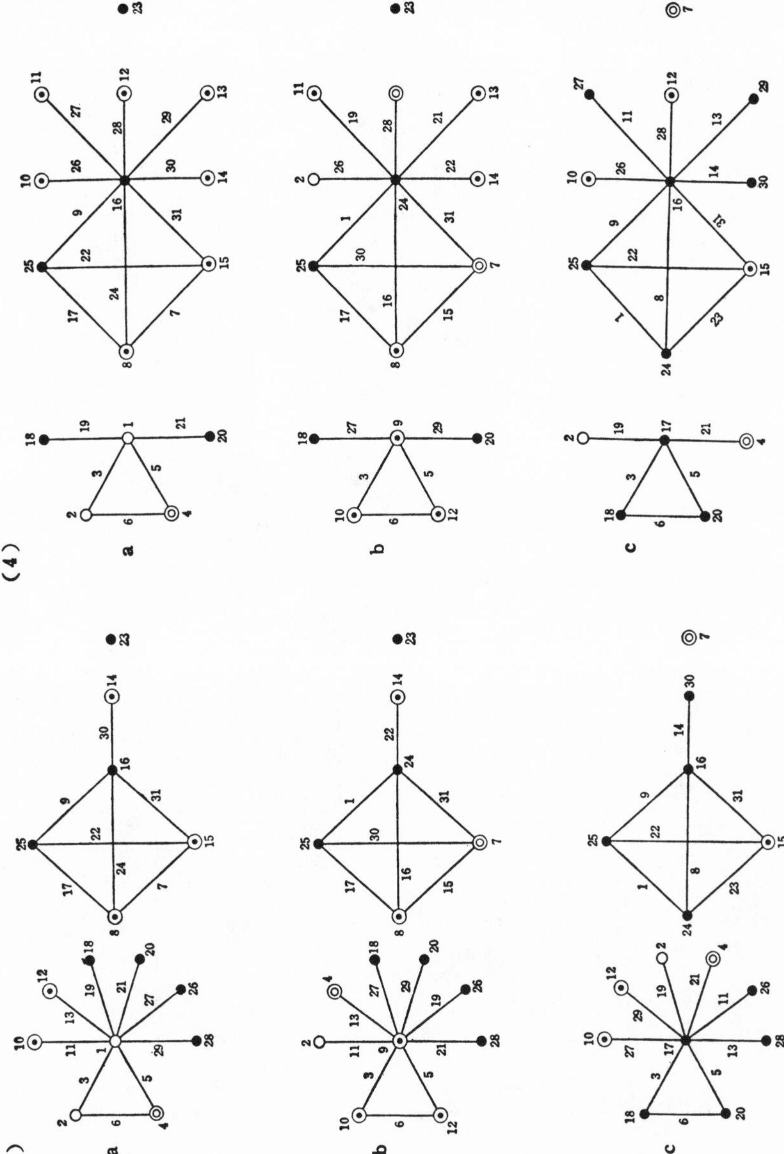

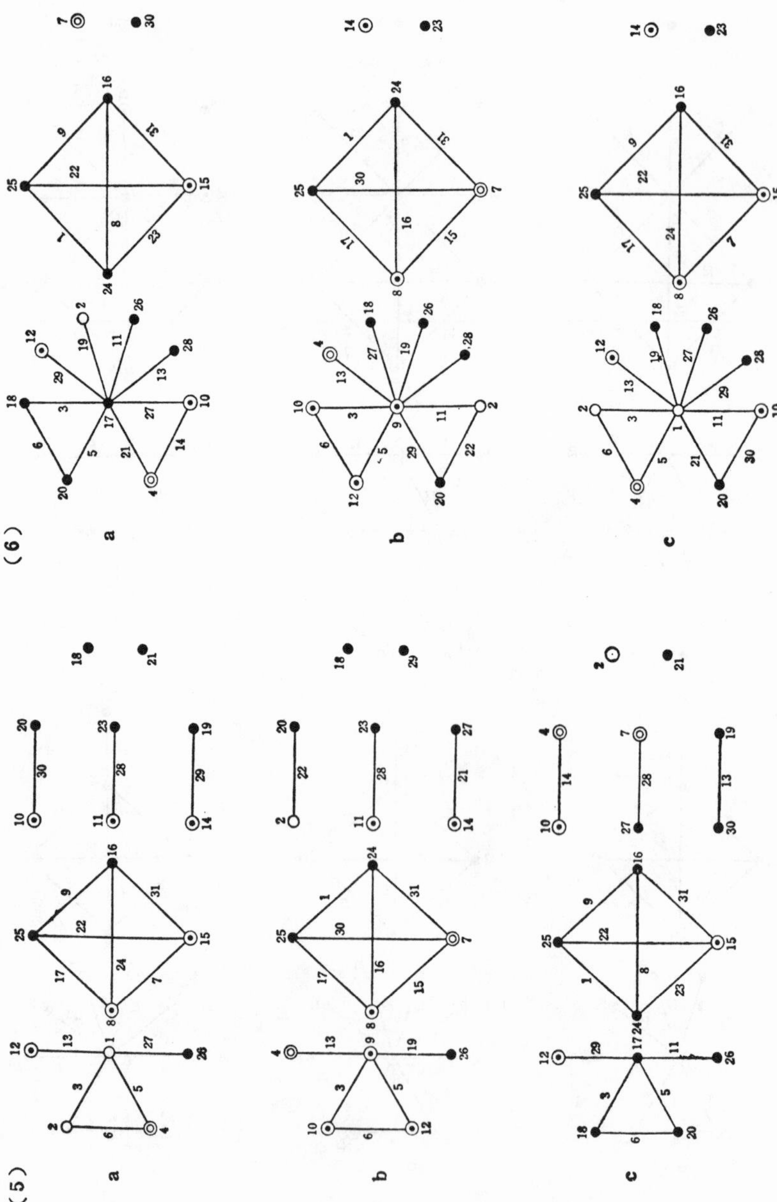
(5)
a
b
c
(6)
a
b
c

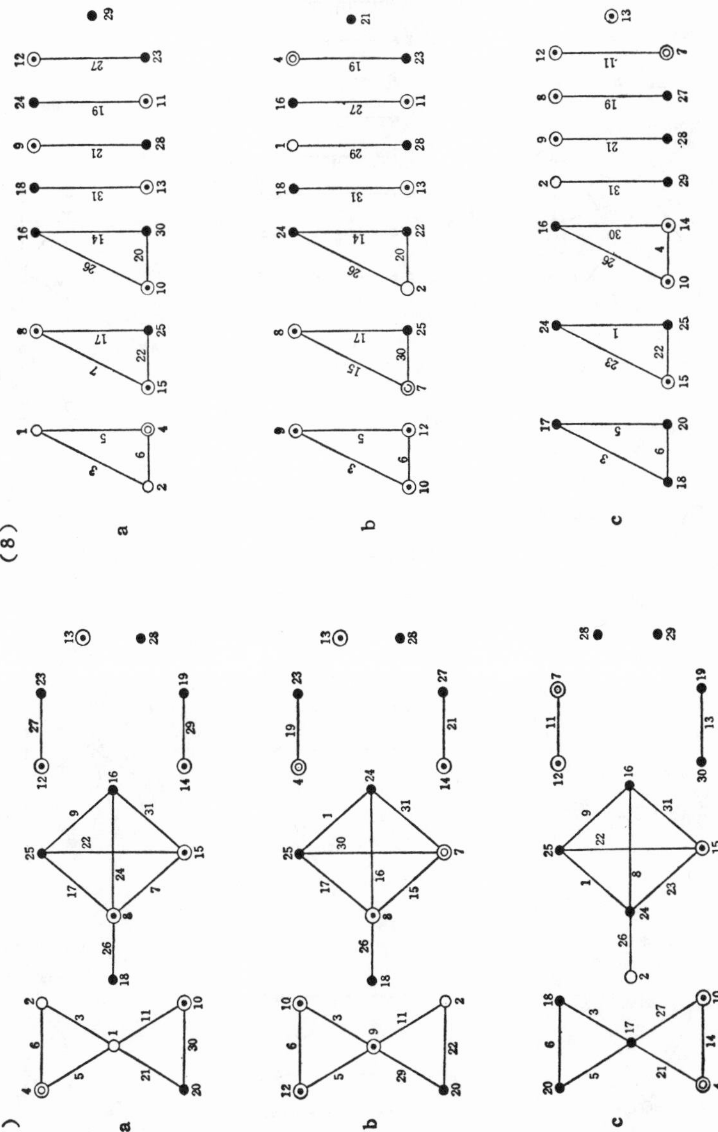

(9) a

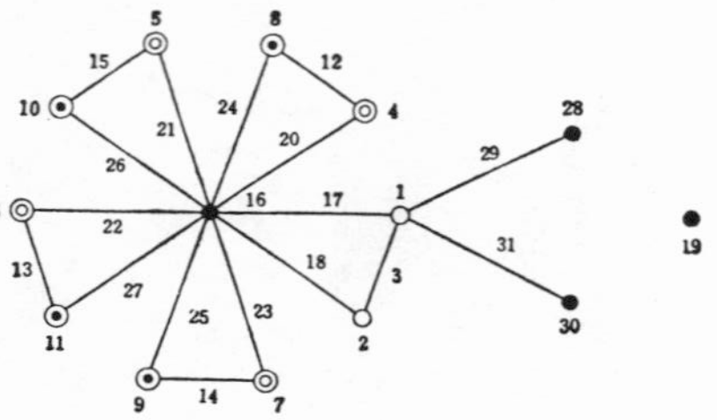

b

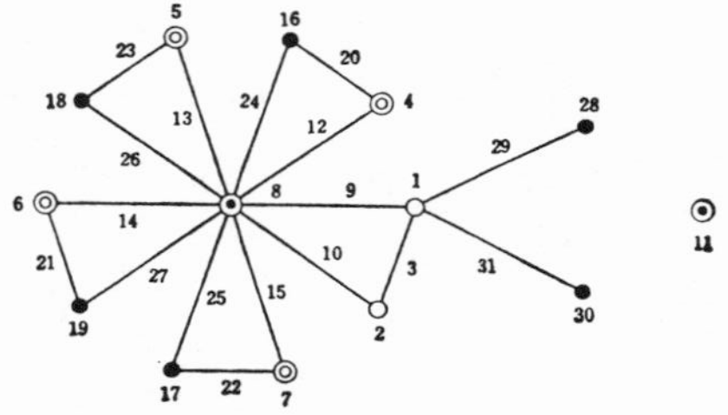

c

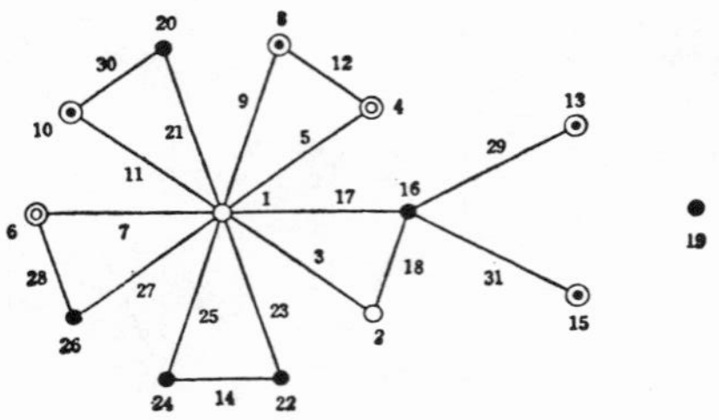

(10) a

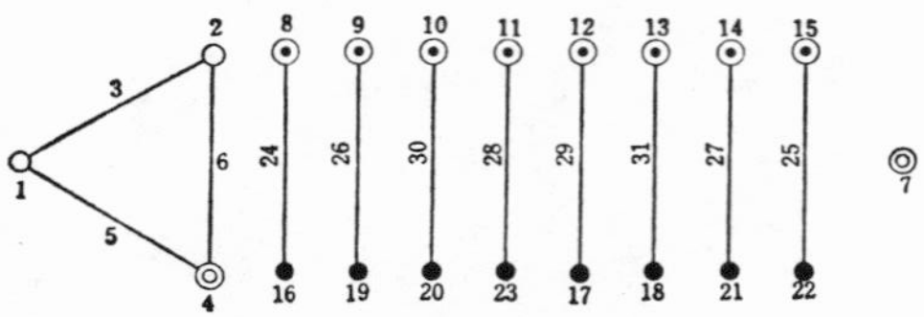

b

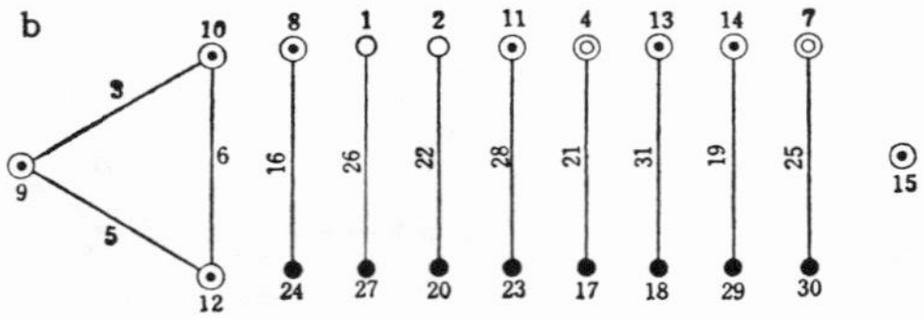

c

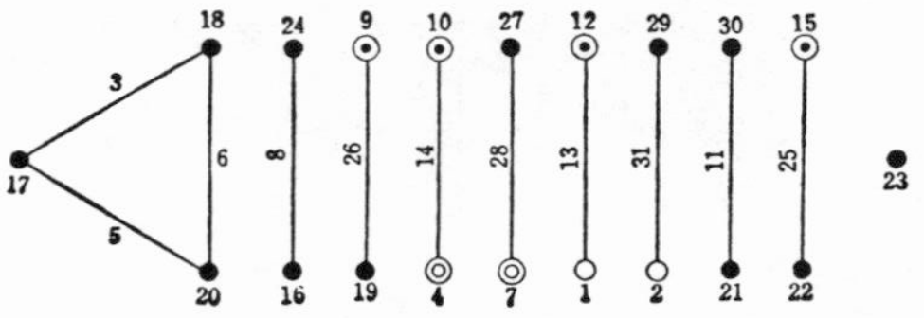

(11) a

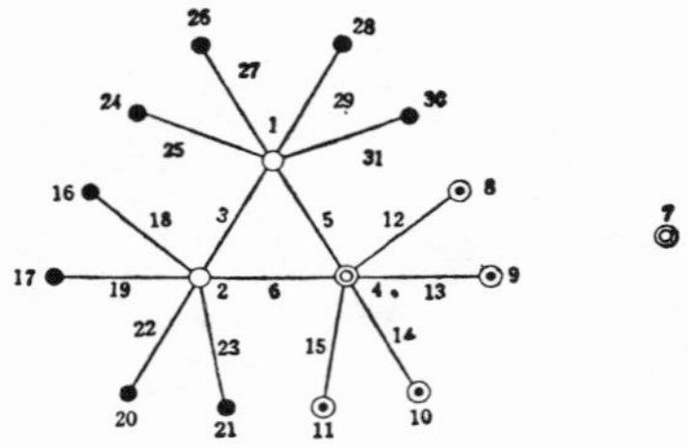

b

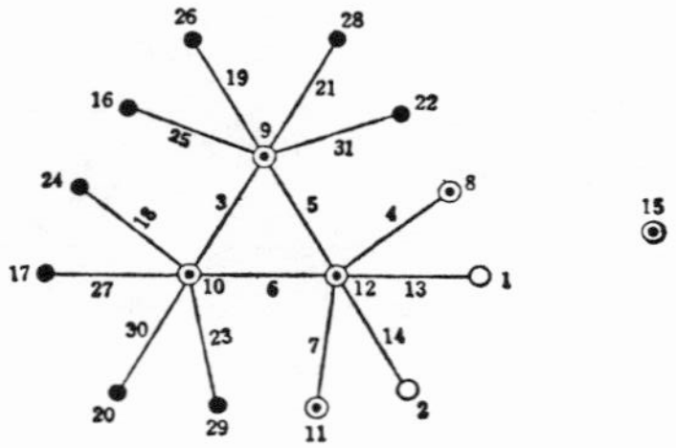

c

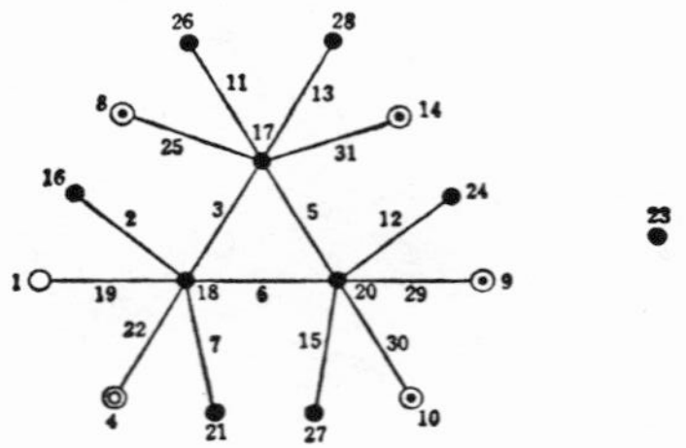

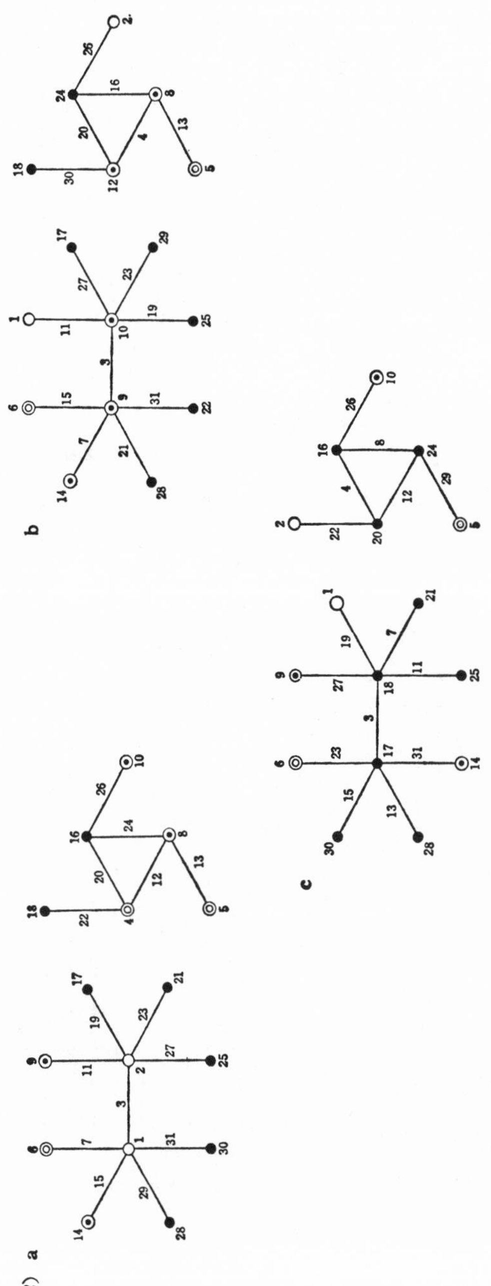
(12) a
b
c

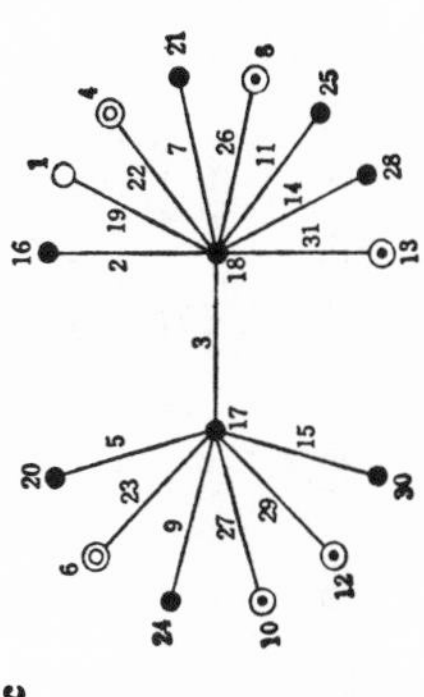
c

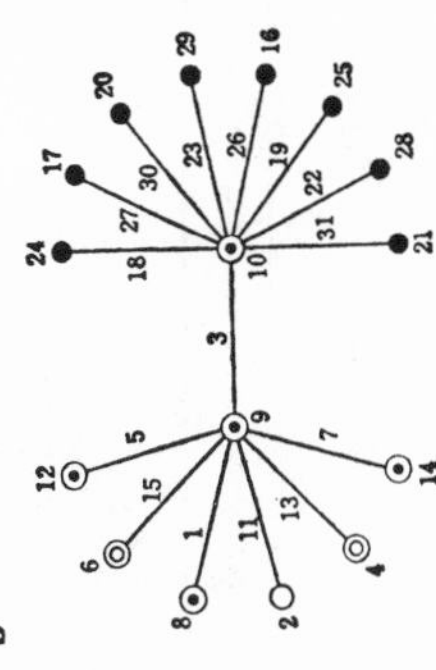
b

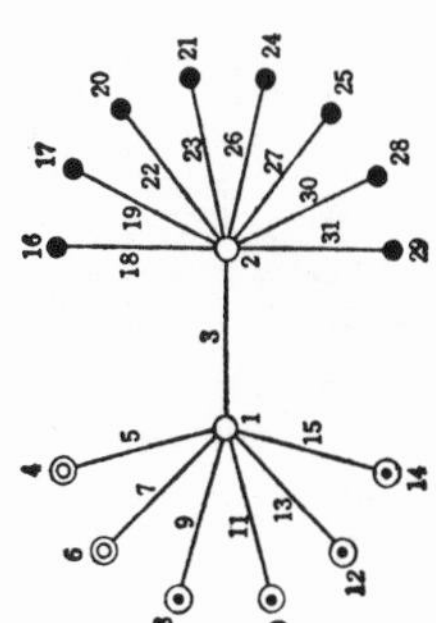
(13) a

$L_{64}(2^{63})$

No. \ Col. no.	1	2	3	4	5	6	7	8	9	10	11	12	13	14	15	16	17	18	19	20	21	22	23	24	25	26	27	28	29	30	31
1	1	1	1	1	1	1	1	1	1	1	1	1	1	1	1	1	1	1	1	1	1	1	1	1	1	1	1	1	1	1	1
2	1	1	1	1	1	1	1	1	1	1	1	1	1	1	1	1	1	1	1	1	1	1	1	1	1	1	1	1	1	1	1
3	1	1	1	1	1	1	1	1	1	1	1	1	1	1	1	2	2	2	2	2	2	2	2	2	2	2	2	2	2	2	2
4	1	1	1	1	1	1	1	1	1	1	1	1	1	1	1	2	2	2	2	2	2	2	2	2	2	2	2	2	2	2	2
5	1	1	1	1	1	1	1	2	2	2	2	2	2	2	2	1	1	1	1	1	1	1	1	2	2	2	2	2	2	2	2
6	1	1	1	1	1	1	1	2	2	2	2	2	2	2	2	1	1	1	1	1	1	1	1	2	2	2	2	2	2	2	2
7	1	1	1	1	1	1	1	2	2	2	2	2	2	2	2	2	2	2	2	2	2	2	2	1	1	1	1	1	1	1	1
8	1	1	1	1	1	1	1	2	2	2	2	2	2	2	2	2	2	2	2	2	2	2	2	1	1	1	1	1	1	1	1
9	1	1	1	2	2	2	2	1	1	1	1	2	2	2	2	1	1	1	1	2	2	2	2	1	1	1	1	2	2	2	2
10	1	1	1	2	2	2	2	1	1	1	1	2	2	2	2	1	1	1	1	2	2	2	2	1	1	1	1	2	2	2	2
11	1	1	1	2	2	2	2	1	1	1	1	2	2	2	2	2	2	2	2	1	1	1	1	2	2	2	2	1	1	1	1
12	1	1	1	2	2	2	2	1	1	1	1	2	2	2	2	2	2	2	2	1	1	1	1	2	2	2	2	1	1	1	1
13	1	1	1	2	2	2	2	2	2	2	2	1	1	1	1	1	1	1	1	2	2	2	2	2	2	2	2	1	1	1	1
14	1	1	1	2	2	2	2	2	2	2	2	1	1	1	1	1	1	1	1	2	2	2	2	2	2	2	2	1	1	1	1
15	1	1	1	2	2	2	2	2	2	2	2	1	1	1	1	2	2	2	2	1	1	1	1	1	1	1	1	2	2	2	2
16	1	1	1	2	2	2	2	2	2	2	2	1	1	1	1	2	2	2	2	1	1	1	1	1	1	1	1	2	2	2	2
17	1	2	2	1	1	2	2	1	1	2	2	1	1	2	2	1	1	2	2	1	1	2	2	1	1	2	2	1	1	2	2
18	1	2	2	1	1	2	2	1	1	2	2	1	1	2	2	1	1	2	2	1	1	2	2	1	1	2	2	1	1	2	2
19	1	2	3	1	1	2	2	1	1	2	2	1	1	2	2	2	2	1	1	2	2	1	1	2	2	1	1	2	2	1	1
20	1	2	2	1	1	2	2	1	1	2	2	1	1	2	2	2	2	1	1	2	2	1	1	2	2	1	1	2	2	1	1
21	1	2	2	1	1	2	2	2	2	1	1	2	2	1	1	1	1	2	2	1	1	2	2	2	2	1	1	2	2	1	1
22	1	2	2	1	1	2	2	2	2	1	1	2	2	1	1	1	1	2	2	1	1	2	2	2	2	1	1	2	2	1	1
23	1	2	2	1	1	2	2	2	2	1	1	2	2	1	1	2	2	1	1	2	2	1	1	1	1	2	2	1	1	2	2
24	1	2	2	1	1	2	2	2	2	1	1	2	2	1	1	2	2	1	1	2	2	1	1	1	1	2	2	1	1	2	2
25	1	2	2	2	2	1	1	1	1	2	2	2	2	1	1	1	1	2	2	2	2	1	1	1	1	2	2	2	2	1	1
26	1	2	2	2	2	1	1	1	1	2	2	2	2	1	1	1	1	2	2	2	2	1	1	1	1	2	2	2	2	1	1
27	1	2	2	2	2	1	1	1	1	2	2	2	2	1	1	2	2	1	1	1	1	2	2	2	2	1	1	1	1	2	2
28	1	2	2	2	2	1	1	1	1	2	2	2	2	1	1	2	2	1	1	1	1	2	2	2	2	1	1	1	1	2	2
29	1	2	2	2	2	1	1	2	2	1	1	1	1	2	2	1	1	2	2	2	2	1	1	2	2	1	1	1	1	2	2
30	1	2	2	2	2	1	1	2	2	1	1	1	1	2	2	1	1	2	2	2	2	1	1	2	2	1	1	1	1	2	2
31	1	2	2	2	2	1	1	2	2	1	1	1	1	2	2	2	2	1	1	1	1	2	2	1	1	2	2	1	1	1	1
32	1	2	2	2	2	1	1	2	2	1	1	1	1	2	2	2	2	1	1	1	1	2	2	1	1	2	2	2	2	1	1
33	2	1	2	1	2	1	2	1	2	1	2	1	2	1	2	1	2	1	2	1	2	1	2	1	2	1	2	1	2	1	2
34	2	1	2	1	2	1	2	1	2	1	2	1	2	1	2	1	2	1	2	1	2	1	2	1	2	1	2	1	2	1	2
35	2	1	2	1	2	1	2	1	2	1	2	1	2	1	2	2	1	2	1	2	1	2	1	2	1	2	1	2	1	2	1
36	2	1	2	1	2	1	2	1	2	1	2	1	2	1	2	2	1	2	1	2	1	2	1	2	1	2	1	2	1	2	1
37	2	1	2	1	2	1	2	2	1	2	1	2	1	2	1	1	2	1	2	1	2	1	2	2	1	2	1	2	1	2	1
38	2	1	2	1	2	1	2	2	1	2	1	2	1	2	1	1	2	1	2	1	2	1	2	2	1	2	1	2	1	2	1
39	2	1	2	1	2	1	2	2	1	2	1	2	1	2	1	2	1	2	1	2	1	2	1	1	2	1	2	1	2	1	2
40	2	1	2	1	2	1	2	2	1	2	1	2	1	2	1	2	1	2	1	2	1	2	1	1	2	1	2	1	2	1	2
41	2	1	2	2	1	2	1	1	2	1	2	2	1	2	1	1	2	1	2	2	1	2	1	1	2	1	2	2	1	2	1
42	2	1	2	2	1	2	1	1	2	1	2	2	1	2	1	1	2	1	2	2	1	2	1	1	2	1	2	2	1	2	1
43	2	1	2	2	1	2	1	1	2	1	2	2	1	2	1	2	1	2	1	1	2	1	2	2	1	2	1	1	2	1	2
44	2	1	2	2	1	2	1	1	2	1	2	2	1	2	1	2	1	2	1	1	2	1	2	2	1	2	1	1	2	1	2
45	2	1	2	2	1	2	1	2	1	2	1	1	2	1	2	1	2	1	2	2	1	2	1	2	1	2	1	1	2	1	2
46	2	1	2	2	1	2	1	2	1	2	1	1	2	1	2	1	2	1	2	2	1	2	1	2	1	2	1	1	2	1	2
47	2	1	2	2	1	2	1	2	1	2	1	1	2	1	2	2	1	2	1	1	2	1	2	1	2	1	2	2	1	2	1
48	2	1	2	2	1	2	1	2	1	2	1	1	2	1	2	2	1	2	1	1	2	1	2	1	2	1	2	2	1	2	1
49	2	2	1	1	2	2	1	1	2	2	1	1	2	2	1	1	2	2	1	1	2	2	1	1	2	2	1	1	2	2	1
50	2	2	1	1	2	2	1	1	2	2	1	1	2	2	1	1	2	2	1	1	2	2	1	1	2	2	1	1	2	2	1
51	2	2	1	1	2	2	1	1	2	2	1	1	2	2	1	2	1	1	2	2	1	1	2	2	1	1	2	2	1	1	2
52	2	2	1	1	2	2	1	1	2	2	1	1	2	2	1	2	1	1	2	2	1	1	2	2	1	1	2	2	1	1	2
53	2	2	1	1	2	2	1	2	1	1	2	2	1	1	2	1	2	2	1	1	2	2	1	2	1	1	2	2	1	1	2
54	2	2	1	1	2	2	1	2	1	1	2	2	1	1	2	1	2	2	1	1	2	2	1	2	1	1	2	2	1	1	2
55	2	2	1	1	2	2	1	2	1	1	2	2	1	1	2	2	1	1	2	2	1	1	2	1	2	2	1	1	2	2	1
56	2	2	1	1	2	2	1	2	1	1	2	2	1	1	2	2	1	1	2	2	1	1	2	1	2	2	1	1	2	2	1
57	2	2	1	2	1	1	2	1	2	2	1	2	1	1	2	1	2	2	1	2	1	1	2	1	2	2	1	2	1	1	2
58	2	2	1	2	1	1	2	1	2	2	1	2	1	1	2	1	2	2	1	2	1	1	2	1	2	2	1	2	1	1	2
59	2	2	1	2	1	1	2	1	2	2	1	2	1	1	2	2	1	1	2	1	2	2	1	2	1	1	2	1	2	2	1
60	2	2	1	2	1	1	2	1	2	2	1	2	1	1	2	2	1	1	2	1	2	2	1	2	1	1	2	1	2	2	1
61	2	2	1	2	1	1	2	2	1	1	2	1	2	2	1	1	2	2	1	2	1	1	2	2	1	1	2	1	2	2	1
62	2	2	1	2	1	1	2	2	1	1	2	1	2	2	1	1	2	2	1	2	1	1	2	2	1	1	2	1	2	2	1
63	2	2	1	2	1	1	2	2	1	1	2	1	2	2	1	2	1	1	2	1	2	2	1	1	2	2	1	2	1	1	2
64	2	2	1	2	1	1	2	2	1	1	2	1	2	2	1	2	1	1	2	1	2	2	1	1	2	2	1	2	1	1	2
	Gp 1	Gp 2		Group 3				Group 4								Group 5															

$L_{64}(2^{63})$ (Continued)

32	33	34	35	36	37	38	39	40	41	42	43	44	45	46	47	48	49	50	51	52	53	54	55	56	57	58	59	60	61	62	63
1	1	1	1	1	1	1	1	1	1	1	1	1	1	1	1	1	1	1	1	1	1	1	1	1	1	1	1	1	1	1	1
2	2	2	2	2	2	2	2	2	2	2	2	2	2	2	2	2	2	2	2	2	2	2	2	2	2	2	2	2	2	2	2
1	1	1	1	1	1	1	1	1	1	1	1	1	1	1	1	2	2	2	2	2	2	2	2	2	2	2	2	2	2	2	2
2	2	2	2	2	2	2	2	2	2	2	2	2	2	2	2	1	1	1	1	1	1	1	1	1	1	1	1	1	1	1	1
1	1	1	1	1	1	1	1	2	2	2	2	2	2	2	2	1	1	1	1	1	1	1	1	2	2	2	2	2	2	2	2
2	2	2	2	2	2	2	2	1	1	1	1	1	1	1	1	2	2	2	2	2	2	2	2	1	1	1	1	1	1	1	1
1	1	1	1	1	1	1	1	2	2	2	2	2	2	2	2	2	2	2	2	2	2	2	2	1	1	1	1	1	1	1	1
2	2	2	2	2	2	2	2	1	1	1	1	1	1	1	1	1	1	1	1	1	1	1	1	2	2	2	2	2	2	2	2
1	1	1	1	2	2	2	2	1	1	1	1	2	2	2	2	1	1	1	1	2	2	2	2	1	1	1	1	2	2	2	2
2	2	2	2	1	1	1	1	2	2	2	2	1	1	1	1	2	2	2	2	1	1	1	1	2	2	2	2	1	1	1	1
1	1	1	1	2	2	2	2	1	1	1	1	2	2	2	2	2	2	2	2	1	1	1	1	2	2	2	2	1	1	1	1
2	2	2	2	1	1	1	1	2	2	2	2	1	1	1	1	1	1	1	1	2	2	2	2	1	1	1	1	2	2	2	2
1	1	1	1	2	2	2	2	2	2	2	2	1	1	1	1	1	1	1	1	2	2	2	2	2	2	2	2	1	1	1	1
2	2	2	2	1	1	1	1	1	1	1	1	2	2	2	2	2	2	2	2	1	1	1	1	1	1	1	1	2	2	2	2
1	1	1	1	2	2	2	2	2	2	2	2	1	1	1	1	2	2	2	2	1	1	1	1	1	1	1	1	2	2	2	2
2	2	2	2	1	1	1	1	1	1	1	1	2	2	2	2	1	1	1	1	2	2	2	2	2	2	2	2	1	1	1	1
1	1	2	2	1	1	2	2	1	1	2	2	1	1	2	2	1	1	2	2	1	1	2	2	1	1	2	2	1	1	2	2
2	2	1	1	2	2	1	1	2	2	1	1	2	2	1	1	2	2	1	1	2	2	1	1	2	2	1	1	2	2	1	1
1	1	2	2	1	1	2	2	1	1	2	2	1	1	2	2	2	2	1	1	2	2	1	1	2	2	1	1	2	2	1	1
2	2	1	1	2	2	1	1	2	2	1	1	2	2	1	1	1	1	2	2	1	1	2	2	1	1	2	2	1	1	2	2
1	1	2	2	1	1	2	2	2	2	1	1	2	2	1	1	1	1	2	2	1	1	2	2	2	2	1	1	2	2	1	1
2	2	1	1	2	2	1	1	1	1	2	2	1	1	2	2	2	2	1	1	2	2	1	1	1	1	2	2	1	1	2	2
1	1	2	2	1	1	2	2	2	2	1	1	2	2	1	1	2	2	1	1	2	2	1	1	1	1	2	2	1	1	2	2
2	2	1	1	2	2	1	1	1	1	2	2	1	1	2	2	1	1	2	2	1	1	2	2	2	2	1	1	2	2	1	1
1	1	2	2	2	2	1	1	1	1	2	2	2	2	1	1	1	1	2	2	2	2	1	1	1	1	2	2	2	2	1	1
2	2	1	1	1	1	2	2	2	2	1	1	1	1	2	2	2	2	1	1	1	1	2	2	2	2	1	1	1	1	2	2
1	1	2	2	2	2	1	1	1	1	2	2	2	2	1	1	2	2	1	1	1	1	2	2	2	2	1	1	1	1	2	2
2	2	1	1	1	1	2	2	2	2	1	1	1	1	2	2	1	1	2	2	2	2	1	1	1	1	2	2	2	2	1	1
1	1	2	2	2	2	1	1	2	2	1	1	1	1	2	2	1	1	2	2	2	2	1	1	2	2	1	1	1	1	2	2
2	2	1	1	1	1	2	2	1	1	2	2	2	2	1	1	2	2	1	1	1	1	2	2	1	1	2	2	2	2	1	1
1	1	2	2	2	2	1	1	2	2	1	1	1	1	2	2	2	2	1	1	1	1	2	2	1	1	2	2	2	2	1	1
2	2	1	1	1	1	2	2	1	1	2	2	2	2	1	1	1	1	2	2	2	2	1	1	2	2	1	1	1	1	2	2
1	2	1	2	1	2	1	2	1	2	1	2	1	2	1	2	1	2	1	2	1	2	1	2	1	2	1	2	1	2	1	2
2	1	2	1	2	1	2	1	2	1	2	1	2	1	2	1	2	1	2	1	2	1	2	1	2	1	2	1	2	1	2	1
1	2	1	2	1	2	1	2	1	2	1	2	1	2	1	2	2	1	2	1	2	1	2	1	2	1	2	1	2	1	2	1
2	1	2	1	2	1	2	1	2	1	2	1	2	1	2	1	1	2	1	2	1	2	1	2	1	2	1	2	1	2	1	2
1	2	1	2	1	2	1	2	2	1	2	1	2	1	2	1	1	2	1	2	1	2	1	2	2	1	2	1	2	1	2	1
2	1	2	1	2	1	2	1	1	2	1	2	1	2	1	2	2	1	2	1	2	1	2	1	1	2	1	2	1	2	1	2
1	2	1	2	1	2	1	2	2	1	2	1	2	1	2	1	2	1	2	1	2	1	2	1	1	2	1	2	1	2	1	2
2	1	2	1	2	1	2	1	1	2	1	2	1	2	1	2	1	2	1	2	1	2	1	2	2	1	2	1	2	1	2	1
1	2	1	2	2	1	2	1	1	2	1	2	2	1	2	1	1	2	1	2	2	1	2	1	1	2	1	2	2	1	2	1
2	1	2	1	1	2	1	2	2	1	2	1	1	2	1	2	2	1	2	1	1	2	1	2	2	1	2	1	1	2	1	2
1	2	1	2	2	1	2	1	1	2	1	2	2	1	2	1	2	1	2	1	1	2	1	2	2	1	2	1	1	2	1	2
2	1	2	1	1	2	1	2	2	1	2	1	1	2	1	2	1	2	1	2	2	1	2	1	1	2	1	2	2	1	2	1
1	2	1	2	2	1	2	1	2	1	2	1	1	2	1	2	1	2	1	2	2	1	2	1	2	1	2	1	1	2	1	2
2	1	2	1	1	2	1	2	1	2	1	2	2	1	2	1	2	1	2	1	1	2	1	2	1	2	1	2	2	1	2	1
1	2	1	2	2	1	2	1	2	1	2	1	1	2	1	2	2	1	2	1	1	2	1	2	1	2	1	2	2	1	2	1
2	1	2	1	1	2	1	2	1	2	1	2	2	1	2	1	1	2	1	2	2	1	2	1	2	1	2	1	1	2	1	2
1	2	2	1	1	2	2	1	1	2	2	1	1	2	2	1	1	2	2	1	1	2	2	1	1	2	2	1	1	2	2	1
2	1	1	2	2	1	1	2	2	1	1	2	2	1	1	2	2	1	1	2	2	1	1	2	2	1	1	2	2	1	1	2
1	2	2	1	1	2	2	1	1	2	2	1	1	2	2	1	2	1	1	2	2	1	1	2	2	1	1	2	2	1	1	2
2	1	1	2	2	1	1	2	2	1	1	2	2	1	1	2	1	2	2	1	1	2	2	1	1	2	2	1	1	2	2	1
1	2	2	1	1	2	2	1	2	1	1	2	2	1	1	2	1	2	2	1	1	2	2	1	2	1	1	2	2	1	1	2
2	1	1	2	2	1	1	2	1	2	2	1	1	2	2	1	2	1	1	2	2	1	1	2	1	2	2	1	1	2	2	1
1	2	2	1	1	2	2	1	2	1	1	2	2	1	1	2	2	1	1	2	2	1	1	2	1	2	2	1	1	2	2	1
2	1	1	2	2	1	1	2	1	2	2	1	1	2	2	1	1	2	2	1	1	2	2	1	2	1	1	2	2	1	1	2
1	2	2	1	2	1	1	2	1	2	2	1	2	1	1	2	1	2	2	1	2	1	1	2	1	2	2	1	2	1	1	2
2	1	1	2	1	2	2	1	2	1	1	2	1	2	2	1	2	1	1	2	1	2	2	1	2	1	1	2	1	2	2	1
1	2	2	1	2	1	1	2	1	2	2	1	2	1	1	2	2	1	1	2	1	2	2	1	2	1	1	2	1	2	2	1
2	1	1	2	1	2	2	1	2	1	1	2	1	2	2	1	1	2	2	1	2	1	1	2	1	2	2	1	2	1	1	2
1	2	2	1	2	1	1	2	2	1	1	2	1	2	2	1	1	2	2	1	2	1	1	2	2	1	1	2	1	2	2	1
2	1	1	2	1	2	2	1	1	2	2	1	2	1	1	2	2	1	1	2	1	2	2	1	1	2	2	1	2	1	1	2
1	2	2	1	2	1	1	2	2	1	1	2	1	2	2	1	2	1	1	2	1	2	2	1	1	2	2	1	2	1	1	2
2	1	1	2	1	2	2	1	1	2	2	1	2	1	1	2	1	2	2	1	2	1	1	2	2	1	1	2	1	2	2	1

Group 6

Interaction Between Two Columns

1	2	3	4	5	6	7	8	9	10	11	12	13	14	15	16	17	18	19	20	21	22	23	24	25	26	27	28	29	30	31	32
(1)	3	2	5	4	7	6	9	8	11	10	13	12	15	14	17	16	19	18	21	20	23	22	25	24	27	26	29	28	31	30	33
	(2)	1	6	7	4	5	10	11	8	9	14	15	12	13	18	19	16	17	22	23	20	21	26	27	24	25	30	31	28	29	34
		(3)	7	6	5	4	11	10	9	8	15	14	13	12	19	18	17	16	23	22	21	20	27	26	25	24	31	30	29	28	35
			(4)	1	2	3	12	13	14	15	8	9	10	11	20	21	22	23	16	17	18	19	28	29	30	31	24	25	26	27	36
				(5)	3	2	13	12	15	14	9	8	11	10	21	20	23	22	17	16	19	18	29	28	31	30	25	24	27	26	37
					(6)	1	14	15	12	13	10	11	8	9	22	23	20	21	18	19	16	17	30	31	28	29	26	27	24	25	38
						(7)	15	14	13	12	11	10	9	8	23	22	21	20	19	18	17	16	31	30	29	28	27	26	25	24	39
							(8)	1	2	3	4	5	6	7	24	25	26	27	28	29	30	31	16	17	18	19	20	21	22	23	40
								(9)	3	2	5	4	7	6	25	24	27	26	29	28	31	30	17	16	19	18	21	20	23	22	41
									(10)	1	6	7	4	5	26	27	24	25	30	31	28	29	18	19	16	17	22	23	20	21	42
										(11)	7	6	5	4	27	26	25	24	31	30	29	28	19	18	17	26	23	22	21	20	43
											(12)	1	2	3	28	29	30	31	24	25	26	27	20	21	22	23	16	17	18	19	44
												(13)	3	2	29	28	31	30	25	24	27	26	21	20	23	22	17	16	19	18	45
													(14)	1	30	31	28	29	26	27	24	25	22	23	20	21	18	19	16	17	46
														(15)	31	30	29	28	27	26	25	24	23	22	21	20	19	18	17	16	47
															(16)	1	2	3	4	5	6	7	8	9	10	11	12	13	14	15	48
																(17)	3	2	5	4	7	6	9	8	11	10	13	12	15	14	49
																	(18)	1	6	7	4	5	10	11	8	9	14	15	12	13	50
																		(19)	7	6	5	4	11	10	9	8	15	14	13	12	51
																			(20)	1	2	3	12	13	14	15	8	9	10	11	52
																				(21)	3	2	13	12	15	14	9	8	11	10	53
																					(22)	1	14	15	12	13	10	11	8	9	54
																						(23)	15	14	13	12	11	10	9	8	55
																							(24)	1	2	3	4	5	6	7	56
																								(25)	3	2	5	4	7	6	57
																									(26)	1	6	7	4	5	58
																										(27)	7	6	5	4	59
																											(28)	1	2	3	60
																												(29)	3	2	61
																													(30)	1	62
																														(31)	63
																															(32)

Interaction Between Two Columns (Continued)

	33	34	35	36	37	38	39	40	41	42	43	44	45	46	47	48	49	50	51	52	53	54	55	56	57	58	59	60	61	62	63
	32	35	34	37	36	39	38	41	40	43	42	45	44	47	46	49	48	51	50	53	52	55	54	57	56	59	58	61	60	63	62
	35	32	33	38	39	36	37	42	43	40	41	46	47	44	45	50	51	48	49	54	55	52	53	58	59	56	57	62	63	60	61
	34	33	32	39	38	37	36	43	42	41	40	47	46	45	44	51	50	49	48	55	54	53	52	59	58	57	56	63	62	61	60
	37	38	39	32	33	34	35	44	45	46	47	40	41	42	43	52	53	54	55	48	49	50	51	60	61	62	63	56	57	58	59
	36	39	38	33	32	35	34	45	44	47	46	41	40	43	42	53	52	55	54	49	48	51	50	61	60	63	62	57	56	59	58
	39	36	37	34	35	32	33	46	47	44	45	42	43	40	41	54	55	52	53	50	51	48	49	62	63	60	61	58	59	56	57
	38	37	36	35	34	33	32	47	46	45	44	43	42	41	40	55	54	53	52	51	50	49	48	63	62	61	60	59	58	57	56
	41	42	43	44	45	46	47	32	33	34	35	36	37	38	39	56	57	58	59	60	61	62	63	48	49	50	51	52	53	54	55
	40	43	42	45	44	47	46	33	32	35	34	37	36	39	38	57	56	59	58	61	60	63	62	49	48	51	50	53	52	55	54
	43	40	41	46	47	44	45	34	35	32	33	38	39	36	37	58	59	56	57	62	63	60	61	50	51	48	49	54	55	52	53
	42	41	40	47	46	45	44	35	34	33	32	39	38	37	36	59	58	57	56	63	62	61	60	51	50	49	48	55	54	53	52
	45	46	47	40	41	42	43	36	37	38	39	32	33	34	35	60	61	62	63	56	57	58	59	52	53	54	55	48	49	50	51
	44	47	46	41	40	43	42	37	36	39	38	33	32	35	34	61	60	63	62	57	56	59	58	53	52	55	54	49	48	51	50
	47	44	45	42	43	40	41	38	39	36	37	34	35	32	33	62	63	60	61	58	59	56	57	54	55	52	53	50	51	48	49
	46	45	44	43	42	41	40	39	38	37	36	35	34	33	32	63	62	61	60	59	58	57	56	55	54	53	52	51	50	49	48
	49	50	51	52	53	54	55	56	57	58	59	60	61	62	63	32	33	34	35	36	37	38	39	40	41	42	43	44	45	46	47
	48	51	50	53	52	55	54	57	56	59	58	61	60	63	62	33	32	35	34	37	36	39	38	41	40	43	42	45	44	47	46
	51	48	49	54	55	52	53	58	59	56	57	62	63	60	61	34	35	32	33	38	39	36	37	42	43	40	41	46	47	44	45
	50	49	48	55	54	53	52	59	58	57	56	63	62	61	60	35	34	33	32	39	38	37	36	43	42	41	40	47	46	45	44
	53	54	55	48	49	50	51	60	61	62	63	56	57	58	59	36	37	38	39	32	33	34	35	44	45	46	47	40	41	42	43
	52	55	54	49	48	51	50	61	60	63	62	57	56	59	58	37	36	39	38	33	32	35	34	45	44	47	46	41	40	43	42
	55	52	53	50	51	48	49	62	63	60	61	58	59	56	57	38	39	36	37	34	35	32	33	46	47	44	45	42	43	40	41
	54	53	52	51	50	49	48	63	62	61	60	59	58	57	56	39	38	37	36	35	34	33	32	47	46	45	44	43	42	41	40
	57	58	59	60	61	62	63	48	49	50	51	52	53	54	55	40	41	42	43	44	45	46	47	32	33	34	35	36	37	38	39
	56	59	58	61	60	63	62	49	48	51	50	53	52	55	54	41	40	43	42	45	44	47	46	33	32	35	34	37	36	39	38
	59	56	57	62	63	60	61	50	51	48	49	54	55	52	53	42	43	40	41	46	47	44	45	34	35	32	33	38	39	36	37
	58	57	56	63	62	61	60	51	50	49	48	55	54	53	52	43	42	41	40	47	46	45	44	35	34	33	32	39	38	37	36
	61	62	63	56	57	58	59	52	53	54	55	48	49	50	51	44	45	46	47	40	41	42	43	36	37	38	39	32	33	34	35
	60	63	62	57	56	59	58	53	52	55	54	49	48	51	50	45	44	47	46	41	40	43	42	37	36	39	38	33	32	35	34
	63	60	61	58	59	56	57	54	55	52	53	50	51	48	49	46	47	44	45	42	43	40	41	38	39	36	37	34	35	32	33
	62	61	60	59	58	57	56	55	54	53	52	51	50	49	48	47	46	45	44	43	42	41	40	39	38	37	36	35	34	33	32
	1	2	3	4	5	6	7	8	9	10	11	12	13	14	15	16	17	18	19	20	21	22	23	24	25	26	27	28	29	30	31
(33)		3	2	5	4	7	6	9	8	11	10	13	12	15	14	17	16	19	18	21	20	23	22	25	24	27	26	29	28	31	30
(34)			1	6	7	4	5	10	11	8	9	14	15	12	13	18	19	16	17	22	23	20	21	26	27	24	25	30	31	28	29
(35)				7	6	5	4	11	10	9	8	15	14	13	12	19	18	17	16	23	22	21	20	27	26	25	24	31	30	29	28
(36)					1	2	3	12	13	14	15	8	9	10	11	20	21	22	23	16	17	18	19	28	29	30	31	24	25	26	27
(37)						3	2	13	12	15	14	9	8	11	10	21	20	23	22	17	16	19	18	29	28	31	30	25	24	27	26
(38)							1	14	15	12	13	10	11	8	9	22	23	20	21	18	19	16	17	30	31	28	29	26	27	24	25
(39)								15	14	13	12	11	10	9	8	23	22	21	20	19	18	17	16	31	30	29	28	27	26	25	24
(40)									1	2	3	4	5	6	7	24	25	26	27	28	29	30	31	16	17	18	19	20	21	22	23
(41)										3	2	5	4	7	6	25	24	27	26	29	28	31	30	17	16	19	18	21	20	23	22
(42)											1	6	7	4	5	26	27	24	25	30	31	28	29	18	19	16	17	22	23	20	21
(43)												7	6	5	4	27	26	25	24	31	30	29	28	19	18	17	16	23	22	21	20
(44)													1	2	3	28	29	30	31	24	25	26	27	20	21	22	23	16	17	18	19
(45)														3	2	29	28	31	30	25	24	27	26	21	20	23	22	17	16	19	18
(46)															1	30	31	28	29	26	27	24	25	22	23	20	21	18	19	16	17
(47)																31	30	29	28	27	26	25	24	23	22	21	20	19	18	17	16
(48)																	1	2	3	4	5	6	7	8	9	10	11	12	13	14	15
(49)																		3	2	5	4	7	6	9	8	11	10	13	12	15	14
(50)																			1	6	7	4	5	10	11	8	9	14	15	12	13
(51)																				7	6	5	4	11	10	9	8	15	14	13	12
(52)																					1	2	3	12	13	14	15	8	9	10	11
(53)																						3	2	13	12	15	14	9	8	11	10
(54)																							1	14	15	12	13	10	11	8	9
(55)																								15	14	13	12	11	10	9	8
(56)																									1	2	3	4	5	6	7
(57)																										3	2	5	4	7	6
(58)																											1	6	7	4	5
(59)																												7	6	5	4
(60)																													1	2	3
(61)																														3	2
(62)																															1

Linear Graphs of L_{64}

(1)

a

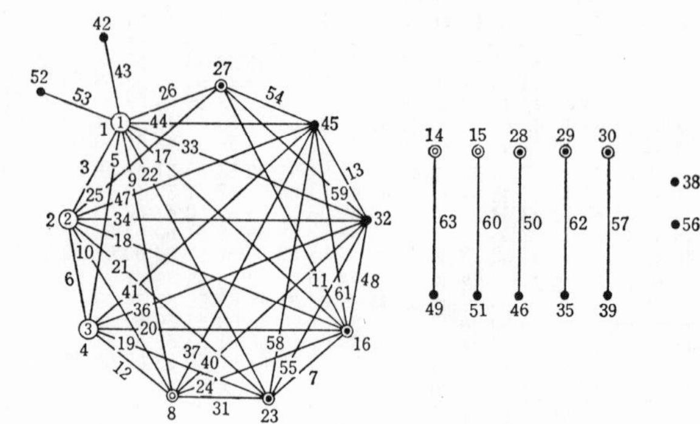

b

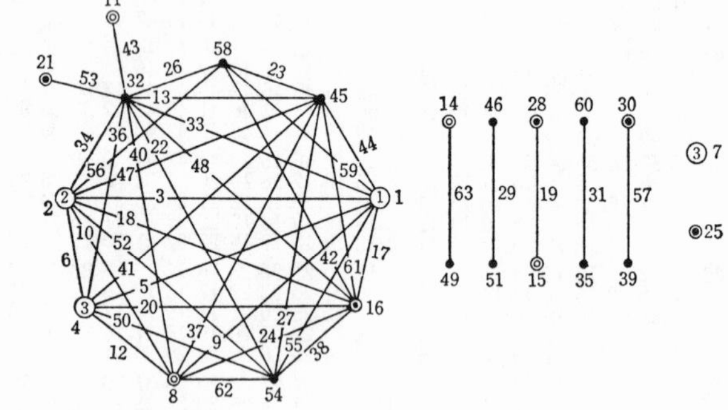

c

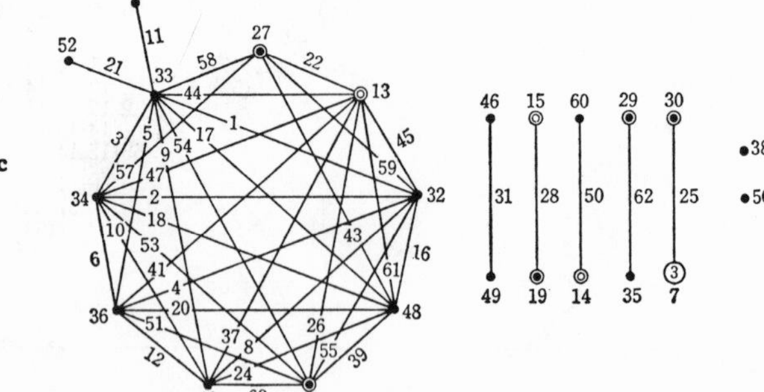

(2)

a

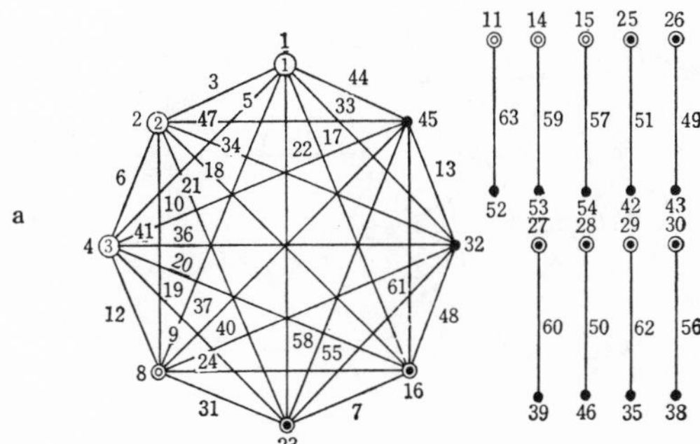

b

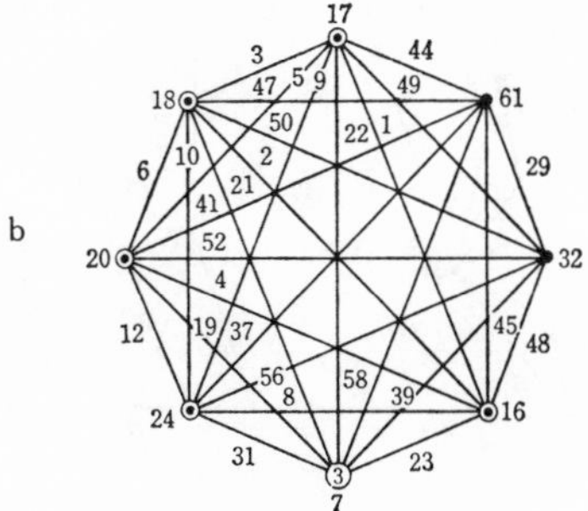

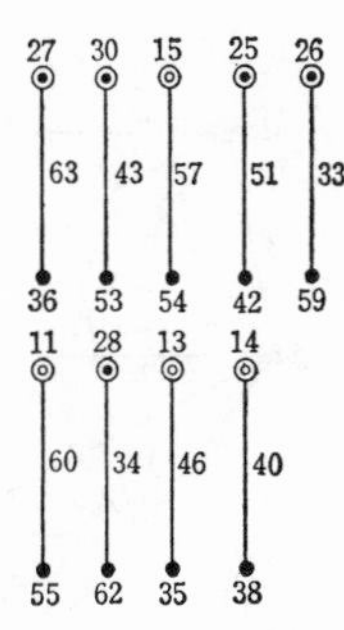

c

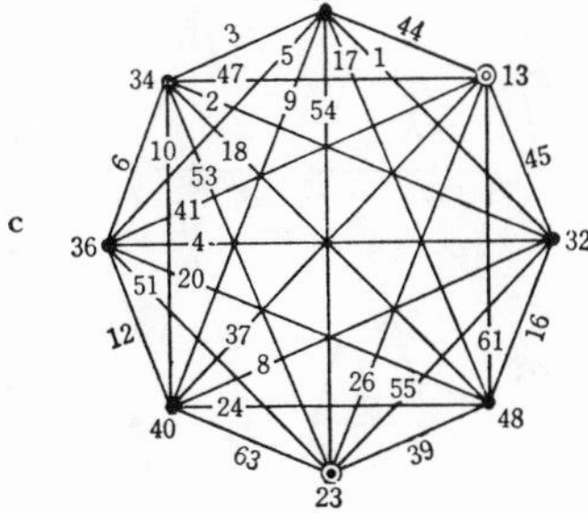

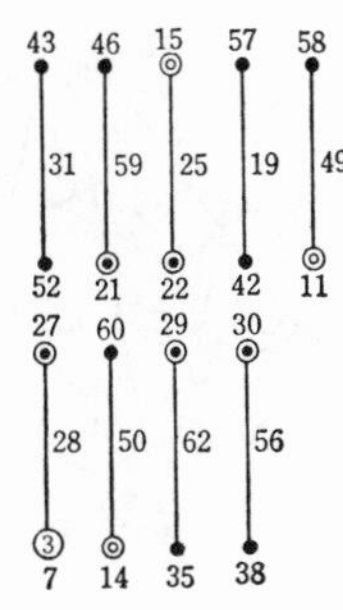

(3)

a

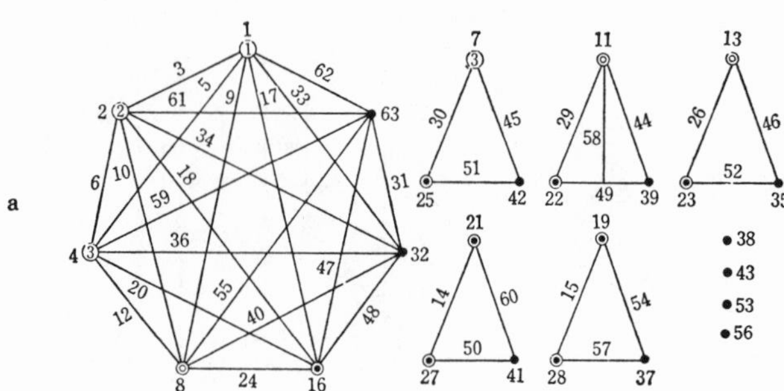
1
3
62
61 5 9 17 33
2
63
34
6 10
18
31
59
4
36
32
47
20
55
40
48
12
8
24
16
7
30
45
51
25
42
11
29
58
44
22
49
39
13
26
46
52
23
35
21
14
60
50
27
41
19
15
54
57
28
37
38
43
53
56

b

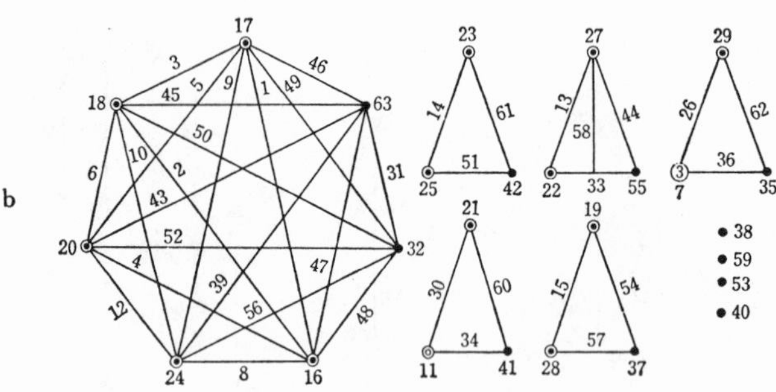
17
3
46
45 5 9 1 49
18
63
50
6 10
2
31
43
52
20
32
4
47
39
12
56
48
24
8
16
23
14
61
51
25
42
27
13
58
44
22
33
55
29
26
62
36
7
35
21
30
60
34
11
41
19
15
54
57
28
37
38
59
53
40

c

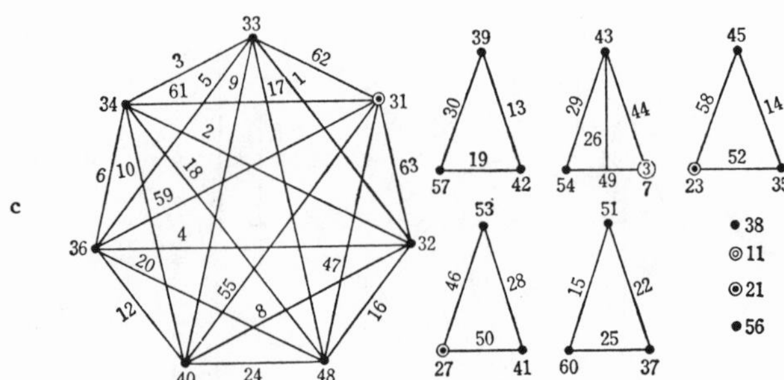
33
3
62
61 5 9 17 1
34
31
2
6 10
18
63
59
4
36
32
20
47
55
12
8
16
40
24
48
39
30
13
19
57
42
43
29
44
26
54
49
7
45
58
14
52
23
35
53
46
28
50
27
41
51
15
22
25
60
37
38
11
21
56

(4)

a

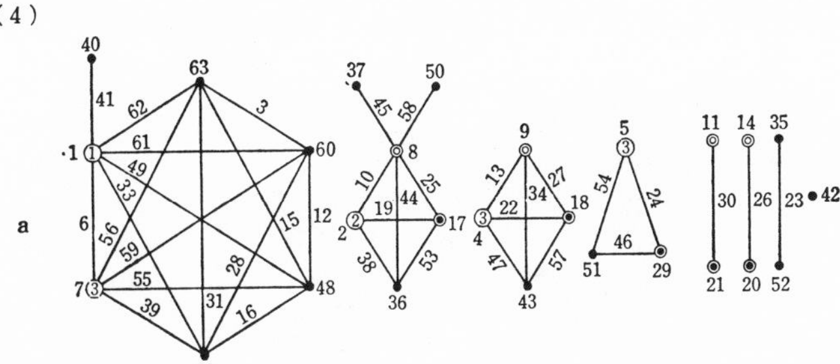
40
63
37 50
41 62 3
45 58
9
5
11 14 35
·1 61 60
8
49 33
10 25
13 27
34
54
6 56 15 12
19 44
2 17
3 22 18
4
24
30 26 23 42
59 28
46
7 55 48
38 53
47 57
51 29
21 20 52
39 31 16
36
43
32

b

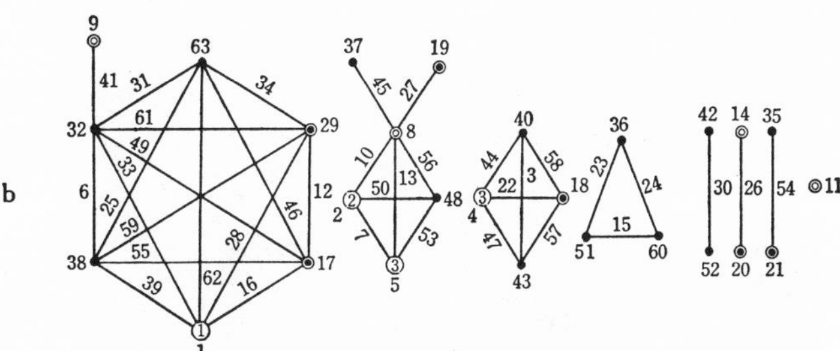
9
63
37 19
41 31 34
45 27
32 61 29
8
40
36
42 14 35
49 33
10 56
13
44 58
3
23
6 25 46 12
50
2 48
22 18
4
24
30 26 54 11
59 28
15
38 55 17
7 53
47 57
51 60
52 20 21
39 62 16
5
43
1

c

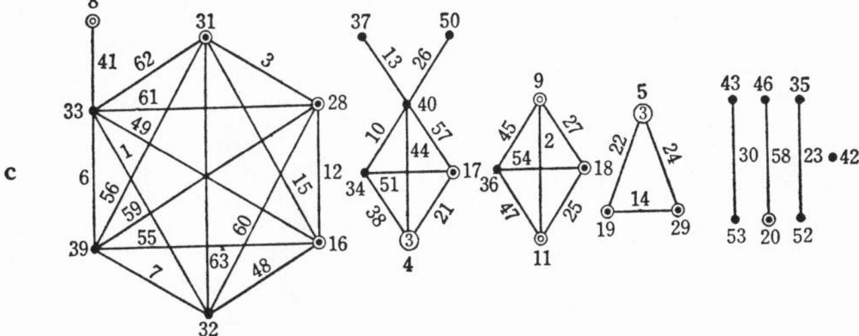
8
31
37 50
41 62 3
13 26
9
5
43 46 35
33 61 28
40
49
10 57
45 27
2
22
1
6 56 15 12
44
34 51 17
54
36 18
24
30 58 23 42
59 60
38 21
47 25
14
19 29
39 55 16
63
7 48
4
11
53 20 52
32

(5)

a

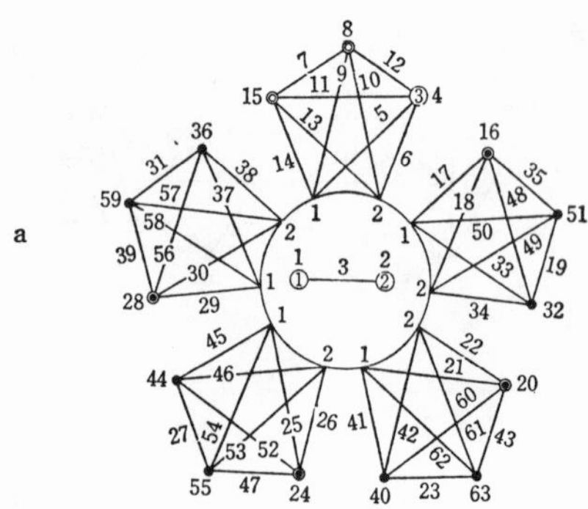

b

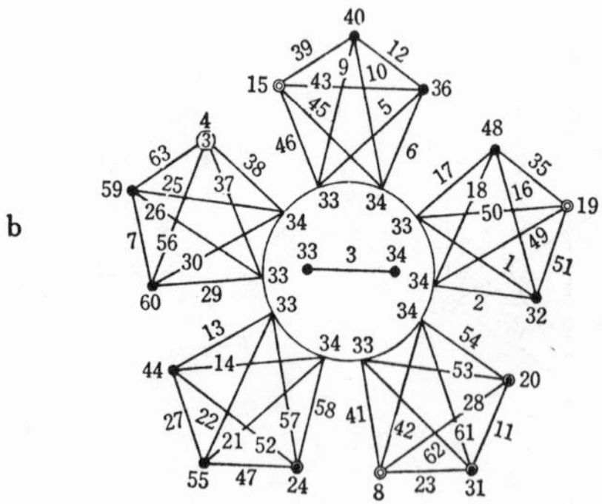

c

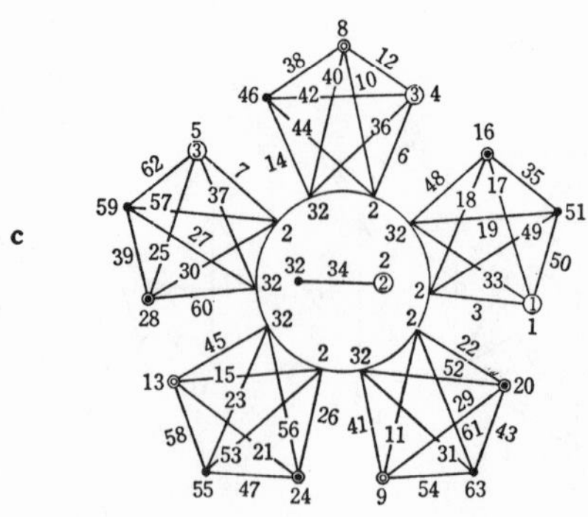

(6)

a

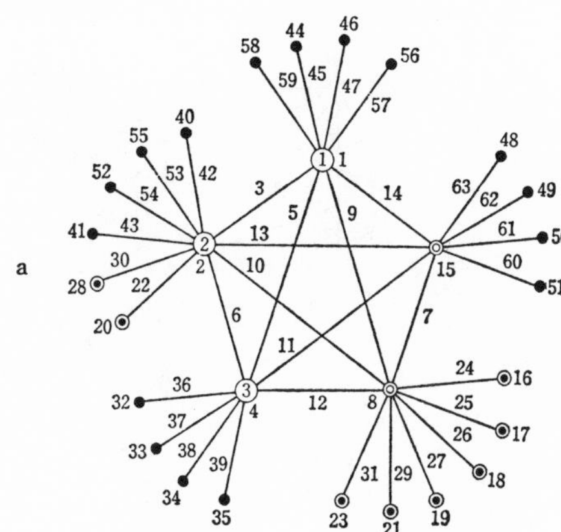

b

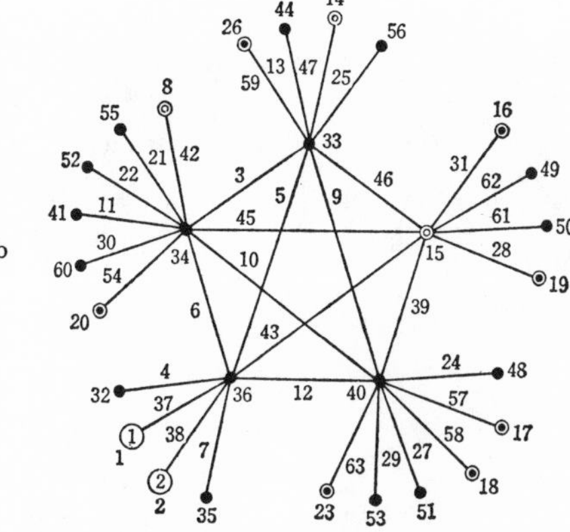

c

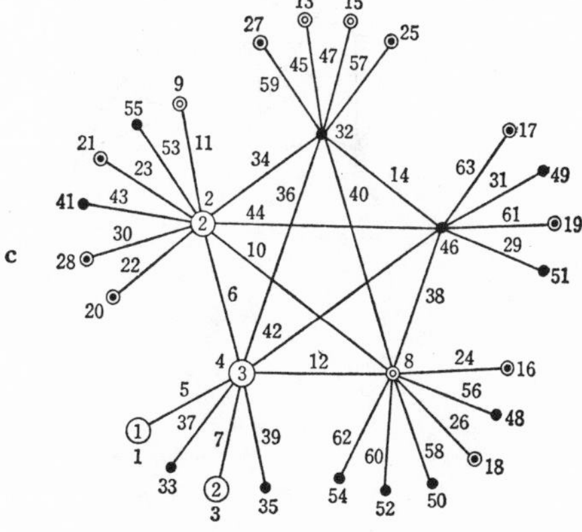

(7)

a

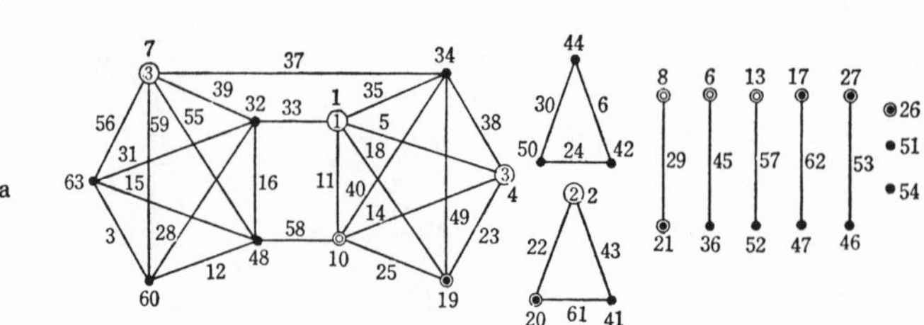

b

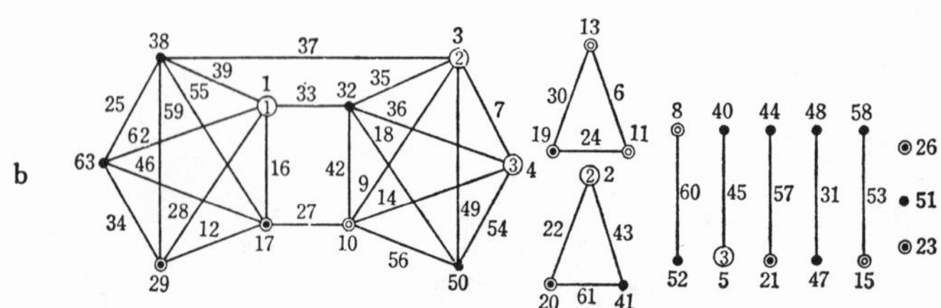

c

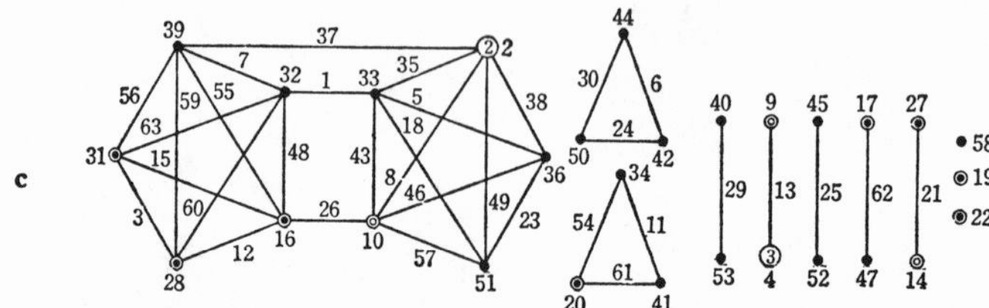

(8)

a

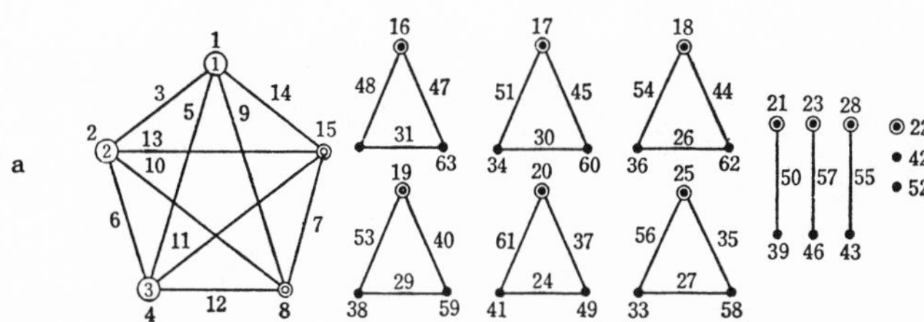

b

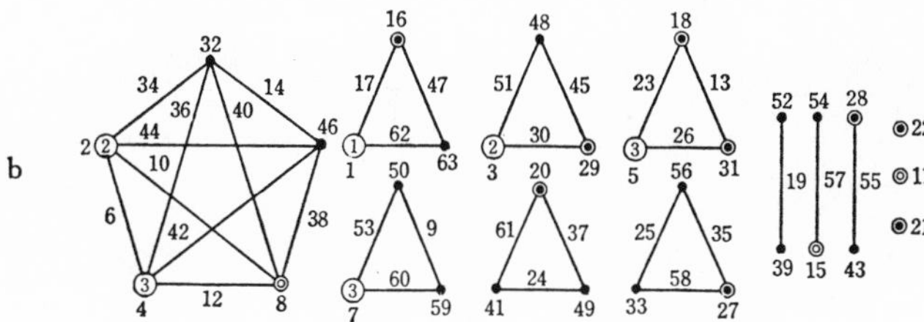

c

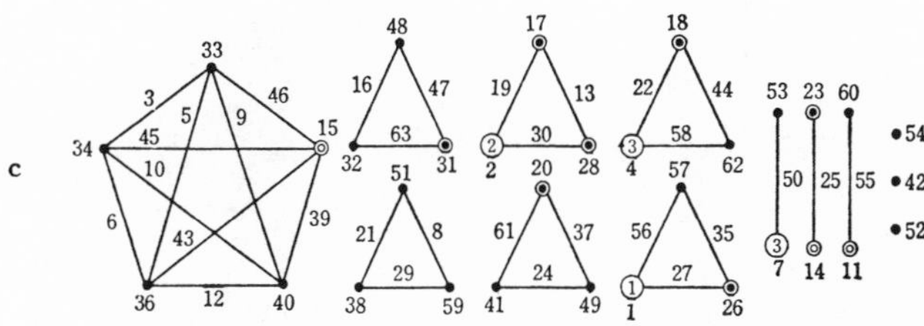

(9)

a

b

c

(10)

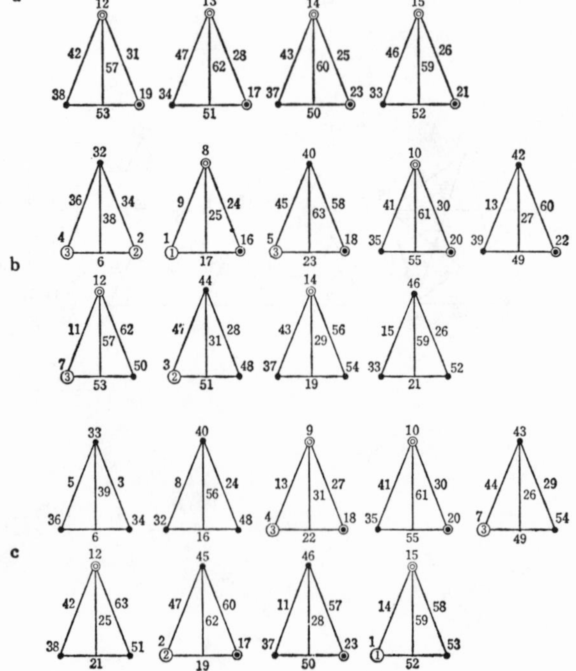

3^n System

$L_9(3^4)$

No. \ Col. no.	1	2	3	4
1	1	1	1	1
2	1	2	2	2
3	1	3	3	3
4	2	1	2	3
5	2	2	3	1
6	2	3	1	2
7	3	1	3	2
8	3	2	1	3
9	3	3	2	1
	Group 1	Group 2		

(1)

1 —3,4— 2

$L_{18}(2^1 \times 3^7)$

No. \ Col. no.	1	2	3	4	5	6	7	8
1	1	1	1	1	1	1	1	1
2	1	1	2	2	2	2	2	2
3	1	1	3	3	3	3	3	3
4	1	2	1	1	2	2	3	3
5	1	2	2	2	3	3	1	1
6	1	2	3	3	1	1	2	2
7	1	3	1	2	1	3	2	3
8	1	3	2	3	2	1	3	1
9	1	3	3	1	3	2	1	2
10	2	1	1	3	3	2	2	1
11	2	1	2	1	1	3	3	2
12	2	1	3	2	2	1	1	3
13	2	2	1	2	3	1	3	2
14	2	2	2	3	1	2	1	3
15	2	2	3	1	2	3	2	1
16	2	3	1	3	2	3	1	2
17	2	3	2	1	3	1	2	3
18	2	3	3	2	1	2	3	1
	Group 1	Group 2	Group 3					

(1)

1 —— 2

(Interactions are found with no sacrifice of columns. They are found from the two-way array of column 1 and column 2)

(Note) Interactions among three levels are partially confounded a little in each of the columns of the remaining three levels. The same can be said as by the Note for L_{12}.

$L_{27}(3^{13})$

No. \ Col. no.	1	2	3	4	5	6	7	8	9	10	11	12	13
1	1	1	1	1	1	1	1	1	1	1	1	1	1
2	1	1	1	1	2	2	2	2	2	2	2	2	2
3	1	1	1	1	3	3	3	3	3	3	3	3	3
4	1	2	2	2	1	1	1	2	2	2	3	3	3
5	1	2	2	2	2	2	2	3	3	3	1	1	1
6	1	2	2	2	3	3	3	1	1	1	2	2	2
7	1	3	3	3	1	1	1	3	3	3	2	2	2
8	1	3	3	3	2	2	2	1	1	1	3	3	3
9	1	3	3	3	3	3	3	2	2	2	1	1	1
10	2	1	2	3	1	2	3	1	2	3	1	2	3
11	2	1	2	3	2	3	1	2	3	1	2	3	1
12	2	1	2	3	3	1	2	3	1	2	3	1	2
13	2	2	3	1	1	2	3	2	3	1	3	1	2
14	2	2	3	1	2	3	1	3	1	2	1	2	3
15	2	2	3	1	3	1	2	1	2	3	2	3	1
16	2	3	1	2	1	2	3	3	1	2	2	3	1
17	2	3	1	2	2	3	1	1	2	3	3	1	2
18	2	3	1	2	3	1	2	2	3	1	1	2	3
19	3	1	3	2	1	3	2	1	3	2	1	3	2
20	3	1	3	2	2	1	3	2	1	3	2	1	3
21	3	1	3	2	3	2	1	3	2	1	3	2	1
22	3	2	1	3	1	3	2	2	1	3	3	2	1
23	3	2	1	3	2	1	3	3	2	1	1	3	2
24	3	2	1	3	3	2	1	1	3	2	2	1	3
25	3	3	2	1	1	3	2	3	2	1	2	1	3
26	3	3	2	1	2	1	3	1	3	2	3	2	1
27	3	3	2	1	3	2	1	2	1	3	1	3	2
	Group 1	Group 2			Group 3								

Table of Interactions Between Two Columns

Col. \ Col. no.	1	2	3	4	5	6	7	8	9	10	11	12	13
	(1)	3 4	2 4	2 3	6 7	5 7	5 6	9 10	8 10	8 9	12 13	11 13	11 12
		(2)	1 4	1 3	8 11	9 12	10 13	5 11	6 12	7 13	5 8	6 9	7 10
			(3)	1 2	9 13	10 11	8 12	7 12	5 13	6 11	6 10	7 8	5 9
				(4)	10 12	8 13	9 11	6 13	7 11	5 12	7 9	5 10	6 8
					(5)	1 7	1 6	2 11	3 13	4 12	2 8	4 10	3 9
						(6)	1 5	4 13	2 12	3 11	3 10	2 9	4 8
							(7)	3 12	4 11	2 13	4 9	3 8	2 10
								(8)	1 10	1 9	2 5	3 7	4 6
									(9)	1 8	4 7	2 6	3 5
										(10)	3 6	4 5	2 7
											(11)	1 13	1 12
												(12)	1 11

Linear Graph of L_{27}

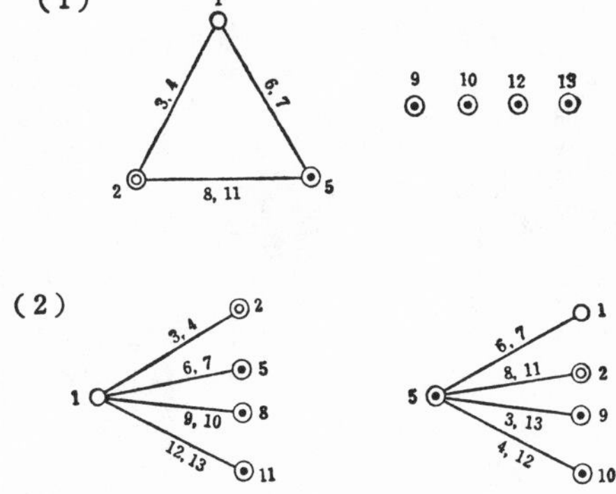

$L_{54}(2^1 \times 3^{25})$

No. \ Col. no.	1	2	3	4	5	6	7	8	9	10	11	12	13	14	15	16	17	18	19	20	21	22	23	24	25	26
1	1	1	1	1	1	1	1	1	1	1	1	1	1	1	1	1	1	1	1	1	1	1	1	1	1	1
2	1	1	1	1	1	1	1	1	2	2	2	2	2	2	2	2	2	2	2	2	2	2	2	2	2	2
3	1	1	1	1	1	1	1	1	3	3	3	3	3	3	3	3	3	3	3	3	3	3	3	3	3	3
4	1	1	2	2	2	2	2	2	1	1	1	1	1	1	2	3	2	3	2	3	2	3	2	3	2	3
5	1	1	2	2	2	2	2	2	2	2	2	2	2	2	3	1	3	1	3	1	3	1	3	1	3	1
6	1	1	2	2	2	2	2	2	3	3	3	3	3	3	1	2	1	2	1	2	1	2	1	2	1	2
7	1	1	3	3	3	3	3	3	1	1	1	1	1	1	3	2	3	2	3	2	3	2	3	2	3	2
8	1	1	3	3	3	3	3	3	2	2	2	2	2	2	1	3	1	3	1	3	1	3	1	3	1	3
9	1	1	3	3	3	3	3	3	3	3	3	3	3	3	2	1	2	1	2	1	2	1	2	1	2	1
10	1	2	1	1	2	2	3	3	1	1	2	2	3	3	1	1	1	1	2	3	2	3	3	2	3	2
11	1	2	1	1	2	2	3	3	2	2	3	3	1	1	2	2	2	2	3	1	3	1	1	3	1	3
12	1	2	1	1	2	2	3	3	3	3	1	1	2	2	3	3	3	3	1	2	1	2	2	1	2	1
13	1	2	2	2	3	3	1	1	1	1	2	2	3	3	2	3	2	3	3	2	3	2	1	1	1	1
14	1	2	2	2	3	3	1	1	2	2	3	3	1	1	3	1	3	1	1	3	1	3	2	2	2	2
15	1	2	2	2	3	3	1	1	3	3	1	1	2	2	1	2	1	2	2	1	2	1	3	3	3	3
16	1	2	3	3	1	1	2	2	1	1	2	2	3	3	3	2	3	2	1	1	1	1	2	3	2	3
17	1	2	3	3	1	1	2	2	2	2	3	3	1	1	1	3	1	3	2	2	2	2	3	1	3	1
18	1	2	3	3	1	1	2	2	3	3	1	1	2	2	2	1	2	1	3	3	3	3	1	2	1	2
19	1	3	1	2	1	3	2	3	1	2	1	3	2	3	1	1	2	3	1	1	3	2	2	3	3	2
20	1	3	1	2	1	3	2	3	2	3	2	1	3	1	2	2	3	1	2	2	1	3	3	1	1	3
21	1	3	1	2	1	3	2	3	3	1	3	2	1	2	3	3	1	2	3	3	2	1	1	2	2	1
22	1	3	2	3	2	1	3	1	1	2	1	3	2	3	2	3	3	2	2	3	1	1	3	2	1	1
23	1	3	2	3	2	1	3	1	2	3	2	1	3	1	3	1	1	3	3	1	2	2	1	3	2	2
24	1	3	2	3	2	1	3	1	3	1	3	2	1	2	1	2	2	1	1	2	3	3	2	1	3	3
25	1	3	3	1	3	2	1	2	1	2	1	3	2	3	3	2	1	1	3	2	2	3	1	1	2	3
26	1	3	3	1	3	2	1	2	2	3	2	1	3	1	1	3	2	2	1	3	3	1	2	2	3	1
27	1	3	3	1	3	2	1	2	3	1	3	2	1	2	2	1	3	3	2	1	1	2	3	3	1	2
28	2	1	1	3	3	2	2	1	1	3	3	2	2	1	1	1	3	2	3	2	2	3	2	3	1	1
29	2	1	1	3	3	2	2	1	2	1	1	3	3	2	2	2	1	3	1	3	3	1	3	1	2	2
30	2	1	1	3	3	2	2	1	3	2	2	1	1	3	3	3	2	1	2	1	1	2	1	2	3	3
31	2	1	2	1	1	3	3	2	1	3	3	2	2	1	2	3	1	1	1	1	3	2	3	2	2	3
32	2	1	2	1	1	3	3	2	2	1	1	3	3	2	3	1	2	2	2	2	1	3	1	3	3	1
33	2	1	2	1	1	3	3	2	3	2	2	1	1	3	1	2	3	3	3	3	2	1	2	1	1	2
34	2	1	3	2	2	1	1	3	1	3	3	2	2	1	3	2	2	3	2	3	1	1	1	1	3	2
35	2	1	3	2	2	1	1	3	2	1	1	3	3	2	1	3	3	1	3	1	2	2	2	2	1	3
36	2	1	3	2	2	1	1	3	3	2	2	1	1	3	2	1	1	2	1	2	3	3	3	3	2	1
37	2	2	1	2	3	1	3	2	1	2	3	1	3	2	1	1	2	3	3	2	1	1	3	2	2	3
38	2	2	1	2	3	1	3	2	2	3	1	2	1	3	2	2	3	1	1	3	2	2	1	3	3	1
39	2	2	1	2	3	1	3	2	3	1	2	3	2	1	3	3	1	2	2	1	3	3	2	1	1	2
40	2	2	2	3	1	2	1	3	1	2	3	1	3	2	2	3	3	2	1	1	2	3	1	1	3	2
41	2	2	2	3	1	2	1	3	2	3	1	2	1	3	3	1	1	3	2	2	3	1	2	2	1	3
42	2	2	2	3	1	2	1	3	3	1	2	3	2	1	1	2	2	1	3	3	1	2	3	3	2	1
43	2	2	3	1	2	3	2	1	1	2	3	1	3	2	3	2	1	1	2	3	3	2	2	3	1	1
44	2	2	3	1	2	3	2	1	2	3	1	2	1	3	1	3	2	2	3	1	1	3	3	1	2	2
45	2	2	3	1	2	3	2	1	3	1	2	3	2	1	2	1	3	3	1	2	2	1	1	2	3	3
46	2	3	1	3	2	3	1	2	1	3	2	3	1	2	1	1	3	2	2	3	3	2	1	1	2	3
47	2	3	1	3	2	3	1	2	2	1	3	1	2	3	2	2	1	3	3	1	1	3	2	2	3	1
48	2	3	1	3	2	3	1	2	3	2	1	2	3	1	3	3	2	1	1	2	2	1	3	3	1	2
49	2	3	2	1	3	1	2	2	1	3	2	3	1	2	2	3	1	1	3	2	1	1	2	3	3	2
50	2	3	2	1	3	1	2	3	2	1	3	1	2	3	3	1	2	2	1	3	2	2	3	1	1	3
51	2	3	2	1	3	1	2	3	3	2	1	2	3	1	1	2	3	3	2	1	3	3	1	2	2	1
52	2	3	3	2	1	2	3	1	1	3	2	3	1	2	3	2	2	3	1	1	2	3	3	2	1	1
53	2	3	3	2	1	2	3	1	2	1	3	1	2	3	1	3	3	1	2	2	3	1	1	3	2	2
54	2	3	3	2	1	2	3	1	3	2	1	2	3	1	2	1	1	2	3	3	1	2	2	1	3	3
	Gp 1	Gp 2	Group 3						Group 4																	

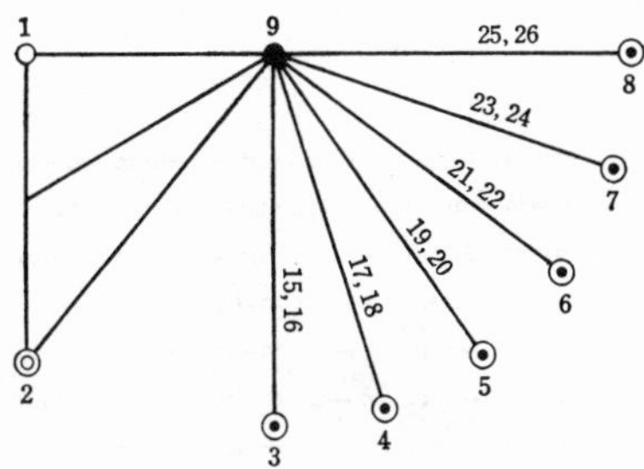

(Note)

i. Column 1 is at two levels.
ii. Interaction of column 1 and column 2 can be found without sacrifice of other columns. Find this interaction from the two-way array of column 1 and column 2.
iii. The three interactions of column 1 × column 9, column 2 × column 9, and column 1 × column 2 × column 9 appear comprehensively in the five columns 10, 11, 12, 13, and 14. The numbers of degrees of freedom are 2, 4, and 4, making a total of 10. The total of the numbers of degrees of freedom of columns 10–14 is also 10. Therefore, if two-level, three-level, and three-level factors are made to correspond to columns 1, 2, and 9, it is possible to find three two-factor interactions and one three-factor interaction from the three-way array.
iv. By letting six levels correspond to the combination of column 1 and column 2, it is also possible to find interactions with column 9.

$L_{81}(3^{40})$

No. \ Col. no.	1	2	3	4	5	6	7	8	9	10	11	12	13	14	15	16	17	18	19	20	21	22	23	24	25	26	27	28	29	30	31	32	33	34	35	36	37	38	39	40
1	1	1	1	1	1	1	1	1	1	1	1	1	1	1	1	1	1	1	1	1	1	1	1	1	1	1	1	1	1	1	1	1	1	1	1	1	1	1	1	1
2	1	1	1	1	1	1	1	1	1	1	1	1	1	2	2	2	2	2	2	2	2	2	2	2	2	2	2	2	2	2	2	2	2	2	2	2	2	2	2	2
3	1	1	1	1	1	1	1	1	1	1	1	1	1	3	3	3	3	3	3	3	3	3	3	3	3	3	3	3	3	3	3	3	3	3	3	3	3	3	3	3
4	1	1	1	1	2	2	2	2	2	2	2	2	2	1	1	1	1	1	1	1	1	1	2	2	2	2	2	2	2	2	2	3	3	3	3	3	3	3	3	3
5	1	1	1	1	2	2	2	2	2	2	2	2	2	2	2	2	2	2	2	2	2	2	3	3	3	3	3	3	3	3	3	1	1	1	1	1	1	1	1	1
6	1	1	1	1	2	2	2	2	2	2	2	2	2	3	3	3	3	3	3	3	3	3	1	1	1	1	1	1	1	1	1	2	2	2	2	2	2	2	2	2
7	1	1	1	1	3	3	3	3	3	3	3	3	3	1	1	1	1	1	1	1	1	1	3	3	3	3	3	3	3	3	3	2	2	2	2	2	2	2	2	2
8	1	1	1	1	3	3	3	3	3	3	3	3	3	2	2	2	2	2	2	2	2	2	1	1	1	1	1	1	1	1	1	3	3	3	3	3	3	3	3	3
9	1	1	1	1	3	3	3	3	3	3	3	3	3	3	3	3	3	3	3	3	3	3	2	2	2	2	2	2	2	2	2	1	1	1	1	1	1	1	1	1
10	1	2	2	2	1	1	1	2	2	2	3	3	3	1	1	1	2	2	2	3	3	3	1	1	1	2	2	2	3	3	3	1	1	1	2	2	2	3	3	3
11	1	2	2	2	1	1	1	2	2	2	3	3	3	2	2	2	3	3	3	1	1	1	2	2	2	3	3	3	1	1	1	2	2	2	3	3	3	1	1	1
12	1	2	2	2	1	1	1	2	2	2	3	3	3	3	3	3	1	1	1	2	2	2	3	3	3	1	1	1	2	2	2	3	3	3	1	1	1	2	2	2
13	1	2	2	2	2	2	2	3	3	3	1	1	1	1	1	1	2	2	2	3	3	3	2	2	2	3	3	3	1	1	1	3	3	3	1	1	1	2	2	2
14	1	2	2	2	2	2	2	3	3	3	1	1	1	2	2	2	3	3	3	1	1	1	3	3	3	1	1	1	2	2	2	1	1	1	2	2	2	3	3	3
15	1	2	2	2	2	2	2	3	3	3	1	1	1	3	3	3	1	1	1	2	2	2	1	1	1	2	2	2	3	3	3	2	2	2	3	3	3	1	1	1
16	1	2	2	2	3	3	3	1	1	1	2	2	2	1	1	1	2	2	2	3	3	3	3	3	3	1	1	1	2	2	2	2	2	2	3	3	3	1	1	1
17	1	2	2	2	3	3	3	1	1	1	2	2	2	2	2	2	3	3	3	1	1	1	1	1	1	2	2	2	3	3	3	3	3	3	1	1	1	2	2	2
18	1	2	2	2	3	3	3	1	1	1	2	2	2	3	3	3	1	1	1	2	2	2	2	2	2	3	3	3	1	1	1	1	1	1	2	2	2	3	3	3
19	1	3	3	3	1	1	1	3	3	3	2	2	2	1	1	1	3	3	3	2	2	2	1	1	1	3	3	3	2	2	2	1	1	1	3	3	3	2	2	2
20	1	3	3	3	1	1	1	3	3	3	2	2	2	2	2	2	1	1	1	3	3	3	2	2	2	1	1	1	3	3	3	2	2	2	1	1	1	3	3	3
21	1	3	3	3	1	1	1	3	3	3	2	2	2	3	3	3	2	2	2	1	1	1	3	3	3	2	2	2	1	1	1	3	3	3	2	2	2	1	1	1
22	1	3	3	3	2	2	2	1	1	1	3	3	3	1	1	1	3	3	3	2	2	2	2	2	2	1	1	1	3	3	3	3	3	3	2	2	2	1	1	1
23	1	3	3	3	2	2	2	1	1	1	3	3	3	2	2	2	1	1	1	3	3	3	3	3	3	2	2	2	1	1	1	1	1	1	3	3	3	2	2	2
24	1	3	3	3	2	2	2	1	1	1	3	3	3	3	3	3	2	2	2	1	1	1	1	1	1	3	3	3	2	2	2	2	2	2	1	1	1	3	3	3
25	1	3	3	3	3	3	3	2	2	2	1	1	1	1	1	1	3	3	3	2	2	2	3	3	3	2	2	2	1	1	1	2	2	2	1	1	1	3	3	3
26	1	3	3	3	3	3	3	2	2	2	1	1	1	2	2	2	1	1	1	3	3	3	1	1	1	3	3	3	2	2	2	3	3	3	2	2	2	1	1	1
27	1	3	3	3	3	3	3	2	2	2	1	1	1	3	3	3	2	2	2	1	1	1	2	2	2	1	1	1	3	3	3	1	1	1	3	3	3	2	2	2
28	2	1	2	3	1	2	3	1	2	3	1	2	3	1	2	3	1	2	3	1	2	3	1	2	3	1	2	3	1	2	3	1	2	3	1	2	3	1	2	3
29	2	1	2	3	1	2	3	1	2	3	1	2	3	2	3	1	2	3	1	2	3	1	2	3	1	2	3	1	2	3	1	2	3	1	2	3	1	2	3	1
30	2	1	2	3	1	2	3	1	2	3	1	2	3	3	1	2	3	1	2	3	1	2	3	1	2	3	1	2	3	1	2	3	1	2	3	1	2	3	1	2
31	2	1	2	3	2	3	1	2	3	1	2	3	1	1	2	3	1	2	3	1	2	3	2	3	1	2	3	1	2	3	1	3	1	2	3	1	2	3	1	2
32	2	1	2	3	2	3	1	2	3	1	2	3	1	2	3	1	2	3	1	2	3	1	3	1	2	3	1	2	3	1	2	1	2	3	1	2	3	1	2	3
33	2	1	2	3	2	3	1	2	3	1	2	3	1	3	1	2	3	1	2	3	1	2	1	2	3	1	2	3	1	2	3	2	3	1	2	3	1	2	3	1
34	2	1	2	3	3	1	2	3	1	2	3	1	2	1	2	3	1	2	3	1	2	3	3	1	2	3	1	2	3	1	2	2	3	1	2	3	1	2	3	1
35	2	1	2	3	3	1	2	3	1	2	3	1	2	2	3	1	2	3	1	2	3	1	1	2	3	1	2	3	1	2	3	3	1	2	3	1	2	3	1	2
36	2	1	2	3	3	1	2	3	1	2	3	1	2	3	1	2	3	1	2	3	1	2	2	3	1	2	3	1	2	3	1	1	2	3	1	2	3	1	2	3
37	2	2	3	1	1	2	3	2	3	1	3	1	2	1	2	3	2	3	1	3	1	2	1	2	3	2	3	1	3	1	2	1	2	3	2	3	1	3	1	2
38	2	2	3	1	1	2	3	2	3	1	3	1	2	2	3	1	3	1	2	1	2	3	2	3	1	3	1	2	1	2	3	2	3	1	3	1	2	1	2	3
39	2	2	3	1	1	2	3	2	3	1	3	1	2	3	1	2	1	2	3	2	3	1	3	1	2	1	2	3	2	3	1	3	1	2	1	2	3	2	3	1

	Gp 1	Gp 2			Group 3									Group 4																										
40	2	2	3	1	2	3	1	3	1	2	1	2	3	1	2	3	2	3	1	3	1	2	2	3	1	3	1	2	1	2	3	3	1	2	1	2	3	2	3	1
41	2	2	3	1	2	3	1	3	1	2	1	2	3	2	3	1	3	1	2	1	2	3	3	1	2	1	2	3	2	3	1	1	2	3	2	3	1	3	1	2
42	2	2	3	1	2	3	1	3	1	2	1	2	3	3	1	2	1	2	3	2	3	1	1	2	3	2	3	1	3	1	2	2	3	1	3	1	2	1	2	3
43	2	2	3	1	3	1	2	1	2	3	2	3	1	1	2	3	2	3	1	3	1	2	3	1	2	1	2	3	2	3	1	2	3	1	3	1	2	1	2	3
44	2	2	3	1	3	1	2	1	2	3	2	3	1	2	3	1	3	1	2	1	2	3	1	2	3	2	3	1	3	1	2	3	1	2	1	2	3	2	3	1
45	2	2	3	1	3	1	2	1	2	3	2	3	1	3	1	2	1	2	3	2	3	1	2	3	1	3	1	2	1	2	3	1	2	3	2	3	1	3	1	2
46	2	3	1	2	1	2	3	3	1	2	2	3	1	1	2	3	3	1	2	2	3	1	1	2	3	3	1	2	2	3	1	1	2	3	3	1	2	2	3	1
47	2	3	1	2	1	2	3	3	1	2	2	3	1	2	3	1	1	2	3	3	1	2	2	3	1	1	2	3	3	1	2	2	3	1	1	2	3	3	1	2
48	2	3	1	2	1	2	3	3	1	2	2	3	1	3	1	2	2	3	1	1	2	3	3	1	2	2	3	1	1	2	3	3	1	2	2	3	1	1	2	3
49	2	3	1	2	2	3	1	1	2	3	3	1	2	1	2	3	3	1	2	2	3	1	2	3	1	1	2	3	3	1	2	3	1	2	2	3	1	1	2	3
50	2	3	1	2	2	3	1	1	2	3	3	1	2	2	3	1	1	2	3	3	1	2	3	1	2	2	3	1	1	2	3	1	2	3	3	1	2	2	3	1
51	2	3	1	2	2	3	1	1	2	3	3	1	2	3	1	2	2	3	1	1	2	3	1	2	3	3	1	2	2	3	1	2	3	1	1	2	3	3	1	2
52	2	3	1	2	3	1	2	2	3	1	1	2	3	1	2	3	3	1	2	2	3	1	3	1	2	2	3	1	1	2	3	2	3	1	1	2	3	3	1	2
53	2	3	1	2	3	1	2	2	3	1	1	2	3	2	3	1	1	2	3	3	1	2	1	2	3	3	1	2	2	3	1	3	1	2	2	3	1	1	2	3
54	2	3	1	2	3	1	2	2	3	1	1	2	3	3	1	2	2	3	1	1	2	3	2	3	1	1	2	3	3	1	2	1	2	3	3	1	2	2	3	1
55	3	1	3	2	1	3	2	1	3	2	1	3	2	1	3	2	1	3	2	1	3	2	1	3	2	1	3	2	1	3	2	1	3	2	1	3	2	1	3	2
56	3	1	3	2	1	3	2	1	3	2	1	3	2	2	1	3	2	1	3	2	1	3	2	1	3	2	1	3	2	1	3	2	1	3	2	1	3	2	1	3
57	3	1	3	2	1	3	2	1	3	2	1	3	2	3	2	1	3	2	1	3	2	1	3	2	1	3	2	1	3	2	1	3	2	1	3	2	1	3	2	1
58	3	1	3	2	2	1	3	2	1	3	2	1	3	1	3	2	1	3	2	1	3	2	2	1	3	2	1	3	2	1	3	3	2	1	3	2	1	3	2	1
59	3	1	3	2	2	1	3	2	1	3	2	1	3	2	1	3	2	1	3	2	1	3	3	2	1	3	2	1	3	2	1	1	3	2	1	3	2	1	3	2
60	3	1	3	2	2	1	3	2	1	3	2	1	3	3	2	1	3	2	1	3	2	1	1	3	2	1	3	2	1	3	2	2	1	3	2	1	3	2	1	3
61	3	1	3	2	3	2	1	3	2	1	3	2	1	1	3	2	1	3	2	1	3	2	3	2	1	3	2	1	3	2	1	2	1	3	2	1	3	2	1	3
62	3	1	3	2	3	2	1	3	2	1	3	2	1	2	1	3	2	1	3	2	1	3	1	3	2	1	3	2	1	3	2	3	2	1	3	2	1	3	2	1
63	3	1	3	2	3	2	1	3	2	1	3	2	1	3	2	1	3	2	1	3	2	1	2	1	3	2	1	3	2	1	3	1	3	2	1	3	2	1	3	2
64	3	2	1	3	1	3	2	2	1	3	3	2	1	1	3	2	2	1	3	3	2	1	1	3	2	2	1	3	3	2	1	1	3	2	2	1	3	3	2	1
65	3	2	1	3	1	3	2	2	1	3	3	2	1	2	1	3	3	2	1	1	3	2	2	1	3	3	2	1	1	3	2	2	1	3	3	2	1	1	3	2
66	3	2	1	3	1	3	2	2	1	3	3	2	1	3	2	1	1	3	2	2	1	3	3	2	1	1	3	2	2	1	3	3	2	1	1	3	2	2	1	3
67	3	2	1	3	2	1	3	3	2	1	1	3	2	1	3	2	2	1	3	3	2	1	2	1	3	3	2	1	1	3	2	3	2	1	1	3	2	2	1	3
68	3	2	1	3	2	1	3	3	2	1	1	3	2	2	1	3	3	2	1	1	3	2	3	2	1	1	3	2	2	1	3	1	3	2	2	1	3	3	2	1
69	3	2	1	3	2	1	3	3	2	1	1	3	2	3	2	1	1	3	2	2	1	3	1	3	2	2	1	3	3	2	1	2	1	3	3	2	1	1	3	2
70	3	2	1	3	3	2	1	1	3	2	2	1	3	1	3	2	2	1	3	3	2	1	3	2	1	1	3	2	2	1	3	2	1	3	3	2	1	1	3	2
71	3	2	1	3	3	2	1	1	3	2	2	1	3	2	1	3	3	2	1	1	3	2	1	3	2	2	1	3	3	2	1	3	2	1	1	3	2	2	1	3
72	3	2	1	3	3	2	1	1	3	2	2	1	3	3	2	1	1	3	2	2	1	3	2	1	3	3	2	1	1	3	2	1	3	2	2	1	3	3	2	1
73	3	3	2	1	1	3	2	3	2	1	2	1	3	1	3	2	3	2	1	2	1	3	1	3	2	3	2	1	2	1	3	1	3	2	3	2	1	2	1	3
74	3	3	2	1	1	3	2	3	2	1	2	1	3	2	1	3	1	3	2	3	2	1	2	1	3	1	3	2	3	2	1	2	1	3	1	3	2	3	2	1
75	3	3	2	1	1	3	2	3	2	1	2	1	3	3	2	1	2	1	3	1	3	2	3	2	1	2	1	3	1	3	2	3	2	1	2	1	3	1	3	2
76	3	3	2	1	2	1	3	1	3	2	3	2	1	1	3	2	3	2	1	2	1	3	2	1	3	1	3	2	3	2	1	3	2	1	2	1	3	1	3	2
77	3	3	2	1	2	1	3	1	3	2	3	2	1	2	1	3	1	3	2	3	2	1	3	2	1	2	1	3	1	3	2	1	3	2	3	2	1	2	1	3
78	3	3	2	1	2	1	3	1	3	2	3	2	1	3	2	1	2	1	3	1	3	2	1	3	2	3	2	1	2	1	3	2	1	3	1	3	2	3	2	1
79	3	3	2	1	3	2	1	2	1	3	1	3	2	1	3	2	3	2	1	2	1	3	3	2	1	2	1	3	1	3	2	2	1	3	1	3	2	3	2	1
80	3	3	2	1	3	2	1	2	1	3	1	3	2	2	1	3	1	3	2	3	2	1	1	3	2	3	2	1	2	1	3	3	2	1	2	1	3	1	3	2
81	3	3	2	1	3	2	1	2	1	3	1	3	2	3	2	1	2	1	3	1	3	2	2	1	3	1	3	2	3	2	1	1	3	2	3	2	1	2	1	3

Interactions Between Two Columns

Col. \ Col.	1	2	3	4	5	6	7	8	9	10	11	12	13	14	15	16	17	18	19	20	21	22	23	24	25	26	27	28	29	30	31	32	33	34	35	36	37	38	39	40
(1)		3 4	2 4	2 3	6 7	5 7	5 6	9 10	8 10	8 9	12 13	11 13	11 12	15 16	14 16	14 15	18 19	17 19	17 18	21 22	20 22	20 21	24 25	23 25	23 24	27 28	26 28	26 27	30 31	29 31	29 30	33 34	32 34	32 33	36 37	35 37	35 36	39 40	38 40	38 39
(2)			1 4	1 3	8 11	9 12	10 13	5 11	6 12	7 13	5 8	6 9	7 10	17 20	18 21	19 22	14 20	15 21	16 22	14 17	15 18	16 19	26 29	27 30	28 31	23 29	24 30	25 31	23 26	24 27	25 28	35 38	36 39	37 40	32 38	33 39	34 40	32 35	33 36	34 37
(3)				1 2	9 13	10 11	8 12	7 12	5 13	6 11	6 10	7 8	5 9	18 22	19 20	17 21	16 21	14 22	15 20	15 19	16 17	14 18	27 31	28 29	26 30	25 30	23 31	24 29	24 28	25 26	23 27	36 40	37 38	35 39	34 39	32 40	33 38	33 37	34 35	32 36
(4)					10 12	8 13	9 11	6 13	7 11	5 12	7 9	5 10	6 8	19 21	17 22	18 20	15 22	16 20	14 21	16 18	14 19	15 17	28 30	26 31	27 29	24 31	25 29	23 30	25 27	23 28	24 26	37 39	35 40	36 38	33 40	34 38	32 39	34 36	32 37	33 35
(5)						1 7	1 6	2 11	3 13	4 12	2 8	4 10	3 9	23 32	24 33	25 34	26 35	27 36	28 37	29 38	30 39	31 40	14 32	15 33	16 34	17 35	18 36	19 37	20 38	21 39	22 40	14 23	15 24	16 25	17 26	18 27	19 28	20 29	21 30	22 31
(6)							1 5	4 13	2 12	3 11	3 10	2 9	4 8	24 34	25 32	23 33	27 37	28 35	26 36	30 40	31 38	29 39	16 33	14 34	15 32	19 36	17 37	18 35	22 39	20 40	21 38	15 25	16 23	14 24	18 28	19 26	17 27	21 31	22 29	20 30
(7)								3 12	4 11	2 13	4 9	3 8	2 10	25 33	23 34	24 32	28 36	26 37	27 35	31 39	29 40	30 38	15 34	16 32	14 33	18 37	19 35	17 36	21 40	22 38	20 39	16 24	14 25	15 23	19 27	17 28	18 26	22 30	20 31	21 29
(8)									1 10	1 9	2 5	3 7	4 6	26 38	27 39	28 40	29 32	30 33	31 34	23 35	24 36	25 37	20 35	21 36	22 37	14 38	15 39	16 40	17 32	18 33	19 34	17 29	18 30	19 31	20 23	21 24	22 25	14 26	15 27	16 28
(9)										1 8	4 7	2 6	3 5	27 40	28 38	26 39	30 34	31 32	29 33	24 37	25 35	23 36	22 36	20 37	21 35	16 39	14 40	15 38	19 33	17 34	18 32	18 31	19 29	17 30	21 25	22 23	20 24	15 28	16 26	14 27
(10)											3 6	4 5	2 7	28 39	26 40	27 38	31 33	29 34	30 32	25 36	23 37	24 35	21 37	22 35	20 36	15 40	16 38	14 39	18 34	19 32	17 33	19 30	17 31	18 29	22 24	20 25	21 23	16 27	14 28	15 26
(11)												1 13	1 12	29 35	30 36	31 37	23 38	24 39	25 40	26 32	27 33	28 34	17 38	18 39	19 40	20 32	21 33	22 34	14 35	15 36	16 37	20 26	21 27	22 28	14 29	15 30	16 31	17 23	18 24	19 25
(12)													1 11	30 37	31 35	29 36	24 40	25 38	23 39	27 34	28 32	26 33	19 39	17 40	18 38	22 33	20 34	21 32	16 36	14 37	15 35	21 28	22 26	20 27	15 31	16 29	14 30	18 25	19 23	17 24
(13)														31 36	29 37	30 35	25 39	23 40	24 38	28 33	26 34	27 32	18 40	19 38	17 39	21 34	22 32	20 33	15 37	16 35	14 36	22 27	20 28	21 26	16 30	14 31	15 29	19 24	17 25	18 23
(14)															1 16	1 15	2 20	3 22	4 21	2 17	4 19	3 18	5 32	6 34	7 33	8 38	9 40	10 39	11 35	12 37	13 36	5 23	7 25	6 24	11 29	13 31	12 30	8 26	10 28	9 27
(15)																1 14	4 22	2 21	3 20	3 19	2 18	4 17	7 34	5 33	6 32	10 40	8 39	9 38	13 37	11 36	12 35	6 25	5 24	7 23	12 31	11 30	13 29	9 28	8 27	10 26
(16)																	3 21	4 20	2 22	4 18	3 17	2 19	6 33	7 32	5 34	9 39	10 38	8 40	12 36	13 35	11 37	7 24	6 23	5 25	13 30	12 29	11 31	10 27	9 26	8 28
(17)																		1 19	1 18	2 14	3 16	4 15	11 38	12 40	13 39	5 35	6 37	7 36	8 32	9 34	10 33	8 29	10 31	9 30	5 26	7 28	6 27	11 23	13 25	12 24
(18)																			1 17	4 16	2 15	3 14	13 40	11 39	12 38	7 37	5 36	6 35	10 34	8 33	9 32	9 31	8 30	10 29	6 28	5 27	7 26	12 25	11 24	13 24

	20	21	22	23	24	25	26	27	28	29	30	31	32	33	34	35	36	37	38	39	40
(19)	3, 15	4, 14	2, 16	12, 39	13, 38	11, 40	6, 36	7, 35	5, 37	9, 33	10, 32	8, 34	10, 30	9, 29	8, 31	7, 27	6, 26	5, 28	13, 24	12, 23	11, 25
(20)		1, 22	1, 21	8, 35	9, 37	10, 36	11, 32	12, 34	13, 33	5, 38	6, 40	7, 39	11, 26	13, 28	12, 27	8, 23	10, 25	9, 24	5, 29	7, 31	6, 30
(21)			1, 20	10, 37	8, 36	9, 35	13, 34	11, 33	12, 32	7, 40	5, 39	6, 38	12, 28	11, 27	13, 26	9, 25	8, 24	10, 23	6, 31	5, 30	7, 29
(22)				9, 36	10, 35	8, 37	12, 33	13, 32	11, 34	6, 39	7, 38	5, 40	13, 27	12, 26	11, 28	10, 24	9, 23	8, 25	7, 30	6, 29	5, 31
(23)					1, 25	1, 24	2, 29	3, 31	4, 30	2, 26	4, 28	3, 27	5, 14	6, 16	7, 15	8, 20	9, 22	10, 21	11, 17	12, 19	13, 18
(24)						1, 23	4, 31	2, 30	3, 29	3, 28	2, 27	4, 26	7, 16	5, 15	6, 14	10, 22	8, 21	9, 20	13, 19	11, 18	12, 17
(25)							3, 30	4, 29	2, 31	4, 27	3, 26	2, 28	6, 15	7, 14	5, 16	9, 21	10, 20	8, 22	12, 18	13, 17	11, 19
(26)								1, 28	1, 27	2, 23	3, 25	4, 24	11, 20	12, 22	13, 21	5, 17	6, 19	7, 18	8, 14	9, 16	10, 15
(27)									1, 26	4, 25	2, 24	3, 23	13, 22	11, 21	12, 20	7, 19	5, 18	6, 17	10, 16	8, 15	9, 14
(28)										3, 24	4, 23	2, 25	12, 21	13, 20	11, 22	6, 18	7, 17	5, 19	9, 15	10, 14	8, 16
(29)											1, 31	1, 30	8, 17	9, 19	10, 18	11, 14	12, 16	13, 15	5, 20	6, 22	7, 21
(30)												1, 29	10, 19	8, 18	9, 17	13, 16	11, 15	12, 14	7, 22	5, 21	6, 20
(31)													9, 18	10, 17	8, 19	12, 15	13, 14	11, 16	6, 21	7, 20	5, 22
(32)														1, 34	1, 33	2, 38	3, 40	4, 39	2, 35	4, 37	3, 36
(33)															1, 32	4, 40	2, 39	3, 38	3, 37	2, 36	4, 35
(34)																3, 39	4, 38	2, 40	4, 36	3, 35	2, 37
(35)																	1, 37	1, 36	2, 32	3, 34	4, 33
(36)																		1, 35	4, 34	2, 33	3, 32
(37)																			3, 33	4, 32	2, 34
(38)																				1, 40	1, 39
(39)																					1, 38

(Note) Refer to the table of $L_{27}(3^{13})$ for interactions between columns 1 – 13.

Linear Graphs of L_{81}

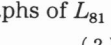

(1)

a

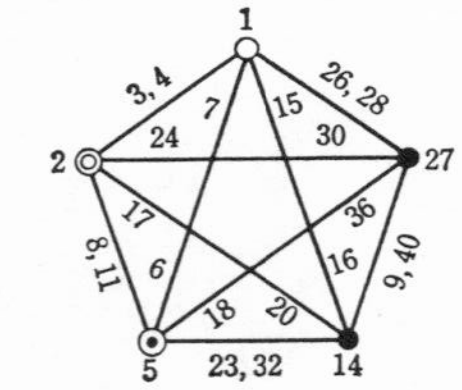

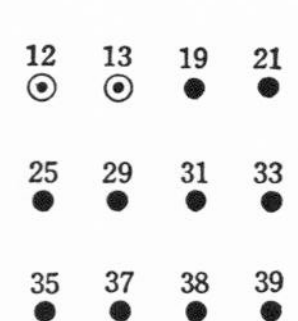

b

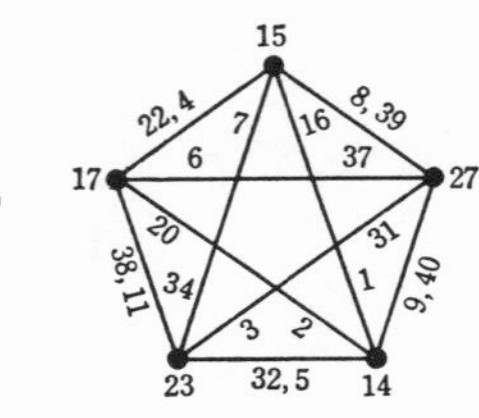

c

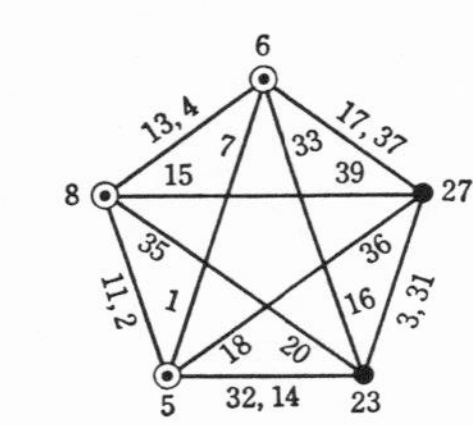

(2)

a

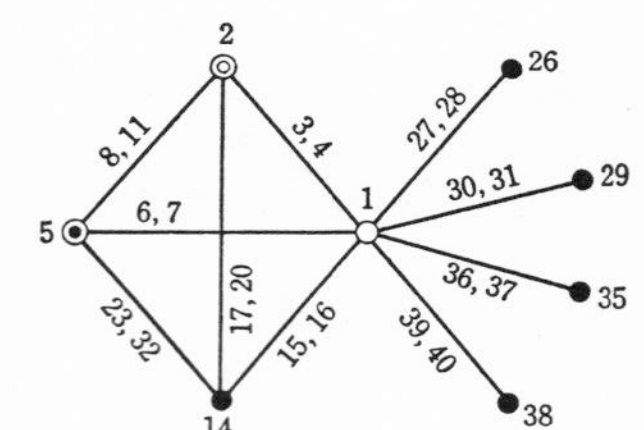

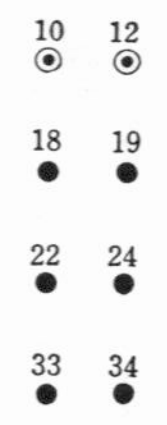

b

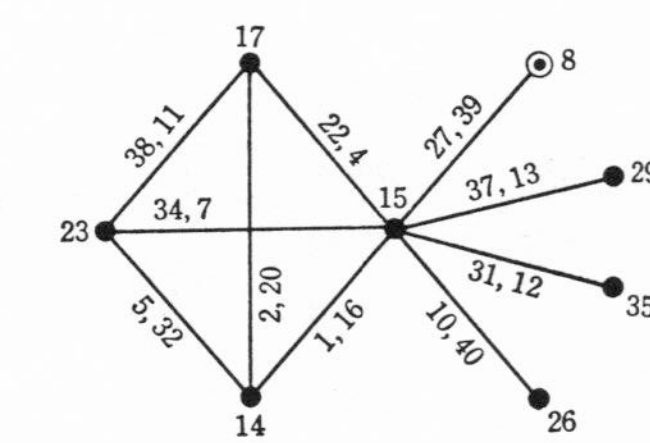

c

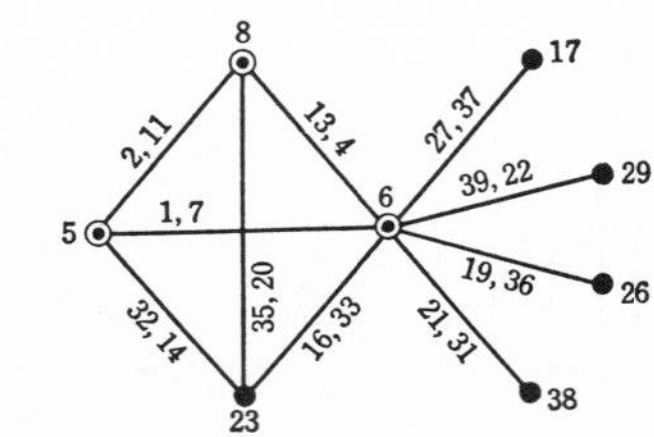

(4)

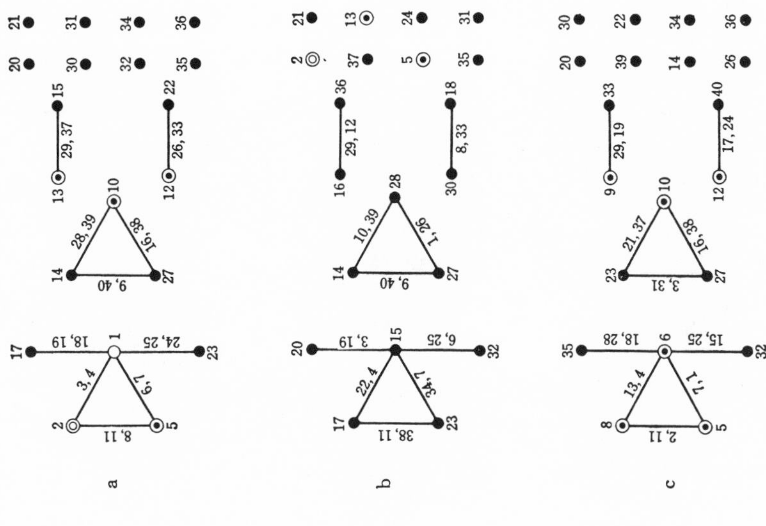

(3)

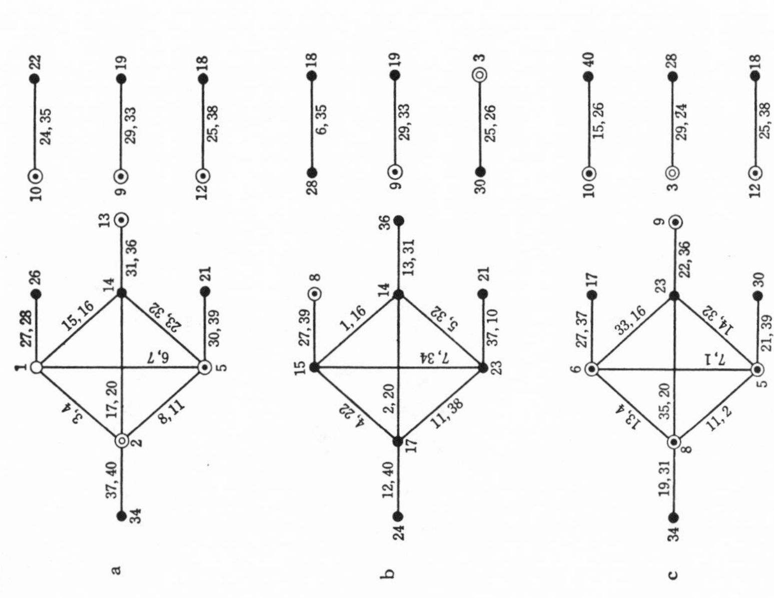

(6)

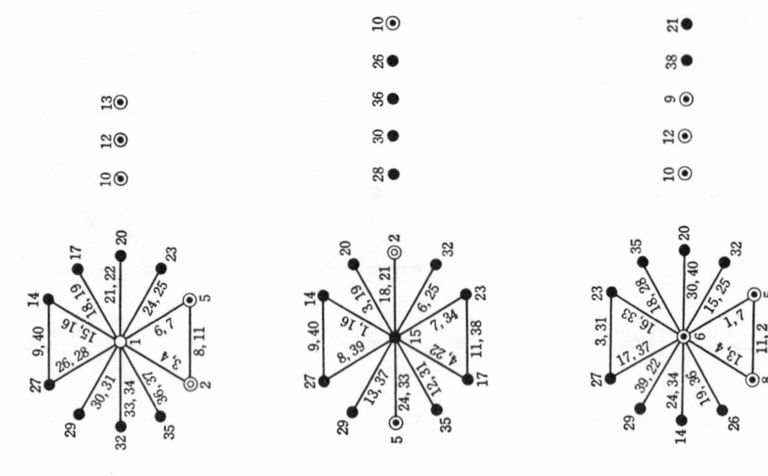

(5)

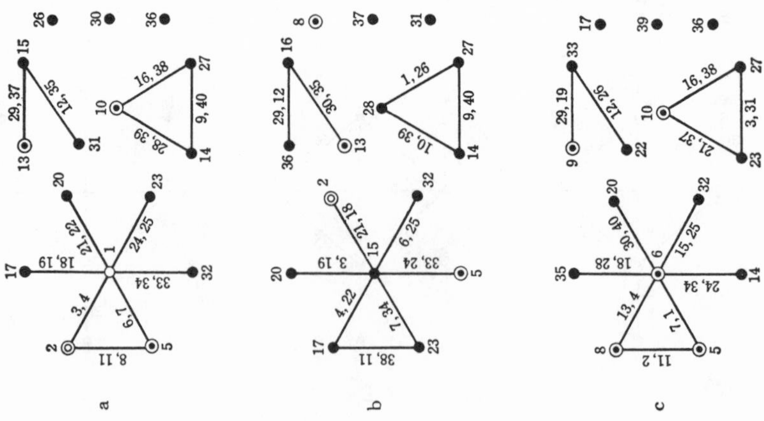

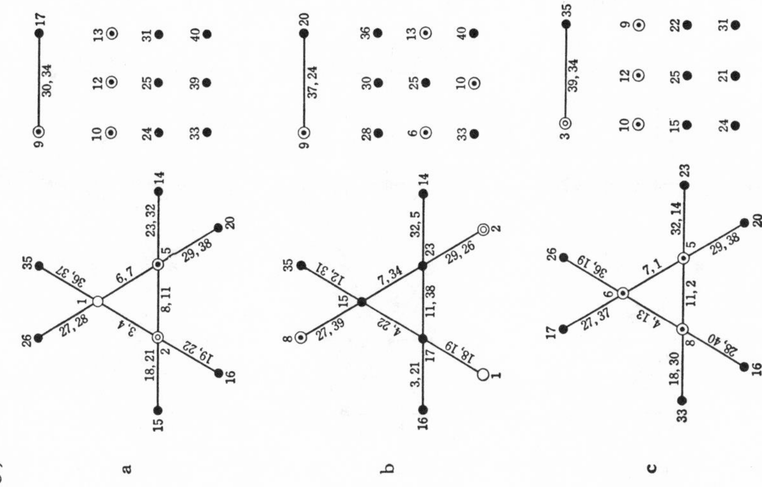

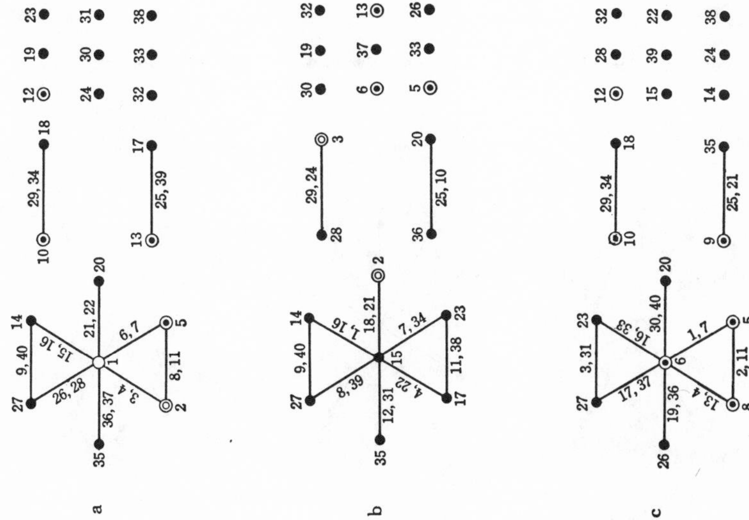

(10)

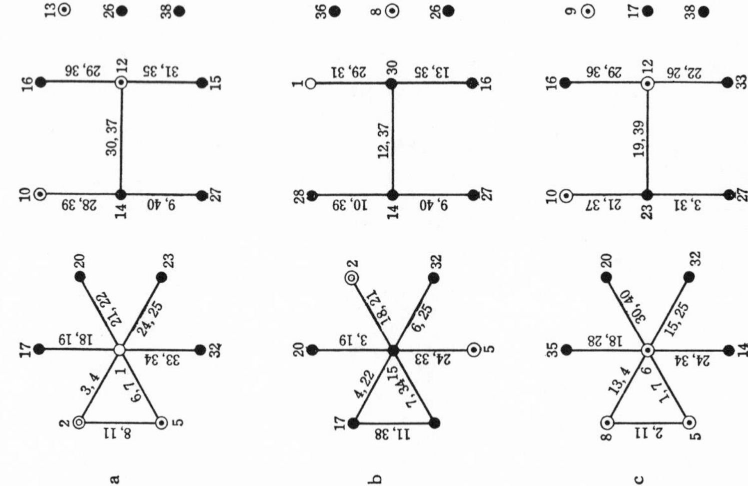

(9)

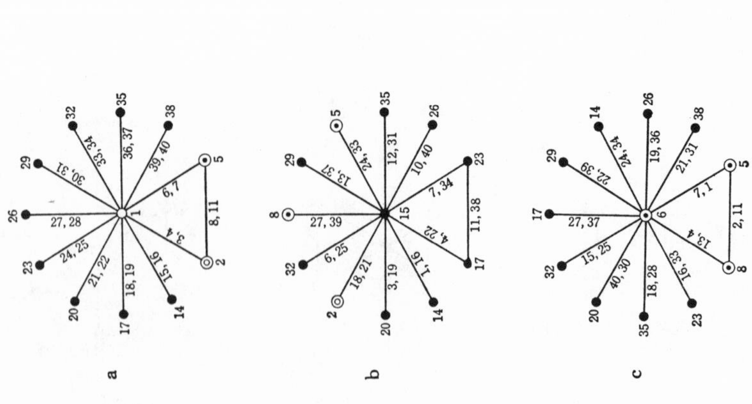

(11)

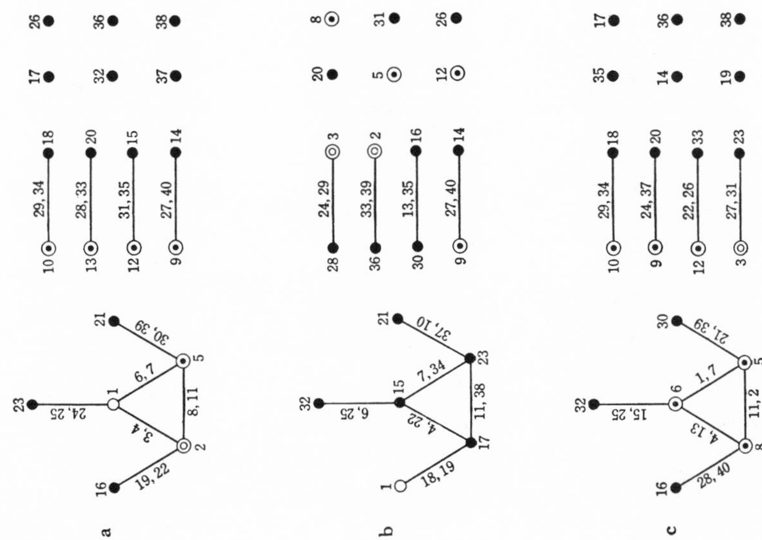

(12)

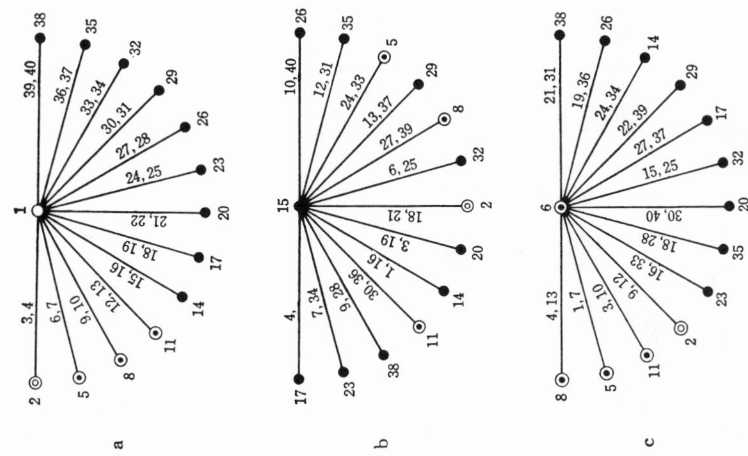

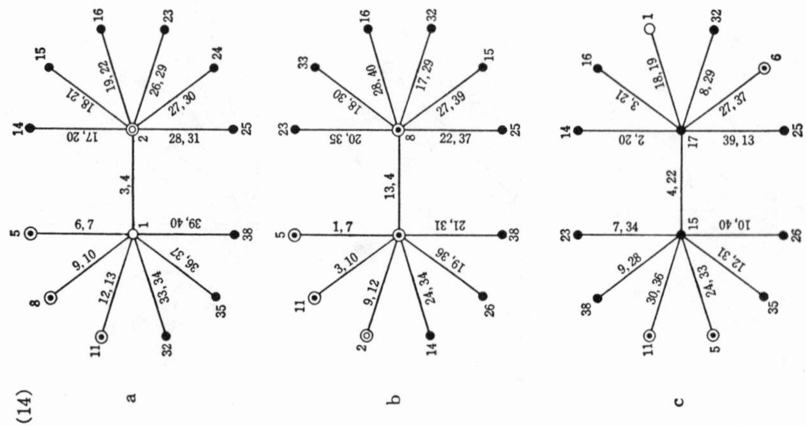

(14)

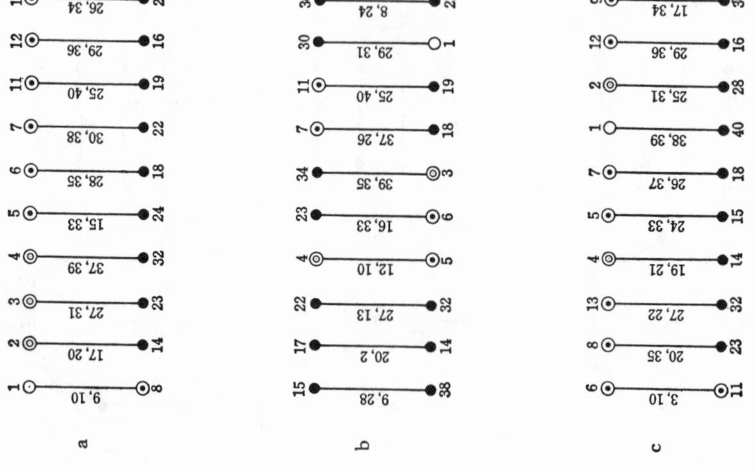

(13)

4^n System

$L_{16}(4^5)$

No. \ Col. no.	1	2	3	4	5
1	1	1	1	1	1
2	1	2	2	2	2
3	1	3	3	3	3
4	1	4	4	4	4
5	2	1	2	3	4
6	2	2	1	4	3
7	2	3	4	1	2
8	2	4	3	2	1
9	3	1	3	4	2
10	3	2	4	3	1
11	3	3	1	2	4
12	3	4	2	1	3
13	4	1	4	2	3
14	4	2	3	1	4
15	4	3	2	4	1
16	4	4	1	3	2
	Gp 1	Group 2			

Linear Graph of L_{16} (1)

1 —— 3, 4, 5 —— 2

$L_{32}(2^1 \times 4^9)$

No. \ Col. no.	1	2	3	4	5	6	7	8	9	10
1	1	1	1	1	1	1	1	1	1	1
2	1	1	2	2	2	2	2	2	2	2
3	1	1	3	3	3	3	3	3	3	3
4	1	1	4	4	4	4	4	4	4	4
5	1	2	1	1	2	2	3	3	4	4
6	1	2	2	2	1	1	4	4	3	3
7	1	2	3	3	4	4	1	1	2	2
8	1	2	4	4	3	3	2	2	1	1
9	1	3	1	2	3	4	1	2	3	4
10	1	3	2	1	4	3	2	1	4	3
11	1	3	3	4	1	2	3	4	1	2
12	1	3	4	3	2	1	4	3	2	1
13	1	4	1	2	4	3	3	4	2	1
14	1	4	2	1	3	4	4	3	1	2
15	1	4	3	4	2	1	1	2	4	3
16	1	4	4	3	1	2	2	1	3	4
17	2	1	1	4	1	4	2	3	2	3
18	2	1	2	3	2	3	1	4	1	4
19	2	1	3	2	3	2	4	1	4	1
20	2	1	4	1	4	1	3	2	3	2
21	2	2	1	4	2	3	4	1	3	2
22	2	2	2	3	1	4	3	2	4	1
23	2	2	3	2	4	1	2	3	1	4
24	2	2	4	1	3	2	1	4	2	3
25	2	3	1	3	3	1	2	4	4	2
26	2	3	2	4	4	2	1	3	3	1
27	2	3	3	1	1	3	4	2	2	4
28	2	3	4	2	2	4	3	1	1	3
29	2	4	1	3	4	2	4	2	1	3
30	2	4	2	4	3	1	3	1	2	4
31	2	4	3	1	2	4	2	4	3	1
32	2	4	4	2	1	3	1	3	4	2
	Gp 1	Gp 2	Group 3							

Linear Graph of L_{32} (1)

1 ——— 2

(Note)

i. Interactions between column 1 (2 levels) and column 2 (4 levels) can be found without sacrifice of other columns.

ii. Interactions between two columns at four levels confound partially a little with each of the remaining four-level columns. It is necessary to use the sequential correction method if one wishes to find the interactions. It is therefore better to use L_{64} if one wishes to find interactions.

iii. This is the same as type (10) of $L_{32}(2^{31})$.

$L_{64}(4^{21})$

No. \ Col. no.	1	2	3	4	5	6	7	8	9	10	11	12	13	14	15	16	17	18	19	20	21
1	1	1	1	1	1	1	1	1	1	1	1	1	1	1	1	1	1	1	1	1	1
2	1	1	1	1	1	2	2	2	2	2	2	2	2	2	2	2	2	2	2	2	2
3	1	1	1	1	1	3	3	3	3	3	3	3	3	3	3	3	3	3	3	3	3
4	1	1	1	1	1	4	4	4	4	4	4	4	4	4	4	4	4	4	4	4	4
5	1	2	2	2	2	1	1	1	1	2	2	2	2	3	3	3	3	4	4	4	4
6	1	2	2	2	2	2	2	2	2	1	1	1	1	4	4	4	4	3	3	3	3
7	1	2	2	2	2	3	3	3	3	4	4	4	4	1	1	1	1	2	2	2	2
8	1	2	2	2	2	4	4	4	4	3	3	3	3	2	2	2	2	1	1	1	1
9	1	3	3	3	3	1	1	1	1	3	3	3	3	4	4	4	4	2	2	2	2
10	1	3	3	3	3	2	2	2	2	4	4	4	4	3	3	3	3	1	1	1	1
11	1	3	3	3	3	3	3	3	3	1	1	1	1	2	2	2	2	4	4	4	4
12	1	3	3	3	3	4	4	4	4	2	2	2	2	1	1	1	1	3	3	3	3
13	1	4	4	4	4	1	1	1	1	4	4	4	4	2	2	2	2	3	3	3	3
14	1	4	4	4	4	2	2	2	2	3	3	3	3	1	1	1	1	4	4	4	4
15	1	4	4	4	4	3	3	3	3	2	2	2	2	4	4	4	4	1	1	1	1
16	1	4	4	4	4	4	4	4	4	1	1	1	1	3	3	3	3	2	2	2	2
17	2	1	2	3	4	1	2	3	4	1	2	3	4	1	2	3	4	1	2	3	4
18	2	1	2	3	4	2	1	4	3	2	1	4	3	2	1	4	3	2	1	4	3
19	2	1	2	3	4	3	4	1	2	3	4	1	2	3	4	1	2	3	4	1	2
20	2	1	2	3	4	4	3	2	1	4	3	2	1	4	3	2	1	4	3	2	1
21	2	2	1	4	3	1	2	3	4	2	1	4	3	3	4	1	2	4	3	2	1
22	2	2	1	4	3	2	1	4	3	1	2	3	4	4	3	2	1	3	4	1	2
23	2	2	1	4	3	3	4	1	2	4	3	2	1	1	2	3	4	2	1	4	3
24	2	2	1	4	3	4	3	2	1	3	4	1	2	2	1	4	3	1	2	3	4
25	2	3	4	1	2	1	2	3	4	3	4	1	2	4	3	2	1	2	1	4	3
26	2	3	4	1	2	2	1	4	3	4	3	2	1	3	4	1	2	1	2	3	4
27	2	3	4	1	2	3	4	1	2	1	2	3	4	2	1	4	3	4	3	2	1
28	2	3	4	1	2	4	3	2	1	2	1	4	3	1	2	3	4	3	4	1	2
29	2	4	3	2	1	1	2	3	4	4	3	2	1	2	1	4	3	3	4	1	2
30	2	4	3	2	1	2	1	4	3	3	4	1	2	1	2	3	4	4	3	2	1
31	2	4	3	2	1	3	4	1	2	2	1	4	3	4	3	2	1	1	2	3	4
32	2	4	3	2	1	4	3	2	1	1	2	3	4	3	4	1	2	2	1	4	3
33	3	1	3	4	2	1	3	4	2	1	3	4	2	1	3	4	2	1	3	4	2
34	3	1	3	4	2	2	4	3	1	2	4	3	1	2	4	3	1	2	4	3	1
35	3	1	3	4	2	3	1	2	4	3	1	2	4	3	1	2	4	3	1	2	4
36	3	1	3	4	2	4	2	1	3	4	2	1	3	4	2	1	3	4	2	1	3
37	3	2	4	3	1	1	3	4	2	2	4	3	1	3	1	2	4	4	2	1	3
38	3	2	4	3	1	2	4	3	1	1	3	4	2	4	2	1	3	3	1	2	4
39	3	2	4	3	1	3	1	2	4	4	1	1	3	1	3	4	2	2	4	3	1
40	3	2	4	3	1	4	2	1	3	3	2	2	4	2	4	3	1	1	3	4	2
41	3	3	1	2	4	1	3	4	2	3	1	2	4	4	2	1	3	2	4	3	1
42	3	3	1	2	4	2	4	3	1	4	2	1	3	3	1	2	4	1	3	4	2
43	3	3	1	2	4	3	1	2	4	1	3	4	2	2	4	3	1	4	2	1	3
44	3	3	1	2	4	4	2	1	3	2	4	3	1	1	3	4	2	3	1	2	4
45	3	4	2	1	3	1	3	4	2	4	2	1	3	2	4	3	1	3	1	2	4
46	3	4	2	1	3	2	4	3	1	3	1	2	4	1	3	4	2	4	2	1	3
47	3	4	2	1	3	3	1	2	4	2	4	3	1	4	2	1	3	1	3	4	2
48	3	4	2	1	3	4	2	1	3	1	3	4	2	3	1	2	4	2	4	3	1
49	4	1	4	2	3	1	4	2	3	1	4	2	3	1	4	2	3	1	4	2	3
50	4	1	4	2	3	2	3	1	4	2	3	1	4	2	3	1	4	2	3	1	4
51	4	1	4	2	3	3	2	4	1	3	2	4	1	3	2	4	1	3	2	4	1
52	4	1	4	2	3	4	1	3	2	4	1	3	2	4	1	3	2	4	1	3	2
53	4	2	3	1	4	1	4	2	3	2	3	1	4	3	2	4	1	4	1	3	2
54	4	2	3	1	4	2	3	1	4	1	4	2	3	4	1	3	2	3	2	4	1
55	4	2	3	1	4	3	2	4	1	4	1	3	2	1	4	2	3	2	3	1	4
56	4	2	3	1	4	4	1	3	2	3	2	4	1	2	3	1	4	1	4	2	3
57	4	3	2	4	1	1	4	2	3	3	2	4	1	4	1	3	2	2	3	1	4
58	4	3	2	4	1	2	3	1	4	4	1	3	2	3	2	4	1	1	4	2	3
59	4	3	2	4	1	3	2	4	1	1	4	2	3	2	3	1	4	4	1	3	2
60	4	3	2	4	1	4	1	3	2	2	3	1	4	1	4	2	3	3	2	4	1
61	4	4	1	3	2	1	4	2	3	4	1	3	2	2	3	1	4	3	2	4	1
62	4	4	1	3	2	2	3	1	4	3	2	4	1	1	4	2	3	4	1	3	2
63	4	4	1	3	2	3	2	4	1	2	3	1	4	4	1	3	2	1	4	2	3
64	4	4	1	3	2	4	1	3	2	1	4	2	3	3	2	4	1	2	3	1	4
	Gp 1	Group 2				Group 3															

Interactions Between Two Columns

Col. \ Col.	1	2	3	4	5	6	7	8	9	10	11	12	13	14	15	16	17	18	19	20	21
	(1)	3 4 5	2 4 5	2 3 5	2 3 4	7 8 9	6 8 9	6 7 9	6 7 8	11 12 13	10 12 13	10 11 13	10 11 12	15 16 17	14 16 17	14 15 17	14 15 16	19 20 21	18 20 21	18 19 21	18 19 20
		(2)	1 4 5	1 3 5	1 3 4	10 14 18	11 15 19	12 16 20	13 17 21	6 14 18	7 15 19	8 16 20	9 17 21	6 10 18	7 11 19	8 12 20	9 13 21	6 10 14	7 11 15	8 12 16	9 13 17
			(3)	1 2 5	1 2 4	11 16 21	10 17 20	13 14 19	12 15 18	7 17 20	6 16 21	9 15 18	8 14 19	8 13 19	9 12 18	6 11 21	7 10 20	9 12 15	8 13 14	7 10 17	6 11 16
				(4)	1 2 3	12 17 19	13 16 18	10 15 21	11 14 20	8 15 21	9 14 20	6 17 19	7 16 18	9 11 20	8 10 21	7 13 18	6 12 19	7 13 16	6 12 17	9 11 14	8 10 15
					(5)	13 15 20	12 14 21	11 17 18	10 16 19	9 16 19	8 17 18	7 14 21	6 15 20	7 12 21	6 13 20	9 10 19	8 11 18	8 11 17	9 10 16	6 13 15	7 12 14
						(6)	1 8 9	1 7 9	1 7 8	2 14 18	3 16 21	4 17 19	5 15 20	2 10 18	5 13 20	3 11 21	4 12 19	2 10 14	4 12 17	5 13 15	3 11 16
							(7)	1 6 9	1 6 8	3 17 20	2 15 19	5 14 21	4 16 18	5 12 21	2 11 19	4 13 18	3 10 20	4 13 16	2 11 15	3 10 17	5 12 14
								(8)	1 6 7	4 15 11	5 17 18	2 16 20	3 14 19	3 13 19	4 10 21	2 12 20	5 11 18	5 11 17	3 13 14	2 12 16	4 10 15
									(9)	5 16 19	4 14 20	3 15 18	2 17 21	4 11 20	3 12 18	5 10 19	2 13 21	3 12 15	5 10 16	4 11 14	2 13 17
										(10)	1 12 13	1 11 13	1 11 12	2 6 18	4 8 21	5 9 19	3 7 20	2 6 14	5 9 16	3 7 17	4 8 15
											(11)	1 10 13	1 10 12	4 9 20	2 7 19	3 6 21	5 8 18	5 8 17	2 7 15	4 9 14	3 6 16
												(12)	1 10 11	5 7 21	3 9 18	2 8 20	4 6 19	3 9 15	4 6 17	2 8 16	5 7 14
													(13)	3 8 19	5 6 20	4 7 18	2 9 21	4 7 16	3 8 14	5 6 15	2 9 17
														(14)	1 16 17	1 15 17	1 15 16	2 6 10	3 8 13	4 9 11	5 7 12
															(15)	1 14 17	1 14 16	3 9 12	2 7 11	5 6 13	4 8 10
																(16)	1 14 15	4 7 13	5 9 10	2 8 12	3 6 11
																	(17)	5 8 11	4 6 12	3 7 10	2 9 13
																		(18)	1 20 21	1 19 21	1 19 20
																			(19)	1 18 21	1 18 20
																				(20)	1 18 19

Linear Graphs of L_{64}

(1)

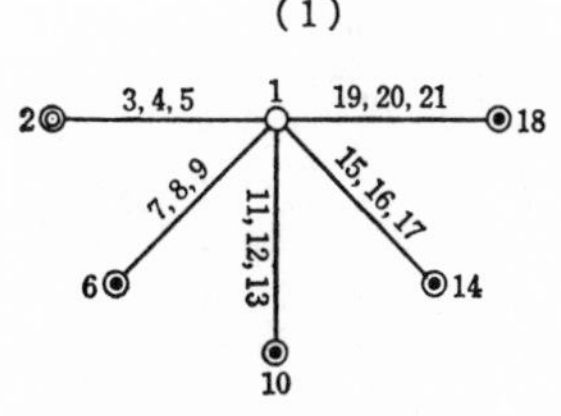

(2)

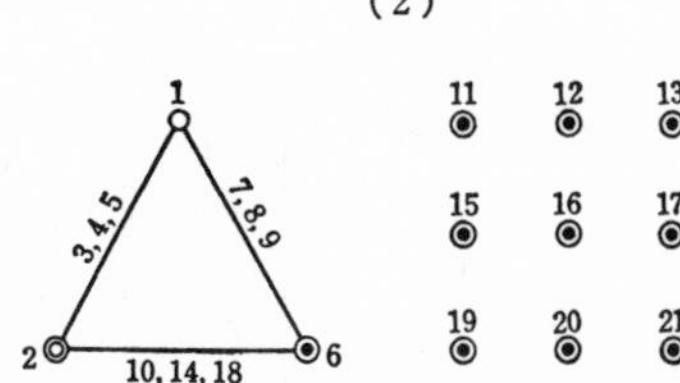

5^n System

$L_{25}(5^6)$

No. \ Col. no.	1	2	3	4	5	6
1	1	1	1	1	1	1
2	1	2	2	2	2	2
3	1	3	3	3	3	3
4	1	4	4	4	4	4
5	1	5	5	5	5	5
6	2	1	2	3	4	5
7	2	2	3	4	5	1
8	2	3	4	5	1	2
9	2	4	5	1	2	3
10	2	5	1	2	3	4
11	3	1	3	5	2	4
12	3	2	4	1	3	5
13	3	3	5	2	4	1
14	3	4	1	3	5	2
15	3	5	2	4	1	3
16	4	1	4	2	5	3
17	4	2	5	3	1	4
18	4	3	1	4	2	5
19	4	4	2	5	3	1
20	4	5	3	1	4	2
21	5	1	5	4	3	2
22	5	2	1	5	4	3
23	5	3	2	1	5	4
24	5	4	3	2	1	5
25	5	5	4	3	2	1
	Gp 1	Group 2				

Linear Graphs of L_{25}

1 —— 3, 4, 5, 6 —— 2

$L_{50}(2^1 \times 5^{11})$

No. \ Col. no.	1	2	3	4	5	6	7	8	9	10	11	12
1	1	1	1	1	1	1	1	1	1	1	1	1
2	1	1	2	2	2	2	2	2	2	2	2	2
3	1	1	3	3	3	3	3	3	3	3	3	3
4	1	1	4	4	4	4	4	4	4	4	4	4
5	1	1	5	5	5	5	5	5	5	5	5	5
6	1	2	1	2	3	4	5	1	2	3	4	5
7	1	2	2	3	4	5	1	2	3	4	5	1
8	1	2	3	4	5	1	2	3	4	5	1	2
9	1	2	4	5	1	2	3	4	5	1	2	3
10	1	2	5	1	2	3	4	5	1	2	3	4
11	1	3	1	3	5	2	4	4	1	3	5	2
12	1	3	2	4	1	3	5	5	2	4	1	3
13	1	3	3	5	2	4	1	1	3	5	2	4
14	1	3	4	1	3	5	2	2	4	1	3	5
15	1	3	5	2	4	1	3	3	5	2	4	1
16	1	4	1	4	2	5	3	5	3	1	4	2
17	1	4	2	5	3	1	4	1	4	2	5	3
18	1	4	3	1	4	2	5	2	5	3	1	4
19	1	4	4	2	5	3	1	3	1	4	2	5
20	1	4	5	3	1	4	2	4	2	5	3	1
21	1	5	1	5	4	3	2	4	3	2	1	5
22	1	5	2	1	5	4	3	5	4	3	2	1
23	1	5	3	2	1	5	4	1	5	4	3	2
24	1	5	4	3	2	1	5	2	1	5	4	3
25	1	5	5	4	3	2	1	3	2	1	5	4
26	2	1	1	1	4	5	4	3	2	5	2	3
27	2	1	2	2	5	1	5	4	3	1	3	4
28	2	1	3	3	1	2	1	5	4	2	4	5
29	2	1	4	4	2	3	2	1	5	3	5	1
30	2	1	5	5	3	4	3	2	1	4	1	2
31	2	2	1	2	1	3	3	2	4	5	5	4
32	2	2	2	3	2	4	4	3	5	1	1	5
33	2	2	3	4	3	5	5	4	1	2	2	1
34	2	2	4	5	4	1	1	5	2	3	3	2
35	2	2	5	1	5	2	2	1	3	4	4	3
36	2	3	1	3	3	1	2	5	5	4	2	4
37	2	3	2	4	4	2	3	1	1	5	3	5
38	2	3	3	5	5	3	4	2	2	1	4	1
39	2	3	4	1	1	4	5	3	3	2	5	2
40	2	3	5	2	2	5	1	4	4	3	1	3
41	2	4	1	4	5	4	1	2	5	2	3	3
42	2	4	2	5	1	5	2	3	1	3	4	4
43	2	4	3	1	2	1	3	4	2	4	5	5
44	2	4	4	2	3	2	4	5	3	5	1	1
45	2	4	5	3	4	3	5	1	4	1	2	2
46	2	5	1	5	2	2	5	3	4	4	3	1
47	2	5	2	1	3	3	1	4	5	5	4	2
48	2	5	3	2	4	4	2	5	1	1	5	3
49	2	5	4	3	5	5	3	1	2	2	1	4
50	2	5	5	4	1	1	4	2	3	3	2	5
	Gp 1	Gp 2	Group 3									

Linear Graph of L_{50}

1 ——— 2

(Note) Interactions of column 1 and column 2 can be found without sacrifice of other columns.

Mixed Systems

$L_{36}(2^{11} \times 3^{12})$

No. \ Col. no.	1	2	3	4	5	6	7	8	9	10	11	12	13	14	15	16	17	18	19	20	21	22	23	1′	2′	3′	4
1	1	1	1	1	1	1	1	1	1	1	1	1	1	1	1	1	1	1	1	1	1	1	1	1	1	1	1
2	1	1	1	1	1	1	1	1	1	1	1	2	2	2	2	2	2	2	2	2	2	2	2	1	1	1	1
3	1	1	1	1	1	1	1	1	1	1	1	3	3	3	3	3	3	3	3	3	3	3	3	1	1	1	1
4	1	1	1	1	1	2	2	2	2	2	2	1	1	1	1	2	2	2	2	3	3	3	3	1	2	2	1
5	1	1	1	1	1	2	2	2	2	2	2	2	2	2	2	3	3	3	3	1	1	1	1	1	2	2	1
6	1	1	1	1	1	2	2	2	2	2	2	3	3	3	3	1	1	1	1	2	2	2	2	1	2	2	1
7	1	1	2	2	2	1	1	1	2	2	2	1	1	2	3	1	2	3	3	1	2	2	3	2	1	2	1
8	1	1	2	2	2	1	1	1	2	2	2	2	2	3	1	2	3	1	1	2	3	3	1	2	1	2	1
9	1	1	2	2	2	1	1	1	2	2	2	3	3	1	2	3	1	2	2	3	1	1	2	2	1	2	1
10	1	2	1	2	2	1	2	2	1	1	2	1	1	3	2	1	3	2	3	2	1	3	2	2	2	1	1
11	1	2	1	2	2	1	2	2	1	1	2	2	2	1	3	2	1	3	1	3	2	1	3	2	2	1	1
12	1	2	1	2	2	1	2	2	1	1	2	3	3	2	1	3	2	1	2	1	3	2	1	2	2	1	1
13	1	2	2	1	2	2	1	2	1	2	1	1	2	3	1	3	2	1	3	3	2	1	2	1	1	1	2
14	1	2	2	1	2	2	1	2	1	2	1	2	3	1	2	1	3	2	1	1	3	2	3	1	1	1	2
15	1	2	2	1	2	2	1	2	1	2	1	3	1	2	3	2	1	3	2	2	1	3	1	1	1	1	2
16	1	2	2	2	1	2	2	1	2	1	1	1	2	3	2	1	1	3	2	3	3	2	1	1	2	2	2
17	1	2	2	2	1	2	2	1	2	1	1	2	3	1	3	2	2	1	3	1	1	3	2	1	2	2	2
18	1	2	2	2	1	2	2	1	2	1	1	3	1	2	1	3	3	2	1	2	2	1	3	1	2	2	2
19	2	1	2	2	1	1	2	2	1	2	1	1	2	1	3	3	3	1	2	2	1	2	3	2	1	2	2
20	2	1	2	2	1	1	2	2	1	2	1	2	3	2	1	1	1	2	3	3	2	3	1	2	1	2	2
21	2	1	2	2	1	1	2	2	1	2	1	3	1	3	2	2	2	3	1	1	3	1	2	2	1	2	2
22	2	1	2	1	2	2	2	1	1	1	2	1	2	2	3	3	1	2	1	1	3	3	2	2	2	1	2
23	2	1	2	1	2	2	2	1	1	1	2	2	3	3	1	1	2	3	2	2	1	1	3	2	2	1	2
24	2	1	2	1	2	2	2	1	1	1	2	3	1	1	2	2	3	1	3	3	2	2	1	2	2	1	2
25	2	1	1	2	2	2	1	2	2	1	1	1	3	2	1	2	3	3	1	3	1	2	2	1	1	1	3
26	2	1	1	2	2	2	1	2	2	1	1	2	1	3	2	3	1	1	2	1	2	3	3	1	1	1	3
27	2	1	1	2	2	2	1	2	2	1	1	3	2	1	3	1	2	2	3	2	3	1	1	1	1	1	3
28	2	2	2	1	1	1	1	2	2	1	2	1	3	2	2	2	1	1	3	2	3	1	3	1	2	2	3
29	2	2	2	1	1	1	1	2	2	1	2	2	1	3	3	3	2	2	1	3	1	2	1	1	2	2	3
30	2	2	2	1	1	1	1	2	2	1	2	3	2	1	1	1	3	3	2	1	2	3	2	1	2	2	3
31	2	2	1	2	1	2	1	1	1	2	2	1	3	3	3	2	3	2	2	1	2	1	1	2	1	2	3
32	2	2	1	2	1	2	1	1	1	2	2	2	1	1	1	3	1	3	3	2	3	2	2	2	1	2	3
33	2	2	1	2	1	2	1	1	1	2	2	3	2	2	2	1	2	1	1	3	1	3	3	2	1	2	3
34	2	2	1	1	2	1	2	1	2	2	1	1	3	1	2	3	2	3	1	2	2	3	1	2	2	1	3
35	2	2	1	1	2	1	2	1	2	2	1	2	1	2	3	1	3	1	2	3	3	1	2	2	2	1	3
36	2	2	1	1	2	1	2	1	2	2	1	3	2	3	1	2	1	2	3	1	1	2	3	2	2	1	3
Columns for response analysis, regression analysis												1	2	3	4	5	6	7	8	9	10	11	12				13
	Gp 1	Group 2										Group 3															

(Notes)

i. By introducing columns 1′, 2′, 3′, and 4′ in place of columns 1, 2, . . . , 11, one obtains $L_{36}(2^3 \times 3^{13})$.

ii. Since interactions are not orthogonal to other columns in the case of $L_{36}(2^{11} \times 3^{12})$, it is best to avoid assignments to find such interactions.

iii. The assignment type is shown here only for $L_{36}(2^3 \times 3^{13})$.

iv. In Chapter 15 and Chapter 16, the column numbers of the bottom row (column 1 - column 13) are used.

Linear Graphs of L_{36}

(1)

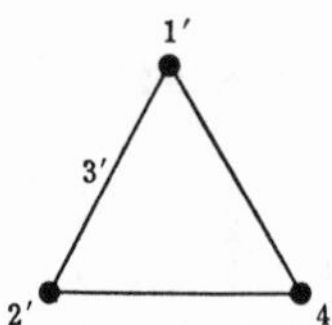

$1' \times 4'$ and $2' \times 4'$ can be found without sacrifice of other columns. Calculate from the three-way array of 1′, 2′, and 4′.

(2)

1′
4′
2′
3′

All interactions can be found without sacrifice of other columns.

$L'_{27}(3^{22})$

No. \ Col. no.	1	2	3	4	5	6	7	8	9	10	11	12	13	14	15	16	17	18	19	20	21	22
1	1	1	1	1	1	1	1	1	1	1	1	1	1	3	3	3	2	2	2	1	1	1
2	1	1	1	1	2	2	2	2	2	2	2	2	2	3	3	3	3	3	3	2	2	2
3	1	1	1	1	3	3	3	3	3	3	3	3	3	1	1	1	2	2	2	2	2	2
4	1	2	2	2	1	1	1	2	2	2	3	3	3	2	2	2	3	3	3	1	1	1
5	1	2	2	2	2	2	2	3	3	3	1	1	1	1	1	1	3	3	3	3	3	3
6	1	2	2	2	3	3	3	1	1	1	2	2	2	1	1	1	1	1	1	1	1	1
7	1	3	3	3	1	1	1	3	3	3	2	2	2	2	2	2	2	2	2	3	3	3
8	1	3	3	3	2	2	2	1	1	1	3	3	3	3	3	3	1	1	1	3	3	3
9	1	3	3	3	3	3	3	2	2	2	1	1	1	2	2	2	1	1	1	2	2	2
10	2	1	2	3	1	2	3	1	2	3	1	2	3	1	2	3	3	1	2	3	1	2
11	2	1	2	3	2	3	1	2	3	1	2	3	1	3	1	2	3	1	2	2	3	1
12	2	1	2	3	3	1	2	3	1	2	3	1	2	3	1	2	1	2	3	3	1	2
13	2	2	3	1	1	2	3	2	3	1	3	1	2	1	2	3	2	3	1	2	3	1
14	2	2	3	1	2	3	1	3	1	2	1	2	3	2	3	1	1	2	3	2	3	1
15	2	2	3	1	3	1	2	1	2	3	2	3	1	1	2	3	1	2	3	1	2	3
16	2	3	1	2	1	2	3	3	1	2	2	3	1	2	3	1	2	3	1	3	1	2
17	2	3	1	2	2	3	1	1	2	3	3	1	2	2	3	1	3	1	2	1	2	3
18	2	3	1	2	3	1	2	2	3	1	1	2	3	3	1	2	2	3	1	1	2	3
19	3	1	3	2	1	3	2	1	3	2	1	3	2	2	1	3	1	3	2	2	1	3
20	3	1	3	2	2	1	3	2	1	3	2	1	3	3	2	1	3	2	1	2	1	3
21	3	1	3	2	3	2	1	3	2	1	3	2	1	2	1	3	3	2	1	1	3	2
22	3	2	1	3	1	3	2	2	1	3	3	2	1	3	2	1	1	3	2	3	2	1
23	3	2	1	3	2	1	3	3	2	1	1	3	2	3	2	1	2	1	3	1	3	2
24	3	2	1	3	3	2	1	1	3	2	2	1	3	1	3	2	1	3	2	1	3	2
25	3	3	2	1	1	3	2	3	2	1	2	1	3	2	1	3	2	1	3	3	2	1
26	3	3	2	1	2	1	3	1	3	2	3	2	1	1	3	2	3	1	3	2	1	3
27	3	3	2	1	3	2	1	2	1	3	1	3	2	1	3	2	3	2	1	3	2	1
	Group 1	Group 2			Group 3									Mixed								

(Note)

i. The four columns 1, 11, 12, and 13 are mutually orthogonal with any of the columns. The nine columns 2, 3, 4, 5, 6, 7, 8, 9, and 10 are mutually orthogonal and are also orthogonal to 1, 11, 12, and 13, but are not orthogonal to the nine columns 14–22. However, even in the worst case, rather than corresponding to the respective levels of certain columns among 14–22 in the proportions 3, 3, and 3, these levels correspond as 2, 2, and 5. The nine columns 14, 15, 16, 17, 18, 19, 20, 21, and 22 are mutually orthogonal and are also orthogonal to 1, 11, 12, and 13.

ii. Columns 14–22 are all partially Group 1, partially Group 2, and partially Group 3. Classed-by-group expression is not given by assignment type, but pay heed in the case of experiments by split-unit design. One need only cause only 1, 11, 12, and 13 to be Group 1.

Linear Graphs of L'_{27}

(1)

a

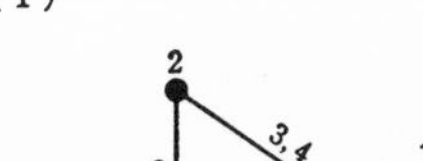

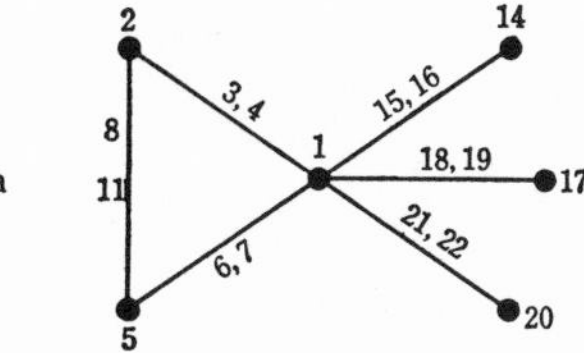

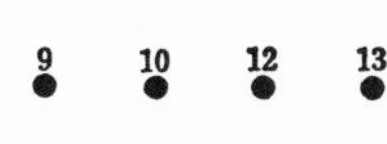

b

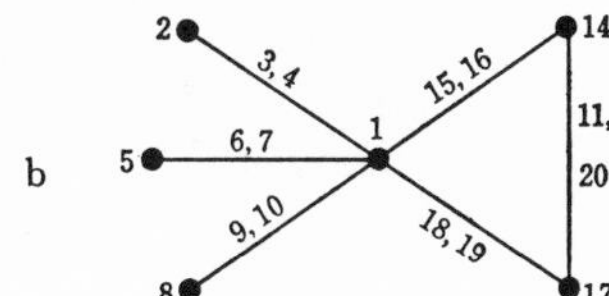

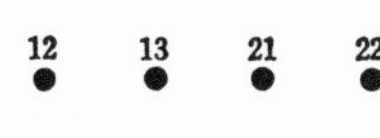

(2)

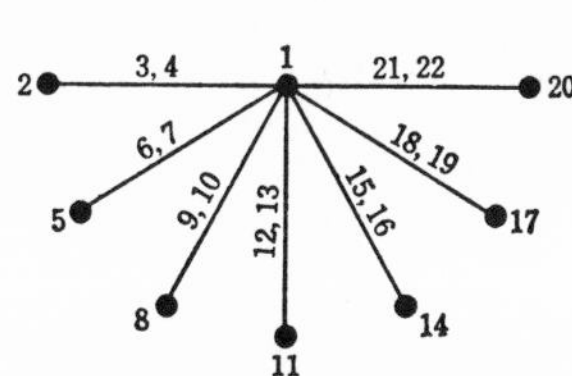

Index

Index

Volume One Contains Pages 1 - 532

Volume Two Contains Pages 533 - 1177